Judith A. Beecher
Judith A. Penna

STUDENT'S
SOLUTIONS MANUAL

ALGEBRA AND TRIGONOMETRY

Bittinger/Beecher

Judith A. Beecher
Judith A. Penna

STUDENT'S SOLUTIONS MANUAL

ALGEBRA AND TRIGONOMETRY

Bittinger/Beecher

ADDISON-WESLEY PUBLISHING COMPANY

Reading, Massachusetts • Menlo Park, California
New York • Don Mills, Ontario • Wokingham, England
Amsterdam • Bonn • Sydney • Singapore • Tokyo • Madrid • San Juan

ISBN 0-201-09156-9

Reproduced by Addison-Wesley from camera-ready copy supplied by the authors.

BCDEFGHIJ-AL-898

TABLE OF CONTENTS

Special thanks are extended to Julie Stephenson for her excellent typing. Her skill, patience, and efficiency made the authors' work much easier.

Judith A. Beecher
Judith A. Penna

STUDENT'S
SOLUTIONS MANUAL

ALGEBRA AND TRIGONOMETRY

Bittinger/Beecher

Exercise Set 1.1

1. - 6.

$-6, 0, 3, -\frac{1}{2}, \sqrt{3}, -2, -\sqrt{7}, \sqrt[3]{2}, \frac{5}{8}, 14, -\frac{9}{4},$

$8.53, 9\frac{1}{2}$

1. Natural numbers: 3, 14

2. Whole numbers: 0, 3, 14

3. Irrational numbers: $\sqrt{3}, -\sqrt{7}, \sqrt[3]{2}$

4. Rational numbers: $-6, 0, 3, -\frac{1}{2}, -2, \frac{5}{8}, 14, -\frac{9}{4},$

$8.53, 9\frac{1}{2}$

5. Integers: -6, 0, 3, -2, 14

6. Real numbers: $-6, 0, 3, -\frac{1}{2}, \sqrt{3}, -2, -\sqrt{7}, \sqrt[3]{2},$

$\frac{5}{8}, 14, -\frac{9}{4}, 8.53, 9\frac{1}{2}$

7. $-\frac{6}{5}$ can be expressed as $\frac{-6}{5}$, a quotient of integers. Thus, $-\frac{6}{5}$ is rational.

8. Rational

9. -9.032 is an ending decimal. Thus, -9.032 is rational.

10. Rational

11. $4.516\overline{516}$ is a repeating decimal. Thus, $4.516\overline{516}$ is rational.

12. Rational

13. 4.303003000300003... has a pattern but not a pattern formed by a repeating block of digits. The numeral is an unending nonrepeating decimal and thus irrational.

14. Irrational

15. $\sqrt{6}$ is irrational because 6 is not a perfect square.

16. Irrational

17. $-\sqrt{14}$ is irrational because 14 is not a perfect square.

18. Irrational

19. $\sqrt{49}$ is rational because 49 is a perfect square.

$\sqrt{49} = \sqrt{7^2} = 7 = \frac{7}{1}$ (Quotient of integers)

20. Rational

21. $\sqrt[3]{5}$ is irrational because 5 is not a perfect cube.

22. Irrational

23. The distance of 12 from 0 is 12, so $|12| = 12$.

24. 2.56

25. The distance of -47 from 0 is 47, so $|-47| = 47$.

26. 0

27. "-x" means "find the additive inverse of x." If x = -7, then we are finding the additive inverse of -7.

$-(-7) = 7$ (The additive inverse of -7 is 7.)

"$-1 \cdot x$" means "find negative one times x." If x = -7, then we are finding -1 times -7.

$-1 \cdot (-7) = 7$ (The product of two negative numbers is positive.)

28. $\frac{10}{3}, \frac{10}{3}$

29. "-x" means "find the additive inverse of x." If x = 57, then we are finding the additive inverse of 57.

$-(57) = -57$ (The additive inverse of 57 is -57.)

"$-1 \cdot x$" means "find negative one times x." If x = 57, then we are finding -1 times 57.

$-1 \cdot 57 = -57$ (The product of a negative number and a positive number is negative.)

30. $-\frac{13}{14}, -\frac{13}{14}$

31. $-(x^2 - 5x + 3)$

$= -(12^2 - 5 \cdot 12 + 3)$ (Substituting 12 for x)

$= -(144 - 60 + 3)$

$= -87$

32. -107

33. $-(7 - y)$

$= -[7 - (-9)]$ (Substituting -9 for y)

$= -[7 + 9]$

$= -16$

34. 12

35. $-3.1 + (-7.2)$

We add their absolute values:

$|-3.1| = 3.1$ $|-7.2| = 7.2$ $3.1 + 7.2 = 10.3$

The sum is negative.

$-3.1 + (-7.2) = -10.3$

36. -416

37. $\frac{9}{2} + \left(-\frac{3}{5}\right)$

Find the absolute values:

$\left|\frac{9}{2}\right| = \frac{9}{2}$ $\left|-\frac{3}{5}\right| = \frac{3}{5}$

Subtract: $\frac{9}{2} - \frac{3}{5} = \frac{45}{10} - \frac{6}{10} = \frac{39}{10}$

The positive number $\left(\frac{9}{2}\right)$ has the greater absolute value, so the answer is positive.

$\frac{9}{2} + \left(-\frac{3}{5}\right) = \frac{39}{10}$

38. -20

39. $-7(-4) = 28$

The product of two negative numbers is positive.

40. 12

41. $(-8.2) \times 6 = -49.2$

The product of a negative number and a positive number is negative.

42. -48

43. $-7(-2)(-3)(-5)$

$= 14 \cdot 15$ (The product of two negative numbers is positive.)

$= 210$

44. 16.33

45. $-\frac{14}{3}\left(-\frac{17}{5}\right)\left(-\frac{21}{2}\right)$

$= -\frac{17}{5}\left[\left(-\frac{14}{3}\right)\left(-\frac{21}{2}\right)\right]$ (Using associativity and commutativity)

$= -\frac{17}{5}\left[\left(-\frac{14}{2}\right)\left(-\frac{21}{3}\right)\right]$

$= -\frac{17}{5}[(-7)(-7)]$

$= -\frac{17}{5} \cdot 49$ (The product of two negative numbers is positive.)

$= -\frac{833}{5}$ (The product of a negative number and a positive number is negative.)

46. $\frac{598}{5}$

47. $\frac{-20}{-4} = 5$ (The quotient of two negative numbers is positive.)

48. -7

49. $\frac{-10}{70}$

$= -\frac{10}{70}$ (The quotient of a negative number and a positive number is negative.)

$= -\frac{1}{7}$ (Simplifying)

50. -5

51. $\frac{2}{7} \div \left(-\frac{14}{3}\right)$

$= \frac{2}{7} \cdot \left(-\frac{3}{14}\right)$ (Multiplying by the reciprocal)

$= \frac{2}{7} \cdot \left(-\frac{3}{2 \cdot 7}\right)$

$= \frac{2}{2} \cdot \left(-\frac{3}{7 \cdot 7}\right)$

$= -\frac{3}{49}$

52. $\frac{7}{10}$

53. $-\frac{10}{3} \div \left(-\frac{2}{15}\right)$

$= -\frac{10}{3} \cdot \left(-\frac{15}{2}\right)$ (Multiplying by the reciprocal)

$= \frac{150}{6}$

$= 25$

54. 8

55. $11 - 15$

$= 11 + (-15)$ $[a - b = a + (-b)]$

$= -4$

56. -29

57. $12 - (-6)$
$= 12 + 6$ $[a - b = a + (-b)]$
$= 18$

58. -9

59. $15.8 - 27.4$
$= 15.8 + (-27.4)$ $[a - b = a + (-b)]$
$= -11.6$

60. -34.8

61. $-\frac{21}{4} - \left(-\frac{7}{4}\right)$
$= -\frac{21}{4} + \frac{7}{4}$ $[a - b = a + (-b)]$
$= -\frac{14}{4}$
$= -\frac{7}{2}$

62. 9

63. a) $(1.4)^2 = 1.96$ b) $\sqrt{2}$
$(1.41)^2 = 1.9881$
$(1.414)^2 = 1.999396$
$(1.4142)^2 = 1.999962$
$(1.41421)^2 = 1.999990$

64. a) 9.261, 9.938375, 9.993948, 9.999517, 9.999935
b) $\sqrt[3]{10}$

65. $k + 0 = k$ Additive Identity Property

66. Commutativity of Multiplication

67. $-1(x + y) = (-1x) + (-1y)$
Distributivity of Multiplication over Addition

68. Associativity of Addition

69. $c + d = d + c$ Commutativity of Addition

70. Multiplicative Identity Property

71. $4(xy) = (4x)y$ Associativity of Multiplication

72. Distributivity of Multiplication over Addition

73. $y\left(\frac{1}{y}\right) = 1,\ y \neq 0$ Multiplicative Inverse Property

74. Additive Inverse Property

75. $7 - 5 \neq 5 - 7$
$7 - 5 = 2$
$5 - 7 = -2$
$2 \neq -2$

76. $16 \div 8 \neq 8 \div 16$
$16 \div 8 = 2$
$8 \div 16 = \frac{1}{2}$

77. $16 \div (4 \div 2) \neq (16 \div 4) \div 2$
$16 \div (4 \div 2) = 16 \div 2 = 8$
$(16 \div 4) \div 2 = 4 \div 2 = 2$
$8 \neq 2$

78. $7 - (5 - 2) \neq (7 - 5) - 2$
$7 - (5 - 2) = 4$
$(7 - 5) - 2 = 0$

79. $|x - 3| = x - 3$ if $x - 3 \geqslant 0$, or $x \geqslant 3$
$|x - 3| = -(x - 3)$ if $x - 3 < 0$, or $x < 3$

80. 1

81. Let $n = 3.7\overline{474}$.

Then $100n = 374.7474...$
$n = 3.7474...$
$99n = 371$
$n = \frac{371}{99}$

82. $\frac{183,062}{9990}$

83. Let $n = 12.34\overline{7652}$.

Then $1,000,000n = 12347652.76527652...$
$100n = 1234.76527652...$
$999,900n = 12346418$
$n = \frac{12,346,418}{999,900}$

Exercise Set 1.2

1. $2^3 \cdot 2^{-4} = 2^{3+(-4)} = 2^{-1}$

2. 3^{-1}

3. $b^2 \cdot b^{-2} = b^{2+(-2)} = b^0 = 1$

4. 1

5. $4^2 \cdot 4^{-5} \cdot 4^6 = 4^{2+(-5)+6} = 4^3$

6. 5^3

7. $2x^3 \cdot 3x^2 = 2 \cdot 3 \cdot x^{3+2} = 6x^5$

8. $12y^7$

9. $(5a^2b)(3a^{-3}b^4) = 5 \cdot 3 \cdot a^{2+(-3)} \cdot b^{1+4} = 15a^{-1}b^5$

10. $12x^{-3}y^7$

11. $(2x)^3(3x)^2 = 2^3x^3 \cdot 3^2x^2 = 8 \cdot 9 \cdot x^{3+2} = 72x^5$

12. $432y^5$

13. $(6x^5y^{-2}z^3)(-3x^2y^3z^{-2})$
$= 6 \cdot (-3) \cdot x^{5+2}y^{-2+3}z^{3+(-2)}$
$= -18x^7yz$

14. $-10x^6yz$

15. $\dfrac{b^{40}}{b^{37}} = b^{40-37} = b^3$

16. a^7

17. $\dfrac{x^2y^{-2}}{x^{-1}y} = x^{2-(-1)}y^{-2-1} = x^3y^{-3}$

18. x^4y^{-5}

19. $\dfrac{9a^2}{(-3a)^2} = \dfrac{9a^2}{9a^2} = 1$

20. 1

21. $\dfrac{24a^5b^3}{8a^4b} = 3a^{5-4}b^{3-1} = 3ab^2$

22. $6x^3y^2$

23. $\dfrac{12x^2y^3z^{-2}}{21xy^2z^3} = \dfrac{12}{21} x^{2-1}y^{3-2}z^{-2-3} = \dfrac{4}{7} xyz^{-5}$

24. $\dfrac{1}{3} x^2y^3z^{-8}$

25. $(2ab^2)^3 = 2^3a^3b^{2 \cdot 3} = 8a^3b^6$

26. $16x^2y^6$

27. $(-2x^3)^4 = (-2)^4x^{3 \cdot 4} = 16x^{12}$

28. $81x^8$

29. $-(2x^3)^4 = -2^4x^{3 \cdot 4} = -16x^{12}$

30. $-81x^8$

31. $(6a^2b^3c)^2 = 6^2a^{2 \cdot 2}b^{3 \cdot 2}c^2 = 36a^4b^6c^2$

32. $25x^6y^4z^2$

33. $(-5c^{-1}d^{-2})^{-2} = (-5)^{-2}c^{-1(-2)}d^{-2(-2)} = \dfrac{1}{25} c^2d^4$

$\left[(-5)^{-2} = \dfrac{1}{(-5)^2} = \dfrac{1}{25}\right]$

34. $\dfrac{1}{16} x^2z^4$

35. $\dfrac{4^{-2} + 2^{-4}}{8^{-1}} = \dfrac{\frac{1}{4^2} + \frac{1}{2^4}}{\frac{1}{8}} = \dfrac{\frac{1}{16} + \frac{1}{16}}{\frac{1}{8}} = \dfrac{\frac{2}{16}}{\frac{1}{8}} = \dfrac{\frac{1}{8}}{\frac{1}{8}} = 1$

36. $\dfrac{119}{72}$

37. $\dfrac{(-2)^4 + (-4)^2}{(-1)^8} = \dfrac{16 + 16}{1} = \dfrac{32}{1} = 32$

38. 25

39. $\dfrac{(3a^2b^{-2}c^4)^3}{(2a^{-1}b^2c^{-3})^2}$
$= \dfrac{3^3a^6b^{-6}c^{12}}{2^2a^{-2}b^4c^{-6}}$
$= \dfrac{27}{4} a^{6-(-2)}b^{-6-4}c^{12-(-6)}$
$= \dfrac{27}{4} a^8b^{-10}c^{18}$

40. $\dfrac{8}{9} a^{11}b^{-3}c^{19}$

41. $\dfrac{6^{-2}x^{-3}y^2}{3^{-3}x^{-4}y} = \dfrac{3^3}{6^2} x^{-3-(-4)}y^{2-1} = \dfrac{27}{36} xy = \dfrac{3}{4} xy$

42. $\dfrac{8}{25} xy^2$

43. $58{,}000{,}000 = 58{,}000{,}000 \times \dfrac{10^7}{10^7}$
$= \dfrac{58{,}000{,}000}{10^7} \times 10^7$
$= 5.8 \times 10^7$

44. 2.7×10^4

45. $365{,}000 = 365{,}000 \times \dfrac{10^5}{10^5}$
$= \dfrac{365{,}000}{10^5} \times 10^5$
$= 3.65 \times 10^5$

46. 3.645×10^3

47. $0.0000027 = 0.0000027 \times \frac{10^6}{10^6}$

$= (0.0000027 \times 10^6) \times \frac{1}{10^6}$

$= 2.7 \times 10^{-6}$

48. 6.58×10^{-5}

49. $0.027 = 0.027 \times \frac{10^2}{10^2}$

$= (0.027 \times 10^2) \times \frac{1}{10^2}$

$= 2.7 \times 10^{-2}$

50. 3.8×10^{-3}

51. $\$910{,}000{,}000{,}000 = \$910{,}000{,}000{,}000 \times \frac{10^{11}}{10^{11}}$

$= \frac{\$910{,}000{,}000{,}000}{10^{11}} \times 10^{11}$

$= \$9.1 \times 10^{11}$

52. 9.3×10^7

53. $4 \times 10^5 = 400{,}000$

54. 0.0005

55. $6.2 \times 10^{-3} = 0.0062$

56. 7,800,000

57. $7.69 \times 10^{12} = 7{,}690{,}000{,}000{,}000$

58. 0.000000854

59. $9.46 \times 10^{12} = 9{,}460{,}000{,}000{,}000$

60. 0.0000000000000000000000017

61. When $x = 5$, $-x^2 = -5^2 = -25$
and $(-x)^2 = (-5)^2 = 25$

62. -49, 49

63. When $x = -1.08$, $-x^2 = -(-1.08)^2 = -1.1664$
and $(-x)^2 = [-(-1.08)]^2 = (1.08)^2 = 1.1664$

64. -3, 3

65. $|9xy| = |9| \cdot |xy| = 9|xy|$ or $9|x||y|$

66. y^4

67. $|3a^2b| = |3| \cdot |a^2| \cdot |b| = 3a^2|b|$

68. $\frac{4|a|}{b^2}$

69. $(x^t \cdot x^{3t})^2 = (x^{4t})^2 = x^{4t \cdot 2} = x^{8t}$

70. 1

71. $(t^{a+x} \cdot t^{x-a})^4 = (t^{2x})^4 = t^{2x \cdot 4} = t^{8x}$

72. m^2yt

73. $(x^ay^b \cdot x^by^a)^c = (x^{a+b}y^{a+b})^c = x^{ac+bc}y^{ac+bc}$

$= (xy)^{ac+bc}$

74. $(mn)^{x^2}$

75. $\left[\frac{(3x^ay^b)^3}{(-3x^ay^b)^2}\right]^2 = \frac{(3x^ay^b)^6}{(-3x^ay^b)^4}$

$= \frac{3^6x^{6a}y^{6b}}{(-3)^4x^{4a}y^{4b}}$

$= \frac{3^6}{3^4} x^{6a-4a}y^{6b-4b}$

$= 3^2x^{2a}y^{2b}$

$= 9x^{2a}y^{2b}$

76. $\frac{x^{6r}}{y^{18t}}$

77. $M = P\left[\frac{\frac{i}{12}\left(1 + \frac{i}{12}\right)^n}{\left(1 + \frac{i}{12}\right)^n - 1}\right]$

$M = 78{,}000\left[\frac{\frac{0.1075}{12}\left(1 + \frac{0.1075}{12}\right)^{300}}{\left(1 + \frac{0.1075}{12}\right)^{300} - 1}\right]$ (Substituting)

$= 78{,}000 \cdot \frac{0.008958333(14.52009644)}{14.52009644 - 1}$

$= 78{,}000(0.009620927)$

$= \$750.43$

78. \$791.88, \$728.12

79. $x^4(x^3)^2 = x^4 \cdot x^5 = x^9$

Exponents were added instead of multiplied.
$x^4(x^3)^2 = x^4 \cdot x^6 = x^{10}$ (Correct)

80. Exponents were added instead of subtracted; $\frac{x^6}{y^{12}}$

81. $(2x^{-4}\ y^6\ z^3)^3 = 6x^{-1}\ y^3\ z^6$
①②③④ ①②③④

Four errors were made.

① $2 \cdot 3$ instead of 2^3

② Exponents were added instead of multiplied.

③ Exponents were subtracted instead of multiplied.

④ Exponents were added instead of multiplied.

$(2x^{-4}y^6z^3)^3 = 2^3x^{-4\cdot3}y^{6\cdot3}z^{3\cdot3} = 8x^{-12}y^{18}z^9$ (Correct)

Exercise Set 1.3

1. $-11x^4 - x^3 + x^2 + 3x - 9$

Term	Degree	
$-11x^4$	4	
$-x^3$	3	
x^2	2	
$3x$	1	$(3x = 3x^1)$
-9	0	$(-9 = -9x^0)$

The degree of a polynomial is the same as that of its term of highest degree.

The term of highest degree is $-11x^4$. Thus, the degree of the polynomial is 4.

2. 3, 2, 1, 0; 3

3. $y^3 + 2y^6 + x^2y^4 - 8$

Term	Degree	
y^3	3	
$2y^6$	6	
x^2y^4	6	$(2 + 4 = 6)$
-8	0	$(-8 = -8x^0)$

The terms of highest degree are $2y^6$ and x^2y^4. Thus, the degree of the polynomial is 6.

4. 2, 5, 7, 0; 7

5. $a^5 + 4a^2b^4 + 6ab + 4a - 3$

Term	Degree	
a^5	5	
$4a^2b^4$	6	$(2 + 4 = 6)$
$6ab$	2	$(1 + 1 = 2)$
$4a$	1	$(4a = 4a^1)$
-3	0	$(-3 = -3a^0)$

The term of highest degree is $4a^2b^4$. Thus, the degree of the polynomial is 6.

6. 6, 8, 4, 2, 0; 8

7. $(5x^2y - 2xy^2 + 3xy - 5)+(-2x^2y - 3xy^2 + 4xy + 7)$
$= (5 - 2)x^2y + (-2 - 3)xy^2 + (3 + 4)xy + (-5 + 7)$
$= 3x^2y - 5xy^2 + 7xy + 2$

8. $2x^2y - 7xy^2 + 8xy + 5$

9. $(-3pq^2 - 5p^2q + 4pq + 3)+(-7pq^2 + 3pq - 4p + 2q)$
$= (-3 - 7)pq^2 - 5p^2q + (4 + 3)pq - 4p + 2q + 3$
$= -10pq^2 - 5p^2q + 7pq - 4p + 2q + 3$

10. $-9pq^2 - 3p^2q + 11pq - 6p + 4q + 5$

11. $(2x+3y+z-7) + (4x-2y-z+8) + (-3x+y-2z-4)$
$= (2+4-3)x + (3-2+1)y + (1-1-2)z + (-7+8-4)$
$= 3x + 2y - 2z - 3$

12. $7x^2 + 12xy - 2x - y - 9$

13. $\left[7x\sqrt{y} - 3y\sqrt{x} + \frac{1}{5}\right] + \left[-2x\sqrt{y} - y\sqrt{x} - \frac{3}{5}\right]$
$= (7 - 2)x\sqrt{y} + (-3 - 1)y\sqrt{x} + \left[\frac{1}{5} - \frac{3}{5}\right]$
$= 5x\sqrt{y} - 4y\sqrt{x} - \frac{2}{5}$

14. $7x\sqrt{y} - 5y\sqrt{x} + 1$

15. The additive inverse of $5x^3 - 7x^2 + 3x - 6$ can be symbolized as

$-(5x^3 - 7x^2 + 3x - 6)$ (With parentheses)

or

$-5x^3 + 7x^2 - 3x + 6$ (Without parentheses)

16. $-(-4y^4 + 7y^2 - 2y - 1)$, $4y^4 - 7y^2 + 2y + 1$

17. $(3x^2 - 2x - x^3 + 2) - (5x^2 - 8x - x^3 + 4)$
$= (3x^2 - 2x - x^3 + 2) + (-5x^2 + 8x + x^3 - 4)$
$= (3 - 5)x^2 + (-2 + 8)x + (-1 + 1)x^3 + (2 - 4)$
$= -2x^2 + 6x - 2$

18. $-4x^2 + 8xy - 5y^2 + 3$

19. $(4a - 2b - c + 3d) - (-2a + 3b + c - d)$
$= (4a - 2b - c + 3d) + (2a - 3b - c + d)$
$= (4 + 2)a + (-2 - 3)b + (-1 - 1)c + (3 + 1)d$
$= 6a - 5b - 2c + 4d$

20. $8a - 8b - 2c + 6d$

21. $(x^4 - 3x^2 + 4x) - (3x^3 + x^2 - 5x + 3)$
$= (x^4 - 3x^2 + 4x) + (-3x^3 - x^2 + 5x - 3)$
$= x^4 - 3x^3 + (-3 - 1)x^2 + (4 + 5)x - 3$
$= x^4 - 3x^3 - 4x^2 + 9x - 3$

22. $2x^4 - 5x^3 - 7x^2 + 10x - 5$

23. $(7x\sqrt{y} - 4y\sqrt{x} + 7.5) - (-2x\sqrt{y} - y\sqrt{x} - 1.6)$
$= (7x\sqrt{y} - 4y\sqrt{x} + 7.5) + (2x\sqrt{y} + y\sqrt{x} + 1.6)$
$= (7 + 2)x\sqrt{y} + (-4 + 1)y\sqrt{x} + (7.5 + 1.6)$
$= 9x\sqrt{y} - 3y\sqrt{x} + 9.1$

24. $13x\sqrt{y} - 5y\sqrt{x} + \frac{5}{3}$

25. $(0.565p^2q - 2.167pq^2 + 16.02pq - 17.1)$
$+ (-1.612p^2q - 0.312pq^2 - 7.141pq - 87.044)$
$= (0.565 - 1.612)p^2q + (-2.167 - 0.312)pq^2 +$
$(16.02 - 7.141)pq + (-17.1 - 87.044)$
$= -1.047p^2q - 2.479pq^2 + 8.879pq - 104.144$

26. $2859.6xy^{-2} - 6153.8xy + 7243.4\sqrt{xy} - 10{,}259.12$

Exercise Set 1.4

1.
$$\begin{array}{rl} 2x^2 + 4x + 16 & \\ 3x - 4 & \\ \hline 6x^3 + 12x^2 + 48x & \text{(Multiplying by } 3x) \\ -8x^2 - 16x - 64 & \text{(Multiplying by } -4) \\ \hline 6x^3 + 4x^2 + 32x - 64 & \text{(Adding)} \end{array}$$

2. $6y^3 + 3y^2 + 9y + 27$

3.
$$\begin{array}{r} 4a^2b - 2ab + 3b^2 \\ ab - 2b + 1 \\ \hline 4a^3b^2 - 2a^2b^2 + 3ab^3 \\ -8a^2b^2 + 4ab^2 - 6b^3 \\ 4a^2b-2ab+3b^2 \\ \hline 4a^3b^2 - 10a^2b^2 + 3ab^3 + 4ab^2 - 6b^3 + 4a^2b-2ab+3b^2 \end{array}$$

4. $2x^4 - x^2y^2 - 4x^3y - 2y^4 + 3xy^3$

5.
$$\begin{array}{r} a^2 + ab + b^2 \\ a - b \\ \hline a^3 + a^2b + ab^2 \\ - a^2b - ab^2 - b^3 \\ \hline a^3 \qquad\qquad - b^3 \end{array}$$
The product is $a^3 - b^3$.

6. $t^3 + 1$

7. $(2x + 3y)(2x + y)$
$= 4x^2 + 2xy + 6xy + 3y^2$ (FOIL)
$= 4x^2 + 8xy + 3y^2$

8. $4a^2 - 8ab + 3b^2$

9. $\left(4x^2 - \frac{1}{2} y\right)\left(3x + \frac{1}{4} y\right)$
$= 12x^3 + x^2y - \frac{3}{2} xy - \frac{1}{8} y^2$ (FOIL)

10. $6y^4 - \frac{1}{2} xy^3 + \frac{3}{5} xy - \frac{1}{20} x^2$

11. $(\sqrt{2}\, x^2 - y^2)(\sqrt{2}\, x - 2y)$
$= 2x^3 - 2\sqrt{2}\, x^2y - \sqrt{2}\, xy^2 + 2y^3$ (FOIL)

12. $3y^3 - \sqrt{3}\, xy^2 - 2\sqrt{3}\, y + 2x$

13. $(2x + 3y)^2$
$= (2x)^2 + 2(2x)(3y) + (3y)^2$
$[(A + B)^2 = A^2 + 2AB + B^2]$
$= 4x^2 + 12xy + 9y^2$

14. $25x^2 + 20xy + 4y^2$

15. $(2x^2 - 3y)^2$
$= (2x^2)^2 - 2(2x^2)(3y) + (3y)^2$
$[(A - B)^2 = A^2 - 2AB + B^2]$
$= 4x^4 - 12x^2y + 9y^2$

16. $16x^4 - 40x^2y + 25y^2$

17. $(2x^3 + 3y^2)^2$
$= (2x^3)^2 + 2(2x^3)(3y^2) + (3y^2)^2$
$[(A + B)^2 = A^2 + 2AB + B^2]$
$= 4x^6 + 12x^3y^2 + 9y^4$

18. $25x^6 + 20x^3y^2 + 4y^4$

19. $\left(\frac{1}{2} x^2 - \frac{3}{5} y\right)^2$
$= \left(\frac{1}{2} x^2\right)^2 - 2\left(\frac{1}{2} x^2\right)\left(\frac{3}{5} y\right) + \left(\frac{3}{5} y\right)^2$
$[(A - B)^2 = A^2 - 2AB + B^2]$
$= \frac{1}{4} x^4 - \frac{3}{5} x^2y + \frac{9}{25} y^2$

20. $\frac{1}{16} x^4 - \frac{1}{3} x^2y + \frac{4}{9} y^2$

21. $(0.5x + 0.7y^2)^2$
$= (0.5x)^2 + 2(0.5x)(0.7y^2) + (0.7y^2)^2$
$[(A + B)^2 = A^2 + 2AB + B^2]$
$= 0.25x^2 + 0.7xy^2 + 0.49y^4$

22. $0.09x^2 + 0.48xy^2 + 0.64y^4$

23. $(3x - 2y)(3x + 2y)$
$= (3x)^2 - (2y)^2$ $[(A - B)(A + B) = A^2 - B^2]$
$= 9x^2 - 4y^2$

24. $9x^2 - 25y^2$

25. $(x^2 + yz)(x^2 - yz)$
$= (x^2)^2 - (yz)^2$ $[(A + B)(A - B) = A^2 - B^2]$
$= x^4 - y^2z^2$

26. $4x^4 - 25x^2y^2$

27. $(3x^2 - \sqrt{2})(3x^2 + \sqrt{2})$
$= (3x^2)^2 - (\sqrt{2})^2$ $[(A - B)(A + B) = A^2 - B^2]$
$= 9x^4 - 2$

28. $25x^4 - 3$

29. $(2x + 3y + 4)(2x + 3y - 4)$
$= (2x + 3y)^2 - 4^2$ $[(A + B)(A - B) = A^2 - B^2]$
$= 4x^2 + 12xy + 9y^2 - 16$

30. $25x^2 + 20xy + 4y^2 - 9$

31. $(x^2 + 3y + y^2)(x^2 + 3y - y^2)$
$= (x^2 + 3y)^2 - (y^2)^2$ $[(A + B)(A - B) = A^2 - B^2]$
$= x^4 + 6x^2y + 9y^2 - y^4$

32. $4x^4 + 4x^2y + y^2 - y^4$

33. $(x + 1)(x - 1)(x^2 + 1)$
$= (x^2 - 1)(x^2 + 1)$
$= x^4 - 1$

34. $y^4 - 16$

35. $(2x + y)(2x - y)(4x^2 + y^2)$
$= (4x^2 - y^2)(4x^2 + y^2)$
$= 16x^4 - y^4$

36. $625x^4 - y^4$

37. $(0.051x + 0.04y)^2$
$= (0.051x)^2 + 2(0.051x)(0.04y) + (0.04y)^2$
$= 0.002601x^2 + 0.00408xy + 0.0016y^2$

38. $1.065024x^2 - 5.184768xy + 6.310144y^2$

39. $(37.86x + 1.42)(65.03x - 27.4)$
$= 2462.0358x^2 - 1037.364x + 92.3426x - 38.908$
$= 2462.0358x^2 - 945.0214x - 38.908$

40. $169.625105x^2 - 711.87827x - 546.525$

41. $(y + 5)^3$
$= y^3 + 3\cdot y^2\cdot 5 + 3\cdot y\cdot 5^2 + 5^3$
$= y^3 + 15y^2 + 75y + 125$

42. $t^3 - 21t^2 + 147t - 343$

43. $(m^2 - 2n)^3$
$= [m^2 + (-2n)]^3$
$= (m^2)^3 + 3(m^2)^2(-2n) + 3(m^2)(-2n)^2 + (-2n)^3$
$= m^6 - 6m^4n + 12m^2n^2 - 8n^3$

44. $27t^6 + 108t^4 + 144t^2 + 64$

45. $(a^n + b^n)(a^n - b^n)$
$= (a^n)^2 - (b^n)^2$
$= a^{2n} - b^{2n}$

46. $t^{2a} - 3t^a - 28$

47. $(x^m - t^n)^3$
$= [x^m + (-t^n)]^3$
$= (x^m)^3 + 3(x^m)^2(-t^n) + 3(x^m)(-t^n)^2 + (-t^n)^3$
$= x^{3m} - 3x^{2m}t^n + 3x^mt^{2n} - t^{3n}$

48. $y^{3n+3}z^{n+3} - 4y^4z^{3n}$

49. $(x - 1)(x^2 + x + 1)(x^3 + 1)$
$= (x^3 - 1)(x^3 + 1)$
$= (x^3)^2 - 1^2$
$= x^6 - 1$

50. $a^{2n} + 2a^nb^n + b^{2n}$

51. $[(2x - 1)^2 - 1]^2$
$= [4x^2 - 4x + 1 - 1]^2$
$= [4x^2 - 4x]^2$
$= (4x^2)^2 - 2(4x^2)(4x) + (4x)^2$
$= 16x^4 - 32x^3 + 16x^2$

52. $-a^4 - 2a^3b + 25a^2 + 2ab^3 - 25b^2 + b^4$

53. $(x^{a-b})^{a+b}$
$= x^{(a-b)(a+b)}$
$= x^{a^2-b^2}$

54. $t^{2m^2+2n^2}$

55. $(a + b + c)^2$
$= (a + b + c)(a + b + c)$
$= a^2 + ab + ac + ba + b^2 + bc + ca + cb + c^2$
$= a^2 + b^2 + c^2 + 2ab + 2ac + 2bc$

56. $a^3 + b^3 + c^3 + 3a^2b + 3a^2c + 3b^2a + 3b^2c + 3c^2a + 3c^2b + 6abc$

57. $(a + b)^4$
$= (a + b)(a + b)^3$
$= (a + b)(a^3 + 3a^2b + 3ab^2 + b^3)$
$= a^4 + 3a^3b + 3a^2b^2 + ab^3 + a^3b + 3a^2b^2 + 3ab^3 + b^4$
$= a^4 + 4a^3b + 6a^2b^2 + 4ab^3 + b^4$

58. $x^5 - y^5$

59. $(m + t)(m^4 - m^3t + m^2t^2 - mt^3 + t^4)$
$= m^5 - m^4t + m^3t^2 - m^2t^3 + mt^4 + m^4t - m^3t^2 + m^2t^3 - mt^4 + t^5$
$= m^5 + t^5$

60. $a^8 - b^8$

61. $(3a + b)^2 = 3a^2 + b^2$

The first term is $9a^2$, not $3a^2$.
The middle term, $2 \cdot 3a \cdot b$, is missing.

$(3a + b)^2 = (3a)^2 + 2 \cdot 3a \cdot b + b^2$
$= 9a^2 + 6ab + b^2$ (Correct)

62. A term is missing, found by taking twice the product of the terms. This is not the product of the sum and difference of two terms. Also $-9y^2$ should by $9y^2$. The correct answer is $4x^2 - 12xy + 9y^2$.

63. $2x(x + 3) + 4(x^2 - 3)$
$= 2x^2 + 3x + 4x^2 - 3$ (1)

The 3x should be 6x; the 2 and 3 were not multiplied. Similarly, -3 should be -12; the 4 and -3 were not multiplied.

$= 6x^2 + x$

3x - 3 is not x since they are not like terms.

$2x(x + 3) + 4(x^2 - 3)$
$= 2x^2 + 6x + 4x^2 - 12$
$= 6x^2 + 6x - 12$ (Correct)

64. FOIL was incorrectly used in step (1); outside and inside products were omitted. In step (2) the 6 was factored out and omitted. The correct answer is $6a^2 - 5ab - 6b^2$.

Exercise Set 1.5

1. Pascal Method: Expand $(m + n)^5$.

The expansion of $(m + n)^5$ has 5 + 1, or 6, terms. The sum of the exponents in each term is 5. The exponents of m start with 5 and decrease to 0. The last term has no factor of m. The first term has no factor of n. The exponents of n start in the second term with 1 and increase to 5. We get the coefficients from the 6th row of Pascal's Triangle.

```
                1
             1     1
          1     2     1
       1     3     3     1
    1     4     6     4     1
 1     5    10    10     5     1
```

$(m + n)^5 = 1 \cdot m^5 + 5 \cdot m^4n^1 + 10 \cdot m^3 \cdot n^2 + 10 \cdot m^2 \cdot n^3 + 5 \cdot m \cdot n^4 + 1 \cdot n^5$
$= m^5 + 5m^4n + 10m^3n^2 + 10m^2n^3 + 5mn^4 + n^5$

2. $a^4 - 4a^3b + 6a^2b^2 - 4ab^3 + b^4$

3. Pascal Method: Expand $(x - y)^6$.

The expansion of $(x - y)^6$ has 6 + 1, or 7 terms. The sum of the exponents in each term is 6. The exponents of x start with 6 and decrease to 0. The last term has no factor of x. The first term has no factor of -y. The exponents of -y start in the second term with 1 and increase to 6. We get the coefficients from the 7th row of Pascal's Triangle.

```
                   1
                1     1
             1     2     1
          1     3     3     1
       1     4     6     4     1
    1     5    10    10     5     1
 1     6    15    20    15     6     1
```

3. (continued)

$$(x - y)^6 = 1\cdot x^6 + 6\cdot x^5\cdot(-y) + 15\cdot x^4\cdot(-y)^2 + 20\cdot x^3\cdot(-y)^3 + 15\cdot x^2\cdot(-y)^4 + 6\cdot x\cdot(-y)^5 + 1\cdot(-y)^6$$
$$= x^6 - 6x^5y + 15x^4y^2 - 20x^3y^3 + 15x^2y^4 - 6xy^5 + y^6$$

4. $p^7 + 7p^6q + 21p^5q^2 + 35p^4q^3 + 35p^3q^4 + 21p^2q^5 + 7pq^6 + q^7$

5. Pascal Method: Expand $(x^2 - 3y)^5$.

The expansion of $(x^2 - 3y)^5$ has 5 + 1, or 6, terms. The sum of the exponents in each term is 5. The exponents of x^2 start with 5 and decrease to 0. The last term has no factor of x^2. The first term has no factor of -3y. The exponents of -3y start in the second term with 1 and increase to 5. We get the coefficients from the 6th row of Pascal's Triangle.

1
1 1
1 2 1
1 3 3 1
1 4 6 4 1
1 5 10 10 5 1

$$(x^2 - 3y)^5 = 1\cdot(x^2)^5 + 5\cdot(x^2)^4\cdot(-3y) + 10\cdot(x^2)^3\cdot(-3y)^2 + 10\cdot(x^2)^2\cdot(-3y)^3 + 5\cdot(x^2)\cdot(-3y)^4 + 1\cdot(-3y)^5$$
$$= x^{10} - 15x^8y + 90x^6y^2 - 270x^4y^3 + 405x^2y^4 - 243y^5$$

6. $2187c^7 - 5103c^6d + 5103c^5d^2 - 2835c^4d^3 + 945c^3d^4 - 189c^2d^5 + 21cd^6 - d^7$

7. Pascal Method: Expand $(3c - d)^6$.

The sum of the exponents in each term is 6. The exponents of 3c start with 6 and decrease to 0. The last term has no factor of 3c. The first term has no factor of -d. The exponents of -d start in the second term with 1 and increase to 6. We get the coefficients from the 7th row of Pascal's Triangle. The expansion has 6 + 1, or 7 terms.

1
1 1
1 2 1
1 3 3 1
1 4 6 4 1
1 5 10 10 5 1
1 6 15 20 15 6 1

$$(3c - d)^6 = 1\cdot(3c)^6 + 6\cdot(3c)^5\cdot(-d) + 15\cdot(3c)^4\cdot(-d)^2 + 20\cdot(3c)^3\cdot(-d)^3 + 15\cdot(3c)^2\cdot(-d)^4 + 6\cdot(3c)\cdot(-d)^5 + 1\cdot(-d)^6$$

7. (continued)

$$= 3^6c^6 - 6\cdot3^5c^5d + 15\cdot3^4c^4d^2 - 20\cdot3^3c^3d^3 + 15\cdot3^2c^2d^4 - 6\cdot3cd^5 + d^6$$
$$= 729c^6 - 6\cdot243c^5d + 15\cdot81c^4d^2 - 20\cdot27c^3d^3 + 15\cdot9c^2d^4 - 6\cdot3cd^5 + d^6$$
$$= 729c^6 - 1458c^5d + 1215c^4d^2 - 540c^3d^3 + 135c^2d^4 - 18cd^5 + d^6$$

Preceding Coefficient Method: Expand $(3c - d)^6$.

We find each term in sequence. There are 7 terms in the expansion. When we get to the 4th coefficient, we can use symmetry to obtain the others.

The 1st term is $(3c)^6$.

The 2nd term is $\frac{1\cdot6}{2-1}(3c)^5(-d)$, or $6(3c)^5(-d)$.

The 3rd term is $\frac{6\cdot5}{3-1}(3c)^4(-d)^2$, or $15(3c)^4(-d)^2$.

The 4th term is $\frac{15\cdot4}{4-1}(3c)^3(-d)^3$, or $20(3c)^3(-d)^3$.

The rest of the coefficients are 15, 6, and 1.

The complete expansion is

$$(3c - d)^6 = (3c)^6 + 6(3c)^5(-d) + 15(3c)^4(-d)^2 + 20(3c)^3(-d)^3 + 15(3c)^2(-d)^4 + 6(3c)(-d)^5 + (-d)^6$$
$$= 729c^6 + 6(243c^5)(-d) + 15(81c^4)(d^2) + 20(27c^3)(-d)^3 + 15(9c^2)(d^4) + 6(3c)(-d)^5 + d^6$$
$$= 729c^6 - 1458c^5d + 1215c^4d^2 - 540c^3d^3 + 135c^2d^4 - 18cd^5 + d^6$$

8. $t^{-12} + 12t^{-10} + 60t^{-8} + 160t^{-6} + 240t^{-4} + 192t^{-2} + 64$

9. Pascal Method: Expand $(x - y)^3$.

The expansion of $(x - y)^3$ has 3 + 1, or 4 terms. The sum of the exponents in each term is 3. The exponents of x start with 3 and decrease to 0. The last term has no factor of x. The first term has no factor of -y. The exponents of -y start in the second term with 1 and increase to 3. We get the coefficients from the 4th row of Pascal's Triangle.

1
1 1
1 2 1
1 3 3 1

$$(x - y)^3 = 1\cdot x^3 + 3\cdot x^2\cdot(-y) + 3\cdot x\cdot(-y)^2 + 1\cdot(-y)^3$$
$$= x^3 - 3x^2y + 3xy^2 - y^3$$

Preceding Coefficient Method: Expand $(x - y)^3$.

We find each term in sequence. There are 4 terms in the expansion.

9. (continued)

The 1st term is x^3.

The 2nd term is $\frac{1\cdot 3}{2-1}x^2(-y)$, or $3x^2(-y)$.

The 3rd term is $\frac{3\cdot 2}{3-1}x(-y)^2$, or $3x(-y)^2$.

The 4th term is $\frac{3\cdot 1}{4-1}(-y)^3$, or $(-y)^3$.

The complete expansion is
$(x - y)^3 = x^3 - 3x^2y + 3xy^2 - y^3$.

10. $x^5 - 5x^4y + 10x^3y^2 - 10x^2y^3 + 5xy^4 - y^5$

11. Preceding Coefficient Method: Expand $\left(\frac{1}{x} + y\right)^7$.

We find each term in sequence. There are 8 terms in the expansion. When we get to the 4th coefficient, we can use symmetry to obtain the others.

The 1st term is $\left(\frac{1}{x}\right)^7$.

The 2nd term is $\frac{1\cdot 7}{2-1}\left(\frac{1}{x}\right)^6 y$, or $7\left(\frac{1}{x}\right)^6 y$.

The 3rd term is $\frac{7\cdot 6}{3-1}\left(\frac{1}{x}\right)^5 y^2$, or $21\left(\frac{1}{x}\right)^5 y^2$.

The 4th term is $\frac{21\cdot 5}{4-1}\left(\frac{1}{x}\right)^4 y^3$, or $35\left(\frac{1}{x}\right)^4 y^3$.

The rest of the coefficients are 35, 21, 7 and 1.

The complete expansion is

$$\left(\frac{1}{x} + y\right)^7 = \left(\frac{1}{x}\right)^7 + 7\left(\frac{1}{x}\right)^6 y + 21\left(\frac{1}{x}\right)^5 y^2 + 35\left(\frac{1}{x}\right)^4 y^3 + 35\left(\frac{1}{x}\right)^3 y^4 + 21\left(\frac{1}{x}\right)^2 y^5 + 7\left(\frac{1}{x}\right) y^6 + y^7$$

$$= x^{-7} + 7x^{-6}y + 21x^{-5}y^2 + 35x^{-4}y^3 + 35x^{-3}y^4 + 21x^{-2}y^5 + 7x^{-1}y^6 + y^7$$

Binomial Theorem Method: Expand $\left(\frac{1}{x} + y\right)^7$.

Note that $a = \frac{1}{x}$, $b = y$, and $n = 7$.

$$\left(\frac{1}{x} + y\right)^7 = \binom{7}{0}\left(\frac{1}{x}\right)^7 + \binom{7}{1}\left(\frac{1}{x}\right)^6 y + \binom{7}{2}\left(\frac{1}{x}\right)^5 y^2 + \binom{7}{3}\left(\frac{1}{x}\right)^4 y^3 + \binom{7}{4}\left(\frac{1}{x}\right)^3 y^4 + \binom{7}{5}\left(\frac{1}{x}\right)^2 y^5 + \binom{7}{6}\left(\frac{1}{x}\right) y^6 + \binom{7}{7} y^7$$

$$= \frac{7!}{7!0!}\left(\frac{1}{x}\right)^7 + \frac{7!}{6!1!}\left(\frac{1}{x}\right)^6 y + \frac{7!}{5!2!}\left(\frac{1}{x}\right)^5 y^2 + \frac{7!}{4!3!}\left(\frac{1}{x}\right)^4 y^3 + \frac{7!}{3!4!}\left(\frac{1}{x}\right)^3 y^4 + \frac{7!}{2!5!}\left(\frac{1}{x}\right)^2 y^5 + \frac{7!}{1!6!}\left(\frac{1}{x}\right) y^6 + \frac{7!}{0!7!} y^7$$

$$= x^{-7} + 7x^{-6}y + 21x^{-5}y^2 + 35x^{-4}y^3 + 35x^{-3}y^4 + 21x^{-2}y^5 + 7x^{-1}y^6 + y^7$$

12. $8s^3 - 36s^2t^2 + 54st^4 - 27t^6$

13. $\left(a - \frac{2}{a}\right)^9$

Let $a = a$, $b = -\frac{2}{a}$, and $n = 9$.

Expand using the Binomial Theorem Method.

$$\left(a - \frac{2}{a}\right)^9 = \binom{9}{0}a^9 + \binom{9}{1}a^8\left(-\frac{2}{a}\right) + \binom{9}{2}a^7\left(-\frac{2}{a}\right)^2 + \binom{9}{3}a^6\left(-\frac{2}{a}\right)^3 + \binom{9}{4}a^5\left(-\frac{2}{a}\right)^4 + \binom{9}{5}a^4\left(-\frac{2}{a}\right)^5 + \binom{9}{6}a^3\left(-\frac{2}{a}\right)^6 + \binom{9}{7}a^2\left(-\frac{2}{a}\right)^7 + \binom{9}{8}a\left(-\frac{2}{a}\right)^8 + \binom{9}{9}\left(-\frac{2}{a}\right)^9$$

$$= \frac{9!}{9!0!}a^9 + \frac{9!}{8!1!}a^8\left(-\frac{2}{a}\right) + \frac{9!}{7!2!}a^7\left(\frac{4}{a^2}\right) + \frac{9!}{6!3!}a^6\left(-\frac{8}{a^3}\right) + \frac{9!}{5!4!}a^5\left(\frac{16}{a^4}\right) + \frac{9!}{4!5!}a^4\left(-\frac{32}{a^5}\right) + \frac{9!}{3!6!}a^3\left(\frac{64}{a^6}\right) + \frac{9!}{2!7!}a^2\left(-\frac{128}{a^7}\right) + \frac{9!}{1!8!}a\left(\frac{256}{a^8}\right) + \frac{9!}{0!9!}\left(-\frac{512}{a^9}\right)$$

$$= a^9 - 9(2a^7) + 36(4a^5) - 84(8a^3) + 126(16a) - 126(32a^{-1}) + 84(64a^{-3}) - 36(128a^{-5}) + 9(256a^{-7}) - 512a^{-9}$$

$$= a^9 - 18a^7 + 144a^5 - 672a^3 + 2016a - 4032a^{-1} + 5376a^{-3} + 4608a^{-5} + 2304a^{-7} - 512a^{-9}$$

14. $512x^9 + 2304x^7 + 4608x^5 + 5376x^3 + 4032x + 2016x^{-1} + 672x^{-3} + 144x^{-5} + 18x^{-7} + x^{-9}$

15. $(1 - 1)^n$

Let $a = 1$ and $b = -1$.

Expand using the Binomial Theorem Method:

$$(1 - 1)^n = \binom{n}{0}1^n + \binom{n}{1}1^{n-1}(-1) + \binom{n}{2}1^{n-2}(-1)^2 + \binom{n}{3}1^{n-3}(-1)^3 + \ldots + \binom{n}{n}(-1)^n$$

$$= 1\cdot 1 + n\cdot 1\cdot(-1) + \binom{n}{2}\cdot 1\cdot 1 + \binom{n}{3}\cdot 1\cdot(-1) + \ldots + \binom{n}{n}(-1)^n$$

$$= 1 - n + \binom{n}{2} - \binom{n}{3} + \ldots + \binom{n}{n}(-1)^n$$

16. $1 + 3n + \binom{n}{2}3^2 + \binom{n}{3}3^3 + \ldots + \binom{n}{n-2}3^{n-2} + \binom{n}{n-1}3^{n-1} + 3^n$

17. $(\sqrt{3} - t)^4$

Let $a = \sqrt{3}$, $b = -t$, and $n = 4$.

Expand using the Binomial Theorem Method.

$$(\sqrt{3} - t)^4 = \binom{4}{0}(\sqrt{3})^4 + \binom{4}{1}(\sqrt{3})^3(-t) + \binom{4}{2}(\sqrt{3})^2(-t)^2 + \binom{4}{3}(\sqrt{3})(-t)^3 + \binom{4}{4}(-t)^4$$

$$= 1\cdot 9 + 4\cdot 3\sqrt{3}\cdot(-t) + 6\cdot 3\cdot t^2 + 4\cdot\sqrt{3}\cdot(-t^3) + 1\cdot t^4$$

$$= 9 - 12\sqrt{3}\,t + 18t^2 - 4\sqrt{3}\,t^3 + t^4$$

18. $125 + 150\sqrt{5}\,t + 375t^2 + 100\sqrt{5}\,t^3 + 75t^4 + 6\sqrt{5}\,t^5 + t^6$

19. $6! = 6\cdot 5\cdot 4\cdot 3\cdot 2\cdot 1 = 720$

20. 24

21. $1! = 1$

22. 1

23. $\binom{5}{0} = \frac{5!}{(5-0)!0!} = \frac{5!}{5!\cdot 1} = 1$

24. 7

25. $\binom{8}{4} = \frac{8!}{(8-4)!4!} = \frac{8!}{4!4!} = \frac{8\cdot 7\cdot 6\cdot 5\cdot 4!}{4\cdot 3\cdot 2\cdot 1\cdot 4!}$

$$= \frac{8\cdot 7\cdot 6\cdot 5}{4\cdot 3\cdot 2\cdot 1}$$

$$= 70$$

26. 9!

27. Find the 3rd term of $(a + b)^6$.

First, we note that $3 = 2 + 1$, $a = a$, $b = b$, and $n = 6$. Then the 3rd term of the expansion of $(a + b)^6$ is

$\binom{6}{2}a^{6-2}b^2$, or $\frac{6!}{4!2!}a^4b^2$, or $15a^4b^2$.

28. $21x^2y^5$

29. Find the 12th term of $(a - 2)^{14}$.

First, we note that $12 = 11 + 1$, $a = a$, $b = -2$, and $n = 14$. Then the 12th term of the expansion of $(a - 2)^{14}$ is

$$\binom{14}{11} a^{14-11}\cdot(-2)^{11} = \frac{14!}{3!11!}a^3(-2048)$$

$$= 364a^3(-2048)$$

$$= -745{,}472a^3$$

30. $3{,}897{,}234x^2$

31. Find the 5th term of $(2x^3 - \sqrt{y})^8$.

First, we note that $5 = 4 + 1$, $a = 2x^3$, $b = -\sqrt{y}$, and $n = 8$. Then the 5th term of the expansion of $(2x^3 - \sqrt{y})^8$ is

$$\binom{8}{4}(2x^3)^{8-4}(-\sqrt{y})^4$$

$$= \frac{8!}{4!4!}(2x^3)^4(-\sqrt{y})^4$$

$$= 70(16x^{12})(y^2)$$

$$= 1120x^{12}y^2$$

32. $\frac{35}{27}b^{-5}$

33. The middle term of the expansion of $(2u - 3v^2)^{10}$ is the 6th term. Note that $6 = 5 + 1$, $a = 2u$, $b = -3v^2$, and $n = 10$. Then the 6th term of the expansion of $(2u - 3v^2)^{10}$ is

$$\binom{10}{5}(2u)^{10-5}(-3v^2)^5$$

$$= \frac{10!}{5!5!}(2u)^5(-3v^2)^5$$

$$= 252(32u^5)(-243v^{10})$$

$$= -1{,}959{,}552u^5v^{10}$$

34. $30x\sqrt{x}$, $30x\sqrt{3}$

35. Expand $(x^{-2} + x^2)^4$ using the Binomial Theorem.

Note that $a = x^{-2}$, $b = x^2$, and $n = 4$.

$$(x^{-2} + x^2)^4 = \binom{4}{0}(x^{-2})^4 + \binom{4}{1}(x^{-2})^3(x^2) + \binom{4}{2}(x^{-2})^2(x^2)^2 + \binom{4}{3}(x^{-2})(x^2)^3 + \binom{4}{4}(x^2)^4$$

$$= \frac{4!}{4!0!}(x^{-8}) + \frac{4!}{3!1!}(x^{-6})(x^2) + \frac{4!}{2!2!}(x^{-4})(x^4) + \frac{4!}{1!3!}(x^{-2})(x^6) + \frac{4!}{0!4!}(x^8)$$

$$= x^{-8} + 4x^{-4} + 6x^0 + 4x^4 + x^8$$

$$= x^{-8} + 4x^{-4} + 6 + 4x^4 + x^8$$

36. $x^{-3} - 6x^{-2} + 15x^{-1} - 20 + 15x - 6x^2 + x^3$

37. $(\sqrt{2} + 1)^6 - (\sqrt{2} - 1)^6$

First, expand $(\sqrt{2} + 1)^6$.

$$(\sqrt{2} + 1)^6 = \binom{6}{0}(\sqrt{2})^6 + \binom{6}{1}(\sqrt{2})^5(1) + \binom{6}{2}(\sqrt{2})^4(1)^2 + \binom{6}{3}(\sqrt{2})^3(1)^3 + \binom{6}{4}(\sqrt{2})^2(1)^4 + \binom{6}{5}(\sqrt{2})(1)^5 + \binom{6}{6}(1)^6$$

$$= \frac{6!}{6!0!} \cdot 8 + \frac{6!}{5!1!} \cdot 4\sqrt{2} + \frac{6!}{4!2!} \cdot 4 + \frac{6!}{3!3!} \cdot 2\sqrt{2} + \frac{6!}{2!4!} \cdot 2 + \frac{6!}{1!5!} \cdot \sqrt{2} + \frac{6!}{0!6!}$$

$$= 8 + 24\sqrt{2} + 60 + 40\sqrt{2} + 30 + 6\sqrt{2} + 1$$

$$= 99 + 70\sqrt{2}$$

Next, expand $(\sqrt{2} - 1)^6$.

$$(\sqrt{2} - 1)^6 = \binom{6}{0}(\sqrt{2})^6 + \binom{6}{1}(\sqrt{2})^5(-1) + \binom{6}{2}(\sqrt{2})^4(-1)^2 + \binom{6}{3}(\sqrt{2})^3(-1)^3 + \binom{6}{4}(\sqrt{2})^2(-1)^4 + \binom{6}{5}(\sqrt{2})(-1)^5 + \binom{6}{6}(-1)^6$$

$$= \frac{6!}{6!0!} \cdot 8 - \frac{6!}{5!1!} \cdot 4\sqrt{2} + \frac{6!}{4!2!} \cdot 4 - \frac{6!}{3!3!} \cdot 2\sqrt{2} + \frac{6!}{2!4!} \cdot 2 - \frac{6!}{1!5!} \cdot \sqrt{2} + \frac{6!}{0!6!}$$

$$= 8 - 24\sqrt{2} + 60 - 40\sqrt{2} + 30 - 6\sqrt{2} + 1$$

$$= 99 - 70\sqrt{2}$$

$$(\sqrt{2} + 1)^6-(\sqrt{2} - 1)^6 = (99 + 70\sqrt{2})-(99 - 70\sqrt{2})$$

$$= 99 + 70\sqrt{2} - 99 + 70\sqrt{2}$$

$$= 140\sqrt{2}$$

38. 34

39. $$\binom{16}{11} = \frac{16!}{5!11!} = \frac{16\cdot15\cdot14\cdot13\cdot12\cdot11!}{5\cdot4\cdot3\cdot2\cdot1\cdot11!}$$

$$= \frac{16\cdot15\cdot14\cdot13\cdot12}{5\cdot4\cdot3\cdot2\cdot1}$$

$$= 4368$$

$$\binom{16}{5} = \frac{16!}{11!5!} = \frac{16!}{5!11!} = 4368$$

40. Use the definition:

$$\binom{n}{r} = \frac{n!}{(n - r)!r!}$$

$$\binom{n}{n-r} = \frac{n!}{[n - (n - r)]!(n - r)!}$$

$$= \frac{n!}{[n - n + r]!(n - r)!}$$

$$= \frac{n!}{r!(n - r)!}$$

$$= \binom{n}{r}$$

41. The $(r + 1)$st term of $\left(\frac{3x^2}{2} - \frac{1}{3x}\right)^{12}$ is $\binom{12}{r}\left(\frac{3x^2}{2}\right)^{12-r}\left(- \frac{1}{3x}\right)^r$. In the term which does not contain x, the exponent of x in the numerator is equal to the exponent of x in the denominator. That is,

$$2(12 - r) = r$$
$$24 - 2r = r$$
$$24 = 3r$$
$$8 = r$$

Find the 9th term:

$$\binom{12}{8}\left(\frac{3x^2}{2}\right)^{12-8}\left(- \frac{1}{3x}\right)^8$$

$$= \frac{12!}{4!8!} \cdot \frac{3^4x^8}{2^4} \cdot \frac{1}{3^8x^8} = \frac{495}{2^4\cdot3^4} = \frac{55}{144}$$

42. $90{,}720x^8y^6$

43. The ratio of the fourth term to the third term is

$$\frac{\binom{5}{3}(a^2)^{5-3}\left(- \frac{1}{2} a \sqrt[3]{c}\right)^3}{\binom{5}{2}(a^2)^{5-2}\left(- \frac{1}{2} a \sqrt[3]{c}\right)^2}$$

$$= \frac{\frac{5!}{2!3!}(a^2)^2\left(- \frac{1}{2} a \sqrt[3]{c}\right)^3}{\frac{5!}{3!2!}(a^2)^3\left(- \frac{1}{2} a \sqrt[3]{c}\right)^2} = \frac{- \frac{1}{2} a \sqrt[3]{c}}{a^2}$$

$$= - \frac{\sqrt[3]{c}}{2a}$$

44. $- \frac{35}{x^{1/6}}$

45. The term of highest degree of $(x^5 + 3)^4$ is the first term, or

$\binom{4}{0}(x^5)^{4-0}3^0 = \frac{4!}{4!0!}x^{20} = x^{20}$.

Therefore, the degree of $(x^5 + 3)^4$ is 20.

Exercise Set 1.6

1. $18a^2b - 15ab^2$
$= 3ab \cdot 6a - 3ab \cdot 5b$
$= 3ab(6a - 5b)$

2. $4xy(x + 3y)$

3. $a(b - 2) + c(b - 2)$
$= (a + c)(b - 2)$

4. $(a - 2)(x^2 - 3)$

5. $x^2 + 3x + 6x + 18$
$= x(x + 3) + 6(x + 3)$
$= (x + 6)(x + 3)$

6. $(x^2 - 6)(3x + 1)$

7. $9x^2 - 25$
$= (3x)^2 - 5^2$
$= (3x + 5)(3x - 5)$

8. $(4x - 3)(4x + 3)$

9. $4xy^4 - 4xz^2$
$= 4x(y^4 - z^2)$
$= 4x[(y^2)^2 - z^2]$
$= 4x(y^2 + z)(y^2 - z)$

10. $5x(y^2 + z^2)(y + z)(y - z)$

11. $y^2 - 6y + 9$
$= y^2 - 2 \cdot y \cdot 3 + 3^2$
$= (y - 3)^2$

12. $(x + 4)^2$

13. $1 - 8x + 16x^2$
$= 1^2 - 2 \cdot 1 \cdot 4x + (4x)^2$
$= (1 - 4x)^2$

14. $(1 + 5x)^2$

15. $4x^2 - 5$
$= (2x)^2 - (\sqrt{5})^2$
$= (2x + \sqrt{5})(2x - \sqrt{5})$

16. $(4x - \sqrt{7})(4x + \sqrt{7})$

17. $x^2y^2 - 14xy + 49$
$= (xy)^2 - 2 \cdot xy \cdot 7 + 7^2$
$= (xy - 7)^2$

18. $(xy - 8)^2$

19. $4ax^2 + 20ax - 56a$
$= 4a(x^2 + 5x - 14)$
$= 4a(x + 7)(x - 2)$

20. $y(7x - 4)(3x + 2)$

21. $a^2 + 2ab + b^2 - c^2$
$= (a + b)^2 - c^2$
$= [(a + b) + c][(a + b) - c]$
$= (a + b + c)(a + b - c)$

22. $(x - y + z)(x - y - z)$

23. $x^2 + 2xy + y^2 - a^2 - 2ab - b^2$
$= (x^2 + 2xy + y^2) - (a^2 + 2ab + b^2)$
$= (x + y)^2 - (a + b)^2$
$= [(x + y) + (a + b)][(x + y) - (a + b)]$
$= (x + y + a + b)(x + y - a - b)$

24. $(r + s + t - v)(r + s - t + v)$

25. $5y^4 - 80x^4$
$= 5(y^4 - 16x^4)$
$= 5(y^2 + 4x^2)(y^2 - 4x^2)$
$= 5(y^2 + 4x^2)(y + 2x)(y - 2x)$

26. $6(y^2 + 4x^2)(y - 2x)(y + 2x)$

27. $x^3 + 8$
$= x^3 + 2^3$
$= (x + 2)(x^2 - 2x + 4)$

28. $(y - 4)(y^2 + 4y + 16)$

29. $3x^3 - \frac{3}{8}$
$= 3\left(x^3 - \frac{1}{8}\right)$
$= 3\left(x - \frac{1}{2}\right)\left(x^2 + \frac{1}{2}x + \frac{1}{4}\right)$

30. $5\left(y + \frac{1}{3}\right)\left(y^2 - \frac{1}{3}y + \frac{1}{9}\right)$

31. $x^3 + 0.001$
$= x^3 + (0.1)^3$
$= (x + 0.1)(x^2 - 0.1x + 0.01)$

32. $(y - 0.5)(y^2 + 0.5y + 0.25)$

33. $3z^3 - 24$
$= 3(z^3 - 8)$
$= 3(z^3 - 2^3)$
$= 3(z - 2)(z^2 + 2z + 4)$

34. $4(t + 3)(t^2 - 3t + 9)$

35. $a^6 - t^6$
$= (a^3)^2 - (t^3)^2$
$= (a^3 + t^3)(a^3 - t^3)$
$= (a + t)(a^2 - at + t^2)(a - t)(a^2 + at + t^2)$

36. $(4m^2 + y^2)(16m^4 - 4m^2y^2 + y^4)$

37. $16a^7b + 54ab^7$
$= 2ab(8a^6 + 27b^6)$
$= 2ab[(2a^2)^3 + (3b^2)^3]$
$= 2ab(2a^2 + 3b^2)(4a^4 - 6a^2b^2 + 9b^4)$

38. $3a^2x(2x - 5a^2)(4x^2 + 10a^2x + 25a^4)$

39. $x^2 - 17.6$
$= x^2 - (\sqrt{17.6})^2$
$= (x + \sqrt{17.6})(x - \sqrt{17.6})$
$= (x + 4.195235)(x - 4.195235)$

40. $(x + 2.8337)(x - 2.8337)$

41. $37x^2 - 14.5y^2$
$= (\sqrt{37}\,x)^2 - (\sqrt{14.5}\,y)^2$
$= (\sqrt{37}\,x + \sqrt{14.5}\,y)(\sqrt{37}\,x - \sqrt{14.5}\,y)$
$= (6.082763x + 3.807887y)(6.082763x - 3.807887y)$

42. $1.96(x + 2.980y)(x - 2.980y)$

43. $(x + h)^3 - x^3$
$= [(x + h) - x][(x + h)^2 + x(x + h) + x^2]$
$= (x + h - x)(x^2 + 2xh + h^2 + x^2 + xh + x^2)$
$= h(3x^2 + 3xh + h^2)$

44. $0.02(x + 0.005)$

45. $y^4 - 84 + 5y^2$
$= y^4 + 5y^2 - 84$
$= (y^2)^2 + 5y^2 - 84$
$= (y^2 + 12)(y^2 - 7)$

46. $(x^2 + 16)(x^2 - 5)$, or $(x^2 + 16)(x + \sqrt{5})(x - \sqrt{5})$

47. $y^2 - \frac{8}{49} + \frac{2}{7}y$
$= y^2 + \frac{2}{7}y - \frac{8}{49}$
$= \left(y + \frac{4}{7}\right)\left(y - \frac{2}{7}\right)$

48. $\left(x + \frac{4}{5}\right)\left(x - \frac{1}{5}\right)$, or $\frac{1}{25}(5x + 4)(5x - 1)$

49. $t^2 - 0.27 + 0.6t$
$= t^2 + 0.6t - 0.27$
$= (t + 0.9)(t - 0.3)$

50. $(m - 0.1)(m + 0.5)$

51. $x^{2n} + 5x^n - 24$
$= (x^n)^2 + 5x^n - 24$
$= (x^n + 8)(x^n - 3)$

52. $(2x^n - 3)(2x^n + 1)$

53. $x^2 + ax + bx + ab$
$= x(x + a) + b(x + a)$
$= (x + b)(x + a)$

54. $(by + a)(dy + c)$

55. $\frac{1}{4}t^2 - \frac{2}{5}t + \frac{4}{25}$
$= \left(\frac{1}{2}t\right)^2 - 2 \cdot \frac{1}{2}t \cdot \frac{2}{5} + \left(\frac{2}{5}\right)^2$
$= \left(\frac{1}{2}t - \frac{2}{5}\right)^2$

56. $\frac{1}{3}\left(\frac{2}{3}r + \frac{1}{2}s\right)^2$, or $\frac{1}{108}(4r + 3s)^2$,
or $\left(\frac{1}{9}r + \frac{1}{12}s\right)\left(\frac{4}{3}r + s\right)$

57. $25y^{2m} - (x^{2n} - 2x^n + 1)$
$= (5y^m)^2 - (x^n - 1)^2$
$= [5y^m + (x^n - 1)][5y^m - (x^n - 1)]$
$= (5y^m + x^n - 1)(5y^m - x^n + 1)$

58. $2(x^{2a} + 3)(2x^{2a} + 5)$

59. $3x^{3n} - 24y^{3m}$

$= 3(x^{3n} - 8y^{3m})$

$= 3[(x^n)^3 - (2y^m)^3]$

$= 3(x^n - 2y^m)(x^{2n} + 2x^ny^m + 4y^{2m})$

60. $(x^{2a} - t^b)(x^{4a} + x^{2a}t^b + t^{2b})$

61. $(y - 1)^4 - (y - 1)^2$

$= (y - 1)^2[(y - 1)^2 - 1]$

$= (y - 1)^2[y^2 - 2y + 1 - 1]$

$= (y - 1)^2(y^2 - 2y)$

$= y(y - 1)^2(y - 2)$

62. $(x + 1)(x^2 + 1)(x - 1)^3$

Exercise Set 1.7

1. $\dfrac{3x - 2}{x(x - 1)}$

To determine the sensible replacements, we set the denominator equal to 0 and solve:

$x(x - 1) = 0$

$x = 0$ or $x - 1 = 0$

$x = 0$ or $x = 1$

Thus 0 and 1 are not sensible replacements. All real numbers except 0 and 1 are sensible replacements.

2. All real numbers except -2, 1, and -1

3. $\dfrac{7y^2 - 2y + 4}{x(x^2 - x - 6)}$

To determine the sensible replacements, we set the denominator equal to 0 and solve:

$x(x^2 - x - 6) = 0$

$x(x - 3)(x + 2) = 0$

$x = 0$ or $x - 3 = 0$ or $x + 2 = 0$

$x = 0$ or $x = 3$ or $x = -2$

Thus 0, 3, and -2 are not sensible replacements. All real numbers except 0, 3, and -2 are sensible replacements.

4. $\dfrac{5x}{2}$; All real numbers

5. $\dfrac{x^2 - 4}{x^2 + 5x + 6} = \dfrac{(x + 2)(x - 2)}{(x + 2)(x + 3)} = \dfrac{x - 2}{x + 3}$

To determine the sensible replacements in the simplified expression, we set the denominator equal to 0 and solve:

$x + 3 = 0$

$x = -3$

Thus -3 is not a sensible replacement. All real numbers except -3 are sensible replacements in the simplified expression.

6. $\dfrac{x - 2}{x + 2}$; All real numbers except -2

7. $\dfrac{x^2 - y^2}{(x - y)^2} \cdot \dfrac{1}{x + y}$

$= \dfrac{(x + y)(x - y)\cdot 1}{(x - y)(x - y)(x + y)}$

$= \dfrac{1}{x - y}$

8. 1

9. $\dfrac{x^2 - 2x - 35}{2x^3 - 3x^2} \cdot \dfrac{4x^3 - 9x}{7x - 49}$

$= \dfrac{(x - 7)(x + 5)(x)(2x + 3)(2x - 3)}{x^2(2x - 3)(7)(x - 7)}$

$= \dfrac{(x + 5)(2x + 3)}{7x}$, or $\dfrac{2x^2 + 13x + 15}{7x}$

10. $\dfrac{(x - 5)(3x + 2)}{7x}$, or $\dfrac{3x^2 - 13x - 10}{7x}$

11. $\dfrac{a^2 - a - 6}{a^2 - 7a + 12} \cdot \dfrac{a^2 - 2a - 8}{a^2 - 3a - 10}$

$= \dfrac{(a - 3)(a + 2)(a - 4)(a + 2)}{(a - 4)(a - 3)(a - 5)(a + 2)}$

$= \dfrac{a + 2}{a - 5}$

12. $\dfrac{(a + 3)^2}{(a - 6)(a + 4)}$, or $\dfrac{a^2 + 6a + 9}{a^2 - 2a - 24}$

13. $\dfrac{m^2 - n^2}{r + s} \div \dfrac{m - n}{r + s}$

$= \dfrac{m^2 - n^2}{r + s} \cdot \dfrac{r + s}{m - n}$

$= \dfrac{(m + n)(m - n)(r + s)}{(r + s)(m - n)}$

$= m + n$

14. $a - b$

15. $\frac{3x + 12}{2x - 8} \div \frac{(x + 4)^2}{(x - 4)^2}$

$= \frac{3x + 12}{2x - 8} \cdot \frac{(x - 4)^2}{(x + 4)^2}$

$= \frac{3(x + 4)(x - 4)(x - 4)}{2(x - 4)(x + 4)(x + 4)}$

$= \frac{3(x - 4)}{2(x + 4)}$, or $\frac{3x - 12}{2x + 8}$

16. $\frac{a + 1}{a - 3}$

17. $\frac{x^2 - y^2}{x^3 - y^3} \cdot \frac{x^2 + xy + y^2}{x^2 + 2xy + y^2}$

$= \frac{(x + y)(x - y)(x^2 + xy + y^2)}{(x - y)(x^2 + xy + y^2)(x + y)(x + y)}$

$= \frac{1}{x + y}$

18. c - 2

19. $\frac{(x - y)^2 - z^2}{(x + y)^2 - z^2} \div \frac{x - y + z}{x + y - z}$

$= \frac{(x - y)^2 - z^2}{(x + y)^2 - z^2} \cdot \frac{x + y - z}{x - y + z}$

$= \frac{(x - y + z)(x - y - z)(x + y - z)}{(x + y + z)(x + y - z)(x - y + z)}$

$= \frac{x - y - z}{x + y + z}$

20. $\frac{a + b - 3}{a - b + 3}$

21. $\frac{3}{2a + 3} + \frac{2a}{2a + 3}$

$= \frac{3 + 2a}{2a + 3}$

$= 1$

22. 2

23. $\frac{y}{y - 1} + \frac{2}{1 - y}$

$= \frac{y}{y - 1} + \frac{-1}{-1} \cdot \frac{2}{1 - y}$

$= \frac{y}{y - 1} + \frac{-2}{y - 1}$

$= \frac{y - 2}{y - 1}$

24. 1

25. $\frac{x}{2x - 3y} - \frac{y}{3y - 2x}$

$= \frac{x}{2x - 3y} - \frac{-1}{-1} \cdot \frac{y}{3y - 2x}$

$= \frac{x}{2x - 3y} - \frac{-y}{2x - 3y}$

$= \frac{x + y}{2x - 3y}$ [x - (-y) = x + y]

26. $\frac{5a}{3a - 2b}$

27. $\frac{3}{x + 2} + \frac{2}{x^2 - 4}$

$= \frac{3}{x + 2} + \frac{2}{(x + 2)(x - 2)}$, LCM = (x + 2)(x - 2)

$= \frac{3}{x + 2} \cdot \frac{x - 2}{x - 2} + \frac{2}{(x + 2)(x - 2)}$

$= \frac{3x - 6}{(x + 2)(x - 2)} + \frac{2}{(x + 2)(x - 2)}$

$= \frac{3x - 4}{(x + 2)(x - 2)}$

28. $\frac{5a + 13}{(a + 3)(a - 3)}$

29. $\frac{y}{y^2 - y - 20} + \frac{2}{y + 4}$

$= \frac{y}{(y + 4)(y - 5)} + \frac{2}{y + 4}$, LCM = (y + 4)(y - 5)

$= \frac{y}{(y + 4)(y - 5)} + \frac{2}{y + 4} \cdot \frac{y - 5}{y - 5}$

$= \frac{y}{(y + 4)(y - 5)} + \frac{2y - 10}{(y + 4)(y - 5)}$

$= \frac{3y - 10}{(y + 4)(y - 5)}$

30. $\frac{-5y - 9}{(y + 3)^2}$

31. $\frac{3}{x + y} + \frac{x - 5y}{x^2 - y^2}$

$= \frac{3}{x + y} + \frac{x - 5y}{(x + y)(x - y)}$, LCM = (x + y)(x - y)

$= \frac{3}{x + y} \cdot \frac{x - y}{x - y} + \frac{x - 5y}{(x + y)(x - y)}$

$= \frac{3x - 3y}{(x + y)(x - y)} + \frac{x - 5y}{(x + y)(x - y)}$

$= \frac{4x - 8y}{(x + y)(x - y)}$

32. $\frac{2a}{(a + 1)(a - 1)}$

33. $\frac{9x + 2}{3x^2 - 2x - 8} + \frac{7}{3x^2 + x - 4}$

$= \frac{9x + 2}{(3x + 4)(x - 2)} + \frac{7}{(3x + 4)(x - 1)}$,

LCM = (3x + 4)(x - 2)(x - 1)

$= \frac{9x + 2}{(3x + 4)(x - 2)} \cdot \frac{x - 1}{x - 1} + \frac{7}{(3x + 4)(x - 1)} \cdot \frac{x-2}{x-2}$

$= \frac{9x^2 - 7x - 2}{(3x + 4)(x - 2)(x - 1)} + \frac{7x - 14}{(3x + 4)(x - 1)(x - 2)}$

$= \frac{9x^2 - 16}{(3x + 4)(x - 2)(x - 1)}$

$= \frac{(3x + 4)(3x - 4)}{(3x + 4)(x - 2)(x - 1)}$

$= \frac{3x - 4}{(x - 2)(x - 1)}$

34. $\dfrac{y}{(y - 2)(y - 3)}$

35. $$\frac{5a}{a - b} + \frac{ab}{a^2 - b^2} + \frac{4b}{a + b}$$
$$= \frac{5a}{a - b} + \frac{ab}{(a + b)(a - b)} + \frac{4b}{a + b},$$
$$\text{LCM} = (a + b)(a - b)$$
$$= \frac{5a}{a - b} \cdot \frac{a + b}{a + b} + \frac{ab}{(a + b)(a - b)} + \frac{4b}{a + b} \cdot \frac{a - b}{a - b}$$
$$= \frac{5a^2 + 5ab}{(a + b)(a - b)} + \frac{ab}{(a + b)(a - b)} + \frac{4ab - 4b^2}{(a + b)(a - b)}$$
$$= \frac{5a^2 + 10ab - 4b^2}{(a + b)(a - b)}$$

36. $\dfrac{6a^2 + 9ab + 3b^2 + 5}{(a + b)(a - b)}$

37. $$\frac{7}{x + 2} - \frac{x + 8}{4 - x^2} + \frac{3x - 2}{4 - 4x + x^2}$$
$$= \frac{7}{x + 2} - \frac{-1}{-1} \cdot \frac{x + 8}{4 - x^2} + \frac{3x - 2}{x^2 - 4x + 4}$$
$$= \frac{7}{x + 2} - \frac{-x - 8}{x^2 - 4} + \frac{3x - 2}{x^2 - 4x + 4}$$
$$= \frac{7}{x + 2} - \frac{-x - 8}{(x + 2)(x - 2)} + \frac{3x - 2}{(x - 2)^2},$$
$$\text{LCM} = (x + 2)(x - 2)^2$$
$$= \frac{7}{x + 2} \cdot \frac{(x - 2)^2}{(x - 2)^2} + \frac{x + 8}{(x + 2)(x - 2)} \cdot \frac{x - 2}{x - 2} + \frac{3x - 2}{(x - 2)^2} \cdot \frac{x + 2}{x + 2}$$
$$= \frac{7x^2 - 28x + 28}{(x + 2)(x - 2)^2} + \frac{x^2 + 6x - 16}{(x + 2)(x - 2)^2} + \frac{3x^2 + 4x - 4}{(x + 2)(x - 2)^2}$$
$$= \frac{11x^2 - 18x + 8}{(x + 2)(x - 2)^2}$$

38. $\dfrac{33 - 32x + 9x^2}{(3 + x)(3 - x)^2}$

39. $$\frac{1}{x + 1} - \frac{x}{x - 2} + \frac{x^2 + 2}{x^2 - x - 2}$$
$$= \frac{1}{x + 1} - \frac{x}{x - 2} + \frac{x^2 + 2}{(x + 1)(x - 2)},$$
$$\text{LCM} = (x + 1)(x - 2)$$
$$= \frac{1}{x + 1} \cdot \frac{x - 2}{x - 2} - \frac{x}{x - 2} \cdot \frac{x + 1}{x + 1} + \frac{x^2 + 2}{(x + 1)(x - 2)}$$
$$= \frac{x - 2}{(x + 1)(x - 2)} - \frac{x^2 + x}{(x + 1)(x - 2)} + \frac{x^2 + 2}{(x + 1)(x - 2)}$$
$$= \frac{x - 2 - x^2 - x + x^2 + 2}{(x + 1)(x - 2)}$$
$$= \frac{0}{(x + 1)(x - 2)}$$
$$= 0$$

40. $\dfrac{3}{x + 2}$

41. $$\frac{\dfrac{x^2 - y^2}{xy}}{\dfrac{x - y}{y}} = \frac{x^2 - y^2}{xy} \cdot \frac{y}{x - y}$$
$$= \frac{(x + y)(x - y)y}{xy(x - y)}$$
$$= \frac{x + y}{x}$$

42. $\dfrac{a}{a + b}$

43. $$\frac{a - a^{-1}}{a + a^{-1}} = \frac{a - \dfrac{1}{a}}{a + \dfrac{1}{a}} = \frac{a \cdot \dfrac{a}{a} - \dfrac{1}{a}}{a \cdot \dfrac{a}{a} + \dfrac{1}{a}}$$
$$= \frac{\dfrac{a^2 - 1}{a}}{\dfrac{a^2 + 1}{a}}$$
$$= \frac{a^2 - 1}{a} \cdot \frac{a}{a^2 + 1}$$
$$= \frac{a^2 - 1}{a^2 + 1}$$

44. $\dfrac{a^2(b - 1)}{b^2(a - 1)}$

45. $$\frac{c + \dfrac{8}{c^2}}{1 + \dfrac{2}{c}} = \frac{c \cdot \dfrac{c^2}{c^2} + \dfrac{8}{c^2}}{1 \cdot \dfrac{c}{c} + \dfrac{2}{c}}$$
$$= \frac{\dfrac{c^3 + 8}{c^2}}{\dfrac{c + 2}{c}}$$
$$= \frac{c^3 + 8}{c^2} \cdot \frac{c}{c + 2}$$
$$= \frac{(c + 2)(c^2 - 2c + 4)c}{c^2(c + 2)}$$
$$= \frac{c^2 - 2c + 4}{c}$$

46. $\dfrac{x^2y^2}{x^2 - xy + y^2}$

47. $$\frac{x^2 + xy + y^2}{\dfrac{x^2}{y} - \dfrac{y^2}{x}} = \frac{x^2 + xy + y^2}{\dfrac{x^2}{y} \cdot \dfrac{x}{x} - \dfrac{y^2}{x} \cdot \dfrac{y}{y}}$$
$$= \frac{x^2 + xy + y^2}{\dfrac{x^3 - y^3}{xy}}$$
$$= (x^2 + xy + y^2) \cdot \frac{xy}{x^3 - y^3}$$
$$= \frac{(x^2 + xy + y^2)xy}{(x - y)(x^2 + xy + y^2)}$$
$$= \frac{xy}{x - y}$$

48. $\dfrac{a + b}{ab}$

49. $\dfrac{\frac{x}{y} - \frac{y}{x}}{\frac{1}{y} + \frac{1}{x}} = \dfrac{\frac{x^2 - y^2}{xy}}{\frac{x + y}{xy}}$

$= \dfrac{x^2 - y^2}{xy} \cdot \dfrac{xy}{x + y}$

$= \dfrac{(x + y)(x - y)xy}{xy(x + y)}$

$= x - y$

50. $-a - b$

51. $\dfrac{x^2y^{-2} - y^2x^{-2}}{xy^{-1} + yx^{-1}} = \dfrac{\frac{x^2}{y^2} - \frac{y^2}{x^2}}{\frac{x}{y} + \frac{y}{x}}$

$= \dfrac{\frac{x^4 - y^4}{x^2y^2}}{\frac{x^2 + y^2}{xy}}$

$= \dfrac{x^4 - y^4}{x^2y^2} \cdot \dfrac{xy}{x^2 + y^2}$

$= \dfrac{(x^2 + y^2)(x + y)(x - y)xy}{x^2y^2(x^2 + y^2)}$

$= \dfrac{(x + y)(x - y)}{xy}$, or $\dfrac{x^2 - y^2}{xy}$

52. $\dfrac{a^2 + b^2}{ab}$

53. $\dfrac{\frac{a}{1 - a} + \frac{1 + a}{a}}{\frac{1 - a}{a} + \frac{a}{1 + a}} = \dfrac{\frac{a}{1 - a} \cdot \frac{a}{a} + \frac{1 + a}{a} \cdot \frac{1 - a}{1 - a}}{\frac{1 - a}{a} \cdot \frac{1 + a}{1 + a} + \frac{a}{1 + a} \cdot \frac{a}{a}}$

$= \dfrac{\frac{a^2 + (1 - a^2)}{a(1 - a)}}{\frac{(1 - a^2) + a^2}{a(1 + a)}}$

$= \dfrac{1}{a(1 - a)} \cdot \dfrac{a(1 + a)}{1}$

$= \dfrac{1 + a}{1 - a}$

54. $\dfrac{1 - x}{1 + x}$

55. $\dfrac{\frac{1}{a^2} + \frac{2}{ab} + \frac{1}{b^2}}{\frac{1}{a^2} - \frac{1}{b^2}} = \dfrac{\frac{1}{a^2} \cdot \frac{b^2}{b^2} + \frac{2}{ab} \cdot \frac{ab}{ab} + \frac{1}{b^2} \cdot \frac{a^2}{a^2}}{\frac{1}{a^2} \cdot \frac{b^2}{b^2} - \frac{1}{b^2} \cdot \frac{a^2}{a^2}}$

$= \dfrac{\frac{b^2 + 2ab + a^2}{a^2b^2}}{\frac{b^2 - a^2}{a^2b^2}}$

$= \dfrac{(b + a)(b + a)}{a^2b^2} \cdot \dfrac{a^2b^2}{(b + a)(b - a)}$

$= \dfrac{b + a}{b - a}$

56. $\dfrac{y + x}{y - x}$

57. $\dfrac{(x + h)^2 - x^2}{h} = \dfrac{x^2 + 2xh + h^2 - x^2}{h}$

$= \dfrac{2xh + h^2}{h}$

$= \dfrac{h(2x + h)}{h}$

$= 2x + h$

58. $\dfrac{-1}{x(x + h)}$

59. $\dfrac{(x + h)^3 - x^3}{h} = \dfrac{x^3 + 3x^2h + 3xh^2 + h^3 - x^3}{h}$

$= \dfrac{3x^2h + 3xh^2 + h^3}{h}$

$= \dfrac{h(3x^2 + 3xh + h^2)}{h}$

$= 3x^2 + 3xh + h^2$

60. $\dfrac{-2x - h}{x^2(x + h)^2}$

61. $\left[\dfrac{\frac{x + 1}{x - 1} + 1}{\frac{x + 1}{x - 1} - 1}\right]^5 = \left[\dfrac{\frac{(x + 1) + (x - 1)}{x - 1}}{\frac{(x + 1) - (x - 1)}{x - 1}}\right]^5$

$= \left[\dfrac{2x}{x - 1} \cdot \dfrac{x - 1}{2}\right]^5$

$= x^5$

62. $\dfrac{5x + 3}{3x + 2}$

63. $\dfrac{a}{b} \div \left(\dfrac{a}{3} + \dfrac{b}{4}\right)$

$= \dfrac{a}{b} \cdot \left(\dfrac{3}{a} + \dfrac{4}{b}\right)$ (1) Step (1) uses the wrong reciprocal. The reciprocal of a sum is not the sum of the reciprocals.

$= \dfrac{a}{b} \cdot \left(\dfrac{3b + 4a}{ab}\right)$ (2) Step (2) would be correct if step (1) had been correct.

$= \dfrac{a(3b + 4a)}{ab^2}$ (3) Step (3) would be correct if steps (1) and (2) had been correct.

$= \dfrac{4a + 3b}{b}$ (4) Step (4) has an improper simplification of the b in the denominator.

$\dfrac{a}{b} \div \left(\dfrac{a}{3} + \dfrac{b}{4}\right) = \dfrac{a}{b} \div \left(\dfrac{4a + 3b}{12}\right)$

$= \dfrac{a}{b} \cdot \dfrac{12}{4a + 3b}$

$= \dfrac{12a}{4ab + 3b^2}$ (Correct answer)

64. In step (1), numerator of first term should be $10x^2$ found by multiplying by $\frac{2x}{2x}$. Step (2) would be correct if step (1) had been correct. Step (3) has an incorrect simplification as does step (4). $\frac{10x^2 + 3y^2}{4xy}$ is the correct answer.

Exercise Set 1.8

1. $\sqrt{x - 3}$

   We substitute -2 for x in x - 3: -2 - 3 = -5. Since the radicand is negative, -2 is not a sensible replacement.

   We substitute 5 for x in x - 3: 5 - 3 = 2. Since the radicand is not negative, 5 is a sensible replacement.

2. Yes, No

3. $\sqrt{3 - 4x}$

   We substitute -1 for x in 3 - 4x: 3 - 4(-1) = 7. Since the radicand is not negative, -1 is a sensible replacement.

   We substitute 1 for x in 3 - 4x: $3 - 4 \cdot 1 = -1$. Since the radicand is negative, 1 is not a sensible replacement.

4. Yes, Yes

5. $\sqrt{1 - x^2}$

   We substitute 1 for x in $1 - x^2$: $1 - 1^2 = 0$. Since the radicand is not negative, 1 is a sensible replacement.

   We substitute 3 for x in $1 - x^2$: $1 - 3^2 = -8$. Since the radicand is negative, 3 is not a sensible replacement.

6. Yes, Yes

7. $\sqrt[3]{2x + 7}$

   We substitute -4 for x in 2x + 7: 2(-4) + 7 = -1. Since every real number, positive, negative, or zero has a cube root, -4 is a sensible replacement.

   We substitute 5 for x in 2x + 7: 2(5) + 7 = 17. Since every real number, positive, negative, or zero has a cube root, 5 is a sensible replacement.

8. No, No

9. $\sqrt{(-11)^2} = |-11| = 11$

10. 1

11. $\sqrt{16x^2} = \sqrt{(4x)^2} = |4x| = 4|x|$

12. $6|t|$

13. $\sqrt{(b + 1)^2} = |b + 1|$

14. $|2c - 3|$

15. $\sqrt[3]{-27x^3} = \sqrt[3]{(-3x)^3} = -3x$

16. -2y

17. $\sqrt{x^2 - 4x + 4} = \sqrt{(x - 2)^2} = |x - 2|$

18. $|y + 8|$

19. $\sqrt[5]{32} = \sqrt[5]{2^5} = 2$

20. -2

21. $\sqrt{180} = \sqrt{36 \cdot 5} = \sqrt{36} \cdot \sqrt{5} = 6\sqrt{5}$

22. $4\sqrt{3}$

23. $\sqrt[3]{54} = \sqrt[3]{27 \cdot 2} = \sqrt[3]{27} \cdot \sqrt[3]{2} = 3\sqrt[3]{2}$

24. $3\sqrt[3]{5}$

25. $\sqrt{128c^2d^{-4}} = \sqrt{\frac{128c^2}{d^4}} = \frac{\sqrt{64c^2 \cdot 2}}{\sqrt{d^4}} = \frac{8|c|\sqrt{2}}{d^2}$

26. $9c^2|d^{-3}|\sqrt{2}$ or $\frac{9c^2\sqrt{2}}{|d^3|}$

27. $\sqrt{3}\sqrt{6} = \sqrt{18} = \sqrt{9 \cdot 2} = 3\sqrt{2}$

28. $4\sqrt{3}$

29. $\sqrt{2x^3y}\sqrt{12xy} = \sqrt{24x^4y^2} = \sqrt{4x^4y^2 \cdot 6} = 2x^2y\sqrt{6}$

30. $2y^2z\sqrt{15}$

31. $\sqrt[3]{3x^2y}\sqrt[3]{36x} = \sqrt[3]{108x^3y} = \sqrt[3]{27x^3 \cdot 4y} = 3x\sqrt[3]{4y}$

32. $2xy\sqrt[5]{x^2}$

33. $\sqrt[3]{2(x+4)}\ \sqrt[3]{4(x+4)^4} = \sqrt[3]{8(x+4)^5}$
$= \sqrt[3]{8(x+4)^3 \cdot (x+4)^2}$
$= 2(x+4)\ \sqrt[3]{(x+4)^2}$

34. $2(x+1)\ \sqrt[3]{9(x+1)}$

35. $\dfrac{\sqrt{21ab^2}}{\sqrt{3ab}} = \sqrt{\dfrac{21ab^2}{3ab}} = \sqrt{7b}$

36. $\dfrac{2\sqrt{2ab}}{a}$

37. $\dfrac{\sqrt[3]{40m}}{\sqrt[3]{5m}} = \sqrt[3]{\dfrac{40m}{5m}} = \sqrt[3]{8} = 2$

38. $\sqrt{5y}$

39. $\dfrac{\sqrt[3]{3x^2}}{\sqrt[3]{24x^5}} = \sqrt[3]{\dfrac{3x^2}{24x^5}} = \sqrt[3]{\dfrac{1}{8x^3}} = \dfrac{1}{2x}$

40. $y\ \sqrt[3]{5}$

41. $\dfrac{\sqrt{a^2-b^2}}{\sqrt{a-b}} = \sqrt{\dfrac{a^2-b^2}{a-b}} = \sqrt{\dfrac{(a+b)(a-b)}{a-b}}$
$= \sqrt{a+b}$

42. $\sqrt{x^2+xy+y^2}$

43. $\sqrt{\dfrac{9a^2}{8b}} = \sqrt{\dfrac{9a^2}{8b} \cdot \dfrac{2b}{2b}} = \sqrt{\dfrac{9a^2}{16b^2} \cdot 2b} = \dfrac{3a}{4b}\sqrt{2b}$

44. $\dfrac{b}{6a}\sqrt{15a}$

45. $\sqrt[3]{\dfrac{2x^2 2y^3}{25z^4}} = \sqrt[3]{\dfrac{2x^2 2y^3}{25z^4} \cdot \dfrac{5z^2}{5z^2}} = \sqrt[3]{\dfrac{y^3}{125z^6} \cdot 20x^2z^2}$
$= \dfrac{y}{5z^2}\ \sqrt[3]{20x^2z^2}$

46. $\dfrac{2x}{y}$

47. $\dfrac{\left(\sqrt[3]{32x^4y}\right)^2}{\left(\sqrt[3]{xy}\right)^2} = \left(\sqrt[3]{\dfrac{32x^4y}{xy}}\right)^2 = \left(\sqrt[3]{32x^3}\right)^2$
$= \left(\sqrt[3]{2^5x^3}\right)^2$
$= \sqrt[3]{2^{10}x^6}$
$= \sqrt[3]{2^9x^6 \cdot 2}$
$= 2^3x^2\ \sqrt[3]{2}$
$= 8x^2\ \sqrt[3]{2}$

48. $4\ \sqrt[3]{4x^2}$

49. $\dfrac{3\sqrt{a^2b^2}\ \sqrt{4xy}}{2\sqrt{a^{-1}b^{-2}}\ \sqrt{9x^{-3}y^{-1}}} = \dfrac{3}{2}\sqrt{\dfrac{a^2b^2}{a^{-1}b^{-2}}}\ \sqrt{\dfrac{4xy}{9x^{-3}y^{-1}}}$
$= \dfrac{3}{2}\sqrt{a^3b^4}\ \sqrt{\dfrac{4}{9}x^4y^2}$
$= \dfrac{3}{2} \cdot ab^2\sqrt{a} \cdot \dfrac{2}{3}x^2y$
$= ab^2x^2y\sqrt{a}$

50. xy^2a^3b

51. $8\sqrt{2} - 6\sqrt{20} - 5\sqrt{8}$
$= 8\sqrt{2} - 6\sqrt{4 \cdot 5} - 5\sqrt{4 \cdot 2}$
$= 8\sqrt{2} - 6 \cdot 2\sqrt{5} - 5 \cdot 2\sqrt{2}$
$= 8\sqrt{2} - 12\sqrt{5} - 10\sqrt{2}$
$= -2\sqrt{2} - 12\sqrt{5}$

52. $4\sqrt{3}$

53. $2\ \sqrt[3]{8x^2} + 5\ \sqrt[3]{27x^2} - 3\ \sqrt[3]{x^3}$
$= 2 \cdot 2\ \sqrt[3]{x^2} + 5 \cdot 3\ \sqrt[3]{x^2} - 3x$
$= 4\ \sqrt[3]{x^2} + 15\ \sqrt[3]{x^2} - 3x$
$= 19\ \sqrt[3]{x^2} - 3x$

54. $(5a^2 - 3b^2)\sqrt{a+b}$

55. $3\sqrt{3y^2} - \dfrac{y\sqrt{48}}{\sqrt{2}} + \sqrt{\dfrac{12}{4y^{-2}}}$
$= 3\sqrt{3y^2} - y\sqrt{24} + \sqrt{3y^2}$
$= 3y\sqrt{3} - 2y\sqrt{6} + y\sqrt{3}$
$= 4y\sqrt{3} - 2y\sqrt{6}$

56. $(3x - 2)\ \sqrt[3]{x^2}$

57. $\left(\sqrt{3} - \sqrt{2}\right)\left(\sqrt{3} + \sqrt{2}\right)$
$= \left(\sqrt{3}\right)^2 - \left(\sqrt{2}\right)^2$
$= 3 - 2$
$= 1$

58. -12

59. $\left(\sqrt{t} - x\right)^2$
$= \left(\sqrt{t}\right)^2 - 2 \cdot \sqrt{t} \cdot x + x^2$
$= t - 2x\sqrt{t} + x^2$

60. $a + 2 + a^{-1}$, or $\dfrac{(a+1)^2}{a}$

61. $5\sqrt{7} + \frac{35}{\sqrt{7}}$

$= 5\sqrt{7} + \frac{35}{\sqrt{7}} \cdot \frac{\sqrt{7}}{\sqrt{7}}$

$= 5\sqrt{7} + \frac{35\sqrt{7}}{7}$

$= 5\sqrt{7} + 5\sqrt{7}$

$= 10\sqrt{7}$

62. $2a^2b + 5a\sqrt{by} - 3y$

63. $\left(\sqrt{x+3} - \sqrt{3}\right)\left(\sqrt{x+3} + \sqrt{3}\right)$

$= \left(\sqrt{x+3}\right)^2 - \left(\sqrt{3}\right)^2$

$= (x + 3) - 3$

$= x$

64. h

65. $T = 2\pi\sqrt{\frac{L}{32}}$

$T = 2(3.14)\sqrt{\frac{2}{32}}$

$= 6.28 \cdot \frac{1}{4}$

$= 1.57$ sec

$T = 2(3.14)\sqrt{\frac{8}{32}}$

$= 6.28 \cdot \frac{1}{2}$

$= 3.14$ sec

$T = 2(3.14)\sqrt{\frac{64}{32}}$

$= 2(3.14)\sqrt{2}$

$\approx 6.28(1.414)$

≈ 8.88 sec

$T = 2(3.14)\sqrt{\frac{100}{32}}$

$= 2(3.14)\sqrt{3.125}$

$\approx 6.28(1.768)$

≈ 11.10 sec

66. $65\sqrt{2} \approx 91.92$ ft

67. $\frac{6}{3 + \sqrt{5}} = \frac{6}{3 + \sqrt{5}} \cdot \frac{3 - \sqrt{5}}{3 - \sqrt{5}}$

$= \frac{6\left(3 - \sqrt{5}\right)}{9 - 5}$

$= \frac{6\left(3 - \sqrt{5}\right)}{4}$

$= \frac{3\left(3 - \sqrt{5}\right)}{2}$

68. $\sqrt{3} + 1$

69. $\sqrt[3]{\frac{16}{9}} = \sqrt[3]{\frac{16}{9} \cdot \frac{3}{3}} = \sqrt[3]{\frac{48}{27}} = \sqrt[3]{\frac{8}{27} \cdot 6} = \frac{2}{3}\sqrt[3]{6}$

70. $\frac{\sqrt[3]{4}}{2}$

71. $\frac{4\sqrt{x} - 3\sqrt{xy}}{2\sqrt{x} + 5\sqrt{y}} = \frac{4\sqrt{x} - 3\sqrt{xy}}{2\sqrt{x} + 5\sqrt{y}} \cdot \frac{2\sqrt{x} - 5\sqrt{y}}{2\sqrt{x} - 5\sqrt{y}}$

$= \frac{8x - 20\sqrt{xy} - 6x\sqrt{y} + 15y\sqrt{x}}{4x - 25y}$

72. $\frac{15x + 10\sqrt{xy} + 6x\sqrt{y} + 4y\sqrt{x}}{9x - 4y}$

73. $\frac{\sqrt{2} + \sqrt{5a}}{6} = \frac{\sqrt{2} + \sqrt{5a}}{6} \cdot \frac{\sqrt{2} - \sqrt{5a}}{\sqrt{2} - \sqrt{5a}}$

$= \frac{2 - 5a}{6(\sqrt{2} - \sqrt{5a})}$

74. $\frac{3 - 5y}{4(\sqrt{3} - \sqrt{5y})}$

75. $\frac{\sqrt{x+1} + 1}{\sqrt{x+1} - 1} = \frac{\sqrt{x+1} + 1}{\sqrt{x+1} - 1} \cdot \frac{\sqrt{x+1} - 1}{\sqrt{x+1} - 1}$

$= \frac{(x + 1) - 1}{(x + 1) - 2\sqrt{x+1} + 1}$

$= \frac{x}{x - 2\sqrt{x+1} + 2}$

76. $\frac{x}{x + 8 + 4\sqrt{x+4}}$

77. $\frac{\sqrt{a+3} - \sqrt{3}}{3} = \frac{\sqrt{a+3} - \sqrt{3}}{3} \cdot \frac{\sqrt{a+3} + \sqrt{3}}{\sqrt{a+3} + \sqrt{3}}$

$= \frac{(a + 3) - 3}{3(\sqrt{a+3} + \sqrt{3})}$

$= \frac{a}{3(\sqrt{a+3} + \sqrt{3})}$

78. $\frac{1}{\sqrt{a+h} + \sqrt{a}}$

79. $\sqrt{8.2x^3y}\,\sqrt{12.5xy} = \sqrt{102.5x^4y^2} \approx 10.124x^2y$

80. $0.1251y^2z$

81. $\sqrt{\frac{6.03a^2}{17.13b}} \approx \sqrt{0.352014\,\frac{a^2}{b} \cdot \frac{b}{b}} \approx \frac{0.5933a\sqrt{b}}{b}$

82. $\frac{0.1974b\sqrt{a}}{a}$

83. $h^2 + \left(\frac{a}{2}\right)^2 = a^2$ (Pythagorean property)

$h^2 + \frac{a^2}{4} = a^2$

$h^2 = \frac{3a^2}{4}$

$h = \sqrt{\frac{3a^2}{4}}$ (The height cannot be negative.)

$h = \frac{a}{2}\sqrt{3}$

84. $A = \frac{a^2}{4}\sqrt{3}$

85.

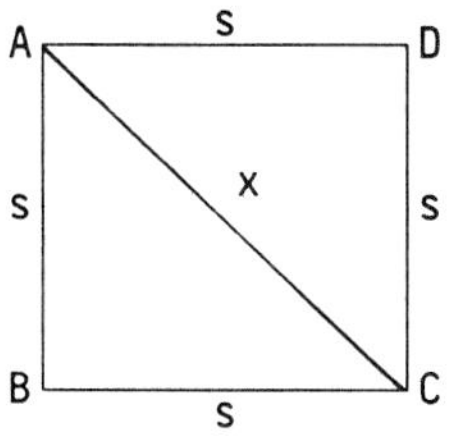

Let x represent the length of $\overline{AC}$. Then using the Pythagorean property we have

$s^2 + s^2 = x^2$

$2s^2 = x^2$

$\sqrt{2s^2} = x$ (The length of a diagonal cannot be negative.)

$s\sqrt{2} = x$

The length of $\overline{AC}$ is $\sqrt{2}\,s$.

86. $\sqrt{2}\,s$

87.

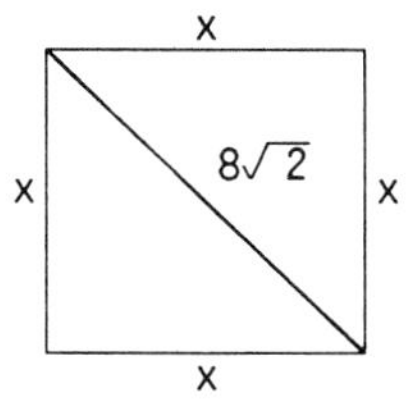

$x^2 + x^2 = (8\sqrt{2})^2$ (Pythagorean property)

$2x^2 = 128$

$x^2 = 64$

$x = 8$ (The length of a side cannot be negative.)

88. 50 ft^2

89. $\sqrt{1 + x^2} + \frac{1}{\sqrt{1 + x^2}}$

$= \sqrt{1 + x^2} \cdot \frac{1 + x^2}{1 + x^2} + \frac{1}{\sqrt{1 + x^2}} \quad \frac{\sqrt{1 + x^2}}{\sqrt{1 + x^2}}$

$= \frac{(1 + x^2)\sqrt{1 + x^2}}{1 + x^2} + \frac{\sqrt{1 + x^2}}{1 + x^2}$

$= \frac{(2 + x^2)\left(\sqrt{1 + x^2}\right)}{1 + x^2}$

90. $\frac{(2 - 3x^2)\sqrt{1 - x^2}}{2(1 - x^2)}$

91. Let a = 16 and b = 9.

Then $\sqrt{a + b} = \sqrt{16 + 9} = \sqrt{25} = 5$

and $\sqrt{a} + \sqrt{b} = \sqrt{16} + \sqrt{9} = 4 + 3 = 7$.

92. $\left(\sqrt{5 + \sqrt{24}}\right)^2 = 5 + \sqrt{24} = 5 + 2\sqrt{6}$

$\left(\sqrt{2} + \sqrt{3}\right)^2 = \left(\sqrt{2}\right)^2 + 2\sqrt{2}\sqrt{3} + \left(\sqrt{3}\right)^2$

$= 2 + 2\sqrt{6} + 3$

$= 5 + 2\sqrt{6}$

Exercise Set 1.9

1. $x^{3/4} = \sqrt[4]{x^3}$

2. $\sqrt[5]{y^2}$

3. $16^{3/4} = (16^{1/4})^3 = \left(\sqrt[4]{16}\right)^3 = 2^3 = 8$

4. 128

5. $125^{-1/3} = \frac{1}{125^{1/3}} = \frac{1}{\sqrt[3]{125}} = \frac{1}{5}$

6. $\frac{1}{16}$

7. $a^{5/4}\,b^{-3/4} = \frac{a^{5/4}}{b^{3/4}} = \frac{\sqrt[4]{a^5}}{\sqrt[4]{b^3}} = \frac{a\sqrt[4]{a}}{\sqrt[4]{b^3}}$, or $\frac{a\sqrt[4]{ab}}{b}$

8. $\sqrt[5]{x^2y^{-1}}$

9. $\sqrt[3]{20^2} = 20^{2/3}$

10. $17^{3/5}$

11. $\left(\sqrt[4]{13}\right)^5 = \sqrt[4]{13^5} = 13^{5/4}$

12. $12^{4/5}$

13. $\sqrt[3]{\sqrt{11}} = \left(\sqrt{11}\right)^{1/3} = (11^{1/2})^{1/3} = 11^{1/6}$

14. $7^{1/12}$

15. $\sqrt{5}\,\sqrt[3]{5} = 5^{1/2} \cdot 5^{1/3} = 5^{1/2 + 1/3} = 5^{5/6}$

16. $2^{5/6}$

17. $\sqrt[5]{32^2} = 32^{2/5} = (32^{1/5})^2 = 2^2 = 4$

18. $\frac{1}{16}$

19. $\sqrt[3]{8y^6} = (8y^6)^{1/3} = (2^3y^6)^{1/3} = 2^{3/3}y^{6/3} = 2y^2$

20. $2c^2d^3$

21. $\sqrt[3]{a^2 + b^2} = (a^2 + b^2)^{1/3}$

22. $(a^3 - b^3)^{1/4}$

23. $\sqrt[3]{27a^3b^9} = (3^3a^3b^9)^{1/3} = 3^{3/3}a^{3/3}b^{9/3} = 3ab^3$

24. $3x^2y^2$

25. $\sqrt[6]{\dfrac{m^{12}n^{24}}{64}} = \left(\dfrac{m^{12}n^{24}}{2^6}\right)^{1/6} = \dfrac{m^{12/6}n^{24/6}}{2^{6/6}} = \dfrac{m^2n^4}{2}$

26. $\dfrac{m^2n^3}{2}$

27. $(2a^{3/2})(4a^{1/2}) = 8a^{3/2 + 1/2} = 8a^2$

28. $24a\sqrt{a}$

29. $\left(\dfrac{x^6}{9b^{-4}}\right)^{-1/2} = \left(\dfrac{x^6}{3^2b^{-4}}\right)^{-1/2} = \dfrac{x^{-3}}{3^{-1}b^2} = \dfrac{3}{x^3b^2}$

30. $\dfrac{2\sqrt[3]{x^2}}{xy}$

31. $\dfrac{x^{2/3}y^{5/6}}{x^{-1/3}y^{1/2}} = x^{2/3-(-1/3)}y^{5/6-1/2} = xy^{1/3} = x\sqrt[3]{y}$

32. $\sqrt[4]{ab}$

33. $\sqrt[3]{6}\sqrt{2} = 6^{1/3}2^{1/2} = 6^{2/6}2^{3/6}$
$= (6^22^3)^{1/6}$
$= \sqrt[6]{36 \cdot 8}$
$= \sqrt[6]{288}$

34. $2\sqrt[4]{2}$

35. $\sqrt[4]{xy}\sqrt[3]{x^2y} = (xy)^{1/4}(x^2y)^{1/3} = (xy)^{3/12}(x^2y)^{4/12}$
$= \left[(xy)^3(x^2y)^4\right]^{1/12}$
$= [x^3y^3x^8y^4]^{1/12}$
$= \sqrt[12]{x^{11}y^7}$

36. $b\sqrt[6]{a^5b}$

37. $\sqrt[3]{a^4\sqrt{a^3}} = \left(a^4\sqrt{a^3}\right)^{1/3} = (a^4a^{3/2})^{1/3}$
$= (a^{11/2})^{1/3}$
$= a^{11/6}$
$= \sqrt[6]{a^{11}}$
$= a\sqrt[6]{a^5}$

38. $a\sqrt[6]{a^5}$

39. $\dfrac{\sqrt{(a + x)^3}\sqrt[3]{(a + x)^2}}{\sqrt[4]{a + x}} = \dfrac{(a + x)^{3/2}(a + x)^{2/3}}{(a + x)^{1/4}}$
$= \dfrac{(a + x)^{26/12}}{(a + x)^{3/12}}$
$= (a + x)^{23/12}$
$= \sqrt[12]{(a + x)^{23}}$
$= (a + x)\sqrt[12]{(a + x)^{11}}$

40. $\dfrac{\sqrt[3]{x + y}}{x + y}$

41. $\left(\sqrt[4]{13}\right)^5 = 13^{5/4} = 13^{1.25} \approx 24.685$

42. 8.372

43. $12.3^{3/2} = 12.3^{1.5} \approx 43.138$

44. 2.098

45. $105.6^{3/4} = 105.6^{0.75} \approx 32.942$

46. 11.671

47. $L = \dfrac{0.000169d^{2.27}}{h}$
$= \dfrac{0.000169(180)^{2.27}}{4}$
≈ 5.563 ft

48. 1.46 ft

49. $L = \dfrac{0.000169d^{2.27}}{h}$
$L = \dfrac{0.000169(200)^{2.27}}{4}$
≈ 7.066 ft

50. 17.74 ft

51. $\left(\sqrt{a\sqrt{a}}\right)^{\sqrt{a}} = \left(a\sqrt{a}/2\right)^{\sqrt{a}} = a^{a/2}$

52. $48 \cdot 2^{1/3} \cdot a^{34/3} \cdot b^{47/9} \cdot c^{34/35}$

53. $T = 34\,x^{-0.41}$

When $x = 1$:
$T = 34(1)^{-0.41}$
$= 34$ hr

When $x = 6$:
$T = 34(6)^{-0.41}$
≈ 16.3 hr

When $x = 8$:
$T = 34(8)^{-0.41}$
≈ 14.5 hr

When $x = 10$:
$T = 34(10)^{-0.41}$
≈ 13.2 hr

When $x = 32$:
$T = 34(32)^{-0.41}$
≈ 8.2 hr

When $x = 64$:
$T = 34(64)^{-0.41}$
≈ 6.2 hr

Exercise Set 2.1

1. $4x + 12 = 60$

$4x + 12 - 12 = 60 - 12$

$4x = 48$

$\frac{1}{4} \cdot 4x = \frac{1}{4} \cdot 48$

$x = 12$

The solution set is {12}.

2. {24}

3. $4 + \frac{1}{2}x = 1$

$4 + \frac{1}{2}x - 4 = 1 - 4$

$\frac{1}{2}x = -3$

$2 \cdot \frac{1}{2}x = 2(-3)$

$x = -6$

The solution set is {-6}.

4. {14}

5. $y + 1 = 2y - 7$ or $y + 1 = 2y - 7$

$1 + 7 = 2y - y$ $y - 2y = -7 - 1$

$8 = y$ $-y = -8$

$y = 8$

The solution set is {8}.

6. $\left\{\frac{18}{5}\right\}$

7. $5x - 2 + 3x = 2x + 6 - 4x$

$8x - 2 = 6 - 2x$

$8x + 2x = 6 + 2$

$10x = 8$

$x = \frac{8}{10}$

$x = \frac{4}{5}$

The solution set is $\left\{\frac{4}{5}\right\}$.

8. {-8}

9. $1.9x - 7.8 + 5.3x = 3.0 + 1.8x$

$7.2x - 7.8 = 3.0 + 1.8x$

$7.2x - 1.8x = 3.0 + 7.8$

$5.4x = 10.8$

$x = \frac{10.8}{5.4}$

$x = 2$

The solution set is {2}.

10. {-3}

11. $7(3x + 6) = 11 - (x + 2)$

$21x + 42 = 11 - x - 2$

$21x + 42 = 9 - x$

$21x + x = 9 - 42$

$22x = -33$

$x = -\frac{33}{22}$

$x = -\frac{3}{2}$

The solution set is $\left\{-\frac{3}{2}\right\}$.

12. $\left\{-\frac{27}{14}\right\}$

13. $2x - (5 + 7x) = 4 - [x - (2x + 3)]$

$2x - 5 - 7x = 4 - x + 2x + 3$

$-5x - 5 = 7 + x$

$-5 - 7 = x + 5x$

$-12 = 6x$

$-\frac{12}{6} = x$

$-2 = x$

The solution set is {-2}.

14. $\emptyset$

15. $(2x - 3)(3x - 2) = 0$

$2x - 3 = 0$ or $3x - 2 = 0$

$2x = 3$ or $3x = 2$

$x = \frac{3}{2}$ or $x = \frac{2}{3}$

The solution set is $\left\{\frac{3}{2}, \frac{2}{3}\right\}$.

16. $\left\{\frac{2}{5}, -\frac{3}{2}\right\}$

17. $x(x - 1)(x + 2) = 0$

$x = 0$ or $x - 1 = 0$ or $x + 2 = 0$

$x = 0$ or $x = 1$ or $x = -2$

The solution set is {0, 1, -2}.

18. $\{0, -2, 3\}$

19. $3x^2 + x - 2 = 0$
$(3x - 2)(x + 1) = 0$

$3x - 2 = 0$ or $x + 1 = 0$
$x = \frac{2}{3}$ or $x = -1$

The solution set is $\left\{\frac{2}{3}, -1\right\}$.

20. $\left\{\frac{3}{5}, 1\right\}$

21. $(x - 1)(x + 1) = 5(x - 1)$
$x^2 - 1 = 5x - 5$
$x^2 - 5x + 4 = 0$
$(x - 4)(x - 1) = 0$

$x - 4 = 0$ or $x - 1 = 0$
$x = 4$ or $x = 1$

The solution set is $\{4, 1\}$.

22. $\{8, 3\}$

23. $x[4(x - 2) - 5(x - 1)] = 2$
$x(4x - 8 - 5x + 5) = 2$
$x(-x - 3) = 2$
$-x^2 - 3x = 2$
$0 = x^2 + 3x + 2$
$0 = (x + 2)(x + 1)$

$x + 2 = 0$ or $x + 1 = 0$
$x = -2$ or $x = -1$

The solution set is $\{-2, -1\}$.

24. $\{5, 10\}$

25. $(3x^2 - 7x - 20)(2x - 5) = 0$
$(3x + 5)(x - 4)(2x - 5) = 0$

$3x + 5 = 0$ or $x - 4 = 0$ or $2x - 5 = 0$
$x = -\frac{5}{3}$ or $x = 4$ or $x = \frac{5}{2}$

The solution set is $\left\{-\frac{5}{3}, 4, \frac{5}{2}\right\}$.

26. $\left\{-\frac{11}{8}, -\frac{1}{4}, \frac{2}{3}\right\}$

27. $16x^3 = x$
$16x^3 - x = 0$
$x(16x^2 - 1) = 0$
$x(4x + 1)(4x - 1) = 0$

$x = 0$ or $4x + 1 = 0$ or $4x - 1 = 0$
$x = 0$ or $x = -\frac{1}{4}$ or $x = \frac{1}{4}$

The solution set is $\left\{0, -\frac{1}{4}, \frac{1}{4}\right\}$.

28. $\left\{0, \frac{1}{3}, -\frac{1}{3}\right\}$

29. $2x^2 = 6x$
$2x^2 - 6x = 0$
$2x(x - 3) = 0$

$2x = 0$ or $x - 3 = 0$
$x = 0$ or $x = 3$

The solution set is $\{0, 3\}$.

30. $\{0, -2\}$

31. $3y^3 - 5y^2 - 2y = 0$
$y(3y^2 - 5y - 2) = 0$
$y(3y + 1)(y - 2) = 0$

$y = 0$ or $3y + 1 = 0$ or $y - 2 = 0$
$y = 0$ $y = -\frac{1}{3}$ or $y = 2$

The solution set is $\left\{0, -\frac{1}{3}, 2\right\}$

32. $\left\{0, 1, \frac{2}{3}\right\}$

33. $(2x - 3)(3x + 2)(x - 1) = 0$

$2x - 3 = 0$ or $3x + 2 = 0$ or $x - 1 = 0$
$x = \frac{3}{2}$ or $x = -\frac{2}{3}$ or $x = 1$

The solution set is $\left\{\frac{3}{2}, -\frac{2}{3}, 1\right\}$.

34. $\left\{4, -3, -\frac{1}{2}\right\}$

35. $(2 - 4y)(y^2 + 3y) = 0$
$2(1 - 2y)y(y + 3) = 0$

$1 - 2y = 0$ or $y = 0$ or $y + 3 = 0$
$y = \frac{1}{2}$ or $y = 0$ or $y = -3$

The solution set is $\left\{\frac{1}{2}, 0, -3\right\}$.

36. {3, -3, 6, -6}

37. $x + 6 < 5x - 6$ or $x + 6 < 5x - 6$

$6 + 6 < 5x - x$ $\quad$ $x - 5x < -6 - 6$

$12 < 4x$ $\quad$ $-4x < -12$

$\frac{12}{4} < x$ $\quad$ $x > \frac{12}{4}$

$3 < x$ $\quad$ $x > 3$

The solution set is $\{x \mid x > 3\}$.

38. $\left\{x \mid x > -\frac{4}{5}\right\}$

39. $3x - 3 + 2x \geqslant 1 - 7x - 9$

$5x - 3 \geqslant -7x - 8$

$5x + 7x \geqslant -8 + 3$

$12x \geqslant -5$

$x \geqslant -\frac{5}{12}$

The solution set is $\left\{x \mid x \geqslant -\frac{5}{12}\right\}$.

40. $\left\{y \mid y \leqslant -\frac{1}{12}\right\}$

41. $14 - 5y \leqslant 8y - 8$ or $14 - 5y \leqslant 8y - 8$

$14 + 8 \leqslant 8y + 5y$ $\quad$ $-5y - 8y \leqslant -8 - 14$

$22 \leqslant 13y$ $\quad$ $-13y \leqslant -22$

$\frac{22}{13} \leqslant y$ $\quad$ $y \geqslant \frac{22}{13}$

The solution set is $\left\{y \mid y \geqslant \frac{22}{13}\right\}$.

42. $\{x \mid x < 5\}$

43. $-\frac{3}{4}x \geqslant -\frac{5}{8} + \frac{2}{3}x$

$\frac{5}{8} \geqslant \frac{3}{4}x + \frac{2}{3}x$

$\frac{5}{8} \geqslant \frac{9}{12}x + \frac{8}{12}x$

$\frac{5}{8} \geqslant \frac{17}{12}x$

$\frac{12}{17} \cdot \frac{5}{8} \geqslant \frac{12}{17} \cdot \frac{17}{12}x$

$\frac{15}{34} \geqslant x$

The solution set is $\left\{x \mid x \leqslant \frac{15}{34}\right\}$.

44. $\left\{x \mid x \geqslant -\frac{3}{14}\right\}$

45. $4x(x - 2) < 2(2x - 1)(x - 3)$

$4x(x - 2) < 2(2x^2 - 7x + 3)$

$4x^2 - 8x < 4x^2 - 14x + 6$

$-8x < -14x + 6$

$-8x + 14x < 6$

$6x < 6$

$x < \frac{6}{6}$

$x < 1$

The solution set is $\{x \mid x < 1\}$.

46. $\{x \mid x > -1\}$

47. $\{x \mid x > 2.5\}$

48. $\{y \mid y \leqslant -7\}$

49. $\{t \mid t^2 = 5\}$

50. $\{m \mid m^3 + 3 = m^2 - 2\}$

51. $\sqrt{x - 3}$

The radicand must be nonnegative. We set $x - 3 \geqslant 0$ and solve for x.

$x - 3 \geqslant 0$

$x \geqslant 3$

The sensible replacements are $\{x \mid x \geqslant 3\}$.

52. $\left\{x \mid x \geqslant \frac{5}{2}\right\}$

53. $\sqrt{3 - 4x}$

The radicand must be nonnegative. We set $3 - 4x \geqslant 0$ and solve for x.

$3 - 4x \geqslant 0$

$-4x \geqslant -3$

$x \leqslant \frac{3}{4}$

The sensible replacements are $\left\{x \mid x \leqslant \frac{3}{4}\right\}$.

54. $\{x \mid x$ is a real number$\}$

55. $2.905x - 3.214 + 6.789x = 3.012 + 1.805x$

$9.694x - 3.214 = 3.012 + 1.805x$

$9.694x - 1.805x = 3.012 + 3.214$

$7.889x = 6.226$

$x = \frac{6.226}{7.889}$

$x \approx 0.7892$

The solution set is {0.7892}.

56. {-1.305, 1.989}

57. $3.12x^2 - 6.715x = 0$
$x(3.12x - 6.715) = 0$

$x = 0$ or $3.12x - 6.715 = 0$
$x = 0$ or $3.12x = 6.715$
$x = 0$ or $x \approx 2.1522$

The solution set is {0, 2.1522}.

58. {0, -1.9492}

59. $1.52(6.51x + 7.3) < 11.2 - (7.2x + 13.52)$
$9.8952x + 11.096 < 11.2 - 7.2x - 13.52$
$9.8952x + 11.096 < -7.2x - 2.32$
$9.8952x + 7.2x < -2.32 - 11.096$
$17.0952x < -13.416$
$x < -\frac{13.416}{17.0952}$
$x < -0.7848$

The solution set is $\{x | x < -0.7848\}$.

60. $\{y | y \geqslant -2.2353\}$

61. $7x^3 + x^2 - 7x - 1 = 0$
$x^2(7x + 1) - (7x + 1) = 0$
$(x^2 - 1)(7x + 1) = 0$
$(x + 1)(x - 1)(7x + 1) = 0$

$x + 1 = 0$ or $x - 1 = 0$ or $7x + 1 = 0$

$x = -1$ or $x = 1$ or $x = -\frac{1}{7}$

The solution set is $\left\{-1, 1, -\frac{1}{7}\right\}$.

62. $\left\{2, -2, -\frac{1}{3}\right\}$

63. $y^3 + 2y^2 - y - 2 = 0$
$y^2(y + 2) - (y + 2) = 0$
$(y^2 - 1)(y + 2) = 0$
$(y + 1)(y - 1)(y + 2) = 0$

$y + 1 = 0$ or $y - 1 = 0$ or $y + 2 = 0$
$y = -1$ or $y = 1$ or $y = -2$

The solution set is {-1, 1, -2}.

64. {-1, 5, -5}

65. $x^2 - x - 20 = x^2 - 25$
$-x - 20 = -25$
$-x = -5$
$x = 5$

The solution set is {5}.

Exercise Set 2.2

1. $3x + 5 = 12$
$3x = 7$
$x = \frac{7}{3}$

The solution set is $\left\{\frac{7}{3}\right\}$.

$3x = 7$
$x = \frac{7}{3}$

The solution set is $\left\{\frac{7}{3}\right\}$.

The equations are equivalent.

2. No

3. $x = 3$

The solution set is {3}.

$x^2 = 9$
$x = \pm 3$

The solution set is {-3, 3}.

The equations are not equivalent.

4. Yes

5. $\frac{(x - 3)(x + 9)}{(x - 3)} = x + 9$, Note: $x \neq 3$
$x + 9 = x + 9$
$9 = 9$

The solution set is the set of all numbers except 3.

$x + 9 = x + 9$
$9 = 9$

The solution set is the set of all numbers. Thus, the equations are not equivalent.

6. No

7. $\frac{1}{4} + \frac{1}{5} = \frac{1}{t}$, LCM = 20t

$20t\left(\frac{1}{4} + \frac{1}{5}\right) = 20t \cdot \frac{1}{t}$

$20t \cdot \frac{1}{4} + 20t \cdot \frac{1}{5} = 20t \cdot \frac{1}{t}$

$5t + 4t = 20$
$9t = 20$
$t = \frac{20}{9}$

The solution set is $\left\{\frac{20}{9}\right\}$.

Check:

$\frac{1}{4} + \frac{1}{5}$	$= \frac{1}{t}$
$\frac{1}{4} + \frac{1}{5}$	$\frac{1}{\frac{20}{9}}$
$\frac{5}{20} + \frac{4}{20}$	$1 \cdot \frac{9}{20}$
$\frac{9}{20}$	$\frac{9}{20}$

8. {-2}

9. $\frac{3}{x - 8} = \frac{x - 5}{x - 8}$, LCM = x - 8

$$(x - 8) \cdot \frac{3}{x - 8} = (x - 8) \cdot \frac{x - 5}{x - 8}$$
$$3 = x - 5$$
$$8 = x$$

The solution set is ∅.

Check:

$$\begin{array}{c|c} \frac{3}{x - 8} & = \frac{x - 5}{x - 8} \\ \hline \frac{3}{8 - 8} & \frac{8 - 5}{8 - 8} \\ \frac{3}{0} & \frac{3}{0} \end{array}$$

Division by zero is undefined.

10. ∅

11. $\frac{x + 2}{4} - \frac{x - 1}{5} = 15$, LCM = 20

$$20\left[\frac{x + 2}{4} - \frac{x - 1}{5}\right] = 20 \cdot 15$$
$$5(x + 2) - 4(x - 1) = 300$$
$$5x + 10 - 4x + 4 = 300$$
$$x + 14 = 300$$
$$x = 286$$

The solution set is {286}.

12. {-1}

13. $x + \frac{6}{x} = 5$, LCM = x

$$x\left[x + \frac{6}{x}\right] = x \cdot 5$$
$$x^2 + 6 = 5x$$
$$x^2 - 5x + 6 = 0$$
$$(x - 3)(x - 2) = 0$$
$$x - 3 = 0 \text{ or } x - 2 = 0$$
$$x = 3 \text{ or } x = 2$$

The solution set is {3, 2}.

Check:

For x = 3:

$$\begin{array}{c|c} x + \frac{6}{x} & = 5 \\ \hline 3 + \frac{6}{3} & 5 \\ 3 + 2 & \\ 5 & \end{array}$$

For x = 2:

$$\begin{array}{c|c} x + \frac{6}{x} & = 5 \\ \hline 2 + \frac{6}{2} & 5 \\ 2 + 3 & \\ 5 & \end{array}$$

14. {4, -3}

15. $\frac{x + 2}{2} + \frac{3x + 1}{5} = \frac{x - 2}{4}$, LCM = 20

$$20\left[\frac{x + 2}{2} + \frac{3x + 1}{5}\right] = 20 \cdot \frac{x - 2}{4}$$
$$10(x + 2) + 4(3x + 1) = 5(x - 2)$$
$$10x + 20 + 12x + 4 = 5x - 10$$
$$22x + 24 = 5x - 10$$
$$22x - 5x = -10 - 24$$
$$17x = -34$$
$$x = -2$$

The solution set is {-2}.

16. {2}

17. $\frac{1}{2} + \frac{2}{x} = \frac{1}{3} + \frac{3}{x}$, LCM = 6x

$$6x\left[\frac{1}{2} + \frac{2}{x}\right] = 6x\left[\frac{1}{3} + \frac{3}{x}\right]$$
$$3x + 12 = 2x + 18$$
$$3x - 2x = 18 - 12$$
$$x = 6$$

The solution set is {6}.

Check:

$$\begin{array}{c|c} \frac{1}{2} + \frac{2}{x} & = \frac{1}{3} + \frac{3}{x} \\ \hline \frac{1}{2} + \frac{2}{6} & \frac{1}{3} + \frac{3}{6} \\ \frac{1}{2} + \frac{1}{3} & \frac{1}{3} + \frac{1}{2} \end{array}$$

18. $\left\{\frac{11}{30}\right\}$

19. $\frac{4}{x^2 - 1} - \frac{2}{x - 1} = \frac{3}{x + 1}$,

LCM = (x + 1)(x - 1)

$$(x+1)(x-1)\left[\frac{4}{(x+1)(x-1)} - \frac{2}{x-1}\right] = (x+1)(x-1) \cdot \frac{3}{x+1}$$
$$4 - 2(x + 1) = 3(x - 1)$$
$$4 - 2x - 2 = 3x - 3$$
$$2 - 2x = 3x - 3$$
$$2 + 3 = 3x + 2x$$
$$5 = 5x$$
$$1 = x$$

Check:

$$\begin{array}{c|c} \frac{4}{x^2 - 1} - \frac{2}{x - 1} & = \frac{3}{x + 1} \\ \hline \frac{4}{1^2 - 1} - \frac{2}{1 - 1} & \frac{3}{1 + 1} \\ \frac{4}{0} - \frac{2}{0} & \frac{3}{2} \end{array}$$

Division by zero is undefined.

The solution set is ∅.

20. ∅

21. $\frac{1}{2t} - \frac{2}{5t} = \frac{1}{10t} - 3$, LCM = 10t

$$10t\left(\frac{1}{2t} - \frac{2}{5t}\right) = 10t\left(\frac{1}{10t} - 3\right)$$
$$5\cdot 1 - 2\cdot 2 = 1\cdot 1 - 10t\cdot 3$$
$$5 - 4 = 1 - 30t$$
$$1 = 1 - 30t$$
$$30t = 0$$
$$t = 0$$

Division by zero is undefined.

The solution set is $\emptyset$.

22. $\emptyset$

23. $1 - \frac{3}{x} = \frac{40}{x^2}$, LCM = x^2

$$x^2\left(1 - \frac{3}{x}\right) = x^2 \cdot \frac{40}{x^2}$$
$$x^2 - 3x = 40$$
$$x^2 - 3x - 40 = 0$$
$$(x - 8)(x + 5) = 0$$
$$x - 8 = 0 \text{ or } x + 5 = 0$$
$$x = 8 \text{ or } x = -5$$

The solution set is {8, -5}.

Check:

For 8:

$1 - \frac{3}{x} = \frac{40}{x^2}$	
$1 - \frac{3}{8}$	$\frac{40}{8^2}$
$\frac{5}{8}$	$\frac{40}{64}$
	$\frac{5}{8}$

For -5:

$1 - \frac{3}{x} = \frac{40}{x^2}$	
$1 - \frac{3}{-5}$	$\frac{40}{(-5)^2}$
$\frac{5}{5} + \frac{3}{5}$	$\frac{40}{25}$
$\frac{8}{5}$	$\frac{8}{5}$

24. {5, -3}

25. $\frac{11 - t^2}{3t^2 - 5t + 2} = \frac{2t + 3}{3t - 2} - \frac{t - 3}{t - 1}$,

LCM = (3t - 2)(t - 1)

$$(3t-2)(t-1)\cdot\frac{11 - t^2}{(3t-2)(t-1)} = (3t-2)(t-1)\left(\frac{2t+3}{3t-2} - \frac{t-3}{t-1}\right)$$
$$11 - t^2 = (t-1)(2t+3)-(3t-2)(t-3)$$
$$11 - t^2 = (2t^2+t-3) - (3t^2-11t+6)$$
$$11 - t^2 = 2t^2+t-3-3t^2+11t-6$$
$$11 - t^2 = -t^2 + 12t - 9$$
$$11 = 12t - 9$$
$$20 = 12t$$
$$\frac{20}{12} = t$$
$$\frac{5}{3} = t$$

The value checks. The solution set is $\left\{\frac{5}{3}\right\}$.

26. {1}

27. $\frac{2.315}{y} - \frac{12.6}{17.4} = \frac{6.71}{7} + 0.763$,

LCM = 7(17.4)y

$$7(17.4)y\left(\frac{2.315}{y} - \frac{12.6}{17.4}\right) = 7(17.4)y\left(\frac{6.71}{7} + 0.763\right)$$
$$281.967 - 88.2y = 116.754y + 92.9334y$$
$$281.967 - 88.2y = 209.6874y$$
$$281.967 = 297.8874y$$
$$0.94656 \approx y$$

The value checks. The solution set is {0.94656}.

28. {0.0855}

29. $\frac{(x - 3)^2}{x - 3} = x - 3$, LCM = x - 3

Note: $x \neq 3$
Division by 0 is undefined.

$$(x - 3)\cdot\frac{(x - 3)^2}{x - 3} = (x - 3)(x - 3)$$
$$(x - 3)^2 = (x - 3)^2$$

The solution set is $\{x \mid x \neq 3\}$.

30. $\{x \mid x \neq 2\}$

31. $\frac{x^3 + 8}{x + 2} = x^2 - 2x + 4$, LCM = x + 2

Note: $x \neq -2$
Division by 0 is undefined.

$$(x + 2)\cdot\frac{x^3 + 8}{x + 2} = (x + 2)(x^2 - 2x + 4)$$
$$x^3 + 8 = x^3 + 8$$

The solution set is $\{x \mid x \neq -2\}$.

32. $\{x \mid x \neq 2\}$

33. $\frac{x + 3}{x} = 3$, LCM = x

$$x\cdot\frac{x + 3}{x} = x\cdot 3$$
$$x + 3 = 3x$$
$$3 = 2x$$
$$\frac{3}{2} = x$$

The solution set is $\left\{\frac{3}{2}\right\}$.

Check:

$\frac{x + 3}{x} = 3$	
$\frac{\frac{3}{2} + 3}{\frac{3}{2}}$	3
$\frac{9}{2}\cdot\frac{2}{3}$	
3	

34. (1) equivalent to (2); (2) not equivalent to (3); (3) equivalent to (4)

35. $$\frac{x^2 + 6x - 16}{x - 2} = x + 8, \quad \text{LCM} = x - 2$$

(The sensible replacements for x are all numbers except 2.)

$$(x - 2) \cdot \frac{x^2 + 6x - 16}{x - 2} = (x - 2)(x + 8)$$
$$x^2 + 6x - 16 = x^2 + 6x - 16$$

The solution set is the set of all numbers except 2. Thus, the equation is an identity.

36. Identity

37. $(x - 1)(x^2 + x + 1) = x^3 - 1$

(All numbers are sensible replacements for x.)

$$x^3 - 1 = x^3 - 1$$

The solution set is the set of all numbers. Thus, the equation is an identity.

38. Identity

39.
$$(x + 7)^2 = x^2 + 49$$
$$x^2 + 14x + 49 = x^2 + 49$$
$$14x = 0$$
$$x = 0$$

All numbers are sensible replacements for the expressions $(x + 7)^2$ and $x^2 + 49$, but only the number 0 makes the equation true. Thus, the equation is not an identity.

40. Not an identity

Exercise Set 2.3

1.
$$P = 2\ell + 2w$$
$$P - 2\ell = 2w$$
$$\frac{P - 2\ell}{2} = w$$

2. $a = \frac{F}{m}$

3. $E = IR$
$$\frac{E}{R} = I$$

4. $m_2 = \frac{Fd^2}{km_1}$

5.
$$\frac{P_1V_1}{T_1} = \frac{P_2V_2}{T_2}$$
$$T_1T_2 \cdot \frac{P_1V_1}{T_1} = T_1T_2 \cdot \frac{P_2V_2}{T_2}$$
$$T_2P_1V_1 = T_1P_2V_2$$
$$\frac{T_2P_1V_1}{P_2V_2} = T_1$$

6. $V_2 = \frac{T_2P_1V_1}{T_1P_2}$

7.
$$S = \frac{H}{m(v_1 - v_2)}$$
$$m(v_1 - v_2)S = H$$
$$mSv_1 - mSv_2 = H$$
$$mSv_1 = H + mSv_2$$
$$v_1 = \frac{H + mSv_2}{mS}$$

or

$$S = \frac{H}{m(v_1 - v_2)}$$
$$m(v_1 - v_2)S = H$$
$$v_1 - v_2 = \frac{H}{mS}$$
$$v_1 = \frac{H}{mS} + v_2$$

8. $V_2 = V_1 - \frac{H}{Sm}$

9.
$$\frac{1}{F} = \frac{1}{m} + \frac{1}{p}$$
$$Fmp \cdot \frac{1}{F} = Fmp\left(\frac{1}{m} + \frac{1}{p}\right)$$
$$mp = Fp + Fm$$
$$mp - Fp = Fm$$
$$p(m - F) = Fm$$
$$p = \frac{Fm}{m - F}$$

10. $F = \frac{mp}{p + m}$

11.
$$(x + a)(x - b) = x^2 + 5$$
$$x^2 - bx + ax - ab = x^2 + 5$$
$$ax - bx = 5 + ab$$
$$x(a - b) = 5 + ab$$
$$x = \frac{5 + ab}{a - b}$$

12. $x = \frac{c^2}{2c}$ or $\frac{c}{2}$

13.
$$10(a + x) = 8(a - x)$$
$$10a + 10x = 8a - 8x$$
$$10x + 8x = 8a - 10a$$
$$18x = -2a$$
$$x = -\frac{2a}{18}$$
$$x = -\frac{a}{9}$$

14. $x = a - 7b$

15. Translate:

79.2 is what percent of 180?

$79.2 = x\% \cdot 180$

Carry out:

$79.2 = x\% \cdot 180$

$79.2 = x \cdot 0.01 \cdot 180$

$79.2 = x \cdot 1.8$

$\frac{79.2}{1.8} = x$

$44 = x$

Check:

$44\% \cdot 180 = 0.44 \cdot 180 = 79.2$

The answer is 44%.

16. 8000

17. Translate:

What percent of 28 is 1.68?

$x\% \cdot 28 = 1.68$

Carry out:

$x\% \cdot 28 = 1.68$

$x \cdot 0.01 \cdot 28 = 1.68$

$x \cdot 0.28 = 1.68$

$x = \frac{1.68}{0.28}$

$x = 6$

Check:

$6\% \cdot 28 = 0.06 \cdot 28 = 1.68$

The answer is 6%.

18. 3.171

19. Translate:

We let x represent the old salary.

11% of the old salary is \$1595.

$11\% \cdot x = 1595$

Carry out:

$11\% \cdot x = 1595$

$0.11x = 1595$

$x = \frac{1595}{0.11}$

$x = 14{,}500$

19. (continued)

Check:

$11\% \cdot 14{,}500 = 0.11 \cdot 14{,}500 = 1595$

The old salary is \$14,500. The new salary is \$14,500 + \$1595, or \$16,095.

20. \$21,000, \$23,520

21. Translate:

$A = P(1 + i)^t$

$702 = P(1 + 0.08)^1$ [Substituting 702 for A, 8% (or 0.08) for i and 1 for t]

Carry out:

$702 = 1.08P$

$\frac{702}{1.08} = P$

$650 = P$

Check:

$A = 650(1 + 0.08)^1 = 650(1.08) = 702$

The original investment was \$650.

22. \$850

23. Familiarize: We make a drawing.

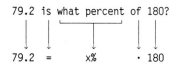

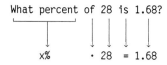

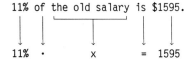

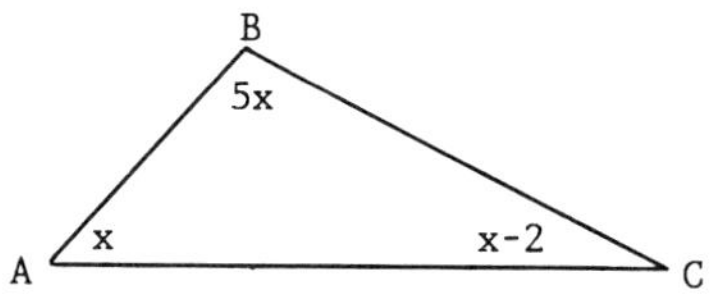

We let x represent the measure of angle A. Then 5x represents angle B, and x - 2 represents angle C. The sum of the angle measures is 180°.

Translate:

Measure of angle A + Measure of angle B + Measure of angle C = 180

$x + 5x + x - 2 = 180$

Carry out:

$x + 5x + x - 2 = 180$

$7x - 2 = 180$

$7x = 182$

$x = 26$

If x = 26, then $5x = 5 \cdot 26$, or 130, and $x - 2 = 26 - 2$, or 24.

23. (continued)

Check:

The measure of angle B, 130°, is five times the measure of angle A, 26°. The measure of angle C, 24°, is 2° less than the measure of angle A, 26°. The sum of the angle measures is 26° + 130° + 24°, or 180°.

Angle A measures 26°. Angle B measures 130°, and angle C measures 24°.

24. 40°, 80°, 60°

25. Familiarize: We make a drawing.

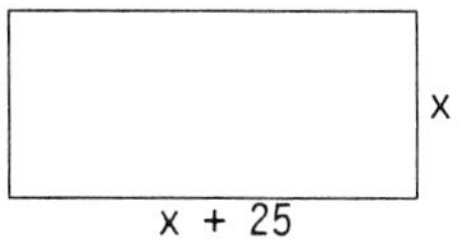

We let x represent the width. Then x + 25 represents the length.

Translate:

Perimeter = 2 × length + 2 × width

$$322 = 2(x + 25) + 2(x)$$

Carry out:

$322 = 2x + 50 + 2x$

$322 = 4x + 50$

$272 = 4x$

$68 = x$

If $x = 68$, then $x + 25 = 68 + 25$, or 93.

Check:

The length is 25 m more than the width: $93 = 68 + 25$. The perimeter is $2 \cdot 93 + 2 \cdot 68$, or $186 + 136$, or 322 m.

The length is 93 m; the width is 68 m.

26. 13 m, 6.5 m

27. Familiarize:

We let x represent the score on the fourth test. Then the average of the four scores is

$$\frac{87 + 64 + 78 + x}{4}.$$

Translate:

The average must be 80.

$$\frac{87 + 64 + 78 + x}{4} = 80$$

Carry out:

$87 + 64 + 78 + x = 4 \cdot 80$

$229 + x = 320$

$x = 91$

27. (continued)

Check:

The average of 87, 64, 78, and 91 is $\frac{87 + 64 + 78 + 91}{4}$, or $\frac{320}{4}$, or 80.

The score on the fourth test must be 91%.

28. 83%

29. Familiarize: We make a drawing.

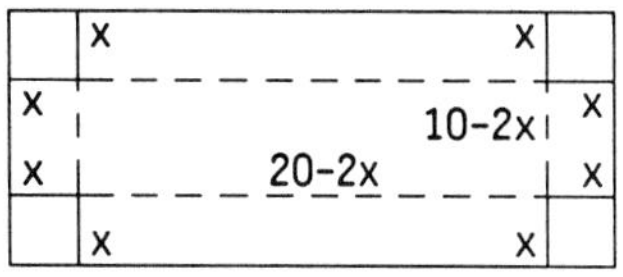

We let x represent the length of a side of the square in each corner. Then the length and width of the resulting base are represented by $20 - 2x$ and $10 - 2x$.

Translate:

The area of the base is 96 cm².

$(20 - 2x)(10 - 2x) = 96$

Carry out:

$200 - 60x + 4x^2 = 96$

$4x^2 - 60x + 104 = 0$

$x^2 - 15x + 26 = 0$

$(x - 13)(x - 2) = 0$

$x - 13 = 0$ or $x - 2 = 0$

$x = 13$ or $x = 2$

Since the length and width of the base cannot be negative, we only consider $x = 2$.

Check:

If $x = 2$, then $20 - 2x = 16$ and $10 - 2x = 6$. The area of the base is $16 \cdot 6$, or 96 cm².

The length of the sides of the squares is 2 cm.

30. 8 cm

31. Translate:

Former population	+	Increase in population	=	Present population
x	+	2%x	=	826,200

Carry out:

$$x + 0.02x = 826{,}200$$
$$1.02x = 826{,}200$$
$$x = \frac{826{,}200}{1.02}$$
$$x = 810{,}000$$

Check:

2%·810,000 = 0.02·810,000 = 16,200. The present population is 810,000 + 16,200, or 826,200.

The former population was 810,000.

32. 720,000

33. Familiarize:

We first make a drawing. We let r represent the speed of the boat in still water. Then the speed downstream is r + 3, and the speed upstream is r - 3. The time to go downstream is the same as the time to go upstream. We call it t.

t hours Downstream →
50 km r + 3 km/h

t hours Upstream ←
30 km r - 3 km/h

We organize the information in a table.

	Distance	Speed	Time
Downstream	50	r + 3	t
Upstream	30	r - 3	t

Translate:

Using t = d/r we can get two expressions for t from the table.

$$t = \frac{50}{r + 3} \qquad t = \frac{30}{r - 3}$$

Thus

$$\frac{50}{r + 3} = \frac{30}{r - 3}$$

33. (continued)

Carry out:

$$(r + 3)(r - 3) \cdot \frac{50}{r + 3} = (r + 3)(r - 3) \cdot \frac{30}{r - 3}$$

(Multiplying by the LCM)

$$50(r - 3) = 30(r + 3)$$
$$50r - 150 = 30r + 90$$
$$50r - 30r = 90 + 150$$
$$20r = 240$$
$$r = 12$$

Check:

If r = 12, the speed downstream is 12 + 3, or 15 km/h, and thus the time downstream is 50/15, or $3.\overline{3}$ hr. If r = 12, the speed upstream is 12 - 3, or 9 km/h, and thus the time upstream is 30/9 or $3.\overline{3}$ hr. The times are equal; the value checks.

The speed of the boat in still water is 12 km/h.

34. 4 km/h

35. Familiarize:

We organize the information in a table. We let r represent the speed of train B. Then the speed of train A is r - 12. The time, t, is the same for each train.

	Distance	Speed	Time
Train A	230	r - 12	t
Train B	290	r	t

Translate:

Using t = d/r we get two expressions for t from the table.

$$t = \frac{230}{r - 12} \qquad t = \frac{290}{r}$$

Thus

$$\frac{230}{r - 12} = \frac{290}{r}$$

Carry out:

$$r(r - 12) \cdot \frac{230}{r - 12} = r(r - 12) \cdot \frac{290}{r}$$
$$230r = 290(r - 12)$$
$$230r = 290r - 3480$$
$$3480 = 60r$$
$$58 = r$$

Check:

If r = 58, then the speed of train A is 58 - 12, or 46 mph. Thus the time for train A is 230/46, or 5 hr. If the speed of train B is 58 mph, the time for train B is 290/58, or 5 hr. The times are the same; the value checks.

The speed of train A is 46 mph; the speed of train B is 58 mph.

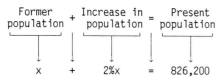

36. Passenger: 80 mph; Freight: 66 mph

37. Familiarize:

We first make a drawing.

Chicago → ← Cleveland

475 mph 500 mph

t hours $t - \frac{1}{3}$ hours

d_1 miles d_2 miles

← 350 miles →

We organize the information in a table.

	Distance	Speed	Time
From Chicago	d_1	475	t
From Cleveland	d_2	500	$t - \frac{1}{3}$

Translate:

Using $d = rt$ we get two equations from the table.

$d_1 = 475t$ $d_2 = 500\left(t - \frac{1}{3}\right)$

We also know that $d_1 + d_2 = 350$. Thus

$475t + 500\left(t - \frac{1}{3}\right) = 350$

Carry out:

$$475t + 500t - \frac{500}{3} = 350$$
$$975t = \frac{1050}{3} + \frac{500}{3}$$
$$975t = \frac{1550}{3}$$
$$2925t = 1550$$
$$t = \frac{1550}{2925}$$
$$t = \frac{62}{117}$$

Substitute $\frac{62}{117}$ for t and solve for d_2.

$$d_2 = 500\left(t - \frac{1}{3}\right)$$
$$d_2 = 500\left(\frac{62}{117} - \frac{1}{3}\right) = 500 \cdot \frac{23}{117} \approx 98.3$$

Check:

If $t = \frac{62}{117}$, then $d_1 = 475 \cdot \frac{62}{117}$, or ≈ 251.7. The sum of the distances is $98.3 + 251.7$, or 350 miles. The value checks.

When the trains meet, they are 98.3 miles from Cleveland.

38. 450 km

39. Familiarize:

A does $\frac{1}{3}$ of the job in 1 hr.

B does $\frac{1}{5}$ of the job in 1 hr.

C does $\frac{1}{7}$ of the job in 1 hr.

Together they do $\frac{1}{3} + \frac{1}{5} + \frac{1}{7}$ of the job in 1 hr.

Translate:

Letting t represent the number of hours the job would take with all working together, then 1/t of the job would be completed in 1 hour. This gives us an equation.

$$\frac{1}{3} + \frac{1}{5} + \frac{1}{7} = \frac{1}{t}$$

Carry out:

$$105t\left(\frac{1}{3} + \frac{1}{5} + \frac{1}{7}\right) = 105t \cdot \frac{1}{t} \quad \text{(Multiplying by the LCM)}$$
$$35t + 21t + 15t = 105$$
$$71t = 105$$
$$t = \frac{105}{71}$$
$$t = 1\frac{34}{71} \text{ hr}$$

The value checks. Thus, when working together, it takes $1\frac{34}{71}$ hr to do the job.

40. $2\frac{4}{13}$ hr

41. Familiarize and Translate:

We let t represent the time B can do a certain job working alone.

A does $\frac{1}{3.15}$ of the job in 1 hr.

B does $\frac{1}{t}$ of the job in 1 hr.

Together they do $\frac{1}{3.15} + \frac{1}{t}$, or $\frac{1}{2.09}$ of the job in 1 hour.

We now have an equation.

$$\frac{1}{3.15} + \frac{1}{t} = \frac{1}{2.09}$$

Carry out:

$$3.15(2.09)t \cdot \left(\frac{1}{3.15} + \frac{1}{t}\right) = 3.15(2.09)t \cdot \frac{1}{2.09}$$

(Multiplying by a common denominator)

$$2.09t + 6.5835 = 3.15t$$
$$6.5835 = 1.06t$$
$$\frac{6.5835}{1.06} = t$$
$$6.21 \approx t$$

41. (continued)

Check:

In 1 hour A does $\frac{1}{3.15}$ of the job and B does $\frac{1}{6.21}$ of the job. Together they do $\frac{1}{3.15} + \frac{1}{6.21}$, or 0.3175 + 0.1610, or 0.4785, of the job in 1 hour. If working together it takes them 2.09 hours, then they do $\frac{1}{2.09}$, or 0.4785, of the job in 1 hour. The value checks.

It would take B, working alone, 6.21 hours to do the job.

42. A: 23.95 hr, B: 51.02 hr

43. a) $A = P(1 + i)^t$

$A = 1000(1 + 0.1375)^1$

$= \$1137.50$

b) $A = P\left(1 + \frac{i}{2}\right)^{2t}$

$A = 1000\left(1 + \frac{0.1375}{2}\right)^2$

$= \$1142.23$

c) $A = P\left(1 + \frac{i}{4}\right)^{4t}$

$A = 1000\left(1 + \frac{0.1375}{4}\right)^4$

$= \$1144.75$

d) $A = P\left(1 + \frac{i}{365}\right)^{365t}$

$A = 1000\left(1 + \frac{0.1375}{365}\right)^{365}$

$= \$1147.37$

e) $A = P\left(1 + \frac{i}{8760}\right)^{8760t}$

$A = 1000\left(1 + \frac{0.1375}{8760}\right)^{8760}$

$= \$1147.40$

44. a) \$2055.46, b) \$2109.47, c) \$2139.05, d) \$2170.24, e) \$2170.58

45. Familiarize:

We organize the information in a table.

	Distance	Speed	Time
Slow trip	144	r	t
Fast trip	144	$r + 4$	$t - \frac{1}{2}$

We let r represent the speed for the slow trip and t the time for the slow trip. Then $r + 4$ and $t - \frac{1}{2}$ represent the speed and time respectively for the fast trip.

Translate:

Using $t = d/r$ we get two equations from the table.

$t = \frac{144}{r}$ $\quad$ $t - \frac{1}{2} = \frac{144}{r + 4}$

or $t = \frac{144}{r + 4} + \frac{1}{2}$

This gives us the following equation:

$\frac{144}{r} = \frac{144}{r + 4} + \frac{1}{2}$

Carry out:

$2r(r + 4) \cdot \frac{144}{r} = 2r(r + 4)\cdot\left(\frac{144}{r + 4} + \frac{1}{2}\right)$

(Multiplying by the LCM)

$288(r + 4) = 288r + r(r + 4)$

$288r + 1152 = 288r + r^2 + 4r$

$0 = r^2 + 4r - 1152$

$0 = (r + 36)(r - 32)$

$r + 36 = 0$ or $r - 32 = 0$

$r = -36$ or $r = 32$

Check:

We only consider $r = 32$ since the speed in this problem cannot be negative. If $r = 32$, then the time for the slow trip is $\frac{144}{32}$, or 4.5 hr. If $r = 32$, then $r + 4 = 36$. Thus, the time for the fast trip is $\frac{144}{36}$, or 4 hr. The time for the fast trip is $\frac{1}{2}$ hr less than the time for the slow trip. The value checks.

The car's speed is 32 mph.

46. 35 mph

47. Freeway: $D = rt$

$D = 55 \cdot 3$, or 165

City: $t = \frac{d}{r}$

$t = \frac{10}{35}$, or $\frac{2}{7}$

We organize the information in a table.

	Distance	Speed	Time
Freeway	165	55	3
City	10	35	$\frac{10}{35}$, or $\frac{2}{7}$
Totals	175		$3\frac{2}{7}$ or $\frac{23}{7}$

$$\text{Average speed} = \frac{\text{Total distance}}{\text{Total time}}$$

$$\text{Average speed} = \frac{175}{\frac{23}{7}}$$

$$= 175 \cdot \frac{7}{23}$$

$$\approx 53.26 \text{ mph}$$

48. 48 km/h

49. Making a drawing is helpful.

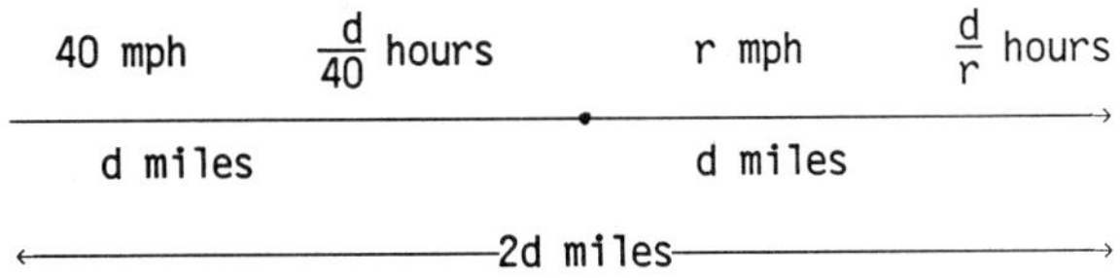

We let d represent the distance of the first half of the trip. Then 2d represents the total distance. We also let r represent the speed for the second half of the trip. The times are represented by $\frac{d}{40}$ and $\frac{d}{r}$. The total time is $\frac{d}{40} + \frac{d}{r}$.

$$\text{Average speed} = \frac{\text{Total distance}}{\text{Total time}}$$

Substituting we get

$$45 = \frac{2d}{\frac{d}{40} + \frac{d}{r}}$$

$$45\left(\frac{d}{40} + \frac{d}{r}\right) = 2d$$

$$45\left(\frac{1}{40} + \frac{1}{r}\right)d = 2d$$

$$\frac{45}{40} + \frac{45}{r} = 2$$

$$\frac{45}{r} = 2 - \frac{45}{40}$$

$$\frac{45}{r} = \frac{35}{40}$$

49. (continued)

$$40 \cdot 45 = r \cdot 35$$

$$1800 = 35r$$

$$\frac{1800}{35} = r$$

$$51\frac{3}{7} = r$$

The value checks. The speed for the second half of the trip is $51\frac{3}{7}$ mph.

50. $4{:}21\frac{9}{11}$

51. It is helpful to make drawings.

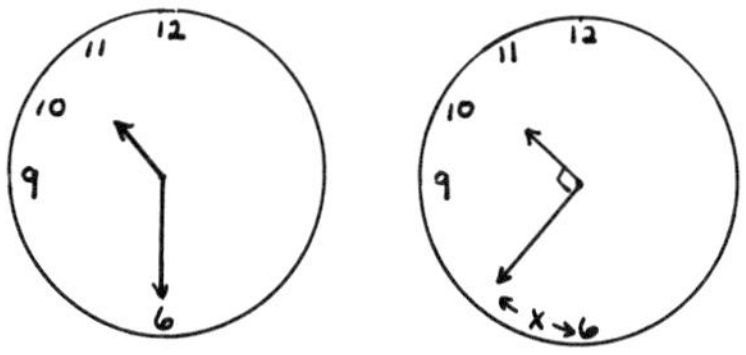

At 10:30 the minute hand is 30 units after the 12. The hour hand is $52\frac{1}{2}$ units after the 12. Let x represent the number of units the minute hand moves before the hands are perpendicular for the first time. When the minute hand moves x units, the hour hand moves $\frac{1}{12}$ x units. Then the minute hand is 30 + x units after the 12, and the hour hand is $52\frac{1}{2} + \frac{1}{12}x$ after the 12. When the hands are perpendicular, they must be 15 units apart. Thus, we have the following equation:

$$\left(52\frac{1}{2} + \frac{1}{12}x\right) - (30 + x) = 15$$

$$52\frac{1}{2} + \frac{1}{12}x - 30 - x = 15$$

$$-\frac{11}{12}x + 22\frac{1}{2} = 15$$

$$\frac{15}{2} = \frac{11}{12}x$$

$$\frac{12}{11} \cdot \frac{15}{2} = x$$

$$8\frac{2}{11} = \frac{90}{11} = x$$

The hands will be perpendicular when the minute hand moves $8\frac{2}{11}$ units. The time will be $10{:}38\frac{2}{11}$.

52. $\frac{2}{3}$ hr

Exercise Set 2.4

1. $\sqrt{-15} = \sqrt{-1\cdot 15} = \sqrt{-1}\,\sqrt{15} = i\sqrt{15}$

2. $i\sqrt{17}$

3. $\sqrt{-81} = \sqrt{-1\cdot 81} = \sqrt{-1}\,\sqrt{81} = 9i$

4. $5i$

5. $-\sqrt{-12} = -\sqrt{-1\cdot 4\cdot 3} = -\sqrt{-1}\,\sqrt{4}\,\sqrt{3} = -2i\sqrt{3}$

6. $-2i\sqrt{5}$

7. $\sqrt{-16} + \sqrt{-25} = i\sqrt{16} + i\sqrt{25} = 4i + 5i = 9i$

8. $4i$

9. $\sqrt{-7} - \sqrt{-10} = i\sqrt{7} - i\sqrt{10} = (\sqrt{7} - \sqrt{10})i$

10. $(\sqrt{5} + \sqrt{7})i$

11. $\sqrt{-5}\,\sqrt{-11} = i\sqrt{5}\cdot i\sqrt{11} = i^2\,\sqrt{5}\,\sqrt{11} = -\sqrt{55}$

12. $-2\sqrt{14}$

13. $-\sqrt{-4}\,\sqrt{-5} = -(i\sqrt{4}\cdot i\sqrt{5}) = -(i^2\cdot 2\cdot\sqrt{5})$
$= -(-1)2\sqrt{5}$
$= 2\sqrt{5}$

14. $3\sqrt{7}$

15. $\dfrac{-\sqrt{5}}{\sqrt{-2}} = \dfrac{-\sqrt{5}}{i\sqrt{2}} = \dfrac{-\sqrt{5}}{i\sqrt{2}}\cdot\dfrac{i}{i} = \dfrac{-i\sqrt{5}}{i^2\sqrt{2}}$
$= \dfrac{-i\sqrt{5}}{-1\sqrt{2}}$
$= \sqrt{\dfrac{5}{2}}\,i$

16. $-\sqrt{\dfrac{7}{5}}\,i$

17. $\dfrac{\sqrt{-9}}{-\sqrt{4}} = \dfrac{3i}{-2} = -\dfrac{3}{2}i$

18. $\dfrac{5}{4}i$

19. $\dfrac{-\sqrt{-36}}{\sqrt{-9}} = \dfrac{-i\sqrt{36}}{i\sqrt{9}} = -\dfrac{6}{3} = -2$

20. $-\dfrac{5}{4}$

21. $(2 + 3i) + (4 + 2i) = 2 + 4 + 3i + 2i = 6 + 5i$

22. $11 + i$

23. $(4 + 3i) + (4 - 3i) = 4 + 4 + 3i - 3i = 8$

24. 0

25. $(8 + 11i) - (6 + 7i) = 8 + 11i - 6 - 7i$
$= 8 - 6 + 11i - 7i$
$= 2 + 4i$

26. $5 - 7i$

27. $2i - (4 + 3i) = 2i - 4 - 3i = -4 - i$

28. $-5 + i$

29. $(1 + 2i)(1 + 3i) = 1 + 3i + 2i + 6i^2$
$= 1 + 5i - 6$ $\quad (i^2 = -1)$
$= -5 + 5i$

30. $13 + i$

31. $(1 + 2i)(1 - 3i) = 1 - 3i + 2i - 6i^2$
$= 1 - i + 6$ $\quad (i^2 = -1)$
$= 7 - i$

32. 13

33. $3i(4 + 2i) = 12i + 6i^2 = 12i - 6$, or $-6 + 12i$

34. $20 + 15i$

35. $(2 + 3i)^2 = 4 + 12i + 9i^2$
$= 4 + 12i - 9$ $\quad (i^2 = -1)$
$= -5 + 12i$

36. $5 - 12i$

37. $i^{13} = i^{12}\cdot i = 1\cdot i = i$ $\quad (i^{12} = (i^4)^3 = 1^3 = 1)$

Note that i^{12} is a power of i^4 and $i^4 = 1$.

38. 1

39. $4x^2 + 25y^2 = (2x + 5yi)(2x - 5yi)$

Check by multiplying:
$(2x + 5yi)(2x - 5yi) = 4x^2 - 25y^2i^2 = 4x^2 + 25y^2$

40. $(4a + 7bi)(4a - 7bi)$

41.

$x^2 - 2x + 5 = 0$	
$(1 + 2i)^2 - 2(1 + 2i) + 5$	0
$1 + 4i + 4i^2 - 2 - 4i + 5$	
$4i^2 + 4$	
$-4 + 4$	
0	

The number $1 + 2i$ is a solution.

42. Yes

43. $4x + 7i = -6 + yi$

We equate the real parts and solve for x.

$4x = -6$

$x = -\frac{3}{2}$

We equate the imaginary parts and solve for y.

$7i = yi$

$7 = y$

44. $x = 3, \quad y = 2$

45. $\frac{4 + 3i}{1 - i} = \frac{4 + 3i}{1 - i} \cdot \frac{1 + i}{1 + i}$

$= \frac{4 + 7i + 3i^2}{1 - i^2} = \frac{1 + 7i}{2} = \frac{1}{2} + \frac{7}{2} i$

46. $\frac{22}{41} - \frac{7}{41} i$

47. $\frac{\sqrt{2} + i}{\sqrt{2} - i} = \frac{\sqrt{2} + i}{\sqrt{2} - i} \cdot \frac{\sqrt{2} + i}{\sqrt{2} + i}$

$= \frac{2 + 2\sqrt{2}\, i + i^2}{2 - i^2} = \frac{1 + 2\sqrt{2}\, i}{3} = \frac{1}{3} + \frac{2\sqrt{2}}{3} i$

48. $\frac{1}{2} + \frac{\sqrt{3}}{2} i$

49. $\frac{3 + 2i}{i} = \frac{3 + 2i}{i} \cdot \frac{-i}{-i}$

$= \frac{-3i - 2i^2}{-i^2} = \frac{2 - 3i}{1} = 2 - 3i$

50. $3 - 2i$

51. $\frac{i}{2 + i} = \frac{i}{2 + i} \cdot \frac{2 - i}{2 - i}$

$= \frac{2i - i^2}{4 - i^2} = \frac{1 + 2i}{5} = \frac{1}{5} + \frac{2}{5} i$

52. $\frac{15}{146} + \frac{33}{146} i$

53. $\frac{1 - i}{(1 + i)^2} = \frac{1 - i}{1 + 2i + i^2} = \frac{1 - i}{2i}$

$= \frac{1 - i}{2i} \cdot \frac{-2i}{-2i} = \frac{-2i + 2i^2}{-4i^2}$

$= \frac{-2 - 2i}{4} = -\frac{1}{2} - \frac{1}{2} i$

54. $-\frac{1}{2} + \frac{1}{2} i$

55. $\frac{3 - 4i}{(2 + i)(3 - 2i)} = \frac{3 - 4i}{6 - i - 2i^2}$

$= \frac{3 - 4i}{8 - i} \cdot \frac{8 + i}{8 + i}$

$= \frac{24 - 29i - 4i^2}{64 - i^2}$

$= \frac{28 - 29i}{65} = \frac{28}{65} - \frac{29}{65} i$

56. $\frac{719}{3233} + \frac{955}{3233} i$

57. $\frac{1 + i}{1 - i} \cdot \frac{2 - i}{1 - i} = \frac{2 + i - i^2}{1 - 2i + i^2}$

$= \frac{3 + i}{-2i} \cdot \frac{2i}{2i}$

$= \frac{6i + 2i^2}{-4i^2}$

$= \frac{-2 + 6i}{4} = -\frac{1}{2} + \frac{3}{2} i$

58. $-\frac{1}{2} - \frac{3}{2} i$

59. $\frac{3 + 2i}{1 - i} + \frac{6 + 2i}{1 - i} = \frac{3 + 6 + 2i + 2i}{1 - i}$

$= \frac{9 + 4i}{1 - i} \cdot \frac{1 + i}{1 + i}$

$= \frac{9 + 13i + 4i^2}{1 - i^2}$

$= \frac{5 + 13i}{2} = \frac{5}{2} + \frac{13}{2} i$

60. $-\frac{1}{2} - \frac{13}{2} i$

61. The reciprocal of $4 + 3i$ is $\frac{1}{4 + 3i}$, or

$\frac{1}{4 + 3i} \cdot \frac{4 - 3i}{4 - 3i} = \frac{4 - 3i}{16 - 9i^2} = \frac{4 - 3i}{25} = \frac{4}{25} - \frac{3}{25} i.$

62. $\frac{4}{25} + \frac{3}{25} i$

63. The reciprocal of $5 - 2i$ is $\frac{1}{5 - 2i}$, or

$\frac{1}{5 - 2i} \cdot \frac{5 + 2i}{5 + 2i} = \frac{5 + 2i}{25 - 4i^2} = \frac{5 + 2i}{29} = \frac{5}{29} + \frac{2}{29} i.$

64. $\frac{2}{29} - \frac{5}{29} i$

65. The reciprocal of i is $\frac{1}{i}$, or
$\frac{1}{i} \cdot \frac{-i}{-i} = \frac{-i}{-i^2} = \frac{-i}{1} = -i.$

66. i

67. The reciprocal of -4i is $\frac{1}{-4i}$, or
$\frac{1}{-4i} \cdot \frac{4i}{4i} = \frac{4i}{-16i^2} = \frac{4i}{16} = \frac{i}{4}.$

68. $-\frac{i}{5}$

69. $(3 + i)x + i = 5i$
$(3 + i)x = 4i$
$x = \frac{4i}{3 + i}$
$x = \frac{4i}{3 + i} \cdot \frac{3 - i}{3 - i}$
$x = \frac{12i - 4i^2}{9 - i^2} = \frac{4 + 12i}{10}$
$x = \frac{2}{5} + \frac{6}{5}i$

70. $\frac{12}{5} - \frac{1}{5}i$

71. $2ix + 5 - 4i = (2 + 3i)x - 2i$
$5 - 4i + 2i = (2 + 3i)x - 2ix$
$5 - 2i = (2 + i)x$
$\frac{5 - 2i}{2 + i} = x$
$\frac{2 - i}{2 - i} \cdot \frac{5 - 2i}{2 + i} = x$
$\frac{8 - 9i}{5} = \frac{10 - 9i + 2i^2}{4 - i^2} = x$
$\frac{8}{5} - \frac{9}{5}i = x$

72. $\frac{8}{29} + \frac{9}{29}i$

73. $(1 + 2i)x + 3 - 2i = 4 - 5i + 3ix$
$(1 + 2i)x - 3ix = 4 - 5i - 3 + 2i$
$(1 - i)x = 1 - 3i$
$x = \frac{1 - 3i}{1 - i}$
$x = \frac{1 - 3i}{1 - i} \cdot \frac{1 + i}{1 + i}$
$x = \frac{1 - 2i - 3i^2}{1 - i^2} = \frac{4 - 2i}{2}$
$x = 2 - i$

74. $-\frac{1}{5} + \frac{7}{5}i$

75. $(5 + i)x + 1 - 3i = (2 - 3i)x + 2 - i$
$(5 + i)x - (2 - 3i)x = 2 - i - 1 + 3i$
$(3 + 4i)x = 1 + 2i$
$x = \frac{1 + 2i}{3 + 4i}$
$x = \frac{1 + 2i}{3 + 4i} \cdot \frac{3 - 4i}{3 - 4i}$
$x = \frac{3 + 2i - 8i^2}{9 - 16i^2} = \frac{11 + 2i}{25}$
$x = \frac{11}{25} + \frac{2}{25}i$

76. $\frac{4}{5} + \frac{3}{5}i$

77. For example, $\sqrt{-1}\sqrt{-1} = i^2 = -1$
but $\sqrt{(-1)(-1)} = \sqrt{1} = 1$.

78. For example, $\sqrt{\frac{4}{-1}} = \sqrt{-4} = 2i$, but $\frac{\sqrt{4}}{\sqrt{-1}} = \frac{2}{i} = -2i$

79. Let $z = a + bi$. Then $z \cdot \overline{z} = (a + bi)(a - bi) = a^2 - b^2i^2 = a^2 + b^2$. Since a and b are real numbers, so is $a^2 + b^2$. Thus $z \cdot \overline{z}$ is real.

80. Let $z = a + bi$. Then $z + \overline{z} = (a + bi) + (a - bi) = 2a$. Since a is a real number, 2a is real. Thus $z + \overline{z}$ is real.

81. Let $z = a + bi$ and $w = c + di$. Then $\overline{z + w} = \overline{(a + bi) + (c + di)} = \overline{(a + c) + (b + d)i}$, by adding.

We now take the conjugate and obtain $(a + c) - (b + d)i$. Now $\overline{z} + \overline{w} = \overline{(a + bi)} + \overline{(c + di)} = (a - bi) + (c - di)$, taking the conjugate. We will now add to obtain $(a + c) - (b + d)i$, the same result as before. Thus $\overline{z + w} = \overline{z} + \overline{w}$.

82. Let $z = a + bi$ and $w = c + di$. Then $\overline{z \cdot w} = \overline{(a + bi)(c + di)} = \overline{(ac - bd) + (ad + bc)i} = (ac + bd) - (ad + bc)i$, taking the conjugate. Now $\overline{z} \cdot \overline{w} = \overline{(a + bi)} \cdot \overline{(c + di)} = (a - bi)(c - di) = (ac + bd) - (ad + bc)i$, the same result as before. Thus $\overline{z \cdot w} = \overline{z} \cdot \overline{w}$.

83. By definition of exponents the conjugate of z^n is the conjugate of the product of n factors of z. Using the result of Exercise 82, the conjugate of n factors of z is the product of n factors of $\overline{z}$. Thus $\overline{z^n} = \overline{z}^n$.

84. If z is a real number, then $z = a + 0i = a$ and $\overline{z} = a - 0i = a$. Thus $\overline{z} = z$.

85. $\overline{3z^5 - 4z^2 + 3z - 5}$

$= \overline{3z^5} - \overline{4z^2} + \overline{3z} - \overline{5}$ By Exercise 81

$= \overline{3}\,\overline{z^5} - \overline{4}\,\overline{z^2} + \overline{3}\,\overline{z} - \overline{5}$ By Exercise 82

$= 3\,\overline{z^5} - 4\,\overline{z^2} + 3\,\overline{z} - 5$ By Exercise 84

$= 3\overline{z}^5 - 4\overline{z}^2 + 3\overline{z} - 5$ By Exercise 83

86. 1

87. Solve $5z - 4\overline{z} = 7 + 8i$ for z.

Let $z = a + bi$. Then $\overline{z} = a - bi$.

$$5z - 4\overline{z} = 7 + 8i$$
$$5(a + bi) - 4(a - bi) = 7 + 8i$$
$$5a + 5bi - 4a + 4bi = 7 + 8i$$
$$a + 9bi = 7 + 8i$$

Equate the real parts.

$$a = 7$$

Equate the imaginary parts.

$$9b = 8$$
$$b = \frac{8}{9}$$

Then $z = a + bi$

$$z = 7 + \frac{8}{9}i$$

88. a

89. Let $z = a + bi$. Then $\overline{z} = a - bi$.

$$\frac{1}{2}i(\overline{z} - z) = \frac{1}{2}i[(a - bi) - (a + bi)]$$
$$= \frac{1}{2}i(-2bi) = -bi^2 = b$$

Exercise Set 2.5

1. $4x^2 = 20$

$x^2 = 5$

$x = \sqrt{5}$ or $x = -\sqrt{5}$

Check:

$4x^2$	$= 20$
$4(\pm\sqrt{5})^2$	20
$4\cdot 5$	
20	

The solutions are $\sqrt{5}$ and $-\sqrt{5}$, or $\pm\sqrt{5}$.

2. $\pm\sqrt{7}$

3. $10x^2 = 0$

$x^2 = 0$

$x = 0$

Check:

$10x^2$	$= 0$
$10\cdot 0^2$	0
$10\cdot 0$	
0	

The solution is 0.

4. 0

5. $2x^2 - 3 = 0$

$2x^2 = 3$

$x^2 = \frac{3}{2}$

$x = \sqrt{\frac{3}{2}}$ or $x = -\sqrt{\frac{3}{2}}$

$x = \sqrt{\frac{3}{2}\cdot\frac{2}{2}}$ or $x = -\sqrt{\frac{3}{2}\cdot\frac{2}{2}}$

$x = \frac{\sqrt{6}}{2}$ or $x = -\frac{\sqrt{6}}{2}$

Check:

$2x^2 - 3$	$= 0$
$2\left(\pm\frac{\sqrt{6}}{2}\right)^2 - 3$	0
$2\cdot\frac{6}{4} - 3$	
$3 - 3$	
0	

The solutions are $\frac{\sqrt{6}}{2}$ and $-\frac{\sqrt{6}}{2}$, or $\pm\frac{\sqrt{6}}{2}$.

6. $\pm\frac{\sqrt{21}}{3}$

7. $2x^2 + 14 = 0$

$2x^2 = -14$

$x^2 = -7$

$x = \sqrt{-7}$ or $x = -\sqrt{-7}$

$x = i\sqrt{7}$ or $x = -i\sqrt{7}$

These numbers check, so the solutions are $i\sqrt{7}$ and $-i\sqrt{7}$, or $\pm i\sqrt{7}$.

8. $\pm i\sqrt{5}$

9. $ax^2 = b$

$x^2 = \frac{b}{a}$

$x = \pm\sqrt{\frac{b}{a}}$, or $\pm\frac{\sqrt{ba}}{a}$

The solution set is $\left\{\sqrt{\frac{b}{a}}, -\sqrt{\frac{b}{a}}\right\}$, or $\left\{\pm\sqrt{\frac{b}{a}}\right\}$.

10. $\pm\sqrt{\frac{k}{\pi}}$

11. $(x - 7)^2 = 5$

$x - 7 = \pm\sqrt{5}$

$x = 7 \pm\sqrt{5}$

The solution set is $\{7 + \sqrt{5}, 7 - \sqrt{5}\}$, or $\{7 \pm\sqrt{5}\}$.

12. $\{-3 \pm \sqrt{2}\}$

13. $\frac{4}{9}x^2 - 1 = 0$

$\frac{4}{9}x^2 = 1$

$x^2 = \frac{9}{4}$

$x = \sqrt{\frac{9}{4}}$ or $x = -\sqrt{\frac{9}{4}}$

$x = \frac{3}{2}$ or $x = -\frac{3}{2}$

Both numbers check, so the solutions are $\frac{3}{2}$ and $-\frac{3}{2}$, or $\pm\frac{3}{2}$.

14. $\left\{\pm \frac{5}{4}\right\}$

15. $(x - h)^2 = a$

$x - h = \pm\sqrt{a}$

$x = h \pm \sqrt{a}$

The solution set is $\{h + \sqrt{a}, h - \sqrt{a}\}$, or $\{h \pm \sqrt{a}\}$.

16. $\left\{h \pm \sqrt{\frac{y - k}{a}}\right\}$

17. $x^2 + 6x + 4 = 0$

$x^2 + 6x + 9 - 9 + 4 = 0$ $\left[\frac{1}{2} \cdot 6 = 3,\ 3^2 = 9;\right.$ thus we add $9 - 9.\left.\right]$

$(x + 3)^2 - 5 = 0$

$(x + 3)^2 = 5$

$x + 3 = \pm\sqrt{5}$

$x = -3 \pm \sqrt{5}$

The solution set is $\{-3 \pm \sqrt{5}\}$.

18. $\{3 \pm \sqrt{13}\}$

19. $y^2 + 7y - 30 = 0$

$y^2 + 7y + \frac{49}{4} - \frac{49}{4} - 30 = 0$ $\left[\frac{1}{2} \cdot 7 = \frac{7}{2},\ \left(\frac{7}{2}\right)^2 = \frac{49}{4};\right.$ thus we add $\left.\frac{49}{4} - \frac{49}{4}\right]$

$\left(y + \frac{7}{2}\right)^2 - \frac{169}{4} = 0$

$\left(y + \frac{7}{2}\right)^2 = \frac{169}{4}$

$y + \frac{7}{2} = \pm\sqrt{\frac{169}{4}} = \pm\frac{13}{2}$

$y = -\frac{7}{2} \pm \frac{13}{2} = \frac{-7 \pm 13}{2}$

$y = \frac{-7 + 13}{2}$ or $y = \frac{-7 - 13}{2}$

$y = 3$ or $y = -10$

The solution set is $\{3, -10\}$.

20. $\{-3, 10\}$

21. $5x^2 - 4x - 2 = 0$

$x^2 - \frac{4}{5}x - \frac{2}{5} = 0$ $\left[\text{Multiplying by } \frac{1}{5}\right]$

$x^2 - \frac{4}{5}x + \frac{4}{25} - \frac{4}{25} - \frac{2}{5} = 0$ $\left[\text{Adding } \frac{4}{25} - \frac{4}{25}\right]$

$\left(x - \frac{2}{5}\right)^2 - \frac{14}{25} = 0$

$\left(x - \frac{2}{5}\right)^2 = \frac{14}{25}$

$x - \frac{2}{5} = \pm\sqrt{\frac{14}{25}} = \pm\frac{\sqrt{14}}{5}$

$x = \frac{2}{5} \pm \frac{\sqrt{14}}{5} = \frac{2 \pm \sqrt{14}}{5}$

The solution set is $\left\{\frac{2 \pm \sqrt{14}}{5}\right\}$.

22. $\left\{\frac{7 \pm \sqrt{13}}{12}\right\}$

23. $2x^2 + 7x - 15 = 0$

$x^2 + \frac{7}{2}x - \frac{15}{2} = 0$ $\left[\text{Multiplying by } \frac{1}{2}\right]$

$x^2 + \frac{7}{2}x + \frac{49}{16} - \frac{49}{16} - \frac{15}{2} = 0$ $\left[\text{Adding } \frac{49}{16} - \frac{49}{16}\right]$

$\left(x + \frac{7}{4}\right)^2 - \frac{169}{16} = 0$

$\left(x + \frac{7}{4}\right)^2 = \frac{169}{16}$

$x + \frac{7}{4} = \pm\sqrt{\frac{169}{16}} = \pm\frac{13}{4}$

$x = -\frac{7}{4} \pm \frac{13}{4} = \frac{-7 \pm 13}{4}$

$x = \frac{-7 + 13}{4}$ or $x = \frac{-7 - 13}{4}$

$x = \frac{3}{2}$ or $x = -5$

The solution set is $\left\{\frac{3}{2}, -5\right\}$.

24. $\left\{\frac{5}{3}\right\}$

25. $x^2 + 4x = 5$

$x^2 + 4x - 5 = 0$

$a = 1, \quad b = 4, \quad c = -5$

$x = \frac{-b \pm \sqrt{b^2 - 4ac}}{2a}$

$x = \frac{-4 \pm \sqrt{4^2 - 4(1)(-5)}}{2(1)} = \frac{-4 \pm \sqrt{16 + 20}}{2}$

$= \frac{-4 \pm \sqrt{36}}{2} = \frac{-4 \pm 6}{2}$

$x = \frac{-4 + 6}{2}$ or $x = \frac{-4 - 6}{2}$

$x = 1$ or $x = -5$

The solution set is $\{1, -5\}$.

26. $\{-3, 5\}$

27. $2y^2 - 3y - 2 = 0$

$a = 2, \quad b = -3, \quad c = -2$

$y = \frac{-b \pm \sqrt{b^2 - 4ac}}{2a}$

$y = \frac{-(-3) \pm \sqrt{(-3)^2 - 4(2)(-2)}}{2(2)} = \frac{3 \pm \sqrt{9 + 16}}{4}$

$= \frac{3 \pm \sqrt{25}}{4} = \frac{3 \pm 5}{4}$

$y = \frac{3 + 5}{4}$ or $y = \frac{3 - 5}{4}$

$y = 2$ or $y = -\frac{1}{2}$

The solution set is $\left\{2, -\frac{1}{2}\right\}$.

28. $\left\{-1, \frac{2}{5}\right\}$

29. $3t^2 + 8t + 3 = 0$

$a = 3, \quad b = 8, \quad c = 3$

$t = \frac{-b \pm \sqrt{b^2 - 4ac}}{2a}$

$t = \frac{-8 \pm \sqrt{8^2 - 4 \cdot 3 \cdot 3}}{2 \cdot 3} = \frac{-8 \pm \sqrt{64 - 36}}{6}$

$= \frac{-8 \pm \sqrt{28}}{6} = \frac{-8 \pm 2\sqrt{7}}{6} = \frac{2(-4 \pm \sqrt{7})}{6}$

$= \frac{-4 \pm \sqrt{7}}{3}$

The solution set is $\left\{\frac{-4 \pm \sqrt{7}}{3}\right\}$.

30. $\{3 \pm \sqrt{7}\}$

31. $3 + u^2 = 12u$

$u^2 - 12u + 3 = 0$

$a = 1, \quad b = -12, \quad c = 3$

$u = \frac{-b \pm \sqrt{b^2 - 4ac}}{2a}$

$u = \frac{-(-12) \pm \sqrt{(-12)^2 - 4 \cdot 1 \cdot 3}}{2 \cdot 1} = \frac{12 \pm \sqrt{144 - 12}}{2}$

$= \frac{12 \pm \sqrt{132}}{2} = \frac{12 \pm 2\sqrt{33}}{2}$

$= \frac{2(6 \pm \sqrt{33})}{2} = 6 \pm \sqrt{33}$

The solution set is $\{6 \pm \sqrt{33}\}$.

32. $\{-2, -4\}$

33. $x^2 - x + 1 = 0$

$a = 1, \quad b = -1, \quad c = 1$

$x = \frac{-b \pm \sqrt{b^2 - 4ac}}{2a}$

$x = \frac{-(-1) \pm \sqrt{(-1)^2 - 4 \cdot 1 \cdot 1}}{2 \cdot 1}$

$= \frac{1 \pm \sqrt{1 - 4}}{2} = \frac{1 \pm \sqrt{-3}}{2}$

$= \frac{1 \pm i\sqrt{3}}{2}$

The solution set is $\left\{\frac{1 \pm \sqrt{3}}{2}\right\}$.

34. $\left\{\frac{-1 \pm i\sqrt{7}}{2}\right\}$

35. $x^2 + 13 = 4x$

$x^2 - 4x + 13 = 0$

$a = 1, \quad b = -4, \quad c = 13$

$x = \frac{-b \pm \sqrt{b^2 - 4ac}}{2a}$

$= \frac{-(-4) \pm \sqrt{(-4)^2 - 4 \cdot 1 \cdot 13}}{2 \cdot 1}$

$= \frac{4 \pm \sqrt{16 - 52}}{2} = \frac{4 \pm \sqrt{-36}}{2}$

$= \frac{4 \pm 6i}{2} = \frac{2(2 \pm 3i)}{2} = 2 \pm 3i$

The solution set is $\{2 \pm 3i\}$.

36. $\left\{\frac{-1 \pm 2i}{5}\right\}$

37. $x^2 - 6x + 9 = 0$

$a = 1, \quad b = -6, \quad c = 9$

We compute the discriminant.

$$\begin{aligned} b^2 - 4ac &= (-6)^2 - 4 \cdot 1 \cdot 9 \\ &= 36 - 36 \\ &= 0 \end{aligned}$$

Since $b^2 - 4ac = 0$, there is just one solution, and it is a real number.

38. One real

39. $x^2 + 7 = 0$

$a = 1, \quad b = 0, \quad c = 7$

We compute the discriminant.

$$\begin{aligned} b^2 - 4ac &= 0^2 - 4 \cdot 1 \cdot 7 \\ &= -28 \end{aligned}$$

Since $b^2 - 4ac < 0$, there are two nonreal solutions.

40. Two nonreal

41. $x^2 - 2 = 0$

$a = 1, \quad b = 0, \quad c = -2$

We compute the discriminant.

$$\begin{aligned} b^2 - 4ac &= 0^2 - 4 \cdot 1 \cdot (-2) \\ &= 8 \end{aligned}$$

Since $b^2 - 4ac > 0$, there are two real solutions.

42. Two real

43. $4x^2 - 12x + 9 = 0$

$a = 4, \quad b = -12, \quad c = 9$

We compute the discriminant.

$$\begin{aligned} b^2 - 4ac &= (-12)^2 - 4 \cdot 4 \cdot 9 \\ &= 144 - 144 \\ &= 0 \end{aligned}$$

Since $b^2 - 4ac = 0$, there is just one solution, and it is a real number.

44. Two real

45. $x^2 - 2x + 4 = 0$

$a = 1, \quad b = -2, \quad c = 4$

We compute the discriminant.

$$\begin{aligned} b^2 - 4ac &= (-2)^2 - 4 \cdot 1 \cdot 4 \\ &= 4 - 16 \\ &= -12 \end{aligned}$$

Since $b^2 - 4ac < 0$, ther are two nonreal solutions.

46. Two nonreal

47. $9t^2 - 3t = 0$

$a = 9, \quad b = -3, \quad c = 0$

We compute the discriminant.

$$\begin{aligned} b^2 - 4ac &= (-3)^2 - 4 \cdot 9 \cdot 0 \\ &= 9 - 0 \\ &= 9 \end{aligned}$$

Since $b^2 - 4ac > 0$, there are two real solutions.

48. Two real

49. $y^2 = \frac{1}{2} y + \frac{3}{5}$

$y^2 - \frac{1}{2} y - \frac{3}{5} = 0$ (Standard form)

$a = 1, \quad b = -\frac{1}{2}, \quad c = -\frac{3}{5}$

We compute the discriminant.

$$\begin{aligned} b^2 - 4ac &= \left(-\frac{1}{2}\right)^2 - 4 \cdot 1 \cdot \left(-\frac{3}{5}\right) \\ &= \frac{1}{4} + \frac{12}{5} \\ &= \frac{53}{20} \end{aligned}$$

Since $b^2 - 4ac > 0$, there are two real solutions.

50. Two real

51. $4x^2 - 4\sqrt{3}\,x + 3 = 0$

$a = 4, \quad b = -4\sqrt{3}, \quad c = 3$

We compute the discriminant.

$$\begin{aligned} b^2 - 4ac &= (-4\sqrt{3})^2 - 4 \cdot 4 \cdot 3 \\ &= 48 - 48 \\ &= 0 \end{aligned}$$

Since $b^2 - 4ac = 0$, there is just one solution, and it is a real number.

52. Two real

53. The solutions are -11 and 9.

$$\begin{aligned} x = -11 \quad &\text{or} \quad x = 9 \\ x + 11 = 0 \quad &\text{or} \quad x - 9 = 0 \end{aligned}$$

$$\begin{aligned} (x + 11)(x - 9) &= 0 \\ x^2 + 2x - 99 &= 0 \end{aligned}$$

54. $x^2 - 16 = 0$

55. The solutions are both 7.

$x = 7$ or $x = 7$

$x - 7 = 0$ or $x - 7 = 0$

$(x - 7)(x - 7) = 0$

$x^2 - 14x + 49 = 0$

56. $x^2 + \frac{4}{3}x + \frac{4}{9} = 0$, or $9x^2 + 12x + 4 = 0$

57. The solutions are $-\frac{2}{5}$ and $\frac{6}{5}$.

$x = -\frac{2}{5}$ or $x = \frac{6}{5}$

$x + \frac{2}{5} = 0$ or $x - \frac{6}{5} = 0$

$\left(x + \frac{2}{5}\right)\left(x - \frac{6}{5}\right) = 0$

$x^2 - \frac{4}{5}x - \frac{12}{25} = 0$

or

$25x^2 - 20x - 12 = 0$

58. $x^2 + \frac{3}{4}x + \frac{1}{8} = 0$, or $8x^2 + 6x + 1 = 0$

59. The solutions are $\frac{c}{2}$ and $\frac{d}{2}$.

$x = \frac{c}{2}$ or $x = \frac{d}{2}$

$x - \frac{c}{2} = 0$ or $x - \frac{d}{2} = 0$

$\left(x - \frac{c}{2}\right)\left(x - \frac{d}{2}\right) = 0$

$x^2 - \left(\frac{c + d}{2}\right)x + \frac{cd}{4} = 0$

or

$4x^2 - 2(c + d)x + cd = 0$

60. $x^2 - \left(\frac{k}{3} + \frac{m}{4}\right)x + \frac{km}{12} = 0$, or

$12x^2 - (4k + 3m)x + km = 0$

61. The solutions are $\sqrt{2}$ and $3\sqrt{2}$.

$x = \sqrt{2}$ or $x = 3\sqrt{2}$

$x - \sqrt{2} = 0$ or $x - 3\sqrt{2} = 0$

$(x - \sqrt{2})(x - 3\sqrt{2}) = 0$

$x^2 - 4\sqrt{2}x + 6 = 0$

62. $x^2 - \sqrt{3}x - 6 = 0$

63. The solutions are $3i$ and $-3i$.

$x = 3i$ or $x = -3i$

$x - 3i = 0$ or $x + 3i = 0$

$(x - 3i)(x + 3i) = 0$

$x^2 - 9i^2 = 0$

$x^2 + 9 = 0$

64. $x^2 + 16 = 0$

65. $x^2 - 0.75x - 0.5 = 0$

$a = 1, \quad b = -0.75, \quad c = -0.5$

$x = \frac{-(-0.75) \pm \sqrt{(-0.75)^2 - 4(1)(-0.5)}}{2(1)}$

$= \frac{0.75 \pm \sqrt{0.5625 + 2}}{2} = \frac{0.75 \pm \sqrt{2.5625}}{2}$

$x = \frac{0.75 + \sqrt{2.5625}}{2}$ or $x = \frac{0.75 - \sqrt{2.5625}}{2}$

$x \approx 1.1754$ or $x \approx -0.4254$

The solution set is $\{1.1754, -0.4254\}$.

66. $\{1.8693, -0.3252\}$

67. $x + \frac{1}{x} = \frac{13}{6}$, LCM = 6x

$6x\left(x + \frac{1}{x}\right) = 6x \cdot \frac{13}{6}$

$6x^2 + 6 = 13x$

$6x^2 - 13x + 6 = 0$

$(3x - 2)(2x - 3) = 0$

$3x - 2 = 0$ or $2x - 3 = 0$

$x = \frac{2}{3}$ or $x = \frac{3}{2}$

The solution set is $\left\{\frac{2}{3}, \frac{3}{2}\right\}$.

68. $\left\{6, \frac{3}{2}\right\}$

69. $t^2 + 0.2t - 0.3 = 0$

$a = 1, \quad b = 0.2, \quad c = -0.3$

$t = \frac{-0.2 \pm \sqrt{(0.2)^2 - 4(1)(-0.3)}}{2\cdot1}$

$= \frac{-0.2 \pm \sqrt{0.04 + 1.2}}{2} = \frac{-0.2 \pm \sqrt{1.24}}{2}$

$= \frac{-0.2 \pm 2\sqrt{0.31}}{2} = \frac{2(-0.1 \pm \sqrt{0.31})}{2}$

$= -0.1 \pm \sqrt{0.31}$

The solution set is $\{-0.1 \pm \sqrt{0.31}\}$.

70. $\left\{\frac{-0.3 \pm \sqrt{0.89}}{2}\right\}$

71. $x^2 + x - \sqrt{2} = 0$

$a = 1, \quad b = 1, \quad c = -\sqrt{2}$

$$x = \frac{-1 \pm \sqrt{1^2 - 4(1)(-\sqrt{2})}}{2\cdot 1} = \frac{-1 \pm \sqrt{1 + 4\sqrt{2}}}{2}$$

The solution set is $\left\{\frac{-1 \pm \sqrt{1 + 4\sqrt{2}}}{2}\right\}$.

72. $\left\{\frac{1 \pm \sqrt{1 + 4\sqrt{3}}}{2}\right\}$

73. $x^2 + \sqrt{5}\,x - \sqrt{3} = 0$

$a = 1, \quad b = \sqrt{5}, \quad c = -\sqrt{3}$

$$x = \frac{-\sqrt{5} \pm \sqrt{(\sqrt{5})^2 - 4(1)(-\sqrt{3})}}{2\cdot 1}$$

$$= \frac{-\sqrt{5} \pm \sqrt{5 + 4\sqrt{3}}}{2}$$

The solution set is $\left\{\frac{-\sqrt{5} \pm \sqrt{5 + 4\sqrt{3}}}{2}\right\}$.

74. $\frac{-\sqrt{3} \pm \sqrt{3 + 8\pi}}{4}$

75. $\sqrt{2}\,x^2 - \sqrt{3}\,x - \sqrt{5} = 0$

$a = \sqrt{2}, \quad b = -\sqrt{3}, \quad c = -\sqrt{5}$

$$x = \frac{-(-\sqrt{3}) \pm \sqrt{(-\sqrt{3})^2 - 4(\sqrt{2})(-\sqrt{5})}}{2\cdot\sqrt{2}}$$

$$= \frac{\sqrt{3} \pm \sqrt{3 + 4\sqrt{10}}}{2\sqrt{2}} \cdot \frac{\sqrt{2}}{\sqrt{2}}$$ (Rationalizing the denominator)

$$= \frac{\sqrt{6} \pm \sqrt{6 + 8\sqrt{10}}}{4}$$

The solution set is $\left\{\frac{\sqrt{6} \pm \sqrt{6 + 8\sqrt{10}}}{4}\right\}$.

76. $\left\{\frac{-5\sqrt{2} \pm \sqrt{34}}{4}\right\}$

77. $(2t - 3)^2 + 17t = 15$

$4t^2 - 12t + 9 + 17t = 15$

$4t^2 + 5t - 6 = 0$

$(4t - 3)(t + 2) = 0$

$4t - 3 = 0$ or $t + 2 = 0$

$t = \frac{3}{4}$ or $t = -2$

The solution set is $\left\{\frac{3}{4}, -2\right\}$.

78. {2, -3}

79. $(x + 3)(x - 2) = 2(x + 11)$

$x^2 + x - 6 = 2x + 22$

$x^2 - x - 28 = 0$

$a = 1, \quad b = -1, \quad c = -28$

$$x = \frac{-(-1) \pm \sqrt{(-1)^2 - 4(1)(-28)}}{2\cdot 1}$$

$$= \frac{1 \pm \sqrt{1 + 112}}{2} = \frac{1 \pm \sqrt{113}}{2}$$

The solution set is $\left\{\frac{1 \pm \sqrt{113}}{2}\right\}$.

80. {-4, 3}

81. $2x^2 + (x - 4)^2 = 5x(x - 4) + 24$

$2x^2 + x^2 - 8x + 16 = 5x^2 - 20x + 24$

$0 = 2x^2 - 12x + 8$

$0 = x^2 - 6x + 4$

$a = 1, \quad b = -6, \quad c = 4$

$$x = \frac{-(-6) \pm \sqrt{(-6)^2 - 4\cdot 1\cdot 4}}{2\cdot 1} = \frac{6 \pm \sqrt{36 - 16}}{2}$$

$$= \frac{6 \pm \sqrt{20}}{2} = \frac{6 \pm 2\sqrt{5}}{2}$$

$$= \frac{2(3 \pm \sqrt{5})}{2} = 3 \pm \sqrt{5}$$

The solution set is $\{3 \pm \sqrt{5}\}$.

82. {-12, 4}

Exercise Set 2.6

1. $F = \frac{kM_1M_2}{d^2}$

$Fd^2 = kM_1M_2$

$d^2 = \frac{kM_1M_2}{F}$

$d = \sqrt{\frac{kM_1M_2}{F}}$

2. $c = \sqrt{\frac{E}{m}}$

3. $S = \frac{1}{2}at^2$

$2S = at^2$

$\frac{2S}{a} = t^2$

$\sqrt{\frac{2S}{a}} = t$

4. $r = \sqrt{\frac{V}{4\pi}}$ or $\frac{1}{2}\sqrt{\frac{V}{\pi}}$

5. $s = -16t^2 + v_0 t$

$0 = -16t^2 + v_0 t - s$

$a = -16, \quad b = v_0, \quad c = -s$

$t = \dfrac{-v_0 \pm \sqrt{v_0^2 - 4(-16)(-s)}}{2(-16)}$

$= \dfrac{-v_0 \pm \sqrt{v_0^2 - 64s}}{-32}$

or

$= \dfrac{v_0 \pm \sqrt{v_0^2 - 64s}}{32}$

6. $r = \dfrac{-3\pi h + \sqrt{9\pi^2 h^2 + 8A\pi}}{4\pi}$

7. $d = \dfrac{n^2 - 3n}{2}$

$2d = n^2 - 3n$

$0 = n^2 - 3n - 2d$

$a = 1, \quad b = -3, \quad c = -2d$

$n = \dfrac{-(-3) \pm \sqrt{(-3)^2 - 4(1)(-2d)}}{2 \cdot 1}$

$= \dfrac{3 \pm \sqrt{9 + 8d}}{2}$

$n = \dfrac{3 + \sqrt{9 + 8d}}{2}$

8. $t = \dfrac{\pi \pm \sqrt{\pi^2 - 12k\sqrt{2}}}{2\sqrt{2}}$

9. $A = P(1 + i)^2$

$\dfrac{A}{P} = (1 + i)^2$

$\pm\sqrt{\dfrac{A}{P}} = 1 + i$

$-1 \pm \sqrt{\dfrac{A}{P}} = i$

$-1 + \sqrt{\dfrac{A}{P}} = i$

10. $i = 2\left[-1 + \sqrt{\dfrac{A}{P}}\right]$

11. $A = P(1 + i)^t$

$7220 = 5120(1 + i)^2$ (Substituting)

$\dfrac{7220}{5120} = (1 + i)^2$

$\pm\sqrt{\dfrac{722}{512}} = 1 + i$

$\pm\sqrt{\dfrac{361}{256}} = 1 + i$

$\pm\dfrac{19}{16} = 1 + i$

$-1 \pm \dfrac{19}{16} = i$

$-1 + \dfrac{19}{16} = i$ or $-1 - \dfrac{19}{16} = i$

$\dfrac{3}{16} = i$ or $-\dfrac{35}{16} = i$

Since the interest rate cannot be negative, $i = \dfrac{3}{16} = 0.1875 = 18.75\%$.

12. 10%

13. $A = P(1 + i)^t$

$9856.80 = 8000(1 + i)^2$ (Substituting)

$\dfrac{9856.80}{8000} = (1 + i)^2$

$\pm\sqrt{\dfrac{9856.80}{8000}} = 1 + i$

$-1 \pm 1.11 = i$

$-1 + 1.11 = i$ or $-1 - 1.11 = i$

$0.11 = i$ or $-2.11 = i$

Since the interest rate cannot be negative, $i = 0.11 = 11\%$.

14. 12.75%

15. $d = \dfrac{n^2 - 3n}{2}$

$27 = \dfrac{n^2 - 3n}{2}$ (Substituting 27 for d)

$54 = n^2 - 3n$

$0 = n^2 - 3n - 54$

$0 = (n - 9)(n + 6)$

$n - 9 = 0$ or $n + 6 = 0$

$n = 9$ or $n = -6$

The number of sides must be positive. Thus the number of sides is 9.

16. 11

17. Familiarize and Translate:

Using the Pythagorean property we determine the length of the other leg.

$10^2 = 6^2 + a^2$

$100 = 36 + a^2$

$64 = a^2$

$8 = a$

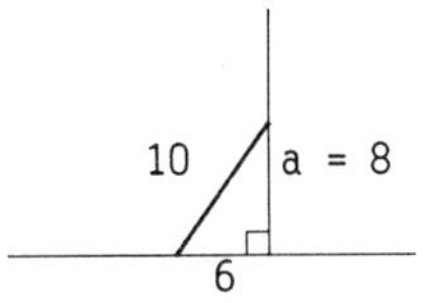

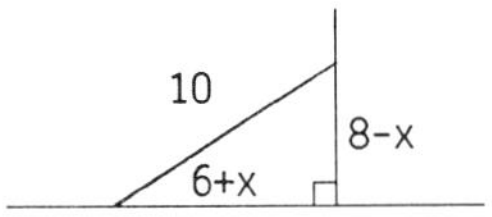

We let x represent the amount the ladder is pulled away and pulled down. The legs then become 6 + x and 8 - x. Then we again use the Pythagorean property.

$10^2 = (6 + x)^2 + (8 - x)^2$

Carry out:

$100 = 36 + 12x + x^2 + 64 - 16x + x^2$

$100 = 2x^2 - 4x + 100$

$0 = 2x^2 - 4x$

$0 = x^2 - 2x$

$0 = x(x - 2)$

$x = 0$ or $x - 2 = 0$

$x = 0$ or $x = 2$

We only consider $x = 2$ since $x \neq 0$ if the ladder is pulled away.

Check:

When $x = 2$, the legs of the triangle become 6 + 2, or 8, and 8 - 2, or 6. The Pythagorean relationship still holds.

$8^2 + 6^2 = 10^2$

$64 + 36 = 100$

The ladder ia pulled away 2 ft and pulled down 2 ft.

18. 7 ft

19. Familiarize:

We make a drawing. We let h represent the height and h + 3 represent the base.

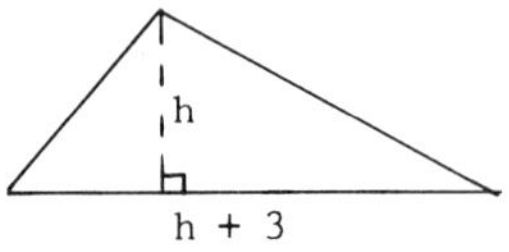

Translate:

$A = \frac{1}{2} \cdot \text{base} \cdot \text{height}$ (Area of triangle)

$18 = \frac{1}{2}(h + 3)h$ (Substituting)

19. (continued)

Carry out:

$36 = h^2 + 3h$

$0 = h^2 + 3h - 36$

$a = 1, \quad b = 3, \quad c = -36$

$$h = \frac{-3 \pm \sqrt{3^2 - 4(1)(-36)}}{2 \cdot 1}$$

$$= \frac{-3 \pm \sqrt{9 + 144}}{2} = \frac{-3 \pm \sqrt{153}}{2}$$

$$h = \frac{-3 + \sqrt{153}}{2}$$ (h cannot be negative)

≈ 4.685

Check:

When $h = 4.685$, $h + 3 = 7.685$ and the area of the triangle is $\frac{1}{2}(7.685)(4.685) \approx 18$.

The height is 4.685 cm.

20. $90\sqrt{2} \approx 127.28$ ft

21. Familiarize: We make a drawing.

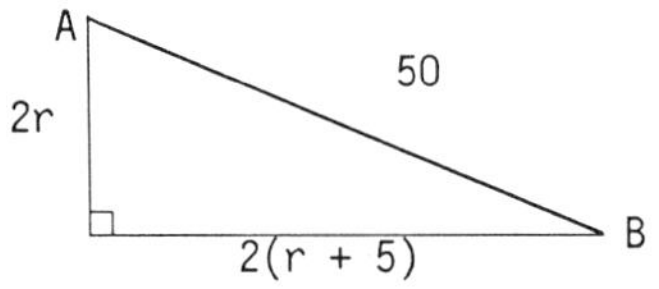

We let r represent the speed of train A. Then r + 5 represents the speed of train B. The distances they travel in 2 hours are 2r and 2(r + 5). After 2 hours they are 50 miles apart.

Translate:

Using the Pythagorean property we have an equation.

$(2r)^2 + [2(r + 5)]^2 = 50^2$

Carry out:

$4r^2 + 4(r + 5)^2 = 2500$

$r^2 + (r + 5)^2 = 625$

$r^2 + r^2 + 10r + 25 = 625$

$2r^2 + 10r - 600 = 0$

$r^2 + 5r - 300 = 0$

$(r + 20)(r - 15) = 0$

$r + 20 = 0$ or $r - 15 = 0$

$r = -20$ or $r = 15$

Check:

Since speed in this problem must be positive, we only check 15. If $r = 15$, then $r + 5 = 20$. In 2 hours train A will travel $2 \cdot 15$, or 30 miles, and train B will travel $2 \cdot 20$, or 40 miles.

21. (continued)

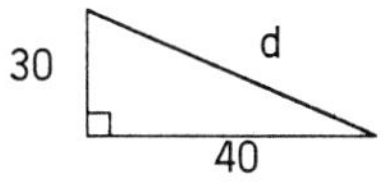

We now calculate how far apart trains A and B will be after 2 hours.

$$30^2 + 40^2 = d^2$$
$$900 + 1600 = d^2$$
$$2500 = d^2$$
$$50 = d$$

The value checks.

The speed of train A is 15 mph; the speed of train B is 20 mph.

22. A: 24 km/h; B: 10 km/h

23. a)

$$s = 4.9t^2 + v_0 t$$
$$75 = 4.9t^2 + 0 \cdot t \quad \text{(Substituting 75 for } s \text{ and 0 for } v_0\text{)}$$
$$75 = 4.9t^2$$
$$\frac{75}{4.9} = t^2$$
$$\sqrt{\frac{75}{4.9}} = t$$
$$3.91 \approx t$$

It takes 3.91 sec to reach the ground.

b)

$$s = 4.9t^2 + v_0 t$$
$$75 = 4.9t^2 + 30t \quad \text{(Substituting 75 for } s \text{ and 30 for } v_0\text{)}$$
$$0 = 4.9t^2 + 30t - 75$$
$$a = 4.9, \quad b = 30, \quad c = -75$$
$$t = \frac{-30 \pm \sqrt{30^2 - 4(4.9)(-75)}}{2(4.9)}$$
$$= \frac{-30 \pm \sqrt{900 + 1470}}{9.8}$$
$$t = \frac{-30 + \sqrt{2370}}{9.8} \quad \text{(t must be positive)}$$
$$\approx 1.906$$

It takes 1.906 sec to reach the ground.

c)

$$s = 4.9t^2 + v_0 t$$
$$s = 4.9(2)^2 + 30(2) \quad \text{(Substituting 2 for } t \text{ and 30 for } v_0\text{)}$$
$$s = 19.6 + 60$$
$$s = 79.6$$

The object will fall 79.6 m.

24. a) 10.1 sec, b) 7.49 sec, c) 272.5 m

25. Familiarize and Translate:

It helps to make a drawing.

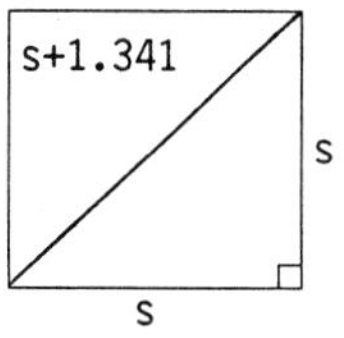

We let s represent the length of the side. Then s + 1.341 represents the diagonal. Using the Pythagorean property we have an equation.

$$s^2 + s^2 = (s + 1.341)^2$$

Carry out:

$$2s^2 = s^2 + 2.682s + 1.798281$$
$$s^2 - 2.682s - 1.798281 = 0$$
$$s = \frac{-(-2.682) \pm \sqrt{(-2.682)^2 - 4(1)(-1.798281)}}{2 \cdot 1}$$
$$= \frac{2.682 \pm \sqrt{7.193124 + 7.193124}}{2}$$
$$s = \frac{2.682 + \sqrt{14.386248}}{2}$$
$$\approx 3.237$$

The value checks. The length of the side is 3.237 cm.

26. 2.2199 cm, 8.0101 cm

27. Familiarize and Translate:

We let x represent the number of students in the group at the outset. Then x - 3 represents the number of students at the last minute.

$$\text{Total cost} = (\text{Cost per person}) \cdot (\text{Number of persons})$$
$$\frac{\text{Total cost}}{\text{Number of persons}} = \text{Cost per person}$$

Using this formula we get the following:

$$\text{Cost per person at the outset} = \frac{140}{x}$$
$$\text{Cost per person at the last minute} = \frac{140}{x - 3}$$

The cost at the last minute was $15 more than the cost at the outset. This gives us an equation.

$$\frac{140}{x - 3} = 15 + \frac{140}{x}$$

27. (continued)

Carry out:

$$x(x - 3) \cdot \frac{140}{x - 3} = x(x - 3)\left[15 + \frac{140}{x}\right]$$

(Multiplying by the LCM)

$$140x = 15x(x - 3) + 140(x - 3)$$
$$140x = 15x^2 - 45x + 140x - 420$$
$$0 = 15x^2 - 45x - 420$$
$$0 = x^2 - 3x - 28$$
$$0 = (x - 7)(x + 4)$$

$x - 7 = 0$ or $x + 4 = 0$

$x = 7$ or $x = -4$

Check:

Since the number of people cannot be negative, we only check x = 7. When x = 7, the cost at the outset is 140/7, or \$20. When x = 7, the cost at the last minute is 140/(7 - 3), or 140/4, or \$35 and \$35 is \$15 more than \$20. The value checks.

Thus, 7 students were in the group at the outset.

28. 12

29. Familiarize and Translate:

Let x represent the number of shares bought. Then $\frac{720}{x}$ represents the cost of each share.

Total cost = (Cost per item)·(Number of items)

$$720 = \left[\frac{720}{x}\right]\cdot(x)$$

When the cost per share is reduced by \$15, the new cost is $\frac{720}{x} - 15$. When the number of shares purchased is increased by 4, the new number of shares is x + 4. The total cost remains the same.

$$720 = \left[\frac{720}{x} - 15\right]\cdot(x + 4)$$

Carry out:

$$720 = 720 + \frac{2880}{x} - 15x - 60$$
$$60 = \frac{2880}{x} - 15x$$
$$60x = 2880 - 15x^2$$
$$15x^2 + 60x - 2880 = 0$$
$$x^2 + 4x - 192 = 0$$
$$(x + 16)(x - 12) = 0$$

$x + 16 = 0$ or $x - 12 = 0$

$x = -16$ or $x = 12$

Check:

The number of shares must be positive. We only check x = 12. The cost per share when x = 12 is 720/12, or \$60. The cost per share when the number is increased by 4 is 720/16, or \$45. Thus, \$45 is \$15 less than \$60, and the value checks.

Thus, 12 shares of stock were bought.

30. \$8

31. $kx^2 + (3 - 2k)x - 6 = 0$

$a = k,\quad b = 3 - 2k,\quad c = -6$

$$x = \frac{-(3 - 2k) \pm \sqrt{(3 - 2k)^2 - 4(k)(-6)}}{2k}$$
$$= \frac{-3 + 2k \pm \sqrt{4k^2 + 12k + 9}}{2k}$$
$$= \frac{-3 + 2k \pm \sqrt{(2k + 3)^2}}{2k}$$
$$= \frac{-3 + 2k \pm (2k + 3)}{2k}$$

$x = \frac{-3 + 2k + 2k + 3}{2k}$ or $x = \frac{-3 + 2k - 2k - 3}{2k}$

$x = \frac{4k}{2k}$ or $x = \frac{-6}{2k}$

$x = 2$ or $x = -\frac{3}{k}$

The solution set is $\left\{2, -\frac{3}{k}\right\}$.

32. $\left\{1, \frac{1}{1 - k}\right\}$

33. $$(m + n)^2x^2 + (m + n)x = 2$$
$$(m + n)^2x^2 + (m + n)x - 2 = 0$$

$a = (m + n)^2,\quad b = (m + n),\quad c = -2$

$$x = \frac{-(m + n) \pm \sqrt{(m + n)^2 - 4(m + n)^2(-2)}}{2(m + n)^2}$$
$$= \frac{-(m + n) \pm \sqrt{9(m + n)^2}}{2(m + n)^2}$$
$$= \frac{-(m + n) \pm 3(m + n)}{2(m + n)^2}$$

$x = \frac{2(m + n)}{2(m + n)^2}$ or $x = \frac{-4(m + n)}{2(m + n)^2}$

$x = \frac{1}{m + n}$ or $x = \frac{-2}{m + n}$

The solution set is $\left\{\frac{1}{m + n}, -\frac{2}{m + n}\right\}$.

34. a) $\{4y, -y\}$, b) $\left\{-x, \frac{x}{4}\right\}$

35. $$A = P(1 + i)^t$$
$$13{,}704 = 9826(1 + i)^3 \quad \text{(Substituting)}$$
$$\frac{13{,}704}{9826} = (1 + i)^3$$
$$\sqrt[3]{\frac{13{,}704}{9826}} = 1 + i$$
$$-1 + \sqrt[3]{\frac{13{,}704}{9826}} = i$$
$$0.117 \approx i$$

The interest rate is 11.7%.

36. $4.5\sqrt{6.75}$, or $\frac{27\sqrt{3}}{4}$

37. Familiarize:

We first make a drawing.

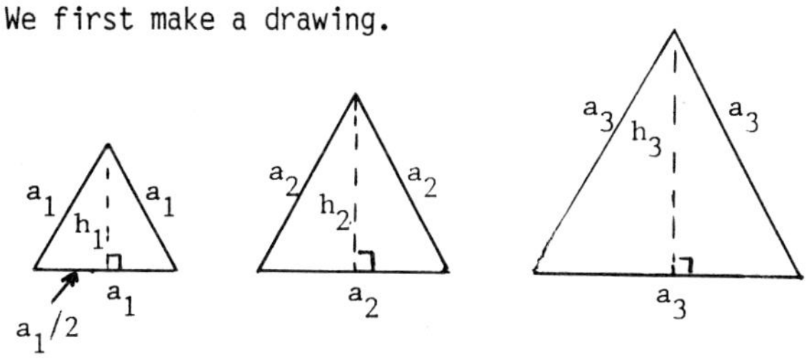

We find the height of the first triangle.

$$\left(\frac{a_1}{2}\right)^2 + (h_1)^2 = a_1^2$$

$$(h_1)^2 = a_1^2 - \frac{a_1^2}{4} = \frac{3a_1^2}{4}$$

$$h_1 = \frac{a_1\sqrt{3}}{4}$$

Next we find the area of the first triangle.

$$A_1 = \frac{1}{2} \cdot a_1 \cdot \frac{a_1\sqrt{3}}{4} = \frac{a_1^2\sqrt{3}}{8}$$

Similarly we find the areas of the other two triangles:

$$A_2 = \frac{a_2^2\sqrt{3}}{8} \text{ and } A_3 = \frac{a_3^2\sqrt{3}}{8}.$$

Translate:

The sum of the areas of the first two triangles equals the area of the third triangle.

$$\frac{a_1^2\sqrt{3}}{8} + \frac{a_2^2\sqrt{3}}{8} = \frac{a_3^2\sqrt{3}}{8}$$

Carry out:

$$\frac{\sqrt{3}}{8}(a_1^2 + a_2^2) = \frac{\sqrt{3}}{8} \cdot a_3^2$$

$$a_1^2 + a_2^2 = a_3^2$$

$$\sqrt{a_1^2 + a_2^2} = a_3$$

Thus the length of a side of the third triangle is $\sqrt{a_1^2 + a_2^2}$.

38. They have the same area: 300 ft².

Exercise Set 2.7

1.

$$\sqrt{3x - 4} = 1$$
$$(\sqrt{3x - 4})^2 = 1^2$$
$$3x - 4 = 1$$
$$3x = 5$$
$$x = \frac{5}{3}$$

Check:

$\sqrt{3x - 4} = 1$	
$\sqrt{3 \cdot \frac{5}{3} - 4}$	1
$\sqrt{5 - 4}$	
$\sqrt{1}$	
1	

The solution is $\frac{5}{3}$.

2. -63

3.

$$\sqrt[4]{x^2 - 1} = 1$$
$$(\sqrt[4]{x^2 - 1})^4 = 1^4$$
$$x^2 - 1 = 1$$
$$x^2 = 2$$
$$x = \pm\sqrt{2}$$

Check:

$\sqrt[4]{x^2 - 1} = 1$	
$\sqrt[4]{(\pm\sqrt{2})^2 - 1}$	1
$\sqrt[4]{2 - 1}$	
$\sqrt[4]{1}$	
1	

The solutions are $\sqrt{2}$ and $-\sqrt{2}$.

4. 168

5.

$$\sqrt{y - 1} + 4 = 0$$
$$\sqrt{y - 1} = -4$$

Note:

The principal root is never negative. Thus, there is no solution.

If we do not observe the above fact, we could continue and reach the same answer.

$$(\sqrt{y - 1})^2 = (-4)^2$$
$$y - 1 = 16$$
$$y = 17$$

Check:

$\sqrt{y - 1} + 4 = 0$	
$\sqrt{17 - 1} + 4$	0
$\sqrt{16} + 4$	
$4 + 4$	
8	

Since 17 does not check, there is no solution.

6. No solution

7. $\sqrt{x - 3} + \sqrt{x + 5} = 4$

$$\sqrt{x + 5} = 4 - \sqrt{x - 3}$$
$$(\sqrt{x + 5})^2 = (4 - \sqrt{x - 3})^2$$
$$x + 5 = 16 - 8\sqrt{x - 3} + (x - 3)$$
$$x + 5 = 13 - 8\sqrt{x - 3} + x$$
$$8\sqrt{x - 3} = 8$$
$$\sqrt{x - 3} = 1$$
$$(\sqrt{x - 3})^2 = 1^2$$
$$x - 3 = 1$$
$$x = 4$$

Check: $\sqrt{x - 3} + \sqrt{x + 5} = 4$

$\sqrt{4 - 3} + \sqrt{4 + 5}$	4
$\sqrt{1} + \sqrt{9}$	
$1 + 3$	
4	

The solution is 4.

8. 9

9. $\sqrt{3x - 5} + \sqrt{2x + 3} + 1 = 0$

$$\sqrt{3x - 5} + \sqrt{2x + 3} = -1$$

Note:

The principal root is never negative. Thus the sum of two principal roots cannot equal -1. There is no solution.

10. 2

11. $\sqrt[3]{6x + 9} + 8 = 5$

$$\sqrt[3]{6x + 9} = -3$$
$$(\sqrt[3]{6x + 9})^3 = (-3)^3$$
$$6x + 9 = -27$$
$$6x = -36$$
$$x = -6$$

Check: $\sqrt[3]{6x + 9} + 8 = 5$

$\sqrt[3]{6(-6) + 9} + 8$	5
$\sqrt[3]{-27} + 8$	
$-3 + 8$	
5	

The solution is -6.

12. $\frac{28}{3}$

13. $\sqrt{6x + 7} = x + 2$

$$(\sqrt{6x + 7})^2 = (x + 2)^2$$
$$6x + 7 = x^2 + 4x + 4$$
$$0 = x^2 - 2x - 3$$
$$0 = (x - 3)(x + 1)$$

$x - 3 = 0$ or $x + 1 = 0$

$x = 3$ or $x = -1$

Both values check. The solutions are 3 and -1.

14. $\frac{1}{3}$, -1

15. $\sqrt{20 - x} = \sqrt{9 - x} + 3$

$$(\sqrt{20 - x})^2 = (\sqrt{9 - x} + 3)^2$$
$$20 - x = (9 - x) + 6\sqrt{9 - x} + 9$$
$$20 - x = 18 - x + 6\sqrt{9 - x}$$
$$2 = 6\sqrt{9 - x}$$
$$1 = 3\sqrt{9 - x}$$
$$1^2 = (3\sqrt{9 - x})^2$$
$$1 = 9(9 - x)$$
$$1 = 81 - 9x$$
$$9x = 80$$
$$x = \frac{80}{9}$$

The value checks. The solution is $\frac{80}{9}$.

16. -1

17. $\sqrt{7.35x + 8.051} = 0.345x + 0.067$

$$(\sqrt{7.35x + 8.051})^2 = (0.345x + 0.067)^2$$
$$7.35x + 8.051 = 0.119025x^2 + 0.04623x + 0.004489$$
$$0 = 0.119025x^2 - 7.30377x - 8.046511$$
$$x = \frac{-(-7.30377) \pm \sqrt{(-7.30377)^2 - 4(0.119025)(-8.046511)}}{2(0.119025)}$$
$$\approx \frac{7.30377 \pm \sqrt{57.176}}{0.23805}$$

$x \approx 62.4459$ or $x \approx -1.0826$

Since -1.0826 does not check and 62.4459 does check, the solution is 62.4459.

18. 0.1444

19. $x^{1/3} = -2$

$$(x^{1/3})^3 = (-2)^3 \qquad (x^{1/3} = \sqrt[3]{x})$$
$$x = -8$$

The value checks. The solution is -8.

20. 32

21. $t^{1/4} = 3$

$(t^{1/4})^4 = 3^4 \qquad (t^{1/4} = \sqrt[4]{t})$

$t = 81$

The value checks. The solution is 81.

22. No solution

23. $8 = \frac{1}{\sqrt{x}}$

$8\sqrt{x} = 1$

$\sqrt{x} = \frac{1}{8}$

$(\sqrt{x})^2 = \left(\frac{1}{8}\right)^2$

$x = \frac{1}{64}$

The value checks. The solution is $\frac{1}{64}$.

24. $\frac{1}{9}$

25. $\sqrt[3]{m} = -5$

$(\sqrt[3]{m})^3 = (-5)^3$

$m = -125$

The value checks. The solution is -125.

26. No solution

27. For L: $T = 2\pi\sqrt{\frac{L}{g}}$

$\frac{T}{2\pi} = \sqrt{\frac{L}{g}}$

$\left(\frac{T}{2\pi}\right)^2 = \left(\sqrt{\frac{L}{g}}\right)^2$

$\frac{T^2}{4\pi^2} = \frac{L}{g}$

$\frac{gT^2}{4\pi^2} = L$

For g: $T = 2\pi\sqrt{\frac{L}{g}}$

$\frac{T}{2\pi} = \sqrt{\frac{L}{g}}$

$\left(\frac{T}{2\pi}\right)^2 = \left(\sqrt{\frac{L}{g}}\right)^2$

$\frac{T^2}{4\pi^2} = \frac{L}{g}$

$gT^2 = 4\pi^2 L$

$g = \frac{4\pi^2 L}{T^2}$

28. $c = \sqrt{H^2 - d^2}$

29. $V = 1.2\sqrt{h}$

$V = 1.2\sqrt{30{,}000} \qquad$ (Substituting)

$\approx 1.2(173.205)$

≈ 207.8

You can see 207.8 miles to the horizon.

30. 10 mi

31. $V = 1.2\sqrt{h}$

$144 = 1.2\sqrt{h} \qquad$ (Substituting)

$\frac{144}{1.2} = \sqrt{h}$

$120 = \sqrt{h}$

$14{,}400 = h$

The airplane is 14,400 ft high.

32. 84 ft

33. $(x - 5)^{2/3} = 2$

$\sqrt[3]{(x - 5)^2} = 2$

$(\sqrt[3]{(x - 5)^2})^3 = 2^3$

$(x - 5)^2 = 2^3$

$x^2 - 10x + 25 = 8$

$x^2 - 10x + 17 = 0$

$a = 1, \quad b = -10, \quad c = 17$

$x = \frac{-(-10) \pm \sqrt{(-10)^2 - 4\cdot 1\cdot 17}}{2\cdot 1}$

$= \frac{10 \pm \sqrt{100 - 68}}{2} = \frac{10 \pm \sqrt{32}}{2}$

$= \frac{10 \pm 4\sqrt{2}}{2} = 5 \pm 2\sqrt{2}$

The solutions are $5 + 2\sqrt{2}$ and $5 - 2\sqrt{2}$.

34. $3 \pm 2\sqrt{2}$

35. $\frac{x + \sqrt{x + 1}}{x - \sqrt{x + 1}} = \frac{5}{11}$

$11(x + \sqrt{x + 1}) = 5(x - \sqrt{x + 1})$

$11x + 11\sqrt{x + 1} = 5x - 5\sqrt{x + 1}$

$6x = -16\sqrt{x + 1}$

$(6x)^2 = (-16\sqrt{x + 1})^2$

$36x^2 = 256(x + 1)$

$36x^2 - 256x - 256 = 0$

$9x^2 - 64x - 64 = 0$

$(9x + 8)(x - 8) = 0$

$9x + 8 = 0 \quad$ or $\quad x - 8 = 0$

$x = -\frac{8}{9} \quad$ or $\quad x = 8$

Only $-\frac{8}{9}$ checks. The solution is $-\frac{8}{9}$.

36. No solution

37. $\sqrt{x+2} - \sqrt{x-2} = \sqrt{2x}$

$$\sqrt{x+2} = \sqrt{2x} + \sqrt{x-2}$$
$$(\sqrt{x+2})^2 = (\sqrt{2x} + \sqrt{x-2})^2$$
$$x + 2 = 2x + 2\sqrt{2x(x-2)} + (x - 2)$$
$$x + 2 = 3x + 2\sqrt{2x(x-2)} - 2$$
$$4 - 2x = 2\sqrt{2x(x-2)}$$
$$(2 - x)^2 = (\sqrt{2x^2 - 4x})^2$$
$$4 - 4x + x^2 = 2x^2 - 4x$$
$$0 = x^2 - 4$$
$$0 = (x + 2)(x - 2)$$

$x + 2 = 0$ or $x - 2 = 0$

$x = -2$ or $x = 2$

Only 2 checks. The solution is 2.

38. 1

39. $\sqrt[4]{x+2} = \sqrt{3x+1}$

$$(\sqrt[4]{x+2})^4 = (\sqrt{3x+1})^4$$
$$x + 2 = (3x + 1)^2$$
$$x + 2 = 9x^2 + 6x + 1$$
$$0 = 9x^2 + 5x - 1$$

$a = 9, \quad b = 5, \quad c = -1$

$$x = \frac{-5 \pm \sqrt{5^2 - 4(9)(-1)}}{2\cdot 9} = \frac{-5 \pm \sqrt{25 + 36}}{18}$$
$$= \frac{-5 \pm \sqrt{61}}{18}$$

Only $\frac{-5 + \sqrt{61}}{18}$ checks. The solution is $\frac{-5 + \sqrt{61}}{18}$.

40. $\frac{5}{4}$

Exercise Set 2.8

1. $x - 10\sqrt{x} + 9 = 0$

We substitute u for $\sqrt{x}$.

$$u^2 - 10u + 9 = 0$$
$$(u - 9)(u - 1) = 0$$

$u - 9 = 0$ or $u - 1 = 0$

$u = 9$ or $u = 1$

Now we substitute $\sqrt{x}$ for u and solve for x.

$\sqrt{x} = 9$ or $\sqrt{x} = 1$

$x = 81$ or $x = 1$

1. (continued)

Check:

For 81:

$$\begin{array}{c|c} x - 10\sqrt{x} + 9 & 0 \\ \hline 81 - 10\sqrt{81} + 9 & 0 \\ 81 - 10\cdot 9 + 9 & \\ 81 - 90 + 9 & \\ 0 & \end{array}$$

For 1:

$$\begin{array}{c|c} x - 10\sqrt{x} + 9 & 0 \\ \hline 1 - 10\sqrt{1} + 9 & 0 \\ 1 - 10 + 9 & \\ 0 & \end{array}$$

The solutions are 81 and 1.

2. 16, $\frac{1}{4}$

3. $x^4 - 10x^2 + 25 = 0$

We substitute u for x^2.

$$u^2 - 10u + 25 = 0$$
$$(u - 5)(u - 5) = 0$$

$u - 5 = 0$ or $u - 5 = 0$

$u = 5$ or $u = 5$

Now we substitute x^2 for u and solve for x.

$x^2 = 5$ or $x^2 = 5$

$x = \pm\sqrt{5}$ or $x = \pm\sqrt{5}$

The solutions are $\pm\sqrt{5}$.

4. 1, -1, $\pm\sqrt{2}$

5. $t^{2/3} + t^{1/3} - 6 = 0$

We substitute u for $t^{1/3}$.

$$u^2 + u - 6 = 0$$
$$(u + 3)(u - 2) = 0$$

$u + 3 = 0$ or $u - 2 = 0$

$u = -3$ or $u = 2$

We now substitute $t^{1/3}$ for u and solve for t.

$t^{1/3} = -3$ or $t^{1/3} = 2$

$t = (-3)^3$ or $t = 2^3$

$t = -27$ or $t = 8$

The solutions are -27 and 8.

6. 64, -8

7. $z^{1/2} = z^{1/4} + 2$

$z^{1/2} - z^{1/4} - 2 = 0$

We substitute u for $z^{1/4}$.

$u^2 - u - 2 = 0$

$(u - 2)(u + 1) = 0$

$u - 2 = 0$ or $u + 1 = 0$

$u = 2$ or $u = -1$

Next we substitute $z^{1/4}$ for u and solve for z.

$z^{1/4} = 2$ or $z^{1/4} = -1$

$z = 2^4$ or $z = (-1)^4$

$z = 16$ or $z = 1$

The number 16 checks, but 1 does not. The solution is 16.

8. 729

9. $(x^2 - 6x)^2 - 2(x^2 - 6x) - 35 = 0$

We substitute u for $x^2 - 6x$.

$u^2 - 2u - 35 = 0$

$(u - 7)(u + 5) = 0$

$u - 7 = 0$ or $u + 5 = 0$

$u = 7$ or $u = -5$

Next we substitute $x^2 - 6x$ for u and solve for x.

$x^2 - 6x = 7$ or $x^2 - 6x = -5$

$x^2 - 6x - 7 = 0$ or $x^2 - 6x + 5 = 0$

$(x - 7)(x + 1) = 0$ or $(x - 5)(x - 1) = 0$

$x - 7 = 0$ or $x + 1 = 0$ or $x - 5 = 0$ or $x - 1 = 0$

$x = 7$ or $x = -1$ or $x = 5$ or $x = 1$

The solutions are 7, -1, 5, and 1.

10. 1

11. $(y^2 - 5y)^2 + (y^2 - 5y) - 12 = 0$

We substitute u for $y^2 - 5y$.

$u^2 + u - 12 = 0$

$(u + 4)(u - 3) = 0$

$u + 4 = 0$ or $u - 3 = 0$

$u = -4$ or $u = 3$

Next we substitute $y^2 - 5y$ for u and solve for y.

$y^2 - 5y = -4$ or $y^2 - 5y = 3$

$y^2 - 5y + 4 = 0$ or $y^2 - 5y - 3 = 0$

$(y - 4)(y - 1) = 0$ or $y = \frac{-(-5) \pm\sqrt{(-5)^2-4(1)(-3)}}{2\cdot1}$

$y - 4 = 0$ or $y - 1 = 0$ or $y = \frac{5 \pm \sqrt{25 + 12}}{2}$

$y = 4$ or $y = 1$ or $y = \frac{5 \pm \sqrt{37}}{2}$

The solutions are 4, 1, and $\frac{5 \pm \sqrt{37}}{2}$.

12. $1, -1, \frac{1}{2}, -\frac{3}{2}$

13. $w^4 - 4w^2 - 2 = 0$

First we substitute u for w^2.

$u^2 - 4u - 2 = 0$

$u = \frac{-(-4) \pm \sqrt{(-4)^2 - 4(1)(-2)}}{2\cdot1}$

$= \frac{4 \pm \sqrt{16 + 8}}{2} = \frac{4 \pm \sqrt{24}}{2}$

$= \frac{4 \pm 2\sqrt{6}}{2} = 2 \pm \sqrt{6}$

Next we substitute w^2 for u and solve for w.

$w^2 = 2 + \sqrt{6}$ or $w^2 = 2 - \sqrt{6}$

$w = \pm\sqrt{2 + \sqrt{6}}$ or Note: $2 - \sqrt{6} < 0$

No solution.

The solutions are $\pm\sqrt{2 + \sqrt{6}}$.

14. $\pm\sqrt{\frac{5 + \sqrt{5}}{2}}, \pm\sqrt{\frac{5 - \sqrt{5}}{2}}$

15. $x^{-2} - x^{-1} - 6 = 0$

We substitute u for x^{-1}.

$u^2 - u - 6 = 0$

$(u - 3)(u + 2) = 0$

$u - 3 = 0$ or $u + 2 = 0$

$u = 3$ or $u = -2$

Then we substitute x^{-1} for u and solve for x.

$x^{-1} = 3$ or $x^{-1} = -2$

$\frac{1}{x} = 3$ or $\frac{1}{x} = -2$

$x = \frac{1}{3}$ or $x = -\frac{1}{2}$

The solutions are $\frac{1}{3}$ and $-\frac{1}{2}$.

16. $-1, \frac{4}{5}$

17. $2x^{-2} + x^{-1} = 1$

$2x^{-2} + x^{-1} - 1 = 0$

We substitute u for x^{-1}.

$2u^2 + u - 1 = 0$

$(2u - 1)(u + 1) = 0$

$2u - 1 = 0$ or $u + 1 = 0$

$u = \frac{1}{2}$ or $u = -1$

17. (continued)

Next we substitute x^{-1} for u and solve for x.

$x^{-1} = \frac{1}{2}$ or $x^{-1} = -1$

$\frac{1}{x} = \frac{1}{2}$ or $\frac{1}{x} = -1$

$x = 2$ or $x = -1$

The solutions are 2 and -1.

18. $-\frac{1}{10}, 1$

19. $x^4 - 24x^2 - 25 = 0$

We substitute u for x^2.

$u^2 - 24u - 25 = 0$

$(u - 25)(u + 1) = 0$

$u - 25 = 0$ or $u + 1 = 0$

$u = 25$ or $u = -1$

Then we substitute x^2 for u and solve for x.

$x^2 = 25$ or $x^2 = -1$

$x = \pm 5$ or $x = \pm i$

The solutions are ± 5 and $\pm i$.

20. $\pm 3, \pm 2i$

21. $\left(\frac{x^2 - 2}{x}\right)^2 - 7\left(\frac{x^2 - 2}{x}\right) - 18 = 0$

We first substitute u for $\frac{x^2 - 2}{x}$.

$u^2 - 7u - 18 = 0$

$(u - 9)(u + 2) = 0$

$u - 9 = 0$ or $u + 2 = 0$

$u = 9$ or $u = -2$

Then we substitute $\frac{x^2 - 2}{x}$ for u and solve for x.

$\frac{x^2 - 2}{x} = 9$ or $\frac{x^2 - 2}{x} = -2$

$x^2 - 2 = 9x$ or $x^2 - 2 = -2x$

$x^2 - 9x - 2 = 0$ or $x^2 + 2x - 2 = 0$

$x = \frac{-(-9)\pm\sqrt{(-9)^2-4(1)(-2)}}{2\cdot 1}$ or $x = \frac{-2\pm\sqrt{2^2-4(1)(-2)}}{2\cdot 1}$

$x = \frac{9 \pm \sqrt{89}}{2}$ or $x = \frac{-2 \pm 2\sqrt{3}}{2}$

or $x = -1 \pm \sqrt{3}$

The solutions are $\frac{9 \pm \sqrt{89}}{2}$ and $-1 \pm \sqrt{3}$.

22. $\frac{5 \pm \sqrt{21}}{2}, \frac{3 \pm \sqrt{5}}{2}$

23. $\frac{x}{x - 1} - 6\sqrt{\frac{x}{x - 1}} - 40 = 0$

We substitute u for $\sqrt{\frac{x}{x - 1}}$.

$u^2 - 6u - 40 = 0$

$(u - 10)(u + 4) = 0$

$u - 10 = 0$ or $u + 4 = 0$

$u = 10$ or $u = -4$

We then substitute $\sqrt{\frac{x}{x - 1}}$ for u and solve for x.

$\sqrt{\frac{x}{x - 1}} = 10$ or $\sqrt{\frac{x}{x - 1}} = -4$

$\frac{x}{x - 1} = 100$ No solution

$x = 100x - 100$

$100 = 99x$

$\frac{100}{99} = x$

The solution is $\frac{100}{99}$.

24. No solution

25. $5\left(\frac{x + 2}{x - 2}\right)^2 = 3\left(\frac{x + 2}{x - 2}\right) + 2$

$5\left(\frac{x + 2}{x - 2}\right)^2 - 3\left(\frac{x + 2}{x - 2}\right) - 2 = 0$

Substitute u for $\frac{x + 2}{x - 2}$.

$5u^2 - 3u - 2 = 0$

$(5u + 2)(u - 1) = 0$

$5u + 2 = 0$ or $u - 1 = 0$

$u = -\frac{2}{5}$ or $u = 1$

Then substitute $\frac{x + 2}{x - 2}$ for u and solve for x.

$\frac{x + 2}{x - 2} = -\frac{2}{5}$ or $\frac{x + 2}{x - 2} = 1$

$5(x + 2) = -2(x - 2)$ or $x + 2 = x - 2$

$5x + 10 = -2x + 4$ or $2 = -2$

$7x = -6$ No solution

$x = -\frac{6}{7}$

The solution is $-\frac{6}{7}$.

26. $-\frac{5}{2}, -5$

27. $\left[\frac{x^2 - 1}{x}\right]^2 - \left[\frac{x^2 - 1}{x}\right] - 2 = 0$

Substitute u for $\frac{x^2 - 1}{x}$.

$$u^2 - u - 2 = 0$$
$$(u - 2)(u + 1) = 0$$

$u - 2 = 0$ or $u + 1 = 0$

$u = 2$ or $u = -1$

Then substitute $\frac{x^2 - 1}{x}$ for u and solve for x.

$\frac{x^2 - 1}{x} = 2$ or $\frac{x^2 - 1}{x} = -1$

$x^2 - 1 = 2x$ or $x^2 - 1 = -x$

$x^2 - 2x - 1 = 0$ or $x^2 + x - 1 = 0$

$x = \frac{-(-2)\pm\sqrt{(-2)^2-4(1)(-1)}}{2\cdot 1}$ or $x = \frac{-1\pm\sqrt{1^2-4(1)(-1)}}{2\cdot 1}$

$x = \frac{2 \pm \sqrt{8}}{2}$ $\qquad x = \frac{-1 \pm \sqrt{5}}{2}$

$x = \frac{2 \pm 2\sqrt{2}}{2}$

$x = 1 \pm \sqrt{2}$

The solutions are $1 \pm \sqrt{2}$ and $\frac{-1 \pm \sqrt{5}}{2}$.

28. $0, \frac{56}{5}$

29. Familiarize and Translate:

We let t_1 represent the time it takes for the object to fall to the ground and t_2 represent the time it takes the sound to get back. The total amount of time, 3 sec, is the sum of t_1 and t_2. This gives us an equation.

$t_1 + t_2 = 3$ (1)

We use the formula $s = 16t^2 + v_0t$ to find an expression for t_1. Since the stone is dropped, v_0 is 0.

$s = 16t_1{}^2$

$\frac{s}{16} = t_1{}^2$

$\sqrt{\frac{s}{16}} = t_1$

$\frac{\sqrt{s}}{4} = t_1$ (2)

We use d = rt (here we use s = rt) to find an expression for t_2.

$s = 1100\cdot t_2$ (Substituting 1100 for r)

$\frac{s}{1100} = t_2$ (3)

We substitute (2) and (3) in (1) and obtain

$$\frac{\sqrt{s}}{4} + \frac{s}{1100} = 3$$

29. (continued)

Carry out:

$$275\sqrt{s} + s = 3300$$

$s + 275\sqrt{s} - 3300 = 0$

Substitute u for $\sqrt{s}$.

$u^2 + 275u - 3300 = 0$

Using the quadratic formula, we get $u \approx 11.5176$. Then $u = \sqrt{s} \approx 11.5176$ and $s \approx 132.66$.

The value checks. The cliff is 132.66 ft high.

30. 229.94 ft

31.
$$6.75x = \sqrt{35x} + 5.36$$

$6.75x - \sqrt{35}\sqrt{x} - 5.36 = 0$ $\quad(\sqrt{35x} = \sqrt{35}\sqrt{x})$

Substitute u for $\sqrt{x}$.

$6.75u^2 - \sqrt{35}u - 5.36 = 0$

$u = \frac{-(-\sqrt{35}) \pm \sqrt{(-\sqrt{35})^2 - 4(6.75)(-5.36)}}{2(6.75)}$

$= \frac{\sqrt{35} \pm \sqrt{35 + 144.72}}{13.5}$

$= \frac{\sqrt{35} \pm \sqrt{179.72}}{13.5}$

$\approx \frac{5.9160798 \pm 13.4059688}{13.5}$

$u \approx 1.4312629$ or $u \approx -0.5548066$

Substitute $\sqrt{x}$ for u and solve for x.

$\sqrt{x} \approx 1.4312629$ or $\sqrt{x} \approx -0.5548066$

$x \approx 2.0485$ $\qquad$ No solution

The solution is 2.0485.

32. $\{\pm 1.9863, \pm 0.8966i\}$

33. $9x^{3/2} - 8 = x^3$

$0 = x^3 - 9x^{3/2} + 8$

$0 = ((\sqrt{x})^2)^3 - 9(\sqrt{x})^3 + 8$

$0 = ((\sqrt{x})^3)^2 - 9(\sqrt{x})^3 + 8$

Substitute u for $(\sqrt{x})^3$.

$0 = u^2 - 9u + 8$

$0 = (u - 8)(u - 1)$

$u - 8 = 0$ or $u - 1 = 0$

$u = 8$ or $u = 1$

33. (continued)

Substitute $(\sqrt{x})^3$ for u and solve for x.

$(\sqrt{x})^3 = 8$ or $(\sqrt{x})^3 = 1$

$\sqrt{x} = 2$ or $\sqrt{x} = 1$

$x = 4$ or $x = 1$

Both values check. The solutions are 4 and 1.

34. $-\frac{3}{2}$, -1

35. $\sqrt{x - 3} - \sqrt[4]{x - 3} = 2$

Substitute u for $\sqrt[4]{x - 3}$.

$u^2 - u - 2 = 0$

$(u - 2)(u + 1) = 0$

$u - 2 = 0$ or $u + 1 = 0$

$u = 2$ or $u = -1$

Substitute $\sqrt[4]{x - 3}$ for u and solve for x.

$\sqrt[4]{x - 3} = 2$ or $\sqrt[4]{x - 3} = -1$

$x - 3 = 16$ No solution

$x = 19$

The solution is 19.

36. 9

Exercise Set 2.9

1. $y = kx$

$0.6 = k(0.4)$ (Substituting)

$\frac{0.6}{0.4} = k$

$\frac{3}{2} = k$

The equation of variation is $y = \frac{3}{2}x$.

2. $y = \frac{0.32}{x}$

3. $y = \frac{k}{x^2}$

$0.15 = \frac{k}{(0.1)^2}$ (Substituting)

$0.15 = \frac{k}{0.01}$

$0.01(0.15) = k$

$0.0015 = k$

The equation of variation is $y = \frac{0.0015}{x^2}$.

4. $y = 0.8xz$

5. $y = k \cdot \frac{xz}{w}$

$\frac{3}{2} = k \cdot \frac{2\cdot 3}{4}$ (Substituting)

$\frac{3}{2} = k \cdot \frac{6}{4}$

$\frac{4}{6} \cdot \frac{3}{2} = k$

$1 = k$

The equation of variation is $y = \frac{xz}{w}$.

6. $y = 0.3xz^2$

7. $y = k \cdot \frac{xz}{w^2}$

$\frac{12}{5} = k \cdot \frac{16\cdot 3}{5^2}$ (Substituting)

$\frac{12}{5} = k \cdot \frac{48}{25}$

$\frac{25}{48} \cdot \frac{12}{5} = k$

$\frac{5}{4} = k$

The equation of variation is $y = \frac{5}{4} \cdot \frac{xz}{w^2}$.

8. $y = \frac{1}{5} \cdot \frac{xz}{wp}$

9. $y = kx$

We double x.

$y = k(2x)$

$y = 2\cdot kx$

Thus, y is doubled.

10. y is multiplied by $\frac{1}{3}$.

11. $y = \frac{k}{x^2}$

We multiply x by n.

$y = \frac{k}{(nx)^2}$

$y = \frac{k}{n^2x^2}$

$y = \frac{1}{n^2} \cdot \frac{k}{x^2}$

Thus, y is mulitplied by $\frac{1}{n^2}$.

12. y is multiplied by n^2.

13. Method 1:

$$A = kN$$
$$42{,}600 = k \cdot 60{,}000 \qquad \text{(Substituting)}$$
$$0.71 = k$$

The equation of variation is A = 0.71N.

$$A = 0.71N$$
$$A = 0.71(750{,}000) \qquad \text{(Substituting)}$$
$$A = 532{,}500$$

Method 2:

$$\frac{A_2}{A_1} = \frac{N_2}{N_1} \qquad \text{(Using a proportion)}$$
$$\frac{A_2}{42{,}600} = \frac{750{,}000}{60{,}000} \qquad \text{(Substituting)}$$
$$A_2 = 42{,}600 \cdot \frac{750{,}000}{60{,}000}$$
$$A_2 = 532{,}500$$

Thus, 532,500 tons will enter the atmosphere.

14. 220 in^3

15. $L = \frac{kwh^2}{\ell}$

The width and height are doubled, and the length is halved.

$$L = \frac{k \cdot 2w \cdot (2h)^2}{\frac{\ell}{2}}$$
$$L = k \cdot 2w \cdot 4h^2 \cdot \frac{2}{\ell}$$
$$L = 16 \cdot \frac{kwh^2}{\ell}$$

Thus, L is multiplied by 16.

16. 256

17. Method 2:

$$\frac{d_2}{d_1} = \frac{\sqrt{h_2}}{\sqrt{h_1}} \qquad \text{(Using a proportion)}$$
$$\frac{54.32}{28.97} = \frac{\sqrt{h_2}}{\sqrt{19.5}} \qquad \text{(Substituting)}$$
$$\sqrt{19.5} \cdot \frac{54.32}{28.97} = \sqrt{h_2}$$
$$\frac{19.5(54.32)^2}{(28.97)^2} = h_2$$
$$68.6 \approx h_2$$

One must be 68.6 m above sea level.

18. 1.263 ohms

19. Method 2:

$$\frac{A_2}{A_1} = \frac{\ell_2^{\,2}}{\ell_1^{\,2}} \qquad \text{(Using a proportion)}$$
$$\frac{A_2}{168.54} = \frac{(10.2)^2}{(5.3)^2} \qquad \text{(Substituting)}$$
$$A_2 = 168.54 \cdot \frac{(10.2)^2}{(5.3)^2}$$
$$A_2 = 624.24$$

The area will be 624.24 m^2.

20. $22.5 \frac{W}{m^2}$

21. Method 1:

$$A = \frac{kR}{I}$$
$$2.92 = \frac{k \cdot 85}{262} \qquad \text{(Substituting)}$$
$$2.92 \cdot \frac{262}{85} = k$$
$$9 \approx k$$

The equation of variation is $A = \frac{9R}{I}$.

$$A = \frac{9R}{I}$$
$$2.92 = \frac{9R}{300} \qquad \text{(Substituting)}$$
$$\frac{2.92(300)}{9} = k$$
$$97 \approx k$$

He would have given up 97 runs.

22. 220 cm^3

23. If p varies directly as q, then p = kq.

Thus, $q = \frac{1}{k}p$, so q varies directly as p.

24. $u = \frac{k}{v}$, so $v = \frac{k}{u}$ and $\frac{1}{u} = \frac{1}{k} \cdot v$

25.

$$A = kd^2$$
$$\pi r^2 = kd^2$$
$$\pi\left(\frac{d}{2}\right)^2 = kd^2$$
$$\frac{\pi}{4} \cdot d^2 = kd^2$$
$$\frac{\pi}{4} = k$$

26. t varies directly as $\sqrt{P}$

Exercise Set 2.10

1. $36\text{ ft}\cdot\frac{1\text{ yd}}{3\text{ ft}}$
$=\frac{36}{3}\cdot\frac{\text{ft}}{\text{ft}}\cdot\text{yd}$
$=12\text{ yd}$

2. 96 oz

3. $6\text{ kg}\cdot 8\,\frac{\text{hr}}{\text{kg}}$
$=6\cdot 8\cdot\frac{\text{kg}}{\text{kg}}\cdot\text{hr}$
$=48\text{ hr}$

4. 27 km

5. $3\text{ cm}\cdot\frac{2\text{g}}{2\text{ cm}}$
$=\frac{3\cdot 2}{2}\cdot\frac{\text{cm}}{\text{cm}}\cdot\text{g}$
$=3\text{ g}$

6. 18 km

7. $6\text{m}+2\text{m}$
$=(6+2)\text{m}$
$=8\text{ m}$

8. 16 tons

9. $5\text{ ft}^3+7\text{ ft}^3$
$=(5+7)\text{ ft}^3$
$=12\text{ ft}^3$

10. 27 yd^3

11. $\frac{3\text{ kg}}{5\text{m}}\cdot\frac{7\text{ kg}}{6\text{m}}$
$=\frac{3\cdot 7}{5\cdot 6}\cdot\frac{\text{kg}}{\text{m}}\cdot\frac{\text{kg}}{\text{m}}$
$=\frac{7\text{ kg}^2}{10\text{ m}^2}$

12. 180

13. $\frac{2000\text{ lb}\cdot(6\text{ mi/hr})^2}{100\text{ ft}}$
$=2000\text{ lb}\cdot\frac{36\text{ mi}^2}{\text{hr}^2}\cdot\frac{1}{100\text{ ft}}$
$=\frac{2000\cdot 36}{100}\cdot\text{lb}\cdot\frac{\text{mi}^2}{\text{hr}^2}\cdot\frac{1}{\text{ft}}$
$=720\,\frac{\text{lb-mi}^2}{\text{hr}^2\text{-ft}}$

14. $14\,\frac{\text{m-kg}}{\text{sec}^2}$

15. $\frac{6\text{ cm}^2\cdot 5\text{ cm/sec}}{2\text{ sec}^2/\text{cm}^2\cdot 2\,\frac{1}{\text{kg}}}$
$=6\text{ cm}^2\cdot\frac{5\text{ cm}}{\text{sec}}\cdot\frac{\text{cm}^2}{2\text{ sec}^2}\cdot\frac{\text{kg}}{2}$
$=\frac{6\cdot 5}{2\cdot 2}\cdot\frac{\text{cm}^2\cdot\text{cm}\cdot\text{cm}^2\cdot\text{kg}}{\text{sec}\cdot\text{sec}^2}$
$=\frac{15}{2}\,\frac{\text{cm}^5\text{-kg}}{\text{sec}^3}$

16. 125 ft-lb

17. $72\text{ in.}=72\text{ in.}\cdot\frac{1\text{ ft}}{12\text{ in.}}$
$=\frac{72}{12}\cdot\frac{\text{in.}}{\text{in.}}\cdot\text{ft}$
$=6\text{ ft}$

18. 1020 min

19. $2\text{ days}=2\text{ days}\cdot\frac{24\text{ hr}}{1\text{ day}}\cdot\frac{60\text{ min}}{1\text{ hr}}\cdot\frac{60\text{ sec}}{1\text{ min}}$
$=2\cdot 24\cdot 60\cdot 60\cdot\frac{\text{day}}{\text{day}}\cdot\frac{\text{hr}}{\text{hr}}\cdot\frac{\text{min}}{\text{min}}\cdot\text{sec}$
$=172{,}800\text{ sec}$

20. 0.1 hr

21. $60\,\frac{\text{kg}}{\text{m}}=60\,\frac{\text{kg}}{\text{m}}\cdot\frac{1000\text{ g}}{1\text{ kg}}\cdot\frac{1\text{ m}}{100\text{ cm}}$
$=\frac{60\cdot 1000}{100}\cdot\frac{\text{kg}}{\text{kg}}\cdot\frac{\text{m}}{\text{m}}\cdot\frac{\text{g}}{\text{cm}}$
$=600\,\frac{\text{g}}{\text{cm}}$

22. $30\,\frac{\text{mi}}{\text{hr}}$

23. $216\text{ m}^2=216\cdot\text{m}\cdot\text{m}$
$=216\cdot 100\text{ cm}\cdot 100\text{ cm}$
$=2{,}160{,}000\text{ cm}^2$

24. $0.81\,\frac{\text{ton}}{\text{yd}^3}$

25. $\frac{\$36}{\text{day}}=\frac{\$36}{\text{day}}\cdot\frac{100¢}{\$1}\cdot\frac{1\text{ day}}{24\text{ hr}}$
$=\frac{36\cdot 100}{24}\cdot\frac{\$}{\$}\cdot\frac{\text{day}}{\text{day}}\cdot\frac{¢}{\text{hr}}$
$=150\,\frac{¢}{\text{hr}}$

26. 60 man-days

27. $1.73\,\frac{\text{mL}}{\text{sec}}=1.73\,\frac{\text{mL}}{\text{sec}}\cdot\frac{1\text{ L}}{1000\text{ mL}}\cdot\frac{60\text{ sec}}{1\text{ min}}\cdot\frac{60\text{ min}}{1\text{ hr}}$
$=\frac{1.73\cdot 60\cdot 60}{1000}\cdot\frac{\text{mL}}{\text{mL}}\cdot\frac{\text{sec}}{\text{sec}}\cdot\frac{\text{min}}{\text{min}}\cdot\frac{\text{L}}{\text{hr}}$
$=6.228\,\frac{\text{L}}{\text{hr}}$

28. $180 \frac{cg}{mL}$

29. $186{,}000 \frac{mi}{sec}$

$= 186{,}000 \frac{mi}{sec} \cdot \frac{60 \text{ sec}}{1 \text{ min}} \cdot \frac{60 \text{ min}}{1 \text{ hr}} \cdot \frac{24 \text{ hr}}{1 \text{ day}} \cdot \frac{365 \text{ days}}{1 \text{ yr}}$

$= 186{,}000 \cdot 60 \cdot 60 \cdot 24 \cdot 365 \frac{sec}{sec} \cdot \frac{min}{min} \cdot \frac{hr}{hr} \cdot \frac{day}{day} \cdot \frac{mi}{yr}$

$= 5{,}865{,}696{,}000{,}000 \frac{mi}{yr}$

30. $6{,}570{,}000 \frac{mi}{yr}$

31. $89.2 \frac{ft}{sec} = 89.2 \frac{ft}{sec} \cdot \frac{1 \text{ m}}{3.3 \text{ ft}} \cdot \frac{60 \text{ sec}}{1 \text{ min}}$

$= \frac{89.2(60)}{3.3} \cdot \frac{ft}{ft} \cdot \frac{sec}{sec} \cdot \frac{m}{min}$

$\approx 1621.8 \frac{m}{min}$

32. 774.5 m^3

33. $640 \text{ mi}^2 = 640 \cdot mi \cdot mi$

$= 640 \cdot 1.6 \text{ km} \cdot 1.6 \text{ km}$

$= 1638.4 \text{ km}^2$

34. $208.13 \frac{lb}{ft}$

35. $\frac{3}{2} = \frac{w}{5}$ (Using a proportion)

$5 \cdot 3 = 2 \cdot w$

$15 = 2w$

$7.5 = w$

A steel rod 3 cm long weighs 7.5 g.

$\frac{500}{2} = \frac{w}{5}$ (Using a proportion)
(5 m = 500 cm)

$5 \cdot 500 = 2 \cdot w$

$2500 = 2w$

$1250 = w$

A steel rod 5 m long weighs 1250 g.

36. 1.72 cm, 0.08 cm

37. 1 mole of oxygen is 32 grams.

$\frac{50}{1} = \frac{x}{32}$ (Using a proportion)

$50 \cdot 32 = x$

$1600 = x$

50 moles of oxygen is 1600 grams of oxygen.

38. 888.8 g

39. 1 mole of neon is 20.2 grams

$\frac{303}{20.2} = \frac{x}{1}$ (Using a proportion)

$15 = x$

303 grams of neon is 15 moles of neon.

40. 11.8 moles

41. $E = mc^2$

Substitute 5000 for m and 2.9979×10^8 for c.

$E = 5000(2.9979 \times 10^8)^2$

$\approx 44{,}937 \times 10^{16} = 4.4937 \times 10^4 \times 10^{16}$

$= 4.4937 \times 10^{20} \text{ g m}^2/\text{sec}^2$

42. 15.48 g

Exercise Set 3.1

1. a) {(Father, Elaine), (Father, Vanessa), (Father, Chuck), (Father, Deron)}

   b) The ordered pair (Father, Chuck) is in the relation because the statement "Father is the father of Chuck" is true.

   c) The ordered pair (Vanessa, Father) is not in the relation because the statement "Vanessa is the father of Father" is not true.

2. a) {(Mother, Elaine), (Mother, Vanessa), (Mother, Chuck), (Mother, Deron)}

   b) Yes, c) No

3. a) The ordered pair (10, \$1270) is in the relation. After 10 years, the value of a \$10,000 whole-life policy is \$1270.

   b) The ordered pair (20, \$3000) is not in the relation. After 20 years, the value of a \$10,000 whole-life policy is \$2790, not \$3000.

4. a) No, b) Yes

5. a) A = {0, 2} and B = {a, b, c}

   A × B = {(0, a), (0, b), (0, c), (2, a), (2, b), (2, c)}

   b) B = {a, b, c} and A = {0, 2}

   B × A = {(a, 0), (a, 2), (b, 0), (b, 2), (c, 0), (c, 2)}

   c) A × A consists of the ordered pairs in the following list.

2	(2, 0)	(2, 2)
0	(0, 0)	(0, 2)
	0	2

   d) B × B consists of the ordered pairs in the following list.

c	(c, a)	(c, b)	(c, c)
b	(b, a)	(b, b)	(b, c)
a	(a, a)	(a, b)	(a, c)
	a	b	c

6. a) {(1, d), (1, e), (1, f), (3, d), (3, e), (3, f), (5, d), (5, e), (5, f), (9, d), (9, e), (9, f)}

   b) {(d, 1), (d, 3), (d, 5), (d, 9), (e, 1), (e, 3), (e, 5), (e, 9), (f, 1), (f, 3), (f, 5), (f, 9)}

   c) {(1, 1), (1, 3), (1, 5), (1, 9), (3, 1), (3, 3), (3, 5), (3, 9), (5, 1), (5, 3), (5, 5), (5, 9), (9, 1), (9, 3), (9, 5), (9, 9)}

   d) {(d, d), (d, e), (d, f), (e, d), (e, e), (e, f), (f, d), (f, e), (f, f)}

7.

2	(2, -1)	(2, 0)	(2, 1)	(2, 2)
1	(1, -1)	(1, 0)	(1, 1)	(1, 2)
0	(0, -1)	(0, 0)	(0, 1)	(0, 2)
-1	(-1, -1)	(-1, 0)	(-1, 1)	(-1, 2)
	-1	0	1	2

A × A

We know that an ordered pair (a, b) is in the relation if a < b.

The set of ordered pairs in the relation < is {(1, 2), (0, 1), (0, 2), (-1, 0), (-1, 1), (-1, 2)}

8. {(2, -1), (2, 0), (2, 1), (1, -1), (1, 0), (0, -1)}

9.

2	(2, -1)	(2, 0)	(2, 1)	(2, 2)
1	(1, -1)	(1, 0)	(1, 1)	(1, 2)
0	(0, -1)	(0, 0)	(0, 1)	(0, 2)
-1	(-1, -1)	(-1, 0)	(-1, 1)	(-1, 2)
	-1	0	1	2

A × A

We know that an ordered pair (a, b) is in the relation if a ⩽ b.

The set of ordered pairs in the relation ⩽ is {(2, 2), (1, 1), (1, 2), (0, 0), (0, 1), (0, 2), (-1, -1), (-1, 0), (-1, 1), (-1, 2)}

10. {(2, -1), (2, 0), (2, 1), (2, 2), (1, -1), (1, 0), (1, 1), (0, -1), (0, 0), (-1, -1)}

11.

2	(2, -1)	(2, 0)	(2, 1)	(2, 2)
1	(1, -1)	(1, 0)	(1, 1)	(1, 2)
0	(0, -1)	(0, 0)	(0, 1)	(0, 2)
-1	(-1, -1)	(-1, 0)	(-1, 1)	(-1, 2)
	-1	0	1	2

A × A

We know that an ordered pair (a, b) is in the relation if a = b.

The set of ordered pairs in the relation = is {(-1, -1), (0, 0), (1, 1), (2, 2)}

12. {(2, -1), (2, 0), (2, 1), (1, -1), (1, 0), (1, 2), (0, -1), (0, 1), (0, 2), (-1, 0), (-1, 1), (-1, 2)}

13. The set of all first members in a relation is called the domain. The set of all second members in a relation is called its range.

Domain: {0, 5, 10, 15, 20}

Range: {0, 490, 1270, 2000, 2790}

14. Domain: {1959, 1969, 1979, 1989}

Range: {209, 231, 255, 282}

15. Domain: {-1, 0, 1}

Range: {0, 1, 2}

16. Domain: {0, 1, 2}

Range: {-1, 0, 1}

17. Domain: {-1, 0, 1, 2}

Range: {-1, 0, 1, 2}

18. Domain: {-1, 0, 1, 2}

Range: {-1, 0, 1, 2}

19. a) b) D = {-1, 0, 1, 2}

2	(2, -1)	(2, 0)	(2, 1)	(2, 2)
1	(1, -1)	(1, 0)	(1, 1)	(1, 2)
0	(0, -1)	(0, 0)	(0, 1)	(0, 2)
-1	(-1, -1)	(-1, 0)	(-1, 1)	(-1, 2)
	-1	0	1	2

c) Domain: {0, 1}

Range: {0, 1, 2}

20. a) b) E = {-1, 1, 3, 5}

5	(5, -1)	(5, 1)	(5, 3)	(5, 5)
3	(3, -1)	(3, 1)	(3, 3)	(3, 5)
1	(1, -1)	(1, 1)	(1, 3)	(1, 5)
-1	(-1, -1)	(-1, 1)	(-1, 3)	(-1, 5)
	-1	1	3	5

c) Domain: {-1, 1}

Range: {1, 3}

Exercise Set 3.2

1. We substitute -1 for x and -3 for y.

y = 5x + 2	
-3	5(-1) + 2
	-5 + 2
	-3

The equation becomes true: (-1, -3) is a solution.

2. No

3. We substitute -2 for x and 7 for y.

4x + 3y = 12	
4(-2) + 3(7)	12
-8 + 21	
13	

The equation becomes false: (-2, 7) is not a solution.

4. Yes

5. We substitute 3 for x and 0 for y.

$x^2 - 2y = 6$	
$3^2 - 2\cdot 0$	6
9 - 0	
9	

The equation becomes false: (3, 0) is not a solution.

6. Yes

7. y = x + 3

We select numbers for x and find the corresponding values for y.

When x = 0, y = 0 + 3 = 3.
When x = 2, y = 2 + 3 = 5.
When x = -3, y = -3 + 3 = 0.
When x = -5, y = -5 + 3 = -2.

x	y
0	3
2	5
-3	0
-5	-2

Plot these solutions and draw the line.

y = x + 3

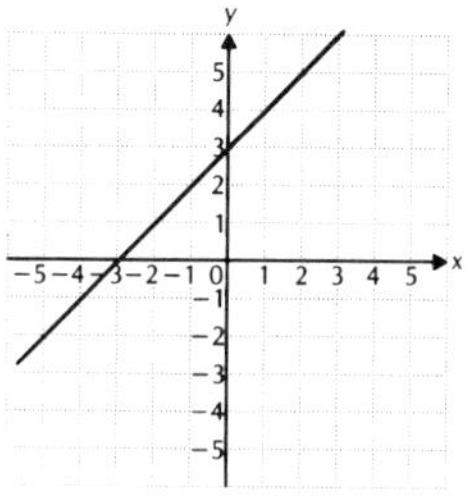

The graph shows the relation {(x, y)|y = x + 3}.

8. y = x - 2

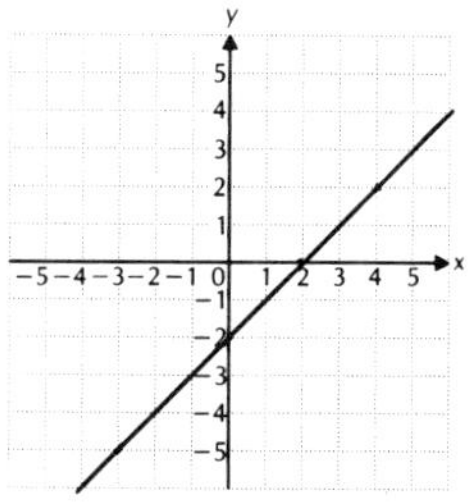

9. $y = 3x - 2$

We select numbers for x and find the corresponding values for y.

When $x = 2$, $y = 3 \cdot 2 - 2 = 4$.

When $x = 0$, $y = 3 \cdot 0 - 2 = -2$.

When $x = -1$, $y = 3(-1) - 2 = -5$.

x	y
2	4
0	-2
-1	-5

Plot these solutions and draw the line.

$y = 3x - 2$

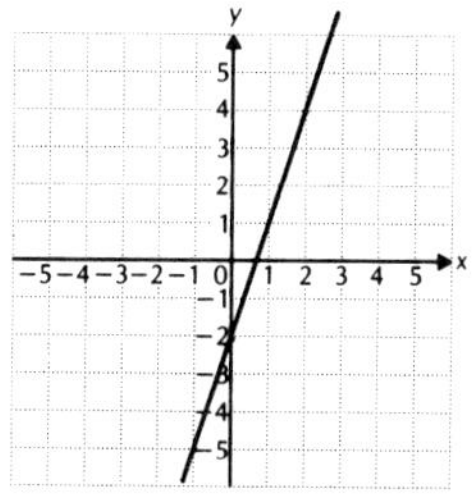

The graph shows the relation $\{(x, y) \mid y = 3x - 2\}$.

10. $y = -4x + 1$

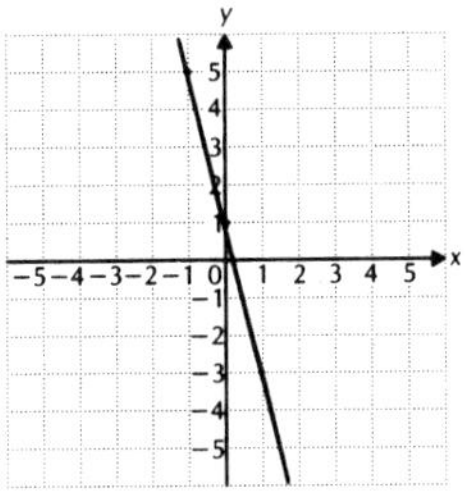

11. $y = x^2$

We select numbers for x and find the corresponding values for y.

When $x = -2$, $y = (-2)^2 = 4$.

When $x = -1$, $y = (-1)^2 = 1$.

When $x = 0$, $y = 0^2 = 0$.

When $x = 1$, $y = 1^2 = 1$.

When $x = 2$, $y = 2^2 = 4$.

x	y
-2	4
-1	1
0	0
1	1
2	4

We plot these ordered pairs and connect the points with a smooth curve.

$y = x^2$

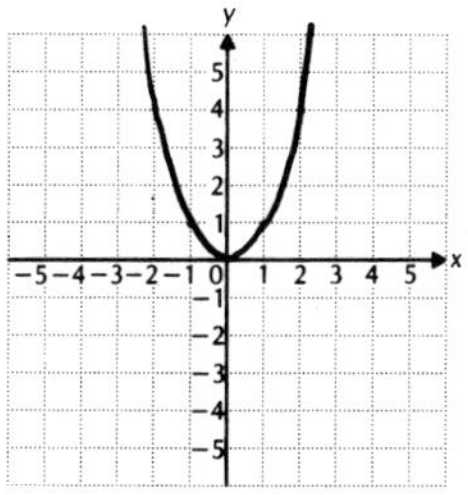

The graph shows the relation $\{(x, y) \mid y = x^2\}$.

12. $y = -x^2$

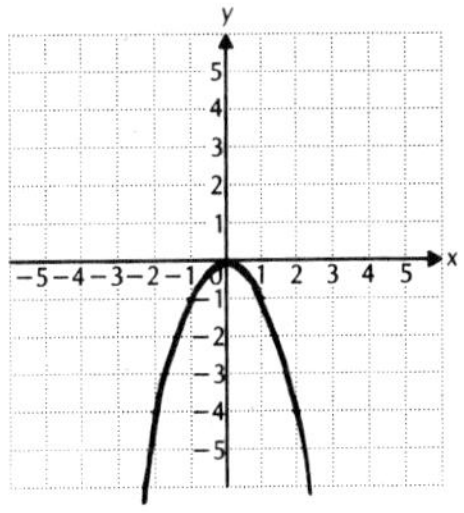

13. $y = x^2 + 2$

We choose numbers for x and find the corresponding values for y.

When $x = -2$, $y = (-2)^2 + 2 = 6$.

When $x = -1$, $y = (-1)^2 + 2 = 3$.

When $x = 0$, $y = 0^2 + 2 = 2$.

When $x = 1$, $y = 1^2 + 2 = 3$.

When $x = 2$, $y = 2^2 + 2 = 6$.

x	y
-2	6
-1	3
0	2
1	3
2	6

We plot these ordered pairs and connect the points with a smooth curve.

$y = x^2 + 2$

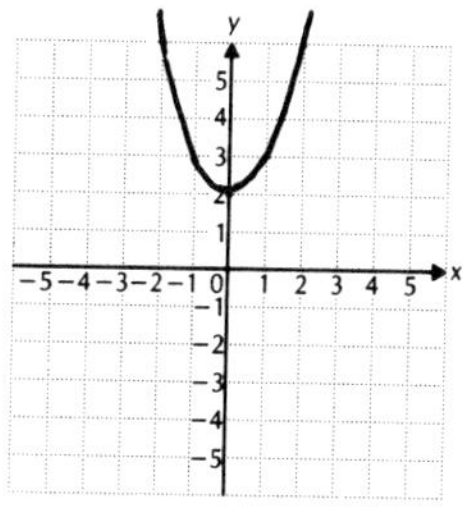

The graph shows the relation $\{(x, y) \mid y = x^2 + 2\}$.

14. $y = x^2 - 2$

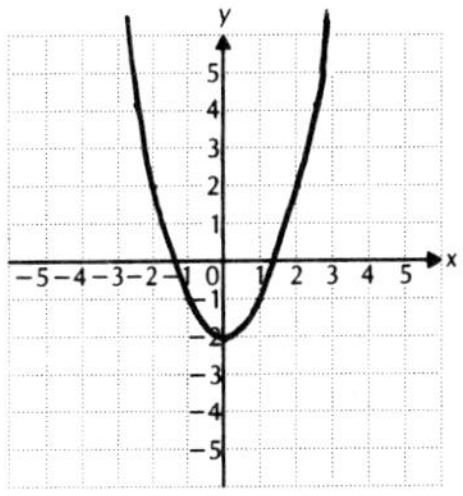

15. $x = y^2 + 2$

Here we select numbers for y and find the corresponding values for x.

When $y = -2$, $x = (-2)^2 + 2 = 6$.

When $y = -1$, $x = (-1)^2 + 2 = 3$.

When $y = 0$, $x = 0^2 + 2 = 2$.

When $y = 1$, $x = 1^2 + 2 = 3$.

When $y = 2$, $y = 2^2 + 2 = 6$.

x	y
6	-2
3	-1
2	0
3	1
6	2

We plot these points and connect them with a smooth curve.

15. (continued)

$x = y^2 + 2$

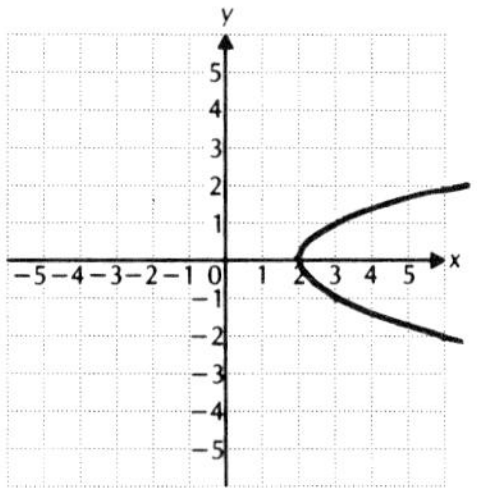

The graph shows the relation $\{(x, y) | x = y^2 + 2\}$.

16. $x = y^2 - 2$

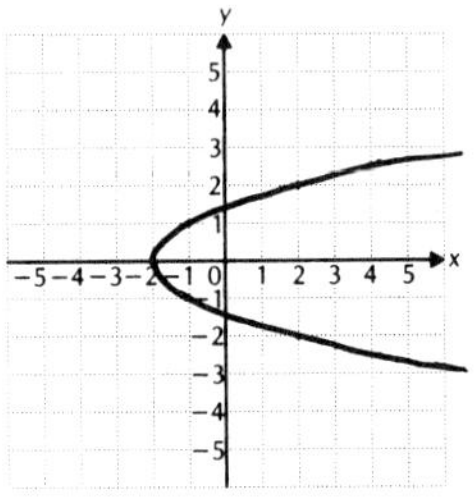

17. $y = |x + 1|$

We find numbers that satisfy the equation.

When $x = -5$, $y = |-5 + 1| = 4$.
When $x = -3$, $y = |-3 + 1| = 2$.
When $x = -1$, $y = |-1 + 1| = 0$.
When $x = 2$, $y = |2 + 1| = 3$.
When $x = 4$, $y = |4 + 1| = 5$.

x	y
-5	4
-3	2
-1	0
2	3
4	5

We plot these points and connect them.

$y = |x + 1|$

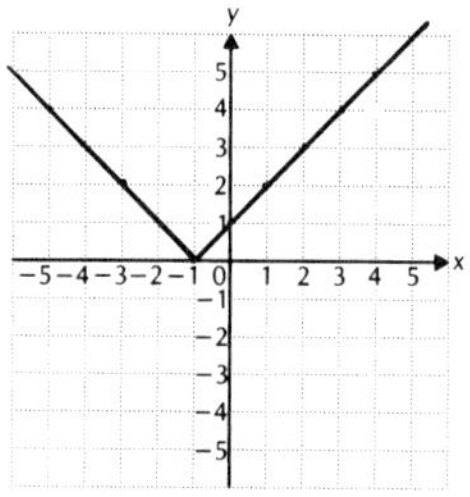

The graph shows the relation $\{(x, y) | y = |x + 1|\}$.

18. $y = |x - 1|$

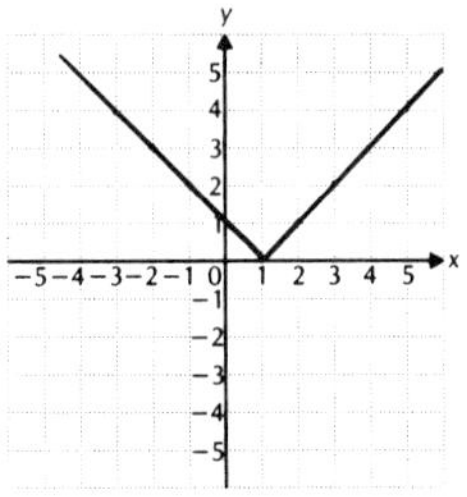

19. $x = |y + 1|$

Here we choose numbers for y and find the corresponding values of x.

When $y = -5$, $x = |-5 + 1| = 4$.
When $y = -2$, $x = |-2 + 1| = 1$.
When $y = -1$, $x = |-1 + 1| = 0$.
When $y = 0$, $x = |0 + 1| = 1$.
When $y = 3$, $x = |3 + 1| = 4$.
When $y = 4$, $x = |4 + 1| = 5$.

x	y
4	-5
1	-2
0	-1
1	0
4	3
5	4

We plot these points and connect them.

$x = |y + 1|$

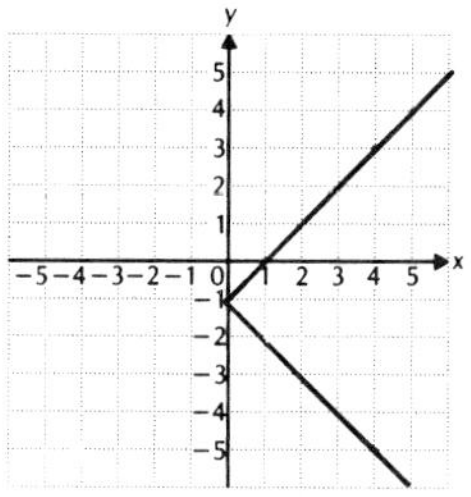

The graph shows the relation $\{(x, y) | x = |y + 1|\}$.

20. $x = |y - 1|$

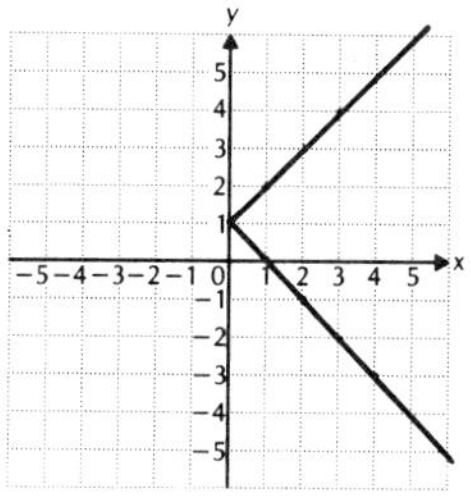

21. $xy = 10$

We find numbers that satisfy the equation. We then plot these points and connect them. Note that neither x nor y can be 0.

x	y
-6	$-\frac{5}{3}$, or $-1\frac{2}{3}$
-5	-2
-3	$-\frac{10}{3}$, or $-3\frac{1}{3}$
-2	-5

x	y
-1	-10
1	10
2	5
3	$3\frac{1}{3}$
5	2
6	$\frac{5}{3}$

$xy = 10$

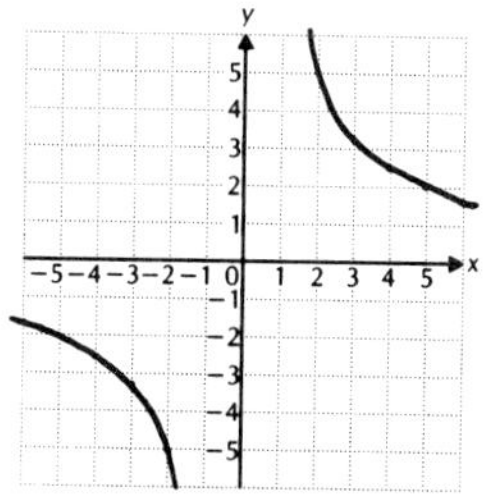

The graph shows the relation $\{(x, y) | xy = 10\}$.

22. $xy = -18$

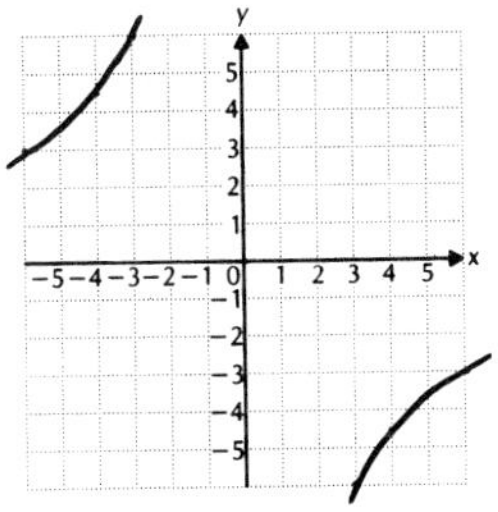

23. $y = \frac{1}{x}$

We find numbers that satisfy the equation. We then plot these points and connect them. Note that neither x nor y can be 0.

x	y
-5	$-\frac{1}{5}$
-3	$-\frac{1}{3}$
-2	$-\frac{1}{2}$
-1	-1
$-\frac{1}{2}$	-2
$-\frac{1}{4}$	-4

x	y
$\frac{1}{4}$	4
$\frac{1}{2}$	2
1	1
2	$\frac{1}{2}$
3	$\frac{1}{3}$
5	$\frac{1}{5}$

$y = \frac{1}{x}$

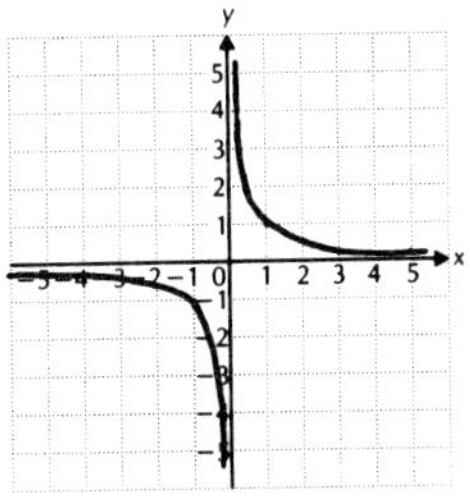

The graph shows the relation $\left\{(x, y) \middle| y = \frac{1}{x}\right\}$.

24. $y = -\frac{2}{x}$

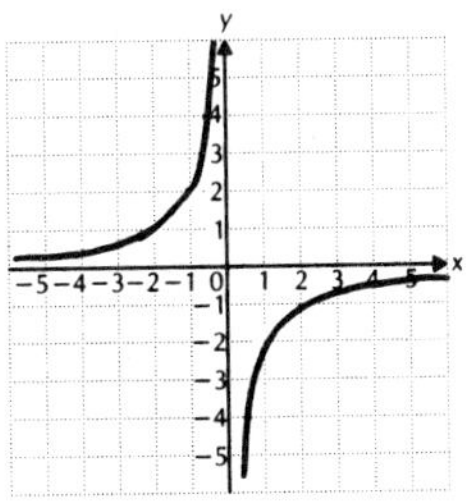

25. $y = \frac{1}{x^2}$

We find numbers that satisfy the equation. We plot these points and connect them. Note that neither x nor y can be 0.

x	y
-3	$\frac{1}{9}$
-2	$\frac{1}{4}$
-1	1
$-\frac{1}{2}$	4
$-\frac{1}{4}$	16

x	y
3	$\frac{1}{9}$
2	$\frac{1}{4}$
1	1
$\frac{1}{2}$	4
$\frac{1}{4}$	16

$y = \frac{1}{x^2}$

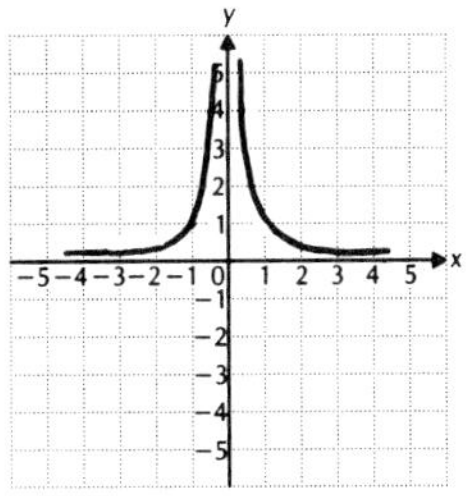

The graph shows the relation $\left\{(x, y) \middle| y = \frac{1}{x^2}\right\}$.

26. $y = x^3$

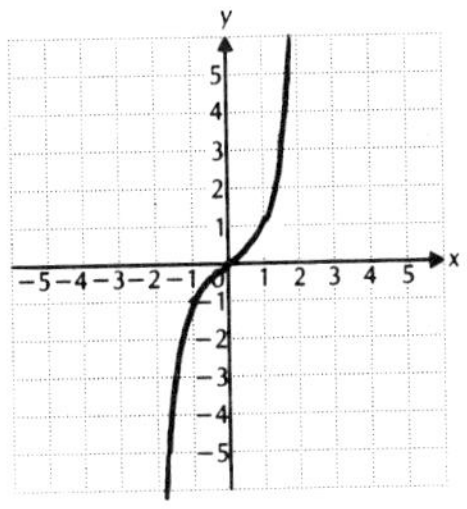

27. $y = \sqrt{x}$

The sensible replacements for x are $x \geqslant 0$. We choose sensible replacements for x and find corresponding values of y.

When $x = 0$, $y = \sqrt{0} = 0$.

When $x = 2$, $y = \sqrt{2} \approx 1.414$.

When $x = 4$, $y = \sqrt{4} = 2$.

When $x = 6$, $y = \sqrt{6} \approx 2.449$.

When $x = 9$, $y = \sqrt{9} = 3$.

x	y
0	0
2	1.414
4	2
6	2.449
9	3

We plot these points and connect them.

$y = \sqrt{x}$

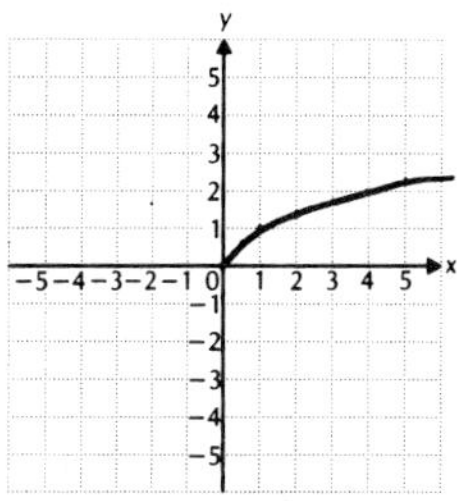

The graph shows the relation $\{(x, y) | y = \sqrt{x}\}$.

28. $y = x^{1/3}$

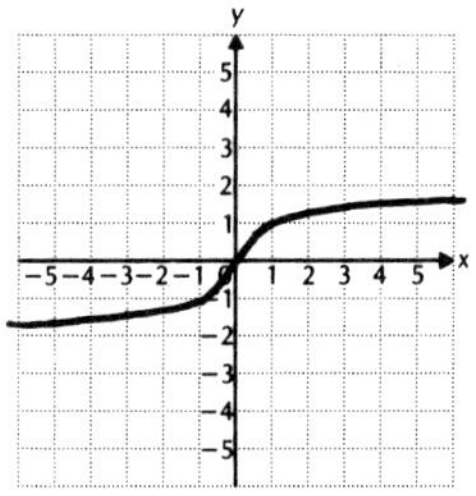

29. $y = 8 - x^2$

We choose numbers for x and find the corresponding values of y.

When $x = -3$, $y = 8 - (-3)^2 = 8 - 9 = -1$.

When $x = -2$, $y = 8 - (-2)^2 = 8 - 4 = 4$.

When $x = -1$, $y = 8 - (-1)^2 = 8 - 1 = 7$.

When $x = 0$, $y = 8 - 0^2 = 8 - 0 = 8$.

When $x = 1$, $y = 8 - 1^2 = 8 - 1 = 7$.

When $x = 2$, $y = 8 - 2^2 = 8 - 4 = 4$.

When $x = 3$, $y = 8 - 3^2 = 8 - 9 = -1$.

x	y
-3	-1
-2	4
-1	7
0	8
1	7
2	4
3	-1

We plot these ordered pairs and connect the points with a smooth curve.

$y = 8 - x^2$

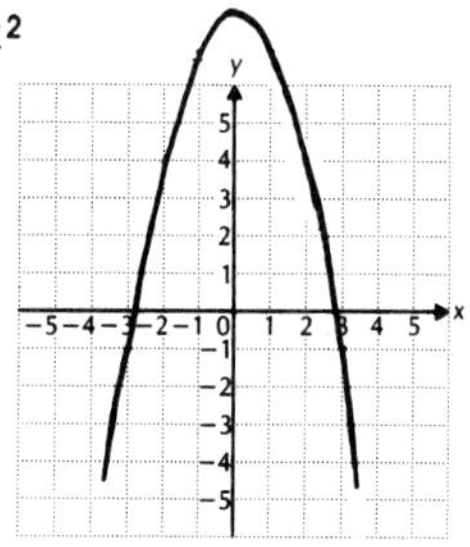

The graph shows the relation $\{(x, y) | y = 8 - x^2\}$.

30. $x = 4 - y^2$

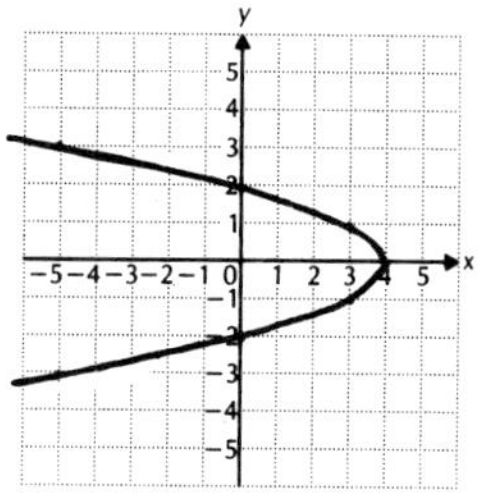

31. $y = x^2 + 1$

x	y
0	1
1	2
-1	2
2	5
-2	5

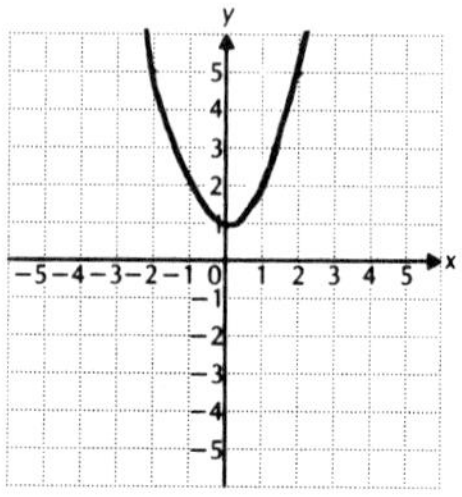

31. (continued)

$y = (-x)^2 + 1$

x	y
0	1
1	2
-1	2
2	5
-2	5

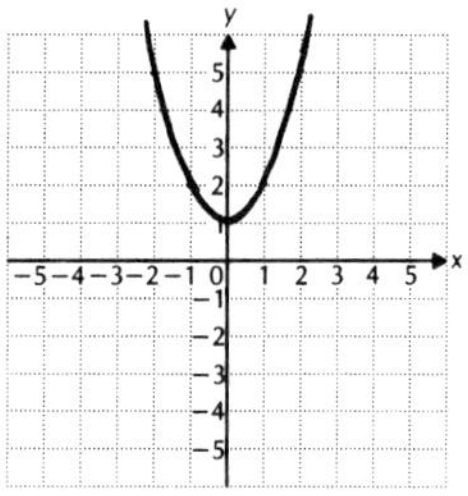

The graphs are the same. The equations are equivalent.

32. $y = x^2 - 2$ $y = 2 - x^2$

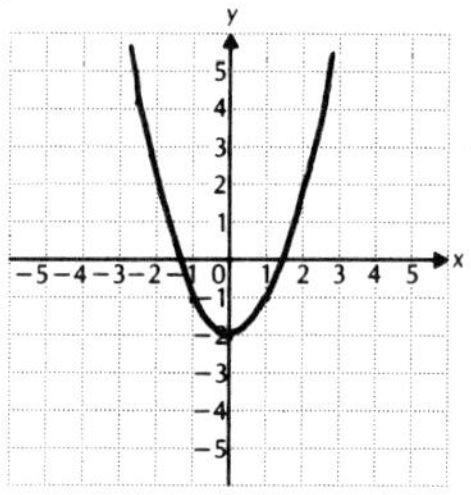

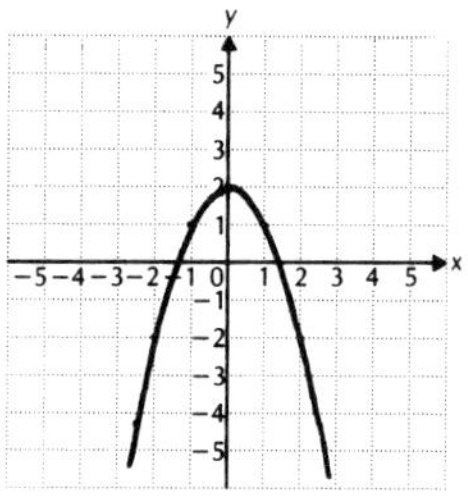

The graphs are not the same. The equations are not equivalent.

33. a)

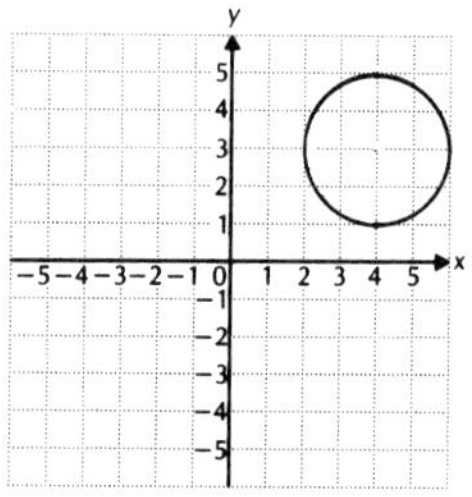

b) Domain: $\{x | 2 \leqslant x \leqslant 6\}$

c) Range: $\{y | 1 \leqslant y \leqslant 5\}$

34. Domain: $\{x | 1 \leqslant x \leqslant 4\}$

Range: $\{y | 1 \leqslant y \leqslant 6\}$

35. $y = -x^2$

Domain: The set of real numbers

Range: $\{y | y \leqslant 0\}$

36. Domain: The set of real numbers

Range: The set of real numbers

37. $x = |y + 1|$

Domain: $\{x | x \geqslant 0\}$

Range: The set of real numbers

38. Domain: $\{x | x \neq 0\}$

Range: $\{y | y \neq 0\}$

39. $\{(x, y) \mid y = 2\}$ is the relation in which the second coordinate is always 2 and the first coordinate may be any real number. Thus, all ordered pairs (x, 2) are solutions. For example,

x	y
-4	2
0	2
3	2

(y must be 2 and x can be any number)

Plot these solutions and complete the graph. The graph is a horizontal line.

$y = 2$

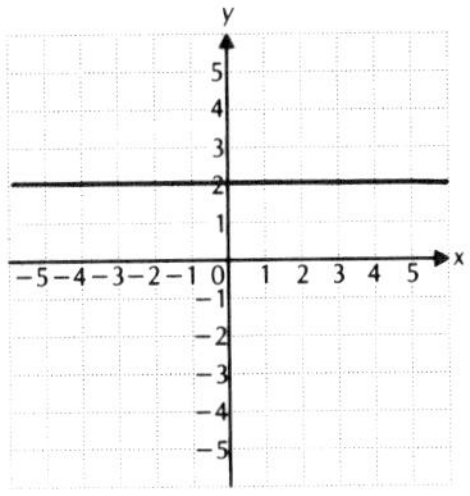

40. $x = -3$

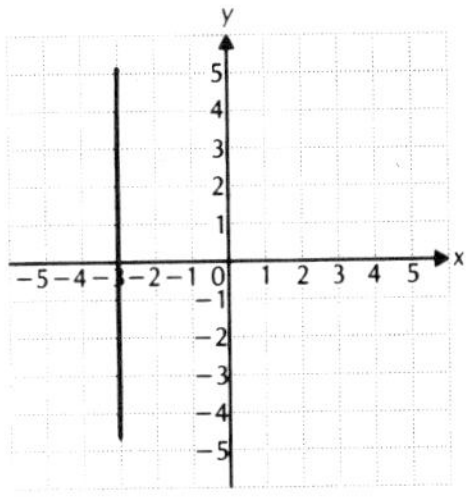

41. $\{(x, y) \mid y = x + 1\}$ is the relation in which the second coordinate is always 1 more than the first coordinate and the first coordinate may be any real number.

When $x = -5$, $y = -5 + 1 = -4$.
When $x = -2$, $y = -2 + 1 = -1$.
When $x = 0$, $y = 0 + 1 = 1$.
When $x = 1$, $y = 1 + 1 = 2$.
When $x = 4$, $y = 4 + 1 = 5$.

x	y
-5	-4
-2	-1
0	1
1	2
4	5

We plot these solutions and draw the graph.

$y = x + 1$

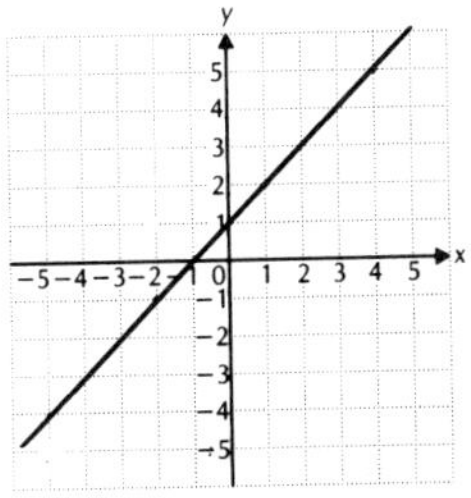

42. $y = x - 1$

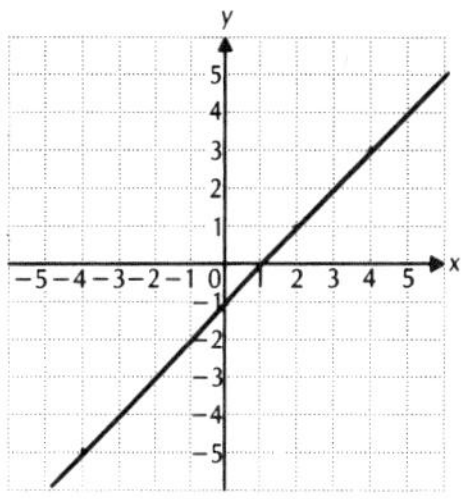

43. $\{(x, y) \mid y = 2x\}$ is the relation in which the second coordinate is always twice the first coordinate and the first coordinate may be any real number.

When $x = -3$, $y = 2(-3) = -6$.
When $x = 0$, $y = 2\cdot 0 = 0$.
When $x = 2$, $y = 2\cdot 2 = 4$.

x	y
-3	-6
0	0
2	4

We plot these solutions and draw the graph.

$y = 2x$

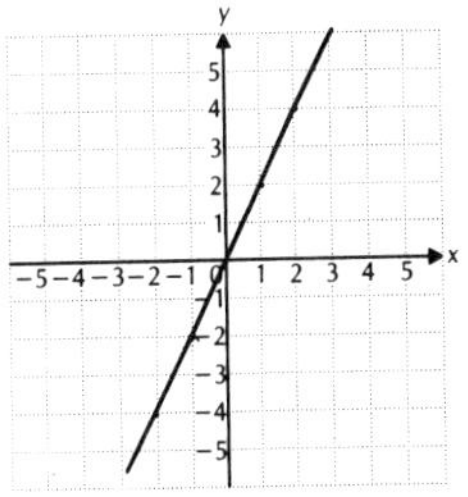

44. $y = \frac{x}{2}$

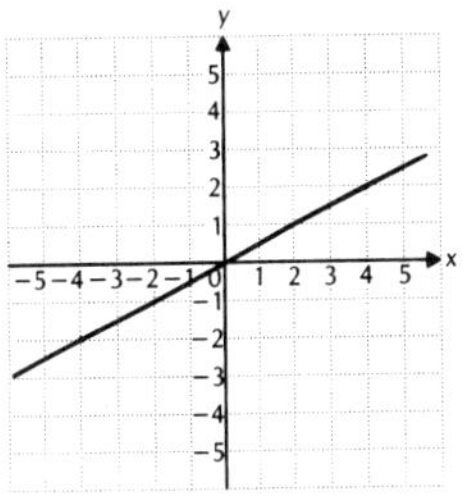

45. $\{(x, y) \mid y = x^2\}$ is the relation in which the second coordinate is always the square of the first coordinate and the first coordinate may be any real number. See Exercise 11.

46. $x = y^2$

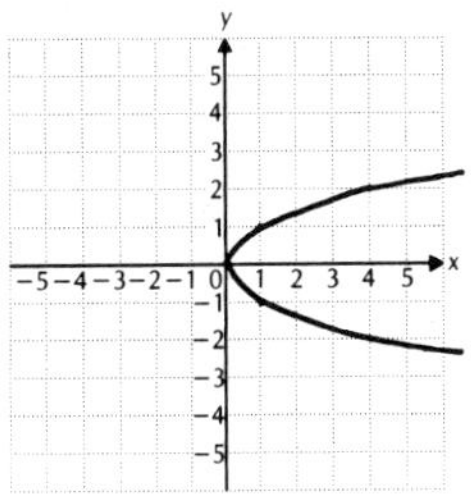

47. $y = |x| + x$

When $x = -5$, $y = |-5| + (-5) = 0$.
When $x = -3$, $y = |-3| + (-3) = 0$.
When $x = -1$, $y = |-1| + (-1) = 0$.
When $x = 0$, $y = |0| + 0 = 0$.
When $x = 1$, $y = |1| + 1 = 2$.
When $x = 3$, $y = |3| + 3 = 6$.
When $x = 5$, $y = |5| + 5 = 10$.

x	y
-5	0
-3	0
-1	0
0	0
1	2
3	6
5	10

$y = |x| + x$

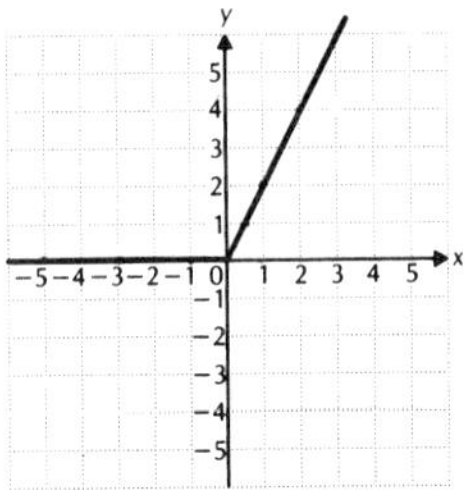

48. $y = x|x|$

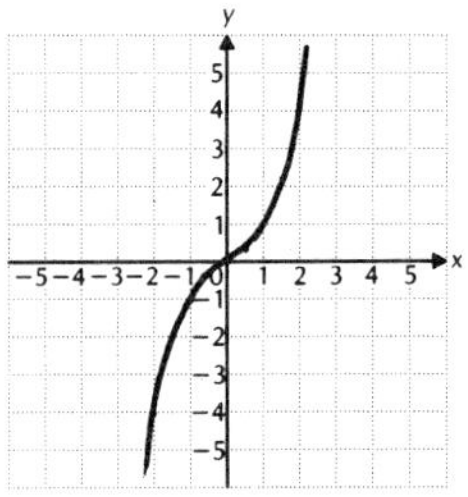

49. $y = |x^2 - 4|$

When $x = 3$ or $x = -3$, $y = |9 - 4| = 5$.
When $x = 2$ or $x = -2$, $y = |4 - 4| = 0$.
When $x = 1$ or $x = -1$, $y = |1 - 4| = 3$.
When $x = 0$, $y = |0 - 4| = 4$.

x	y
3	5
-3	5
2	0
-2	0
1	3
-1	3
0	4

$y = |x^2 - 4|$

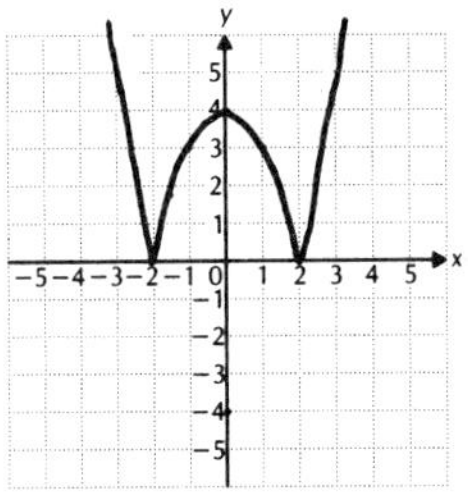

50. $y = x^{2/3}$

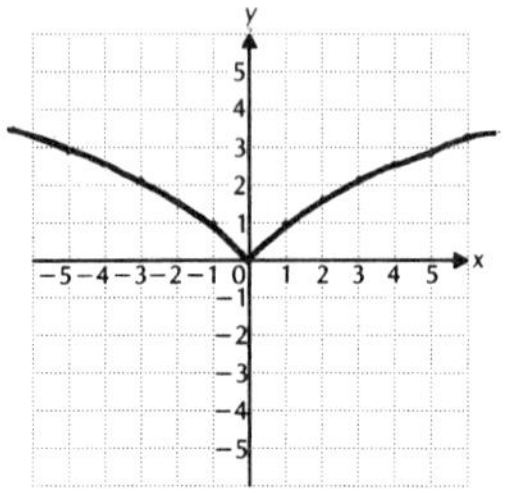

51. $|y| = x + 1$, or $x = |y| - 1$

Here we choose values for y and find corresponding values of x.

When $y = 4$ or -4, $x = 4 - 1 = 3$.
When $y = 3$ or -3, $x = 3 - 1 = 2$.
When $y = 2$ or -2, $x = 2 - 1 = 1$.
When $y = 1$ or -1, $x = 1 - 1 = 0$.
When $y = 0$, $x = 0 - 1 = -1$.

x	y
3	4
3	-4
2	3
2	-3
1	2
1	-2
0	1
0	-1
-1	0

$|y| = x + 1$

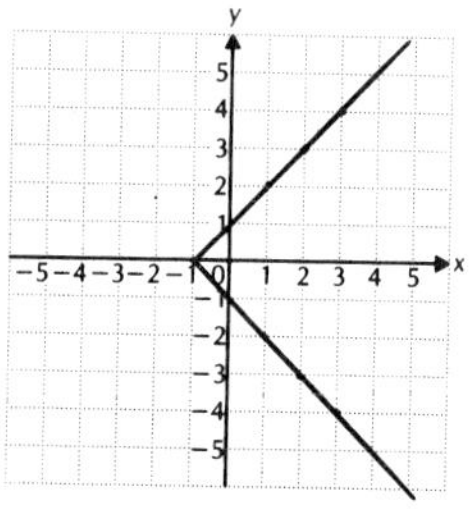

52. $y = |x^3|$

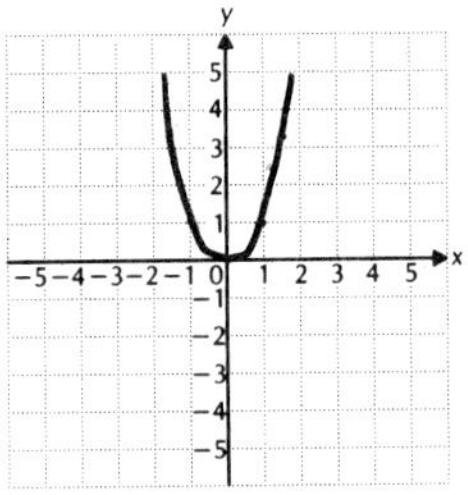

53. $|y| = |x|$

When $x = 0$, $y = 0$.
When $x = 1$, $y = 1$ or -1.
When $x = -1$, $y = 1$ or -1.
When $x = 4$, $y = 4$ or -4.
When $x = -4$, $y = 4$ or -4.

x	y
0	0
1	1
1	-1
-1	1
-1	-1
4	4
4	-4
-4	4
-4	-4

$|y| = |x|$

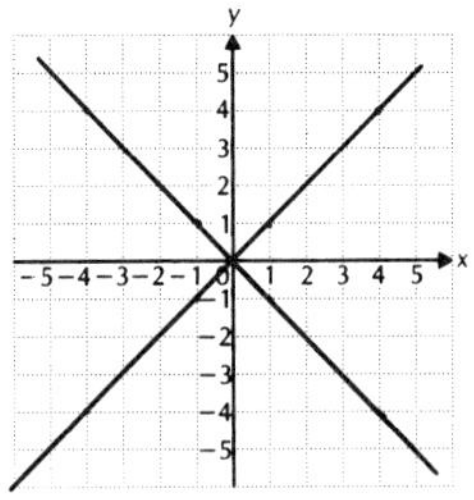

54. Graph consists only of the origin (0, 0).

55. $|xy| = 1$

$xy = 1$ or $xy = -1$

x	y
4	$\frac{1}{4}$
2	$\frac{1}{2}$
1	1
$\frac{1}{2}$	2
$\frac{1}{4}$	4
-4	$-\frac{1}{4}$
-2	$-\frac{1}{2}$
-1	-1
$-\frac{1}{2}$	-2
$-\frac{1}{4}$	-4

x	y
4	$-\frac{1}{4}$
2	$-\frac{1}{2}$
1	-1
$\frac{1}{2}$	-2
$\frac{1}{4}$	-4
-4	$\frac{1}{4}$
-2	$\frac{1}{2}$
-1	1
$-\frac{1}{2}$	2
$-\frac{1}{4}$	4

$|xy| = 1$

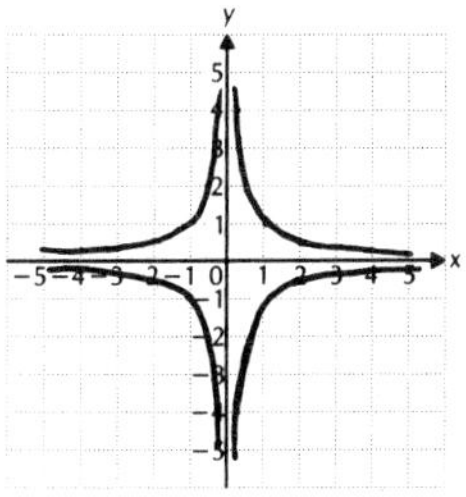

56. $\{(x, y)|1 < x < 4 \text{ and } -3 < y < -1\}$

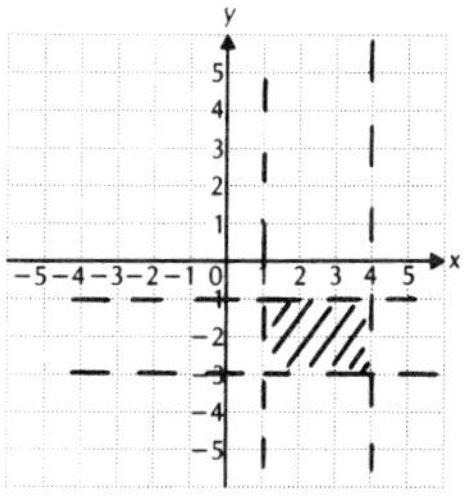

Exercise Set 3.3

1. The graphs b, c, and d are graphs of functions. No vertical line crosses the graph more than once. The graph in a) is not the graph of a function because it fails the vertical line test. We can find a vertical line which meets the graph in more than one point.

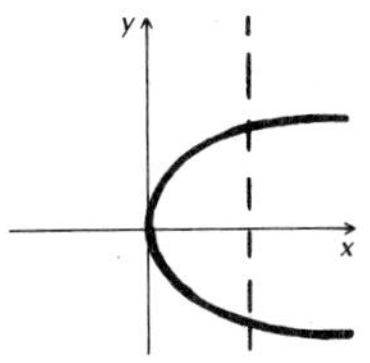

2. b, c, d

3. $f(x) = 5x^2 + 4x$

a) $f(0) = 5\cdot 0^2 + 4\cdot 0 = 0 + 0 = 0$

b) $f(-1) = 5(-1)^2 + 4(-1) = 5 - 4 = 1$

c) $f(3) = 5\cdot 3^2 + 4\cdot 3 = 45 + 12 = 57$

d) $f(t) = 5t^2 + 4t$

e) $f(t - 1) = 5(t - 1)^2 + 4(t - 1)$
$= 5(t^2 - 2t + 1) + 4(t - 1)$
$= 5t^2 - 10t + 5 + 4t - 4$
$= 5t^2 - 6t + 1$

f) $f(a + h) = 5(a + h)^2 + 4(a + h)$
$= 5(a^2 + 2ah + h^2) + 4(a + h)$
$= 5a^2 + 10ah + 5h^2 + 4a + 4h$

$f(a) = 5a^2 + 4a$

$$\frac{f(a + h) - f(a)}{h} = \frac{(5a^2+10ah+5h^2+4a+4h)-(5a^2+4a)}{h}$$
$$= \frac{10ah + 5h^2 + 4h}{h}$$
$$= \frac{h(10a + 5h + 4)}{h}$$
$$= 10a + 5h + 4$$

4. a) 1, b) 6, c) 22, d) $3t^2 - 2t + 1$, e) $3a^2 + 6ah + 3h^2 - 2a - 2h + 1$, f) $6a + 3h - 2$

5. $f(x) = 2|x| + 3x$

a) $f(1) = 2|1| + 3\cdot 1 = 2 + 3 = 5$

b) $f(-2) = 2|-2| + 3(-2) = 4 - 6 = -2$

c) $f(-4) = 2|-4| + 3(-4) = 8 - 12 = -4$

d) $f(2y) = 2|2y| + 3(2y) = 4|y| + 6y$

e) $f(a + h) = 2|a + h| + 3(a + h)$
$= 2|a + h| + 3a + 3h$

f) $$\frac{f(a + h) - f(a)}{h}$$
$$= \frac{(2|a + h| + 3a + 3h) - (2|a| + 3a)}{h}$$
$$= \frac{2|a + h| - 2|a| + 3h}{h}$$

6. a) -1, b) -4, c) -56, d) $27y^3 - 6y$, e) $h^3 + 6h^2 + 10h + 4$, f) $h^2 + 6h + 10$

7. $f(x) = 4.3x^2 - 1.4x$

a) $f(1.034) = 4.3(1.034)^2 - 1.4(1.034)$
$= 4.5973708 - 1.4476 = 3.1497708$

b) $f(-3.441) = 4.3(-3.441)^2 - 1.4(-3.441)$
$= 50.9140683 + 4.8174 = 55.7314683$

c) $f(27.35) = 4.3(27.35)^2 - 1.4(27.35)$
$= 3216.49675 - 38.29 = 3178.20675$

7. (continued)

d) $f(-16.31) = 4.3(-16.31)^2 - 1.4(-16.31)$
$= 1143.86923 + 22.834 = 1166.70323$

8. a) 6.4467, b) 12.0308, c) 1.9259, d) 4.1744

9. $f(x) = \frac{x^2 - x - 2}{2x^2 - 5x - 3}$

a) $f(0) = \frac{0^2 - 0 - 2}{2 \cdot 0^2 - 5 \cdot 0 - 3} = \frac{-2}{-3} = \frac{2}{3}$

b) $f(4) = \frac{4^2 - 4 - 2}{2 \cdot 4^2 - 5 \cdot 4 - 3} = \frac{10}{9}$

c) $f(-1) = \frac{(-1)^2 - (-1) - 2}{2(-1)^2 - 5(-1) - 3} = \frac{0}{4} = 0$

d) $f(3) = \frac{3^2 - 3 - 2}{2 \cdot 3^2 - 5 \cdot 3 - 3} = \frac{4}{0}$ (Does not exist)

We cannot divide by 0. This function value does not exist.

10. a) $\frac{\sqrt{26}}{5}$, b) $\frac{\sqrt{2}}{3}$, c) Does not exist as a real number,
d) Does not exist as a real number

11. $f(x) = 7x + 4$

There are no restrictions on the numbers we can substitute into this formula. Thus, the domain is the entire set of real numbers.

12. All real numbers

13. $f(x) = 4 - \frac{2}{x}$

Division by 0 is undefined. Thus, $x \neq 0$. The domain is all real numbers except 0.

14. $\{x | x \geqslant 3\}$

15. $f(x) = \sqrt{7x + 4}$

Since this formula only makes sense if the radicand is nonnegative, we want the replacements for x which make the following inequality true.

$7x + 4 \geqslant 0$

$7x \geqslant -4$

$x \geqslant -\frac{4}{7}$

The domain is $\left\{x \middle| x \geqslant -\frac{4}{7}\right\}$.

16. $\{x | x \neq 3, -3\}$

17. $f(x) = \frac{1}{x^2 - 4}$

This formula makes sense as long as a replacement for x does not make the denominator 0. To find those replacements which do make the denominator 0, we solve $x^2 - 4 = 0$.

$x^2 - 4 = 0$

$(x + 2)(x - 2) = 0$

$x = -2$ or $x = 2$

Thus the domain consists of all real numbers except -2 and 2. This set can be named $\{x | x \neq -2, 2\}$.

18. $\{x | x \neq 0, 2, -2\}$

19. $f(x) = \frac{4x^3 + 4}{4x^2 - 5x - 6}$

This formula makes sense as long as a replacement for x does not make the denominator 0. To find those replacements which do make the denominator 0, we solve $4x^2 - 5x - 6 = 0$.

$4x^2 - 5x - 6 = 0$

$(4x + 3)(x - 2) = 0$

$4x + 3 = 0$ or $x - 2 = 0$

$x = -\frac{3}{4}$ or $x = 2$

Thus the domain consists of all real numbers except $-\frac{3}{4}$ and 2. This set can be named $\left\{x \middle| x \neq -\frac{3}{4}, 2\right\}$.

20. $\{x | x \neq 2, -2\}$

21. $f(x) = 3x^2 + 2$ $g(x) = 2x - 1$

$f \circ g(x) = f(g(x)) = f(2x - 1)$
$= 3(2x - 1)^2 + 2$
$= 3(4x^2 - 4x + 1) + 2$
$= 12x^2 - 12x + 3 + 2$
$= 12x^2 - 12x + 5$

$g \circ f(x) = g(f(x)) = g(3x^2 + 2)$
$= 2(3x^2 + 2) - 1$
$= 6x^2 + 4 - 1$
$= 6x^2 + 3$

22. $f \circ g(x) = 8x^2 - 17$, $g \circ f(x) = 32x^2 + 48x + 13$

23. $f(x) = 4x^2 - 1$ $\qquad g(x) = \frac{2}{x}$

$f \circ g(x) = f(g(x)) = f\left(\frac{2}{x}\right)$
$= 4\left(\frac{2}{x}\right)^2 - 1$
$= 4 \cdot \frac{4}{x^2} - 1$
$= \frac{16}{x^2} - 1$

$g \circ f(x) = g(f(x)) = g(4x^2 - 1)$
$= \frac{2}{4x^2 - 1}$

24. $f \circ g(x) = \frac{3}{2x^2 + 3}$, $\quad g \circ f(x) = \frac{18}{x^2} + 3$

25. $f(x) = x^2 + 1$ $\qquad g(x) = x^2 - 1$

$f \circ g(x) = f(g(x)) = f(x^2 - 1)$
$= (x^2 - 1)^2 + 1$
$= x^4 - 2x^2 + 1 + 1$
$= x^4 - 2x^2 + 2$

$g \circ f(x) = g(f(x)) = g(x^2 + 1)$
$= (x^2 + 1)^2 - 1$
$= x^4 + 2x^2 + 1 - 1$
$= x^4 + 2x^2$

26. $f \circ g(x) = \frac{1}{x^2 + 4x + 4}$, $\quad g \circ f(x) = \frac{1}{x^2} + 2$

27. $f(z) = z^2 - 4z + i$

$f(3 + i) = (3 + i)^2 - 4(3 + i) + i$
$= 9 + 6i + i^2 - 12 - 4i + i$
$= i^2 + 3i - 3$
$= -1 + 3i - 3$
$= -4 + 3i$

28. $8 - 11i$

29. $g(-2) = 0$, $g(-3) = -3$, $g(0) = 3$, $g(2) = 2$

30. 0, -2, 0, 3

31. $f(x) = \frac{1}{x}$ $\qquad f(x + h) = \frac{1}{x + h}$

$$\frac{f(x + h) - f(x)}{h} = \frac{\frac{1}{x + h} - \frac{1}{x}}{h}$$
$$= \frac{\frac{1}{x + h} \cdot \frac{x}{x} - \frac{1}{x} \cdot \frac{x + h}{x + h}}{h}$$
$$= \frac{\frac{x - (x + h)}{x(x + h)}}{h}$$
$$= \frac{-h}{x(x + h)} \cdot \frac{1}{h}$$
$$= \frac{-1}{x(x + h)}$$

32. $\frac{-2x - h}{x^2(x + h)^2}$

33. $f(x) = \sqrt{x}$ $\qquad f(x + h) = \sqrt{x + h}$

$$\frac{f(x + h) - f(x)}{h} = \frac{\sqrt{x + h} - \sqrt{x}}{h}$$
$$= \frac{\sqrt{x + h} - \sqrt{x}}{h} \cdot \frac{\sqrt{x + h} + \sqrt{x}}{\sqrt{x + h} + \sqrt{x}}$$
$$= \frac{(x + h) - x}{h(\sqrt{x + h} + \sqrt{x})}$$
$$= \frac{h}{h(\sqrt{x + h}) + \sqrt{x})}$$
$$= \frac{1}{\sqrt{x + h} + \sqrt{x}}$$

34. $\left\{x \middle| x \neq \frac{5}{2} \text{ and } x \geqslant 0\right\}$

35. $f(x) = \frac{\sqrt{x + 3}}{x^2 - x - 2}$

The replacements for x cannot make the denominator 0. To find those replacements which do make the denominator 0, we solve $x^2 - x - 2 = 0$.

$x^2 - x - 2 = 0$
$(x - 2)(x + 1) = 0$

$x - 2 = 0$ or $x + 1 = 0$
$x = 2$ or $\quad x = -1$

Thus the domain cannot include 2 and -1.

The radicand in the numerator must be nonnegative, thus we want the replacements for x which make the following inequality true.

$x + 3 \geqslant 0$
$x \geqslant -3$

The domain for f(x) is the set of all reals greater than or equal to -3 except 2 and -1. This set can be named $\{x | x \neq 2, -1 \text{ and } x \geqslant -3\}$.

36. $\{x | x > 0\}$

37. $f(x) = \sqrt{x^2 + 1}$

Since $x^2 + 1 \geqslant 0$ for all x, the domain is the set of all real numbers.

38. $f \circ f(x) = \frac{x - 1}{x}$, $f \circ [f \circ f(x)] = x$

39. $f \circ g(x) = \frac{16}{x^2} - 1$

The only nonsensible replacements are those which make the denominator, x^2, equal to 0.

$x^2 = 0$

$x = 0$

The domain is $\{x | x \neq 0\}$.

$g \circ f(x) = \frac{2}{4x^2 - 1}$

The only nonsensible replacements are those which make the denominator, $4x^2 - 1$, equal to 0.

$4x^2 - 1 = 0$

$(2x + 1)(2x - 1) = 0$

$2x + 1 = 0$ or $2x - 1 = 0$

$x = -\frac{1}{2}$ or $x = \frac{1}{2}$

The domain is $\left\{x \middle| x \neq -\frac{1}{2}, \frac{1}{2}\right\}$

40. No

Exercise Set 3.4

1. Test for symmetry with respect to the y-axis.

$3y = x^2 + 4$ (Original equation)

$3y = (-x)^2 + 4$ (Replacing x by -x)

$3y = x^2 + 4$ (Simplifying)

Since the resulting equation is equivalent to the original, the graph is symmetric with respect to the y-axis.

Test for symmetry with respect to the x-axis.

$3y = x^2 + 4$ (Original equation)

$3(-y) = x^2 + 4$ (Replacing y by -y)

$-3y = x^2 + 4$ (Simplifying)

Since the resulting equation is not equivalent to the original, the graph is not symmetric with respect to the x-axis.

Test for symmetry with respect to the origin.

$3y = x^2 + 4$ (Original equation)

$3(-y) = (-x)^2 + 4$ (Replacing x by -x and y by -y)

$-3y = x^2 + 4$ (Simplifying)

Since the resulting equation is not equivalent to the original, the graph is not symmetric with respect to the origin.

2. x-axis: No, y-axis: Yes, origin: No

3. Test for symmetry with respect to the y-axis.

$y^3 = 2x^2$ (Original equation)

$y^3 = 2(-x)^2$ (Replacing x by -x)

$y^3 = 2x^2$ (Simplifying)

Since the resulting equation is equivalent to the original, the graph is symmetric with respect to the y-axis.

Test for symmetry with respect to the x-axis.

$y^3 = 2x^2$ (Original equation)

$(-y)^3 = 2x^2$ (Replacing y by -y)

$-y^3 = 2x^2$ (Simplifying)

Since the resulting equation is not equivalent to the original, the graph is not symmetric with respect to the x-axis.

Test for symmetry with respect to the origin.

$y^3 = 2x^2$ (Original equation)

$(-y)^3 = 2(-x)^2$ (Replacing x by -x and y by -y)

$-y^3 = 2x^2$ (Simplifying)

Since the resulting equation is not equivalent to the original, the graph is not symmetric with respect to the origin.

4. x-axis: No, y-axis: Yes, origin: No

5. Test for symmetry with respect to the y-axis.

$2x^4 + 3 = y^2$ (Original equation)

$2(-x)^4 + 3 = y^2$ (Replacing x by -x)

$2x^4 + 3 = y^2$ (Simplifying)

Since the resulting equation is equivalent to the original, the graph is symmetric with respect to the y-axis.

Test for symmetry with respect to the x-axis.

$2x^4 + 3 = y^2$ (Original equation)

$2x^4 + 3 = (-y)^2$ (Replacing y by -y)

$2x^4 + 3 = y^2$ (Simplifying)

Since the resulting equation is equivalent to the original, the graph is symmetric with respect to the x-axis.

Test for symmetry with respect to the origin.

$2x^4 + 3 = y^2$ (Original equation)

$2(-x)^4 + 3 = (-y)^2$ (Replacing x by -x and y by -y)

$2x^4 + 3 = y^2$ (Simplifying)

Since the resulting equation is equivalent to the original, the graph is symmetric with respect to the origin.

6. x-axis: Yes, y-axis: Yes, origin: Yes

7. Test for symmetry with respect to the y-axis.

$2y^2 = 5x^2 + 12$ (Original equation)

$2y^2 = 5(-x)^2 + 12$ (Replacing x by -x)

$2y^2 = 5x^2 + 12$ (Simplifying)

Since the resulting equation is equivalent to the original, the graph is symmetric with respect to the y-axis.

Test for symmetry with respect to the x-axis.

$2y^2 = 5x^2 + 12$ (Original equation)

$2(-y)^2 = 5x^2 + 12$ (Replacing y by -y)

$2y^2 = 5x^2 + 12$ (Simplifying)

Since the resulting equation is equivalent to the original, the graph is symmetric with respect to the x-axis.

Test for symmetry with respect to the origin.

$2y^2 = 5x^2 + 12$ (Original equation)

$2(-y)^2 = 5(-x)^2 + 12$ (Replacing x by -x and y by -y)

$2y^2 = 5x^2 + 12$ (Simplifying)

Since the resulting equation is equivalent to the original, the graph is symmetric with respect to the origin.

8. x-axis: Yes, y-axis: Yes, origin: Yes

9. Test for symmetry with respect to the y-axis.

$2x - 5 = 3y$ (Original equation)

$2(-x) - 5 = 3y$ (Replacing x by -x)

$-2x - 5 = 3y$ (Simplifying)

Since the resulting equation is not equivalent to the original, the graph is not symmetric with respect to the y-axis.

Test for symmetry with respect to the x-axis.

$2x - 5 = 3y$ (Original equation)

$2x - 5 = 3(-y)$ (Replacing y by -y)

$2x - 5 = -3y$ (Simplifying)

Since the resulting equation is not equivalent to the original, the graph is not symmetric with respect to the x-axis.

Test for symmetry with respect to the origin.

$2x - 5 = 3y$ (Original equation)

$2(-x) - 5 = 3(-y)$ (Replacing x by -x and y by -y)

$-2x - 5 = -3y$ (Simplifying)

or

$2x + 5 = 3y$ (Multiplying by -1)

Since the resulting equation is not equivalent to the original, the graph is not symmetric with respect to the origin.

10. x-axis: No, y-axis: No, origin: No

11. Test for symmetry with respect to the b-axis.

$3b^3 = 4a^3 + 2$ (Original equation)

$3b^3 = 4(-a)^3 + 2$ (Replacing a by -a)

$3b^3 = -4a^3 + 2$ (Simplifying)

Since the resulting equation is not equivalent to the original, the graph is not symmetric with respect to the b-axis.

Test for symmetry with respect to the a-axis.

$3b^3 = 4a^3 + 2$ (Original equation)

$3(-b)^3 = 4a^3 + 2$ (Replacing b by -b)

$-3b^3 = 4a^3 + 2$ (Simplifying)

Since the resulting equation is not equivalent to the original, the graph is not symmetric with respect to the a-axis.

Test for symmetry with respect to the origin.

$3b^3 = 4a^3 + 2$ (Original equation)

$3(-b)^3 = 4(-a)^3 + 2$ (Replacing a by -a and b by -b)

$-3b^3 = -4a^3 + 2$ (Simplifying)

or

$3b^3 = 4a^3 - 2$ (Multiplying by -1)

Since the resulting equation is not equivalent to the original, the graph is not symmetric with respect to the origin.

12. p-axis: No, q-axis: No, origin: No

13. $3x^2 - 2y^2 = 3$ (Original equation)

$3(-x)^2 - 2(-y)^2 = 3$ (Replacing x by -x and y by -y)

$3x^2 - 2y^2 = 3$ (Simplifying)

Since the resulting equation is equivalent to the original equation, the graph is symmetric with respect to the origin.

14. Yes

15. $5x - 5y = 0$ (Original equation)

$5(-x) - 5(-y) = 0$ (Replacing x by -x and y by -y)

$-5x + 5y = 0$ (Simplifying)

$5x - 5y = 0$ (Multiplying by -1)

Since the resulting equation is equivalent to the original equation, the graph is symmetric with respect to the origin.

16. Yes

17. $3x + 3y = 0$ (Original equation)

$3(-x) + 3(-y) = 0$ (Replacing x by -x and y by -y)

$-3x - 3y = 0$ (Simplifying)

$3x + 3y = 0$ (Multiplying by -1)

Since the resulting equation is equivalent to the original equation, the graph is symmetric with respect to the origin.

18. Yes

19.

$$3x = \frac{5}{y} \quad \text{(Original equation)}$$

$$3(-x) = \frac{5}{-y} \quad \text{(Replacing x by -x and y by -y)}$$

$$-3x = -\frac{5}{y} \quad \text{(Simplifying)}$$

$$3x = \frac{5}{y} \quad \text{(Multiplying by -1)}$$

Since the resulting equation is equivalent to the original, the graph is symmetric with respect to the origin.

20. Yes

21.

$y = |2x|$ (Original equation)

$-y = |2(-x)|$ (Replacing x by -x and y by -y)

$-y = |-2x|$

$-y = |2x|$

Since the resulting equation is not equivalent to the original, the graph is not symmetric with respect to the origin.

22. No

23.

$$3a^2 + 4a = 2b \quad \text{(Original equation)}$$

$$3(-a)^2 + 4(-a) = 2(-b) \quad \text{(Replacing a by -a and b by -b)}$$

$$3a^2 - 4a = -2b$$

or

$$-3a^2 + 4a = 2b$$

Since the resulting equation is not equivalent to the original, the graph is not symmetric with respect to the origin.

24. No

25. The vertices of the original polygon are (0, 4), (4, 4), (-2, -2), and (1, -2).

When the graph is reflected across the x-axis the vertices will be (0, -4), (4, -4), (-2, 2), and (1, 2).

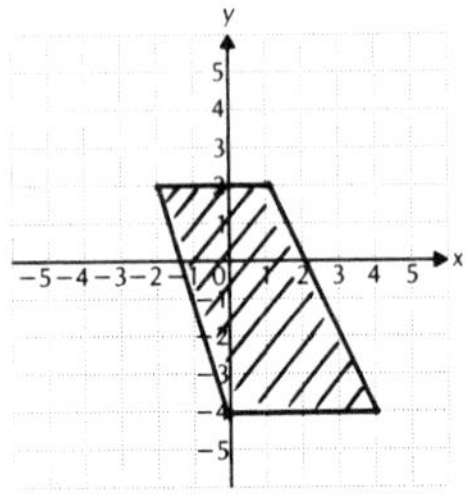

26.

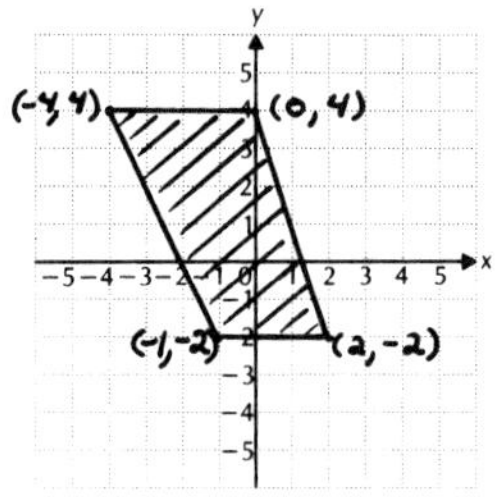

27. The vertices of the original polygon are (0, 4), (4, 4), (1, -2), and (-2, -2).

When the graph is reflected across the line y = x, the vertices will be (4, 0), (4, 4), (-2, 1), and (-2, -2).

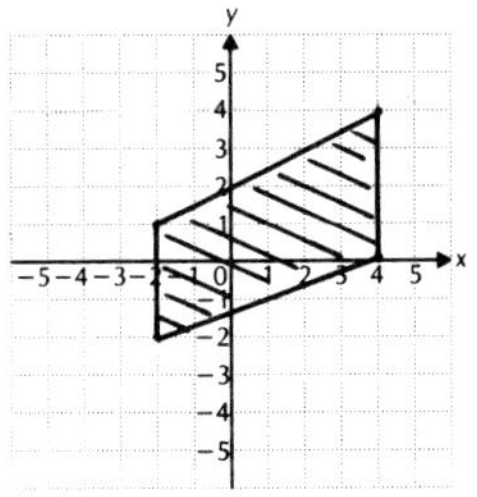

28.

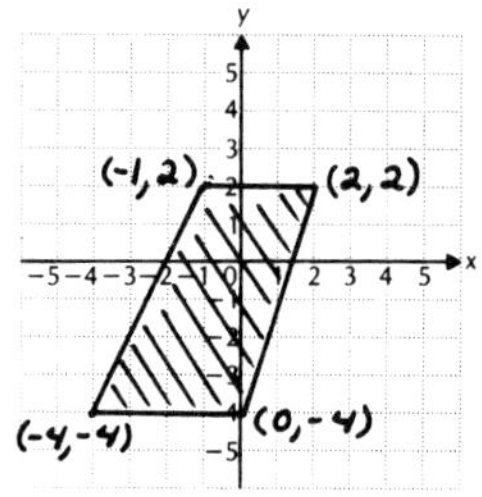

Exercise Set 3.5

1. Graph $y + 3 = |x|$ by transforming the graph of $y = |x|$.

$y = |x|$ (Dashed graph below)

$y + 3 = |x|$ (Replacing y by y + 3)

$y - (-3) = |x|$ $[y + 3 = y - (-3)]$

Using Theorem 1, the -3 tells us that the graph of $y - (-3) = |x|$ is a vertical translation, downward 3 units, of the graph of $y = |x|$. The shapes of the graphs are the same.

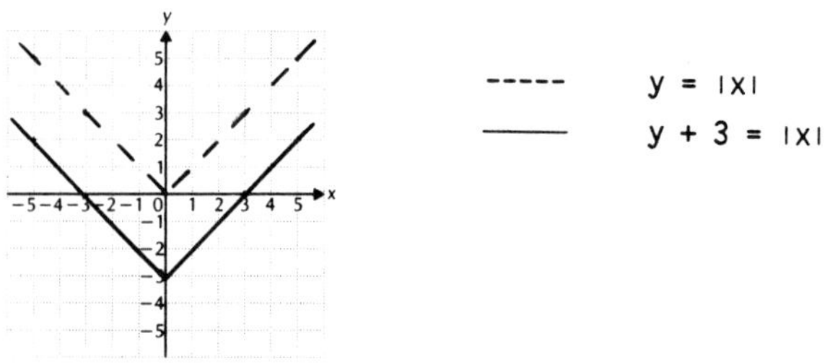

2. $y = 2 + |x|$

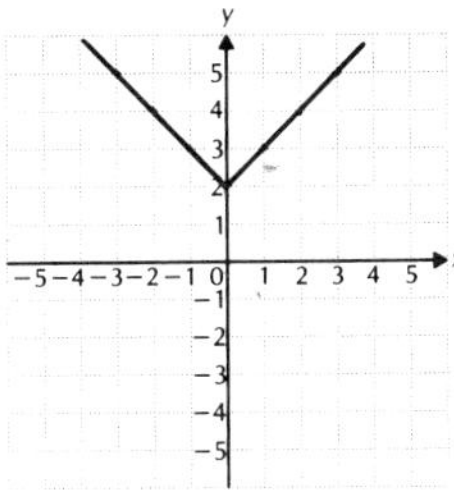

3. Graph $y = |x - 1|$ by transforming the graph of $y = |x|$.

$y = |x|$ (Dashed graph below)

$y = |x - 1|$ (Replacing x by x - 1)

Using Theorem 2, the 1 tells us that the graph of $y = |x - 1|$ is a horizontal translation, right 1 unit, of the graph of $y = |x|$. The shapes of the graphs are the same.

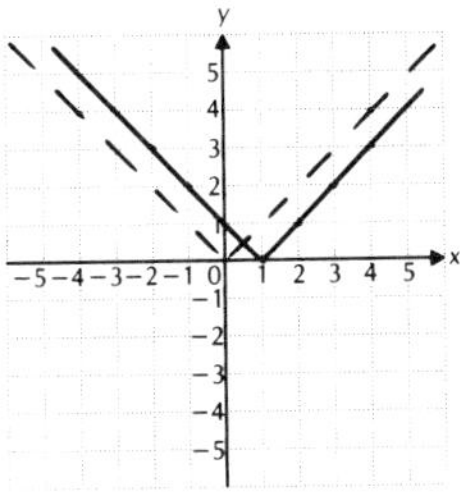

----- $y = |x|$

——— $y = |x - 1|$

4. $y = |x + 2|$

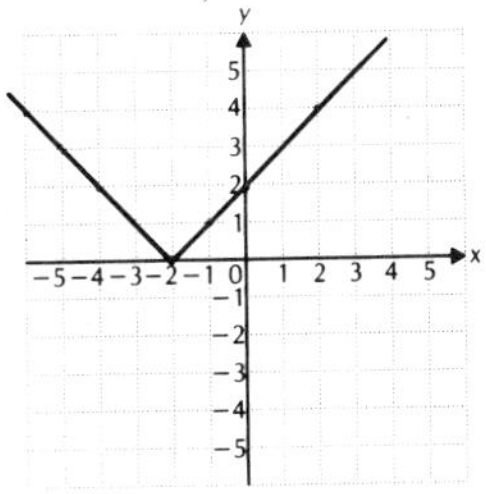

5. Graph $y = -4|x|$ by transforming the graph of $y = |x|$. Since $y = -4|x|$ is equivalent to $\frac{y}{-4} = |x|$, we can obtain the new equation by replacing y in $y = |x|$ by $\frac{y}{-4}$. Using Theorem 3, the -4 tells us that the graph of $y = |x|$ is stretched vertically (each function value multiplied by 4) and reflected across the x-axis. The result is the graph of $y = -4|x|$.

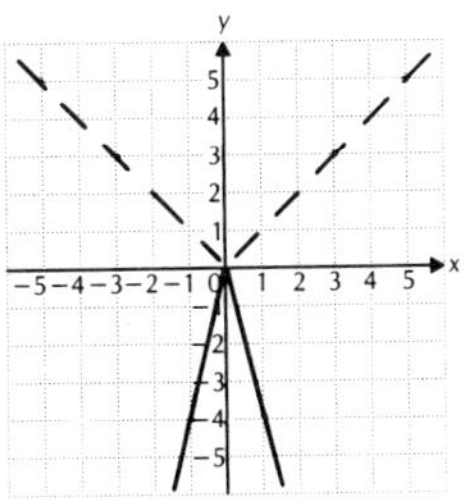

----- $y = |x|$

——— $y = -4|x|$

6. $\frac{y}{3} = |x|$

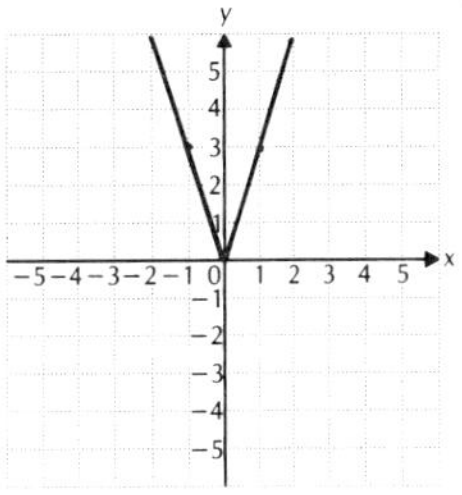

7. Graph $y = \frac{1}{3}|x|$ by transforming the graph of $y = |x|$. Since $y = \frac{1}{3}|x|$ is equivalent to $\frac{y}{\frac{1}{3}} = |x|$, we can obtain the new equation by replacing y in $y = |x|$ by $\frac{y}{\frac{1}{3}}$. Using Theorem 3, the $\frac{1}{3}$ tells us that the graph of $y = |x|$ is shrunk vertically $\left(\text{each function value multiplied by } \frac{1}{3}\right)$ to obtain the graph of $y = \frac{1}{3}|x|$.

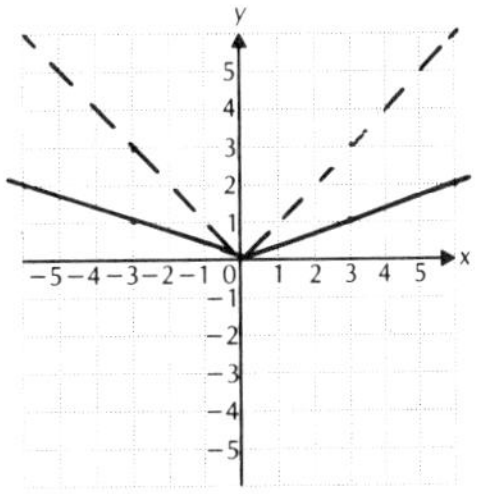

----- $y = |x|$

——— $y = \frac{1}{3}|x|$

8. $y = -\frac{1}{4}|x|$

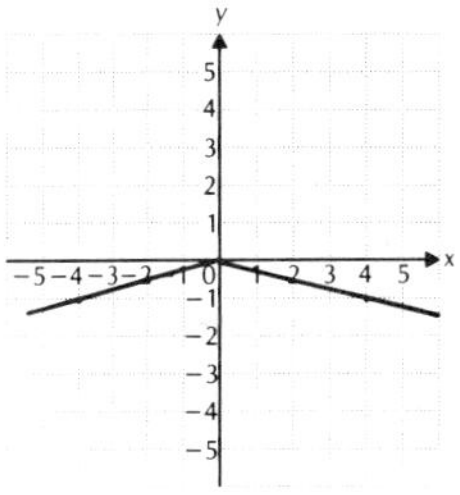

9. Graph $y = |2x|$ by transforming the graph of $y = |x|$.

$y = |2x|$

$y = 2|x|$ (Simplifying)

$\frac{y}{2} = |x|$ $\left(\text{Multiplying by } \frac{1}{2}\right)$

Since $y = |2x|$ is equivalent to $\frac{y}{2} = |x|$, we can obtain the new equation by replacing the y in $y = |x|$ by $\frac{y}{2}$. Using Theorem 3, the 2 tells us that the graph of $y = |x|$ is stretched vertically (each function value multiplied by 2) to obtain the graph of $y = |2x|$.

9. (continued)

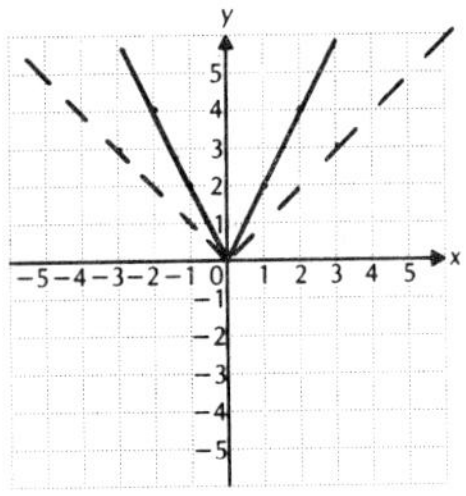

----- $y = |x|$

—— $y = |2x|$

10. $y = \left|\frac{x}{3}\right|$

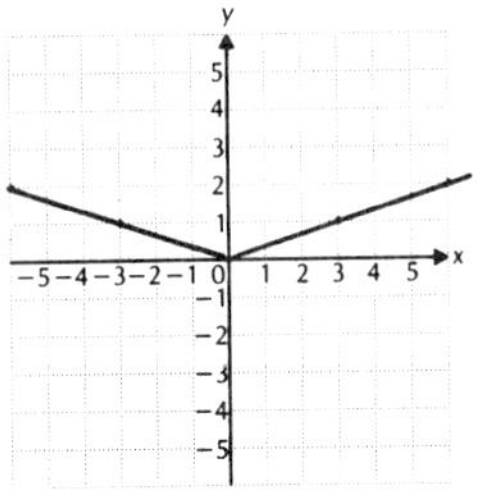

11. Graph $y = |x - 2| + 3$ by transforming the graph of $y = |x|$. Since $y = |x - 2| + 3$ is equivalent to $y - 3 = |x - 2|$, we can obtain the new equation by replacing x by x - 2 and y by y - 3 in $y = |x|$. Using Theorem 2, the 2 tells us that the graph of $y = |x|$ is translated 2 units to the right. Using Theorem 1, the 3 tells us that the graph of $y = |x|$ is also translated 3 units upward. The result is the graph of $y = |x - 2| + 3$

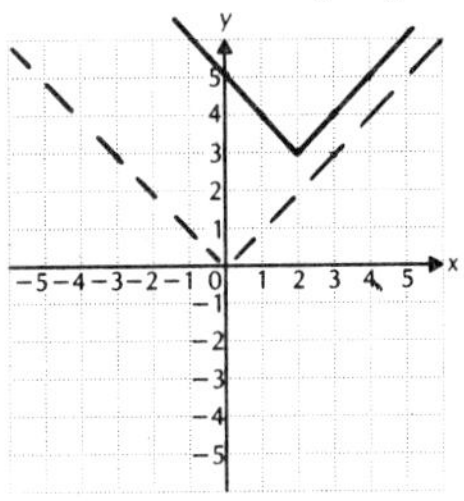

----- $y = |x|$

—— $y = |x - 2| + 3$

12. $y = 2|x + 1| - 3$

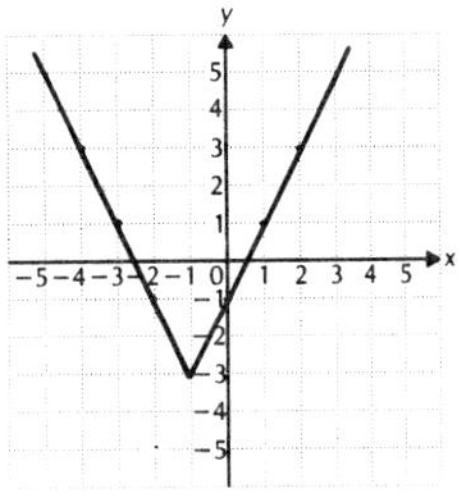

13. Graph $y = -3|x - 2|$ by transforming the graph of $y = |x|$. Since $y = -3|x - 2|$ is equivalent to $\frac{y}{-3} = |x - 2|$, we can obtain the new equation by replacing y by $\frac{y}{-3}$ and x by x - 2 in $y = |x|$. Using Theorem 2, the 2 tells us that the graph of $y = |x|$ is translated two units to the right. Using Theorem 3, the -3 tells us that the graph of $y = |x|$ is also stretched vertically (each function value multiplied by 3) and reflected across the x-axis. The result is the graph of $y = -3|x - 2|$.

13. (continued)

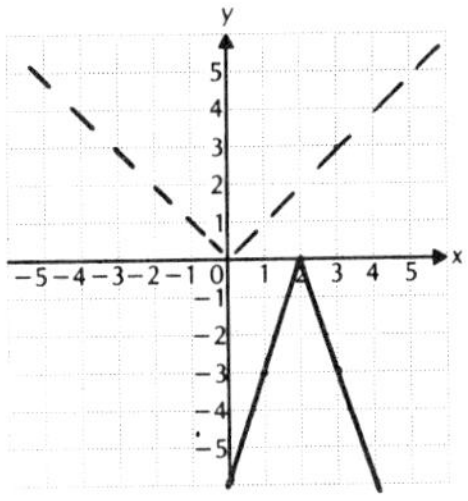

----- $y = |x|$

—— $y = -3|x - 2|$

14. $y = \frac{1}{3}|x + 2| + 1$

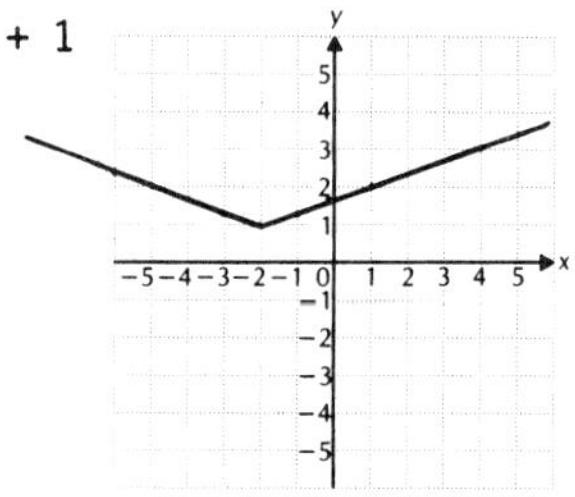

15. Graph $y = 2 + f(x)$ by transforming the graph of $y = f(x)$. Since $y = 2 + f(x)$ is equivalent to $y - 2 = f(x)$, we can obtain the new equation by replacing y in $y = f(x)$ by y - 2. Using Theorem 1, the 2 tells us that the graph of $y = f(x)$ is translated 2 units upward to obtain the graph of $y = 2 + f(x)$.

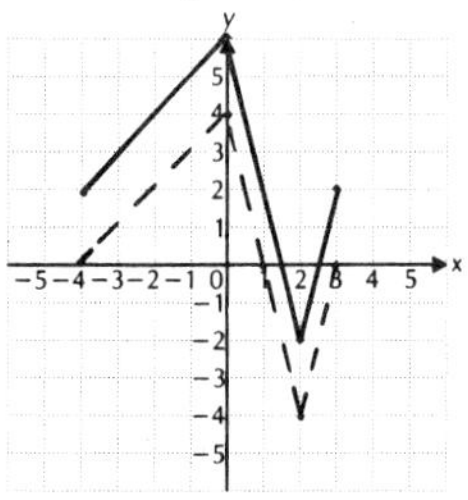

----- $y = f(x)$

—— $y = 2 + f(x)$

16. $y + 1 = f(x)$

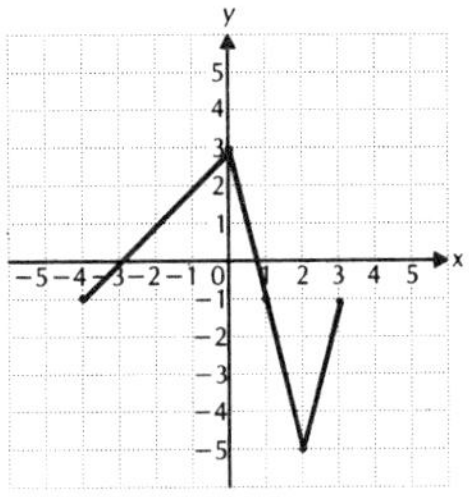

17. Graph $y = f(x - 1)$ by transforming the graph of $y = f(x)$. We can obtain the new equation by replacing x in $y = f(x)$ by x - 1. Using Theorem 2, the 1 tells us that the graph of $y = f(x)$ is translated 1 unit to the right to obtain the graph of $y = f(x - 1)$.

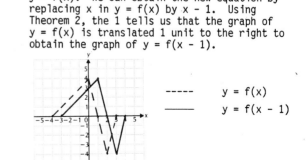

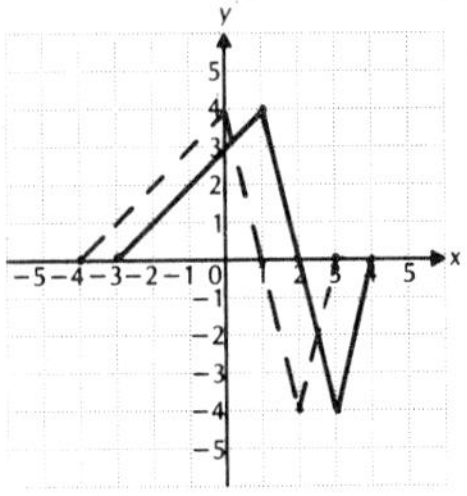

----- $y = f(x)$

—— $y = f(x - 1)$

18. $y = f(x + 2)$

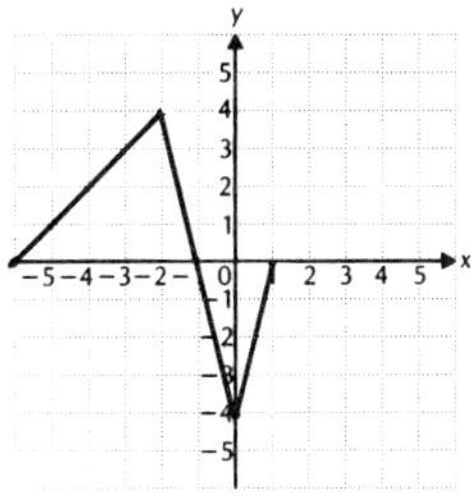

21. (continued)

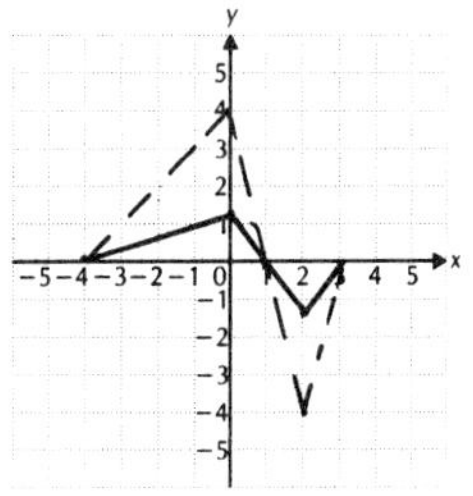

----- $y = f(x)$

——— $y = \frac{1}{3} f(x)$

19. Graph $\frac{y}{-2} = f(x)$ by transforming the graph of $y = f(x)$. We can obtain the new equation by replacing y in $y = f(x)$ by $\frac{y}{-2}$. Using Theorem 3, the -2 tells us that the graph of $y = f(x)$ is stretched vertically (each function value multiplied by 2) and reflected across the x-axis. The result is the graph of $\frac{y}{-2} = f(x)$.

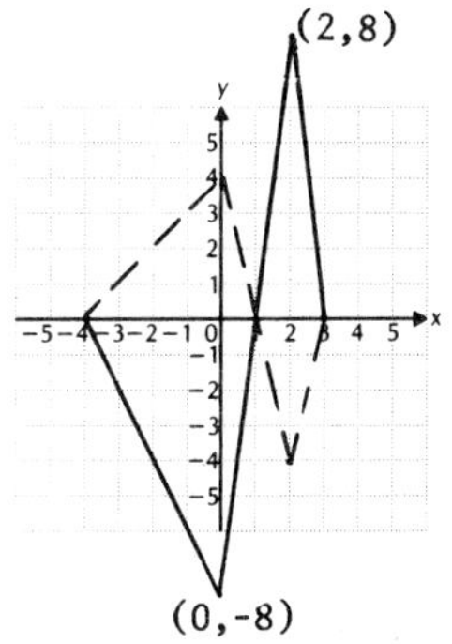

----- $y = f(x)$

——— $\frac{y}{-2} = f(x)$

22. $y = -\frac{1}{2} f(x)$

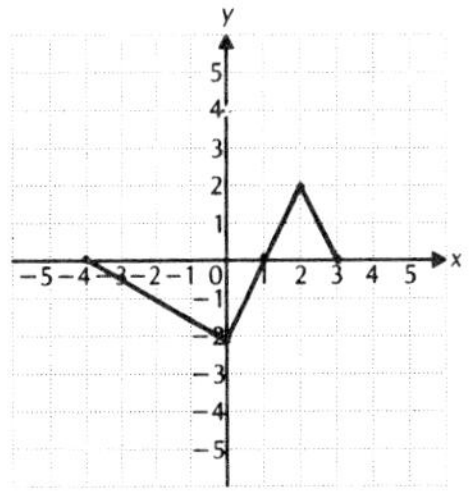

23. Graph $y = f(2x)$ by transforming the graph of $y = f(x)$. Since $y = f(2x)$ is equivalent to $y = f\left(\frac{x}{\frac{1}{2}}\right)$, we can obtain the new equation by replacing x in $y = f(x)$ by $\frac{x}{\frac{1}{2}}$. Using Theorem 4, the $\frac{1}{2}$ tells us that the graph of $y = f(x)$ is shrunk horizontally (each x-value is cut in half) to obtain the graph of $y = f(2x)$.

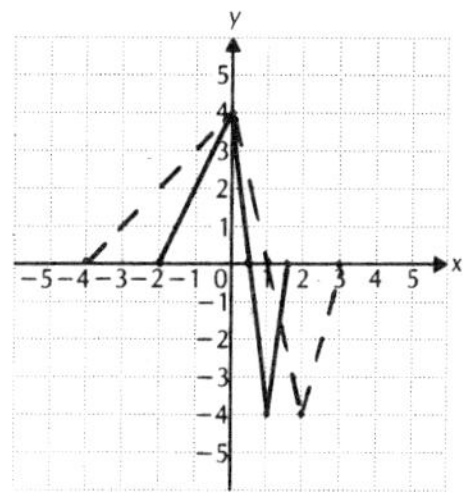

----- $y = f(x)$

——— $y = f(2x)$

20. $y = 3f(x)$

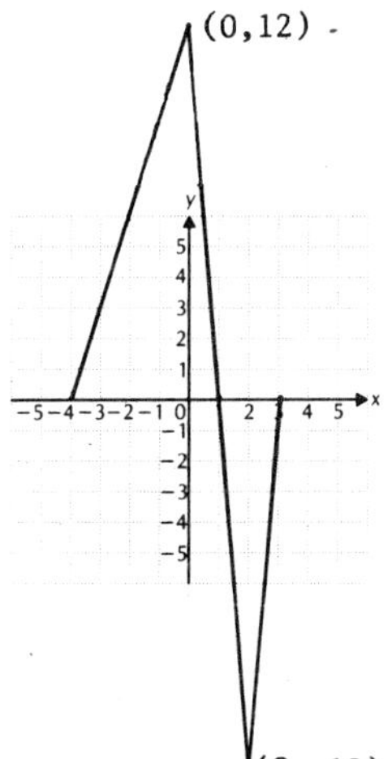

24. $y = f(3x)$

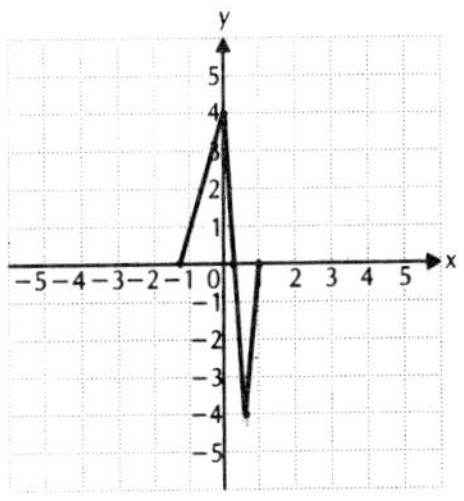

21. Graph $y = \frac{1}{3} f(x)$ by transforming the graph of $y = f(x)$. Since $y = \frac{1}{3} f(x)$ is equivalent to $\frac{y}{\frac{1}{3}} = f(x)$, we can obtain the new equation by replacing y in $y = f(x)$ by $\frac{y}{\frac{1}{3}}$. Using Theorem 3, the $\frac{1}{3}$ tells us that the graph of $y = f(x)$ is shrunk vertically $\left(\text{each function value multiplied by } \frac{1}{3}\right)$ to obtain the graph of $y = \frac{1}{3} f(x)$.

25. Graph $y = f(-2x)$ by transforming the graph of $y = f(x)$. Since $y = f(-2x)$ is equivalent to $y = f\left[\frac{x}{-\frac{1}{2}}\right]$, we can obtain the new equation by replacing x in $y = f(x)$ by $\frac{x}{-\frac{1}{2}}$. Using Theorem 4, the $-\frac{1}{2}$ tells us that the graph of $y = f(x)$ is shrunk horizontally (each x-value is cut in half) and reflected across the y-axis. The result is the graph of $y = f(-2x)$.

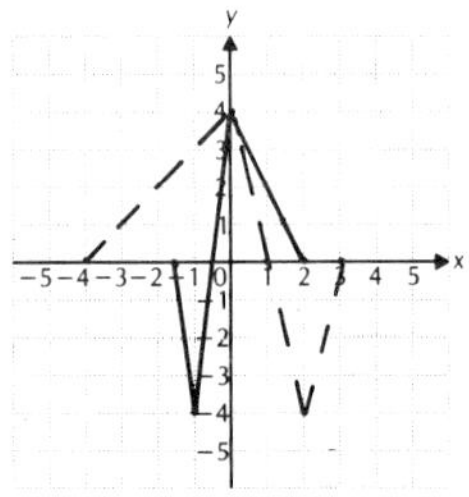

----- $y = f(x)$

—— $y = f(-2x)$

26. $y = f(-3x)$

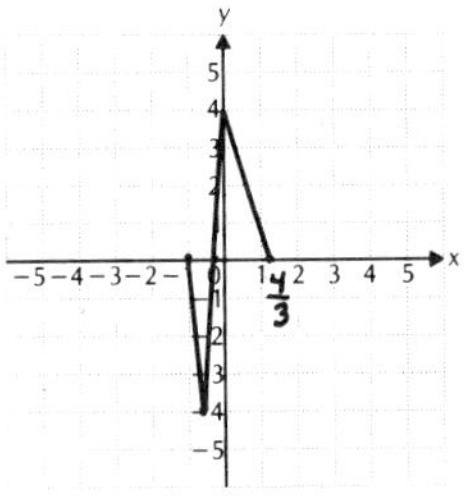

27. Graph $y = f\left[\frac{x}{-2}\right]$ by transforming the graph of $y = f(x)$. Using Theorem 4, the -2 tells us that the graph of $y = f(x)$ is stretched horizontally (each x-value is doubled) and reflected across the y-axis. The result is the graph of $y = f\left[\frac{x}{-2}\right]$.

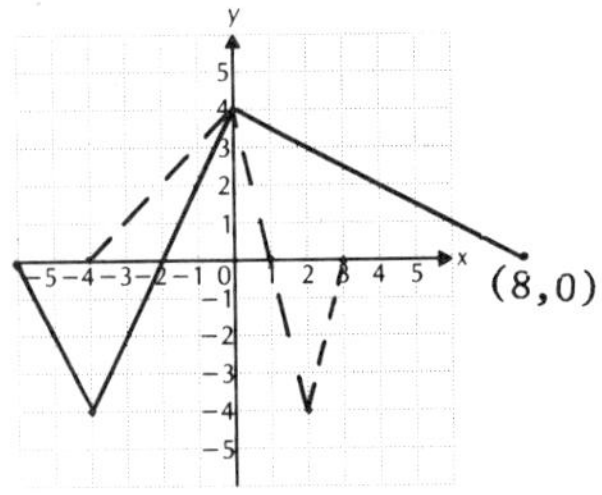

----- $y = f(x)$

—— $y = f\left[\frac{x}{-2}\right]$

28. $y = f\left[\frac{1}{3}x\right]$

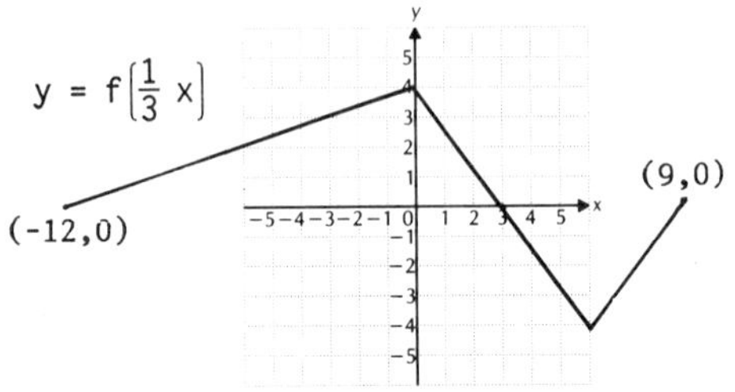

29. Graph $y = f(x - 2) + 3$ by transforming the graph of $y = f(x)$. Since $y = f(x - 2) + 3$ is equivalent to $y - 3 = f(x - 2)$, we can obtain the new equation by replacing x by $x - 2$ and y by $y - 3$ in $y = f(x)$. Using Theorem 2, the 2 tells us that the graph of $y = f(x)$ is translated 2 units to the right. Using Theorem 1, the 3 tells us that the graph of $y = f(x)$ is also translated 3 units upward. The result is the graph of $y = f(x - 2) + 3$.

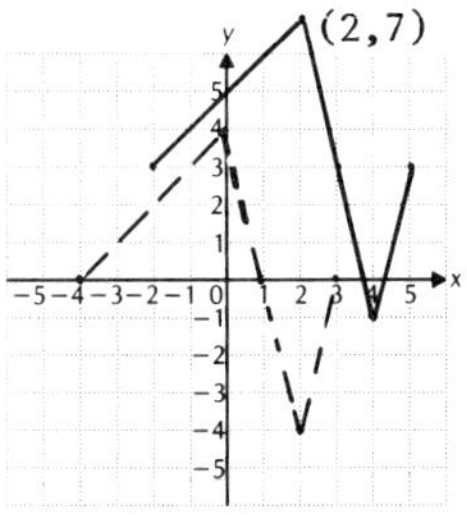

----- $y = f(x)$

—— $y = f(x - 2) + 3$

30. $y = -3f(x - 2)$

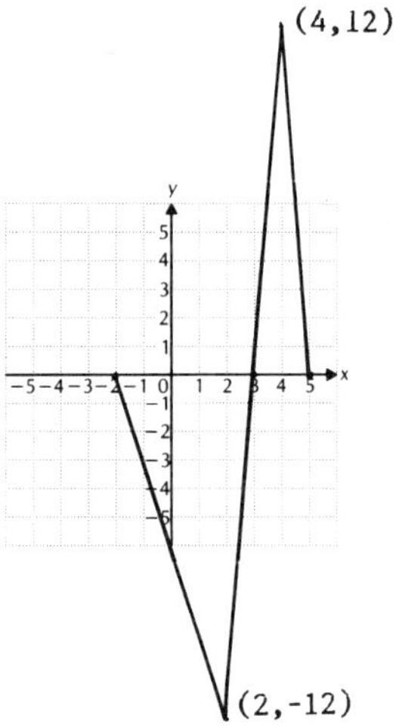

31. Graph $y = 2\cdot f(x + 1) - 2$ by transforming the graph of $y = f(x)$.

$$y = 2\cdot f(x + 1) - 2$$
$$y + 2 = 2f(x + 1)$$
$$\frac{y + 2}{2} = f(x + 1)$$
$$\frac{y - (-2)}{2} = f[x - (-1)]$$

Using Theorem 1, the -2 tells us that the graph of $y = f(x)$ is translated downward 2 units. Using Theorem 2, the -1 tells us that the graph of $y = f(x)$ is translated 1 unit to the left. Using Theorem 3, the 2 tells us that the graph of $y = f(x)$ is stretched vertically (each function value multiplied by 2). The result is the graph of $y = 2\cdot f(x + 1) - 2$.

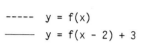

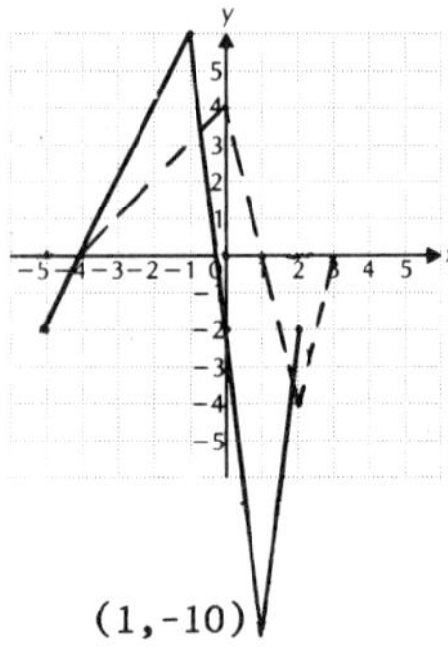

----- $y = f(x)$

—— $y = 2\cdot f(x + 1) - 2$

32. $y = \frac{1}{2} f(x + 2) - 1$

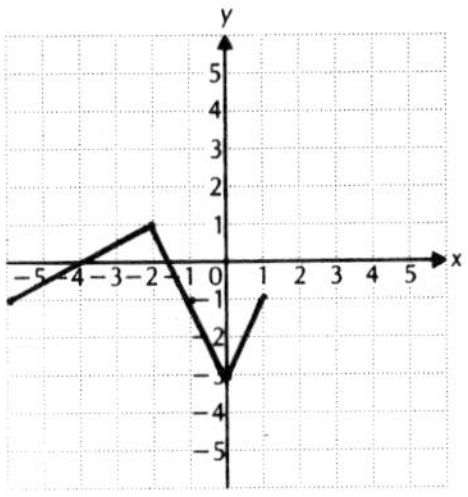

33. Graph $y = -\frac{1}{2} f(x - 3) + 2$ by transforming the graph of $y = f(x)$.

$$y = -\frac{1}{2} f(x - 3) + 2$$

$$y - 2 = -\frac{1}{2} f(x - 3)$$

$$\frac{y - 2}{-\frac{1}{2}} = f(x - 3)$$

Using Theorem 1, the 2 tells us that the graph of $y = f(x)$ is translated upward 2 units. Using Theorem 2, the 3 tells us that the graph of $y = f(x)$ is translated 3 units to the right. Using Theorem 3, the $-\frac{1}{2}$ tells us that the graph of $y = f(x)$ is shrunk vertically $\left[\text{each function value multiplied by } \frac{1}{2}\right]$ and reflected across the x-axis. The result is the graph of $y = -\frac{1}{2} f(x - 3) + 2$.

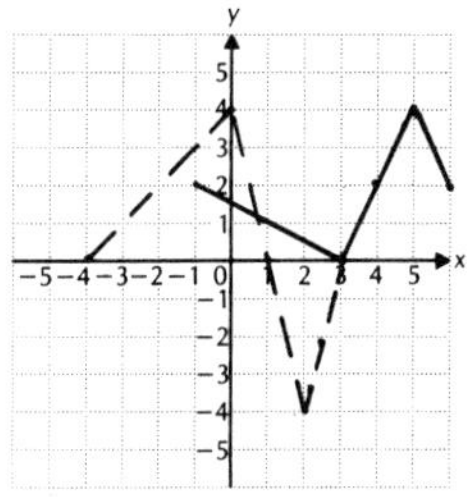

----- $y = f(x)$

——— $y = -\frac{1}{2} f(x - 3) + 2$

34. $y = -2f(x + 1) - 1$

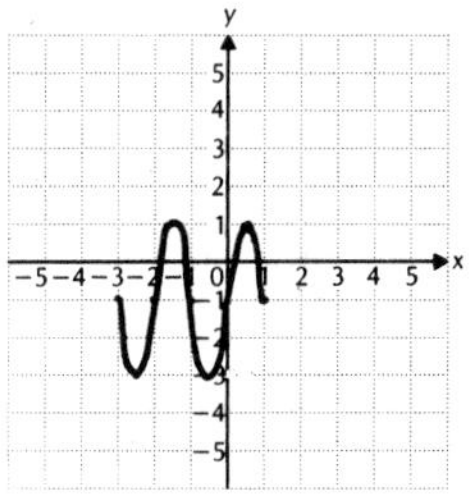

35. Graph $y = 3f(x + 2) + 1$ by transforming the graph of $y = f(x)$.

$$y = 3f(x + 2) + 1$$

$$y - 1 = 3f(x + 2)$$

$$\frac{y - 1}{3} = f[x - (-2)]$$

35. (continued)

Using Theorem 1, the 1 tells us that the graph of $y = f(x)$ is translated upward 1 unit. Using Theorem 2, the -2 tells us that the graph of $y = f(x)$ is translated 2 units to the left. Using Theorem 3, the 3 tells us that the graph of $y = f(x)$ is stretched vertically (each function value multiplied by 3). The result is the graph of $y = 3f(x + 2) + 1$.

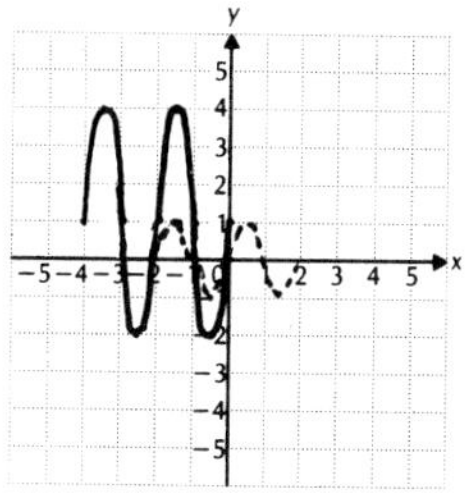

----- $y = f(x)$

——— $y = 3f(x + 2) + 1$

36. $y = \frac{5}{2} f(x - 3) - 2$

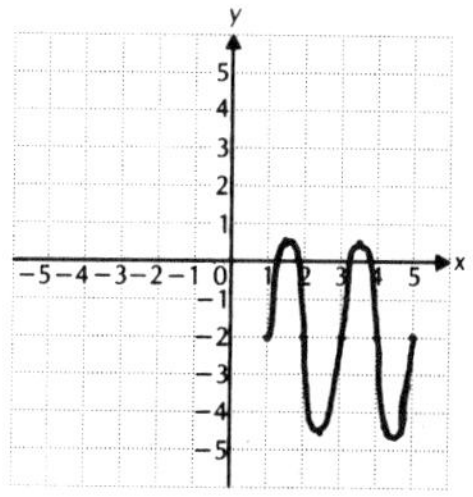

37. See the answer section in the text.

38. A horizontal translation 2.5 units to the right, a vertical stretching by a factor of $\sqrt{3}/2$, followed by a vertical translation 5.3 units down.

Exercise Set 3.6

1. a) The graph is symmetric with respect to the y-axis. Thus the function is even.

Reflect the graph across the origin. Are the graphs the same? No.

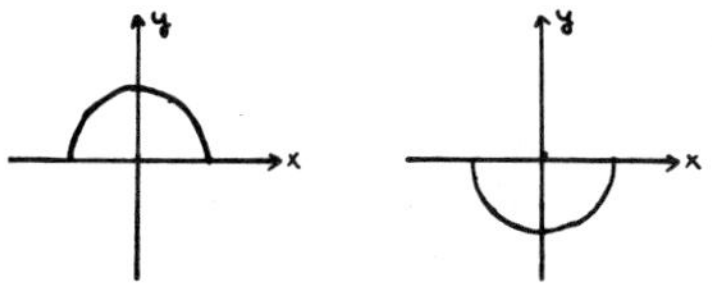

The graph is not symmetric with respect to the origin. Thus the function is not odd.

b) The graph is symmetric with respect to the y-axis. Thus the function is even.

Reflect the graph across the origin. Are the graphs the same? No.

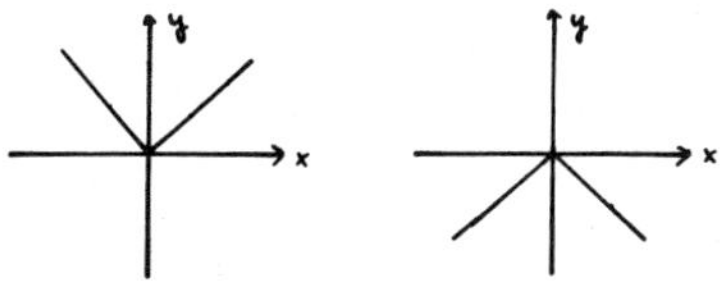

1. (continued)

The graph is not symmetric with respect to the origin. Thus the function is not odd.

c) Reflect the graph across the y-axis. Are the graphs the same? No.

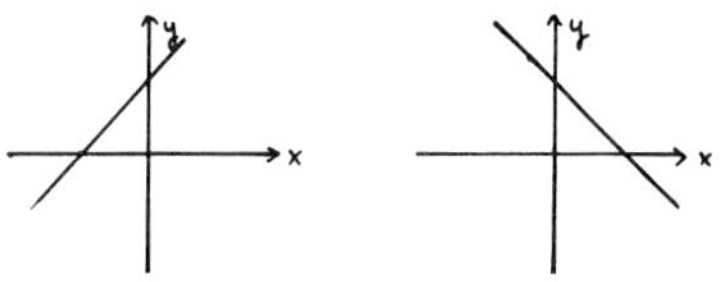

The graph is not symmetric with respect to the y-axis. Thus the function is not even.

Reflect the graph across the origin. Are the graphs the same? No.

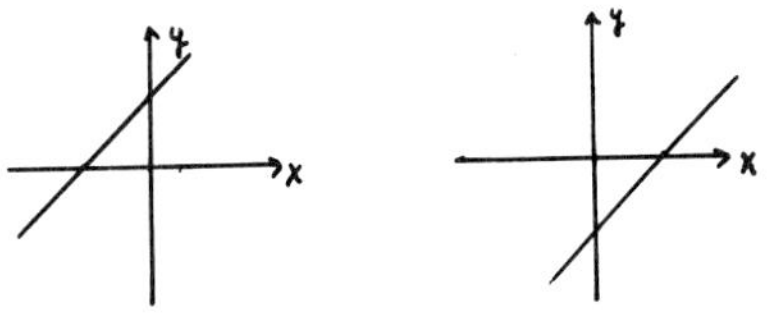

The graph is not symmetric with respect to the origin. Thus the function is not odd.

Therefore the function is neither even nor odd.

d) Reflect the graph across the y-axis. Are the graphs the same? No.

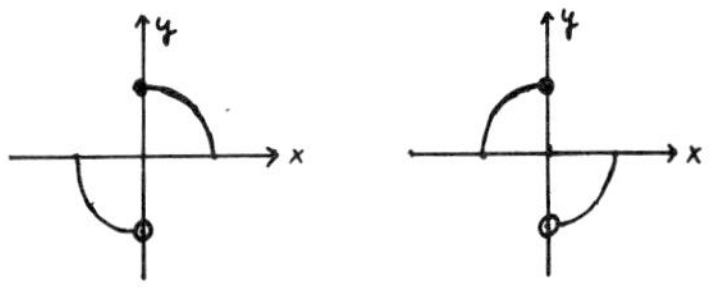

The graph is not symmetric with respect to the y-axis. Thus the function is not even.

Reflect the graph across the origin. Are the graphs the same? No.

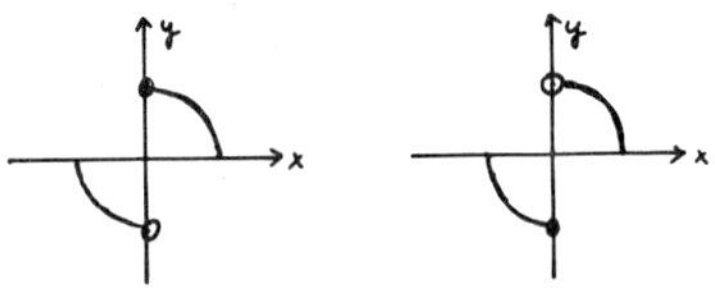

The graph is not symmetric with respect to the origin. Thus the function is not odd.

Therefore the function is neither even nor odd.

2. a) Odd, b) Even, c) Neither, d) Neither

3. Determine whether $f(x) = 2x^2 + 4x$ is even.

Find $f(-x)$.

$f(-x) = 2(-x)^2 + 4(-x) = 2x^2 - 4x$

Since $f(x)$ and $f(-x)$ are not the same for all x in the domain, f is not an even function.

Determine whether $f(x) = 2x^2 + 4x$ is odd.

Find $f(-x)$.

$f(-x) = 2(-x)^2 + 4(-x) = 2x^2 - 4x$

Find $-f(x)$.

$-f(x) = -(2x^2 + 4x) = -2x^2 - 4x$

Since $f(-x)$ and $-f(x)$ are not the same for all x in the domain, f is not an odd function.

Thus, $f(x) = 2x^2 + 4x$ is neither even nor odd.

4. Odd

5. Determine whether $f(x) = 3x^4 - 4x^2$ is even.

Find $f(-x)$.

$f(-x) = 3(-x)^4 - 4(-x)^2 = 3x^4 - 4x^2$

Since $f(x)$ and $f(-x)$ are the same for all x in the domain, f is an even function.

Determine whether $f(x) = 3x^4 - 4x^2$ is odd.

Find $f(-x)$.

$f(-x) = 3(-x)^4 - 4(-x)^2 = 3x^4 - 4x^2$

Find $-f(x)$.

$-f(x) = -(3x^4 - 4x^2) = -3x^4 + 4x^2$

Since $f(-x)$ and $-f(x)$ are not the same for all x in the domain, f is not an odd function.

6. Even

7. Determine whether $f(x) = 7x^3 + 4x - 2$ is even.

Find $f(-x)$.

$f(-x) = 7(-x)^3 + 4(-x) - 2 = -7x^3 - 4x - 2$

Since $f(x)$ and $f(-x)$ are not the same for all x in the domain, f is not an even function.

Determine whether $f(x) = 7x^3 + 4x - 2$ is odd.

Find $f(-x)$.

$f(-x) = 7(-x)^3 + 4(-x) - 2 = -7x^3 - 4x - 2$

Find $-f(x)$.

$-f(x) = -(7x^3 + 4x - 2) = -7x^3 - 4x + 2$

Since $f(-x)$ and $-f(x)$ are not the same for all x in the domain, f is not an odd function.

Thus, $f(x) = 7x^3 + 4x - 2$ is neither even nor odd.

8. Odd

9. Determine whether $f(x) = |3x|$ is even.

Find $f(-x)$.

$f(-x) = |3(-x)| = |-3x| = |3x|$

Since $f(x)$ and $f(-x)$ are the same for all x in the domain, f is an even function.

Determine whether $f(x) = |3x|$ is odd.

Find $f(-x)$.

$f(-x) = |3(-x)| = |-3x| = |3x|$

Find $-f(x)$.

$-f(x) = -|3x|$

Since $f(-x)$ and $-f(x)$ are not the same for all x in the domain, f is not an odd function.

10. Even

11. Determine whether $f(x) = x^{17}$ is even.

Find $f(-x)$.

$f(-x) = (-x)^{17} = -x^{17}$

Find $f(x)$ and $f(-x)$ are not the same for all x in the domain, f is not an even function.

Determine whether $f(x) = x^{17}$ is odd.

Find $f(-x)$.

$f(-x) = (-x)^{17} = -x^{17}$

Find $-f(x)$.

$-f(x) = -(x^{17}) = -x^{17}$

Since $f(-x)$ and $-f(x)$ are the same for all x in the domain, f is an odd function.

12. Odd

13. Determine whether $f(x) = x - |x|$ is even.

Find $f(-x)$.

$f(-x) = (-x) - |(-x)| = -x - |x|$

Since $f(x)$ and $f(-x)$ are not the same for all x in the domain, f is not an even function.

Determine whether $f(x) = x - |x|$ is odd.

Find $f(-x)$.

$f(-x) = (-x) - |(-x)| = -x - |x|$

Find $-f(x)$.

$-f(x) = -(x - |x|) = -x + |x|$

Since $f(-x)$ and $-f(x)$ are not the same for all x in the domain, f is not an odd function.

Therefore, $f(x) = x - |x|$ is neither odd nor even.

14. Neither

15. Determine whether $f(x) = \sqrt[3]{x}$ is even.

Find $f(-x)$.

$f(-x) = \sqrt[3]{-x} = -\sqrt[3]{x}$

Since $f(x)$ and $f(-x)$ are not the same for all x in the domain, f is not an even function.

15. (continued)

Determine whether $f(x) = \sqrt[3]{x}$ is odd.

Find $f(-x)$.

$f(-x) = \sqrt[3]{-x} = -\sqrt[3]{x}$

Find $-f(x)$.

$-f(x) = -(\sqrt[3]{x}) = -\sqrt[3]{x}$

Since $f(-x)$ and $-f(x)$ are the same for all x in the domain, f is an odd function.

16. Even

17. Determine whether $f(x) = 0$ is even.

Find $f(-x)$.

$f(-x) = 0$

Since $f(x)$ and $f(-x)$ are the same for all x in the domain, f is an even function.

Determine whether $f(x) = 0$ is odd.

Find $f(-x)$.

$f(-x) = 0$

Find $-f(x)$.

$-f(x) = -(0) = 0$

Since $f(-x)$ and $-f(x)$ are the same for all x in the domain, f is an odd function.

18. Neither

19. Determine whether $f(x) = \sqrt{x^2 + 1}$ is even.

Find $f(-x)$.

$f(-x) = \sqrt{(-x)^2 + 1} = \sqrt{x^2 + 1}$

Since $f(x)$ and $f(-x)$ are the same for all x in the domain, f is an even function.

Determine whether $f(x) = \sqrt{x^2 + 1}$ is odd.

Find $f(-x)$.

$f(-x) = \sqrt{(-x)^2 + 1} = \sqrt{x^2 + 1}$

Find $-f(x)$.

$-f(x) = -\sqrt{x^2 + 1}$

Since $f(-x)$ and $-f(x)$ are not the same for all x in the domain, f is not an odd function.

20. Odd

21. a) No, b) Yes, c) Yes, d) No

22. a) Yes, b) No, c) No, d) Yes

23. The length of the shortest recurring interval is 4. Thus, the period is 4.

24. 2

25. a) Since neither -2 nor 2 is included in the interval, we use a parenthesis before -2 and a parenthesis after 2. The notation is (-2, 2).

b) Since -5 is not included in the interval, we use a parenthesis before -5. Since -1 is included in the interval we use a bracket after -1. The notation is (-5, -1].

c) Since both endpoints, c and d, are included in the interval, we use a bracket before c and a bracket after d. The notation is [c, d].

d) Since -5 is included in the interval, we use a bracket before -5. Since 1 is not included in the interval, we use a parenthesis after 1. The notation is [-5, 1).

26. a) (-3, 1), b) [c, d), c) $\left[\frac{1}{2}, 3\right]$, d) $\left[-2, 1\frac{1}{2}\right]$

27. a) $\{x \mid -2 < x < 4\}$

The interval does not included either -2 or 4. Interval notation is (-2, 4).

b) $\left\{x \middle| -\frac{1}{4} < x \leqslant \frac{1}{4}\right\}$

The interval does not include $-\frac{1}{4}$ and does include $\frac{1}{4}$. Interval notation is $\left(-\frac{1}{4}, \frac{1}{4}\right]$.

c) $\{x \mid 7 \leqslant x < 10\pi\}$

The interval does include 7 and does not included 10π. Interval notation is $[7, 10\pi)$.

d) $\{x \mid -9 \leqslant x \leqslant -6\}$

The interval does include both -9 and -6. Interval notation is [-9, -6].

28. a) (-5, 0), b) $[-\sqrt{2}, \sqrt{2})$, c) $\left[-\frac{\pi}{2}, \frac{\pi}{2}\right]$,

d) $\left[-12, -\frac{1}{2}\right]$

29. a) Yes, b) Yes, c) No, d) Yes, e) Yes

30. a) Yes, b) Yes, c) No, d) No, e) Yes

31. The function has discontinuities where x = -3 and x = 2.

32. x = 0

33. a) Increasing, b) Neither
c) Decreasing, d) Neither

34. a) Neither, b) Decreasing
c) Increasing, d) Neither

35. Graph $f(x) = \begin{cases} 1 & \text{for } x < 0 \\ -1 & \text{for } x \geqslant 0. \end{cases}$

We graph f(x) = 1 for inputs less than 0. Note that f(x) = 1 <u>only</u> for numbers less than 0. We use an open circle at the point (0, 1).

We graph f(x) = -1 for inputs greater than or equal to 0. Note that f(x) = -1 <u>only</u> for numbers greater than or equal to 0. We use a solid circle at the point (0, -1).

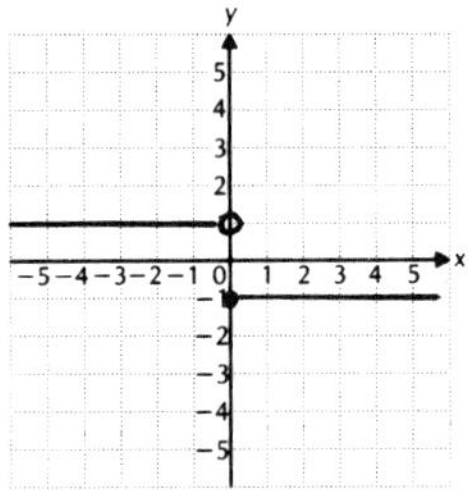

36.

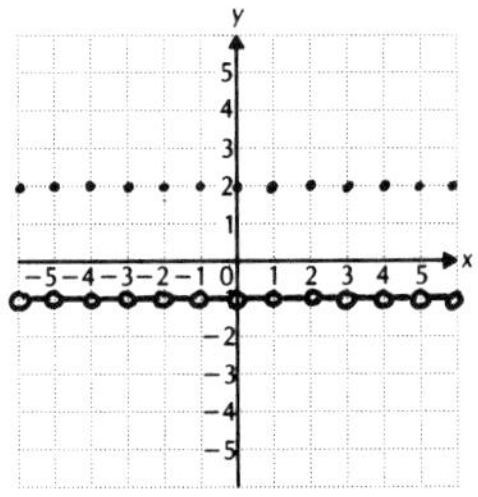

37. $f(x) = \begin{cases} 3 & \text{for } x \leqslant -3 \\ |x| & \text{for } -3 < x \leqslant 3 \\ -3 & \text{for } x > 3 \end{cases}$

We graph f(x) = 3 for inputs less than or equal to -3. Note that f(x) = 3 <u>only</u> for numbers less than or equal to -3.

We graph f(x) = |x| for inputs greater than -3 and less than or equal to 3. Note that f(x) = |x| <u>only</u> on the interval (-3, 3]. We use a solid circle at the point (3, 3).

We graph f(x) = -3 for inputs greater than 3. Note that f(x) = -3 <u>only</u> for numbers greater than 3. We use an open circle at the point (3, -3).

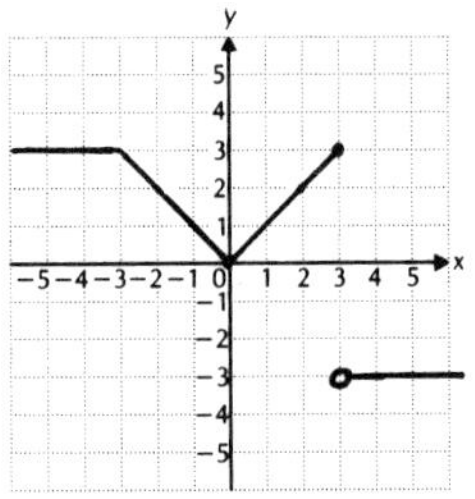

38.

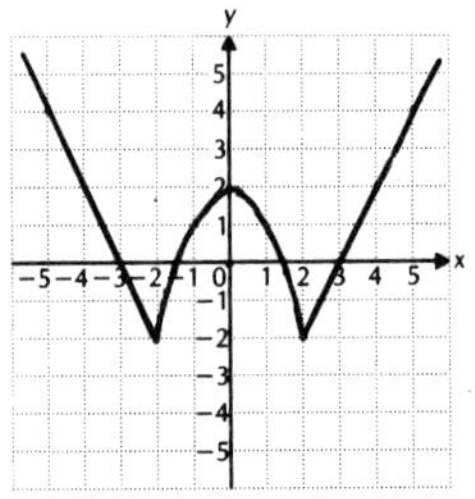

39.

$$p(x) = \begin{cases} 22¢ \text{ if } 0 < x \leqslant 1 \\ 39¢ \text{ if } 1 < x \leqslant 2 \\ 56¢ \text{ if } 2 < x \leqslant 3 \\ 73¢ \text{ if } 3 < x \leqslant 4 \\ 90¢ \text{ if } 4 < x \leqslant 5 \\ \$1.07 \text{ if } 5 < x \leqslant 6 \\ \$1.24 \text{ if } 6 < x \leqslant 7 \\ \$1.41 \text{ if } 7 < x \leqslant 8 \\ \$1.58 \text{ if } 8 < x \leqslant 9 \\ \$1.75 \text{ if } 9 < x \leqslant 10 \\ \$1.92 \text{ if } 10 < x \leqslant 11 \\ \$2.09 \text{ if } 11 < x \leqslant 12 \end{cases}$$

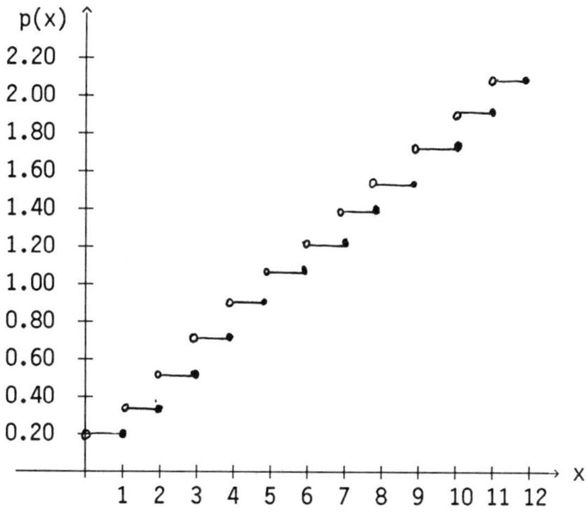

40.

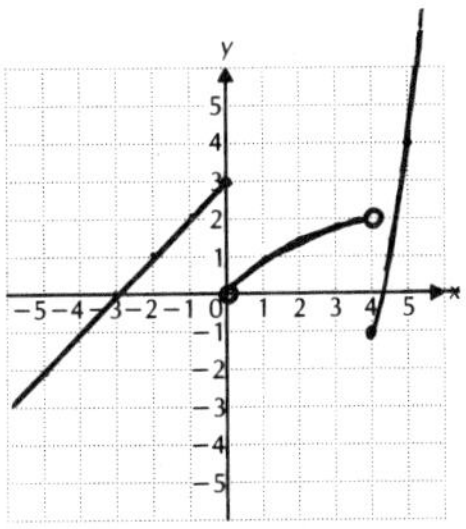

41. a) The function is increasing on the interval [0, 1].

b) The function is decreasing on the interval [-1, 0].

42. a) Increasing: [1, 3]

b) Decreasing: [-3, 1]

43. a)

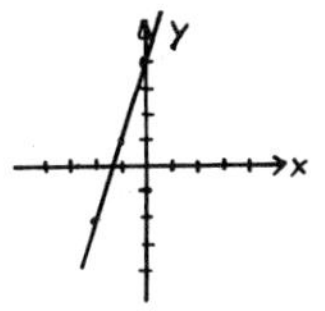

Increasing

b)

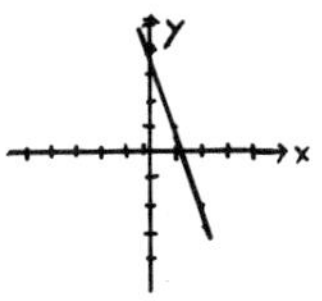

Decreasing

43. (continued)

c)

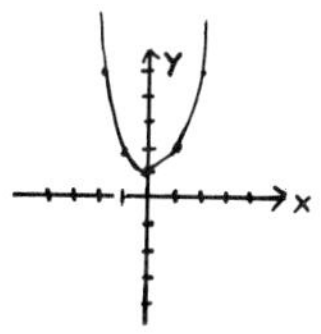

Neither

d)

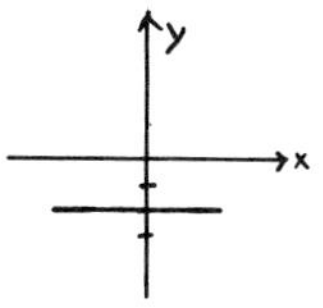

Neither

e)

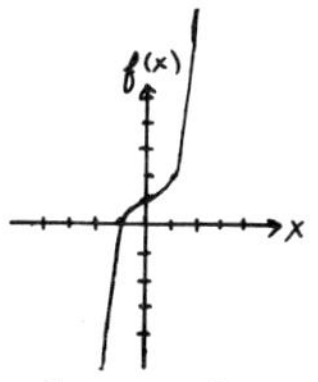

Increasing

f)

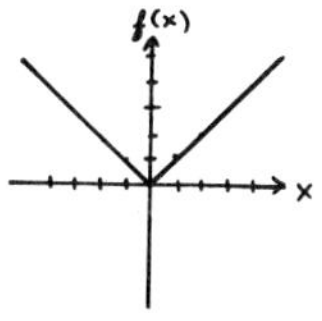

Neither

44. Increasing: b), f)
Decreasing: a)
Neither: c), d), e)

45. a) If $x = 0$, then $x' = 0 + 2 = 2$.

If $x = 1$, then $x' = 1 + 2 = 3$.

The interval [0, 1] is mapped to the interval [2, 3].

b) If $x = -2$, then $x' = -2 + 2 = 0$.

If $x = 7$, then $x' = 7 + 2 = 9$.

The interval (-2, 7] is mapped to the interval (0, 9].

c) If $x = -8$, then $x' = -8 + 2 = -6$.

If $x = -1$, then $x' = -1 + 2 = 1$.

The interval (-8, -1) is mapped to the interval (-6, 1).

46. a) [-3, -2.6], b) (-4.6, -11), c) [-0.6, 1.4)

Exercise Set 4.1

1. a) The equation $3y = 2x - 5$ is linear because it is equivalent to $2x - 3y = 5$. Here $A = 2$, $B = -3$, and $C = 5$.

   b) The equation $5x + 3 = 4y$ is linear because it is equivalent to $5x - 4y = -3$. Here $A = 5$, $B = -4$, and $C = -3$.

   c) The equation $3y = x^2 + 2$ is not linear because the x is squared.

   d) The equation $y = 3$ is linear because it is equivalent to $0x + y = 3$. Here $A = 0$, $B = 1$, and $C = 3$.

   e) The equation $xy = 5$ is not linear because the product xy occurs.

   f) The equation $3x^2 + 2y = 4$ is not linear because the x is squared.

   g) The equation $3x + \frac{1}{y} = 4$ is not linear because y is in a denominator.

   h) The equation $5x - 2 = 4y$ is linear because it is equivalent to $5x - 4y = 2$.

2. a, b, g, h

3. Graph: $8x - 3y = 24$

   First find two of its points. Here the easiest points are the intercepts.

   y-intercept: Set $x = 0$ and find y.

   $8x - 3y = 24$
   $8 \cdot 0 - 3y = 24$
   $-3y = 24$
   $y = -8$

   The y-intercept is $(0, -8)$.

   x-intercept: Set $y = 0$ and find x.

   $8x - 3y = 24$
   $8x - 3 \cdot 0 = 24$
   $8x = 24$
   $x = 3$

   The x-intercept is $(3, 0)$.

   Plot these two points and draw a line through them.

   $8x - 3y = 24$

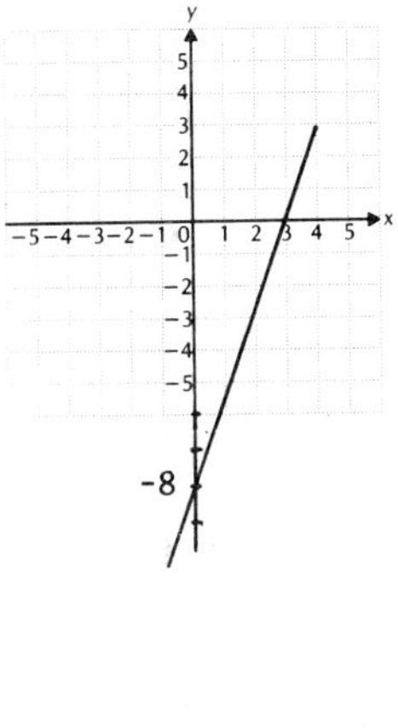

3. (continued)

   A third point should be used as a check.

   Set $x = 2$ and find y.

   $8x - 3y = 24$
   $8 \cdot 2 - 3y = 24$
   $16 - 3y = 24$
   $-3y = 8$
   $y = -\frac{8}{3}$

   The ordered pair $\left(2, -\frac{8}{3}\right)$ is also a point on the line.

4. $5x - 10y = 50$

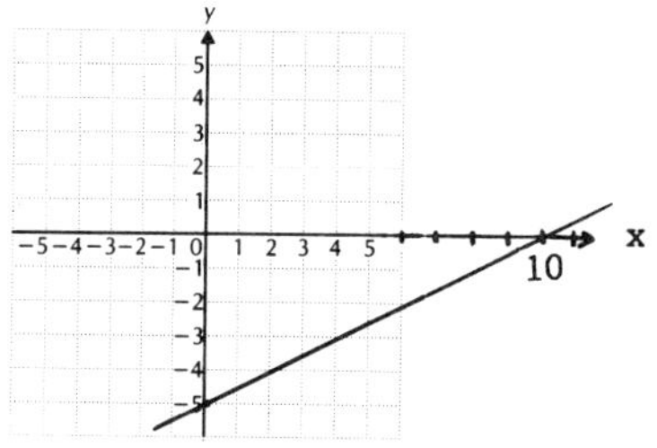

5. Graph: $3x + 12 = 4y$

   First find two of its points. Here the easiest points to find are the intercepts.

   y-intercept: Set $x = 0$ and find y.

   $3x + 12 = 4y$
   $3 \cdot 0 + 12 = 4y$
   $12 = 4y$
   $3 = y$

   The y-intercept is $(0, 3)$.

   x-intercept: Set $y = 0$ and find x.

   $3x + 12 = 4y$
   $3x + 12 = 4 \cdot 0$
   $3x + 12 = 0$
   $3x = -12$
   $x = -4$

   The x-intercept is $(-4, 0)$.

   Plot these points and draw a line through them. A third point should be used as a check.

   $3x + 12 = 4y$

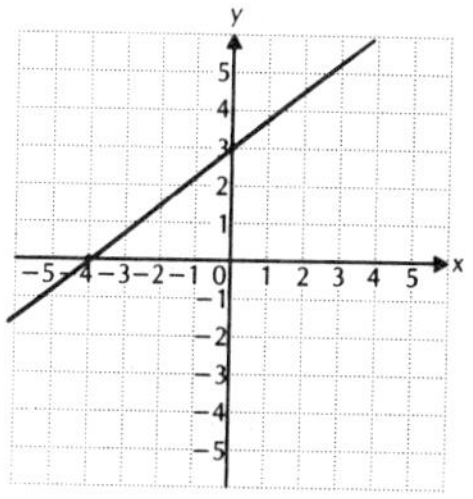

6. $4x - 20 = 5y$

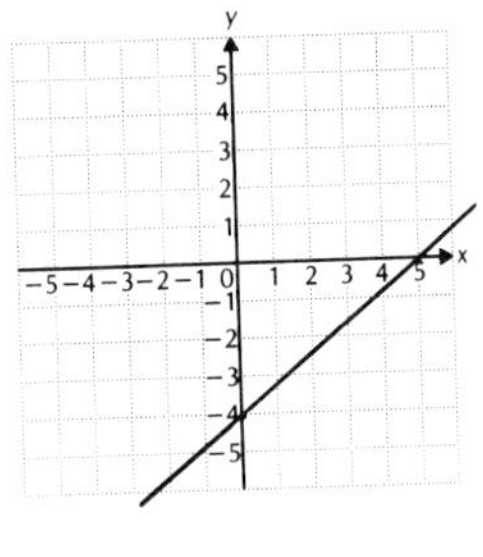

7. Graph: $y = -2$

The equation $y = -2$ says that the second coordinate of every ordered pair of the graph is -2. Below is a table of a few ordered pairs that are solutions of $y = -2$. (It might help to think of $y = -2$ as $0 \cdot x + y = -2$.)

x	y
-3	-2
0	-2
1	-2
4	-2

(x can be any number, but y must be -2)

Plot these points and draw the line through them.

$y = -2$

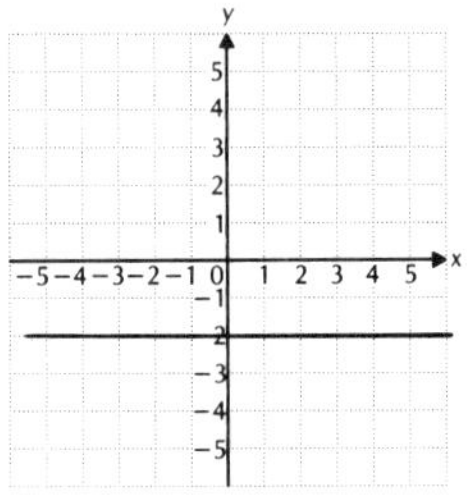

8. $2y - 3 = 9$

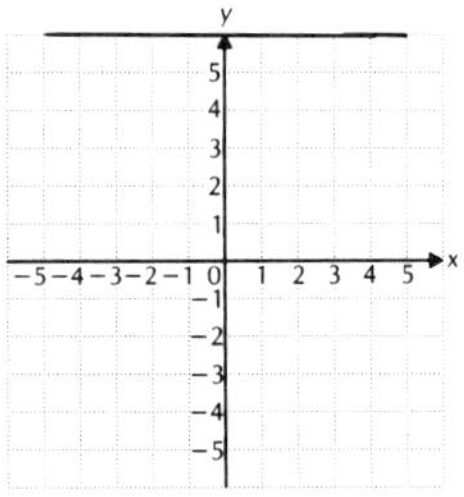

9. Graph: $5x + 2 = 17$

$$5x = 15$$
$$x = 3$$

The equation $5x + 2 = 17$ is equivalent to $x = 3$. The equation $x = 3$ says that the first coordinate of every ordered pair of the graph is 3. Below is a table of a few ordered pairs that are solutions of $x = 3$. (It might help to think of $x = 3$ as $x + 0 \cdot y = 3$.)

x	y
3	0
3	-2
3	4
3	-1

(y can be any number, but x must be 3)

Plot these point and draw a line through them.

9. (continued)

$x = 3$

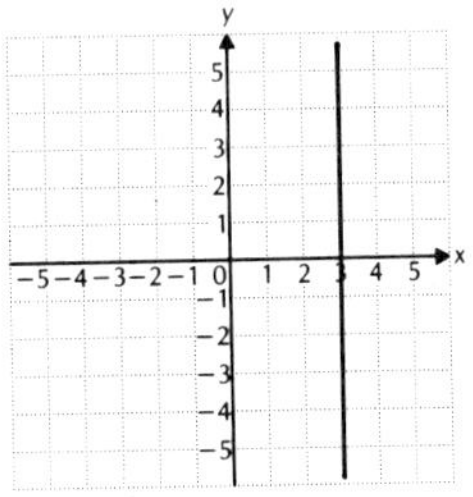

10. $19 = 5 - 2x$

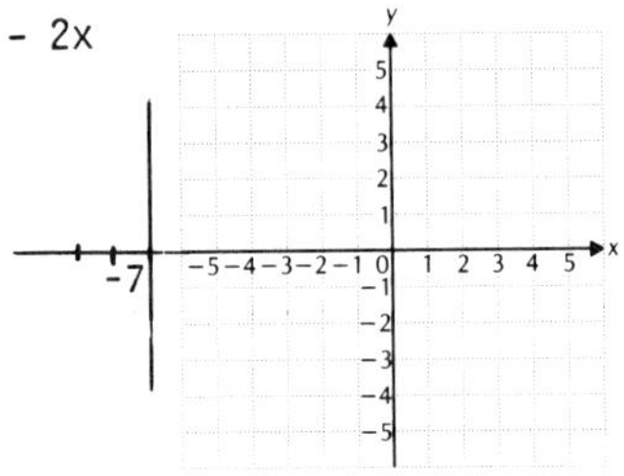

11. Find the slope of the line containing (6, 2) and (-2, 1).

$$m = \frac{\text{change in } y}{\text{change in } x} = \frac{2 - 1}{6 - (-2)} = \frac{1}{8}$$

$$\text{or } = \frac{1 - 2}{-2 - 6} = \frac{-1}{-8} = \frac{1}{8}$$

Note that it does not matter in which order we choose the points, so long as we take the differences in the same order. We get the same slope either way.

12. $\frac{3}{2}$

13. Find the slope of the line containing (2, -4) and (4, -3).

$$m = \frac{\text{change in } y}{\text{change in } x} = \frac{-4 - (-3)}{2 - 4} = \frac{-1}{-2} = \frac{1}{2}$$

$$\text{or } = \frac{-3 - (-4)}{4 - 2} = \frac{1}{2}$$

14. $-\frac{11}{10}$

15. Find the slope of the line containing (2π, 5) and (π, 4).

$$m = \frac{\text{change in } y}{\text{change in } x} = \frac{5 - 4}{2\pi - \pi} = \frac{1}{\pi}$$

16. 0

17. Find an equation of the line through (3, 2) with $m = 4$.

$(y - y_1) = m(x - x_1)$ (Point-slope equation)

$(y - 2) = 4(x - 3)$ (Substituting)

$y - 2 = 4x - 12$

$y = 4x - 10$

18. $y = -2x + 15$

19. Find an equation of the line with y-intercept -5 and m = 2.

$y = mx + b$ (Slope-intercept equation)

$y = 2x + (-5)$ (Substituting)

$y = 2x - 5$

20. $y = \frac{1}{4}x + \pi$

21. Find an equation of the line containing (1, 4) and (5, 6). Let $(x_1, y_1) = (1, 4)$ and $(x_2, y_2) = (5, 6)$.

$y - y_1 = \frac{y_2 - y_1}{x_2 - x_1}(x - x_1)$ (Two-point equation)

$y - 4 = \frac{6 - 4}{5 - 1}(x - 1)$ (Substituting)

$y - 4 = \frac{2}{4}(x - 1)$

$y - 4 = \frac{1}{2}x - \frac{1}{2}$

$y = \frac{1}{2}x + \frac{7}{2}$

22. $y = \frac{3}{4}x + \frac{3}{2}$

23. $y = 2x + 3$

$y = mx + b$ (Slope-intercept equation)

The slope is 2, and the y-intercept is 3, or (0, 3).

24. m = -1, b = 6

25. $2y = -6x + 10$

$y = -3x + 5$

$y = mx + b$ (Slope-intercept equation)

The slope is -3, and the y-intercept is 5, or (0, 5).

26. m = 4, b = -3

27. $3x - 4y = 12$

$-4y = -3x + 12$

$y = \frac{3}{4}x - 3$

$y = mx + b$ (Slope-intercept equation)

The slope is $\frac{3}{4}$, and the y-intercept is -3, or (0, -3).

28. $m = -\frac{5}{2}$, $b = -\frac{7}{2}$

29. $3y + 10 = 0$

$3y = -10$

$y = -\frac{10}{3}$

or

$y = 0 \cdot x + \left(-\frac{10}{3}\right)$

$y = mx + b$ (Slope-intercept equation)

The slope is 0, and the y-intercept is $-\frac{10}{3}$, or $\left(0, -\frac{10}{3}\right)$.

30. m = 0, b = 7

31. $y - y_1 = m(x - x_1)$ (Point-slope equation)

$y - (-2.563) = 3.516(x - 3.014)$ (Substituting)

$y + 2.563 = 3.516x - 10.597224$

$y = 3.516x - 13.160224$

32. $y = -0.00014x - 17.624$

33. $y - y_1 = \frac{y_2 - y_1}{x_2 - x_1}(x - x_1)$ (Two-point equation)

$y - 2.443 = \frac{11.012 - 2.443}{8.114 - 1.103}(x - 1.103)$ (Substituting)

$y - 2.443 = \frac{8.569}{7.011}(x - 1.103)$

$y - 2.443 = 1.222(x - 1.103)$

$y - 2.443 = 1.222x - 1.348$

$y = 1.222x + 1.095$

34. $y = -1.4213x + 1583.6$

35. $f(x) = mx + b$ (Linear function)

$f(3x) = 3f(x)$ (Given)

$m(3x) + b = 3(mx + b)$

$3mx + b = 3mx + 3b$

$b = 3b$

$0 = 2b$

$0 = b$

Thus, if $f(3x) = 3f(x)$, then

$f(x) = mx + 0$

$f(x) = mx$

36. $f(x) = mx$ or $f(x) = mx + b$

37. $f(x) = mx + b$ (Linear function)
$f(x + 2) = f(x) + 2$ (Given)
$m(x + 2) + b = mx + b + 2$
$mx + 2m + b = mx + b + 2$
$2m = 2$
$m = 1$

Thus, if $f(x + 2) = f(x) + 2$, then
$f(x) = 1 \cdot x + b$
$f(x) = x + b$

38. $f(x) = b$

39. $f(x) = mx + b$ (Linear function)
$f(c + d) = m(c + d) + b = mc + md + b$
$f(c) + f(d) = (mc + b) + (md + b) = mc + md + 2b$

$mc + md + b \neq mc + md + 2b$

Thus, $f(c + d) = f(c) + f(d)$ is false.

40. False

41. $f(x) = mx + b$ (Linear function)
$f(kx) = m(kx) + b = mkx + b$
$kf(x) = k(mx + b) = mkx + kb$

$mkx + b \neq mkx + kb$

Thus, $f(kx) = kf(x)$ is false.

42. False

43. If A(9, 4), B(-1, 2), and C(4, 3) are on the same line, then the slope of the segment $\overline{AB}$ must be the same as the slope of the segment $\overline{BC}$. (Note that B is a point of each segment.)

Slope of $\overline{AB} = \frac{4 - 2}{9 - (-1)} = \frac{2}{10} = \frac{1}{5}$

Slope of $\overline{BC} = \frac{2 - 3}{-1 - 4} = \frac{-1}{-5} = \frac{1}{5}$

Since the slopes are the same, and B is on both lines, A, B, and C are on the same line.

44. No

45.

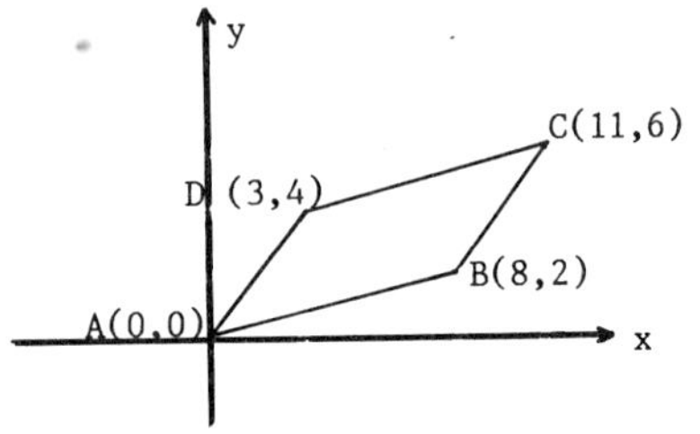

Slope of $\overline{AB} = \frac{2 - 0}{8 - 0} = \frac{2}{8} = \frac{1}{4}$

Slope of $\overline{CD} = \frac{6 - 4}{11 - 3} = \frac{2}{8} = \frac{1}{4}$

45. (continued)

Slope of $\overline{BC} = \frac{6 - 2}{11 - 8} = \frac{4}{3}$

Slope of $\overline{DA} = \frac{4 - 0}{3 - 0} = \frac{4}{3}$

$\overline{AB}$ and $\overline{CD}$ have the same slope.
$\overline{BC}$ and $\overline{DA}$ have the same slope.

46. $\overline{EG}$: $m = \frac{1}{3}$; $\overline{FH}$: $m = -3$

47. Consider a line containing the ordered pairs (0, 32) and (100, 212).

The slope of the line is $\frac{212 - 32}{100 - 0}$, or $\frac{9}{5}$.

$F(C) = mC + b$ (Linear function)

$F(C) = \frac{9}{5} C + b$ [Substituting $\frac{9}{5}$ for m]

We now determine b.

$32 = \frac{9}{5} \cdot 0 + b$ (Substituting 0 for C and 32 for F)

$32 = b$

The linear function is

$F(C) = \frac{9}{5} C + 32$

48. $C = \frac{5}{9} (F - 32)$, or $\frac{5}{9} F - \frac{5}{9} \cdot 32$

49. $P = mQ + b$, $m \neq 0$ (Linear function)

We solve for Q.

$P - b = mQ$

$\frac{P - b}{m} = Q$

$\frac{P}{m} - \frac{b}{m} = Q$ (Linear function)

50. By definition, $y = kx$. Thus, y is linear with $b = 0$.

51. Road grade $= \frac{\text{vertical distance}}{\text{horizontal distance}}$

$= \frac{50}{1250} = 0.04 = 4\%$

The equation of the line containing (0, 0) and (1250, 50) is $y = 4\%x$. (Note: The y-intercept is 0).

52. Grade = 6.7%; $y = 6.7\%x$

Exercise Set 4.2

1. Solve each equation for y.

$2x - 5y = -3$ $\qquad$ $2x + 5y = 4$

$-5y = -2x - 3$ $\qquad$ $5y = -2x + 4$

$y = \frac{2}{5}x + \frac{3}{5}$ $\qquad$ $y = -\frac{2}{5}x + \frac{4}{5}$

The slopes are $\frac{2}{5}$ and $-\frac{2}{5}$. The slopes are not equal. The product of the slopes is not -1. Thus, the lines are neither parallel nor perpendicular.

2. Parallel

3. Solve each equation for y.

$y = 4x - 5$ $\qquad$ $4y = 8 - x$

$y = -\frac{1}{4}x + 2$

The slopes are 4 and $-\frac{1}{4}$. Their product is -1, so the lines are perpendicular.

4. Perpendicular

5. We first solve for y.

$3x - y = 7$

$-y = -3x + 7$

$y = 3x - 7$

The slope of the given line is 3. The line parallel to $y = 3x - 7$ and containing the point (0, 3) must also have slope 3.

$y - y_1 = m(x - x_1)$ (Point-slope equation)

$y - 3 = 3(x - 0)$ (Substituting 0 for x_1, 3 for y_1, and 3 for m)

$y - 3 = 3x$

$y = 3x + 3$

6. $y = -2x - 13$

7. The line x = 2 is vertical, so any line parallel to it must be vertical. The line we seek has one x-coordinate which is 3, so all x-coordinates on the line must be 3. The equation is x = 3.

8. x = 3

9. The line y = 4 is horizontal, so any line parallel to it must be horizontal. The line we seek has one y-coordinate which is -3, so all y-coordinates on the line must be -3. The equation is y = -3.

10. y = 2

11. We first solve for y.

$5x - 2y = 4$

$-2y = -5x + 4$

$y = \frac{5}{2}x - 2$

The slope of the given line is $\frac{5}{2}$. The slope of the line perpendicular to $y = \frac{5}{2}x - 2$ and containing (-3, -5) must be $-\frac{2}{5}$.

$y - y_1 = m(x - x_1)$ (Point-slope equation)

$y - (-5) = -\frac{2}{5}[x - (-3)]$ [Substituting -3 for x_1, -5 for y_1, and $-\frac{2}{5}$ for m]

$y + 5 = -\frac{2}{5}(x + 3)$

$y + 5 = -\frac{2}{5}x - \frac{6}{5}$

$y = -\frac{2}{5}x - \frac{31}{5}$

12. $y = \frac{4}{3}x - 6$

13. The line x = 1 is vertical, so any line perpendicular to it must be horizontal. The line we seek has one y-coordinate which is 3, so all y-coordinates on the line must be 3. The equation is y = 3.

14. y = -2

15. The line y = 2 is horizontal, so any line perpendicular to it must be vertical. The line we seek has one x-coordinate which is -3, so all x-coordinates on the line must be -3. The equation is x = -3.

16. x = 4

17. We first solve for y.

$4.323x - 7.071y = 16.61$

$-7.071y = -4.323x + 16.61$

$y = 0.611x - 2.349$

The slope of the given line is 0.611. The slope of the line parallel to $y = 0.611x - 2.349$ and containing (-2.603, 1.818) must also be 0.611.

$y - y_1 = m(x - x_1)$ (Point-slope equation)

$y - 1.818 = 0.611[x - (-2.603)]$

(Substituting -2.603 for x_1, 1.818 for y_1, and 0.611 for m)

$y - 1.818 = 0.611x + 1.590$

$y = 0.611x + 3.408$

18. $y = 0.642x - 4.930$

19. Let $(x_1, y_1) = (-3, -2)$ and $(x_2, y_2) = (1, 1)$.

Then use the distance formula.

$d = \sqrt{(x_1 - x_2)^2 + (y_1 - y_2)^2}$

$d = \sqrt{(-3 - 1)^2 + (-2 - 1)^2} = \sqrt{(-4)^2 + (-3)^2}$

$= \sqrt{16 + 9} = \sqrt{25} = 5$

20. $3\sqrt{5}$

21. Let $(x_1, y_1) = (0, -7)$ and $(x_2, y_2) = (3, -4)$.

Then use the distance formula.

$d = \sqrt{(x_1 - x_2)^2 + (y_1 - y_2)^2}$

$d = \sqrt{(0 - 3)^2 + [-7 - (-4)]^2} = \sqrt{(-3)^2 + (-3)^2}$

$= \sqrt{9 + 9} = \sqrt{18} = 3\sqrt{2}$

22. $4\sqrt{2}$

23. Let $(x_1, y_1) = (a, -3)$ and $(x_2, y_2) = (2a, 5)$.

$d = \sqrt{(x_1 - x_2)^2 + (y_1 - y_2)^2}$

$d = \sqrt{(a - 2a)^2 + (-3 - 5)^2}$

$= \sqrt{(-a)^2 + (-8)^2} = \sqrt{a^2 + 64}$

24. $\sqrt{64 + k^2}$

25. Let $(x_1, y_1) = (0, 0)$ and $(x_2, y_2) = (a, b)$.

$d = \sqrt{(x_1 - x_2)^2 + (y_1 - y_2)^2}$

$d = \sqrt{(0 - a)^2 + (0 - b)^2}$

$= \sqrt{(-a)^2 + (-b)^2} = \sqrt{a^2 + b^2}$

26. $\sqrt{5}$

27. Let $(x_1, y_1) = (\sqrt{a}, \sqrt{b})$ and $(x_2, y_2) = (-\sqrt{a}, \sqrt{b})$.

$d = \sqrt{(x_1 - x_2)^2 + (y_1 - y_2)^2}$

$d = \sqrt{[\sqrt{a} - (-\sqrt{a})]^2 + (\sqrt{b} - \sqrt{b})^2}$

$= \sqrt{(2\sqrt{a})^2 + 0^2} = \sqrt{4a} = 2\sqrt{a}$

28. $2\sqrt{c^2 + d^2}$

29. Let $(x_1, y_1) = (7.3482, -3.0991)$ and $(x_2, y_2) = (18.9431, -17.9054)$.

$d = \sqrt{(x_1 - x_2)^2 + (y_1 - y_2)^2}$

$d = \sqrt{(7.3482 - 18.9431)^2 + [-3.0991-(-17.9054)]^2}$

$= \sqrt{(-11.5949)^2 + (14.8063)^2}$

$= \sqrt{134.4417 + 219.2265}$

$= \sqrt{353.6682} = 18.8061$

30. 404.4729

31. First we find the square of the distances between the points:

Between (9, 6) and (-1, 2)

$d^2 = [9 - (-1)]^2 + (6 - 2)^2 = 100 + 16 = 116$

Between (-1, 2) and (1, -3)

$d^2 = (-1 - 1)^2 + [2 - (-3)]^2 = 4 + 25 = 29$

Between (1, -3) and (9, 6)

$d^2 = (1 - 9)^2 + (-3 - 6)^2 = 64 + 81 = 145$

Since the sum of 116 and 29 is 145, it follows from the Pythagorean theorem that the points are vertices of a right triangle.

32. Yes

33. Let $(x_1, y_1) = (-4, 7)$ and $(x_2, y_2) = (3, -9)$.

$\left(\frac{x_1 + x_2}{2}, \frac{y_1 + y_2}{2}\right)$ (Midpoint formula)

$\left(\frac{-4 + 3}{2}, \frac{7 + (-9)}{2}\right)$ (Substituting)

$\left(\frac{-1}{2}, \frac{-2}{2}\right)$

The midpoint is $\left(-\frac{1}{2}, -1\right)$.

34. (5, -1)

35. Let $(x_1, y_1) = (a, b)$ and $(x_2, y_2) = (a, -b)$.

$\left(\frac{x_1 + x_2}{2}, \frac{y_1 + y_2}{2}\right)$ (Midpoint formula)

$\left(\frac{a + a}{2}, \frac{b + (-b)}{2}\right)$ (Substituting)

$\left(\frac{2a}{2}, \frac{0}{2}\right)$

The midpoint is (a, 0).

36. (0, d)

37. Let $(x_1, y_1) = (-3.895, 8.1212)$ and $(x_2, y_2) = (2.998, -8.6677)$.

$\left[\frac{x_1 + x_2}{2}, \frac{y_1 + y_2}{2}\right]$ (Midpoint formula)

$\left[\frac{-3.895 + 2.998}{2}, \frac{8.1212 + (-8.6677)}{2}\right]$ (Substituting)

$\left[\frac{-0.897}{2}, \frac{-0.5465}{2}\right]$

The midpoint is (-0.4485, 0.27325).

38. (4.652, 0.03775)

39. The slope of the line containing (-1, 4) and (2, -3) is $\frac{4 - (-3)}{-1 - 2}$, or $-\frac{7}{3}$. The slope of the line parallel to this line and containing (4, -2) is also $-\frac{7}{3}$.

$y - y_1 = m(x - x_1)$ (Point-slope equation)

$y - (-2) = -\frac{7}{3}(x - 4)$ [Substituting 4 for x_1, -2 for y_1, and $-\frac{7}{3}$ for m]

$y + 2 = -\frac{7}{3}x + \frac{28}{3}$

$y = -\frac{7}{3}x + \frac{22}{3}$

40. $y = \frac{2}{5}x + \frac{17}{5}$

41.

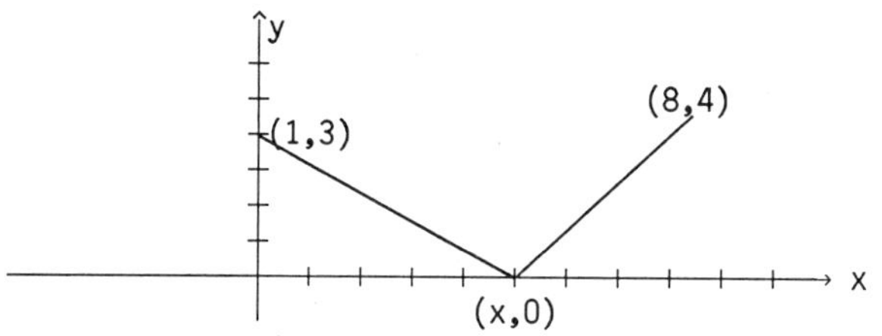

Between (1, 3) and (x, 0):

$d = \sqrt{(1 - x)^2 + (3 - 0)^2}$

$= \sqrt{1 - 2x + x^2 + 9} = \sqrt{x^2 - 2x + 10}$

Between (8, 4) and (x, 0):

$d = \sqrt{(8 - x)^2 + (4 - 0)^2}$

$= \sqrt{64 - 16x + x^2 + 16} = \sqrt{x^2 - 16x + 80}$

Thus,

$\sqrt{x^2 - 2x + 10} = \sqrt{x^2 - 16x + 80}$

$x^2 - 2x + 10 = x^2 - 16x + 80$

$14x = 70$

$x = 5$

The point on the x-axis equidistant from the points (1, 3) and (8, 4) is (5, 0).

42. (0, 4)

Exercise Set 4.3

1. $f(x) = x^2$

$f(x) = (x - 0)^2 + 0$ [In the form $f(x) = a(x - h)^2 + k$]

a) The vertex is (0, 0).

b) The line of symmetry goes through the vertex; it is the line x = 0.

c) Since the coefficient of x^2 is positive (1 > 0), the parabola opens upward. Thus we have a minimum value which is 0.

2. a) (0, 0), b) x = 0, c) 0 is a maximum

3. $f(x) = -2(x - 9)^2$

$f(x) = -2(x - 9)^2 + 0$ [In the form $f(x) = a(x - h)^2 + k$]

a) The vertex is (9, 0).

b) The line of symmetry goes through the vertex; it is the line x = 9.

c) Since the coefficient of x^2 is negative (-2 < 0), the parabola opens downward. Thus we have a maximum value which is 0.

4. a) (7, 0), b) x = 7, c) 0 is a minimum

5. $f(x) = 2(x - 1)^2 - 4$

$f(x) = 2(x - 1)^2 + (-4)$ [In the form $f(x) = a(x - h)^2 + k$]

a) The vertex is (1, -4)

b) The line of symmetry goes through the vertex; it is the line x = 1.

c) Since the coefficient of x^2 is positive (2 > 0), the parabola opens upward. Thus we have a minimum value which is -4.

6. a) (-4, -3), b) x = -4, c) -3 is a maximum

7. a) $f(x) = -x^2 + 2x + 3$

$= -(x^2 - 2x) + 3$

$= -(x^2 - 2x + 1 - 1) + 3$ (Adding 1 - 1 inside parentheses)

$= -(x^2 - 2x + 1) + 1 + 3$

$= -(x - 1)^2 + 4$ [In the form $f(x) = a(x - h)^2 + k$]

b) The vertex is (1, 4).

c) Since the coefficient of x^2 is negative (-1 < 0), the parabola opens downward. Thus we have a maximum value which is 4.

8. a) $f(x) = -(x - 4)^2 + 9$

b) (4, 9)

c) 9 is a maximum

9. a) $f(x) = x^2 + 3x$

$= x^2 + 3x + \frac{9}{4} - \frac{9}{4}$ $\left(\text{Adding } \frac{9}{4} - \frac{9}{4}\right)$

$= \left(x + \frac{3}{2}\right)^2 - \frac{9}{4}$

$= \left[x - \left(-\frac{3}{2}\right)\right]^2 + \left(-\frac{9}{4}\right)$

[In the form $f(x) = a(x - h)^2 + k$]

b) The vertex is $\left(-\frac{3}{2}, -\frac{9}{4}\right)$.

c) Since the coefficient of x^2 is positive $(1 > 0)$, the parabola opens upward. Thus we have a minimum value which is $-\frac{9}{4}$.

10. a) $f(x) = \left(x - \frac{9}{2}\right)^2 - \frac{81}{4}$

b) $\left(\frac{9}{2}, -\frac{81}{4}\right)$

c) $-\frac{81}{4}$ is a minimum

11. a) $f(x) = -\frac{3}{4}x^2 + 6x$

$= -\frac{3}{4}(x^2 - 8x)$

$= -\frac{3}{4}(x^2 - 8x + 16 - 16)$

(Adding 16 - 16 inside parentheses)

$= -\frac{3}{4}(x^2 - 8x + 16) + 12$

$= -\frac{3}{4}(x - 4)^2 + 12$

[In the form $f(x) = a(x - h)^2 + k$]

b) The vertex is (4, 12).

c) Since the coefficient of x^2 is negative $\left(-\frac{3}{4} < 0\right)$, the parabola opens downward. Thus we have a maximum value which is 12.

12. a) $f(x) = \frac{3}{2}(x + 1)^2 - \frac{3}{2}$

b) $\left(-1, -\frac{3}{2}\right)$

c) $-\frac{3}{2}$ is a minimum

13. a) $f(x) = 3x^2 + x - 4$

$= 3\left(x^2 + \frac{1}{3}x\right) - 4$

$= 3\left(x^2 + \frac{1}{3}x + \frac{1}{36} - \frac{1}{36}\right) - 4$

$\left(\text{Adding } \frac{1}{36} - \frac{1}{36} \text{ inside parentheses}\right)$

$= 3\left(x^2 + \frac{1}{3}x + \frac{1}{36}\right) - \frac{1}{12} - 4$

$= 3\left(x + \frac{1}{6}\right)^2 - \frac{49}{12}$

$= 3\left[x - \left(-\frac{1}{6}\right)\right]^2 + \left(-\frac{49}{12}\right)$

[In the form $f(x) = a(x - h)^2 + k$]

b) The vertex is $\left(-\frac{1}{6}, -\frac{49}{12}\right)$.

c) Since the coefficient of x^2 is positive $(3 > 0)$, the parabola opens upward. Thus we have a minimum value which is $-\frac{49}{12}$.

14. a) $f(x) = -2\left(x - \frac{1}{4}\right)^2 - \frac{7}{8}$

b) $\left(\frac{1}{4}, -\frac{7}{8}\right)$

c) $-\frac{7}{8}$ is a maximum

15. $f(x) = -x^2 + 2x + 3$

$= -(x^2 - 2x) + 3$

$= -(x^2 - 2x + 1 - 1) + 3$ (Adding 1 - 1)

$= -(x^2 - 2x + 1) + 1 + 3$

$= -(x - 1)^2 + 4$

It is not necessary to rewrite the function in the form $f(x) = a(x - h)^2 + k$, but it can be helpful. Since the coefficient of x^2 is negative, the parabola opens downward. The vertex is (1, 4). The axis of symmetry is $x = 1$. We calculate several input-output pairs and plot these points.

x	f(x)
1	4
0	3
2	3
-1	0
3	0

$f(x) = -x^2 + 2x + 3$

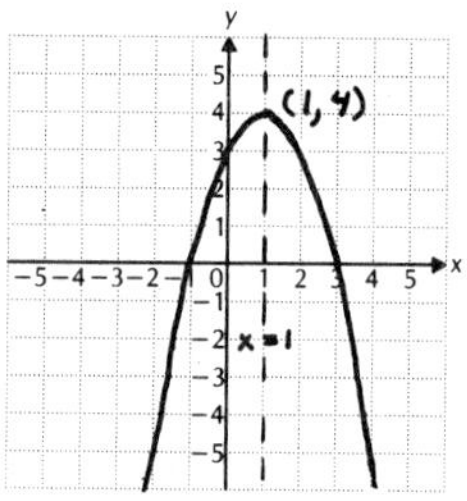

16. $f(x) = x^2 - 3x - 4$

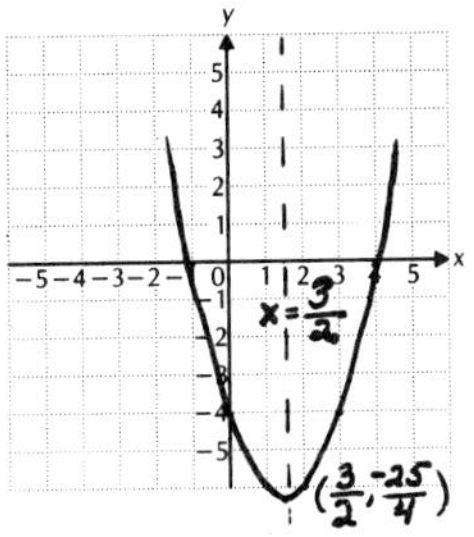

17. $f(x) = x^2 - 8x + 19$

$= (x^2 - 8x + 16 - 16) + 19$ (Adding 16 - 16)

$= (x^2 - 8x + 16) - 16 + 19$

$= (x - 4)^2 + 3$

Since the coefficient of x^2 is positive, the parabola opens upward. The vertex is (4, 3). The axis of symmetry is x = 4. We calculate several input-output pairs and plot these points.

x	f(x)
4	3
3	4
5	4
2	7
6	7

$f(x) = x^2 - 8x + 19$

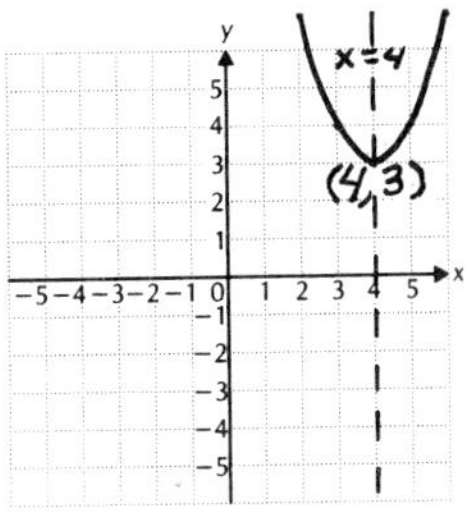

18. $f(x) = -x^2 - 8x - 17$

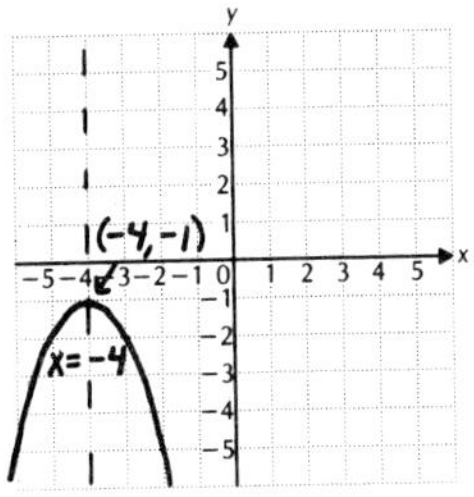

19. $f(x) = -\frac{1}{2}x^2 - 3x + \frac{1}{2}$

$= -\frac{1}{2}(x^2 + 6x) + \frac{1}{2}$

$= -\frac{1}{2}(x^2 + 6x + 9 - 9) + \frac{1}{2}$ (Adding 9 - 9)

$= -\frac{1}{2}(x^2 + 6x + 9) + \frac{9}{2} + \frac{1}{2}$

$= -\frac{1}{2}(x + 3)^2 + 5$

$= -\frac{1}{2}[x - (-3)]^2 + 5$

Since the coefficient of x^2 is negative, the parabola opens downward. The vertex is (-3, 5). The axis of symmetry is x = -3. We calculate several input-output pairs and plot these points.

19. (continued)

x	f(x)
-3	5
-4	$\frac{9}{2}$
-2	$\frac{9}{2}$
-5	3
-1	3

$f(x) = -\frac{1}{2}x^2 - 3x + \frac{1}{2}$

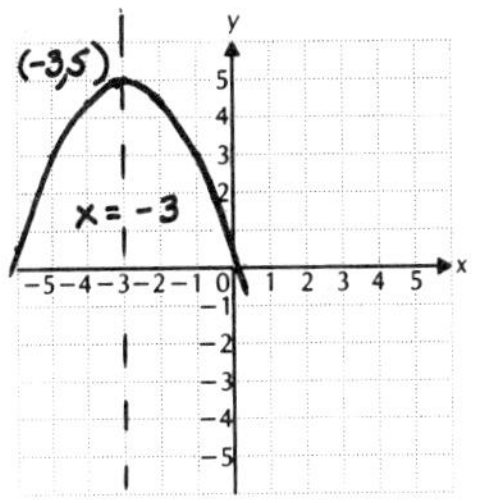

20. $f(x) = 2x^2 - 4x - 2$

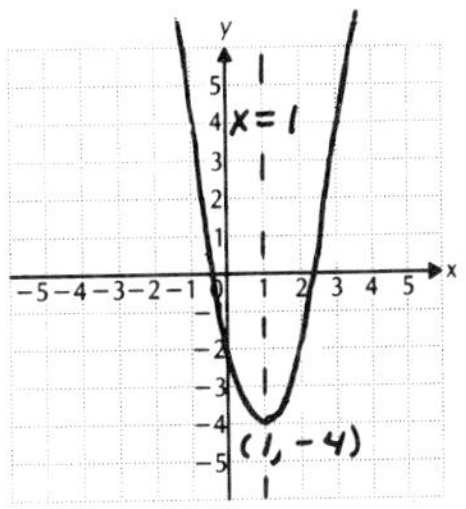

21. $f(x) = 3x^2 - 24x + 50$

$= 3(x^2 - 8x) + 50$

$= 3(x^2 - 8x + 16 - 16) + 50$ (Adding 16 - 16)

$= 3(x^2 - 8x + 16) - 48 + 50$

$= 3(x - 4)^2 + 2$

Since the coefficient of x^2 is positive, the parabola opens upward. The vertex is (4, 2). The axis of symmetry is x = 4. We calculate several input-output pairs and plot these points.

x	f(x)
4	2
3	5
5	5
2	14
6	14

$f(x) = 3x^2 - 24x + 50$

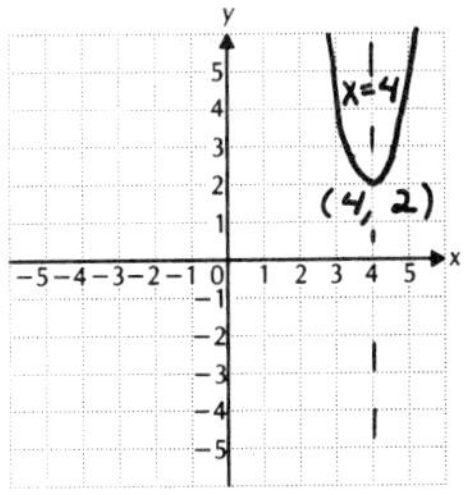

22. $f(x) = -2x^2 + 2x + 1$

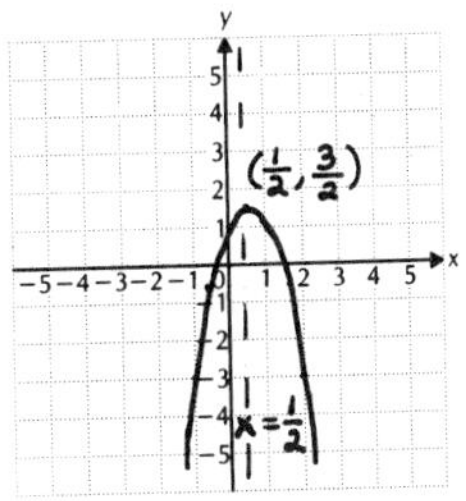

23. $f(x) = -x^2 + 2x + 3$

We solve the equation

$0 = -x^2 + 2x + 3$

$0 = x^2 - 2x - 3$ (Muliplying by -1)

$0 = (x + 1)(x - 3)$

$x + 1 = 0$ or $x - 3 = 0$

$x = -1$ or $x = 3$

Thus the x-intercepts are (-1, 0) and (3, 0).

24. (-1, 0), (4, 0)

25. $f(x) = x^2 - 8x + 5$

We solve the equation

$0 = x^2 - 8x + 5$

$a = 1, \quad b = -8, \quad c = 5$

$$x = \frac{-(-8) \pm \sqrt{(-8)^2 - 4\cdot1\cdot5}}{2\cdot1} = \frac{8 \pm \sqrt{64 - 20}}{2}$$

$$= \frac{8 \pm \sqrt{44}}{2} = \frac{8 \pm 2\sqrt{11}}{2} = 4 \pm \sqrt{11}$$

Thus the x-intercepts are $(4 - \sqrt{11}, 0)$ and $(4 + \sqrt{11}, 0)$.

26. None

27. $f(x) = -5x^2 + 6x - 5$

We solve the equation

$0 = -5x^2 + 6x - 5$

$0 = 5x^2 - 6x + 5$

$a = 5, \quad b = -6, \quad c = 5$

$b^2 - 4ac$ (the discriminant) $= -64 < 0$

Thus, there are no x-intercepts.

28. $\left(\frac{-1 - \sqrt{41}}{4}, 0\right), \quad \left(\frac{-1 + \sqrt{41}}{4}, 0\right)$

29. $f(x) = ax^2 + bx + c$

$$= a\left(x^2 + \frac{b}{a}x\right) + c$$

$$= a\left(x^2 + \frac{b}{a}x + \frac{b^2}{4a^2} - \frac{b^2}{4a^2}\right) + c$$

$$= a\left(x^2 + \frac{b}{a}x + \frac{b^2}{4a^2}\right) - \frac{b^2}{4a} + c$$

$$= a\left(x + \frac{b}{2a}\right)^2 + \frac{-b^2 + 4ac}{4a}$$

$$= a\left[x - \left(-\frac{b}{2a}\right)\right]^2 + \frac{4ac - b^2}{4a}$$

30. $f(x) = 3\left(x + \frac{m}{6}\right)^2 + \frac{11m^2}{12}$

31. $f(x) = |x^2 - 1|$

First graph $f(x) = x^2 - 1$.

x	f(x)
0	-1
-1	0
1	0
2	3
-2	3

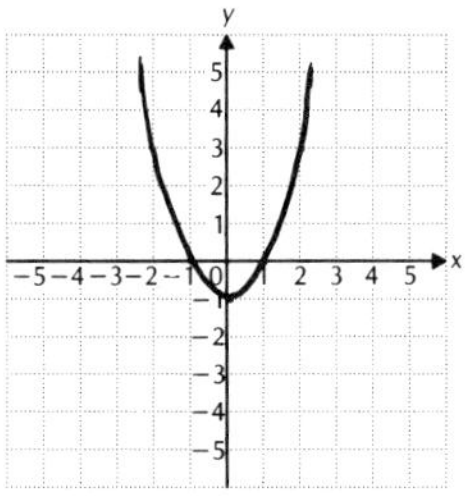

Then reflect the negative values across the x-axis. The point (0, -1) is now (0, 1).

$f(x) = |x^2 - 1|$

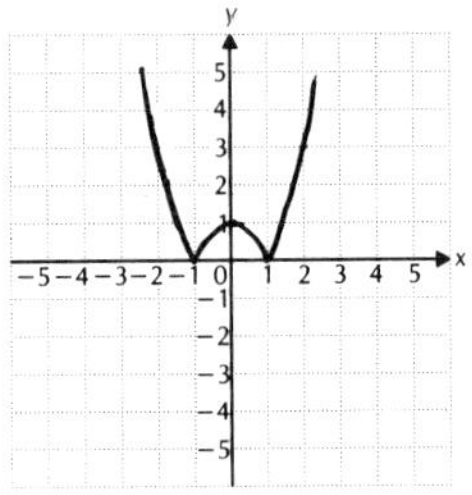

32. $f(x) = |3 - 2x - x^2|$

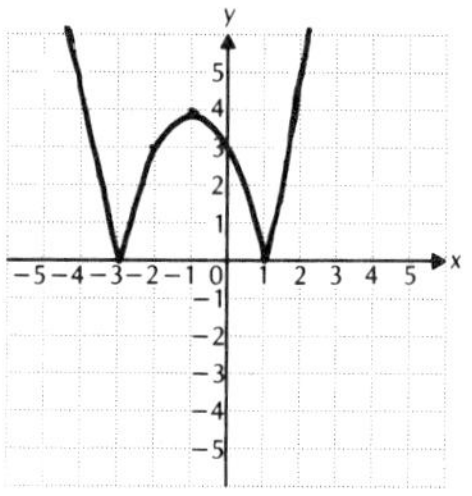

33. $f(x) = 2.31x^2 - 3.135x - 5.89$

$= 2.31(x^2 - 1.3571x) - 5.89$

$= 2.31(x^2 - 1.3571x + 0.4604 - 0.4604) - 5.89$

$= 2.31(x^2 - 1.3571x + 0.4604) - 1.0635 - 5.89$

$= 2.31(x - 0.6785)^2 - 6.9535$

The vertex is (0.6785, -6.9535). The coefficient of x^2 is positive. Thus, the parabola opens upward and has a minimum value of -6.9535.

34. Maximum, 7.014

35. Let x and y represent the numbers.

$x - y = 4.932$, or $y = x - 4.932$

The function we want to minimize is

$P(x) = x\cdot y$

$= x(x - 4.932)$

$= x^2 - 4.932x$

35. (continued)

We complete the square.

$P(x) = x^2 - 4.932x$

$= x^2 - 4.932x + 6.081 - 6.081$

$= (x - 2.466)^2 - 6.081$

The vertex is (2.466, -6.081). The coefficient of x^2 is positive. Thus the parabola opens upward and has a minimum value of -6.081 when x = 2.466.

If x = 2.466, then y = 2.466 - 4.932, or -2.466.

The numbers are 2.466 and -2.466. The minimum product is 2.466(-2.466), or -6.081.

36. Maximum, 114.009; Both 10.678

Exercise Set 4.4

1. a) E is a linear function of t. The ordered pairs are in the form (t, E). To find the equation, we use the two known ordered pairs (0, 72) and (20, 75) and the two-point equation.

$$E - E_1 = \frac{E_2 - E_1}{t_2 - t_1}(t - t_1)$$

$$E - 72 = \frac{75 - 72}{20 - 0}(t - 0) \quad \text{(Substituting)}$$

$$E - 72 = \frac{3}{20}t$$

$$E(t) = 0.15t + 72$$

b) In 1990, t = 1990 - 1950 = 40.

$E(t) = 0.15t + 72$

$E(40) = 0.15(40) + 72$ (Substituting)

$= 6 + 72$

$= 78$

The life expectancy of females in 1990 will be 78.

In 1995, t = 1995 - 1950 = 45.

$E(t) = 0.15t + 72$

$E(45) = 0.15(45) + 72$ (Substituting)

$= 6.75 + 72$

$= 78.75$

The life expectancy of females in 1995 will be 78.75.

2. a) $E(t) = 0.15t + 65$, b) 1990: 71, 1985: 71.75

3. a) R is a linear function of t. The ordered pairs are in the form (t, R). To find the equation, we use the two known ordered pairs (0, 10.43) and (50, 9.93) and the two-point equation.

$$R - R_1 = \frac{R_2 - R_1}{t_2 - t_1}(t - t_1)$$

$$R - 10.43 = \frac{9.93 - 10.43}{50 - 0}(t - 0)$$

(Substituting)

$$R - 10.43 = \frac{-0.5}{50}t$$

$$R(t) = -0.01t + 10.43$$

b) In 1990, t = 1990 - 1920 = 70.

$R(t) = -0.01t + 10.43$

$R(70) = -0.01(70) + 10.43$ (Substituting)

$= -0.7 + 10.43$

$= 9.73$

The record in 1990 will be 9.73 sec.

In 1992, t = 1992 - 1920 = 72.

$R(t) = -0.01t + 10.43$

$R(72) = -0.01(72) + 10.43$

$= -0.72 + 10.43$

$= 9.71$

The record in 1992 will be 9.71 sec.

c) $R(t) = -0.01t + 10.43$

$9.0 = -0.01t + 10.43$

[Substituting 9.0 for R(t)]

$0.01t = 1.43$

$$t = \frac{1.43}{0.01} = 143$$

In the year 2063 (1920 + 143), the record will be 9.0 sec.

4. a) $D(t) = 0.2t + 19$,

b) 1992: 27.4 quadrillion BTU

2000: 29 quadrillion BTU

5. a) C(x) = Fixed costs + Variable costs

$C(x) = 22,500 + 40x$

b) $R(x) = 85x$

c) $P(x) = R(x) - C(x)$

$P(x) = 85x - (22,500 + 40x)$

$= 45x - 22,500$

d) $P(x) = 45x - 22,500$

$P(3000) = 45(3000) - 22,500$

$= 135,000 - 22,500$

$= 112,500$

A profit of \$112,500 will be realized if 3000 pairs of skis are sold.

5. (continued)

e) At the break-even point $P(x) = 0$.

$$P(x) = 45x - 22{,}500$$
$$0 = 45x - 22{,}500$$
$$22{,}500 = 45x$$
$$500 = x$$

The break-even value of x is 500.

f) $x > 500$

g) $x < 500$

6. a) $C(x) = 20x + 10{,}000$
b) $R(x) = 100x$
c) $P(x) = 80x - 10{,}000$
d) $x = 125$
e) $x < 125$
f) $x > 125$

7. $s = -4.9t^2 + v_0t + h$

$s = -4.9t^2 + 147t + 560$
(Substituting 147 for v_0 and 560 for h)

a) We now complete the square.

$$s = -4.9(t^2 - 30t) + 560$$
$$= -4.9(t^2 - 30t + 225 - 225) + 560$$
$$= -4.9(t^2 - 30t + 225) + 1102.5 + 560$$
$$= -4.9(t - 15)^2 + 1662.5$$

The vertex of the graph is the point (15, 1662.5). The maximum height reached is 1662.5 m and it is attained 15 sec after the end of the burn.

b) To find when the rocket reaches the ground, we set s = 0 in our equation and solve for t.

$$0 = -4.9(t - 15)^2 + 1662.5$$
$$4.9(t - 15)^2 = 1662.5$$
$$(t - 15)^2 = \frac{1662.5}{4.9}$$
$$t - 15 = \sqrt{\frac{1662.5}{4.9}} \approx 18.4$$
$$t \approx 33.4$$

The rocket will reach the ground about 33.4 sec after the end of the burn.

8. a) 4302.5 m after 25 sec
b) After 54.6 sec

9. 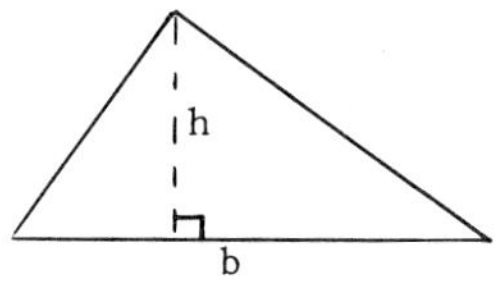

$b + h = 20$
$b = 20 - h$

$$A = \frac{1}{2}bh$$
$$A = \frac{1}{2}(20 - h)h \quad \text{(Substituting 20 - h for b)}$$
$$= \frac{1}{2}(20h - h^2)$$
$$= -\frac{1}{2}h^2 + 10h$$
$$= -\frac{1}{2}(h^2 - 20h)$$
$$= -\frac{1}{2}(h^2 - 20h + 100 - 100) \quad \text{(Adding 100 - 100)}$$
$$= -\frac{1}{2}(h^2 - 20h + 100) + 50$$
$$= -\frac{1}{2}(h - 10)^2 + 50$$

The function has a maximum output of 50 when h = 10. If h = 10, then b = 20 - 10, or 10. The triangle has a maximum area of 50 when the base is 10 and the height is 10.

10. b = 7, h = 7

11. Let x = the number of additional trees per acre which should be planted. Then the number of trees planted per acre is represented by (20 + x) and the yield per tree by (40 - x).

Total yield per acre = (Yield per tree)·(Number of trees per acre)

$$Y(x) = (40 - x)\cdot(20 + x)$$
$$= 800 + 20x - x^2$$

We first complete the square.

$$Y(x) = -x^2 + 20x + 800$$
$$= -(x^2 - 20x) + 800$$
$$= -(x^2 - 20x + 100 - 100) + 800$$
$$= -(x^2 - 20x + 100) + 100 + 800$$
$$= -(x - 10)^2 + 900$$

The function has a maximum output of 900 when x = 10. Thus, 20 + 10, or 30 trees per acre should be planted for maximum yield.

12. \$6

13. Let x = the number of 20¢ decreases (if x is negative, the price would be increased).

The attendance is increased by 30 for each decrease of 20¢ in the price. If x represents the number of 20¢ decreases, the new average attendance can be expressed as

$(500 + 30x)$

and the new admission price can be expressed as

$(4 - 0.20x)$.

Note that if 4 is in dollars, then 20¢ must be in dollars, 0.20.

Now express the revenue R as a function of x and complete the square.

$R(x) = (\text{Number of people})\cdot(\text{Ticket price})$

$R(x) = (500 + 30x)(4 - 0.20x)$

$= 2000 - 100x + 120x - 6x^2$ (FOIL)

$= -6x^2 + 20x + 2000$

$= -6\left(x^2 - \frac{10}{3}x\right) + 2000$

$= -6\left(x^2 - \frac{10}{3}x + \frac{25}{9} - \frac{25}{9}\right) + 2000$

(Completing the square)

$= -6\left(x^2 - \frac{10}{3}x + \frac{25}{9}\right) + \frac{50}{3} + 2000$

$= -6\left(x - \frac{5}{3}\right)^2 + \frac{6050}{3}$

The revenue function has a maximum output of $\frac{6050}{3}$ when x is $\frac{5}{3}$. This means that the maximum revenue will be about \$2016.67 when there is a $\frac{5}{3}$ decrease of 20¢ in the admission price. Since $\frac{5}{3}\cdot 20 = \frac{100}{3} \approx 33¢$, the admission price of \$4 should be decreased 33¢ to achieve the maximum revenue.

The new admission price is \$4 - \$0.33, or \$3.67.

14. 300 m^2

Exercise Set 4.5

1. The intersection of two sets consists of those element common to the sets.

$\{3, 4, 5, 8, 10\} \cap \{1, 2, 3, 4, 5, 6, 7\}$

$= \{3, 4, 5\}$

Only the elements 3, 4, and 5 are in both sets.

2. {1, 2, 3, 4, 5, 6, 7, 8, 10}

3. The union of two sets consists of the elements that are in one or both of the sets.

$\{0, 2, 4, 6, 8\} \cup \{4, 6, 9\} = \{0, 2, 4, 6, 8, 9\}$

Note that the elements 4 and 6 are in both sets, but each is listed only once in the union.

4. {4, 6}

5. $\{a, b, c\} \cap \{c, d\} = \{c\}$

Since the element c is the only element which is in both sets, the intersection only contains c.

6. {a, b, c, d}

7. Graph: $\{x|7 \le x\} \cup \{x|x < 9\}$

Graph the solution sets separately, and then find the union.

Graph: $\{x|7 \le x\}$ $(7 \le x \text{ means } x \ge 7)$

The solid circle at 7 indicates that 7 is in the solution set.

Graph: $\{x|x < 9\}$

The open circle at 9 indicates that 9 is not in the solution set.

Now graph the union:

$\{x|7 \le x\} \cup \{x|x < 9\}$ = The set of all real numbers

8. 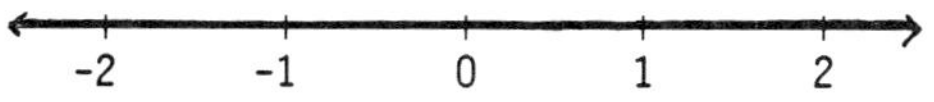

9. Graph: $\left\{x\middle|-\frac{1}{2} \le x\right\} \cap \left\{x\middle|x < \frac{1}{2}\right\}$

Graph the solution sets separately, and then find the intersection.

Graph: $\left\{x\middle|-\frac{1}{2} \le x\right\}$ $\left(-\frac{1}{2} \le x \text{ means } x \ge -\frac{1}{2}\right)$

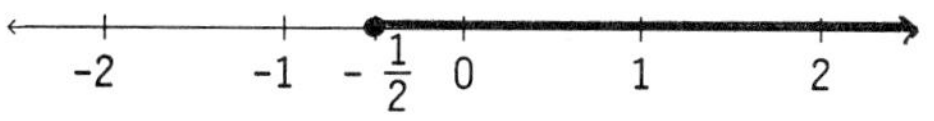

The solid circle indicates that $-\frac{1}{2}$ is in the solution set.

Graph: $\left\{x\middle|x < \frac{1}{2}\right\}$

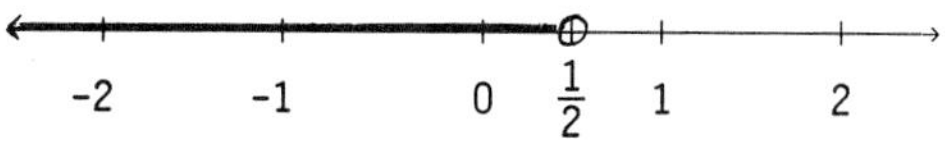

The open circle at $\frac{1}{2}$ indicates that $\frac{1}{2}$ is not in the solution set.

Now graph the intersection:

$\left\{x\middle|-\frac{1}{2} \le x\right\} \cap \left\{x\middle|x < \frac{1}{2}\right\} = \left\{x\middle|-\frac{1}{2} \le x < \frac{1}{2}\right\}$

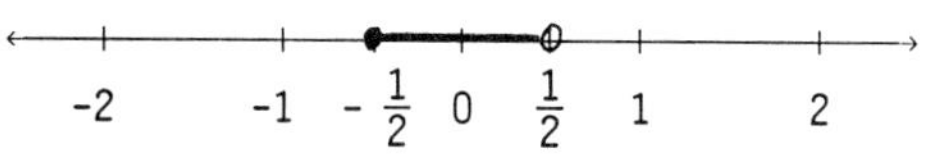

10. 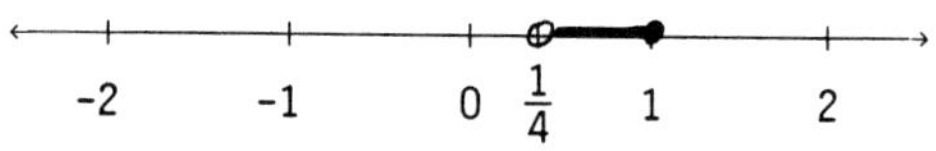

11. Graph: $\{x|x < -\pi\} \cup \{x|x > \pi\}$

Graph the solution sets separately, and then find the union.

Graph: $\{x|x < -\pi\}$

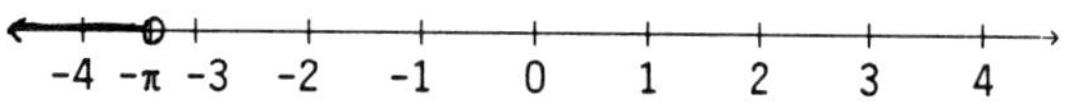

The open circle at $-\pi$ indicates that $-\pi$ is not in the solution set.

Graph: $\{x|x > \pi\}$

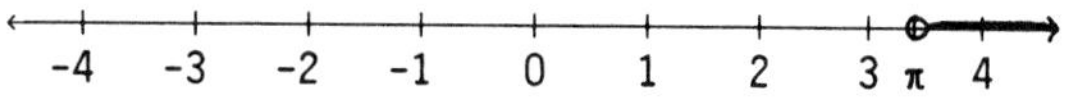

The open circle at π indicates that π is not in the solution set.

Now graph the union:

$\{x|x < -\pi\} \cup \{x|x > \pi\} = \{x|x < -\pi \text{ or } x > \pi\}$

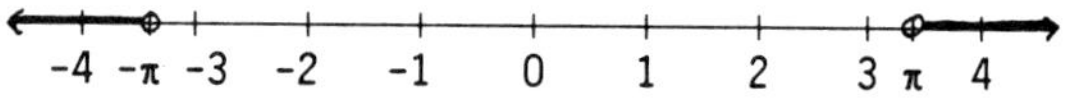

12. 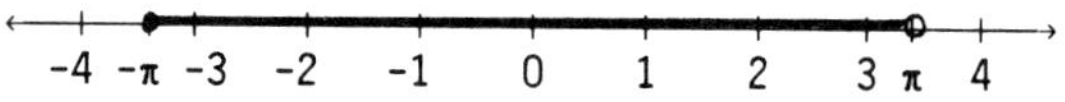

13. Graph: $\{x|x < -7\} \cup \{x|x = -7\}$

Graph the solution sets separately, and then find the union.

Graph: $\{x|x < -7\}$

The open circle at -7 indicates that -7 is not in the solution set.

Graph: $\{x|x = -7\}$

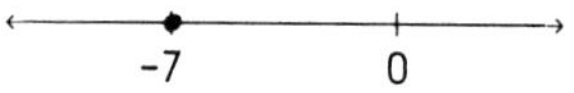

The solid circle at -7 indicates that -7 is in the solution set.

Now graph the union:

$\{x|x < -7\} \cup \{x|x = -7\} = \{x|x \leqslant -7\}$

14. 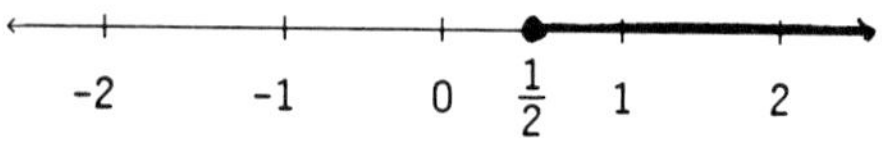

15. Graph: $\{x|x \geqslant 5\} \cap \{x|x \leqslant -3\}$

Graph the solution sets separately, and then find the intersection.

Graph: $\{x|x \geqslant 5\}$

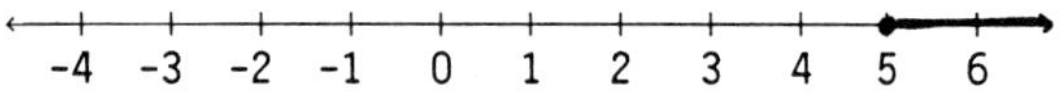

The solid circle at 5 indicates that 5 is in the solution set.

15. (continued)

Graph: $\{x|x \leqslant -3\}$

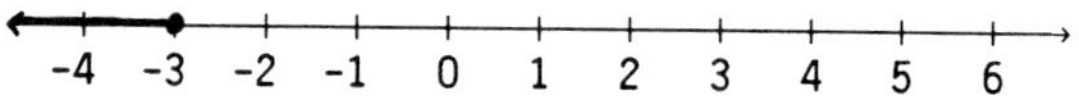

The solid circle at -3 indicates that -3 is in the solution set.

There are no elements in the intersection.

$\{x|x \geqslant 5\} \cap \{x|x \leqslant -3\} = \emptyset$

16. 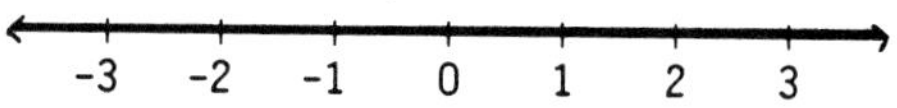

17. $-2 \leqslant x + 1 < 4$

$-3 \leqslant x < 3$ (Adding -1)

The solution set is $\{x|-3 \leqslant x < 3\}$.

18. $\{x|-5 < x \leqslant 3\}$

19. $5 \leqslant x - 3 \leqslant 7$

$8 \leqslant x \leqslant 10$ (Adding 3)

The solution set is $\{x|8 \leqslant x \leqslant 10\}$.

20. $\{x|3 < x < 11\}$

21. $-3 \leqslant x + 4 \leqslant -3$

$-7 \leqslant x \leqslant -7$

The solution set is $\{-7\}$.

22. $\emptyset$

23. $-2 < 2x + 1 < 5$

$-3 < 2x < 4$ (Adding -1)

$-\frac{3}{2} < x < 2$ $\left(\text{Multiplying by } \frac{1}{2}\right)$

The solution set is $\left\{x \middle| -\frac{3}{2} < x < 2\right\}$.

24. $\left\{x \middle| -\frac{4}{5} \leqslant x \leqslant \frac{2}{5}\right\}$

25. $-4 \leqslant 6 - 2x < 4$

$-10 \leqslant -2x < -2$ (Adding -6)

$5 \geqslant x > 1$ $\left(\text{Multiplying by } -\frac{1}{2}\right)$

or $1 < x \leqslant 5$

The solution set is $\{x|1 < x \leqslant 5\}$.

26. $\{x|-1 \leqslant x < 2\}$

27. $-5 < \frac{1}{2}(3x + 1) \leqslant 7$

$-10 < 3x + 1 \leqslant 14$ (Multiplying by 2)

$-11 < 3x \leqslant 13$ (Adding -1)

$-\frac{11}{3} < x \leqslant \frac{13}{3}$ $\left[\text{Multiplying by } \frac{1}{3}\right]$

The solution set is $\left\{x \middle| -\frac{11}{3} < x \leqslant \frac{13}{3}\right\}$.

28. $\left\{x \middle| \frac{7}{4} < x \leqslant \frac{13}{6}\right\}$.

29. $3x \leqslant -6$ or $x - 1 > 0$

$x \leqslant -2$ or $x > 1$

The solution set is $\{x | x \leqslant -2 \text{ or } x > 1\}$.

30. The set of real numbers

31. $2x + 3 \leqslant -4$ or $2x + 3 \geqslant 4$

$2x \leqslant -7$ or $2x \geqslant 1$

$x \leqslant -\frac{7}{2}$ or $x \geqslant \frac{1}{2}$

The solution set is $\left\{x \middle| x \leqslant -\frac{7}{2} \text{ or } x \geqslant \frac{1}{2}\right\}$.

32. $\left\{x \middle| x < -\frac{4}{3} \text{ or } x > 2\right\}$

33. $2x - 20 < -0.8$ or $2x - 20 > 0.8$

$2x < 19.2$ or $2x > 20.8$

$x < 9.6$ or $x > 10.4$

The solution set is $\{x | x < 9.6 \text{ or } x > 10.4\}$.

34. $\left\{x \middle| x \leqslant -3 \text{ or } x \geqslant -\frac{7}{5}\right\}$

35. $x + 14 \leqslant -\frac{1}{4}$ or $x + 14 \geqslant \frac{1}{4}$

$x \leqslant -\frac{57}{4}$ or $x \geqslant -\frac{55}{4}$

The solution set is $\left\{x \middle| x \leqslant -\frac{57}{4} \text{ or } x \geqslant -\frac{55}{4}\right\}$.

36. $\left\{x \middle| x < \frac{17}{2} \text{ or } x > \frac{19}{2}\right\}$

37. $x \leqslant 3x - 2 \leqslant 2 - x$

$x \leqslant 3x - 2$ and $3x - 2 \leqslant 2 - x$

$-2x \leqslant -2$ and $4x \leqslant 4$

$x \geqslant 1$ and $x \leqslant 1$

The word "and" corresponds to set intersection.

The solution set is

$\{x | x \geqslant 1\} \cap \{x | x \leqslant 1\}$, or $\{1\}$.

38. $\left\{x \middle| -\frac{1}{4} < x \leqslant \frac{5}{9}\right\}$

39. $(x + 1)^2 > x(x - 3)$

$x^2 + 2x + 1 > x^2 - 3x$

$2x + 1 > -3x$

$5x > -1$

$x > -\frac{1}{5}$

The solution set is $\left\{x \middle| x > -\frac{1}{5}\right\}$

40. $\left\{x \middle| x > \frac{13}{5}\right\}$

41. $(x + 1)^2 \leqslant (x + 2)^2 \leqslant (x + 3)^2$

$x^2 + 2x + 1 \leqslant x^2 + 4x + 4 \leqslant x^2 + 6x + 9$

$2x + 1 \leqslant 4x + 4 \leqslant 6x + 9$

$2x + 1 \leqslant 4x + 4$ and $4x + 4 \leqslant 6x + 9$

$-2x \leqslant 3$ and $-2x \leqslant 5$

$x \geqslant -\frac{3}{2}$ and $x \geqslant -\frac{5}{2}$

The word "and" corresponds to set intersection.

The solution set is

$\left\{x \middle| x \geqslant -\frac{3}{2}\right\} \cap \left\{x \middle| x \geqslant -\frac{5}{2}\right\}$, or $\left\{x \middle| x \geqslant -\frac{3}{2}\right\}$.

42. $\{x | -1 < x \leqslant 1\}$

43. We first make a drawing.

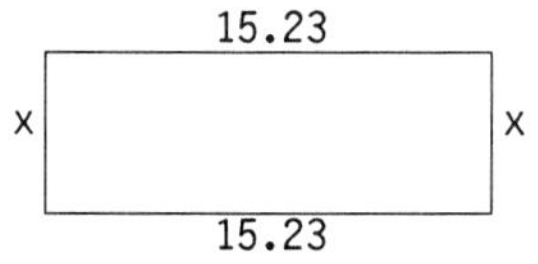

Let x represent the width. Then $2x + 2(15.23)$, or $2x + 30.46$, represents the perimeter. The perimeter is greater than 40.23 cm and less than 137.8 cm. We now have the following inequality.

$40.23 < 2x + 30.46 < 137.8$

$9.77 < 2x < 107.34$ (Adding -30.46)

$4.885 < x < 53.67$ $\left[\text{Multiplying by } \frac{1}{2}\right]$

Widths greater than 4.885 cm and less than 53.67 cm will give a perimeter greater than 40.23 cm and less than 137.8 cm.

44. $\{b | 0 \text{ m} < b \leqslant 40.72 \text{ m}\}$

45. Let x represent the score on the fourth test.

The average of the four scores is $\frac{83 + 87 + 93 + x}{4}$. This average must be greater than or equal to 90. We solve the following inequality.

$$\frac{83 + 87 + 93 + x}{4} \geqslant 90$$

$$83 + 87 + 93 + x \geqslant 360$$

$$263 + x \geqslant 360$$

$$x \geqslant 97$$

The score on the fourth test must be greater than or equal to 97. It will of course be less than or equal to 100%. Yes, an A is possible.

46. $\{x \mid 132\% \leqslant x \leqslant 172\%\}$; No

47. $f(x) = \frac{\sqrt{x + 2}}{\sqrt{x - 2}}$

The radicand in the denominator must be greater than 0. Thus, $x - 2 > 0$, or $x > 2$. Also the radicand in the numerator must be greater than or equal to 0. Thus, $x + 2 \geqslant 0$, or $x \geqslant -2$. The domain of f(x) is the intersection of the two solution sets.

$\{x \mid x > 2\} \cap \{x \mid x \geqslant -2\} = \{x \mid x > 2\}$.

The domain of f(x) is $\{x \mid x > 2\}$.

48. $\{x \mid -5 < x \leqslant 3\}$

Exercise Set 4.6

1. $|x| = 7$

To solve we look for all numbers x whose distance from 0 is 7. There are two of them, so there are two solutions, -7 and 7. The graph is as follows.

$\{-7, 7\}$

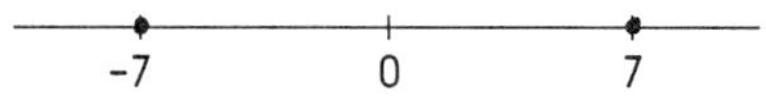

2. $\{-\pi, \pi\}$

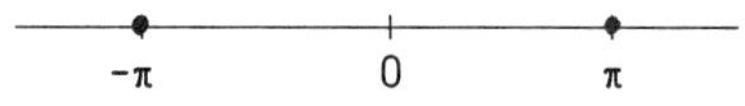

3. $|x| < 7$

To solve we look for all numbers x whose distance from 0 is less than 7. These are the numbers between -7 and 7. The solution set and its graph are as follows.

$\{x \mid -7 < x < 7\}$

4. $\{x \mid -\pi \leqslant x \leqslant \pi\}$

5. $|x| \geqslant \pi$

To solve we look for all numbers x whose distance from 0 is greater than or equal to π. The solution set and its graph are as follows.

$\{x \mid x \leqslant -\pi \text{ or } x \geqslant \pi\}$

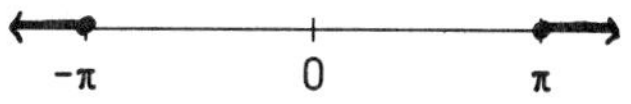

6. $\{x \mid x < -7 \text{ or } x > 7\}$

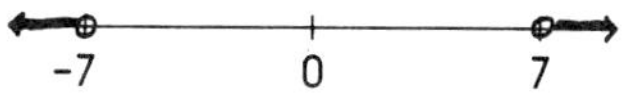

7. $|x - 1| = 4$

Method 1:

We translate the graph of $|x| = 4$ to the right 1 unit.

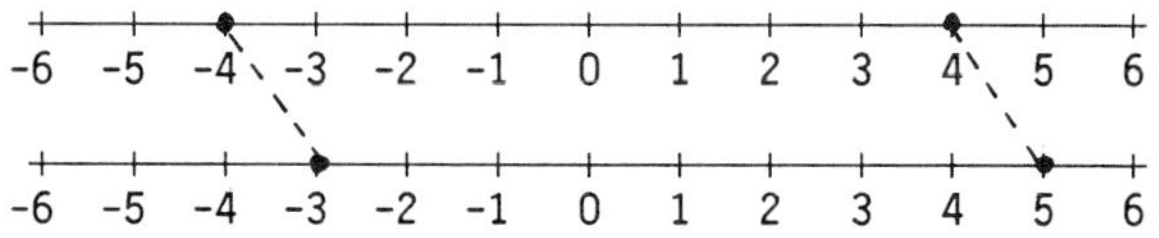

The solution set is { -3, 5}

Method 2:

The solutions are those numbers x whose distance from 1 is 4. Thus to find the solutions graphically we locate 1. Then we locate those numbers 4 units to the left and 4 units to the right.

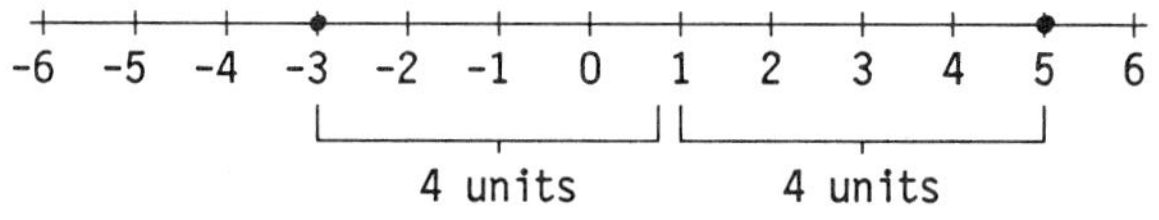

The solution set is {-3, 5}.

Method 3:

$|x - 1| = 4$

$x - 1 = -4$ or $x - 1 = 4$ (Property i)

$x = -3$ or $x = 5$ (Adding 1)

The solution set is {-3, 5}.

8. {2, 12}

9. $|x + 8| < 9$, or $|x - (-8)| < 9$

Method 1:

We translate the graph of $|x| < 9$ to the left 8 units.

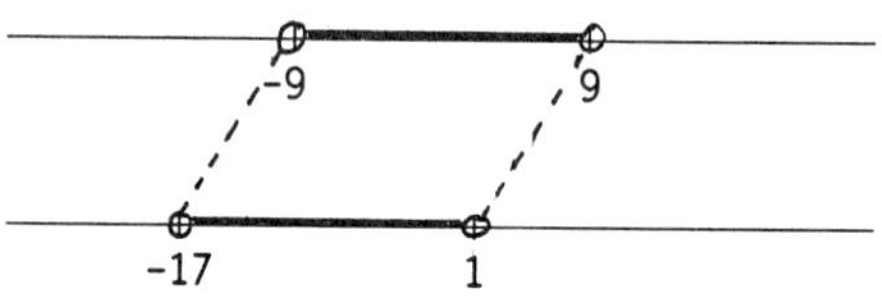

The solution set is $\{x \mid -17 < x < 1\}$.

9. (continued)

Method 2:

The solutions are those numbers x whose distance from -8 is less than 9. Thus to find the solutions graphically we locate -8. Then we locate those numbers that are less than 9 units to the left and less than 9 units to the right.

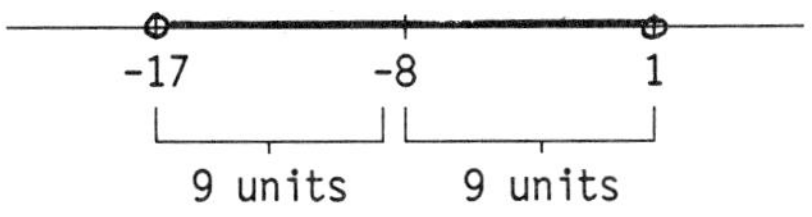

The solution set is $\{x \mid -17 < x < 1\}$.

Method 3:

$|x + 8| < 9$

$-9 < x + 8 < 9$ (Property ii)

$-17 < x < 1$ (Adding -8)

The solution set is $\{x \mid -17 < x < 1\}$.

10. $\{x \mid -16 \leqslant x \leqslant 4\}$

11. $|x + 8| \geqslant 9$, or $|x - (-8)| \geqslant 9$

Method 1:

We translate the graph of $|x| \geqslant 9$ to the left 8 units.

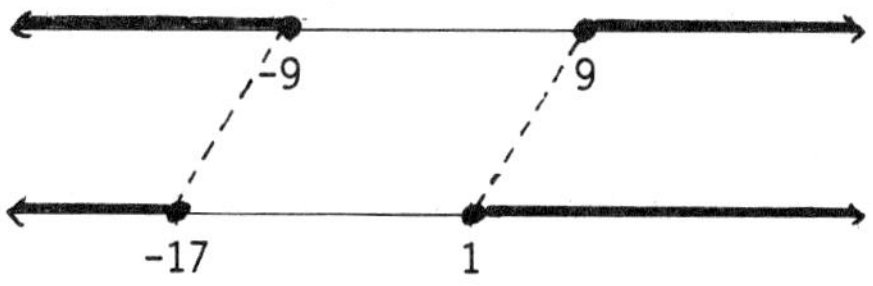

The solution set is $\{x \mid x \leqslant -17 \text{ or } x \geqslant 1\}$.

Method 2:

The solutions are those numbers x whose distance from -8 is greater than or equal to 9. Thus to find the solutions graphically we locate -8. Then we locate those numbers that are greater than or equal to 9 units to the left and greater than or equal to 9 units to the right.

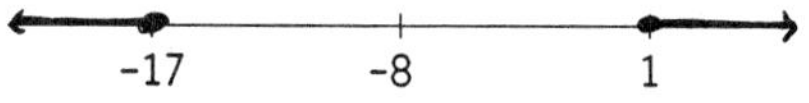

The solution set is $\{x \mid x \leqslant -17 \text{ or } x \geqslant 1\}$.

Method 3:

$|x + 8| \geqslant 9$

$x + 8 \leqslant -9$ or $x + 8 \geqslant 9$ (Property iii)

$x \leqslant -17$ or $x \geqslant 1$ (Adding -8)

The solution set is $\{x \mid x \leqslant -17 \text{ or } x \geqslant 1\}$.

12. $\{x \mid x < -16 \text{ or } x > 4\}$

13. $\left|x - \frac{1}{4}\right| < \frac{1}{2}$

$-\frac{1}{2} < x - \frac{1}{4} < \frac{1}{2}$ (Property ii)

$-\frac{1}{4} < x < \frac{3}{4}$ $\left(\text{Adding } \frac{1}{4}\right)$

The solution set is $\left\{x \middle| -\frac{1}{4} < x < \frac{3}{4}\right\}$.

14. $\{x \mid 0.3 \leqslant x \leqslant 0.7\}$

15. $|3x| = 1$

$3x = -1$ or $3x = 1$ (Property i)

$x = -\frac{1}{3}$ or $x = \frac{1}{3}$ $\left(\text{Multiplying by } \frac{1}{3}\right)$

The solution set is $\left\{-\frac{1}{3}, \frac{1}{3}\right\}$.

16. $\left\{-\frac{4}{5}, \frac{4}{5}\right\}$

17. $|3x + 2| = 1$

$3x + 2 = -1$ or $3x + 2 = 1$ (Property i)

$3x = -3$ or $3x = -1$ (Adding -2)

$x = -1$ or $x = -\frac{1}{3}$ $\left(\text{Multiplying by } \frac{1}{3}\right)$

The solution set is $\left\{-1, -\frac{1}{3}\right\}$.

18. $\left\{-\frac{4}{7}, \frac{12}{7}\right\}$

19. $|3x| < 1$

$-1 < 3x < 1$ (Property ii)

$-\frac{1}{3} < x < \frac{1}{3}$ $\left(\text{Multiplying by } \frac{1}{3}\right)$

The solution set is $\left\{x \middle| -\frac{1}{3} < x < \frac{1}{3}\right\}$.

20. $\left\{x \middle| -\frac{4}{5} \leqslant x \leqslant \frac{4}{5}\right\}$

21. $|2x + 3| \leqslant 9$

$-9 \leqslant 2x + 3 \leqslant 9$ (Property ii)

$-12 \leqslant 2x \leqslant 6$ (Adding -3)

$-6 \leqslant x \leqslant 3$ $\left(\text{Multiplying by } \frac{1}{2}\right)$

The solution set is $\{x \mid -6 \leqslant x \leqslant 3\}$.

22. $\{x \mid -8 < x < 5\}$

23. $|x - 5| > 0.1$

$x - 5 < -0.1$ or $x - 5 > 0.1$ (Property iii)

$x < 4.9$ or $x > 5.1$ (Adding 5)

The solution set is $\{x | x < 4.9 \text{ or } x > 5.1\}$.

24. $\{x | x \leqslant 6.6 \text{ or } x \geqslant 7.4\}$

25. $\left|x + \frac{2}{3}\right| \leqslant \frac{5}{3}$

$-\frac{5}{3} \leqslant x + \frac{2}{3} \leqslant \frac{5}{3}$ (Property ii)

$-\frac{7}{3} \leqslant x \leqslant \frac{3}{3}$ $\left(\text{Adding } -\frac{2}{3}\right)$

The solution set is $\left\{x \middle| -\frac{7}{3} \leqslant x \leqslant 1\right\}$.

26. $\left\{x \middle| -1 < x < -\frac{1}{2}\right\}$

27. $|6 - 4x| \leqslant 8$

$-8 \leqslant 6 - 4x \leqslant 8$ (Property ii)

$-14 \leqslant -4x \leqslant 2$ (Adding -6)

$\frac{14}{4} \geqslant x \geqslant -\frac{2}{4}$ $\left(\text{Multiplying by } -\frac{1}{4}\right)$

$\frac{7}{2} \geqslant x \geqslant -\frac{1}{2}$

The solution set is $\left\{x \middle| -\frac{1}{2} \leqslant x \leqslant \frac{7}{2}\right\}$.

28. $\left\{x \middle| x > \frac{15}{2} \text{ or } x < -\frac{5}{2}\right\}$

29. $\left|\frac{2x + 1}{3}\right| > 5$

$\frac{2x + 1}{3} < -5$ or $\frac{2x + 1}{3} > 5$ (Property iii)

$2x + 1 < -15$ or $2x + 1 > 15$ (Multiplying by 3)

$2x < -16$ or $2x > 14$ (Adding -1)

$x < -8$ or $x > 7$ $\left(\text{Multiplying by } \frac{1}{2}\right)$

The solution set is $\{x | x < -8 \text{ or } x > 7\}$.

30. $\left\{x \middle| -\frac{22}{3} \leqslant x \leqslant 6\right\}$

31. $\left|\frac{13}{4} + 2x\right| > \frac{1}{4}$

$\frac{13}{4} + 2x < -\frac{1}{4}$ or $\frac{13}{4} + 2x > \frac{1}{4}$ (Property iii)

$2x < -\frac{14}{4}$ or $2x > -\frac{12}{4}$

$2x < -\frac{7}{2}$ or $2x > -3$

$x < -\frac{7}{4}$ or $x > -\frac{3}{2}$

The solution set is $\left\{x \middle| x < -\frac{7}{4} \text{ or } x > -\frac{3}{2}\right\}$.

32. $\left\{x \middle| -\frac{2}{3} < x < \frac{1}{9}\right\}$

33. $\left|\frac{3 - 4x}{2}\right| \leqslant \frac{3}{4}$

$-\frac{3}{4} \leqslant \frac{3 - 4x}{2} \leqslant \frac{3}{4}$ (Property ii)

$-3 \leqslant 6 - 8x \leqslant 3$ (Multiplying by 4)

$-9 \leqslant -8x \leqslant -3$ (Adding -6)

$\frac{9}{8} \geqslant x \geqslant \frac{3}{8}$ $\left(\text{Multiplying by } -\frac{1}{8}\right)$

The solution set is $\left\{x \middle| \frac{3}{8} \leqslant x \leqslant \frac{9}{8}\right\}$.

34. $\left\{x \middle| x \leqslant -\frac{3}{4} \text{ or } x \geqslant \frac{7}{4}\right\}$

35. $|x| = -3$

Since $|x| \geqslant 0$ for all x, there is no x such that $|x|$ would be negative. There is no solution.

36. $\emptyset$

37. $|2x - 4| < -5$

Since $|2x - 4| \geqslant 0$ for all x, there is no x such that $|2x - 4|$ would be less than -5. There is no solution.

38. $\emptyset$

39. $|x + 17.217| > 5.0012$

$x + 17.217 < -5.0012$ or $x + 17.217 > 5.0012$ (Property iii)

$x < -22.2182$ or $x > -12.2158$ (Adding -17.217)

The solution set is $\{x | x < -22.2182 \text{ or } x > -12.2158\}$.

40. $\{x | 1.9234 < x < 2.1256\}$

41. $|3.0147x - 8.9912| \leqslant 6.0243$

$-6.0243 \leqslant 3.0147x - 8.9912 \leqslant 6.0243$ (Property ii)

$2.9669 \leqslant 3.0147x \leqslant 15.0155$ (Adding 8.9912)

$0.9841 \leqslant x \leqslant 4.9808$ $\left(\text{Multiplying by } \frac{1}{3.0147}\right)$

The solution set is $\{x | 0.9841 \leqslant x \leqslant 4.9808\}$.

42. $\{x | x \geqslant 7.9277 \text{ or } x \leqslant -0.5746\}$

43. $|4x - 5| = x + 1$

$4x - 5 = -(x + 1)$ or $4x - 5 = x + 1$ (Property i)

$4x - 5 = -x - 1$ or $3x = 6$

$5x = 4$ or $x = 2$

$x = \frac{4}{5}$ or $x = 2$

The solution set is $\left\{\frac{4}{5}, 2\right\}$.

44. {-11, 5}

45. $\big||x| - 1\big| = 3$

$|x| - 1 = -3$ or $|x| - 1 = 3$ (Property i)

$|x| = -2$ or $|x| = 4$

The solution set of $|x| = -2$ is $\emptyset$.

The solution set of $|x| = 4$ is $\{-4, 4\}$.

46. The set of real numbers

47. $|x + 2| \leqslant |x - 5|$

Divide the set of reals into three intervals:

$x \geqslant 5$

$-2 \leqslant x < 5$

$x < -2$

Find the solution set of $|x + 2| \leqslant |x - 5|$ for each interval. Then take the union of the three solution sets.

If $x \geqslant 5$, then $|x + 2| = x + 2$ and $|x - 5| = x - 5$.

Solve: $x + 2 \leqslant x - 5$

$2 \leqslant -5$ (False)

The solution set for this interval is $\emptyset$.

If $-2 \leqslant x < 5$, $|x + 2| = x + 2$ and $|x - 5| = -(x - 5)$.

Solve: $x + 2 \leqslant -(x - 5)$

$x + 2 \leqslant -x + 5$

$2x \leqslant 3$

$x \leqslant \frac{3}{2}$

The solution set for this interval is $\left\{x \middle| x \leqslant \frac{3}{2}\right\}$.

If $x < -2$, then $|x + 2| = -(x + 2)$ and $|x - 5| = -(x - 5)$.

Solve: $-(x + 2) \leqslant -(x - 5)$

$-x - 2 \leqslant -x + 5$

$-2 \leqslant 5$ (True for any x such that $x < -2$)

The solution set for this interval is $\{x|x < -2\}$.

The <u>union</u> of the above three solution sets is $\left\{x \middle| x \leqslant \frac{3}{2}\right\}$. This set is the solution of $|x + 2| \leqslant |x - 5|$.

48. $\left\{x \middle| x < \frac{1}{2}\right\}$

49. $|x| + |x - 1| < 10$

Divide the set of reals into threee intervals:

$x \geqslant 1$

$0 \leqslant x < 1$

$x < 0$

Find the solution set of $|x| + |x - 1| < 10$ for each interval. Then take the union of the three solution sets.

If $x \geqslant 1$, then $|x| = x$ and $|x - 1| = x - 1$.

Solve: $x + (x - 1) < 10$

$2x - 1 < 10$

$2x < 11$

$x < \frac{11}{2}$

The solution set for his interval is $\left\{x \middle| 1 \leqslant x < \frac{11}{2}\right\}$.

If $0 \leqslant x < 1$, then $|x| = x$ and $|x - 1| = -(x - 1)$.

Solve: $x + [-(x - 1)] < 10$

$x - x + 1 < 10$

$1 < 10$ (True for any x such that $0 \leqslant x < 1$)

The solution set for this interval is $\{x|0 \leqslant x < 1\}$.

If $x < 0$, then $|x| = -x$ and $|x - 1| = -(x - 1)$.

Solve: $-x + [-(x - 1)] < 10$

$-x - x + 1 < 10$

$-2x + 1 < 10$

$-2x < 9$

$x > -\frac{9}{2}$

The solution set for this interval is $\left\{x \middle| -\frac{9}{2} < x < 0\right\}$.

The <u>union</u> of the above three solution sets is $\left\{x \middle| -\frac{9}{2} < x < \frac{11}{2}\right\}$. This set is the solution set of $|x| + |x - 1| < 10$.

50. The set of real numbers

Exercise Set 4.7

1. Solve: $x^2 - x - 2 < 0$

Consider the function $f(x) = x^2 - x - 2$. The inputs that produce outputs that are less than 0 are the solutions of the inequality.

Set $f(x) = 0$ and factor to find the x-intercepts.

$$x^2 - x - 2 = 0$$
$$(x + 1)(x - 2) = 0$$
$$x + 1 = 0 \quad \text{or} \quad x - 2 = 0$$
$$x = -1 \quad \text{or} \quad x = 2$$

The x-intercepts are (-1, 0) and (2, 0).

The table below lists a few ordered pairs that are also solutions of the function. Plot these points and draw the graph.

x	f(x)
-2	4
0	-2
1	-2
3	4
$\frac{1}{2}$	$-\frac{9}{4}$

$f(x) = x^2 - x - 2$

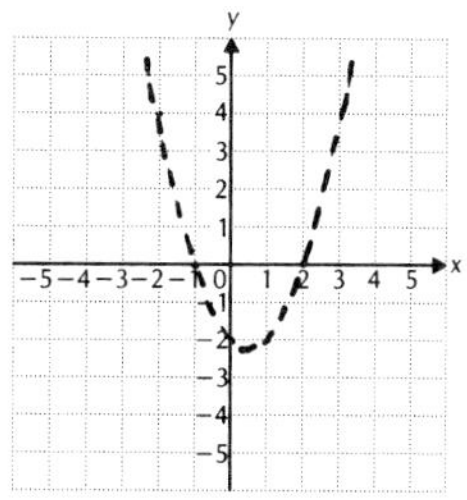

From the graph we can easily see that the solution set is $\{x \mid -1 < x < 2\}$.

2. $\{x \mid x < -2 \text{ or } x > 2\}$

3. Solve: $x^2 \geqslant 1$, or $x^2 - 1 \geqslant 0$

Consider the function $f(x) = x^2 - 1$. The inputs that produce outputs that are greater than or equal to 0 are the solutions of the inequality.

Set $f(x) = 0$ and factor to find the x-intercpets.

$$x^2 - 1 = 0$$
$$(x + 1)(x - 1) = 0$$
$$x + 1 = 0 \quad \text{or} \quad x - 1 = 0$$
$$x = -1 \quad \text{or} \quad x = 1$$

The x-intercepts are (-1, 0) and (1, 0).

The table below lists a few ordered pairs that are also solutions of the function. Plot these points and draw the graph.

x	f(x)
-2	3
0	-1
2	3

$f(x) = x^2 - 1$

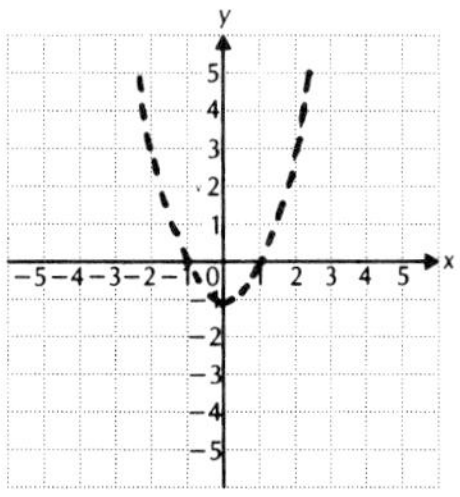

3. (continued)

From the graph we can easily see that the solution set is $\{x \mid x \leqslant -1 \text{ or } x \geqslant 1\}$.

4. $\{x \mid 1 - \sqrt{6} < x < 1 + \sqrt{6}\}$

5. Solve: $x^2 - 2x + 1 \geqslant 0$

Consider the function $f(x) = x^2 - 2x + 1$. The inputs that produce outputs that are greater than or equal to 0 are the solutions of the inequality.

Set $f(x) = 0$ and factor to find the x-intercepts.

$$x^2 - 2x + 1 = 0$$
$$(x - 1)(x - 1) = 0$$
$$x - 1 = 0 \quad \text{or} \quad x - 1 = 0$$
$$x = 1 \quad \text{or} \quad x = 1$$

The only x-intercept is (1, 0).

The table below lists a few ordered pairs that are also solutions of the function. Plot these points and draw the graph.

x	f(x)
-1	4
0	1
2	1
3	4

$f(x) = x^2 - 2x + 1$

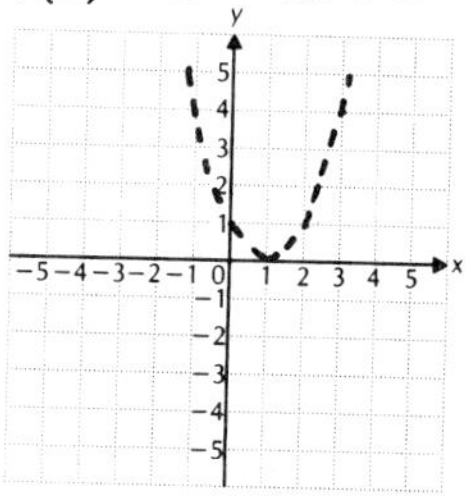

From the graph we can easily see that all values of x produce outputs greater than or equal to 0. Thus the solution set is the entire set of real numbers.

6. $\emptyset$

7. Solve: $x^2 < 6x - 4$, or $x^2 - 6x + 4 < 0$

Consider the function $f(x) = x^2 - 6x + 4$. The inputs that produce outputs that are less than 0 are the solutions of the inequality.

Set $f(x) = 0$ and use the quadratic formula to find the x-intercepts.

$$x^2 - 6x + 4 = 0$$
$$a = 1, \quad b = -6, \quad c = 4$$
$$x = \frac{-(-6) \pm \sqrt{(-6)^2 - 4\cdot1\cdot4}}{2\cdot1} = \frac{6 \pm \sqrt{36 - 16}}{2}$$
$$= \frac{6 \pm \sqrt{20}}{2} = \frac{6 \pm 2\sqrt{5}}{2} = 3 \pm \sqrt{5}$$

The x-intercepts are $(3 - \sqrt{5}, 0)$ and $(3 + \sqrt{5}, 0)$.

7. (continued)

The table below lists a few ordered pairs that are also solutions of the function. Plot these points and draw the graph.

x	f(x)
0	4
1	-1
2	-4
3	-5
4	-4
5	-1
6	4

$f(x) = x^2 - 6x + 4$

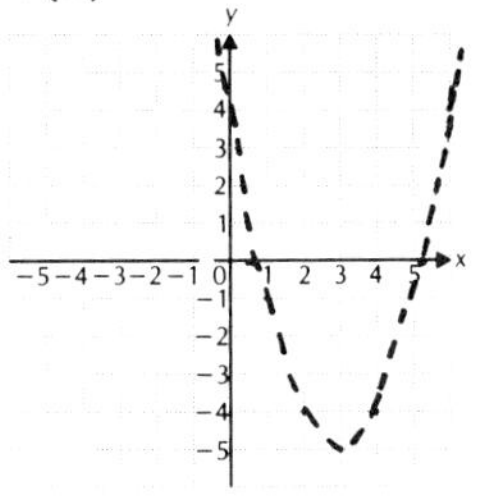

From the graph we can easily see that the solution set is

$\{x \mid 3 - \sqrt{5} < x < 3 + \sqrt{5}\}$.

8. $\left\{x \middle| \frac{5 - \sqrt{17}}{2} < x < \frac{5 + \sqrt{17}}{2}\right\}$

9. Solve: $x^2 - 8x - 20 \leqslant 0$

Consider the function $f(x) = x^2 - 8x - 20$. The inputs that produce outputs that are less than or equal to 0 are the solutions of the inequality.

Set $f(x) = 0$ and factor to find the x-intercepts.

$x^2 - 8x - 20 = 0$

$(x + 2)(x - 10) = 0$

$x + 2 = 0$ or $x - 10 = 0$

$x = -2$ or $x = 10$

The x-intercepts are (-2, 0) and (10, 0).

The graph of the function opens upwards. Function values will be nonpositive between and including its intercepts. The solution set is

$\{x \mid -2 \leqslant x \leqslant 10\}$.

10. $\{x \mid x \leqslant -2 \text{ or } x \geqslant 10\}$

11. Solve: $x^2 - 2x > 8$, or $x^2 - 2x - 8 > 0$

Consider the function $f(x) = x^2 - 2x - 8$. The inputs that produce outputs that are greater than 0 are the solutions of the inequality.

Set $f(x) = 0$ and factor to find the x-intercepts.

$x^2 - 2x - 8 = 0$

$(x + 2)(x - 4) = 0$

$x + 2 = 0$ or $x - 4 = 0$

$x = -2$ or $x = 4$

The x-intercepts are (-2, 0) and (4, 0).

The graph of the function opens upwards. Function values will be greater than 0 when x is less than -2 and when x is greater than 4. The solution set is

$\{x \mid x < -2 \text{ or } x > 4\}$.

12. $\{x \mid -2 < x < 4\}$

13. Solve: $4x^2 + 7x < 15$, or $4x^2 + 7x - 15 < 0$

Consider the function $f(x) = 4x^2 + 7x - 15$. The inputs that produce outputs that are less than 0 are the solutions of the inequality.

Set $f(x) = 0$ and factor to find the x-intercepts.

$4x^2 + 7x - 15 = 0$

$(x + 3)(4x - 5) = 0$

$x + 3 = 0$ or $4x - 5 = 0$

$x = -3$ or $4x = 5$

$x = -3$ or $x = \frac{5}{4}$

The x-intercepts are (-3, 0) and $\left(\frac{5}{4}, 0\right)$.

The graph of the function opens upwards. Function values will be less than 0 between its intercepts. The solution set is

$\left\{x \middle| -3 < x < \frac{5}{4}\right\}$.

14. $\left\{x \middle| x \leqslant -3 \text{ or } x \geqslant \frac{5}{4}\right\}$

15. Solve: $2x^2 + x > 5$, or $2x^2 + x - 5 > 0$

Consider the function $f(x) = 2x^2 + x - 5$. The inputs that produce outputs that are greater than 0 are the solutions of the inequality.

Set $f(x) = 0$ and use the quadratic formula to find the x-intercepts.

$2x^2 + x - 5 = 0$

$a = 2, \quad b = 1, \quad c = -5$

$$x = \frac{-1 \pm \sqrt{1^2 - 4(2)(-5)}}{2 \cdot 2} = \frac{-1 \pm \sqrt{1 + 40}}{4}$$

$$= \frac{-1 \pm \sqrt{41}}{4}$$

The x-intercepts are $\left(\frac{-1 - \sqrt{41}}{4}, 0\right)$ and $\left(\frac{-1 + \sqrt{41}}{4}, 0\right)$.

The graph of the function opens upwards. Function values will be greater than 0 when x is less than $\frac{-1 - \sqrt{41}}{4}$ and when x is greater than $\frac{-1 + \sqrt{41}}{4}$. The solution set is

$\left\{x \middle| x < \frac{-1 - \sqrt{41}}{4} \text{ or } x > \frac{-1 + \sqrt{41}}{4}\right\}$.

16. $\left\{x \middle| \frac{-1 - \sqrt{17}}{4} \leqslant x \leqslant \frac{-1 + \sqrt{17}}{4}\right\}$.

17. Solve: $5x(x + 1)(x - 1) > 0$

The solutions of $5x(x + 1)(x - 1) = 0$ are -1, 0, and 1. They divide the real-number line as pictured below.

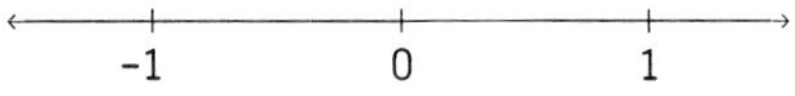

The product $5x(x + 1)(x - 1)$ is positive or negative depending on the signs of the factors $5x$, $x + 1$, and $x - 1$. We tabulate signs in these intervals.

Interval	5x	x + 1	x - 1	Product
$x < -1$	-	-	-	-
$-1 < x < 0$	-	+	-	+
$0 < x < 1$	+	+	-	-
$x > 1$	+	+	+	+

The product of three numbers is positive when it has an even number of negative factors. We see from the table that the solution set is

$\{x \mid -1 < x < 0 \text{ or } x > 1\}$.

18. $\{x \mid x < -2 \text{ or } 0 < x < 2\}$

19. Solve: $(x + 3)(x + 2)(x - 1) < 0$

The solutions of $(x + 3)(x + 2)(x - 1) = 0$ are -3, -2, and 1. They divide the real-number line as pictured below.

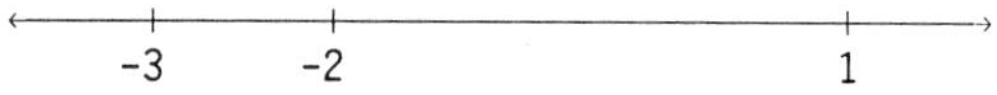

The product $(x + 3)(x + 2)(x - 1)$ is positive or negative depending on the signs of the factors $x + 3$, $x + 2$, and $x - 1$. We tabulate signs in these intervals.

Interval	x + 3	x + 2	x - 1	Product
$x < -3$	-	-	-	-
$-3 < x < -2$	+	-	-	+
$-2 < x < 1$	+	+	-	-
$x > 1$	+	+	+	+

The product of three numbers is negative when there is an odd number of negative factors. We see from the table that the solution set is

$\{x \mid x < -3 \text{ or } -2 < x < 1\}$.

20. $\{x \mid -1 < x < 2 \text{ or } x > 3\}$

21. $\frac{1}{4 - x} < 0$

The solution of $4 - x = 0$ is 4. It divides the number line in a natural way. The quotient is positive or negative depending on the sign of $4 - x$. We tabulate signs in these intervals.

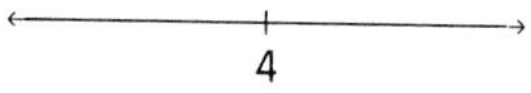

Interval	4 - x	Quotient
$x < 4$	+	+
$x > 4$	-	-

We see from the table that the solution set is

$\{x \mid x > 4\}$.

22. $\left\{x \middle| x < -\frac{5}{2}\right\}$

23. $\frac{3x + 2}{x - 3} > 0$

The solutions of $3x + 2 = 0$ and $x - 3 = 0$ are $-\frac{2}{3}$ and 3. They divide the number line in a natural way. The quotient is positive or negative depending on the signs of $3x + 2$ and $x - 3$. We tabulate signs in these intervals.

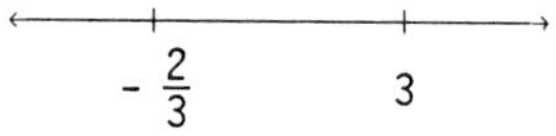

Interval	3x + 2	x - 3	Quotient
$x < -\frac{2}{3}$	-	-	+
$-\frac{2}{3} < x < 3$	+	-	-
$x > 3$	+	+	+

We see from the table that the solution set is

$\left\{x \middle| x < -\frac{2}{3} \text{ or } x > 3\right\}$.

24. $\left\{x \middle| x < -\frac{3}{4} \text{ or } x > \frac{5}{2}\right\}$

25.
$$\frac{x + 1}{2x - 3} \geqslant 1$$
$$\frac{x + 1}{2x - 3} - 1 \geqslant 0$$
$$\frac{x + 1}{2x + 3} - \frac{2x - 3}{2x - 3} \geqslant 0$$
$$\frac{x + 1 - 2x + 3}{2x - 3} \geqslant 0$$
$$\frac{-x + 4}{2x - 3} \geqslant 0$$

Let us consider the equality portion:

$$\frac{-x + 4}{2x - 3} = 0$$

25. (continued)

This has the solution 4. Thus 4 is in the solution set. Next consider the inequality portion:

$$\frac{-x + 4}{2x - 3} > 0$$

The solutions of $-x + 4 = 0$ and $2x - 3 = 0$ are 4 and $\frac{3}{2}$. They divide the number line in a natural way. The quotient is positive or negative depending on the signs of $-x + 4$ and $2x - 3$. We tabulate signs in these intervals.

Interval	$-x + 4$	$2x - 3$	Quotient
$x < \frac{3}{2}$	+	-	-
$\frac{3}{2} < x < 4$	+	+	+
$x > 4$	-	+	-

We see that from the table that the solution set of $\frac{-x + 4}{2x - 3} > 0$ is $\left\{x \middle| \frac{3}{2} < x < 4\right\}$. Thus the solution set of the inequality in question is $\left\{x \middle| \frac{3}{2} < x \leqslant 4\right\}$.

26. $\left\{x \middle| 2 < x \leqslant \frac{5}{2}\right\}$

27.
$$\frac{x + 1}{x + 2} \leqslant 3$$
$$\frac{x + 1}{x + 2} - 3 \leqslant 0$$
$$\frac{x + 1}{x + 2} - 3 \cdot \frac{x + 2}{x + 2} \leqslant 0$$
$$\frac{x + 1 - 3x - 6}{x + 2} \leqslant 0$$
$$\frac{-2x - 5}{x + 2} \leqslant 0$$

Let us consider the equality portion:

$$\frac{-2x - 5}{x + 2} = 0$$

This has the solution $-\frac{5}{2}$. Thus $-\frac{5}{2}$ is in the solution set. Next consider the inequality portion:

$$\frac{-2x - 5}{x + 2} < 0$$

27. (continued)

The solutions of $-2x - 5 = 0$ and $x + 2 = 0$ are $-\frac{5}{2}$ and -2. They divide the number line in a natural way. The quotient is positive or negative depending on the signs of $-2x - 5$ and $x + 2$. We tabulate signs in these intervals.

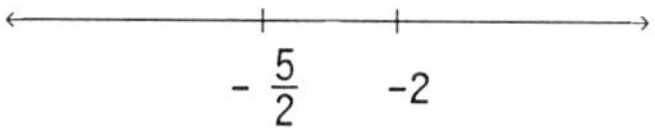

Interval	$-2x - 5$	$x + 2$	Quotient
$x < -\frac{5}{2}$	+	-	-
$-\frac{5}{2} < x < -2$	-	-	+
$x > -2$	-	+	-

We see from the table that the solution set of $\frac{-2x - 5}{x + 2} < 0$ is $\left\{x \middle| x < -\frac{5}{2} \text{ or } x > -2\right\}$. Thus the solution set of the inequality in question is $\left\{x \middle| x \leqslant -\frac{5}{2} \text{ or } x > -2\right\}$.

28. $\left\{x \middle| x < \frac{3}{2} \text{ or } x \geqslant 4\right\}$

29. $(x + 1)(x - 2) > (x + 3)^2$
$$x^2 - x - 2 > x^2 + 6x + 9$$
$$-x - 2 > 6x + 9$$
$$-11 > 7x$$
$$-\frac{11}{7} > x, \text{ or } x < -\frac{11}{7}$$

The solution set is $\left\{x \middle| x < -\frac{11}{7}\right\}$.

30. $\{x | x \leqslant 13\}$

31. $x^3 - x^2 > 0$
$x^2(x - 1) > 0$

The solutions of $x^2(x - 1) = 0$ are 0 and 1. They divide the number line in a natural way.

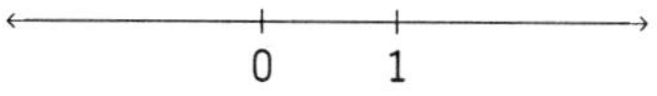

The product $x^2(x - 1)$ is positive or negative depending on the signs of the factors x^2 and $x - 1$. We tabulate signs in these intervals.

Interval	x^2	$x - 1$	Product
$x < 0$	+	-	-
$0 < x < 1$	+	-	-
$x > 1$	+	+	+

31. (continued)

The product of two numbers is positive when either both are positive or both are negative. We see from the table that the solution set is $\{x \mid x > 1\}$.

32. $\{x \mid x > 2 \text{ or } -2 < x < 0\}$

33.

$$x + \frac{4}{x} > 4$$

$$\frac{x^2 + 4}{x} > 4$$

$$\frac{x^2 + 4}{x} - 4 > 0$$

$$\frac{x^2 - 4x + 4}{x} > 0$$

The solutions of $x^2 - 4x + 4 = 0$ and $x = 0$ are 2 and 0. They divide the number line in a natural way. The quotient is positive or negative depending on the signs of $x^2 - 4x + 4$, or $(x - 2)^2$, and x. We tabulate signs in these intervals.

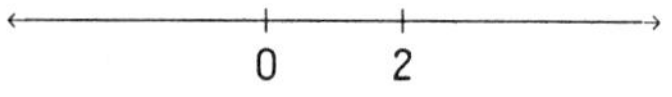

Interval	$(x - 2)^2$	x	Quotient
$x < 0$	+	-	-
$0 < x < 2$	+	+	+
$x > 2$	+	+	+

We see from the table that the solution set is $\{x \mid x > 0 \text{ and } x \neq 2\}$

34. $\{x \mid 0 < x \leqslant 1\}$

35.

$$\frac{1}{x^3} \leqslant \frac{1}{x^2}$$

$$\frac{1}{x^3} - \frac{1}{x^2} \leqslant 0$$

$$\frac{1 - x}{x^3} \leqslant 0$$

Let us consider the equality portion:

$$\frac{1 - x}{x^3} = 0$$

This has the solution 1. Thus 1 is in the solution set. Next consider the inequality portion:

$$\frac{1 - x}{x^3} < 0$$

The solutions of $1 - x = 0$ and $x^3 = 0$ are 1 and 0. They divide the real number line in a natural way. The quotient is positive or negative depending on the signs of $1 - x$ and x^3. We tabulate signs in these intervals.

35. (continued)

Interval	$1 - x$	x^3	Quotient
$x < 0$	+	-	-
$0 < x < 1$	+	+	+
$x > 1$	-	+	-

We see from the table that the solution set of $\frac{1 - x}{x^3} < 0$ is $\{x \mid x < 0 \text{ or } x > 1\}$. Thus the solution set of the inequality in question is $\{x \mid x < 0 \text{ or } x \geqslant 1\}$.

36. $\{x \mid x > 0 \text{ and } x \neq 1\}$

37.

$$\frac{2 + x - x^2}{x^2 + 5x + 6} < 0$$

$$\frac{(2 - x)(1 + x)}{(x + 3)(x + 2)} < 0$$

The solutions of $(2 - x)(1 + x) = 0$ are 2 and -1. The solutions of $(x + 3)(x + 2) = 0$ are -3 and -2. These solutions divide the number line in a natural way.

The quotient is positive or negative depending on the signs of $2 - x$, $1 + x$, $x + 3$, and $x + 2$. We tabulate signs in these intervals.

Interval	$2 - x$	$1 + x$	$(2 - x)(1 + x)$
$x < -3$	+	-	-
$-3 < x < -2$	+	-	-
$-2 < x < -1$	+	-	-
$-1 < x < 2$	+	+	+
$x > 2$	-	+	-

Interval	$x + 3$	$x + 2$	$(x + 3)(x + 2)$
$x < -3$	-	-	+
$-3 < x < -2$	+	-	-
$-2 < x < -1$	+	+	+
$-1 < x < 2$	+	+	+
$x > 2$	+	+	+

37. (continued)

Interval	Quotient
$x < -3$	-
$-3 < x < -2$	+
$-2 < x < -1$	-
$-1 < x < 2$	+
$x > 2$	-

We see from the table that the solution set is $\{x \mid x < -3 \text{ or } -2 < x < -1 \text{ or } x > 2\}$.

38. $\{x \mid -2 < x < 2 \text{ and } x \neq 0\}$

39. $\left|\dfrac{x + 3}{x - 4}\right| < 2$

$-2 < \dfrac{x + 3}{x - 4} < 2$ (Property ii)

$-2 < \dfrac{x + 3}{x - 4}$ and $\dfrac{x + 3}{x - 4} < 2$

$0 < 2 + \dfrac{x + 3}{x - 4}$ and $\dfrac{x + 3}{x - 4} - 2 < 0$

$0 < \dfrac{3x - 5}{x - 4}$ and $\dfrac{-x + 11}{x - 4} < 0$

First solve $0 < \dfrac{3x - 5}{x - 4}$, or $\dfrac{3x - 5}{x - 4} > 0$.

The solutions of $3x - 5 = 0$ and $x - 4 = 0$ are $\frac{5}{3}$ and 4. They divide the number line in a natural way. The quotient is positive or negative depending on the signs of $3x - 5$ and $x - 4$. We tabulate signs in these intervals.

Interval	$3x - 5$	$x - 4$	Quotient
$x < \frac{5}{3}$	-	-	+
$\frac{5}{3} < x < 4$	+	-	-
$x > 4$	+	+	+

From the table we see that the solution set of $0 < \dfrac{3x - 5}{x - 4}$ is $\left\{x \middle| x < \frac{5}{3} \text{ or } x > 4\right\}$.

Next solve $\dfrac{-x + 11}{x - 4} < 0$.

39. (continued)

The solutions of $-x + 11 = 0$ and $x - 4 = 0$ are 11 and 4. They divide the number line in a natural way. The quotient is positive or negative depending on the signs of $-x + 11$ and $x - 4$. We tabulate signs in these intervals.

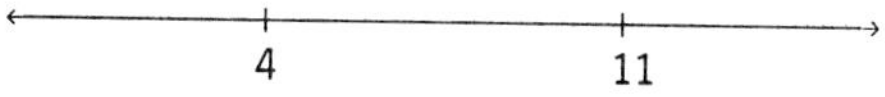

Interval	$-x + 11$	$x - 4$	Quotient
$x < 4$	+	-	-
$4 < x < 11$	+	+	+
$x > 11$	-	+	-

From the table we see that the solution set of $\dfrac{-x + 11}{x - 4} < 0$ is $\{x \mid x < 4 \text{ or } x > 11\}$.

The intersection of the two solution sets is the solution set of $\left|\dfrac{x + 3}{x - 4}\right| < 2$.

$\left\{x \middle| x < \frac{5}{3} \text{ or } x > 4\right\} \cap \{x \mid x < 4 \text{ or } x > 11\}$

$= \left\{x \middle| x < \frac{5}{3} \text{ or } x > 11\right\}$

40. $\{x \mid -\sqrt{5} \leqslant x \leqslant \sqrt{5}\}$

41. $(7 - x)^{-2} < 0$

$\dfrac{1}{(7 - x)^2} < 0$

Since 1 is positive and $(7 - x)^2$ is positive for all values of x, the quotient $\dfrac{1}{(7 - x)^2}$ is always positive. Thus the solution set for the inequality is $\emptyset$.

42. $\{x \mid x < 1\}$

43. $\left|1 + \dfrac{1}{x}\right| < 3$

$-3 < 1 + \dfrac{1}{x} < 3$ (Property ii)

$-3 < \dfrac{x + 1}{x} < 3$

$-3 < \dfrac{x + 1}{x}$ and $\dfrac{x + 1}{x} < 3$

$0 < 3 + \dfrac{x + 1}{x}$ and $\dfrac{x + 1}{x} - 3 < 0$

$0 < \dfrac{4x + 1}{x}$ and $\dfrac{-2x + 1}{x} < 0$

First solve $0 < \dfrac{4x + 1}{x}$, or $\dfrac{4x + 1}{x} > 0$.

43. (continued)

The solutions of $4x + 1 = 0$ and $x = 0$ are $-\frac{1}{4}$ and 0. They divide the number line in a natural way. The quotient is positive or negative depending on the signs of $4x + 1$ and x. We tabulate signs in these intervals.

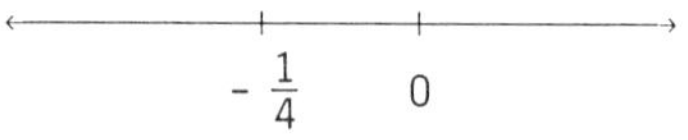

Interval	4x + 1	x	Quotient
$x < -\frac{1}{4}$	-	-	+
$-\frac{1}{4} < x < 0$	+	-	-
$x > 0$	+	+	+

From the table we see that the solution set of $0 < \frac{4x + 1}{x}$ is $\left\{x \middle| x < -\frac{1}{4} \text{ or } x > 0\right\}$.

Next solve $\frac{-2x + 1}{x} < 0$.

The solutions of $-2x + 1 = 0$ and $x = 0$ are $\frac{1}{2}$ and 0. They divide the number line in a natural way. The quotient is positive or negative depending on the signs of $-2x + 1$ and x. We tabulate signs in these intervals.

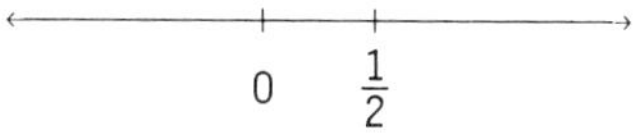

Interval	-2x + 1	x	Quotient
$x < 0$	+	-	-
$0 < x < \frac{1}{2}$	+	+	+
$x > \frac{1}{2}$	-	+	-

From the table we see that the solution set of $\frac{-2x + 1}{x} < 0$ is $\left\{x \middle| x < 0 \text{ or } x > \frac{1}{2}\right\}$.

The intersection of the two solution sets is the solution set of $\left|1 + \frac{1}{x}\right| < 3$.

$$\left\{x \middle| x < -\frac{1}{4} \text{ or } x > 0\right\} \cap \left\{x \middle| x < 0 \text{ or } x > \frac{1}{2}\right\}$$

$$= \left\{x \middle| x < -\frac{1}{4} \text{ or } x > \frac{1}{2}\right\}$$

44. $\{x \mid x \neq -5\}$

45. $\left|2 - \frac{1}{x}\right| \leqslant 2 + \left|\frac{1}{x}\right|$

Note that $\frac{1}{x}$ is not defined when $x = 0$. Thus $x \neq 0$.

Divide the set of reals into three intervals:

$x < 0$

$0 < x < \frac{1}{2}$

$x \geqslant \frac{1}{2}$

Find the solution set of $\left|2 - \frac{1}{x}\right| \leqslant 2 + \left|\frac{1}{x}\right|$ for each interval. Then take the union of the three solution sets.

If $x < 0$, then $\left|2 - \frac{1}{x}\right| = 2 - \frac{1}{x}$ and $\left|\frac{1}{x}\right| = -\frac{1}{x}$.

Solve: $2 - \frac{1}{x} \leqslant 2 - \frac{1}{x}$

$2 \leqslant 2$

True for any $x < 0$. The solution set for this interval is $\{x \mid x < 0\}$.

If $0 < x < \frac{1}{2}$, then $\left|2 - \frac{1}{x}\right| = -\left(2 - \frac{1}{x}\right)$ and $\left|\frac{1}{x}\right| = \frac{1}{x}$.

Solve: $-\left(2 - \frac{1}{x}\right) \leqslant 2 + \frac{1}{x}$

$-2 + \frac{1}{x} \leqslant 2 + \frac{1}{x}$

$-2 \leqslant 2$

True for all x such that $0 < x < \frac{1}{2}$. The solution set for this interval is $\left\{x \middle| 0 < x < \frac{1}{2}\right\}$.

If $x \geqslant \frac{1}{2}$, then $\left|2 - \frac{1}{x}\right| = 2 - \frac{1}{x}$ and $\left|\frac{1}{x}\right| = \frac{1}{x}$.

Solve: $2 - \frac{1}{x} \leqslant 2 + \frac{1}{x}$

$-\frac{1}{x} \leqslant \frac{1}{x}$

$-1 \leqslant 1$

True for all x such that $x \geqslant \frac{1}{2}$. The solution set for this interval is $\left\{x \middle| x \geqslant \frac{1}{2}\right\}$.

The <u>union</u> of the above three solution sets is the set of all reals except 0.

46. $\{x \mid -2 < x \leqslant -1 \text{ or } 3 \leqslant x < 4 \text{ or } x = 2\}$

47. $|x^2 + 3x - 1| < 3$

$-3 < x^2 + 3x - 1 < 3$ (Property ii)

$-3 < x^2 + 3x - 1$ and $x^2 + 3x - 1 < 3$

$0 < x^2 + 3x + 2$ and $x^2 + 3x - 4 < 0$

$0 < (x + 2)(x + 1)$ and $(x + 4)(x - 1) < 0$

Find the solution set of each of these inequalities. Then find the intersection of those solution sets.

47. (continued)

First solve $0 < (x + 2)(x + 1)$.

The solutions of $(x + 2)(x + 1) = 0$ are -2 and -1. They divide the real-number line as pictured below. The product is positive or negative depending on the signs of the factors $x + 2$ and $x + 1$. We tabulate signs in these intervals.

Interval	x + 2	x + 1	Product
$x < -2$	-	-	+
$-2 < x < -1$	+	-	-
$x > -1$	+	+	+

From the table we can see that the solution set of the inequality $0 < (x + 2)(x + 1)$ is $\{x \mid x < -2 \text{ or } x > -1\}$.

Next solve $(x + 4)(x - 1) < 0$.

The solutions of $(x + 4)(x - 1) = 0$ are -4 and 1. They divide the real-number line as pictured below. The product is positive or negative depending on the signs of the factors $x + 4$ and $x - 1$. We tabulate the signs in these intervals.

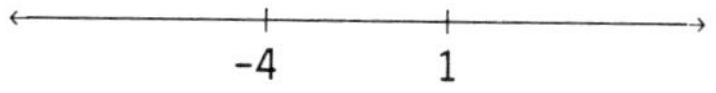

Interval	x + 4	x - 1	Product
$x < -4$	-	-	+
$-4 < x < 1$	+	-	-
$x > 1$	+	+	+

From the table we can see that the solution set of the inequality $(x + 4)(x - 1) < 0$ is $\{x \mid -4 < x < 1\}$.

The intersection of the two solution sets is the solution set of $|x^2 + 3x - 1| < 3$.

$$\{x \mid x < -2 \text{ or } x > -1\} \cap \{x \mid -4 < x < 1\}$$
$$= \{x \mid -4 < x < -2 \text{ or } -1 < x < 1\}$$

Graphs are helpful when finding the intersection.

48. $\{x \mid x \leqslant -1 \text{ or } 1 \leqslant x \leqslant 4 \text{ or } x \geqslant 6\}$

49. We first make a drawing.

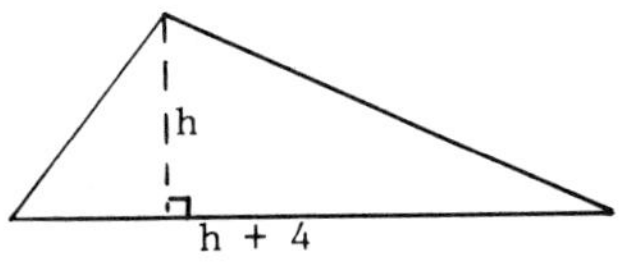

49. (continued)

We let h represent the height and h + 4 the base. The area is $\frac{1}{2}(h + 4)h$. We now have an inequality:

$$\frac{1}{2}h(h + 4) > 10$$
$$h(h + 4) > 20$$
$$h^2 + 4h > 20$$
$$h^2 + 4h - 20 > 0$$

Consider the function $f(h) = h^2 + 4h - 20$. The inputs that produce outputs that are greater than 0 are the solutions of the inequality.

Set $f(h) = 0$ and use the quadratic formula to find the x-intercepts.

$h^2 + 4h - 20 = 0$

$$h = \frac{-4 \pm \sqrt{4^2 - 4(1)(-20)}}{2 \cdot 1} = \frac{-4 \pm \sqrt{16 + 80}}{2}$$
$$= \frac{-4 \pm \sqrt{96}}{2} = \frac{-4 \pm 4\sqrt{6}}{2} = -2 \pm 2\sqrt{6}$$

The x-intercepts are $(-2 - 2\sqrt{6}, 0)$ and $(-2 + 2\sqrt{6}, 0)$.

The graph of the function opens upwards. Function values will be greater than 0 when h is less than $-2 - 2\sqrt{6}$ and when h is greater than $-2 + 2\sqrt{6}$. Since the height can only be positive, we only consider $h > -2 + 2\sqrt{6}$.

Thus, $\{h \mid h > -2 + 2\sqrt{6} \text{ cm}\}$.

50. $\left\{w \middle| w > \frac{-3 + \sqrt{69}}{2} \text{ m}\right\}$

51. $R(x) = 50x - x^2$, $C(x) = 5x + 350$

a) $P(x) = R(x) - C(x)$ (Profit function)
$P(x) = (50x - x^2) - (5x + 350)$
$= -x^2 + 45x - 350$

When the profit is 0, the company breaks even. Set $P(x) = 0$ and find the x-intercepts.

$$0 = -x^2 + 45x - 350$$
$$0 = x^2 - 45x + 350$$
$$0 = (x - 10)(x - 35)$$

$x - 10 = 0$ or $x - 35 = 0$
$x = 10$ or $x = 35$

The x-intercepts are (10, 0) and (35, 0). The break-even values are 10 and 35.

b) The parabola opens downward. The values of x that produce a profit are between the x-intercepts: $\{x \mid 10 < x < 35\}$.

c) The values of x that result in a loss are those less than 10 and those greater than 35: $\{x \mid x < 10 \text{ or } x > 35\}$.

52. a) 10, 60

b) $\{x \mid 10 < x < 60\}$

c) $\{x \mid x < 10 \text{ or } x > 60\}$

53. The discriminant of $ax^2 + bx + c$ is $b^2 - 4ac$.
The discriminant of $x^2 + kx + 1$ is $k^2 - 4$.

a) If the discriminant is positive, there are two real-number solutions for the quadratic equation.

Set $k^2 - 4 > 0$ and solve for k.

$(k + 2)(k - 2) > 0$

The solutions of $(k + 2)(k - 2) = 0$ are -2 and 2. They divide the number line as pictured below. The product is positive or negative depending on the signs of the factors $k + 2$ and $k - 2$. We tabulate signs in these intervals.

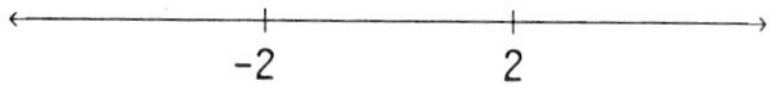

Interval	k + 2	k - 2	Product
$x < -2$	-	-	+
$-2 < x < 2$	+	-	-
$x > 2$	+	+	+

From the table we can see that the solution set is $\{k \mid k < -2 \text{ or } k > 2\}$. Thus, $x^2 + kx + 1 = 0$ will have two real-number solutions if the values for k are in the set $\{k \mid k < -2 \text{ or } k > 2\}$.

b) If the discriminant is negative, there will be no real-number solutions for the quadratic equation.

Set $k^2 - 4 < 0$ and solve for k.

$(k + 2)(k - 2) < 0$

From the table in part a) we see that the solution set is $\{k \mid -2 < k < 2\}$. Thus, $x^2 + kx + 1 = 0$ will have <u>no</u> real-number solutions if the values for k are in the set $\{k \mid -2 < k < 2\}$.

54. a) $\{k \mid k < -2\sqrt{2} \text{ or } k > 2\sqrt{2}\}$

b) $\{k \mid -2\sqrt{2} < k < 2\sqrt{2}\}$

55. $f(x) = \sqrt{1 - x^2}$

The radicand must be greater than or equal to 0

$1 - x^2 \geqslant 0$

Consider the function $g(x) = 1 - x^2$. The inputs that produce outputs that are greater than or equal to 0 are the solutions of the inequality.

Set $g(x) = 0$ and factor to find the x-intercepts.

$1 - x^2 = 0$

$(1 + x)(1 - x) = 0$

$1 + x = 0$ or $1 - x = 0$

$x = -1$ or $x = 1$

The x-intercepts are (-1, 0) and (1, 0).

The graph of the function opens downward. Function values will be greater than or equal to 0 when x is between and including the x-intercepts.

The domain is $\{x \mid -1 \leqslant x \leqslant 1\}$.

56. $\{x \mid -1 < x < 1\}$

57. $g(x) = \sqrt{x^2 + 2x - 3}$

The radicand must be greater than or equal to 0.

$x^2 + 2x - 3 \geqslant 0$

Consider the function $f(x) = x^2 + 2x - 3$. The inputs that produce outputs greater than or equal to 0 are the solutions of the inequality.

Set $f(x) = 0$ and factor to find the x-intercepts.

$x^2 + 2x - 3 = 0$

$(x + 3)(x - 1) = 0$

$x + 3 = 0$ or $x - 1 = 0$

$x = -3$ or $x = 1$

The x-intercepts are (-3, 0) and (1, 0).

The graph of the function opens upward. Function values will be greater than or equal to 0 when x is less than or equal to -3 and when x is greater than or equal to 1.

The domain is $\{x \mid x \leqslant -3 \text{ or } x \geqslant 1\}$.

58. $\{n \mid 10 \leqslant n \leqslant 20\}$

59. Solve $\frac{n(n - 1)}{2} \geqslant 105$.

$n(n - 1) \geqslant 210$

$n^2 - n - 210 \geqslant 0$

Consider the function $f(x) = n^2 - n - 210$. The inputs that produce outputs that are greater than or equal to 0 are solutions of the inequality.

59. (continued)

Set $f(x) = 0$ and factor to find the x-intercepts.

$$n^2 - n - 210 = 0$$

$$(n - 15)(n + 14) = 0$$

$$n - 15 = 0 \quad \text{or} \quad n + 14 = 0$$

$$n = 15 \quad \text{or} \quad n = -14$$

The x-intercepts are (15, 0) and (-14, 0).

The graph of the function opens upwards. Function values will be greater than 0 when x is less than -14 or when x is greater than 15.

The solution set of the inequality is $\{n | n \leqslant -14 \text{ or } n \geqslant 15\}$. However, since there cannot be a negative number of teams, the solution of the given problem is $\{n | n \geqslant 15\}$.

Exercise Set 5.1

1. The degree of the polynomial, $x^4 - 3x^2 + 1$, is 4.

2. 5

3. The degree of the polynomial, $-2x + 5$ (or $-2x^1 + 5$), is 1.

4. 1

5. The degree of the polynomial, $2x^2 - 3x + 4$, is 2.

6. 2

7. The degree of the polynomial, 3 (or $3x^0$), is 0.

8. No degree

9. $P(x) = 4x^2 - 3x + 2$

$$P(4) = 4\cdot 4^2 - 3\cdot 4 + 2 = 64 - 12 + 2 = 54$$

$$P(0) = 4\cdot 0^2 - 3\cdot 0 + 2 = 0 - 0 + 2 = 2$$

10. $Q(3) = -84$, $Q(-1) = 0$

11. $P(y) = 8y^3 - 12y - 5$

$$P(-2) = 8(-2)^3 - 12(-2) - 5 = -64 + 24 - 5 = -45$$

$$P\left(\frac{1}{3}\right) = 8\left(\frac{1}{3}\right)^3 - 12\cdot\frac{1}{3} - 5 = 8\cdot\frac{1}{27} - 4 - 5 = \frac{8}{27} - 9 = \frac{8}{27} - \frac{243}{27} = -\frac{235}{27}$$

12. $Q(-3) = -168$, $Q(0) = -9$

13. $P(x) = -4x^3 + 2x^2 - 7x + 5$

$P(a) = -4a^3 + 2a^2 - 7a + 5$

14. $-6b^3 - 12b^2 + 8b - 3$

15. $P(x) = x^3 - 4x^2 + 3x - 7$

$$P(2a) = (2a)^3 - 4(2a)^2 + 3(2a) - 7 = 8a^3 - 4(4a^2) + 3(2a) - 7 = 8a^3 - 16a^2 + 6a - 7$$

16. $-\frac{7}{8}c^3 + \frac{5}{2}c^2 + 6$

17. $P(x) = -3x^3 + 6x^2 - 4$

$P(x + a) = -3(x + a)^3 + 6(x + a)^2 - 4$

We find $(x + a)^3$ and $(x + a)^2$.

$$(x + a)^3 = (x + a)^2(x + a) = (x^2 + 2xa + a^2)(x + a) = x^3 + 3x^2a + 3xa^2 + a^3$$

$$(x + a)^2 = x^2 + 2xa + a^2$$

We now substitute and simplify.

$$P(x + a) = -3(x^3 + 3x^2a + 3xa^2 + a^3) + 6(x^2 + 2xa + a^2) - 4 = -3x^3 - 9x^2a - 9xa^2 - 3a^3 + 6x^2 + 12xa + 6a^2 - 4$$

18. $48x^3 + 72x^2 + 52x + 3$

19. $P(x) = x^3 + 6x^2 - x - 30$

$$P(2) = 2^3 + 6(2)^2 - 2 - 30 \quad \text{(Substituting)} = 8 + 24 - 2 - 30 = 0$$

Since $P(2) = 0$, 2 is a root, or zero, of the polynomial.

$$P(3) = 3^3 + 6(3)^2 - 3 - 30 \quad \text{(Substituting)} = 27 + 54 - 3 - 30 = 48$$

Since $P(3) \neq 0$, 3 is not a root, or zero, of the polynomial.

20. 2 No, 3 No, -1 No

21. a)

$$\begin{array}{r|l} & x^2 + 8x + 15 \\ \hline x - 2 & x^3 + 6x^2 - x - 30 \\ & \underline{x^3 - 2x^2} \\ & 8x^2 - x \\ & \underline{8x^2 - 16x} \\ & 15x - 30 \\ & \underline{15x - 30} \\ & 0 \end{array}$$

Since the remainder is 0, we know that $x - 2$ is a factor of $x^3 + 6x^2 - x - 30$.

21. (continued)

b)
$$\begin{array}{r} x^2 + 9x + 26 \\ x - 3 \overline{\smash{)}\, x^3 + 6x^2 - x - 30} \\ \underline{x^3 - 3x^2} \\ 9x^2 - x \\ \underline{9x^2 - 27x} \\ 26x - 30 \\ \underline{26x - 78} \\ 48 \end{array}$$

Since the remainder is not 0, we know that $x - 3$ is not a factor of $x^3 + 6x^2 - x - 30$.

c)
$$\begin{array}{r} x^2 + 5x - 6 \\ x + 1 \overline{\smash{)}\, x^3 + 6x^2 - x - 30} \\ \underline{x^3 + x^2} \\ 5x^2 - x \\ \underline{5x^2 + 5x} \\ -6x - 30 \\ \underline{-6x - 6} \\ -24 \end{array}$$

Since the remainder is not 0, we know that $x + 1$ is not a factor of $x^3 + 6x^2 - x - 30$.

22. a) No, b) No, c) No

23. See work in Exercise 21 a).
$P(x) = d(x) \cdot Q(x) + R(x)$

$x^3 + 6x^2 - x - 30 = (x - 2)(x^2 + 8x + 15) + 0$

24. $Q(x) = 2x^2 + x + 3$, $R(x) = 5$
$P(x) = (x - 2)(2x^2 + x + 3) + 5$

25. See work in Exercise 21 b).
$P(x) = d(x) \cdot Q(x) + R(x)$

$x^3 + 6x^2 - x - 30 = (x - 3)(x^2 + 9x + 26) + 48$

26. $Q(x) = 2x^2 + 3x + 10$, $R(x) = 29$
$P(x) = (x - 3)(2x^2 + 3x + 10) + 29$

27.
$$\begin{array}{r} x^2 - 2x + 4 \\ x + 2 \overline{\smash{)}\, x^3 + 0x^2 + 0x - 8} \\ \underline{x^3 + 2x^2} \\ -2x^2 + 0x \\ \underline{-2x^2 - 4x} \\ 4x - 8 \\ \underline{4x + 8} \\ -16 \end{array}$$

$P(x) = d(x) \cdot Q(x) + R(x)$

$x^3 - 8 = (x + 2)(x^2 - 2x + 4) - 16$

28. $Q(x) = x^2 - x + 1$, $R(x) = 26$
$P(x) = (x + 1)(x^2 - x + 1) + 26$

29.
$$\begin{array}{r} x^2 + 5 \\ x^2 + 4 \overline{\smash{)}\, x^4 + 9x^2 + 20} \\ \underline{x^4 + 4x^2} \\ 5x^2 + 20 \\ \underline{5x^2 + 20} \\ 0 \end{array}$$

$P(x) = d(x) \cdot Q(x) + R(x)$

$x^4 + 9x^2 + 20 = (x^2 + 4)(x^2 + 5) + 0$

30. $Q(x) = x^2 - x + 1$, $R(x) = 1$
$P(x) = (x^2 + x + 1)(x^2 - x + 1) + 1$

31.
$$\begin{array}{r} \frac{5}{2}x^5 + \frac{5}{4}x^4 - \frac{5}{8}x^3 - \frac{39}{16}x^2 - \frac{29}{32}x + \frac{113}{64} \\ 2x^2 - x + 1 \overline{\smash{)}\, 5x^7 + 0x^6 + 0x^5 - 3x^4 + 0x^3 + 2x^2 + 0x - 3} \\ \underline{5x^7 - \frac{5}{2}x^6 + \frac{5}{2}x^5} \\ \frac{5}{2}x^6 - \frac{5}{2}x^5 - 3x^4 \\ \underline{\frac{5}{2}x^6 - \frac{5}{4}x^5 + \frac{5}{4}x^4} \\ -\frac{5}{4}x^5 + \frac{17}{4}x^4 + 0x^3 \\ \underline{-\frac{5}{4}x^5 - \frac{5}{8}x^4 - \frac{5}{8}x^3} \\ -\frac{39}{8}x^4 + \frac{5}{8}x^3 + 2x^2 \\ \underline{-\frac{39}{8}x^4 + \frac{39}{16}x^3 - \frac{39}{16}x^2} \\ -\frac{29}{16}x^3 + \frac{71}{16}x^2 + 0x \\ \underline{-\frac{29}{16}x^3 + \frac{29}{32}x^2 - \frac{29}{32}x} \\ \frac{113}{32}x^2 + \frac{29}{32}x - 3 \\ \underline{\frac{113}{32}x^2 - \frac{113}{64}x + \frac{113}{64}} \\ \frac{171}{64}x - \frac{305}{64} \end{array}$$

$P(x) = d(x) \cdot Q(x) + R(x)$

$5x^7 - 3x^4 + 2x^2 - 3$

$= (2x^2 - x + 1)\left(\frac{5}{2}x^5 + \frac{5}{4}x^4 - \frac{5}{8}x^3 - \frac{39}{16}x^2 - \frac{29}{32}x + \frac{113}{64}\right) + \left(\frac{171}{64}x - \frac{305}{64}\right)$

32. $P(x) = (3x^2 + 2x - 1)\left(2x^3 + \frac{2}{3}x - \frac{13}{9}\right) + \frac{41x - 31}{9}$

33. $P(x) = x^5 - 64$

a) $P(2) = 2^5 - 64 = 32 - 64 = -32$

b)
$$
\begin{array}{r}
x^4 + 2x^3 + 4x^2 + 8x + 16 \\
x - 2 \overline{\smash{)}\, x^5 + 0x^4 + 0x^3 + 0x^2 + 0x - 64} \\
\underline{x^5 - 2x^4} \qquad\qquad\qquad\qquad \\
2x^4 + 0x^3 \qquad\qquad\qquad \\
\underline{2x^4 - 4x^3} \qquad\qquad\qquad \\
4x^3 + 0x^2 \qquad\qquad \\
\underline{4x^3 - 8x^2} \qquad\qquad \\
8x^2 + 0x \qquad \\
\underline{8x^2 - 16x} \qquad \\
16x - 64 \\
\underline{16x - 32} \\
-32
\end{array}
$$

The remainder is -32.

c) $P(-1) = (-1)^5 - 64 = -1 - 64 = -65$

d)
$$
\begin{array}{r}
x^4 - x^3 + x^2 - x + 1 \\
x + 1 \overline{\smash{)}\, x^5 + 0x^4 + 0x^3 + 0x^2 + 0x - 64} \\
\underline{x^5 + x^4} \qquad\qquad\qquad\qquad \\
-x^4 + 0x^3 \qquad\qquad\qquad \\
\underline{-x^4 - x^3} \qquad\qquad\qquad \\
x^3 + 0x^2 \qquad\qquad \\
\underline{x^3 + x^2} \qquad\qquad \\
-x^2 + 0x \qquad \\
\underline{-x^2 - x} \qquad \\
x - 64 \\
\underline{x + 1} \\
-65
\end{array}
$$

The remainder is -65.

34. a) 0, b) 0, c) 12, d) 12

35. $f(n) = \frac{1}{2}(n^2 - n)$

When $n = 8$:

$f(8) = \frac{1}{2}(8^2 - 8) = \frac{1}{2}(64 - 8) = \frac{1}{2}(56) = 28$

When $n = 20$:

$f(20) = \frac{1}{2}(20^2 - 20) = \frac{1}{2}(400 - 20) = \frac{1}{2}(380) = 190$

36. 4.034 ppm, 5.792 ppm, 6.998 ppm, 7.376 ppm, 6.65 ppm, 4.544 ppm, 0.782 ppm

37. a) $5'7'' = 5 \text{ ft} + 7 \text{ in.} = 5 \text{ ft} \cdot \frac{12 \text{ in.}}{1 \text{ ft}} + 7 \text{ in.} =$

$60 \text{ in.} + 7 \text{ in.} = 67 \text{ in.}$

$W(67) = (67/12.3)^3 \approx 162$

The threshold weight of a person who is 5'7" is approximately 162 lb.

37. (continued)

$5'10'' = 5 \text{ ft} + 10 \text{ in.} = 5 \text{ ft} \cdot \frac{12 \text{ in.}}{1 \text{ ft}} + 10 \text{ in.}$

$= 60 \text{ in.} + 10 \text{ in.} = 70 \text{ in.}$

$W(70) = (70/12.3)^3 \approx 184$

The threshold weight of a person who is 5'10" is approximately 184 lb.

b) $6'1'' = 6 \text{ ft} + 1 \text{ in.} = 6 \text{ ft} \cdot \frac{12 \text{ in.}}{1 \text{ ft}} + 1 \text{ in.} =$

$72 \text{ in.} + 1 \text{ in.} = 73 \text{ in.}$

$W(73) = (73/12.3)^3 \approx 209$

The threshold weight of a person who is 6'1" is approximately 209 lb. The author should watch his weight.

38. 0.0055 units, 0.4725 units, 1.8626 units, 16.1868 units, 38.8132 units, 55.4341 units, 76.2088 units

Exercise Set 5.2

1. $(2x^4 + 7x^3 + x - 12) \div (x + 3)$

$= (2x^4 + 7x^3 + 0x^2 + x - 12) \div [x - (-3)]$

-3	2	7	0	1	-12
		-6	-3	9	-30
	2	1	-3	10	-42

The quotient is $2x^3 + x^2 - 3x + 10$.
The remainder is -42.

2. $Q(x) = x^2 - 5x + 3$, $R(x) = 9$

3. $(x^3 - 2x^2 - 8) \div (x + 2)$

$= (x^3 - 2x^2 + 0x - 8) \div [x - (-2)]$

-2	1	-2	0	-8
		-2	8	-16
	1	-4	8	-24

The quotient is $x^2 - 4x + 8$.
The remainder is -24.

4. $Q(x) = x^2 + 2x + 1$, $R(x) = 12$

5. $(x^4 - 1) \div (x - 1)$

$= (x^4 + 0x^3 + 0x^2 + 0x - 1) \div (x - 1)$

1	1	0	0	0	-1
		1	1	1	1
	1	1	1	1	0

The quotient is $x^3 + x^2 + x + 1$.
The remainder is 0.

6. $Q(x) = x^4 - 2x^3 + 4x^2 - 8x + 16$, $R(x) = 0$

7. $(2x^4 + 3x^2 - 1) \div \left(x - \frac{1}{2}\right)$

$= (2x^4 + 0x^3 + 3x^2 + 0x - 1) \div \left(x - \frac{1}{2}\right)$

$$\begin{array}{r|rrrr|r} \frac{1}{2} & 2 & 0 & 3 & 0 & -1 \\ & & 1 & \frac{1}{2} & \frac{7}{4} & \frac{7}{8} \\ \hline & 2 & 1 & \frac{7}{2} & \frac{7}{4} & -\frac{1}{8} \end{array}$$

The quotient is $2x^3 + x^2 + \frac{7}{2}x + \frac{7}{4}$.

The remainder is $-\frac{1}{8}$.

8. $Q(x) = 3x^3 + \frac{3}{4}x^2 - \frac{29}{16}x - \frac{29}{64}$, $R(x) = \frac{483}{256}$

9. $(x^4 - y^4) \div (x - y)$
$= (x^4 + 0x^3 + 0x^2 + 0x - y^4) \div (x - y)$

$$\begin{array}{r|rrrr|r} y & 1 & 0 & 0 & 0 & -y^4 \\ & & y & y^2 & y^3 & y^4 \\ \hline & 1 & y & y^2 & y^3 & 0 \end{array}$$

The quotient is $x^3 + x^2y + xy^2 + y^3$.
The remainder is 0.

10. $Q(x) = x^2 + 2ix + (2 - 4i)$, $R(x) = -6 - 2i$

11. $P(x) = x^3 - 6x^2 + 11x - 6$

Find $P(1)$.

$$\begin{array}{r|rrr|r} 1 & 1 & -6 & 11 & -6 \\ & & 1 & -5 & 6 \\ \hline & 1 & -5 & 6 & 0 \end{array}$$

$P(1) = 0$

Find $P(-2)$.

$$\begin{array}{r|rrr|r} -2 & 1 & -6 & 11 & -6 \\ & & -2 & 16 & -54 \\ \hline & 1 & -8 & 27 & -60 \end{array}$$

$P(-2) = -60$

Find $P(3)$.

$$\begin{array}{r|rrr|r} 3 & 1 & -6 & 11 & -6 \\ & & 3 & -9 & 6 \\ \hline & 1 & -3 & 2 & 0 \end{array}$$

$P(3) = 0$

12. $P(-3) = 69$, $P(-2) = 41$, $P(1) = -7$

13. $P(x) = 2x^5 - 3x^4 + 2x^3 - x + 8$

Find $P(20)$.

$$\begin{array}{r|rrrrr|r} 20 & 2 & -3 & 2 & 0 & -1 & 8 \\ & & 40 & 740 & 14{,}840 & 296{,}800 & 5{,}935{,}980 \\ \hline & 2 & 37 & 742 & 14{,}840 & 296{,}799 & 5{,}935{,}988 \end{array}$$

$P(20) = 5{,}935{,}988$

Find $P(-3)$.

$$\begin{array}{r|rrrrr|r} -3 & 2 & -3 & 2 & 0 & -1 & 8 \\ & & -6 & 27 & -87 & 261 & -780 \\ \hline & 2 & -9 & 29 & -87 & 260 & -772 \end{array}$$

$P(-3) = -772$

14. $P(-10) = -220{,}050$, $P(5) = -750$

15. $P(x) = 3x^3 + 5x^2 - 6x + 18$

If -3 is a root of $P(x)$, then $P(-3) = 0$.
Find $P(-3)$ using synthetic division.

$$\begin{array}{r|rrr|r} -3 & 3 & 5 & -6 & 18 \\ & & -9 & 12 & -18 \\ \hline & 3 & -4 & 6 & 0 \end{array}$$

Since $P(-3) = 0$, -3 is a root of $P(x)$.

If 2 is a root of $P(x)$, then $P(2) = 0$.
Find $P(2)$ using synthetic division.

$$\begin{array}{r|rrr|r} 2 & 3 & 5 & -6 & 18 \\ & & 6 & 22 & 32 \\ \hline & 3 & 11 & 16 & 50 \end{array}$$

Since $P(2) \neq 0$, 2 is not a root of $P(x)$.

16. -4 Yes, 2 No

17. $P(x) = x^3 - \frac{7}{2}x^2 + x - \frac{3}{2}$

If -3 is a root of $P(x)$, then $P(-3) = 0$.
Find $P(-3)$ using synthetic division.

$$\begin{array}{r|rrr|r} -3 & 1 & -\frac{7}{2} & 1 & -\frac{3}{2} \\ & & -3 & \frac{39}{2} & -\frac{123}{2} \\ \hline & 1 & -\frac{13}{2} & \frac{41}{2} & -63 \end{array}$$

Since $P(-3) \neq 0$, -3 is not a root of $P(x)$.

If $\frac{1}{2}$ is a root of $P(x)$, then $P\left(\frac{1}{2}\right) = 0$.

Find $P\left(\frac{1}{2}\right)$ using synthetic division.

17. (continued)

$$\begin{array}{r|rrrr} \frac{1}{2} & 1 & -\frac{7}{2} & 1 & -\frac{3}{2} \\ & & \frac{1}{2} & -\frac{3}{2} & -\frac{1}{4} \\ \hline & 1 & -3 & -\frac{1}{2} & -\frac{7}{4} \end{array}$$

Since $P\left(\frac{1}{2}\right) \neq 0$, $\frac{1}{2}$ is not a root of $P(x)$.

18. i Yes, $-i$ Yes, -2 Yes

19. $P(x) = x^3 + 4x^2 + x - 6$

Try $x - 1$. Use synthetic division to see whether $P(1) = 0$.

$$\begin{array}{r|rrrr} 1 & 1 & 4 & 1 & -6 \\ & & 1 & 5 & 6 \\ \hline & 1 & 5 & 6 & 0 \end{array}$$

Since $P(1) = 0$, $x - 1$ is a factor of $P(x)$. Thus $P(x) = (x - 1)(x^2 + 5x + 6)$.

Factoring the trinomial we get

$P(x) = (x - 1)(x + 2)(x + 3)$.

To solve the equation $P(x) = 0$, use the principle of zero products.

$(x - 1)(x + 2)(x + 3) = 0$

$x - 1 = 0$ or $x + 2 = 0$ or $x + 3 = 0$

$x = 1$ or $x = -2$ or $x = -3$

The solutions are 1, -2, and -3.

20. $P(x) = (x - 2)(x + 3)(x + 4)$; 2, -3, -4

21. $P(x) = x^3 - 6x^2 + 3x + 10$

Try $x - 1$. Use synthetic division to see whether $P(1) = 0$.

$$\begin{array}{r|rrrr} 1 & 1 & -6 & 3 & 10 \\ & & 1 & -5 & -2 \\ \hline & 1 & -5 & -2 & 8 \end{array}$$

Since $P(1) \neq 0$, $x - 1$ is not a factor of $P(x)$.

Try $x + 1$. Use synthetic division to see whether $P(-1) = 0$.

$$\begin{array}{r|rrrr} -1 & 1 & -6 & 3 & 10 \\ & & -1 & 7 & -10 \\ \hline & 1 & -7 & 10 & 0 \end{array}$$

Since $P(-1) = 0$, $x + 1$ is a factor of $P(x)$. Thus $P(x) = (x + 1)(x^2 - 7x + 10)$.

Factoring the trinomial we get

$P(x) = (x + 1)(x - 2)(x - 5)$.

21. (continued)

To solve the equation $P(x) = 0$, use the principle of zero products.

$(x + 1)(x - 2)(x - 5) = 0$

$x + 1 = 0$ or $x - 2 = 0$ or $x - 5 = 0$

$x = -1$ or $x = 2$ or $x = 5$

The solutions are -1, 2, and 5.

22. $P(x) = (x - 1)(x - 2)(x + 5)$; 1, 2, -5

23. $P(x) = x^3 - x^2 - 14x + 24$

Try $x + 1$, $x - 1$, and $x + 2$. Using synthetic division we find that $P(-1) \neq 0$, $P(1) \neq 0$, and $P(-2) \neq 0$. Thus $x + 1$, $x - 2$, and $x + 2$ are not factors of $P(x)$.

Try $x - 2$. Use synthetic division to see whether $P(2) = 0$.

$$\begin{array}{r|rrrr} 2 & 1 & -1 & -14 & 24 \\ & & 2 & 2 & -24 \\ \hline & 1 & 1 & -12 & 0 \end{array}$$

Since $P(2) = 0$, $x - 2$ is a factor of $P(x)$. Thus $P(x) = (x - 2)(x^2 + x - 12)$.

Factoring the trinomial we get

$P(x) = (x - 2)(x + 4)(x - 3)$.

To solve the equation $P(x) = 0$, use the principle of zero products.

$(x - 2)(x + 4)(x - 3) = 0$

$x - 2 = 0$ or $x + 4 = 0$ or $x - 3 = 0$

$x = 2$ or $x = -4$ or $x = 3$

The solutions are 2, -4, and 3.

24. $P(x) = (x - 2)(x - 4)(x + 3)$; 2, 4, -3

25. $P(x) = x^4 - x^3 - 19x^2 + 49x - 30$

Try $x - 1$. Use synthetic division to see whether $P(1) = 0$.

$$\begin{array}{r|rrrrr} 1 & 1 & -1 & -19 & 49 & -30 \\ & & 1 & 0 & -19 & 30 \\ \hline & 1 & 0 & -19 & 30 & 0 \end{array}$$

Since $P(1) = 0$, $x - 1$ is a factor of $P(x)$. Thus $P(x) = (x - 1)(x^3 - 19x + 30)$.

We continue to use synthetic division to factor $x^3 - 19x + 30$. Trying $x - 1$, $x + 1$, and $x + 2$ we find that $P(1) \neq 0$, $P(-1) \neq 0$, and $P(-2) \neq 0$. Thus $x - 1$, $x + 1$, and $x + 2$ are not factors of $x^3 - 19x + 30$. Try $x - 2$.

$$\begin{array}{r|rrrr} 2 & 1 & 0 & -19 & 30 \\ & & 2 & 4 & -30 \\ \hline & 1 & 2 & -15 & 0 \end{array}$$

25. (continued)

Since $P(2) = 0$, $x - 2$ is a factor of $x^3 - 19x + 30$.

Thus $P(x) = (x - 1)(x - 2)(x^2 + 2x - 15)$.

Factoring the trinomial we get

$P(x) = (x - 1)(x - 2)(x - 3)(x + 5)$.

To solve the equation $P(x) = 0$, use the principle of zero products.

$(x - 1)(x - 2)(x - 3)(x + 5) = 0$

$x - 1 = 0$ or $x - 2 = 0$ or $x - 3 = 0$ or $x + 5 = 0$

$x = 1$ or $x = 2$ or $x = 3$ or $x = -5$

The solutions are 1, 2, 3, and -5.

26. $P(x) = (x + 1)(x + 2)(x + 3)(x + 5)$;

-1, -2, -3, -5

27. $\frac{6x^2}{x^2 + 11} + \frac{60}{x^3 - 7x^2 + 11x - 77} = \frac{1}{x - 7}$

LCM $= (x^2 + 11)(x - 7)$

$$(x^2+11)(x-7)\left[\frac{6x^2}{x^2+11}+\frac{60}{(x^2+11)(x-7)}\right]=(x^2+11)(x-7)\cdot\frac{1}{x-7}$$

$$6x^2(x - 7) + 60 = x^2 + 11$$

$$6x^3 - 42x^2 + 60 = x^2 + 11$$

$$6x^3 - 43x^2 + 49 = 0$$

Use synthetic division to find factors of $P(x) = 6x^3 - 43x^2 + 49$. Try $x + 1$. Use synthetic division to see whether $P(-1) = 0$.

-1	6	-43	0	49
		-6	49	-49
	6	-49	49	0

Since $P(-1) = 0$, $x + 1$ is a factor of $P(x)$. Thus $P(x) = (x + 1)(6x^2 - 49x + 49)$. Factoring the trinomial we get $P(x) = (x + 1)(6x - 7)(x - 7)$.

To solve $P(x) = 0$, use the principle of zero products.

$(x + 1)(6x - 7)(x - 7) = 0$

$x + 1 = 0$ or $6x - 7 = 0$ or $x - 7 = 0$

$x = -1$ or $6x = 7$ or $x = 7$

$x = -1$ or $x = \frac{7}{6}$ or $x = 7$

The value $x = 7$ does not check, but $x = -1$ and $x = \frac{7}{6}$ do. The solutions are -1 and $\frac{7}{6}$.

28. $-1 \pm \sqrt{7}$

29. To solve $x^3 + 2x^2 - 13x + 10 > 0$ we first factor and solve $x^3 + 2x^2 - 13x + 10 = 0$. Using synthetic division we find the solutions to be -5, 1, and 2. They are not solutions of the inequality, but they divide the number line in a natural way. The product is positive or negative, for values other than -5, 1, and 2, depending on the signs of the factors $x + 5$, $x - 1$, and $x - 2$. We tabulate signs in these intervals.

$x^3 + 2x^2 - 13x + 10 > 0$

$(x + 5)(x - 1)(x - 2) > 0$

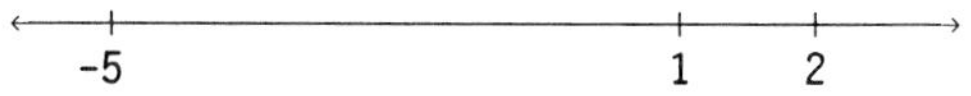

Interval	x + 5	x - 1	x - 2	Product
$x < -5$	-	-	-	-
$-5 < x < 1$	+	-	-	+
$1 < x < 2$	+	+	-	-
$x > 2$	+	+	+	+

We see from the table that the product $(x + 5)(x - 1)(x - 2)$ is positive only in the intervals $-5 < x < 1$ and $x > 2$. The solution set is

$\{x \mid -5 < x < 1 \text{ or } x > 2\}$.

30. $\{x \mid -5 < x < 1 \text{ or } 2 < x < 3\}$

31. $P(x) = x^3 - kx^2 + 3x + 7k$

Think of $x + 2$ as $x - (-2)$.

Find $P(-2)$.

-2	1	-k	3	7k
		-2	2k + 4	-4k - 14
	1	-k - 2	2k + 7	3k - 14

Thus $P(-2) = 3k - 14$.

We know that if $x + 2$ is a factor of $P(x)$, then $P(-2) = 0$.

We solve $0 = 3k - 14$ for k.

$0 = 3k - 14$

$14 = 3k$

$\frac{14}{3} = k$

32. a) -85.1587, b) -485.1587

33. Divide $x^2 + kx + 4$ by $x - 1$.

1	1	k	4
		1	k + 1
	1	k + 1	k + 5

The remainder is $k + 5$.

33. (continued)

Divide $x^2 + kx + 4$ by $x + 1$.

$$
\begin{array}{r|rrr}
-1 & 1 & k & 4 \\
 & & -1 & -k + 1 \\
\hline
 & 1 & k - 1 & -k + 5
\end{array}
$$

The remainder is $-k + 5$.

Set $k + 5 = -k + 5$ and solve for k.

$k + 5 = -k + 5$

$2k = 0$

$k = 0$

34. $k = -\frac{3}{2}$

Exercise Set 5.3

1. $(x + 3)^2(x - 1) = (x + 3)(x + 3)(x - 1)$

The factor $x + 3$ occurs twice. Thus the root -3 has a multiplicity of two.

The factor $x - 1$ occurs only one time. Thus the root 1 has a mulitplicity of one.

2. 3, Multiplicity 2; -4, Multiplicity 3; 0, Multiplicity 4

3. $x^3(x - 1)^2(x + 4) = 0$

$x \cdot x \cdot x(x - 1)(x - 1)(x + 4) = 0$

The factor x occurs three times. Thus the root 0 has a multiplicity of three.

The factor $x - 1$ occurs twice. Thus the root 1 has a multiplicity of two.

The factor $x + 4$ occurs only one time. Thus the root -4 has a multiplicity of one.

4. 3, Multiplicity 2; 2, Multiplicity 2

5. Find a polynomial of degree 3 with -2, 3, and 5 as roots.

By Theorem 2 such a polynomial has factors $x + 2$, $x - 3$, and $x - 5$, so we have

$P(x) = a_n(x + 2)(x - 3)(x - 5)$.

The number a_n can be any nonzero number. The simplest polynomial will be obtained if we let it be 1. Multiplying the factors, we obtain

$P(x) = (x + 2)(x - 3)(x - 5)$

$= (x^2 - x - 6)(x - 5)$

$= x^3 - 6x^2 - x + 30$

6. $x^3 - 2x^2 + x - 2$

7. Find a polynomial of degree 3 with -3, 2i, and -2i as roots.

By Theorem 2 such a polynomial has factors $x + 3$, $x - 2i$, and $x + 2i$, so we have

$P(x) = a_n(x + 3)(x - 2i)(x + 2i)$.

The number a_n can be any nonzero number. The simplest polynomial will be obtained if we let it be 1. Multiplying the factors, we obtain

$P(x) = (x + 3)(x - 2i)(x + 2i)$

$= (x + 3)(x^2 + 4)$

$= x^3 + 3x^2 + 4x + 12$

8. $x^3 - x^2 + 15x + 17$

9. Find a polynomial of degree 3 with $\sqrt{2}$, $-\sqrt{2}$, and $\sqrt{3}$ as roots.

By Theorem 2 such a polynomial has factors $x - \sqrt{2}$, $x + \sqrt{2}$, and $x - \sqrt{3}$, so we have

$P(x) = a_n(x - \sqrt{2})(x + \sqrt{2})(x - \sqrt{3})$.

The number a_n can be any nonzero number. The simplest polynomial will be obtained if we let it be 1. Multiplying the factors, we obtain

$P(x) = (x - \sqrt{2})(x + \sqrt{2})(x - \sqrt{3})$

$= (x^2 - 2)(x - \sqrt{3})$

$= x^3 - \sqrt{3}\,x^2 - 2x + 2\sqrt{3}$

The coefficients are 1, $-\sqrt{3}$, -2, and $2\sqrt{3}$; $-\sqrt{3}$ and $2\sqrt{3}$ are not rational. Using Theorem 7, we know that all coefficients could not be rational since $\sqrt{3}$ is a root and $-\sqrt{3}$ is not a root.

10. $x^4 - 3x^3 - 7x^2 + 15x + 18$

11. A polynomial or polynomial equation of degree 5 has at most 5 roots. Three of the roots are 6, $-3 + 4i$, and $4 - \sqrt{5}$. Using Theorems 6 and 7 we know that $-3 - 4i$ and $4 + \sqrt{5}$ are also roots.

12. $1 + i$

13. Find a polynomial of lowest degree with rational coefficients that has $1 + i$ and 2 as some of its roots.

$1 - i$ is also a root. (Theorem 6)

Thus the polynomial is

$a_n(x - 2)[x - (1 + i)][x - (1 - i)]$.

If we let $a_n = 1$, we obtain

$(x - 2)[(x - 1) - i][(x - 1) + i]$

$= (x - 2)[(x - 1)^2 - i^2]$

$= (x - 2)(x^2 - 2x + 1 + 1)$

$= (x - 2)(x^2 - 2x + 2)$

$= x^3 - 4x^2 + 6x - 4$

14. $x^3 - 3x^2 + x + 5$

15. Find a polynomial of lowest degree with rational coefficients that has $-4i$ and 5 as some of its roots.

$4i$ is also a root. (Theorem 6)

Thus the polynomial is

$a_n(x - 5)(x + 4i)(x - 4i)$.

If we let $a_n = 1$, we obtain

$(x - 5)[x^2 - (4i)^2]$

$= (x - 5)(x^2 + 16)$

$= x^3 - 5x^2 + 16x - 80$

16. $x^4 - 6x^3 + 11x^2 - 10x + 2$

17. Find a polynomial of lowest degree with rational coefficients that has $\sqrt{5}$ and $-3i$ as some of its roots.

$-\sqrt{5}$ is also a root. (Theorem 7)

$3i$ is also a root. (Theorem 6)

Thus the polynomial is

$a_n(x - \sqrt{5})(x + \sqrt{5})(x + 3i)(x - 3i)$.

If we let $a_n = 1$, we obtain

$(x^2 - 5)(x^2 + 9)$

$= x^4 + 4x^2 - 45$

18. $x^4 + 14x^2 - 32$

19. If $-i$ is a root of $x^4 - 5x^3 + 7x^2 - 5x + 6$, i is also a root (Theorem 6). Thus $x + i$ and $x - i$ or $(x + i)(x - i)$ which is $x^2 + 1$ are factors of the polynomial. Divide $x^4 - 5x^3 + 7x^2 - 5x + 6$ by $x^2 + 1$ to find the other factors.

$$\begin{array}{r|l} & x^2 - 5x + 6 \\ \hline x^2 + 1 & x^4 - 5x^3 + 7x^2 - 5x + 6 \\ & \underline{x^4 \qquad\quad + x^2} \\ & -5x^3 + 6x^2 - 5x \\ & \underline{-5x^3 \qquad\quad - 5x} \\ & 6x^2 \qquad + 6 \\ & \underline{6x^2 \qquad + 6} \\ & 0 \end{array}$$

Thus

$x^4 - 5x^3 + 7x^2 - 5x + 6 = (x + i)(x - i)(x^2 - 5x + 6)$

$= (x + i)(x - i)(x - 2)(x - 3)$

Using the principle of zero products we find the other roots to be i, 2, and 3.

20. $-2i$, 2, -2

21. $x^3 - 6x^2 + 13x - 20 = 0$

If 4 is a root, then $x - 4$ is a factor. Use synthetic division to find another factor.

$$\begin{array}{r|rrrr} 4 & 1 & -6 & 13 & -20 \\ & & 4 & -8 & 20 \\ \hline & 1 & -2 & 5 & 0 \end{array}$$

$(x - 4)(x^2 - 2x + 5) = 0$

$x - 4 = 0$ or $x^2 - 2x + 5 = 0$ (Principle of zero products)

$x = 4$ or $x = \dfrac{2 \pm \sqrt{4 - 20}}{2}$ (Quadratic formula)

$x = 4$ or $x = \dfrac{2 \pm 4i}{2} = 1 \pm 2i$

The other roots are $1 + 2i$ and $1 - 2i$.

22. $-1 + \sqrt{3}i$, $-1 - \sqrt{3}i$

23. $x^3 - 4x^2 + x - 4 = 0$

Using synthetic division we find that i is a root.

$$\begin{array}{r|rrrr} i & 1 & -4 & 1 & -4 \\ & & i & -4i - 1 & 4 \\ \hline & 1 & -4 + i & -4i & 0 \end{array}$$

Using Theorem 6 we know that if i is a root then $-i$ is also a root. Thus $(x - i)(x + i)$, or $x^2 + 1$, is a factor of $x^3 - 4x^2 + x - 4$. Using division we find that $x - 4$ is also a factor.

$$\begin{array}{r|l} & x - 4 \\ \hline x^2 + 1 & x^3 - 4x^2 + x - 4 \\ & \underline{x^3 \qquad\quad + x} \\ & -4x^2 \qquad - 4 \\ & \underline{-4x^2 \qquad - 4} \\ & 0 \end{array}$$

Therefore, $(x - i)(x + i)(x - 4) = 0$

Using the principle of zero products we find that the solutions are i, $-i$, and 4.

24. -3, $2 + i$, $2 - i$

25. $x^4 - 2x^3 - 2x - 1 = 0$

Using synthetic division we find that i is a root.

$$\begin{array}{r|rrrrr} i & 1 & -2 & 0 & -2 & -1 \\ & & i & -2i - 1 & 2 - i & 1 \\ \hline & 1 & -2 + i & -2i - 1 & -i & 0 \end{array}$$

Using Theorem 6 we know that if i is a root then $-i$ is also a root. Thus $(x - i)(x + i)$, or $x^2 + 1$, is a factor of $x^4 - 2x^3 - 2x - 1$. Using division we find that $x^2 - 2x - 1$ is also a factor.

25. (continued)

$$\begin{array}{r|l} & x^2 - 2x - 1 \\ \hline x^2 + 1 & x^4 - 2x^3 + 0x^2 - 2x - 1 \\ & \underline{x^4 \qquad\quad + x^2} \\ & -2x^3 - x^2 - 2x \\ & \underline{-2x^3 \qquad - 2x} \\ & -x^2 \qquad - 1 \\ & \underline{-x^2 \qquad - 1} \\ & 0 \end{array}$$

Therefore, $(x - i)(x + i)(x^2 - 2x - 1) = 0$. Using the principle of zero products and the quadratic formula, we find that the roots are i, $-i$, $1 + \sqrt{2}$, and $1 - \sqrt{2}$.

26. $-a$

27. For $P(x) = a_nx^n + a_{n-1}x^{n-1} + \ldots + a_0$ with a_i positive, consider any positive x. Every term will be positive, hence $P(x) > 0$.

Exercise Set 5.4

1. $P(x) = x^5 - 3x^2 + 1$

Since the leading coefficient is 1, the only possibilities for rational roots (Theorem 8) are the factors of the last coefficient 1: 1 and -1.

2. $\pm$ (1, 2, 3, 4, 6, 12)

3. $P(x) = 15x^6 + 47x^2 + 2$

By Theorem 8, if c/d, a rational number in lowest terms, is a root of P(x), then c must be a factor of 2 and d must be a factor of 15.

The possibilities for c and d are

c: 1, -1, 2, -2 d: 1, -1, 3, -3, 5, -5, 15, -15

The resulting possibilities for c/d are

$\frac{c}{d}$: $1, -1, \frac{1}{3}, -\frac{1}{3}, \frac{1}{5}, -\frac{1}{5}, \frac{1}{15}, -\frac{1}{15}$

$2, -2, \frac{2}{3}, -\frac{2}{3}, \frac{2}{5}, -\frac{2}{5}, \frac{2}{15}, -\frac{2}{15}$

4. $\pm \left\{1, 2, 3, 6, \frac{1}{10}, \frac{1}{5}, \frac{3}{10}, \frac{3}{5}, \frac{1}{2}, \frac{3}{2}, \frac{2}{5}, \frac{6}{5}\right\}$

5. $P(x) = x^3 + 3x^2 - 2x - 6$

Using Theorem 8, the only possible rational roots are 1, -1, 2, -2, 3, -3, 6 and -6. Of these eight possibilities we know that at most three of them could be roots because P(x) is of degree 3. Use synthetic division to determine which are roots.

1	1	3	-2	-6
		1	4	2
	1	4	2	-4

$P(1) \neq 0$, so 1 is not a root.

5. (continued)

-1	1	3	-2	-6
		-1	-2	4
	1	2	-4	-2

$P(-1) \neq 0$, so -1 is not a root.

2	1	3	-2	-6
		2	10	16
	1	5	8	10

$P(2) \neq 0$, so 2 is not a root.

-2	1	3	-2	-6
		-2	-2	8
	1	1	-4	2

$P(-2) \neq 0$, so -2 is not a root.

3	1	3	-2	-6
		3	18	48
	1	6	16	42

$P(3) \neq 0$, so 3 is not a root.

-3	1	3	-2	-6
		-3	0	6
	1	0	-2	0

$P(-3) = 0$, so -3 is a root.

We can now express P(x) as follows:

$P(x) = (x + 3)(x^2 - 2)$.

To find the other roots we solve the quadratic equation $x^2 - 2 = 0$. The other roots are $\sqrt{2}$ and $-\sqrt{2}$. These are irrational numbers. Thus the only rational root is -3.

6. $1, \sqrt{3}, -\sqrt{3}$; $(x - 1)(x^2 - 3) = 0$

7. $P(x) = 5x^4 - 4x^3 + 19x^2 - 16x - 4$

Using Theorem 8, the only possible rational roots are $\pm \left\{1, 2, 4, \frac{1}{5}, \frac{2}{5}, \frac{4}{5}\right\}$. Of these twelve possibilities we know that at most four of them could be roots because P(x) is of degree 4. Using synthetic division we determine that the only rational roots are 1 and $-\frac{1}{5}$.

1	5	-4	19	-16	-4
		5	1	20	4
	5	1	20	4	0

$P(1) = 0$, so 1 is a root.

We can now express P(x) as follows:
$P(x) = (x - 1)(5x^3 + x^2 + 20x + 4)$.

We now use $5x^3 + x^2 + 20x + 4$ and check to see if 1 is a double root and check for other possible rational roots. The only other rational root is $-\frac{1}{5}$.

$-\frac{1}{5}$	5	1	20	4
		-1	0	-4
	5	0	20	0

7. (continued)

$P\left(-\frac{1}{5}\right) = 0$, so $-\frac{1}{5}$ is a root.

We can now express P(x) as follows:

$P(x) = (x - 1)\left(x + \frac{1}{5}\right)(5x^2 + 20)$

To find the other roots we solve the quadratic equation $5x^2 + 20 = 0$.

$5x^2 + 20 = 0$

$5x^2 = -20$

$x^2 = -4$

$x = \pm 2i$

The other roots are 2i and -2i, neither of which is rational. The only rational roots are 1 and $-\frac{1}{5}$.

8. $\frac{1}{3}$, -1, 1 - i, 1 + i;

$P(x) = (x + 1)(3x - 1)(x^2 - 2x + 2)$

9. $P(x) = x^4 - 3x^3 - 20x^2 - 24x - 8$

Using Theorem 8, the only possible rational roots are 1, -1, 2, -2, 4, -4, 8, and -8. Of these eight possibilities, we know that at most four of them could be roots because P(x) is of degree 4. Using synthetic division we determine that the only rational roots are -1 and -2.

-1	1	-3	-20	-24	-8
		-1	4	16	8
	1	-4	-16	-8	0

P(-1) = 0, so -1 is a root.

We can now express P(x) as follows:

$P(x) = (x + 1)(x^3 - 4x^2 - 16x - 8)$.

We now use $x^3 - 4x^2 - 16x - 8$ and check to see if -1 is a double root and check for other possible rational roots. The only other rational root is -2.

-2	1	-4	-16	-8
		-2	12	8
	1	-6	-4	0

P(-2) = 0, so -2 is a root.

We can now express P(x) as follows:

$P(x) = (x + 1)(x + 2)(x^2 - 6x - 4)$.

Since the factor $x^2 - 6x - 4$ is quadratic, use the quadratic formula to find the other roots, $3 + \sqrt{13}$ and $3 - \sqrt{13}$. These are irrational numbers. The only rational roots are -1 and -2.

10. 1, 2, $-4 \pm \sqrt{21}$;

$P(x) = (x - 1)(x - 2)(x^2 + 8x - 5)$

11. $P(x) = x^3 - 4x^2 + 2x + 4$

Using Theorem 8, the only possible rational roots are 1, -1, 2, -2, 4, and -4. Of these six possibilities, we know that at most three of them could be roots, because P(x) is of degree 3. Use synthetic division to determine which are roots.

1	1	-4	2	4
		1	-3	-1
	1	-3	-1	3

$P(1) \neq 0$, so 1 is not a root.

-1	1	-4	2	4
		-1	5	-7
	1	-5	7	-3

$P(-1) \neq 0$, so -1 is not a root.

2	1	-4	2	4
		2	-4	-4
	1	-2	-2	0

P(2) = 0, so 2 is a root.

We can now express P(x) as

$P(x) = (x - 2)(x^2 - 2x - 2)$.

To find the other roots we use the quadratic formula to solve the equation $x^2 - 2x - 2 = 0$. The other roots are $1 + \sqrt{3}$ and $1 - \sqrt{3}$. These are irrational roots.

The only rational root is 2.

12. 4, $2 \pm \sqrt{3}$; $P(x) = (x - 4)(x^2 - 4x + 1)$

13. $P(x) = x^3 + 8$

Using Theorem 8, the only possible rational roots are 1, -1, 2, -2, 4, -4, 8, and -8. Of these eight possibilities we know that at most three of them could be roots because P(x) is of degree 3. Using synthetic division we determine that -2 is the only rational root.

-2	1	0	0	8
		-2	4	-8
	1	-2	4	0

P(-2) = 0, so -2 is a root.

We can now express P(x) as follows:

$P(x) = (x + 2)(x^2 - 2x + 4)$

Since the factor $x^2 - 2x + 4$ is quadratic, use the quadratic formula to find the other roots, $1 + \sqrt{3}i$ and $1 - \sqrt{3}i$. These are irrational numbers. Thus the only rational root is -2.

14. 2, $-1 + i\sqrt{3}$, $-1 - i\sqrt{3}$;

$P(x) = (x - 2)(x^2 + 2x + 4) = 0$

15. $P(x) = \frac{1}{3}x^3 - \frac{1}{2}x^2 - \frac{1}{6}x + \frac{1}{6}$

$6P(x) = 2x^3 - 3x^2 - x + 1$

We multiplied by 6, the LCM of the denominators. This equation is equivalent to the first, and all coefficients on the right are integers. Thus any root of $6P(x)$ is a root of $P(x)$.

The possible rational roots of $2x^3 - 3x^2 - x + 1$ are $1, -1, \frac{1}{2}$, and $-\frac{1}{2}$. Of these four possibilities we know that at most three of them could be roots because the degree of $6P(x)$ is 3. Using synthetic division we determine that the only rational root is $\frac{1}{2}$.

$\frac{1}{2}$	2	-3	-1	1
		1	-1	-1
	2	-2	-2	0

Thus $6P\left(\frac{1}{2}\right) = 0$, so $\frac{1}{2}$ is a root.

We can now express $6P(x)$ as follows:

$$6P(x) = \left(x - \frac{1}{2}\right)(2x^2 - 2x - 2)$$
$$= 2\left(x - \frac{1}{2}\right)(x^2 - x - 2)$$

Since the factor $x^2 - x - 2$ is quadratic, use the quadratic formula to find the other roots, $\frac{1 + \sqrt{5}}{2}$ and $\frac{1 - \sqrt{5}}{2}$. These are irrational numbers. Thus the only rational root is $\frac{1}{2}$.

16. $\frac{3}{4}$, i, -i; $(4x - 3)(x^2 + 1) = 0$

17. $P(x) = x^4 + 32$

Using Theorem 8, the only possible rational roots are $\pm$ (1, 2, 4, 8, 16, 32). Of these twelve possibilities we know that at most four of them could be roots because P(x) is of degree 4. Use synthetic division to check each possibility.

1	1	0	0	0	32
		1	1	1	1
	1	1	1	1	33

-1	1	0	0	0	32
		-1	1	-1	1
	1	-1	1	-1	33

2	1	0	0	0	32
		2	4	8	16
	1	2	4	8	48

-2	1	0	0	0	32
		-2	4	-8	16
	1	-2	4	-8	48

$P(1) \neq 0$, $P(-1) \neq 0$, $P(2) \neq 0$, $P(-2) \neq 0$; therefore 1, -1, 2, and -2 are not roots. Similarly we can show that $\pm$ (4, 8, 16, 32) are not roots. Thus there are no rational roots.

18. None

19. $x^3 - x^2 - 4x + 3 = 0$

The possible rational roots are 1, -1, 3, and -3. Of these four possibilities, at most three of them could be roots because the polynomial equation is of degree 3. Use synthetic division to check each possibility.

1	1	-1	-4	3
		1	0	-4
	1	0	-4	-1

-1	1	-1	-4	3
		-1	2	2
	1	-2	-2	5

3	1	-1	-4	3
		3	6	6
	1	2	2	9

-3	1	-1	-4	3
		-3	12	-24
	1	-4	8	-21

$P(1) \neq 0$, $P(-1) \neq 0$, $P(3) \neq 0$, $P(-3) \neq 0$; therefore 1, -1, 3, and -3 are not roots. Thus there are no rational roots.

20. $-\frac{3}{2}$

21. $x^4 + 2x^3 + 2x^2 - 4x - 8 = 0$

The possible rational roots are $\pm$ (1, 2, 4, 8). Of these eight possibilities, at most four of them could be roots because the polynomial equation is of degree 4. Use synthetic division to check each possibility.

1	1	2	2	-4	-8
		1	3	5	1
	1	3	5	1	-7

-1	1	2	2	-4	-8
		-1	-1	-1	5
	1	1	1	-5	-3

2	1	2	2	-4	-8
		2	8	20	32
	1	4	10	16	24

-2	1	2	2	-4	-8
		-2	0	-4	16
	1	0	2	-8	8

$P(1) \neq 0$, $P(-1) \neq 0$, $P(2) \neq 0$, $P(-2) \neq 0$; therefore 1, -1, 2, and -2 are not roots. Similarly we can show that 4, -4, 8, and -8 are not roots. Thus there are no rational roots.

22. None

23. $P(x) = x^5 - 5x^4 + 5x^3 + 15x^2 - 36x + 20$

The possible rational roots are $\pm$ (1, 2, 4, 5, 10, 20). Of these twelve possibilities at most five of them could be roots because P(x) is of degree 5. Using synthetic division we determine the only rational roots are -2, 1, and 2.

-2	1	-5	5	15	-36	20
		-2	14	-38	46	-20
	1	-7	19	-23	10	0

P(-2) = 0, so -2 is a root.

1	1	-5	5	15	-36	20
		1	-4	1	16	-20
	1	-4	1	16	-20	0

P(1) = 0, so 1 is a root.

2	1	-5	5	15	-36	20
		2	-6	-2	26	-20
	1	-3	-1	13	-10	0

P(2) = 0, so 2 is a root.

Similarly we can show that $P(-1) \neq 0$, $P(-4) \neq 0$, $P(4) \neq 0$, $P(-5) \neq 0$, $P(5) \neq 0$, $P(-10) \neq 0$, $P(10) \neq 0$, $P(-20) \neq 0$, and $P(20) \neq 0$. Thus -1, -4, 4, -5, 5, -10, 10, -20, and 20 are not roots. The only rational roots are -2, 1, and 2.

24. 2, -2, 3

25. $x^3 - 64 = 0$

$x^3 = 64$

$x = 4$

The length of a side is 4 cm.

26. 5 cm

27.

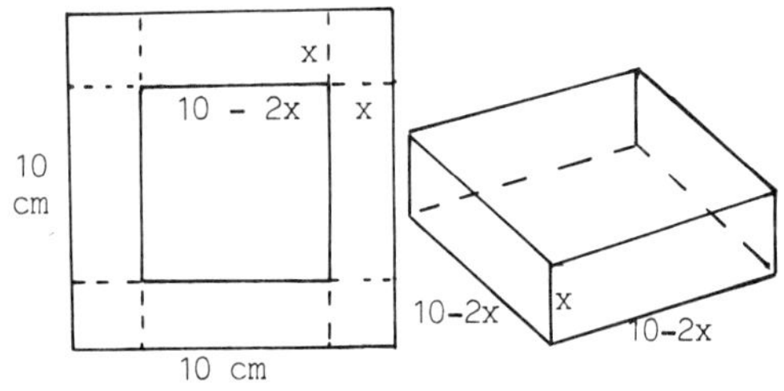

$V = \ell wh$ (Volume formula for rectangular prism)

$48 = (10 - 2x)(10 - 2x)(x)$

(Substituting 10 - 2x for ℓ and w, x for h, and 48 for V)

$48 = 100x - 40x^2 + 4x^3$

$12 = 25x - 10x^2 + x^3$

$0 = x^3 - 10x^2 + 25x - 12$

The possible rational roots of this equation are $\pm$ (1, 2, 3, 4, 6, 12). Using synthetic division we find that 3 is the only rational root.

27. (continued)

3	1	-10	25	-12
		3	-21	12
	1	-7	4	0

We can now express the polynomial equation as follows:

$0 = (x - 3)(x^2 - 7x + 4)$

Since the factor $x^2 - 7x + 4$ is quadratic, we use the quadratic formula to find the other roots, $\frac{7 + \sqrt{33}}{2}$ and $\frac{7 - \sqrt{33}}{2}$. Since the length, x, must be positive and less then 5, $\frac{7 + \sqrt{33}}{2}$ is not a possible solution. The length of a side of the square can be 3 cm or $\frac{7 - \sqrt{33}}{2}$ cm.

28. 5 cm, $\frac{15 - 5\sqrt{5}}{2}$ cm

Exercise Set 5.5

1. $3x^5 - 2x^2 + x - 1$

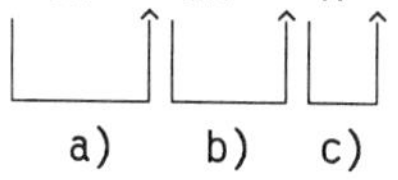

a) From positive to negative: a variation
b) From negative to positive: a variation
c) From positive to negative: a variation

The number of variations of sign is three. Therefore, the number of positive real roots is either 3 or 1 (Theorem 9).

2. 3 or 1

3. $6x^7 + 2x^2 + 5x + 4 = 0$

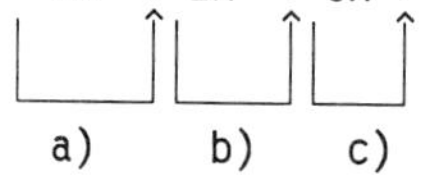

a) From positive to positive: no variation
b) From positive to positive: no variation
c) From positive to positive: no variation

There are no variations of sign. Thus, there are no positive real roots (Theorem 9).

4. None

5. $3p^{18} + 2p^4 - 5p^2 + p + 3$

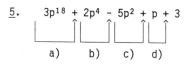

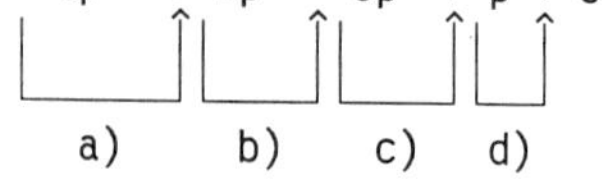

a) From positive to positive: no variation
b) From positive to negative: a variation
c) From negative to positive: a variation
d) From positive to positive: no variation

5. (continued)

The number of variations of sign is two. Therefore, the number of positive real roots is either 2 or 0 (Theorem 9).

6. 2 or 0

7. $P(x) = 3x^5 - 2x^2 + x - 1$

Replace x by -x.

$P(-x) = 3(-x)^5 - 2(-x)^2 + (-x) - 1$

$= -3x^5 - 2x^2 - x - 1$

a) b) c)

a) From negative to negative: no variation
b) From negative to negative: no variation
c) From negative to negative: no variation

There are no variations of sign. Thus, there are no negative real roots of P(x) (Theorem 10).

8. None

9. $6x^7 + 2x^2 + 5x + 4 = 0$

Replace x by -x.

$6(-x)^7 + 2(-x)^2 + 5(-x) + 4 = 0$

$-6x^7 + 2x^2 - 5x + 4 = 0$

a) b) c)

a) From negative to positive: a variation
b) From positive to negative: a variation
c) From negative to positive: a variation

The number of variations of signs is three. Therefore, the number of negative real roots is 3 or 1 (Theorem 10).

10. 1

11. $P(p) = 3p^{18} + 2p^3 - 5p^2 + p + 3$

Replace p with -p.

$P(-p) = 3(-p)^{18} + 2(-p)^3 - 5(-p)^2 + (-p) + 3$

$= 3p^{18} - 2p^3 - 5p - p + 3$

a) b) c) d)

a) From positive to negative: a variation
b) From negative to negative: no variation
c) From negative to negative: no variation
d) From negative to positive: a variation

The number of variations of sign is two. Therefore, the number of negative real roots of P(p) is 2 or 0 (Theorem 10).

12. 3 or 1

13. Determine an upper bound to the roots of

$3x^4 - 15x^2 + 2x - 3$

Remember when using Theorem 11 that the upper bound we are finding is positive. Also note that the upper bound is not unique. Answers may vary.

Try 1:

1	3	0	-15	2	-3
		3	3	-12	-10
	3	3	-12	-10	-13

Since some of the coefficients of the quotient and the remainder are negative, there is no guarantee that 1 is an upper bound (Theorem 11).

Try 2:

2	3	0	-15	2	-3
		6	12	-6	-8
	3	6	-3	-4	-11

Again there is no guarantee that 2 is an upper bound.

Try 3:

3	3	0	-15	2	-3
		9	27	36	114
	3	9	12	38	111

Since all of the numbers in the bottom row are nonnegative, 3 is an upper bound (Theorem 11).

14. 2 (Answers may vary.)

15. Determine an upper bound to the roots of

$6x^3 - 17x^2 - 3x - 1$

Remember when using Theorem 11 that the upper bound we are finding is positive. Also note that the upper bound is not unique. Answers may vary.

Try 2:

2	6	-17	-3	-1
		12	-10	-26
	6	-5	-13	-27

Since some of the coefficients of the quotient and the remainder are negative, there is no guarantee that 2 is an upper bound.

Try 3:

3	6	-17	-3	-1
		18	3	0
	6	1	0	-1

Since the remainder is negative, there is no guarantee that 3 is an upper bound.

Try 4:

4	6	-17	-3	-1
		24	28	100
	6	7	25	99

15. (continued)

Since all of the numbers in the bottom row are nonnegative, 4 is an upper bound.

16. 3 (Answers may vary.)

17. Determine a lower bound to the roots of
$3x^4 - 15x^3 + 2x - 3$

Here we use Theorem 13. Remember that the lower bound we are finding is negative. Also note that the lower bound is not unique. Answers may vary.

Try -1:

-1	3	-15	0	2	-3
		-3	18	-18	16
	3	-18	18	-16	13

Since the odd-numbered coefficients (from left to right) are nonnegative and the even-numbered ones are nonpositive, -1 is a lower bound.

18. -1 (Answers may vary.)

19. Determine a lower bound to the roots of
$6x^3 + 15x^2 + 3x - 1$

Here we use Theorem 13. Remember that the lower bound we are finding is negative. Also note that the lower bound is not unique. Answers may vary.

Try -2:

-2	6	15	3	-1
		-12	-6	6
	6	3	-3	5

We do not know whether or not -2 is a lower bound.

Try -3:

-3	6	15	3	-1
		-18	9	-36
	6	-3	12	-37

Since the odd-numbered coefficients (from left to right) are nonnegative and the even-numbered ones are nonpositive, -3 is a lower bound.

20. -2 (Answers may vary.)

21. $P(x) = x^4 - 2x^2 + 12x - 8$

There are three variations of sign, so the number of positive real roots is 3 or 1 (Theorem 9).

$P(-x) = (-x)^4 - 2(-x)^2 + 12(-x) - 8$
$= x^4 - 2x^2 - 12x - 8$

There is one variation of sign, so the number of negative real roots of P(x) is 1 (Theorem 10).

Look for an upper bound (Use Theorem 11).

21. (continued)

Try 1:

1	1	0	-2	12	-8
		1	1	-1	11
	1	1	-1	11	3

Since one of the coefficients is negative, we do not know whether or not 1 is an upper bound.

Try 2:

2	1	0	-2	12	-8
		2	4	4	32
	1	2	2	16	24

Since all of the numbers in the bottom row are nonnegative, 2 is an upper bound.

Look for a lower bound (Use Theorem 13).

Try -2:

-2	1	0	-2	12	-8
		-2	4	-4	-16
	1	-2	2	8	-24

We do not know whether or not -2 is a lower bound.

Try -3:

-3	1	0	-2	12	-8
		-3	9	-21	27
	1	-3	7	-9	19

Since the odd-numbered coefficients (from left to right) are nonnegative and the even-numbered ones are nonpositive, -3 is a lower bound.

22. 3 or 1 positive, 1 negative,
Upper bound: 3, Lower bound: -4

23. $x^4 - 2x^2 - 8 = 0$

There is one variation of sign, so the number of positive real roots is 1 (Theorem 9).

$(-x)^4 - 2(-x)^2 - 8 = 0$ (Replacing x by -x)
$x^4 - 2x^2 - 8 = 0$

There is one variation of sign, so the number of negative real roots is 1 (Theorem 10). Since the degree of the equation is 4 and there is only one positive real root and one negative real root, there must be two nonreal roots.

Look for an upper bound (Use Theorem 11).

Try 1:

1	1	0	-2	0	-8
		1	1	-1	-1
	1	1	-1	-1	-9

Since some of the coefficients of the quotient and the remainder are negative, we do not know whether or not 1 is an upper bound.

23. (continued)

Try 2:

2	1	0	-2	0	-8
		2	4	4	8
	1	2	2	4	0

Since all of the numbers in the bottom row are nonnegative, 2 is an upper bound. Also note that 2 is a root, P(2) = 0.

Look for a lower bound (Use Theorem 13).

Try -1:

-1	1	0	-2	0	-8
		-1	1	1	-1
	1	-1	-1	1	-9

We do not know whether or not -1 is a lower bound.

Try -2:

-2	1	0	-2	0	-8
		-2	4	-4	8
	1	-2	2	-4	0

Since the odd-numbered coefficients (from left to right) are nonnegative and the even-numbered ones are nonpositive, -2 is a lower bound.

24. 1 positive, 1 negative,
Upper bound: 2, Lower bound: -2

25. $P(x) = x^4 - 9x^2 - 6x + 4$

There are two variations of sign, so the number of positive real roots is 2 or 0 (Theorem 9).

$$P(-x) = (-x)^4 - 9(-x)^2 - 6(-x) + 4$$
$$= x^4 - 9x^2 + 6x + 4$$

There are two variations of sign, so the number of negative real roots of P(x) is 2 or 0 (Theorem 10).

Look for an upper bound (Use Theorem 11).

Try 3:

3	1	0	-9	-6	4
		3	9	0	-18
	1	3	0	-6	-14

Since one of the coefficients and the remainder are negative, we do not know whether or not 3 is an upper bound.

Try 4:

4	1	0	-9	-6	4
		4	16	28	88
	1	4	7	22	92

Since all of the numbers in the bottom row are nonnegative, 4 is an upper bound.

Look for a lower bound (Use Theorem 13).

25. (continued)

Try -2:

-2	1	0	-9	-6	4
		-2	4	10	-8
	1	-2	-5	4	-4

We do not know whether or not -2 is a lower bound.

Try -3:

-3	1	0	-9	-6	4
		-3	9	0	18
	1	-3	0	-6	22

Since the odd-numbered coefficients (from left to right) are nonnegative and the even-numbered ones are nonpositive, -3 is a lower bound.

26. 2 or 0 positive, 2 or 0 negative,
Upper bound: 5, Lower bound: -5

27. $x^4 + 3x^2 + 2 = 0$

There are no variations of sign. Thus there are no negative real roots (Theorem 10).

Since there are no real roots, ther is no upper bound and no lower bound.

28. No positive or negative roots

29. See answer section in text.

30. Let $P(x) = x^n - 1$. There is one variation of sign, hence one positive root.

$P(-x) = (-x)^n - 1$ has no variation, so there is no negative root.

Exercise Set 5.6

1. Graph: $P(x) = x^3 - 3x^2 - 2x - 6$

The polynomial is of degree 3 with leading coefficient positive. Thus as we move far to the right, function values will increase beyond bound; and as we move far to the left, function values will decrease beyond bound.

Make a table of values. The first entry is the y-intercept. The other rows contain the bottom line in synthetic division.

1. (continued)

	1	-3	-2	-6	(x, y)
0				-6	(0, -6)*
1	1	-2	-4	-10	(1, -10)
2	1	-1	-4	-14	(2, -14)
3	1	0	-2	-12	(3, -12)
4	1	1	2	2	(4, 2)**
5	1	2	8	34	(5, 34)
6	1	3	16	90	(6, 90)
-1	1	-4	2	-8	(-1, -8)***
-2	1	-5	8	-22	(-2, -22)
-3	1	-6	16	-54	(-3, -54)

* (0, -6) (The y-intercept)

** (4, 2) (4 is an upper bound to the roots)

*** (-1, -8) (-1 is a lower bound to the roots)

By Descartes' rule of signs we know that there is one positive real root and either 2 or 0 negative real roots. From the table we see that the graph will cross the x-axis between 3 and 4. This is because P(3) is negative and P(4) is positive. Somewhere between 3 and 4, then, P(x) must be 0. Since -1 is a lower bound, negative real roots, if they exist, must occur between -1 and 0. We find a few more values using synthetic division.

	1	-3	-2	-6	(x, y)
$-\frac{3}{4}$	1	$-\frac{15}{4}$	$\frac{13}{16}$	$-6\frac{39}{64}$	$\left(-\frac{3}{4}, -6\frac{39}{64}\right)$
$-\frac{1}{2}$	1	$-\frac{7}{2}$	$-\frac{1}{4}$	$-5\frac{7}{8}$	$\left(-\frac{1}{2}, -5\frac{7}{8}\right)$
$-\frac{1}{4}$	1	$-\frac{13}{4}$	$-\frac{19}{16}$	$-5\frac{45}{64}$	$\left(-\frac{1}{4}, -5\frac{45}{64}\right)$

We now conclude that P(x) < 0 between -1 and 0. Thus there are no negative real roots. If there are no negative real roots, there will be two nonreal ones, and they will be conjugates of each other. We now plot points and connect them. See the graph in the text.

2. $P(x) = x^3 + 4x^2 - 3x - 12$

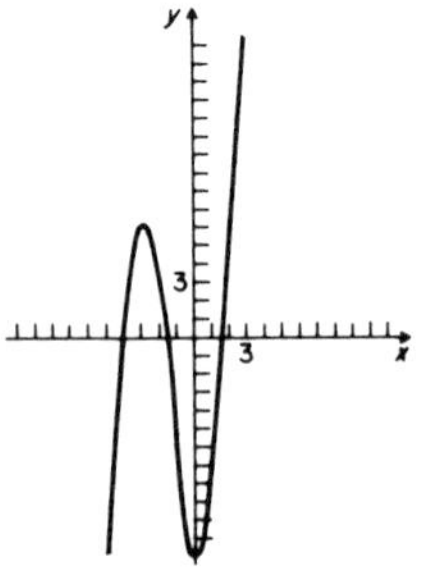

3. Graph: $P(x) = 2x^4 + x^3 - 7x^2 - x + 6$

This polynomial is of degree 4 with leading coefficient positive. Thus as we move far to the right or far to the left, function values will increase beyond bound.

Make a table of values. The first entry is the y-intercept. The other rows contain the bottom line in synthetic division.

	2	1	-7	-1	6	(x, y)
0					6	(0, 6)*
1	2	3	-4	-5	1	(1, 1)
2	2	5	3	5	16	(2, 16)**
3	2	7	14	41	129	(3, 129)
-1	2	-1	-6	5	1	(-1, 1)
-2	2	-3	-1	1	4	(-2, 4)
-3	2	-5	8	-25	81	(-3, 81)***

* (0, 6) (The y-intercept)

** (2, 16) (2 is an upper bound to the roots)

*** (-3, 81) (-3 is a lower bound to the roots)

After plotting the above points, we realize it would be helpful to also know the values listed below.

	2	1	-7	-1	6	(x, y)
$-\frac{3}{2}$	2	-2	-4	5	$-\frac{3}{2}$	$\left(-\frac{3}{2}, -\frac{3}{2}\right)$
$-\frac{1}{2}$	2	0	-7	$\frac{5}{2}$	$\frac{19}{4}$	$\left(-\frac{1}{2}, \frac{19}{4}\right)$
$\frac{1}{2}$	2	2	-6	-4	4	$\left(\frac{1}{2}, 4\right)$
$\frac{3}{2}$	2	4	-1	$-\frac{5}{2}$	$\frac{9}{4}$	$\left(\frac{3}{2}, \frac{9}{4}\right)$

By Descartes' rule of signs we know that there are either 2 or 0 negative real roots. From the tables we see that the graph will cross the x-axis between -2 and $-\frac{3}{2}$. This is because P(-2) is positive and $P\left(-\frac{3}{2}\right)$ is negative. Somewhere between -2 and $-\frac{3}{2}$, then, P(x) must be 0. We also see that the graph will cross the x-axis between $-\frac{3}{2}$ and -1. This is because $P\left(-\frac{3}{2}\right)$ is negative and P(-1) is positive. Somewhere between $-\frac{3}{2}$ and -1, P(x) must be 0. Thus there will be two negative real roots.

By Descartes' rule of signs, we also know that there are either 2 or 0 positive real roots. From the tables we see that if there are positive real roots, they must be between 1 and $\frac{3}{2}$.

Make a table of values.

3. (continued)

	2	1	-7	-1	6	(x, y)
$\frac{9}{8}$	2	$\frac{13}{4}$	$-\frac{107}{32}$	$-\frac{1219}{256}$	$\frac{1317}{2048}$	$\left(\frac{9}{8}, \frac{1317}{2048}\right)$
$\frac{5}{4}$	2	$\frac{7}{2}$	$-\frac{21}{8}$	$-\frac{137}{32}$	$\frac{83}{128}$	$\left(\frac{5}{4}, \frac{83}{128}\right)$
$\frac{11}{8}$	2	$\frac{15}{4}$	$-\frac{59}{32}$	$-\frac{905}{256}$	$\frac{2333}{2048}$	$\left(\frac{11}{8}, \frac{2333}{2048}\right)$

We now conclude that $P(x) > 0$ for $x > 0$, thus there will be no positive real roots. If there are no positive roots, there will be two nonreal ones, and they will be conjugates of each other. We now plot points and connect them. See the graph in the text.

4. $P(x) = 3x^4 + 5x^3 + 5x^2 - 5x - 6$

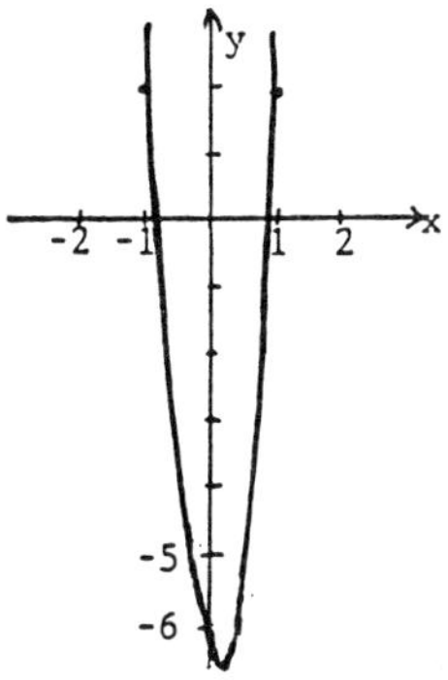

5. Graph: $P(x) = x^5 - 2x^4 - x^3 + 2x^2$

The polynomial is of degree 5 with leading coefficient positive. Thus as we move far to the right, function values will increase beyond bound; and as we move far to the left, function values will decrease beyond bound.

Make a table of values. The first entry is the y-intercept. The other rows contain the bottom line in synthetic division.

	1	-2	-1	2	0	0	(x, y)
0						0	(0, 0)
1	1	-1	-2	0	0	0	(1, 0)
2	1	0	-1	0	0	0	(2, 0)
3	1	1	2	8	24	72	(3, 72)
-1	1	-3	2	0	0	0	(-1, 0)
-2	1	-4	7	-12	24	-48	(-2, -48)

Note:

(0, 0) (y-intercept, 0 is a double root)

(1, 0) (1 is a root)

(2, 0) (2 is a root)

(-1, 0) (-1 is a root)

Since the polynomial is of degree 5, it will have at most 5 roots. Factoring the polynomial we can determine that 0 is a double root:

$P(x) = x^2(x^3 - 2x^2 - x + 2)$. From the table above we see that the five roots are -1, 0 (double), 1 and 2.

5. (continued)

After plotting the points in the table above, we realize it would be helpful to also know the values listed below.

	1	-2	-1	2	0	0	(x, y)
$-\frac{1}{2}$	1	$-\frac{5}{2}$	$\frac{1}{4}$	$\frac{15}{8}$	$-\frac{15}{16}$	$\frac{15}{32}$	$\left(-\frac{1}{2}, \frac{15}{32}\right)$
$\frac{1}{2}$	1	$-\frac{3}{2}$	$-\frac{7}{4}$	$\frac{9}{8}$	$\frac{9}{16}$	$\frac{9}{32}$	$\left(\frac{1}{2}, \frac{9}{32}\right)$
$\frac{3}{2}$	1	$-\frac{1}{2}$	$-\frac{7}{4}$	$-\frac{5}{8}$	$-\frac{15}{16}$	$-\frac{45}{32}$	$\left(\frac{3}{2}, -\frac{45}{32}\right)$

We plot these values and complete the graph. See the graph in the text.

6. $P(x) = x^5 + 4x^4 - 5x^3 - 14x^2 - 8x$

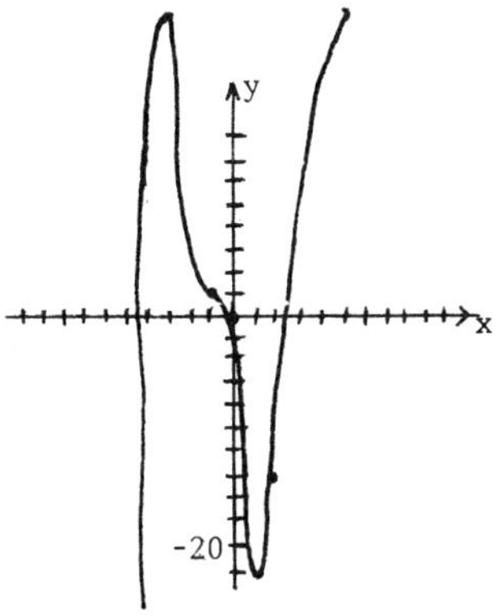

7. Graph $P(x) = x^3 - 3x - 2$, the corresponding polynomial function for the equation $x^3 - 3x - 2 = 0$. This polynomial is of degree 3 with leading coefficient positive. Thus as we move far to the right, function values will increase beyond bound; and as we move far to the left, function values will decrease beyond bound. The function is not odd or even. However, $x^3 - 3x$ is an odd function, with the origin as a point of symmetry. The function $P(x)$ is a translation of the function $x^3 - 3x$ downward 2 units, hence the point (0, -2) is a point of symmetry.

Make a table of values. The first entry is the y-intercept. The other rows contain the bottom line in synthetic division.

	1	0	-3	-2	(x, y)
0				-2	(0, -2)*
1	1	1	-2	-4	(1, -4)
2	1	2	1	0	(2, 0)**
3	1	3	6	16	(3, 16)
-1	1	-1	-2	0	(-1, 0)
-2	1	-2	1	-4	(-2, -4)***
-3	1	-3	6	-20	(-3, -20)

* (0, -2) (The y-intercept)

** (2, 0) (2 is an upper bound to the roots)

*** (-2, -4) (-2 is a lower bound to the roots)

7. (continued)

Because we know the curve is symmetric with respect to the point (0, -2), we do not really need to calculate P(x) for negative values of x. By Descartes' rule of signs we know that there will be 1 positive real root, which is 2, and 2 or 0 negative real roots. Plot the points listed in the table and draw the graph. First draw the part of the graph where $x > 0$. By symmetry it is easy to complete the graph where $x < 0$. See the graph in the text. From the graph we see that -1 is a double root. This can be checked with synthetic division.

8. $x^3 - 3x^2 + 3 = 0$

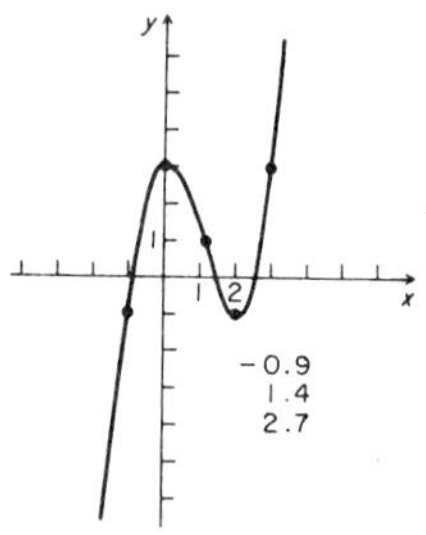

9. Graph $P(x) = x^3 - 3x - 4$, the corresponding polynomial function for the equation $x^3 - 3x - 4 = 0$. This polynomial is of degree 3 with leading coefficient positive. Thus as we move far to the right, function values will increase beyond bound; and as we move far to the left, function values will decrease beyond bound. The function is not odd or even. However, $x^3 - 3x$ is an odd function, with the origin as a point of symmetry. The function P(x) is a translation of the function $x^3 - 3x$ downward 4 units, hence the point (0, -4) is a point of symmetry.

Make a table of values. The first entry is the y-intercept. The other rows contain the bottom line in synthetic division.

	1	0	-3	-4	(x, y)
0				-4	(0, -4)*
1	1	1	-2	-6	(1, -6)
2	1	2	1	-2	(2, -2)
3	1	3	6	14	(3, 14)**
-1	1	-1	-2	-2	(-1, -2)
-2	1	-2	1	-6	(-2, -6)***
-3	1	-3	6	-22	(-3, -22)

* (0, -4) (The y-intercept)

** (3, 14) (3 is an upper bound to the roots)

*** (-2, -6) (-2 is a lower bound to the roots)

9. (continued)

Because we know the curve is symmetric with respect to the point (0, -4), we do not really need to calculate P(x) for negative values of x. By Descartes' rule of signs we know that there will be 1 positive real root and 2 or 0 negative real roots. From the table we see that the graph will cross the x-axis between 2 and 3. This is because P(2) is negative and P(3) is positive. Somewhere between 2 and 3, then, P(x) must be 0. Thus there is a real root between 2 and 3.

Plot the points and draw the graph. First draw the part of the graph where $x > 0$. Since the curve is symmetric with respect to (0, -4), it is easy to complete the graph where $x < 0$. See the graph in the text. From the graph we see that there are no negative real roots and the one positive real root is approximately 2.2. We also know that if there are no negative real roots, there will be two nonreal ones, and they will be conjugates of each other.

10. $x^3 - 3x^2 + 5 = 0$

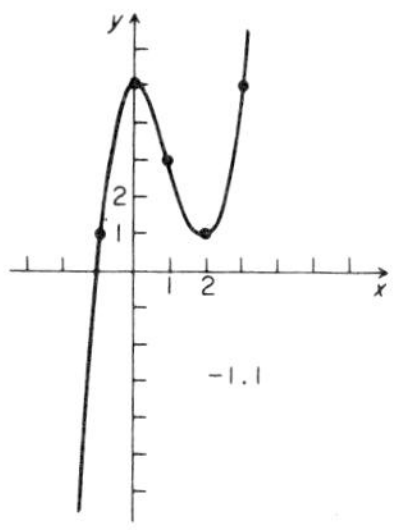

11. Graph $P(x) = x^4 + x^2 + 1$, the corresponding polynomial function for the equation $x^4 + x^2 + 1 = 0$. The polynomial is of degree 4 with leading coefficient positive. Thus as we move far to the right or far to the left, function values will increase beyond bound. Since the function is even, the graph is symmetric with respect to the y-axis. Thus we only need to make a table of values of $x \geqslant 0$. The first entry is the y-intercept. The other rows contain the bottom line in synthetic division.

	1	0	1	0	1	(x, y)
0					1	(0, 1)
1	1	1	2	2	3	(1, 3)
2	1	2	5	10	21	(1, 21)

We also observe, using Descartes' rule of signs, that there are no real roots. Plot the points and complete the graph. See the graph in the text.

12. $x^4 + 2x^2 + 2 = 0$

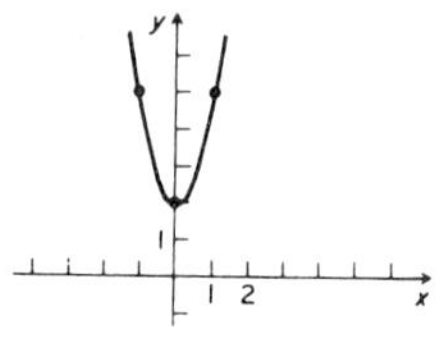

No real roots

13. Graph $P(x) = x^4 - 6x^2 + 8$, the corresponding polynomial function for the equation $x^4 - 6x^2 + 8 = 0$. The polynomial is of degree 4 with leading coefficient positive. Thus as we move far to the right or far to the left, function values will increase beyond bound. The function is even, thus the graph is symmetric with respect to the y-axis. We only need to make a table of values for $x \geqslant 0$. At most the function will have four roots. Factoring the polynomial, $P(x) = (x + 2)(x - 2)(x + \sqrt{2})(x - \sqrt{2})$, we can easily determine that the roots are -2, 2, $-\sqrt{2}$, and $\sqrt{2}$.

Make a table of values. The first entry is the y-intercept. The other rows contain the bottom line in synthetic division.

	1	0	-6	0	8	(x, y)	
0					8	(0, 8)	
1	1	1	-5	-5	3	(1, 3)	
2	1	2	-2	-4	0	(2, 0)	(Root)
3	1	3	3	9	35	(3, 35)	(Upper bound)

We plot these points, their reflections across the y-axis, and the roots. See the graph in the text.

14. $x^4 - 4x^2 + 2 = 0$

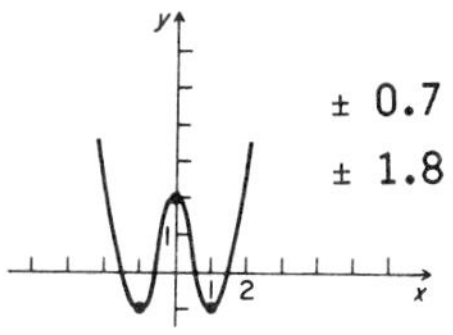

15. Graph $P(x) = x^5 + x^4 - x^3 - x^2 - 2x - 2$, the corresponding polynomial function for the equation $x^5 + x^4 - x^3 - x^2 - 2x - 2 = 0$. The polynomial is of degree 5 with leading coefficient positive. Thus as we move far to the right, function values will increase beyond bound; and as we move far to the left, function values will decrease beyond bound. By Descartes' rule of signs we know that there will be 1 positive real root and either 4, 2, or 0 negative real roots.

Make a table of values. The first entry is the y-intercept. The other rows contain the bottom line in synthetic division.

	1	1	-1	-1	-2	-2	(x, y)
0						-2	(0, -2)*
1	1	2	1	0	-2	-4	(1, -4)
2	1	3	5	9	16	30	(2, 30)**
-1	1	0	-1	0	-2	0	(-1, 0)
-2	1	-1	1	-3	4	-10	(-2, -10)***

* (0, -2) (The y-intercept)

** (2, 30) (2 is an upper bound to the roots)

*** (-2, -10) (-2 is a lower bound to the roots)

From the table we see that the graph will cross the x-axis between 1 and 2. This is because P(1) is negative and P(2) is positive. Somewhere between 1 and 2, then, P(x) must be 0. We also note that P(x) = 0 when x is -1.

15. (continued)

It is helpful to find a few more values using synthetic division.

	1	1	-1	-1	-2	-2	(x,y)
$-\frac{3}{2}$	1	$-\frac{1}{2}$	$-\frac{1}{4}$	$-\frac{5}{8}$	$-\frac{17}{16}$	$-\frac{13}{32}$	$\left(-\frac{3}{2}, -\frac{13}{32}\right)$
$-\frac{1}{2}$	1	$\frac{1}{2}$	$-\frac{5}{4}$	$-\frac{3}{8}$	$-\frac{29}{16}$	$-\frac{35}{32}$	$\left(-\frac{1}{2}, -\frac{35}{32}\right)$
$\frac{1}{2}$	1	$\frac{3}{2}$	$-\frac{1}{4}$	$-\frac{9}{8}$	$-\frac{41}{16}$	$-\frac{105}{32}$	$\left(\frac{1}{2}, -\frac{105}{32}\right)$
$\frac{3}{2}$	1	$\frac{5}{2}$	$\frac{11}{4}$	$\frac{25}{8}$	$\frac{43}{16}$	$\frac{65}{32}$	$\left(\frac{3}{2}, \frac{65}{32}\right)$
$-\frac{5}{4}$	1	$-\frac{1}{4}$	$-\frac{11}{16}$	$-\frac{9}{64}$	$-\frac{467}{256}$	$\frac{287}{1024}$	$\left(-\frac{5}{4}, \frac{287}{1024}\right)$
$\frac{5}{4}$	1	$\frac{9}{4}$	$\frac{29}{16}$	$\frac{81}{64}$	$-\frac{107}{256}$	$-2\frac{535}{1024}$	$\left(\frac{5}{4}, -2\frac{535}{1024}\right)$

From the tables we see that the graph will also cross the x-axis between $-\frac{3}{2}$ and $-\frac{5}{4}$ because $P\left(-\frac{3}{2}\right)$ is negative and $P\left(-\frac{5}{4}\right)$ is positive. Thus there are only two negative real roots. Plot these points and draw the graph. The solutions appear to be approximately -1.4, -1, and 1.4. See the graph in the text.

16. $x^5 - 2x^4 - 2x^3 + 4x^2 - 3x + 6 = 0$

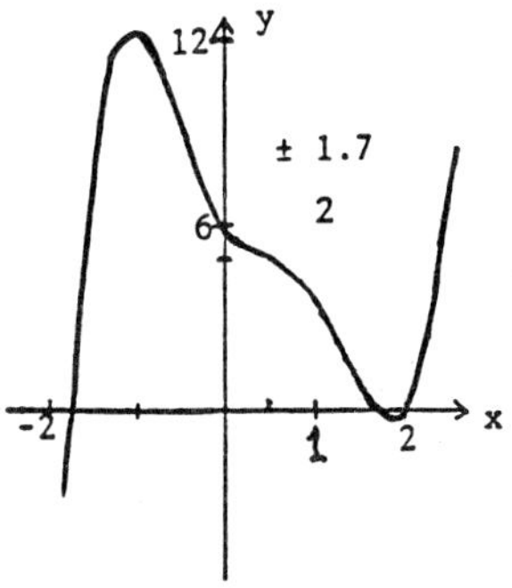

17. $2x^5 + 2x^3 - x^2 - 1 = 0$

Since P(0) = -1 and P(1) = 2, we know that the equation has a solution between 0 and 1. The solution is in the interval (0, 1). Use synthetic division to narrow the interval.

	2	0	2	-1	0	-1
0.7	2	1.4	2.98	1.09	0.76	-0.47
0.8	2	1.6	3.28	1.62	1.30	0.04

Since P(0.7) = -0.47 and P(0.8) = 0.04, we know that the solution is between 0.7 and 0.8. The solution is in the interval (0.7, 0.8). Use synthetic division to further narrow the interval.

	2	0	2	-1	0	-1
0.79	2	1.58	3.25	1.57	1.24	-0.02
0.80	2	1.6	3.28	1.62	1.30	0.04

17. (continued)

Since P(0.79) = -0.02 and P(0.80) = 0.04, we know that the solution is between 0.79 and 0.80. The solution is the interval (0.79, 0.80). To the nearest hundredth the solution is 0.79.

18. 1.41

19. From the graph we see that P(x) has one negative real root between -2 and -1. We use synthetic division to narrow the interval.

	1	-2	-1	4
-1.3	1	-3.3	3.29	-0.277
-1.2	1	-3.2	2.84	0.592

Since P(-1.3) is negative and P(-1.2) is positive we know there is a solution between -1.3 and -1.2. Again use synthetic division to narrow the interval.

	1	-2	-1	4
-1.27	1	-3.27	3.1529	-0.004183
-1.26	1	-3.26	3.1076	0.084424

Since P(-1.27) is negative and P(-1.26) is positive, we know there is a solution between -1.27 and -1.26. The solution is in the interval (-1.27, -1.26). To the nearest hundredth the solution is -1.27.

20. -0.7, 1.24, 3.46

21. Consider, for example

P(x) = $3x^4 - 5x^3 + 4x^2 - 5$

a	3	-5	4	-5
		3a	a(3a-5)	a(a(a-5)+4)
	3	3a-5	a(3a-5)+4	a(a(a-5)+4)-5
				R

The expression for R, which is P(a), is the same expression obtained by factoring to find nested form. Generalizing completes the proof.

22. P(x) = $5.8x^4 - 2.3x^2 - 6.1$

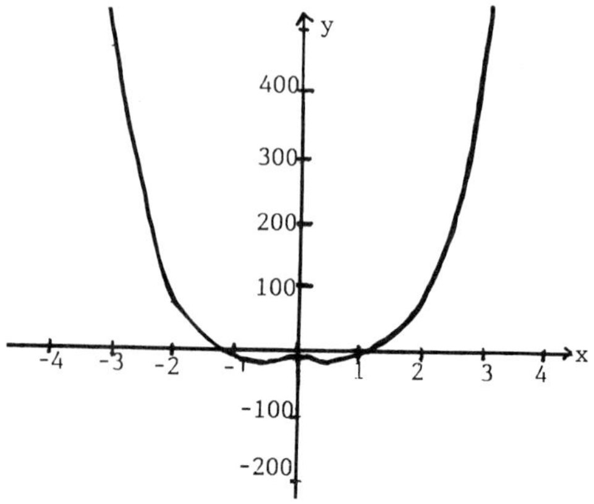

23. f(x) = $x^3 - 9x^2 + 27x + 50$

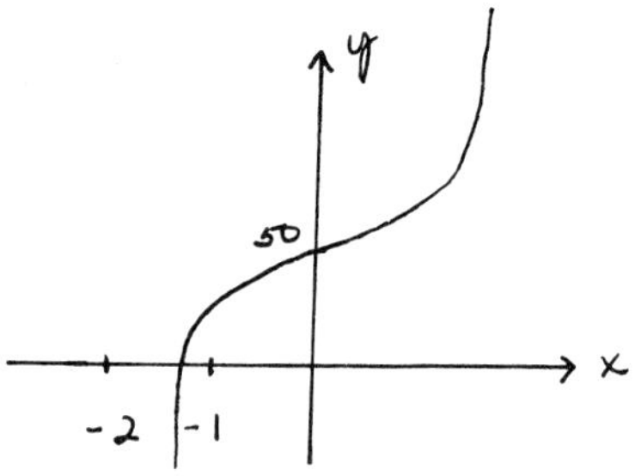

Root: -1.25433

24. f(x) = $x^3 - 3x + 1$

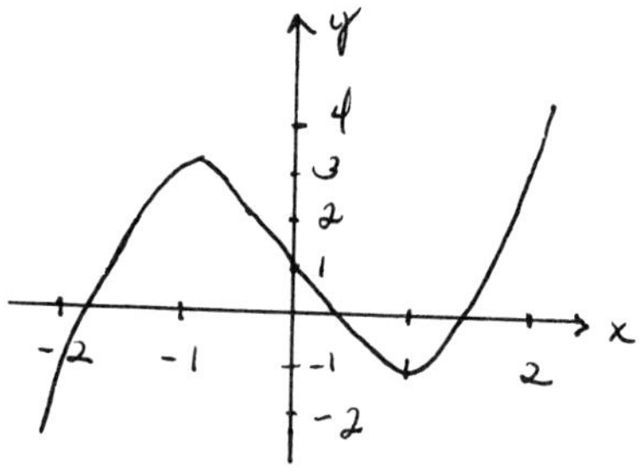

Roots: -1.87939, 0.34731, 1.53209

25. f(x) = $x^4 + 4x^3 - 36x^2 - 160x + 300$

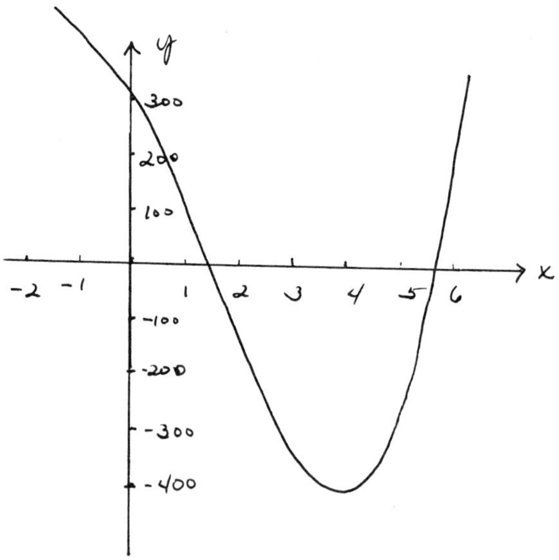

Roots: 1.48928, 5.67342

26. $f(x) = x^6 + 4x^5 - 54x^4 - 160x^3 + 641x^2 + 828x - 1260$

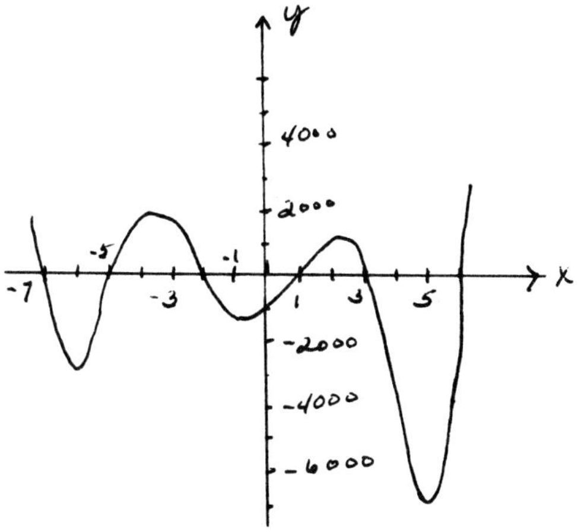

Roots: -7, -5, -2, 1, 3, 6

27. Use a computer or calculator to graph $N(t) = -0.046t^3 + 2.08t + 2$.

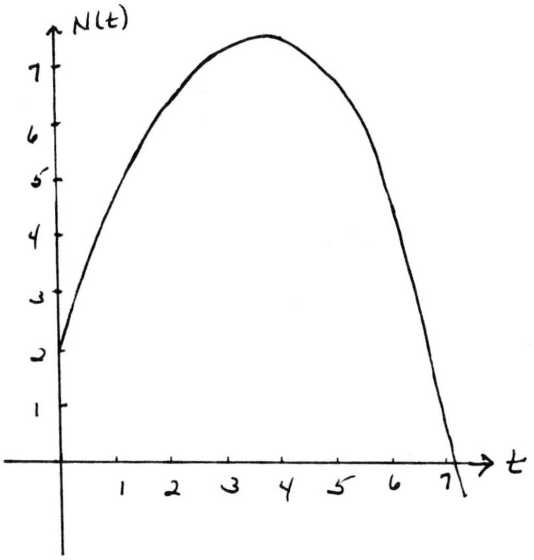

The concentration will be 0 after approximately 7.1616 hours.

28. 8%

Exercise Set 5.7

1. $y = \frac{1}{x - 3}$

The function is neither even nor odd and is not symmetric with respect to the origin or the line y = x. The degree of the denominator is greater than that of the numerator. Thus the x-axis is an asymptote. The denominator tells us that the function is defined for all x except 3. Therefore the line x = 3 is a vertical asymptote. There are no zeros of the function. The vertical asymptote divides the x-axis into two intervals as illustrated below.

Tabulate signs in these intervals.

Interval	1	x - 3	y
x < 3	+	-	-
x > 3	+	+	+

Next make a table of values and plot them.

x	$3\frac{1}{8}$	$3\frac{1}{2}$	4	5	8	$2\frac{3}{4}$	2	1	0	-2
y	8	2	1	$\frac{1}{2}$	$\frac{1}{5}$	-4	-1	$-\frac{1}{2}$	$-\frac{1}{3}$	$-\frac{1}{5}$

Using all available information, draw the graph. See the graph in the text.

2.

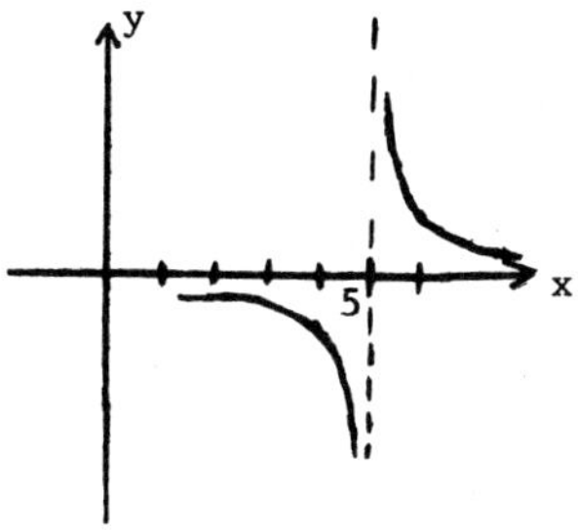

3. $y = \frac{-2}{x - 5}$

The function is neither even nor odd and is not symmetric with respect to the origin or the line y = x. The degree of the denominator is greater than that of the numerator. Thus the x-axis is an asymptote. The denominator tells us that the function is defined for all x except 5. Therefore the line x = 5 is a vertical asymptote. There are no zeros of the function. The vertical asymptote divides the x-axis into two intervals as illustrated below.

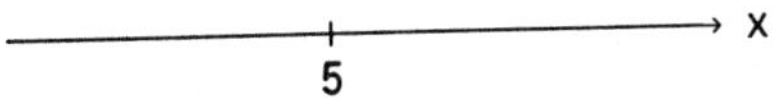

Tabulate signs in these intervals.

3. (continued)

Interval	-2	x - 5	y
x < 5	-	-	+
x < 5	-	+	-

Make a table of values and plot them.

x	-3	0	2	4	$4\frac{2}{3}$	$5\frac{1}{3}$	6	8	11
y	$\frac{1}{4}$	$\frac{2}{5}$	$\frac{2}{3}$	2	6	-6	-2	$-\frac{2}{3}$	$-\frac{1}{3}$

Using all available information, draw the graph. See the graph in the text.

4.

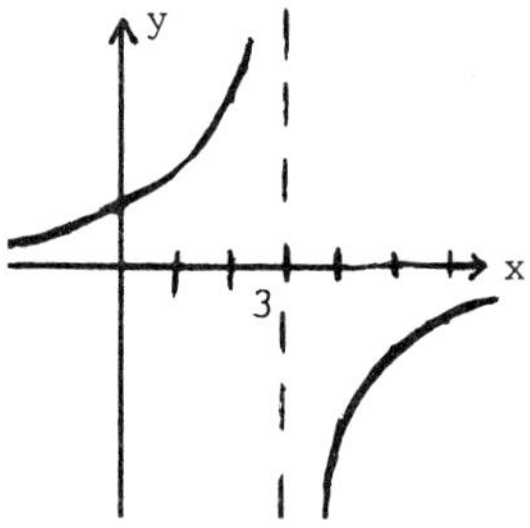

5. $y = \frac{2x + 1}{x}$

The function is neither even nor odd and no symmetries are apparent.

Since the numerator and denominator have the same degree, the horizontal asymptote can be determined by dividing the leading coefficients of the two polynomials. Thus $2 \div 1 = 2$, and we know that $y = 2$ is a horizontal asymptote.

The value making the numerator zero is $-\frac{1}{2}$. Since this value does not make the denominator zero, it is a zero of the function. The denominator tells us that the function is defined for all x except 0. Therefore the line x = 0 is a vertical asymptote. The zero and the vertical asymptote divide the x-axis into intervals as illustrated below.

$$\underset{-\frac{1}{2}}{+} \qquad \underset{0}{+} \qquad \rightarrow x$$

Tabulate the signs in these intervals.

Interval	2x + 1	x	y
$x < -\frac{1}{2}$	-	-	+
$-\frac{1}{2} < x < 0$	+	-	-
$0 < x$	+	+	+

Next make a table of function values and plot them.

5. (continued)

x	-100	-10	-3	$-\frac{1}{2}$	$-\frac{1}{10}$	1	4	10	100
y	$1\frac{99}{100}$	$1\frac{9}{10}$	$1\frac{2}{3}$	0	-8	3	$2\frac{1}{4}$	$2\frac{1}{10}$	$2\frac{1}{100}$

Using all available information, draw the graph. See the graph in the text.

6.

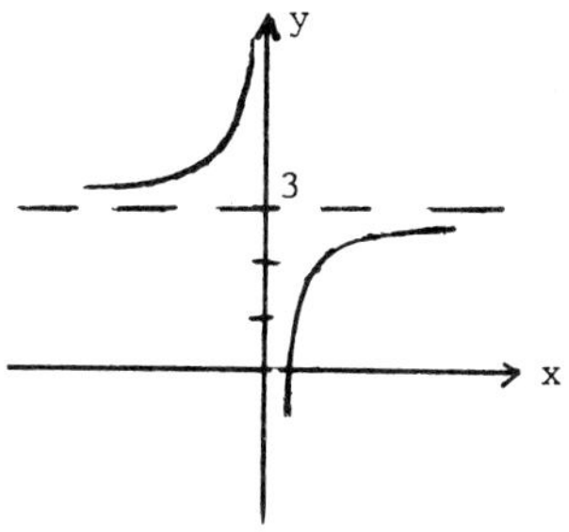

7. $y = \frac{1}{(x - 2)^2}$

The function is neither even nor odd and is not symmetric with respect to the origin or the line y = x. The degree of the denominator is greater than that of the numerator. Thus the x-axis is an asymptote. The denominator tells us that the function is defined for all x except 2. Therefore the line x = 2 is a vertical asymptote. There are no zeros of the function, and all function values are positive.

Make a table of values, plot them, and draw the graph.

x	$2\frac{1}{2}$	3	4	5	$1\frac{1}{2}$	1	0	-1
y	4	1	$\frac{1}{4}$	$\frac{1}{9}$	4	1	$\frac{1}{4}$	$\frac{1}{9}$

See the graph in the text.

8.

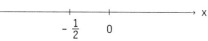

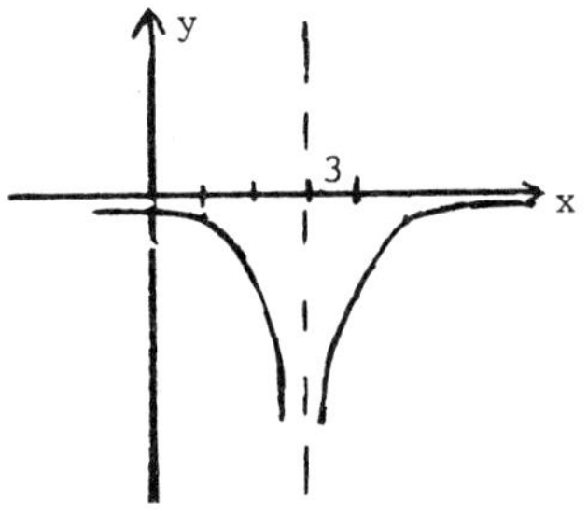

9. $y = \frac{2}{x^2}$

Note that the function is defined for all x except 0. Therefore the line x = 0 is an asymptote. As |x| gets very large, y approaches 0. Therefore the x-axis is also an asymptote, for both positive and negative values of x. The function is even. Therefore, it is symmetric with respect to the y-axis. All function values are positive. Therefore, the entire graph is above the x-axis.

Make a table of values, plot them, and draw the graph.

x	$\pm\frac{1}{2}$	± 1	± 2	± 3	± 4
y	8	2	$\frac{1}{2}$	$\frac{2}{9}$	$\frac{1}{8}$

See the graph in the text.

10.

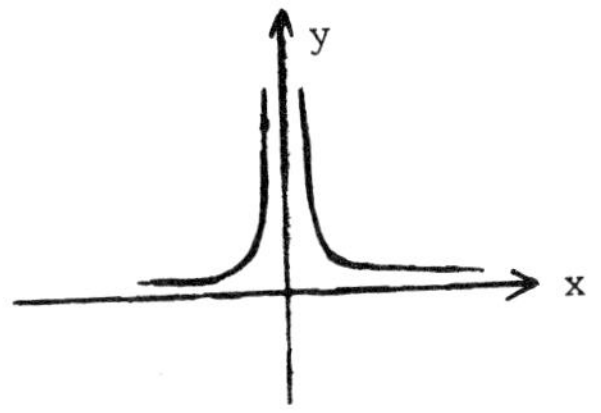

11. $y = \frac{1}{x^2 + 3}$

The function is even, so the graph is symmetric with respect to the y-axis. The degree of the denominaotr is greater than that of the numerator. Thus the x-axis is an asymptote. The denominator is not factorable. Neither is the numerator. Therefore, there are no vertical asymptotes, and there are no zeros. The numerator is a positive constant and the denominator can never be negative, so this function has only positive values.

Make a table of values, plot them, and draw the graph.

x	0	± 1	± 2	± 3
y	$\frac{1}{3}$	$\frac{1}{4}$	$\frac{1}{7}$	$\frac{1}{12}$

See the graph in the text.

12.

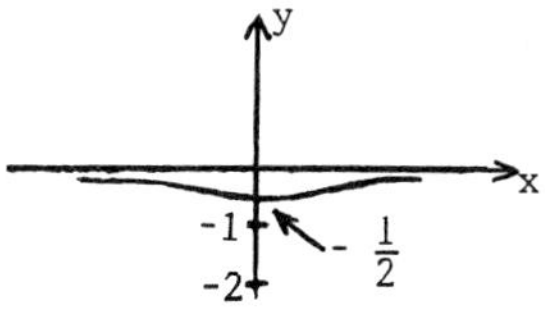

13. $y = \frac{x - 1}{x + 2}$

The function is neither even nor odd and no symmetries are apparent.

Since the numerator and denominator have the same degree, the horizontal asymptote can be determined by dividing the leading coefficients of the two polynomials. Thus $1 \div 1 = 1$, and we know that y = 1 is a horizontal asymptote.

The value making the numerator zero is 1. Since this value does not make the denominator zero, it is a zero of the function. The denominator tells us that the function is defined for all x except -2. Therefore the line x = -2 is a vertical asymptote. THe zero and the vertical asymptote divide the x-axis into intervals as illustrated below.

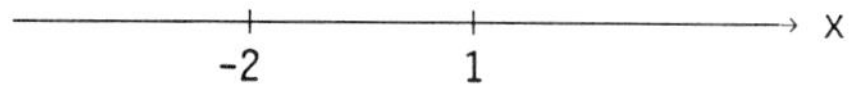

Tabulate signs in these intervals.

Interval	x - 1	x + 2	y
x < -2	-	-	+
-2 < x < 1	-	+	-
1 < x	+	+	+

Next make a table of function values and plot them.

x	-6	-4	-3	$-2\frac{1}{2}$	$-1\frac{1}{2}$	-1	0	1	2	4	6	20
y	$\frac{7}{4}$	$\frac{5}{2}$	4	7	-5	-2	$-\frac{1}{2}$	0	$\frac{1}{4}$	$\frac{1}{2}$	$\frac{5}{8}$	$\frac{19}{22}$

Using all available information, draw the graph. See the graph in the text.

14.

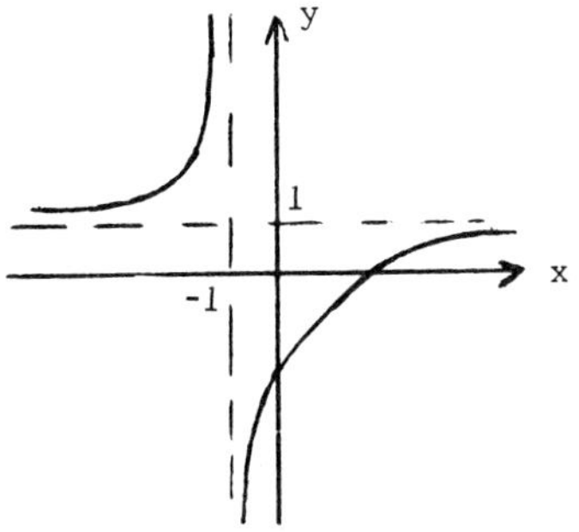

15. $y = \frac{3x}{x^2 + 5x + 4} = \frac{3x}{(x + 4)(x + 1)}$

The function is neither even nor odd and no symmetries are apparent.

The degree of the denominator is greater than that of the numerator. Thus the x-axis is an asymptote.

15. (continued)

The value making the numerator zero is 0. Since this value does not make the denominator zero, it is a zero of the function. The denominator tells us that the function is defined for all x except -4 and -1. Therefore the lines x = -4 and x = -1 are vertical asymptotes. The zero and the vertical asymptotes divide the x-axis into intervals as illustrated below.

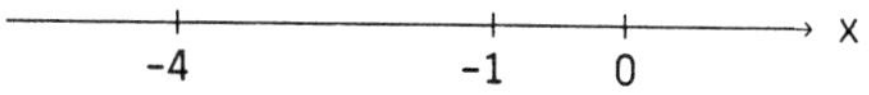

Tabulate signs in these intervals.

Interval	3x	x + 4	x + 1	(x+4)(x+1)	y
x < -4	-	-	-	+	-
-4 < x < -1	-	+	-	-	+
-1 < x < 0	-	+	+	+	-
0 < x	+	+	+	+	+

Make a table of function values and plot them.

x	-6	-5	$-\frac{9}{2}$	$-\frac{7}{2}$	-3	-2	$-\frac{3}{2}$
y	$-\frac{9}{5}$	$-\frac{15}{4}$	$-\frac{54}{7}$	$\frac{42}{5}$	$\frac{9}{2}$	3	$\frac{18}{5}$

x	$-\frac{3}{4}$	$-\frac{1}{2}$	0	1	2	3
y	$-\frac{36}{13}$	$-\frac{6}{7}$	0	$\frac{3}{10}$	$\frac{1}{3}$	$\frac{7}{28}$

Using all the information, draw the graph. See the graph in the text.

16.

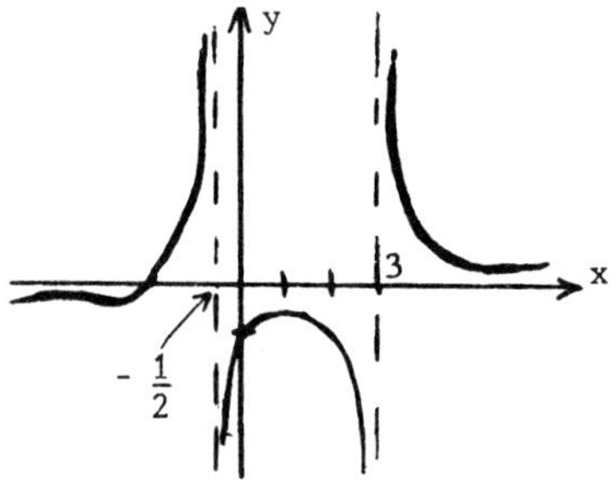

17. $y = \frac{x^2 - 4}{x - 1} = \frac{(x + 2)(x - 2)}{x - 1}$

The function is neither even nor odd and no symmetries are apparent.

The degree of the numerator is one greater than that of the denominator. Dividing numerator by denominator will show that y = x + 1 is an oblique asymptote.

17. (continued)

The values making the numerator zero are 2 and -2. Since these values do not make the denominator zero, they are zeros of the function. The denominator tells us that x = 1 is a vertical asymptote. The zeros and the vertical asymptote divide the x-axis into intervals as illustrated below.

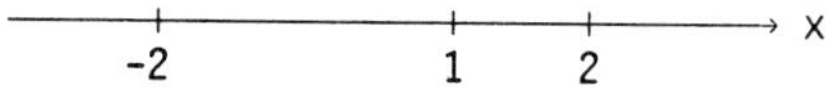

Tabulate signs in these intervals.

Interval	x-2	x+2	(x-2)(x+2)	x-1	y
x < -2	-	-	+	-	-
-2 < x < 1	-	+	-	-	+
1 < x < 2	-	+	-	+	-
2 < x	+	+	+	+	+

Next make a table of function values and plot them.

x	-3	-2	-1	0	$\frac{1}{2}$	$\frac{3}{2}$	2	3
y	$-1\frac{1}{4}$	0	$1\frac{1}{2}$	4	$7\frac{1}{2}$	$-3\frac{1}{2}$	9	$2\frac{1}{2}$

Using all available information, draw the graph. See the graph in the text.

18.

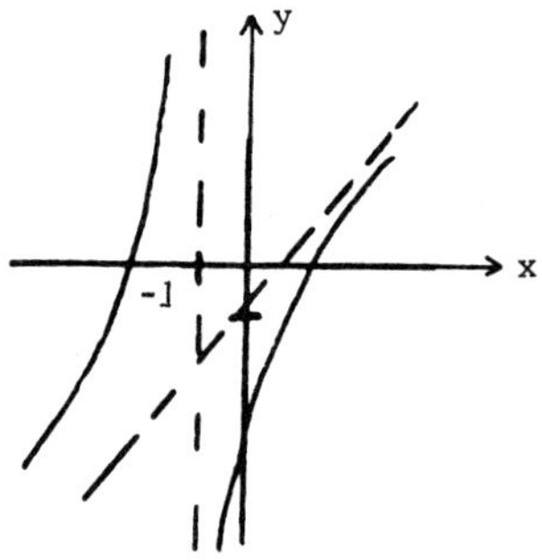

19. $y = \frac{x^2 + x - 2}{2x^2 + 1} = \frac{(x + 2)(x - 1)}{2x^2 + 1}$

The function is neither even nor odd and no symmetries are apparent.

Since the numerator and denominator have the same degree, the horizontal asymptote can be determined by dividing the leading coefficients of the two polynomials. Thus 1 ÷ 2 = 1/2, and we know that y = 1/2 is a horizontal asymptote.

The values making the numerator zero are -2 and 1. Since these values do not make the denominator zero, they are zeros of the function. The denominator tells us that the function is defined for all x. Therefore there are no vertical asymptotes. The zeros divide the x-axis into intervals as illustrated below.

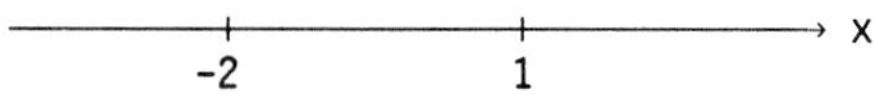

19. (continued)

Tabulate signs in these intervals.

Interval	x+2	x-1	(x+2)(x-1)	$2x^2+1$	y
x < -2	-	-	+	+	+
-2 < x < 1	+	-	-	+	-
1 < x	+	+	+	+	+

Make a table of function values and plot them.

x	-5	-3	-2	-1	0	1	2	3
y	$\frac{6}{17}$	$\frac{4}{19}$	0	$-\frac{2}{3}$	-2	0	$\frac{4}{9}$	$\frac{10}{19}$

Using all the information, draw the graph. See the graph in the text.

20.

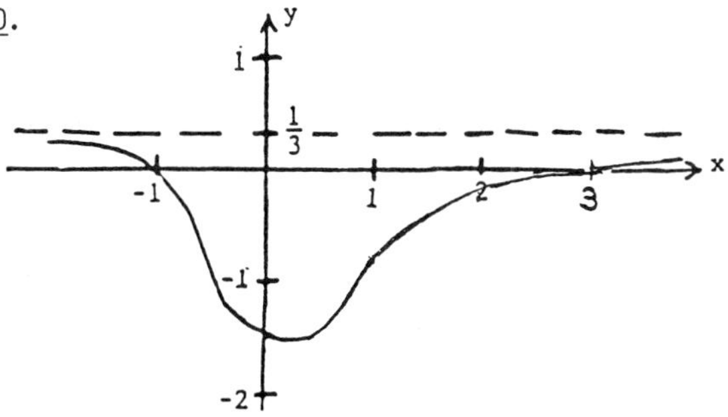

21. $y = \frac{x - 1}{x^2 - 2x - 3} = \frac{x - 1}{(x - 3)(x + 1)}$

The function is neither even nor odd and no symmetries are apparent.

The degree of the denominator is greater than the numerator. Thus the x-axis is an asymptote.

The value making the numerator zero is 1. Since this value does not make the denominator zero, it is a zero of the function. The denominator tells us that the function is defined for all x except 3 and -1. Therefore the lines x = 3 and x = -1 are vertical asymptotes. The zero and the vertical asymptotes divide the x-axis into intervals as illustrated below.

(number line: -1, 1, 3, x)

Tabulate signs in these intervals.

Interval	x-1	x-3	x+1	(x-3)(x+1)	y
x < -1	-	-	-	+	-
-1 < x < 1	-	-	+	-	+
1 < x < 3	+	-	+	-	-
3 < x	+	+	+	+	+

Make a table of function values and plot them.

21. (continued)

x	-5	-3	-2	$-\frac{3}{2}$	$-\frac{1}{2}$	0	1
y	$-\frac{3}{16}$	$-\frac{1}{3}$	$-\frac{3}{5}$	$-\frac{10}{9}$	$\frac{6}{7}$	$\frac{1}{3}$	0

x	2	$\frac{5}{2}$	4	5
y	$-\frac{1}{3}$	$-\frac{6}{7}$	$\frac{3}{5}$	$\frac{1}{3}$

Using all the information, draw the graph. See the graph in the text.

22.

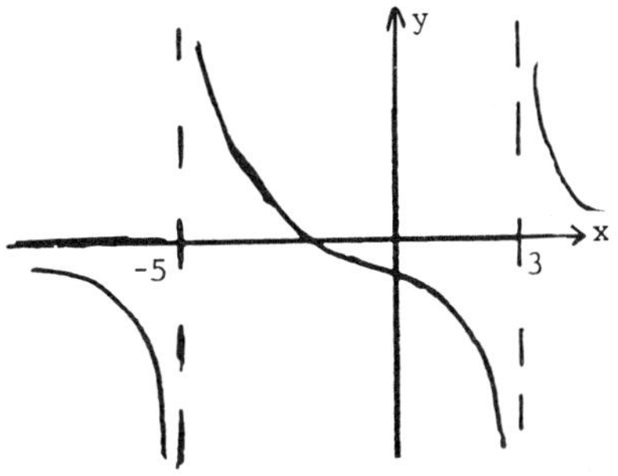

23. $y = \frac{x + 2}{(x - 1)^3}$

The function is neither even nor odd and no symmetries are apparent.

The degree of the denominator is greater than the numerator. Thus the x-axis is an asymptote.

The value making the numerator zero is -2. Since the value does not make the denominator zero, it is a zero of the function. The denominator tells us that the function is defined for all x except 1. Therefore the line x = 1 is a vertical asymptote. The zero and the vertical asymptote divide the x-axis into intervals as illustrated below.

(number line: -2, 1, x)

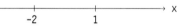

Tabulate signs in these intervals.

Interval	x + 2	$(x - 1)^3$	y
x < -2	-	-	+
-2 < x < 1	+	-	-
1 < x	+	+	+

Make a table of function values and plot them.

x	-5	-4	-3	-2	-1	0	$\frac{1}{2}$
y	$\frac{1}{72}$	$\frac{2}{125}$	$\frac{1}{64}$	0	$-\frac{1}{8}$	-2	-20

23. (continued)

x	$\frac{3}{2}$	2	3
y	28	4	$\frac{5}{8}$

Using all the information, draw the graph.
See the graph in the text.

24.

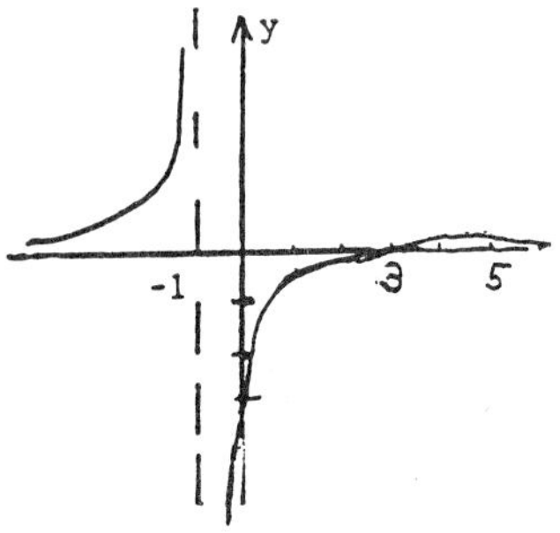

25. $y = \frac{x^3 + 1}{x} = \frac{(x + 1)(x^2 - x + 1)}{x}$

The function is neither even nor odd and no symmetries are apparent.

The degree of the numerator is two greater than the denominator. There are no horizontal or oblique asymptotes.

The only real root of the numerator is -1. Since this value does not make the denominator zero, it is a zero of the function. The denominator tells us that x = 0 is a vertical asymptote. The zero and vertical asymptote divide the x-axis into intervals as illustrated below.

$$\xrightarrow[\quad -1 \qquad 0 \quad]{} x$$

Tabulate signs in these intervals.

Interval	$x^3 + 1$	x	y
$x < -1$	-	-	+
$-1 < x < 0$	+	-	-
$0 < x$	+	+	+

Make a table of function values and plot them.

x	-3	-2	-1	$-\frac{1}{2}$	$\frac{1}{2}$	1	2	3
y	$\frac{26}{3}$	$\frac{7}{2}$	0	$-\frac{7}{4}$	$\frac{9}{4}$	2	$\frac{9}{2}$	$\frac{28}{3}$

Draw the graph using all the known information.
See the graph in the text.

26.

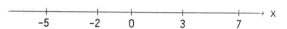

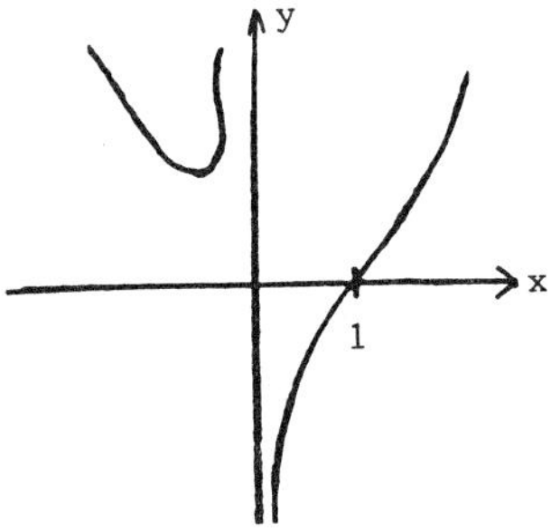

27. $y = \frac{x^3 + 2x^2 - 15x}{x^2 - 5x - 14} = \frac{x(x + 5)(x - 3)}{(x + 2)(x - 7)}$

The function is neither even nor odd and no symmetries are apparent.

The degree of the numerator is one greater than that of the denominator. Dividing numerator by denominator will show that y = x + 7 is an oblique asymptote.

The values making the numerator zero are 0, -5, and 3. Since these values do not make the denominator zero, they are zeros of the function. The denominator tells us that x = -2 and x = 7 are vertical asymptotes. The zeros and vertical asymptotes divide the x-axis into intervals as illustrated below.

$$\xrightarrow[\quad -5 \qquad -2 \qquad 0 \qquad 3 \qquad 7 \quad]{} x$$

Tabulate signs in these intervals.

Interval	x	x+5	x-3	x+2	x-7	y
$x < -5$	-	-	-	-	-	-
$-5 < x < -2$	-	+	-	-	-	+
$-2 < x < 0$	-	+	-	+	-	-
$0 < x < 3$	+	+	-	+	-	+
$3 < x < 7$	+	+	+	+	-	-
$7 < x$	+	+	+	+	+	+

Make a table of function values and plot them.

x	-100	-10	-8	-7	-6	-5	-4	-3	-2.5
y	-93.2	-4.8	-2.9	-2	-1.0	0	1.3	3.6	7.2

x	-2.1	-1.9	-1.5	-1	0	1	2	3	4	5
y	11.1	-32.4	-5.6	-2	0	0.7	0.7	0	-2	-7.1

x	6	7.1	7.5	8	10	20	100
y	-24.8	394	88.8	42	29.2	29.7	107

Draw the graph using all available information.
See the graph in the text.

28.

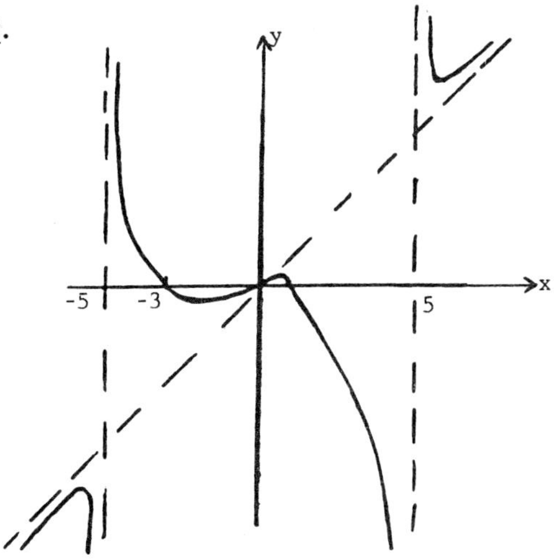

29. $y = \frac{1}{x^2 + 3} - 5 = \frac{1}{x^2 + 3} - 5 \cdot \frac{x^2 + 3}{x^2 + 3}$

$= \frac{1 - 5x^2 - 15}{x^2 + 3} = \frac{-5x^2 - 14}{x^2 + 3}$

The function is even, so the graph is symmetric with respect to the y-axis.

Since the numerator and denominator have the same degree, the horizontal asymptote can be determined by dividing the leading coefficients of the two polynomials. Thus $-5 \div 1 = -5$ and we know that $y = -5$ is a horizontal asymptote.

The numerator tells us that there are no zeros of the function. The denominator tells us that the function is defined for all x, thus there are no vertical asymptotes.

Make a table of function values and plot them.

x	0	± 1	± 2	± 3
y	-4.67	-4.75	-4.86	-4.92

See the graph in the text.

30.

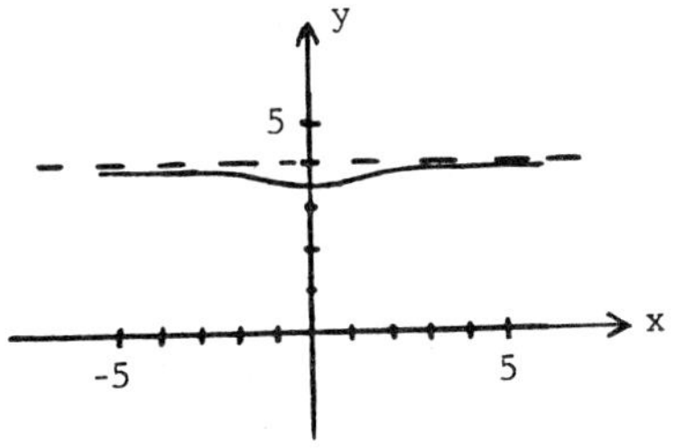

31. $y = \frac{1}{|x + 2|}$

The function is neither even nor odd and is not symmetric with respect to the origin or the line $y = x$. The degree of the denominator is greater than that of the numerator. Thus the x-axis is an asymptote. The denominator tells us that the function is defined for all x except -2. Therefore the line $x = -2$ is a vertical asymptote. There are no zeros of the function, and all function values are positive.

Make a table of values, plot them, and draw the graph.

x	$-2\frac{1}{2}$	-3	-4	-5	$-1\frac{1}{2}$	-1	0	1
y	2	1	$\frac{1}{2}$	$\frac{1}{3}$	2	1	$\frac{1}{2}$	$\frac{1}{3}$

See the graph in the text.

32.

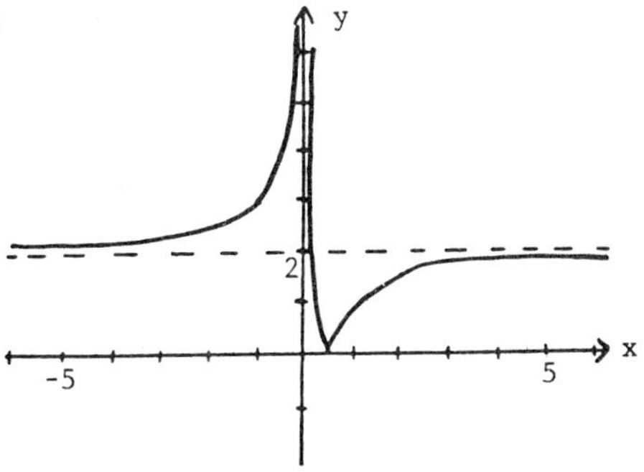

33. $y = \left|\frac{1}{x - 2} - 3\right| = \left|\frac{7 - 3x}{x - 2}\right|$

The function is neither even nor odd and no symmetries are apparent. The function is defined for all x except 2. Therefore, the line $x = 2$ is an asymptote. As $|x|$ gets very large, y approaches 3. Therefore, the line $y = 3$ is also an asymptote, for both positive and negative values of x. All function values are positive. The value making the numerator zero is 7/3. Since the value does not make the denominator zero, it is a zero of the function.

Make a table of function values and plot them.

x	-2	0	1	$\frac{3}{2}$	$\frac{13}{6}$	$\frac{7}{3}$	$\frac{5}{2}$	3	8
y	$\frac{13}{4}$	$\frac{7}{2}$	4	5	3	0	1	2	$\frac{7}{3}$

Using all known information, draw the graph. See the grpah in the text.

34.

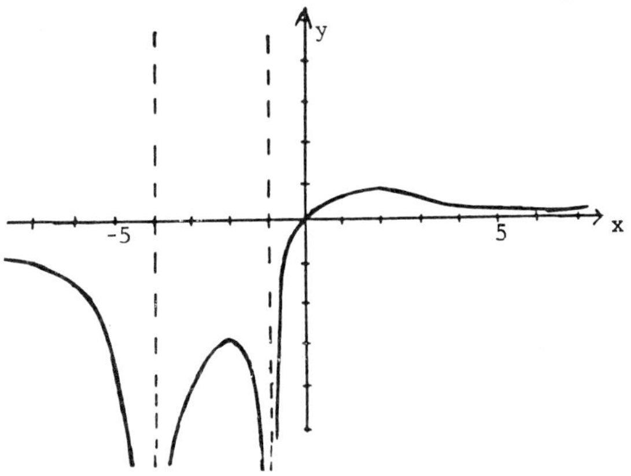

Exercise Set 5.8

1. $\frac{x+7}{(x-3)(x+2)} = \frac{A}{x-3} + \frac{B}{x+2}$ (Theorem 14)

$\frac{x+7}{(x-3)(x+2)} = \frac{A(x+2) + B(x-3)}{(x-3)(x+2)}$ (Adding)

Equate the numerators:

$x + 7 = A(x + 2) + B(x - 3)$

Let $x + 2 = 0$, or $x = -2$. Then we get

$-2 + 7 = 0 + B(-2 - 3)$

$5 = -5B$

$-1 = B$

Next let $x - 3 = 0$, or $x = 3$. Then we get

$3 + 7 = A(3 + 2) + 0$

$10 = 5A$

$2 = A$

The decomposition is as follows:

$\frac{2}{x-3} - \frac{1}{x+2}$

2. $\frac{1}{x+1} + \frac{1}{x-1}$

3. $\frac{7x-1}{6x^2-5x+1}$

$= \frac{7x-1}{(3x-1)(2x-1)}$ (Factoring the denominator)

$= \frac{A}{3x-1} + \frac{B}{2x-1}$ (Theorem 14)

$= \frac{A(2x-1) + B(3x-1)}{(3x-1)(2x-1)}$ (Adding)

Equate the numerators:

$7x - 1 = A(2x - 1) + B(3x - 1)$

Let $2x - 1 = 0$, or $x = \frac{1}{2}$. Then we get

$7\left(\frac{1}{2}\right) - 1 = 0 + B\left(3 \cdot \frac{1}{2} - 1\right)$

$\frac{5}{2} = \frac{1}{2}B$

$5 = B$

3. (continued)

Next let $3x - 1 = 0$, or $x = \frac{1}{3}$. We get

$7\left(\frac{1}{3}\right) - 1 = A\left(2 \cdot \frac{1}{3} - 1\right) + 0$

$\frac{7}{3} - 1 = A\left(\frac{2}{3} - 1\right)$

$\frac{4}{3} = -\frac{1}{3}A$

$-4 = A$

The decomposition is as follows:

$\frac{-4}{3x-1} + \frac{5}{2x-1}$

4. $\frac{3x^2 - 11x - 26}{(x^2-4)(x+1)}$

5. $\frac{3x^2 - 11x - 26}{(x^2-4)(x+1)}$

$= \frac{3x^2 - 11x - 26}{(x+2)(x-2)(x+1)}$ (Factoring the denominator)

$= \frac{A}{x+2} + \frac{B}{x-2} + \frac{C}{x+1}$ (Theorem 14)

$= \frac{A(x-2)(x+1) + B(x+2)(x+1) + C(x+2)(x-2)}{(x+2)(x-2)(x+1)}$ (Adding)

Equate the numerators:

$3x^2 - 11x - 26 = A(x - 2)(x + 1) + B(x + 2)(x + 1) + C(x + 2)(x - 2)$

Let $x + 2 = 0$, or $x = -2$. Then, we get

$3(-2)^2 - 11(-2) - 26 = A(-2 - 2)(-2 + 1) + 0 + 0$

$12 + 22 - 26 = A(-4)(-1)$

$8 = 4A$

$2 = A$

Next let $x - 2 = 0$, or $x = 2$. Then, we get

$3 \cdot 2^2 - 11 \cdot 2 - 26 = 0 + B(2 + 2)(2 + 1) + 0$

$12 - 22 - 26 = B \cdot 4 \cdot 3$

$-36 = 12B$

$-3 = B$

Finally let $x + 1 = 0$, or $x = -1$. We get

$3(-1)^2 - 11(-1) - 26 = 0 + 0 + C(-1 + 2)(-1 - 2)$

$3 + 11 - 26 = C(1)(-3)$

$-12 = -3C$

$4 = C$

The decomposition is as follows:

$\frac{2}{x+2} - \frac{3}{x-2} + \frac{4}{x+1}$

6. $\frac{6}{x-4} + \frac{3}{x-2} - \frac{4}{x+1}$

7. $\dfrac{9}{(x + 2)^2(x - 1)}$

$= \dfrac{A}{x + 2} + \dfrac{B}{(x + 2)^2} + \dfrac{C}{x - 1}$ (Theorem 14)

$= \dfrac{A(x + 2)(x - 1) + B(x - 1) + C(x + 2)^2}{(x + 2)^2(x - 1)}$ (Adding)

Equate the numerators:

$9 = A(x + 2)(x - 1) + B(x - 1) + C(x + 2)^2$

Let $x - 1 = 0$, or $x = 1$. Then, we get

$9 = 0 + 0 + C(1 + 2)^2$

$9 = 9C$

$1 = C$

Next let $x + 2 = 0$, or $x = -2$. Then, we get

$9 = 0 + B(-2 - 1) + 0$

$9 = -3B$

$-3 = B$

To find A we equate the coefficients of x^2. Consider $9 = 0x^2 + 0x + 9$. Then

$0 = A + C$

Substituting 1 for C, we get $A = -1$.

The decomposition is as follows:

$\dfrac{-1}{x + 2} - \dfrac{3}{(x + 2)^2} + \dfrac{1}{x - 1}$

8. $\dfrac{1}{x - 2} + \dfrac{3}{(x - 2)^2} - \dfrac{2}{(x - 2)^3}$

9. $\dfrac{2x^2 + 3x + 1}{(x^2 - 1)(2x - 1)}$

$= \dfrac{2x^2 + 3x + 1}{(x + 1)(x - 1)(2x - 1)}$ (Factoring the denominator)

$= \dfrac{A}{x + 1} + \dfrac{B}{x - 1} + \dfrac{C}{2x - 1}$ (Theorem 14)

$= \dfrac{A(x-1)(2x-1) + B(x+1)(2x-1) + C(x+1)(x-1)}{(x + 1)(x -1)(2x - 1)}$ (Adding)

Equate the numerators:

$2x^2 + 3x + 1 = A(x - 1)(2x - 1) + B(x + 1)(2x - 1) + C(x + 1)(x - 1)$

Let $x + 1 = 0$, or $x = -1$. Then, we get

$2(-1)^2 + 3(-1) + 1 = A(-1 - 1)[2(-1) - 1] + 0 + 0$

$2 - 3 + 1 = A(-2)(-3)$

$0 = 6A$

$0 = A$

Next let $x - 1 = 0$, or $x = 1$. Then, we get

$2\cdot 1^2 + 3\cdot 1 + 1 = 0 + B(1 + 1)(2\cdot 1 - 1) + 0$

$2 + 3 + 1 = B\cdot 2\cdot 1$

$6 = 2B$

$3 = B$

9. (continued)

Finally we let $2x -1 = 0$, or $x = \frac{1}{2}$. We get

$2\left(\frac{1}{2}\right)^2 + 3\left(\frac{1}{2}\right) + 1 = 0 + 0 + C\left(\frac{1}{2} + 1\right)\left(\frac{1}{2} - 1\right)$

$\frac{1}{2} + \frac{3}{2} + 1 = C \cdot \frac{3}{2} \cdot \left(-\frac{1}{2}\right)$

$3 = -\frac{3}{4}C$

$-4 = C$

The decomposition is as follows:

$\dfrac{3}{x - 1} - \dfrac{4}{2x - 1}$

10. $\dfrac{-4}{x - 3} + \dfrac{3}{x - 2} + \dfrac{2}{x - 1}$

11. $\dfrac{x^4 - 3x^3 - 3x^2 + 10}{(x + 1)^2(x - 3)}$

$= \dfrac{x^4 - 3x^3 - 3x^2 + 10}{x^3 - x^2 - 5x - 3}$ (Multiplying the denominator)

Since the degree of the numerator is greater than the degree of the denominator, we divide and then use Theorem 14.

$$\begin{array}{r} x - 2 \\ x^3 - x^2 - 5x - 3 \;\big)\; x^4 - 3x^3 - 3x^2 + 0x + 10 \\ \underline{x^4 - x^3 - 5x^2 - 3x} \\ -2x^3 + 2x^2 + 3x + 10 \\ \underline{-2x^3 + 2x^2 + 10x + 6} \\ -7x + 4 \end{array}$$

The original expression is thus equivalent to the following:

$x - 2 + \dfrac{-7x + 4}{x^3 - x^2 - 5x - 3}$

We proceed to decompose the fraction.

$\dfrac{-7x + 4}{(x + 1)^2(x - 3)}$

$= \dfrac{A}{x + 1} + \dfrac{B}{(x + 1)^2} + \dfrac{C}{x - 3}$ (Theorem 14)

$= \dfrac{A(x + 1)(x - 3) + B(x - 3) + C(x + 1)^2}{(x + 1)^2(x - 3)}$ (Adding)

Equate the numerators:

$-7x + 4 = A(x + 1)(x - 3) + B(x - 3) + C(x + 1)^2$

Let $x - 3 = 0$, or $x = 3$. Then, we get

$-7\cdot 3 + 4 = 0 + 0 + C(3 + 1)^2$

$-17 = 16C$

$-\frac{17}{16} = C$

Let $x + 1 = 0$, or $x = -1$. Then, we get

$-7(-1) + 4 = 0 + B(-1 - 3) + 0$

$11 = -4B$

$-\frac{11}{4} = B$

To find A we equate the coefficients of x^2.

11. (continued)

Consider $-7x + 4 = 0x^2 - 7x + 4$.

$0 = A + C$

Substituting $-\frac{17}{16}$ for C, we get $A = \frac{17}{16}$.

The decomposition is as follows:

$$\frac{17/16}{x + 1} - \frac{11/4}{(x + 1)^2} - \frac{17/16}{x - 3}$$

The original expression is equivalent to the following:

$$x - 2 + \frac{17/16}{x + 1} - \frac{11/4}{(x + 1)^2} - \frac{17/16}{x - 3}$$

12. $10x - 5 + \frac{6}{x - 3} + \frac{14}{x + 2}$

13. $\frac{-x^2 + 2x - 13}{(x^2 + 2)(x - 1)}$

$= \frac{Ax + B}{x^2 + 2} + \frac{C}{x - 1}$ (Theorem 14)

$= \frac{(Ax + B)(x - 1) + C(x^2 + 2)}{(x^2 + 2)(x - 1)}$ (Adding)

Equate the numerators:

$-x^2 + 2x - 13 = (Ax + B)(x - 1) + C(x^2 + 2)$

Let $x - 1 = 0$, or $x = 1$. Then we get

$-1 + 2\cdot 1 - 13 = 0 + C(1^2 + 2)$

$-1 + 2 - 13 = C(1 + 2)$

$-12 = 3C$

$-4 = C$

Equate the coefficients of x^2:

$-1 = A + C$

Substituting -4 for C, we get A = 3.

Equte the constant terms:

$-13 = -B + 2C$

Substituting -4 for C, we get B = 5.

The decomposition is as follows:

$$\frac{3x + 5}{x^2 + 2} + \frac{-4}{x - 1}$$

14. $\frac{41x + 3}{x^2 + 1} - \frac{15}{x + 5}$

15. $\frac{6 + 26x - x^2}{(2x - 1)(x + 2)^2}$

$= \frac{A}{2x - 1} + \frac{B}{x + 2} + \frac{C}{(x + 2)^2}$ (Theorem 14)

$= \frac{A(x + 2)^2 + B(2x - 1)(x + 2) + C(2x - 1)}{(2x - 1)(x + 2)^2}$ (Adding)

Equate the numerators:

$6 + 26x - x^2 = A(x + 2)^2 + B(2x - 1)(x + 2) + C(2x - 1)$

15. (continued)

Let $2x - 1 = 0$, or $x = \frac{1}{2}$. Then, we get

$6 + 26\cdot\frac{1}{2} - \left(\frac{1}{2}\right)^2 = A\left(\frac{1}{2} + 2\right)^2 + 0 + 0$

$6 + 13 - \frac{1}{4} = A\left(\frac{5}{2}\right)^2$

$\frac{75}{4} = \frac{25}{4}A$

$3 = A$

Let $x + 2 = 0$, or $x = -2$. We get

$6 + 26(-2) - (-2)^2 = 0 + 0 + C[2(-2) - 1]$

$6 - 52 - 4 = -5C$

$-50 = -5C$

$10 = C$

Equate the coefficients of x^2:

$-1 = A + 2B$

Substituting 3 for A, we obtain B = -2.

The decomposition is as follows:

$$\frac{3}{2x - 1} - \frac{2}{x + 2} + \frac{10}{(x + 2)^2}$$

16. $\frac{1}{x - 1} - \frac{1}{x + 1} + \frac{3}{(x + 1)^2} - \frac{3}{(x + 1)^3} + \frac{2}{(x + 1)^4}$

17. $\frac{6x^3 + 5x^2 + 6x - 2}{2x^2 + x - 1}$

Since the degree of the numerator is greater than the degree of the denominator, we divide and then use Theorem 14.

$$\begin{array}{r|l} & 3x + 1 \\ 2x^2 + x - 1 & 6x^3 + 5x^2 + 6x - 2 \\ & \underline{6x^3 + 3x^2 - 3x} \\ & 2x^2 + 9x - 2 \\ & \underline{2x^2 + x - 1} \\ & 8x - 1 \end{array}$$

The original expression is equivalent to

$$3x + 1 + \frac{8x - 1}{2x^2 + x - 1}$$

We proceed to decompose the fraction.

$\frac{8x - 1}{2x^2 + x - 1} = \frac{8x - 1}{(2x - 1)(x + 1)}$ (Factoring the denominator)

$= \frac{A}{2x - 1} + \frac{B}{x + 1}$ (Theorem 14)

$= \frac{A(x + 1) + B(2x - 1)}{(2x - 1)(x + 1)}$ (Adding)

Equate the numerators:

$8x - 1 = A(x + 1) + B(2x - 1)$

Let $x + 1 = 0$, or $x = -1$. Then we get

$8(-1) - 1 = 0 + B[2(-1) - 1]$

$-8 - 1 = B(-2 - 1)$

$-9 = -3B$

$3 = B$

17. (continued)

Next let $2x - 1 = 0$, or $x = \frac{1}{2}$. We get

$$8\left(\frac{1}{2}\right) - 1 = A\left(\frac{1}{2} + 1\right) + 0$$

$$4 - 1 = A\left(\frac{3}{2}\right)$$

$$3 = \frac{3}{2} A$$

$$2 = A$$

The decomposition is

$$\frac{2}{2x - 1} + \frac{3}{x + 1}.$$

The original expression is equivalent to

$$3x + 1 + \frac{2}{2x - 1} + \frac{3}{x + 1}.$$

18. $2x - 1 + \frac{1}{x + 3} + \frac{-4}{x - 1}$

19. $\frac{2x^2 - 11x + 5}{(x - 3)(x^2 + 2x - 5)}$

$= \frac{A}{x - 3} + \frac{Bx + C}{x^2 + 2x - 5}$ (Theorem 14)

$= \frac{A(x^2 + 2x - 5) + (Bx + C)(x - 3)}{(x - 3)(x^2 + 2x - 5)}$ (Adding)

Equate the numerators:

$2x^2 - 11x + 5 = A(x^2 + 2x - 5) + (Bx + C)(x - 3)$

Let $x - 3 = 0$, or $x = 3$. Then, we get

$$2 \cdot 3^2 - 11 \cdot 3 + 5 = A(3^2 + 2 \cdot 3 - 5) + 0$$

$$18 - 33 + 5 = A(9 + 6 - 5)$$

$$-10 = 10A$$

$$-1 = A$$

Equate the coefficients of x^2:

$2 = A + B$

Substituting -1 for A, we get $B = 3$.

Equate the constant terms:

$5 = -5A - 3C$

Substituting -1 for A, we get $C = 0$.

The decomposition is as follows:

$$\frac{-1}{x - 3} + \frac{3x}{x^2 + 2x - 5}$$

20. $\frac{2}{x - 5} + \frac{x}{x^2 + x - 4}$

21. $\frac{x}{x^4 - a^4}$

$= \frac{x}{(x^2 + a^2)(x + a)(x - 1)}$ (Factoring the denominator)

$= \frac{Ax + B}{x^2 + a^2} + \frac{C}{x + a} + \frac{D}{x - a}$ (Theorem 14)

$= \frac{(Ax-B)(x+a)(x-a) + C(x^2+a^2)(x-a) + D(x^2+a^2)(x+a)}{(x^2 + a^2)(x + a)(x - a)}$

Equate the numerators:

$x = (Ax + B)(x + a)(x - a) + C(x^2 + a^2)(x - a) + D(x^2 + a^2)(x + a)$

Let $x - a = 0$, or $x = a$. Then, we get

$$a = 0 + 0 + D(a^2 + a^2)(a + a)$$

$$a = D(2a^2)(2a)$$

$$a = 4a^3D$$

$$\frac{1}{4a^3} = D$$

Let $x + a = 0$, or $x = -a$. We get

$$-a = 0 + C[(-a)^2 + a^2](-a - a) + 0$$

$$-a = C(2a^2)(-2a)$$

$$-a = -4a^3C$$

$$\frac{1}{4a^2} = C$$

Equate the coefficients of x^3:

$0 = A + C + D$

Substituting $\frac{1}{4a^2}$ for C and for D, we get $A = -\frac{1}{2a^2}$.

Equate the constant terms:

$0 = -Ba^2 - Ca^3 + Da^3$

Substitute $\frac{1}{4a^2}$ for C and for D. Then solve for B.

$$0 = -Ba^2 - \frac{1}{4a^2} \cdot a^3 + \frac{1}{4a^2} \cdot a^3$$

$$0 = -Ba^2$$

$$0 = B$$

The decomposition is as follows:

$$\frac{-\frac{1}{2a^2} x}{x^2 + a^2} + \frac{\frac{1}{4a^2}}{x + a} + \frac{\frac{1}{4a^2}}{x - a}$$

22. $P(x) = x^3 + 16$;

$$\frac{-1}{x + 1} + \frac{1}{x - 2} + \frac{6}{(x - 2)^2} + \frac{12}{(x - 2)^3} + \frac{24}{(x - 2)^4}$$

23. $y = \frac{3x}{x^2 + 5x + 4} = \frac{3x}{(x + 4)(x + 1)}$

Using Theorem 14 and adding, we get

$$\frac{3x}{(x + 4)(x + 1)} = \frac{A}{x + 4} + \frac{B}{x + 1}$$

$$= \frac{A(x + 1) + B(x + 4)}{(x + 4)(x + 1)}$$

23. (continued)

Equate the numerators:

$3x = A(x + 1) + B(x + 4)$

Let $x + 1 = 0$, or $x = -1$. We get

$3(-1) = 0 + B(-1 + 4)$

$-3 = 3B$

$-1 = B$

Next let $x + 4 = 0$, or $x = -4$. Then we get

$3(-4) = A(-4 + 1) + 0$

$-12 = -3A$

$4 = A$

The decomposition is $y = \frac{4}{x + 4} - \frac{1}{x + 1}$.

If $x = 4$, $y = \frac{4}{4 + 4} - \frac{1}{4 + 1} = \frac{1}{2} - \frac{1}{5} = \frac{3}{10}$.

If $x = 2$, $y = \frac{4}{2 + 4} - \frac{1}{2 + 1} = \frac{2}{3} - \frac{1}{3} = \frac{1}{3}$.

If $x = 0$, $y = \frac{4}{0 + 4} - \frac{1}{0 + 1} = 1 - 1 = 0$.

If $x = -2$, $y = \frac{4}{-2 + 4} - \frac{1}{-2 + 1} = 2 + 1 = 3$.

Continue finding ordered pairs as above. Then plot these points and draw the graph. Note that $x \neq -4$ and $x \neq -1$.

24. $\frac{\frac{1}{2}}{x - 3} + \frac{\frac{1}{2}}{x + 1}$

Exercise Set 6.1

1. Interchange x and y.

$y = 4x - 5$

$\downarrow \quad \downarrow$

$x = 4y - 5$ (Equation of the inverse relation)

2. $x = 3y + 5$

3. Interchange x and y.

$x^2 - 3y^2 = 3$

$\downarrow \quad \downarrow$

$y^2 - 3x^2 = 3$ (Equation of the inverse relation)

4. $2y^2 + 5x^2 = 4$

5. Interchange x and y.

$y = 3x^2 + 2$

$\downarrow \quad \downarrow$

$x = 3y^2 + 2$ (Equation of the inverse relation)

6. $x = 5y^2 - 4$

7. Interchange x and y.

$xy = 7$

$\downarrow\downarrow$

$yx = 7$ (Equation of the inverse relation)

8. $yx = -5$

9. Graph $y = x^2 + 1$. A few solutions of the equation are (0, 1), (1, 2), (-1, 2), (2, 5), and (-2, 5). Plot these points and draw the curve. Then reflect the graph of $y = x^2 + 1$ across the line $y = x$ (----). A few solutions of the inverse relation, $x = y^2 + 1$, are (1, 0), (2, 1), (2, -1), (5, 2), and (5, -2). Both graphs are shown below.

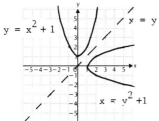

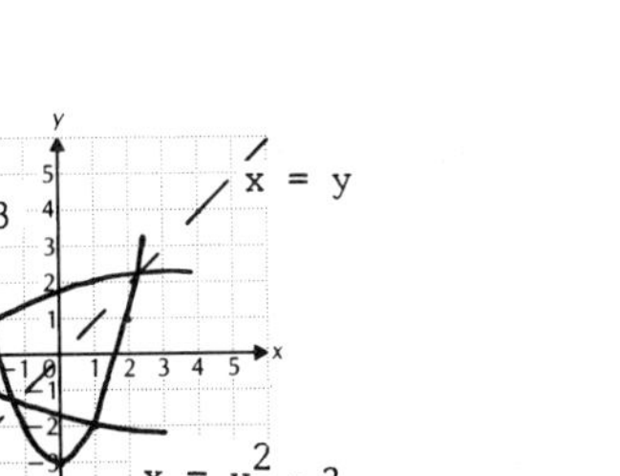

10.

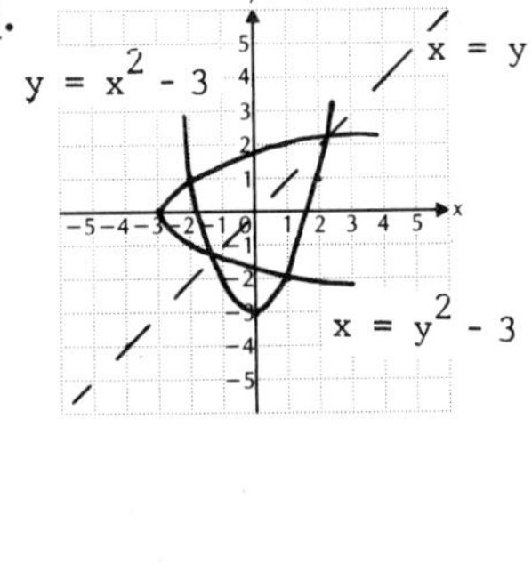

11. Graph $y = |x|$. A few solutions of the equation are (0, 0), (2, 2), (-2, 2), (5, 5), and (-5, 5). Plot these points and draw the graph. Then reflect the graph across the line $y = x$ (----). A few solutions of the inverse relation, $x = |y|$, are (0, 0), (2, 2), (2, -2), (5, 5), and (5, -5). Both graphs are shown below.

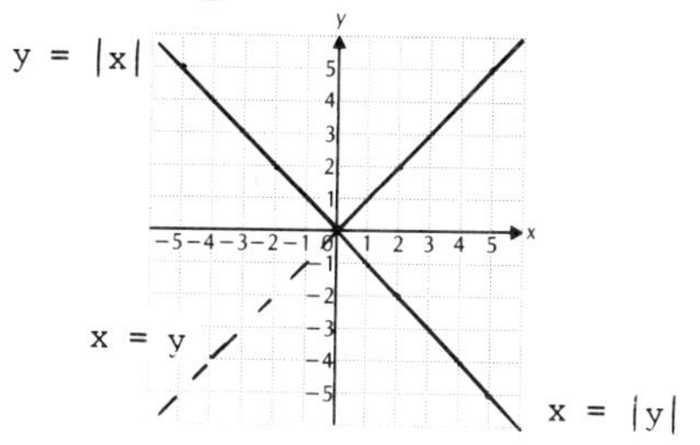

12.

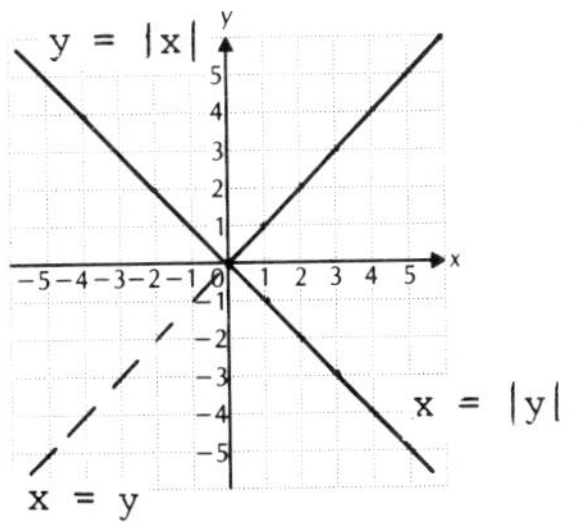

13. $3x + 2y = 4$

$\downarrow \quad \downarrow$

$3y + 2x = 4$ (Interchanging x and y)

Is the resulting equation equivalent to the original? Note that (2, -1) is a solution of $3x + 2y = 4$, but it is not a solution of $3y + 2x = 4$.

$3x + 2y = 4$	
$3(2) + 2(-1)$	4
$6 + -2$	
4	

$3y + 2x = 4$	
$3(-1) + 2(2)$	4
$-3 + 4$	
1	

Thus, the equations are not equivalent, so the graph of $3x + 2y = 4$ is not symmetric with respect to the line $y = x$.

14. No

15. $4x + 4y = 3$

$\downarrow \quad \downarrow$

$4y + 4x = 3$ (Interchanging x and y)

The commutative law of addition guarantees that the resulting equation is equivalent to the original. Thus the graph is symmetric with respect to the line $y = x$.

16. Yes

17. $xy = 10$

$\downarrow\downarrow$

$yx = 10$ (Interchanging x and y)

The commutative law of multiplication guarantees that the resulting equation is equivalent to the original. Thus the graph is symmetric with respect to the line $y = x$.

18. Yes

19. $3x = \frac{4}{y}$

$\downarrow \quad \downarrow$

$3y = \frac{4}{x}$ (Interchanging x and y)

Is $3y = \frac{4}{x}$ equivalent to $3x = \frac{4}{y}$?

$$3y = \frac{4}{x}$$

$$3y \cdot x = \frac{4}{x} \cdot x \quad \text{(Multiplying by x)}$$

$$3yx = 4$$

$$3yx \cdot \frac{1}{y} = 4 \cdot \frac{1}{y} \quad \left(\text{Multiplying by } \frac{1}{y}\right)$$

$$3x = \frac{4}{y}$$

Since $3y = \frac{4}{x}$ is equivalent to $3x = \frac{4}{y}$, the graph of $3x = \frac{4}{y}$ is symmetric with respect to the line $y = x$.

20. Yes

21. $y = |2x|$

$\downarrow \quad \downarrow$

$x = |2y|$ (Interchanging x and y)

Is the resulting equation equivalent to the original? Note that (-3, 6) is a solution of $y = |2x|$, but it is not a solution of $x = |2y|$.

$y = \lvert 2x\rvert$	
6	$\lvert 2(-3)\rvert$
	$\lvert -6\rvert$
	6

$x = \lvert 2y\rvert$	
-3	$\lvert 2\cdot 6\rvert$
	$\lvert 12\rvert$
	12

Thus, the equations are not equivalent, so the graph of $y = |2x|$ is not symmetric with respect to the line $y = x$.

22. No

23. $4x^2 + 4y^2 = 3$

$\downarrow \quad \downarrow$

$4y^2 + 4x^2 = 3$ (Interchanging x and y)

The commutative law of addition guarantees that the resulting equation is equivalent to the original. Thus the graph is symmetric with respect to the line $y = x$.

24. Yes

25. Below are graphs of the inverses. The inverses in parts a) and c) represent functions because they pass the vertical line test.

The inverses in parts b) and d) do not represent functions becuase they fail the vertical line test.

a)

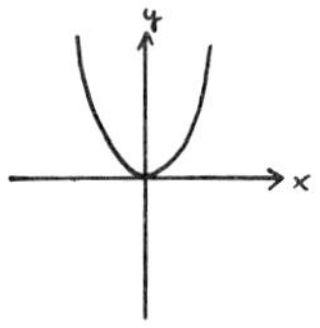

b)

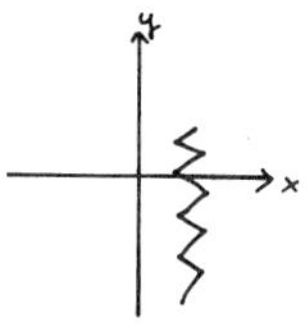

c)

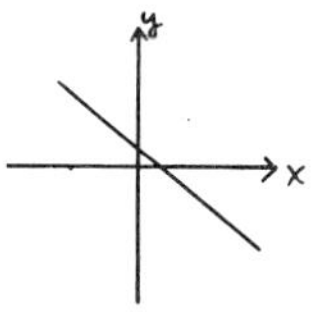

d)

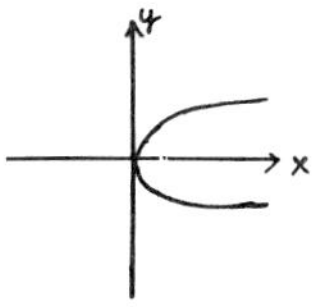

26. a, d

27. $f(x) = 2x + 5$

Replace f(x) with y: $y = 2x + 5$

Interchange x and y: $x = 2y + 5$

Solve for y:

$$x = 2y + 5$$

$$x - 5 = 2y$$

$$\frac{x - 5}{2} = y$$

Thus $f^{-1}(x) = \frac{x - 5}{2}$.

28. $g^{-1}(x) = \frac{x + 1}{3}$

29. $f(x) = \sqrt{x + 1}$

Replace f(x) with y: $y = \sqrt{x + 1}$

Interchange x and y: $x = \sqrt{y + 1}$

Solve for y:

$$x = \sqrt{y + 1}$$

$$x^2 = y + 1 \quad \text{(Principle of squaring)}$$

$$x^2 - 1 = y$$

Thus $f^{-1}(x) = x^2 - 1$.

30. $g^{-1}(x) = x^2 + 1$

31. $f^{-1}(f(3)) = 3$ [Theorem 1: $f^{-1}(f(a)) = a$]

$f(f^{-1}(-125)) = -125$ [Theorem 1: $f(f^{-1}(a)) = a$]

32. $g^{-1}(g(5)) = 5$, $g(g^{-1}(-12)) = -12$

33. $f^{-1}(f(12{,}053)) = 12{,}053$

[Theorem 1: $f^{-1}(f(a)) = a$]

$f(f^{-1}(-17{,}243)) = -17{,}243$

[Theorem 1: $f(f^{-1}(a)) = a$]

34. $g^{-1}(g(489)) = 489$, $g(g^{-1}(-17{,}422)) = -17{,}422$

35. $y = x^2$

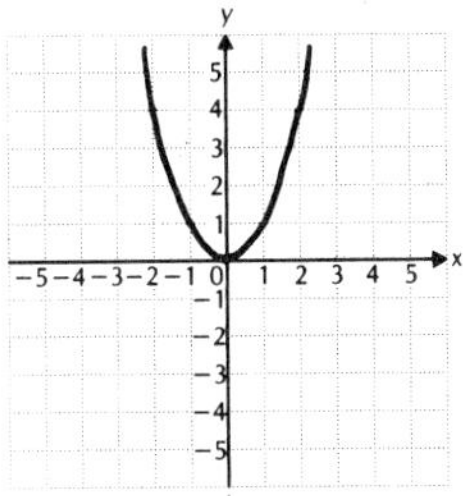

Let $y = 3.1$, then $x = 1.8$ or -1.8.
Thus, $\sqrt{3.1}$ (the principle square root) ≈ 1.8.

36. -1.7

37. Graph $y = \frac{1}{x^2}$. Graph $x = \frac{1}{y^2}$.

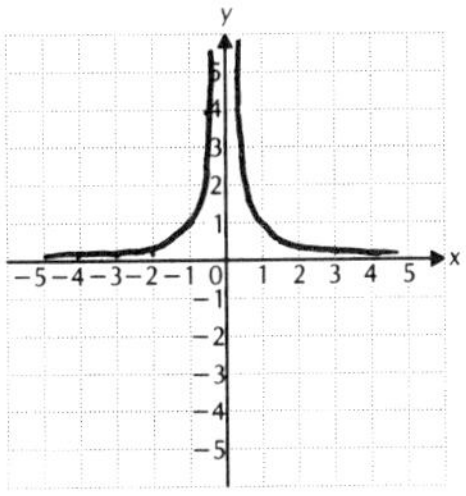

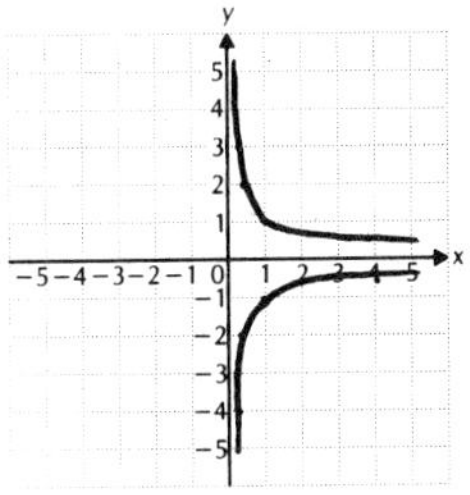

Test $y = \frac{1}{x^2}$ for symmetry with respect to the x-axis.

Replace y by -y: $-y = \frac{1}{x^2}$

Since $y = \frac{1}{x^2}$ is not equivalent to $-y = \frac{1}{x^2}$, $y = \frac{1}{x^2}$ is not symmetric with respect to the x-axis.

Test $y = \frac{1}{x^2}$ for symmetry with respect to the y-axis.

Replace x by -x: $y = \frac{1}{(-x)^2} = \frac{1}{x^2}$

Since the equations are equivalent, $y = \frac{1}{x^2}$ is symmetric with respect to the y-axis.

Test $y = \frac{1}{x^2}$ for symmetry with respect to the origin.

Replace x by -x and y by -y: $-y = \frac{1}{(-x)^2} = \frac{1}{x^2}$

Since $y = \frac{1}{x^2}$ is not equivalent to $-y = \frac{1}{x^2}$, $y = \frac{1}{x^2}$ is not symmetric with respect to the origin.

37. (continued)

Test $y = \frac{1}{x^2}$ for symmetry with respect to the line $y = x$.

Interchange x and y: $x = \frac{1}{y^2}$

Since $y = \frac{1}{x^2}$ is not equivalent to $x = \frac{1}{y^2}$, $y = \frac{1}{x^2}$ is not symmetric with respect to the line $y = x$.

Exercise Set 6.2

1. a) Graph: $y = 4^x$

x	y
0	1
1	4
2	16
-1	$\frac{1}{4}$
-2	$\frac{1}{16}$

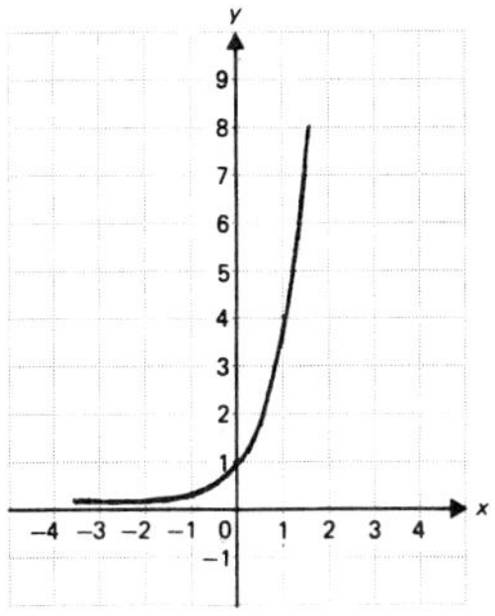

b) Graph: $y = \left(\frac{1}{4}\right)^x$

x	y
0	1
1	$\frac{1}{4}$
2	$\frac{1}{16}$
-1	4
-2	16

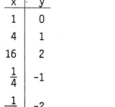

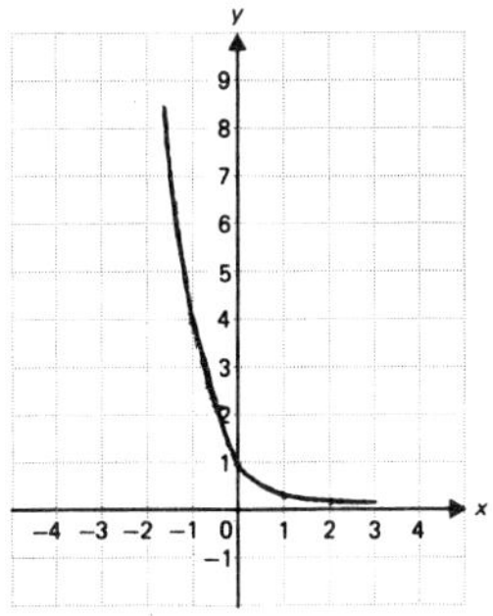

Note that $y = \left(\frac{1}{4}\right)^x = 4^{-x}$, so we could have found this graph by reflecting the graph in a), $y = 4^x$, across the y-axis.

c) The equation $y = \log_4 x$ is equivalent to $x = 4^y$. The graph is a reflection of $y = 4^x$ across the line $y = x$.

For $y = 4^x$:

x	y
0	1
1	4
2	16
-1	$\frac{1}{4}$
-2	$\frac{1}{16}$

For $y = \log_4 x$ (or $x = 4^y$):

x	y
1	0
4	1
16	2
$\frac{1}{4}$	-1
$\frac{1}{16}$	-2

We could also have found these values directly:

$\log_4 1 = 0$ $\log_4 4 = 1$ $\log_4 16 = 2$

$\log_4 \frac{1}{4} = -1$ $\log_4 \frac{1}{16} = -2$

1. (continued)

$y = \log_4 x$

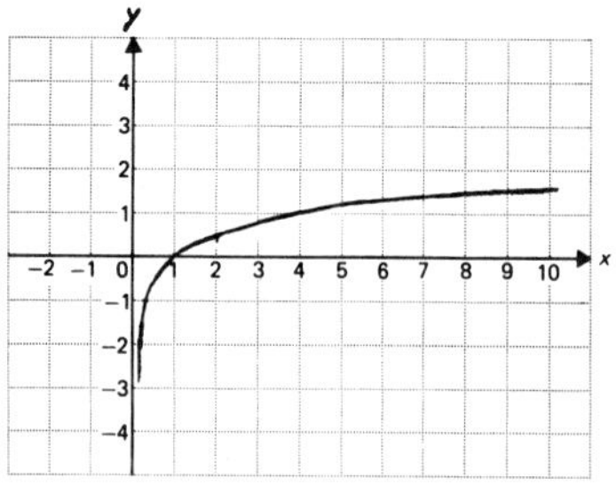

2. a) $y = 3^x$

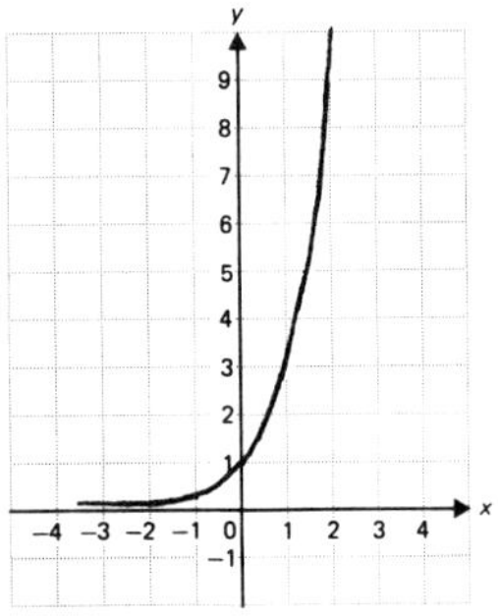

b) $y = \left(\frac{2}{3}\right)^x$

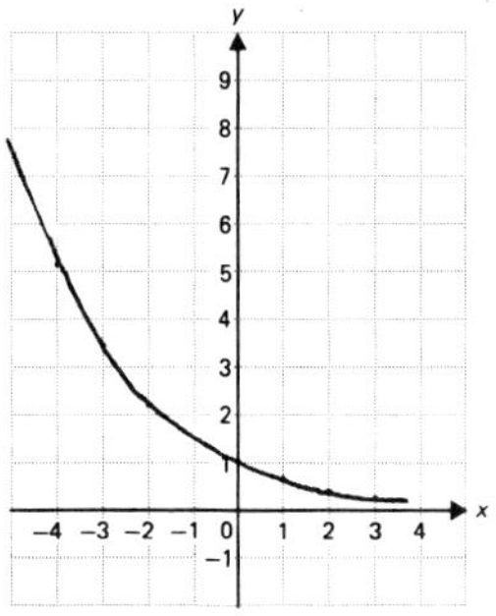

c) $y = \log_3 x$

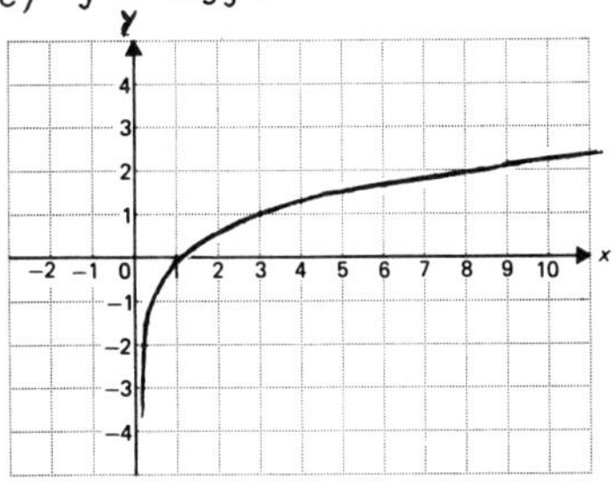

3. Graph: $y = 5^x$

x	y
0	1
1	5
2	25
-1	$\frac{1}{5}$
-2	$\frac{1}{25}$

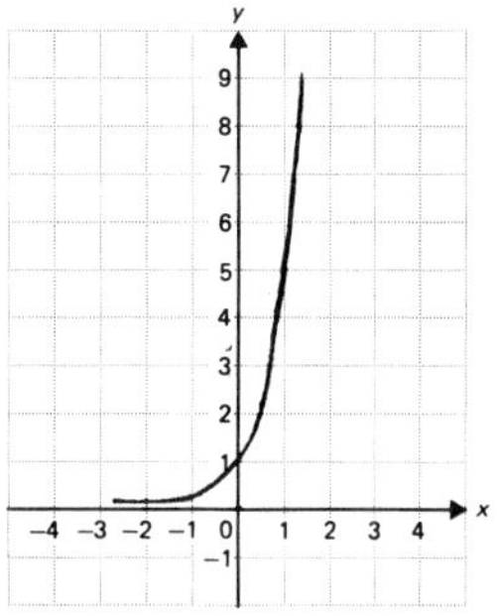

4. $y = 2^x$

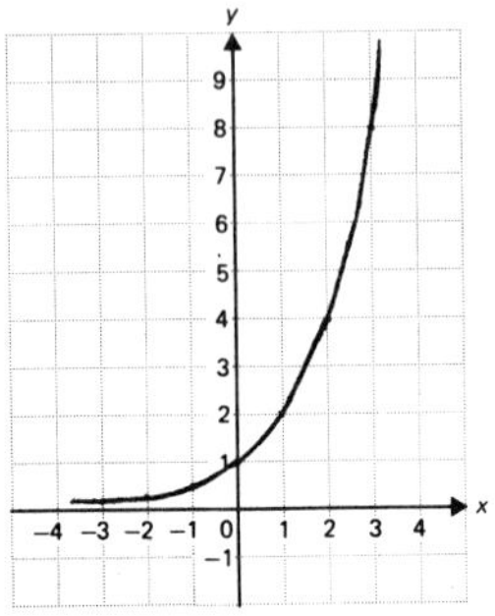

5. Graph: $y = \log_2 x$ (or $x = 2^y$)

x	y
1	0
2	1
4	2
$\frac{1}{2}$	-1
$\frac{1}{4}$	-2

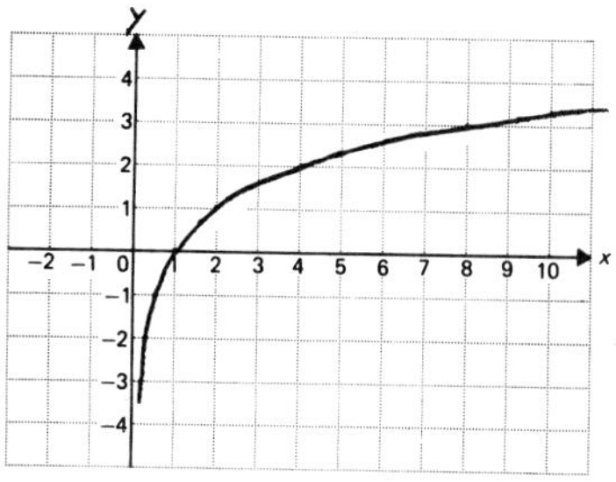

6. $y = \log_{10} x$

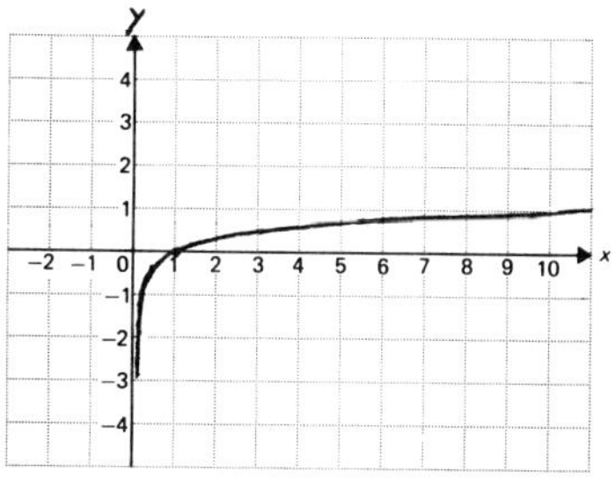

7. $\log_2 32 = 5$ (Logarithmic equation)
$2^5 = 32$ (Exponential equation)
Remember: The logarithm is the exponent.

8. $10^3 = 1000$

9. $\log_{10} 0.01 = -2$ (Logarithmic equation)
$10^{-2} = 0.01$ (Exponential equation)

10. $(\sqrt{5})^2 = 5$

11. $\log_6 6 = 1$ (Logarithmic equation)
$6^1 = 6$ (Exponential equation)

12. $b^N = M$

13. $6^0 = 1$ (Exponential equation)
$\log_6 1 = 0$ (Logarithmic equation)
Remember: The logarithm is the exponent.

14. $\log_{10} 0.001 = -3$

15. $\left(\frac{4}{3}\right)^{-2} = \frac{9}{16}$ (Exponential equation)

$\log_{\frac{4}{3}} \frac{9}{16} = -2$ (Logarithmic equation)

16. $\log_5 625 = 4$

17. $5^{-2} = \frac{1}{25}$ (Exponential equation)

$\log_5 \frac{1}{25} = -2$ (Logarithmic equation)

18. $\log_8 2 = \frac{1}{3}$

19. $3^{0.08} = 1.0833$ (Exponential equation)

$\log_e 1.0833 = 0.08$ (Logarithmic equation)

20. $\log_{10} 3 = 0.4771$

21. $\log_{10} x = 3$ (Logarithmic equation)

$10^3 = x$ (Exponential equation)

$1000 = x$

22. 9

23. $\log_x \frac{1}{32} = 5$ (Logarithmic equation)

$x^5 = \frac{1}{32}$ (Exponential equation)

Since $\left(\frac{1}{2}\right)^5 = \frac{1}{32}$, $x = \frac{1}{2}$.

24. 2

25. $\log_2 16 = x$ (Logarithmic equation)

$2^x = 16$ (Exponential equation)

$2^x = 2^4$

$x = 4$

26. 1

27. $\log_4 2 = x$ (Logarithmic equation)

$4^x = 2$ (Exponential equation)

Since $4^{1/2} = 2$, $x = \frac{1}{2}$. $(4^{1/2} = \sqrt{4} = 2)$

28. 6

29. $\log_{10} 10^2 = x$ (Logarithmic equation)

$10^x = 10^2$ (Exponential equation)

$x = 2$

30. 4

31. $\log_\pi \pi = x$ (Logarithmic equation)

$\pi^x = \pi$ (Exponential equation)

$x = 1$ $(\pi = \pi^1)$

32. 1

33. $\log_{10} 0.001 = x$ (Logarithmic equation)

$10^x = 0.001$ (Exponential equation)

$10^x = 10^{-3}$

$x = -3$

34. 3

35. $3^{\log_3 4x} = 4x$ [Theorem 3: $a^{\log_a x} = x$]

36. $4x - 5$

37. $\log_Q Q^{\sqrt{5}} = \sqrt{5}$ [Theorem 3: $\log_a a^x = x$]

38. $|x - 4|$

39. Note that $y = \log_2 (x + 3) = \log_2 [x - (-3)]$. That is, the graph of $y = \log_2 (x + 3)$ is a translation of the graph of $y = \log_2 x$, 3 units to the left. Thus, we first graph $y = \log_2 x$, then we translate it 3 units to the left.

a) First graph $y = \log_2 x$.

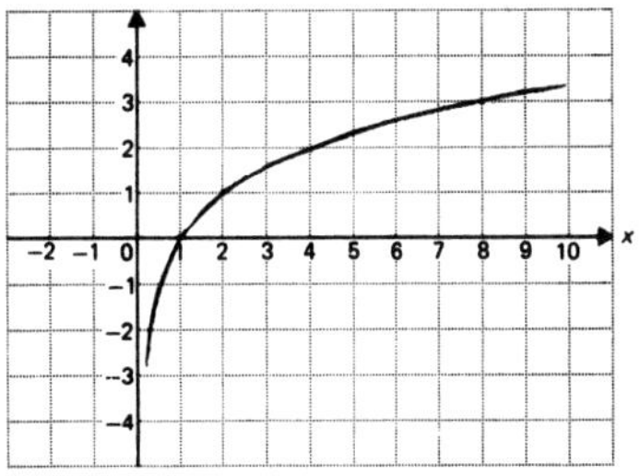

b) Now we translate $y = \log_2 x$ 3 units to the left to obtain the graph of $y = \log_2 (x + 3)$.

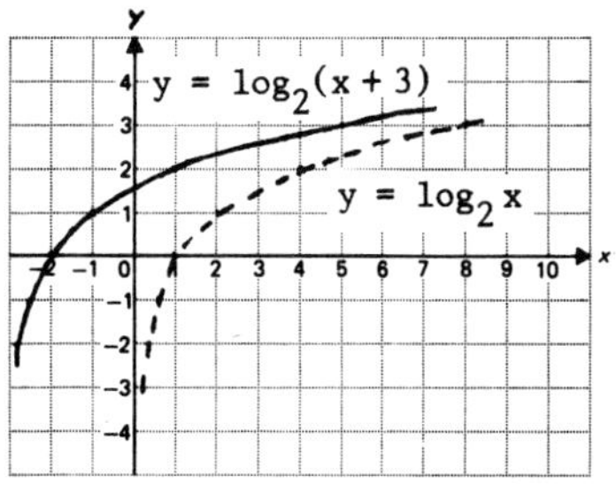

40. $y = \log_3 (x - 2)$

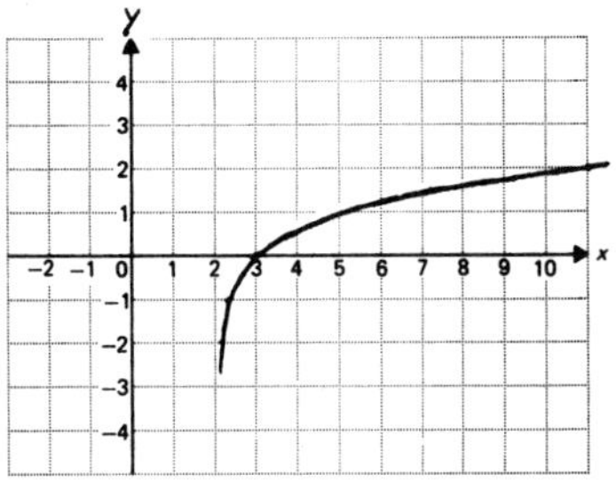

41. $y = 2^x - 1$

x	y
0	0
1	1
2	3
3	7
-1	$-\frac{1}{2}$
-2	$-\frac{3}{4}$
-3	$-\frac{7}{8}$

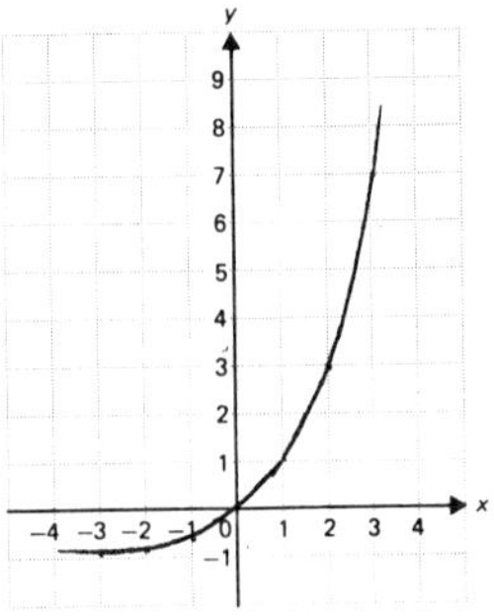

42. $y = 2^{x-3}$

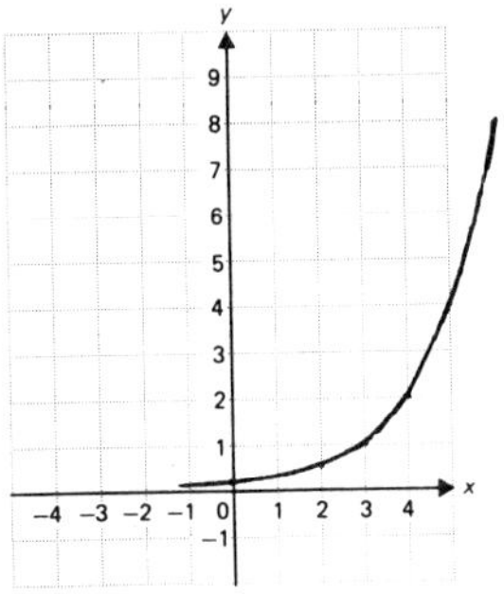

43. $f(x) = 2^{|x|}$

x	f(x)
0	1
1	2
-1	2
2	4
-2	4
3	8
-3	8

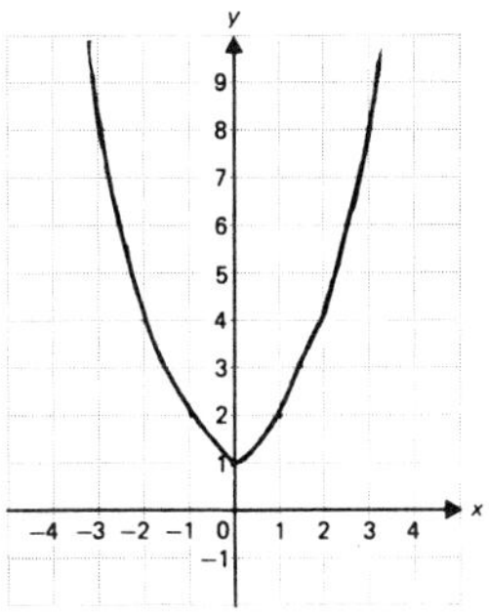

44. $f(x) = \log_3 |x|$

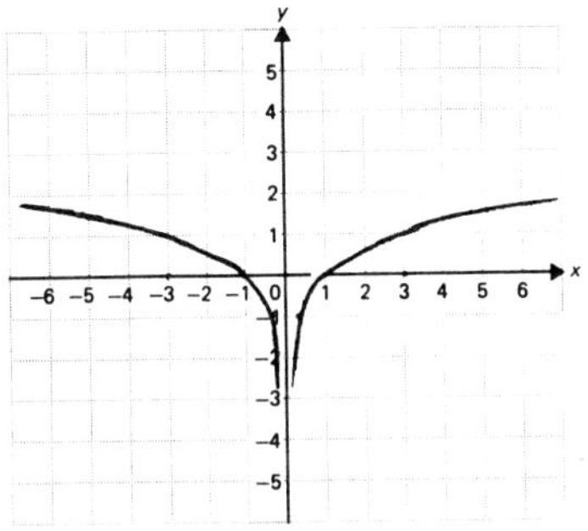

45. $f(x) = 2^x + 2^{-x}$

We select numbers for x and find corresponding values for f(x).

$f(0) = 2^0 + 2^{-0} = 1 + 1 = 2$

$f(1) = 2^1 + 2^{-1} = 2 + \frac{1}{2} = 2\frac{1}{2}$

$f(-1) = 2^{-1} + 2^1 = \frac{1}{2} + 2 = 2\frac{1}{2}$

$f(2) = 2^2 + 2^{-2} = 4 + \frac{1}{4} = 4\frac{1}{4}$

$f(-2) = 2^{-2} + 2^2 = \frac{1}{4} + 4 = 4\frac{1}{4}$

x	f(x)
0	2
1	$2\frac{1}{2}$
-1	$2\frac{1}{2}$
2	$4\frac{1}{4}$
-2	$4\frac{1}{4}$

$f(x) = 2^x + 2^{-x}$

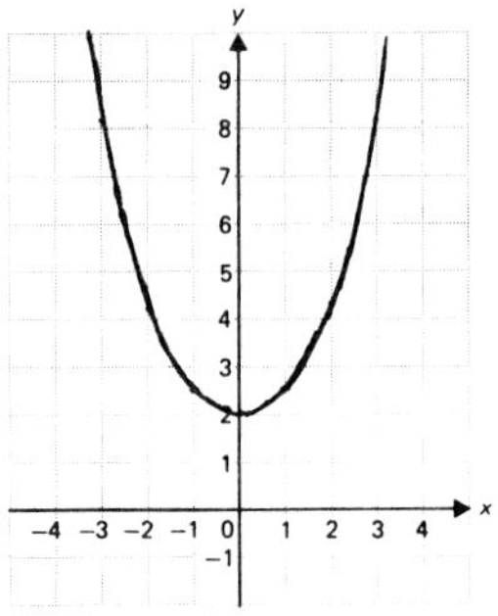

46. $f(x) = 2^{-(x-1)}$

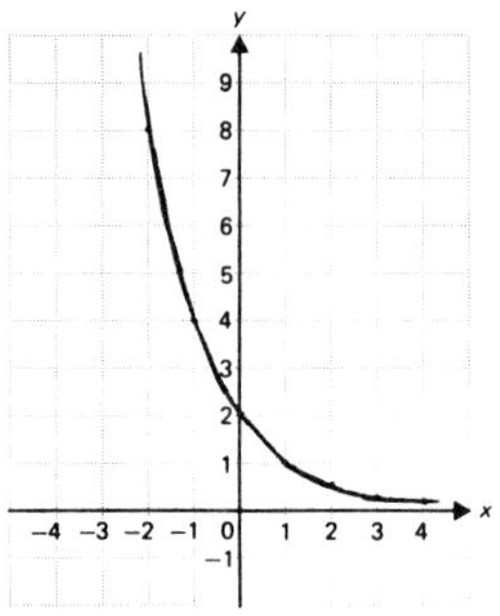

47. $f(x) = 3^x$

3^x is defined for all reals. Thus the domain is the set of all real numbers.

48. $\{x | x > 0\}$

49. $f(x) = \log_a x^2$

The domain of a logarithmic function is the set of all positive real numbers. Since $x^2 > 0$ for all reals, except 0, the domain is $\{x | x \neq 0\}$.

50. $\{x | x > 0\}$

51. $f(x) = \log_{10} (3x - 4)$

The domain of a logarithmic function is the set of all positive real numbers. Thus,

$$3x - 4 > 0$$
$$3x > 4$$
$$x > \frac{4}{3}$$

The domain is $\left\{x \middle| x > \frac{4}{3}\right\}$.

52. $\{x | x \neq 0\}$

53. $f(x) = \log_6 (x^2 - 9)$

The domain of a logarithmic function is the set of all positive real numbers. For $x^2 - 9$ to be greater than 0, x must be less than -3 or greater than 3. The domain is $\{x | x < -3 \text{ or } x > 3\}$.

54. $\{x | x > 0\}$

55. Solve $3^x \leqslant 1$ by graphing. To do this we first graph $y = 3^x$.

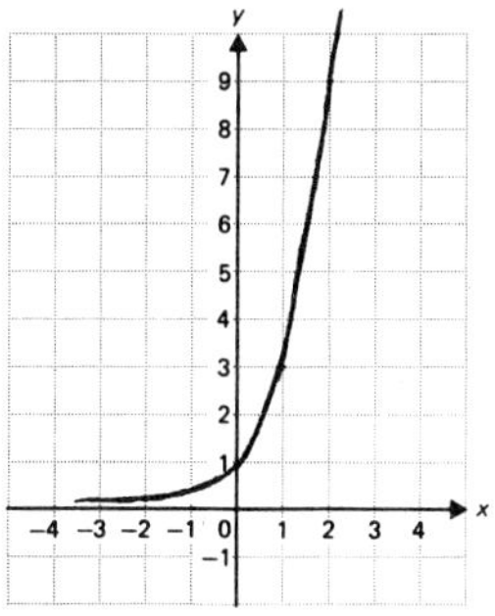

Next we study the graph to see for which inputs the outputs are less than or equal to 1.

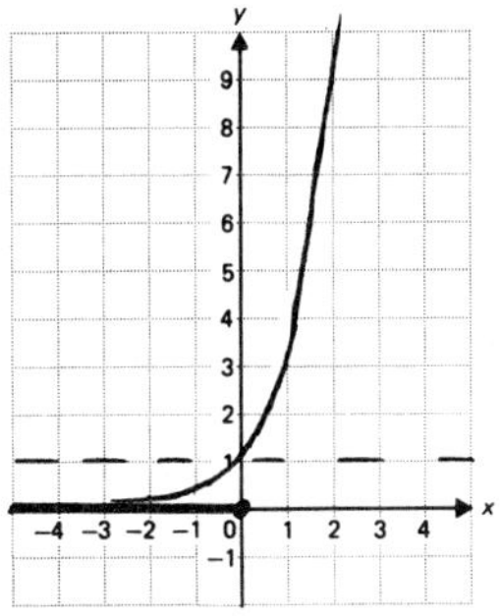

We see that $3^x \leqslant 1$ for those values of x for which $x \leqslant 0$. Thus the solution set of $3^x \leqslant 1$ is $\{x | x \leqslant 0\}$.

56. $\{x | 0 < x < 1\}$

57. Solve $\log_2 x \geqslant 4$ by graphing. To do this we first graph $y = \log_2 x$ (or $2^y = x$).

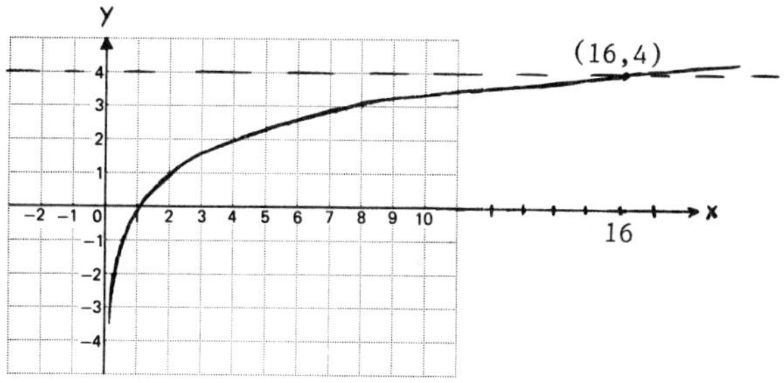

Next we study the graph to see for which inputs the outputs are greater than or equal to 4. We see that $\log_2 x \geqslant 4$ for those values of x for which $x \geqslant 16$. Thus the solution set of $\log_2 x \geqslant 4$ is $\{x | x \geqslant 16\}$.

58. $\{x | 3 < x \leqslant 35\}$

59. Solve $2^{x+3} > 1$ by graphing. To do this we first graph $y = 2^{x+3}$.

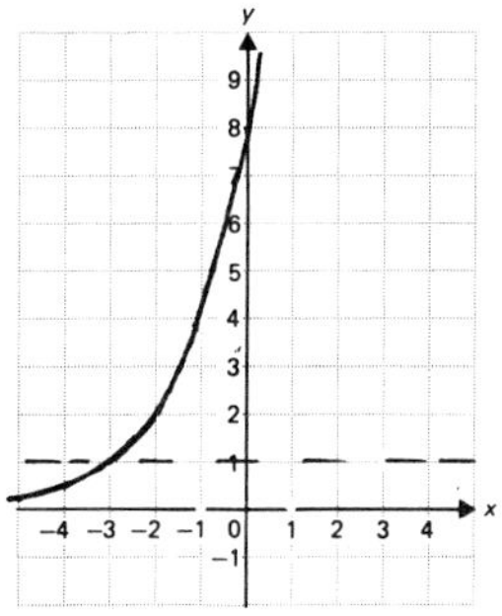

Next we study the graph to see for which inputs the outputs are greater than 1. We see that $2^{x+3} > 1$ for those values of x for which $x > -3$. Thus the solution set is $\{x | x > -3\}$.

60. 27, 30.135326, 31.489136, 31.523749, 31.541070, 31.544189

61. $5^\pi \approx 156.9925$

$\pi^5 \approx 306.0197$

π^5 is larger than 5^π.

62. $8^{\sqrt{3}}$

63. Graph: $y = (0.745)^x$

A few solutions are (0, 1), (1, 0.745), (2, 0.555), (3, 0.413), (4, 0.308), (-1, 1.342), (-2, 1.802), (-3, 2.418), and (-4, 3.246). Plot these points and draw the curve.

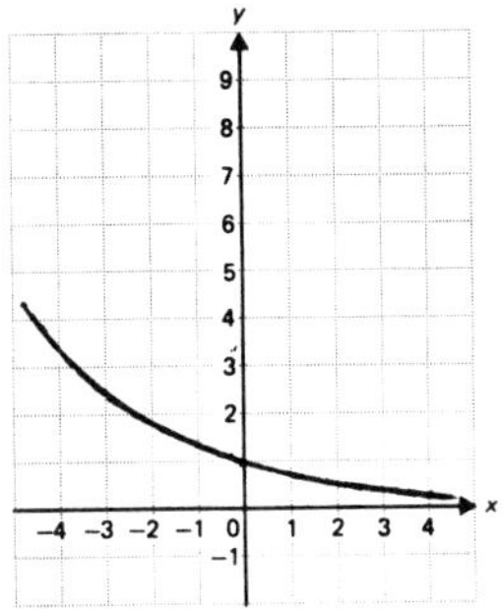

Exercise Set 6.3

1. $\log_a x^2y^3z$

$= \log_a x^2 + \log_a y^3 + \log_a z$ (Theorem 4)

$= 2 \log_a x + 3 \log_a y + \log_a z$ (Theorem 5)

2. $\log_a 5 + \log_a x + 4 \log_a y + 3 \log_a z$

3. $\log_b \frac{xy^2}{z^3}$

$= \log_b xy^2 - \log_b z^3$ (Theorem 6)

$= \log_b x + \log_b y^2 - \log_b z^3$ (Theorem 4)

$= \log_b x + 2 \log_b y - 3 \log_b z$ (Theorem 5)

4. $\frac{1}{3} [4 \log_c x - 3 \log_c y - 2 \log_c z]$

5. $\frac{2}{3} \log_a 64 - \frac{1}{2} \log_a 16$

$= \log_a 64^{2/3} - \log_a 16^{1/2}$ (Theorem 5)

$= \log_a (4^3)^{2/3} - \log_a (4^2)^{1/2}$

$= \log_a 4^2 - \log_a 4$

$= \log_a \frac{4^2}{4}$ (Theorem 6)

$= \log_a 4$

6. $\log_a \frac{y^3 \sqrt{x}}{x^2}$

7. $\log_a 2x + 3(\log_a x - \log_a y)$

$= \log_a 2x + 3 \log_a x - 3 \log_a y$ (Distributive Law)

$= \log_a 2x + \log_a x^3 - \log_a y^3$ (Theorem 5)

$= \log_a 2x^4 - \log_a y^3$ (Theorem 4)

$= \log_a \frac{2x^4}{y^3}$ (Theorem 6)

8. $\log_a x$

9. $\log_a \frac{a}{\sqrt{x}} - \log_a \sqrt{ax}$

$= \log_a ax^{-1/2} - \log_a a^{1/2}x^{1/2}$

$= \log_a \frac{ax^{-1/2}}{a^{1/2}x^{1/2}}$ (Theorem 6)

$= \log_a a^{1/2}x^{-1}$, or $\log_a \frac{\sqrt{a}}{x}$

$= \log_a a^{1/2} + \log_a x^{-1}$ (Theorem 4)

$= \frac{1}{2} - \log_a x$ (Theorems 7 and 5)

10. $\log_a (x + 2)$

11. $\log_a (x^3 + y^3) - \log_a (x + y)$

$= \log_a \frac{x^3 + y^3}{x + y}$ (Theorem 6)

$= \log_a \frac{(x + y)(x^2 - xy + y^2)}{x + y}$

$= \log_a (x^2 - xy + y^2)$

12. $\log_a (x^3 - y^3)$

13. $\log_a \sqrt{1 - x^2}$

$= \log_a (1 - x^2)^{1/2}$

$= \frac{1}{2} \log_a (1 - x^2)$ (Theorem 5)

$= \frac{1}{2} \log_a [(1 - x)(1 + x)]$

$= \frac{1}{2} [\log_a (1 - x) + \log_a (1 + x)]$ (Theorem 4)

$= \frac{1}{2} \log_a (1 - x) + \frac{1}{2} \log_a (1 + x)$

14. $\frac{1}{2} \log_a (x + t) - \frac{1}{2} \log_a (x - t)$

15. $\log_{10} 4$

$= \log_{10} 2^2$ ($4 = 2^2$)

$= 2 \log_{10} 2$ (Theorem 5)

$= 2(0.301)$ (Substituting 0.301 for $\log_{10} 2$)

$= 0.602$

16. 0.699

17. $\log_{10} 50$

$= \log_{10} \frac{100}{2}$ $\left[50 = \frac{100}{2}\right]$

$= \log_{10} 10^2 - \log_{10} 2$ ($100 = 10^2$) (Theorem 6)

$= 2 \log_{10} 10 - \log_{10} 2$ (Theorem 5)

$= 2 - 0.301$ (Substituting)

$= 1.699$

18. 1.079

19. $\log_{10} 60$
$= \log_{10} (10 \cdot 2 \cdot 3)$ $(60 = 10 \cdot 2 \cdot 3)$
$= \log_{10} 10 + \log_{10} 2 + \log_{10} 3$ (Theorem 4)
$= 1 + 0.301 + 0.477$ (Substituting)
$= 1.778$

20. -0.477

21. $\log_{10} \sqrt{\frac{2}{3}}$
$= \log_{10} \left(\frac{2}{3}\right)^{1/2}$
$= \frac{1}{2} \log_{10} \frac{2}{3}$ (Theorem 5)
$= \frac{1}{2} (\log_{10} 2 - \log_{10} 3)$ (Theorem 6)
$= \frac{1}{2} (0.301 - 0.477)$ (Substituting)
$= -0.088$

22. 0.2158

23. $\log_{10} 90$
$= \log_{10} (3^2 \cdot 10)$
$= \log_{10} 3^2 + \log_{10} 10$ (Theorem 4)
$= 2 \log_{10} 3 + \log_{10} 10$ (Theorem 5)
$= 2(0.477) + 1$ (Substituting)
$= 1.954$

24. 0.051

25. $\log_{10} \frac{9}{10}$
$= \log_{10} 3^2 - \log_{10} 10$ $(9 = 3^2)$ (Theorem 6)
$= 2 \log_{10} 3 - \log_{10} 10$ (Theorem 5)
$= 2(0.477) - 1$ (Substituting)
$= -0.046$

26. -0.602

27. $\frac{\log_a M}{\log_a N} = \log_a M - \log_a N$, False
By Theorem 6,
$\log_a M - \log_a N = \log_a \frac{M}{N}$, not $\frac{\log_a M}{\log_a N}$.

28. False

29. $\frac{\log_a M}{c} = \frac{1}{c} \log_a M = \log_a M^{1/c}$
Thus, true by Theorem 5.

30. True

31. $\log_a 2x = 2 \log_a x$, False
By Theorem 4,
$\log_a 2x = \log_a 2 + \log_a x$, not $2 \log_a x$.

32. True

33. $\log_a (M + N) = \log_a M + \log_a N$, False
By Theorem 4,
$\log_a M + \log_a N = \log_a MN$, not $\log_a (M + N)$.

34. True

35. $\log_\pi \pi^{2x+3} = 4$
$2x + 3 = 4$ (Theorem 3)
$2x = 1$
$x = \frac{1}{2}$

36. $\frac{9}{8}$

37. $4^{2 \log_4 x} = 7$
$4^{\log_4 x^2} = 7$ (Theorem 5)
$x^2 = 7$ (Theorem 3)
$x = \pm \sqrt{7}$

Only $\sqrt{7}$ checks. Since the domain of a logarithmic function is the set of all positive real numbers, $\log_4 x$ is not defined if $x = -\sqrt{7}$. Thus $-\sqrt{7}$ is not a solution.

The solution is $\sqrt{7}$.

38. 3

39. $(x + 3) \cdot \log_a a^x = x$
$\log_a (a^x)^{(x+3)} = x$ (Theorem 5)
$\log_a a^{(x^2+3x)} = x$
$x^2 + 3x = x$ (Theorem 3)
$x^2 + 2x = 0$
$x(x + 2) = 0$

$x = 0$ or $x + 2 = 0$
$x = 0$ or $x = -2$

Both values check. The solution are 0 and -2.

40. $\{x \mid x > 0\}$

41. $\log_a 5x = \log_a 5 + \log_a x$

$\log_a 5x = \log_a 5x$ (Theorem 4)

Since the domain of a logarithmic function is the set of all positive real numbers, the solution set is $\{x | x > 0\}$.

42. $\{x | x > -2\}$

Exercise Set 6.4

1. $\log 3.921 \approx 0.5934$

2. 0.9291

3. $\log 37{,}590 \approx 4.5751$

4. 5.7484

5. $\log 5149 \approx 3.7117$

6. 3.5057

7. $\log 0.1414 \approx -0.8496$

8. -0.4872

9. $\log 0.0005123 \approx -3.2905$

10. -2.0084

11. $\log 0.00001234 \approx -4.9087$

12. -1.9087

13. $\log (5.621 \times 10^5)$
$= \log 5.621 + \log 10^5$
$\approx 0.7498 + 5$
≈ 5.7498

14. 7.7870

15. $\log (8.042 \times 10^{38})$
$= \log 8.042 + \log 10^{38}$
$\approx 0.9054 + 38$
≈ 38.9054

16. 29.1623

17. $\log (4.625 \times 10^{-12})$
$= \log 4.625 + \log 10^{-12}$
$\approx 0.6651 + (-12)$
≈ -11.3349

18. -19.2983

19. Since the domain of a logarithmic function is the set of all positive real numbers, log (-4.923) does not exist.

20. Does not exist

21. $\text{antilog } 4.6524 = 10^{4.6524} \approx 44{,}915.89$

22. 1050.51

23. $10^{5.9231} \approx 83{,}772.15$

24. 2,589,405.11

25. $\text{antilog } 0.003215 = 10^{0.003215} \approx 1.0074$

26. 1.00095

27. $\text{antilog } (-3.0143) = 10^{-3.0143} \approx 0.0009676$

28. 0.00001199

29. $10^{-5.9231} \approx 0.000001194$

30. 0.000000924

31. $\text{antilog } (8.1111 \times 10^{-4})$
$= \text{antilog } 0.00081111 = 10^{0.00081111} \approx 1.0019$

32. 1.0210

33. $10^{(3.8146 \times 10^{-3})} = 10^{0.0038146} \approx 1.0088$

34. 1.1245

35. $\text{antilog } (2.8307 \times 10^{-8})$
$= 10^{0.000000028307} \approx 1.000000065$

36. 1.000004602

37. $10^{(4.0060 \times 10^{-7})} = 10^{0.00000040060} \approx 1.00000092$

38. 1.000025584

39. $\ln 3285 \approx 8.0971$

40. 11.4354

41. $\ln 0.1248 \approx -2.0810$

42. -2.9123

43. $\ln 0.0005127 \approx -7.5758$

44. -10.1706

45. Since the domain of a logarithmic function is the set of all positive real numbers, ln (-8.762) does not exist.

46. Does not exist

47.
$$\begin{aligned}&\ln (8.041 \times 10^{28})\\ &= \ln 8.041 + \ln 10^{28}\\ &= \ln 8.041 + 28 \ln 10\\ &\approx 2.0846 + 28(2.3026)\\ &\approx 2.0846 + 64.4728\\ &\approx 66.5574\end{aligned}$$

48. 28.34

49.
$$\begin{aligned}&\ln (5.043 \times 10^{-14})\\ &= \ln 5.043 + \ln 10^{-14}\\ &= \ln 5.043 - 14 \ln 10\\ &\approx 1.6180 - 14(2.3026)\\ &\approx 1.6180 - 32.2364\\ &\approx -30.6184\end{aligned}$$

50. -33.42

51. $\text{antilog}_e\ 3.4935 = e^{3.4935} \approx 32.9009$

52. 1458.4070

53. $e^{1.0312} \approx 2.8044$

54. 824.8487

55. $\text{antilog}_e\ 31.891 = e^{31.891} \approx 7.0808 \times 10^{13}$

56. 1.0532×10^{31}

57. $e^{17.814} \approx 54{,}515{,}738.62$

58. 3.5268×10^{54}

59. $\text{antilog}_e\ (-17.123) = e^{-17.123} \approx 0.000000036$

60. 0.02191

61. $e^{-12.832} \approx 0.000002673$

62. 0.006454

63. Find $\log_4 100$. We will use common logarithms in the Change of Base Theorem:

$$\log_b M = \frac{\log_a M}{\log_a b}$$

Substitute 10 for a, 4 for b, and 100 for M.

$$\begin{aligned}\log_4 100 &= \frac{\log_{10} 100}{\log_{10} 4}\\ &\approx \frac{2.0000}{0.6021} \quad \text{(Using a calculator)}\\ &\approx 3.3219\end{aligned}$$

64. 2.7268

65. Find $\log_2 12$. We will use common logarithms in the Change of Base Theorem:

$$\log_b M = \frac{\log_a M}{\log_a b}$$

Substitute 10 for a, 2 for b, and 12 for M.

$$\begin{aligned}\log_2 12 &= \frac{\log_{10} 12}{\log_{10} 2}\\ &\approx \frac{1.0792}{0.3010} \quad \text{(Using a calculator)}\\ &\approx 3.5850\end{aligned}$$

66. 2.2920

67. Find $\log_{100} 50$. We will use common logarithms in the Change of Base Theorem:

$$\log_b M = \frac{\log_a M}{\log_a b}$$

Substitute 10 for a, 100 for b, and 50 for M.

$$\begin{aligned}\log_{100} 50 &= \frac{\log_{10} 50}{\log_{10} 100}\\ &\approx \frac{1.6990}{2} \quad \text{(Using a calculator)}\\ &= 0.8495\end{aligned}$$

68. 0.7384

69. Find $\log_{0.5} 7$. We will use common logarithms in the Change of Base Theorem:

$$\log_b M = \frac{\log_a M}{\log_a b}$$

Substitute 10 for a, 0.5 for b and 7 for M.

$$\begin{aligned}\log_{0.5} 7 &= \frac{\log_{10} 7}{\log_{10} 0.5}\\ &\approx \frac{0.8451}{-0.3010} \quad \text{(Using a calculator)}\\ &\approx -2.8074\end{aligned}$$

70. -0.3010

71. Find $\log_3 0.3$. We will use common logarithms in the Change of Base Theorem:

$$\log_b M = \frac{\log_a M}{\log_b b}$$

Substitute 10 for a, 3 for b, and 0.3 for M.

$$\log_3 0.3 = \frac{\log_{10} 0.3}{\log_{10} 3}$$

$$\approx \frac{-0.5229}{0.4771} \quad \text{(Using a calculator)}$$

$$\approx -1.0959$$

72. -4.0589

73. Find $\log_\pi 100$. We will use common logarithms in the Change of Base Theorem:

$$\log_b M = \frac{\log_a M}{\log_a b}$$

Substitute 10 for a, π for b, and 100 for M.

$$\log_\pi 100 = \frac{\log_{10} 100}{\log_{10} \pi}$$

$$\approx \frac{2.0000}{0.4971} \quad \text{(Using a calculator)}$$

$$\approx 4.0229$$

74. 2.8119

75. $y = 10^x$

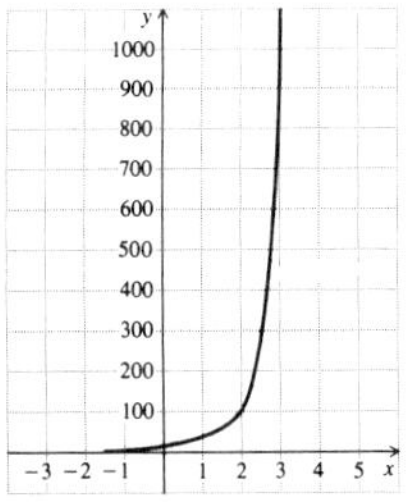

76. $y = \log x$

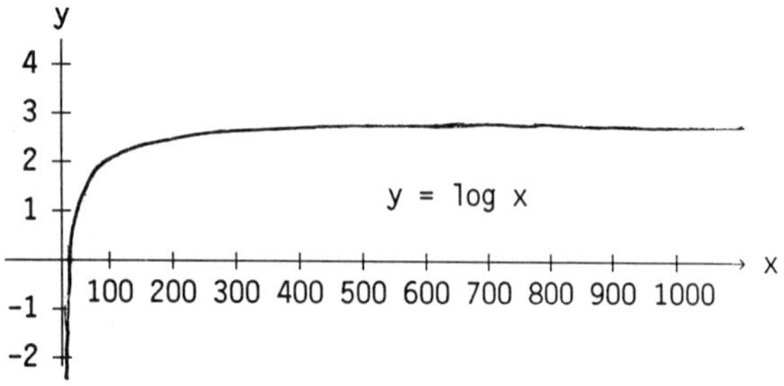

77. $y = e^x$

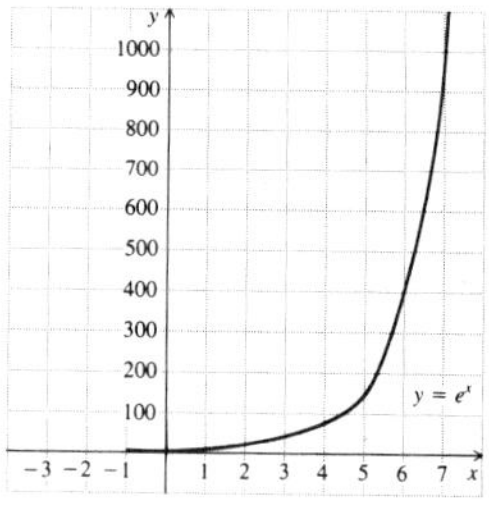

78. $y = \ln x$

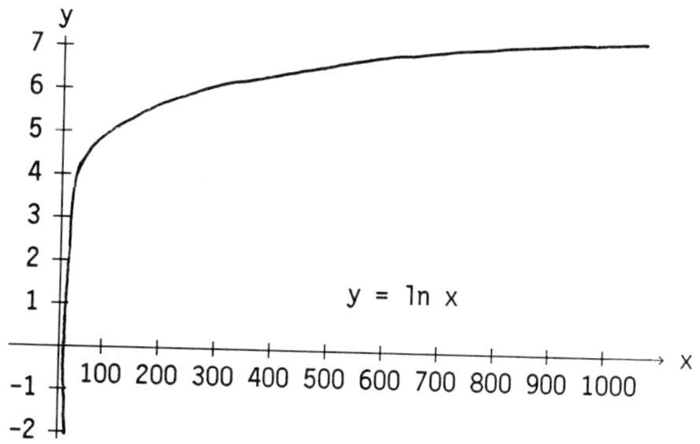

79. a) Let a = e, b = 10, M = e. Substitute in the Change of Base Theorem.

$$\log e = \frac{\ln e}{\ln 10} = \frac{1}{\ln 10}$$

b) Let a = e, b = 10, M = M. Substitute in the Change of Base Theorem.

$$\log M = \frac{\ln M}{\ln 10}$$

c) Let a = M, b = b, M = M. Substitute in the Change of Base Theorem.

$$\log_b M = \frac{\log_M M}{\log_M b} = \frac{1}{\log_M b}$$

d) Let a = 10, b = e, M = M. Substitute in the Change of Base Theorem.

$$\ln M = \frac{\log M}{\log E}$$

80. 3

Exercise Set 6.5

In this section we will use a calculator with an $\boxed{e^x}$ key to find approximate values of e^x. If your calculator does not have an $\boxed{e^x}$ key, you can raise 2.718 ($e \approx 2.718$) to the x power using an exponential key.

1. Graph $f(x) = 2e^x$.

Use a calculator to find approximate values of $2e^x$, and use these values to draw the graph.

1. (continued)

x	$2e^x$
-2	0.3
-1	0.7
0	2
1	5.4
2	14.8

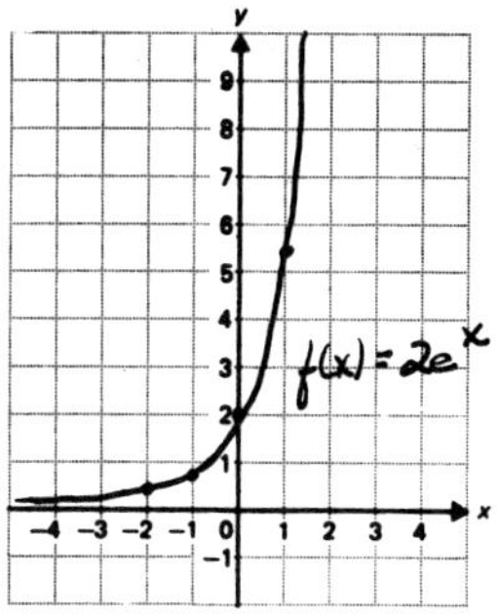

2.

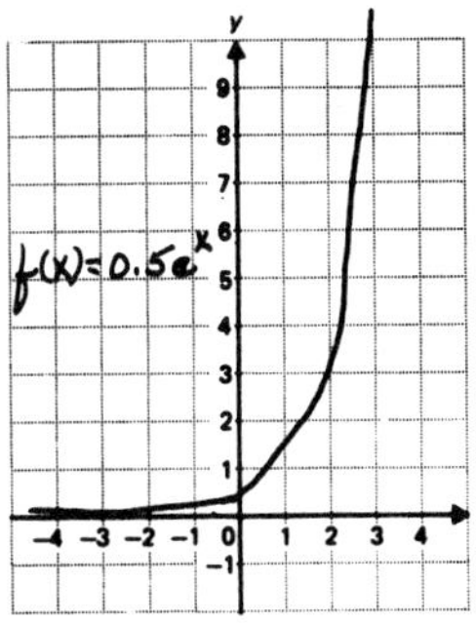

3. Graph $f(x) = e^{(1/2)x}$.

Use a calculator to find approximate values of $e^{(1/2)x}$, and use these values to draw the graph.

x	$e^{(1/2)x}$
-2	0.4
-1	0.6
0	1
1	1.6
2	2.7

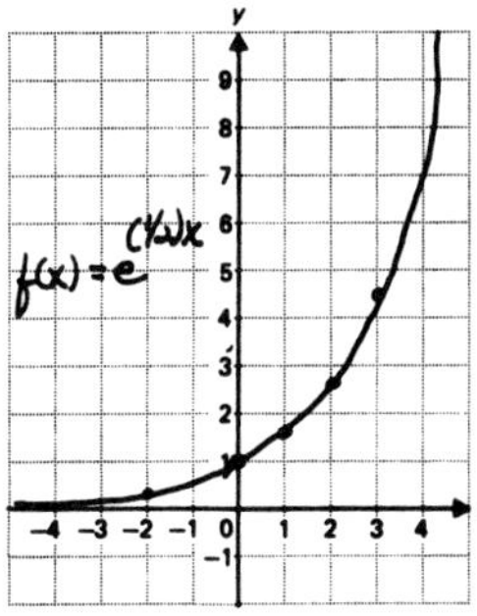

4.

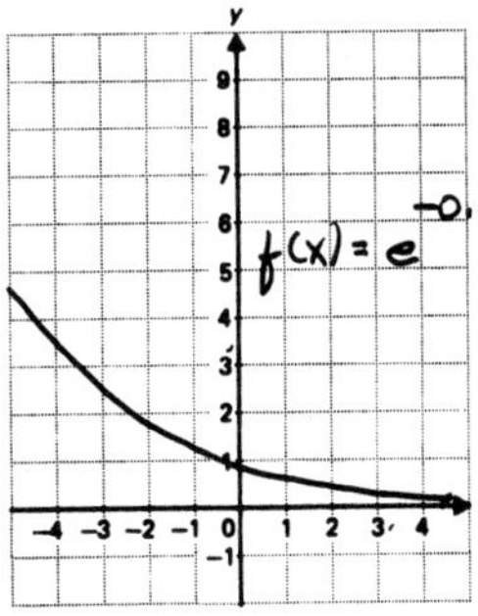

5. Graph $f(x) = e^{x+1}$.

Use a calculator to find approximate values of e^{x+1}, and use these values to draw the graph.

x	e^{x+1}
-2	0.4
-1	1
0	2.7
1	7.4
2	20.1

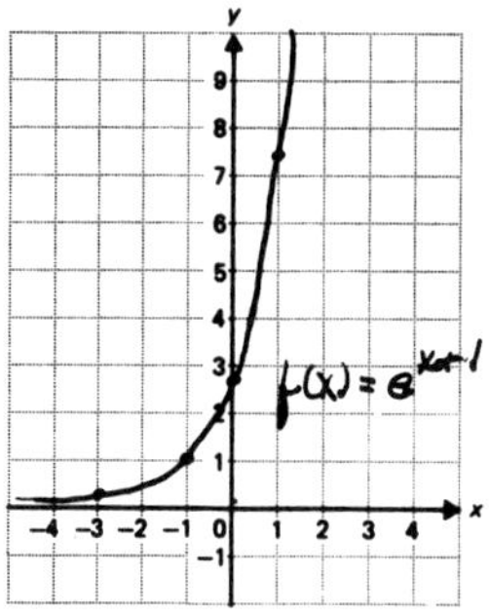

6.

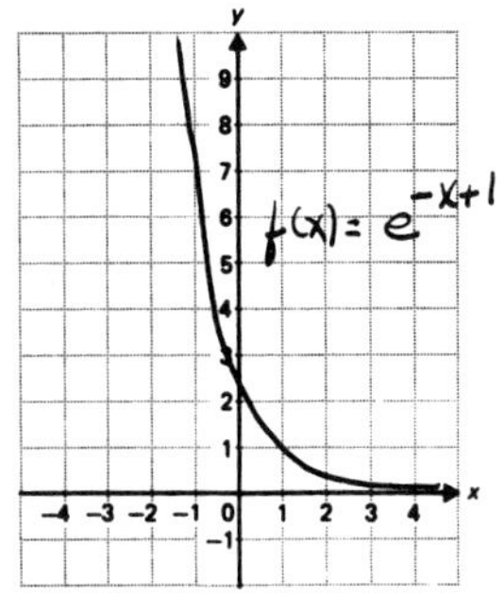

7. Graph $f(x) = e^{2x} + 1$.

Use a calculator to find approximate values of $e^{2x} + 1$, and use these values to draw the graph.

x	$e^{2x} + 1$
-2	1.0 +
-1	1.1
0	2
1	8.4
2	55.6

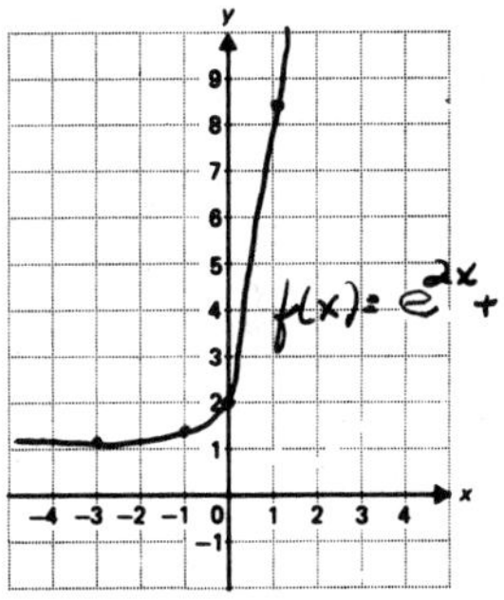

8.

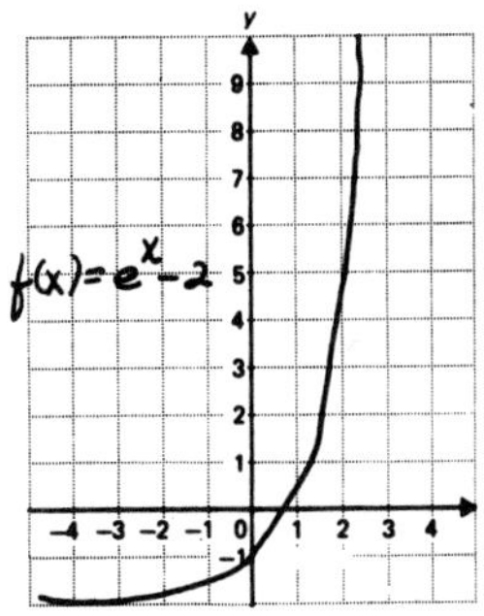

9. Graph $f(x) = 1 - e^{-0.01x}$, for nonnegative values of x. Use a calculator to find approximate values of $1 - e^{-0.01x}$ for nonnegative values of x, and use these values to draw the graph.

x	$1 - e^{-0.01x}$
0	0
100	0.63
200	0.86
300	0.95
400	0.98

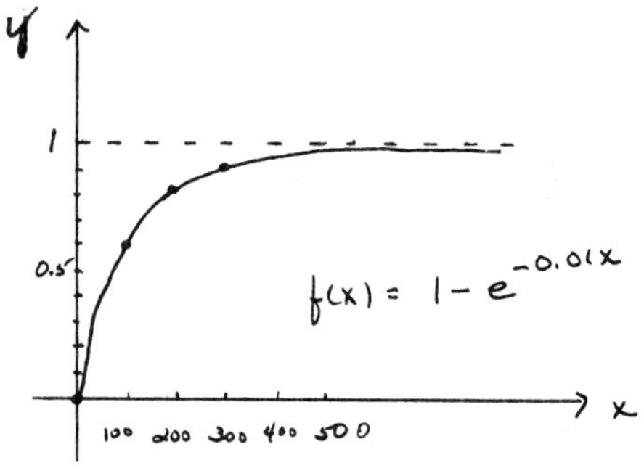

10.

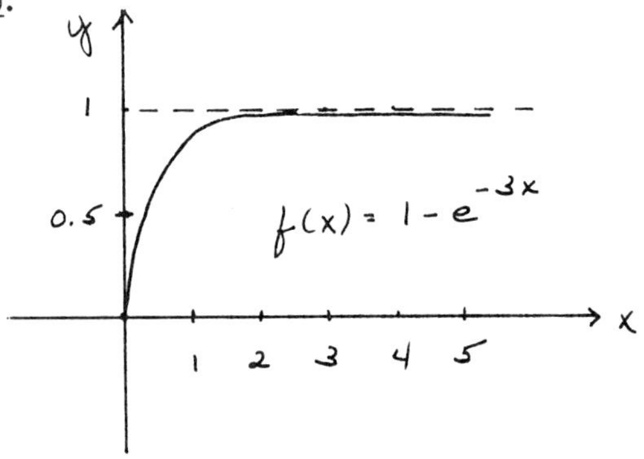

11. Graph $f(x) = 2(1 - e^{-x})$, for nonnegative values of x. Use a calculator to find approximate values of $2(1 - e^{-x})$ for nonnegative values of x, and use these values to draw the grpah.

x	$2(1 - e^{-x})$
0	0
1	1.26
2	1.73
3	1.90
4	1.96

11. (continued)

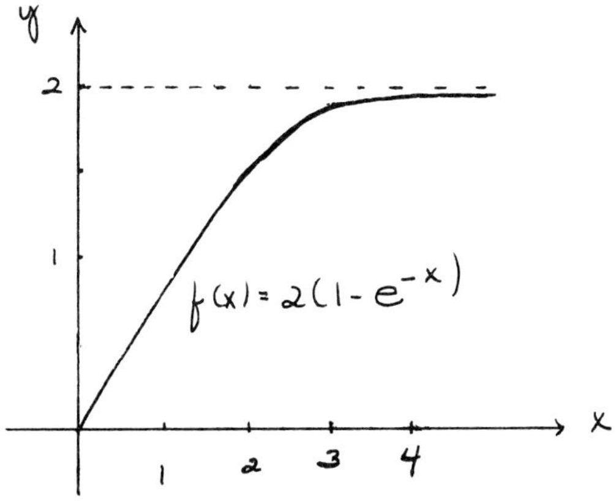

12.

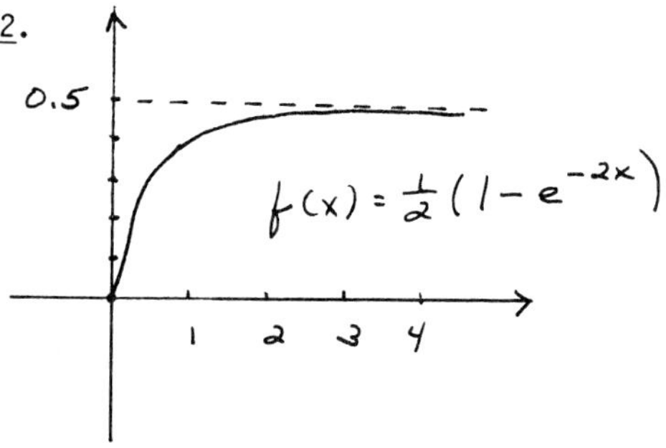

13. Graph $f(x) = 4 \ln x$.

Find some solutions with a calculator, plot them, and draw the graph.

x	0.25	0.5	1	1.5	2	3
4 ln x	-5.55	-2.77	0	1.62	2.77	4.39

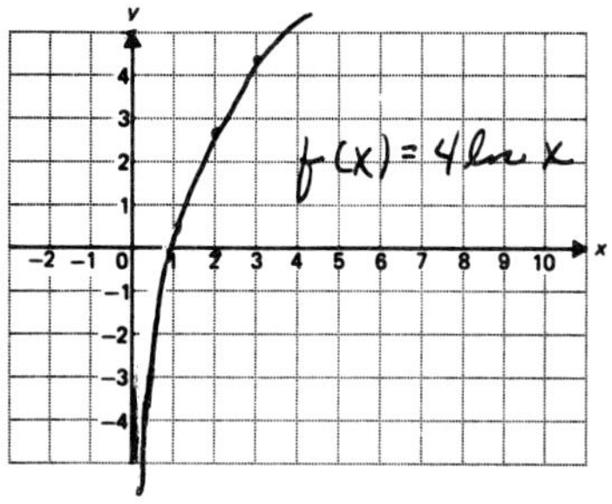

14.

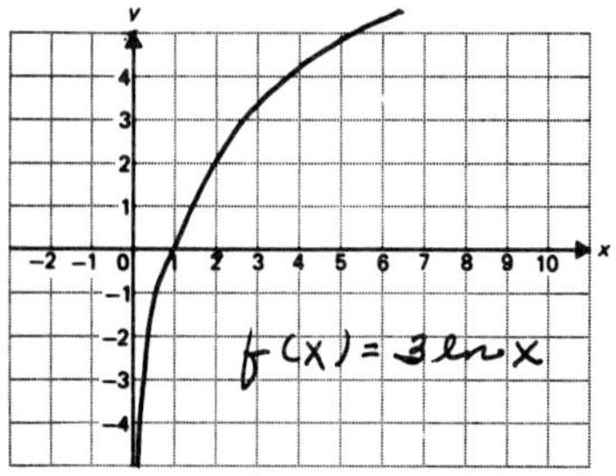

15. Graph $f(x) = \frac{1}{2} \ln x$.

Find some solutions with a calculator, plot them, and draw the graph.

x	0.5	1	2	4	7	8
$\frac{1}{2} \ln x$	-0.35	0	0.35	0.69	0.97	1.04

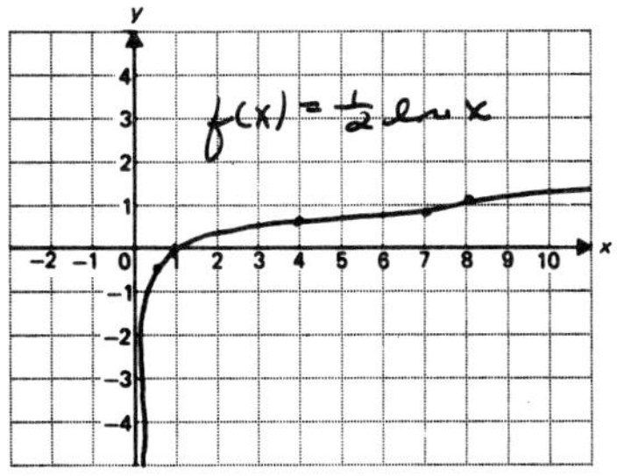

16.

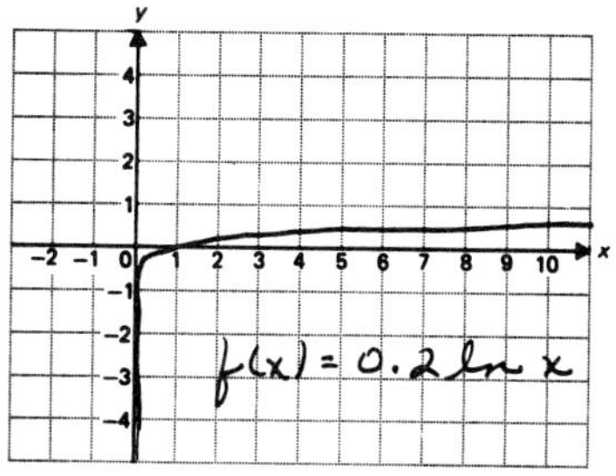

17. Graph $f(x) = \ln (x - 2)$.

Find some solutions using a calculator, plot them, and draw the graph. When $x = 3$, $f(x) = \ln (3 - 2) = \ln 1 = 0$, and so on.

x	2.5	3	4	6	8	10
ln (x-2)	-0.69	0	0.69	1.39	1.79	2.08

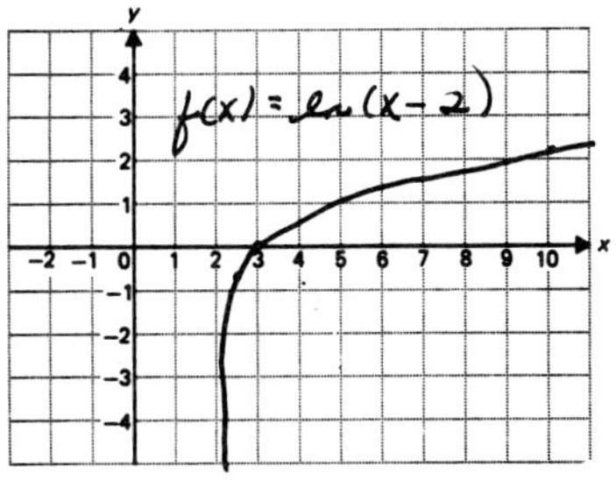

18.

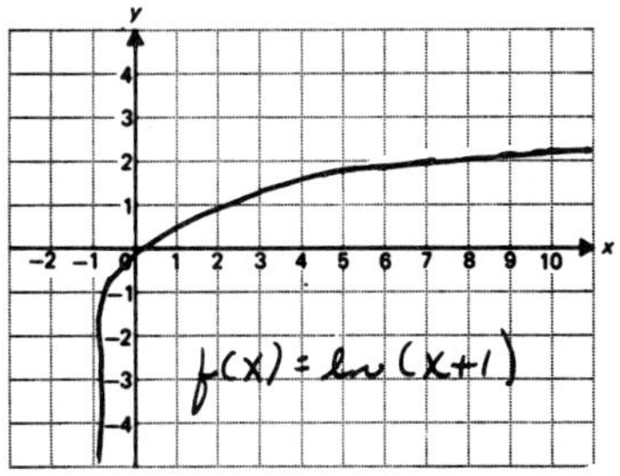

19. Graph $f(x) = 2 - \ln x$.

Find some solutions using a calculator, plot them, and draw the graph.

x	0.25	0.5	1	2	4
2 - ln x	3.39	2.69	2	1.31	0.61

x	7	8	10
2 - ln x	0.05	-0.08	-0.30

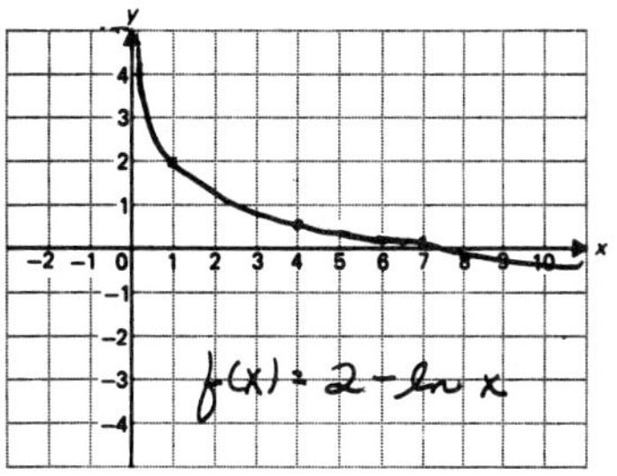

20.

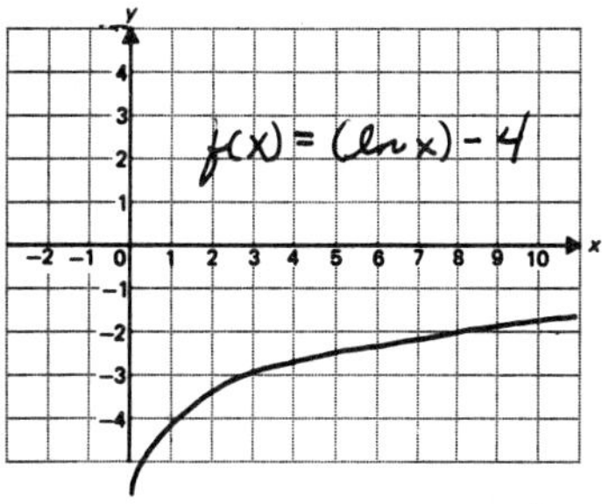

21. $P(t) = 1 - e^{-0.2t}$

a) $P(1) = 1 - e^{-0.2(1)} = 1 - e^{-0.2} \approx 0.181$, or 18.1%

$P(4) = 1 - e^{-0.2(4)} = 1 - e^{-0.8} \approx 0.551$, or 55.1%

$P(6) = 1 - e^{-0.2(6)} = 1 - e^{-1.2} \approx 0.699$, or 69.9%

$P(12) = 1 - e^{-0.2(12)} = 1 - e^{-2.4} \approx 0.909$, or 90.9%

b) Plot the points (0, 0%), (1, 18.1%), (4, 55.1%), (6, 69.9%), and (12, 90.9%) and draw the graph.

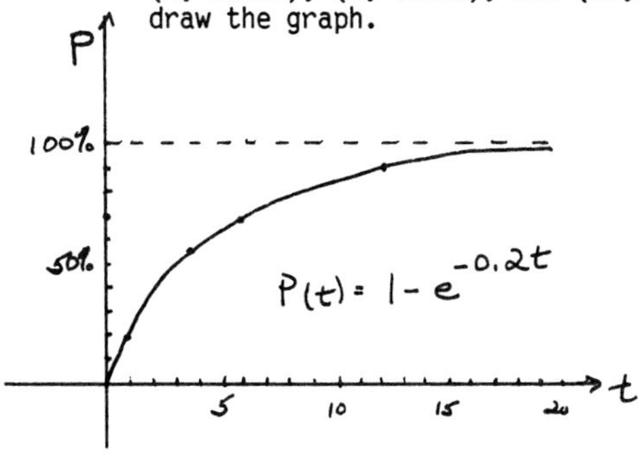

22. a) 26, 78, 91, 95

b)

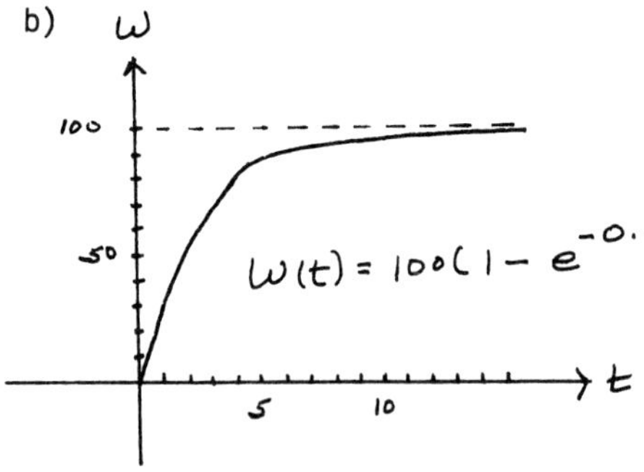

23. $V(t) = \$58(1 - e^{-1.1t}) + \$20$

a) $V(1) = \$58(1 - e^{-1.1(1)}) + \$20$

$= \$58(1 - e^{-1.1}) + \$20$

$\approx \$58.69$

$V(2) = \$58(1 - e^{-1.1(2)}) + \$20$

$= \$58(1 - e^{-2.2}) + \$20$

$\approx \$71.57$

$V(4) = \$58(1 - e^{-1.1(4)}) + \$20$

$= \$58(1 - e^{-4.4}) + \$20$

$\approx \$77.29$

$V(6) = \$58(1 - e^{-1.1(6)}) + \$20$

$= \$58(1 - e^{-6.6}) + \$20$

$\approx \$77.92$

$V(12) = \$58(1 - e^{-1.1(12)}) + \$20$

$= \$58(1 - e^{-13.2}) + \$20$

$\approx \$77.99 +$

b) Plot the points (0, \$20), (1, \$58.69), (2, \$71.57), (4, \$77.29), (6, \$77.92), and (12, \$77.99 +) and draw the graph.

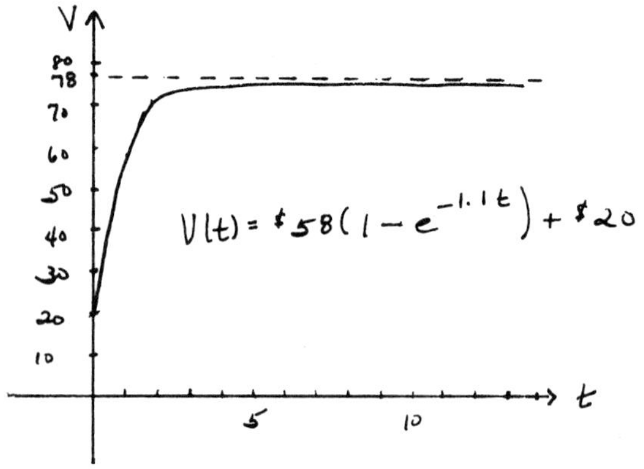

24. a) 25.9%, 59.3%, 83.5%, 97.3%

b)

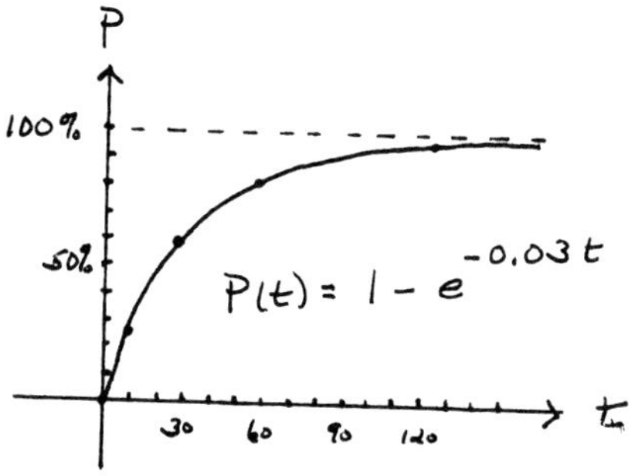

25. Graph $g(x) = e^{|x|}$.

Use a calculator to find some values of $e^{|x|}$, and use these values to draw the graph.

x	$e^{\|x\|}$
-3	20.09
-2	7.39
-1	2.72
0	1
1	2.72
2	7.39

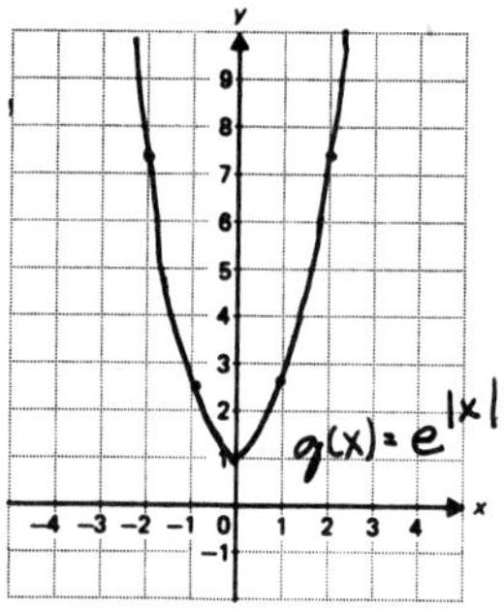

26.

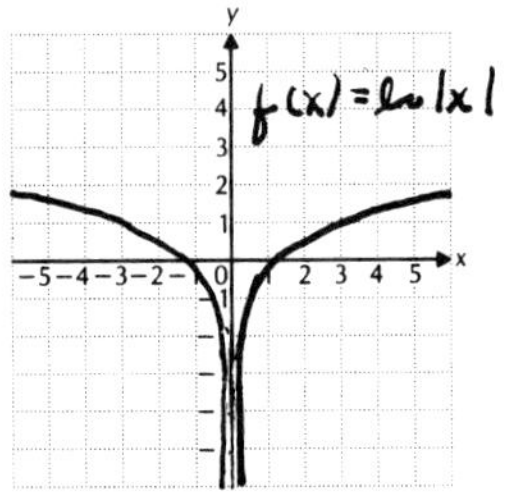

27. Graph $f(x) = |\ln x|$.

Find some solutions using a calculator, plot them, and draw the graph.

x	0.25	0.5	1	3	6	10
$\|\ln x\|$	1.39	0.69	0	1.10	1.79	2.30

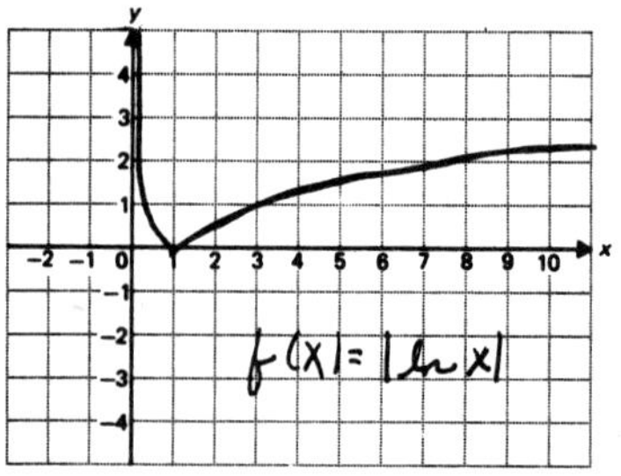

28.

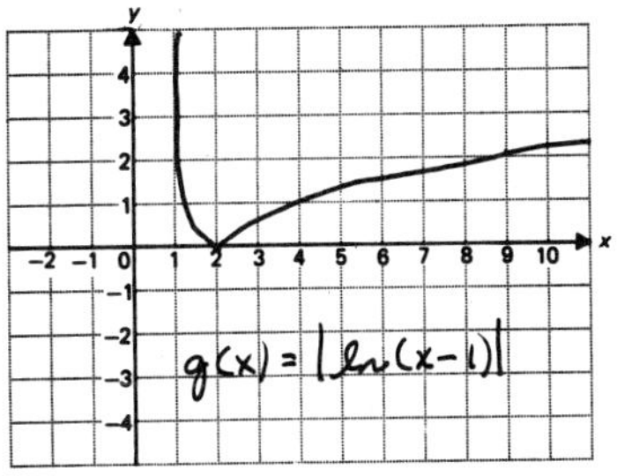

29. Graph $f(x) = \frac{e^x + e^{-x}}{2}$

Use a calculator to find some values of $\frac{e^x + e^{-x}}{2}$, and use these values to draw the graph.

x	$\frac{e^x + e^{-x}}{2}$
-3	13.76
-2	3.76
-1	1.54
0	1
1	1.54
2	3.76

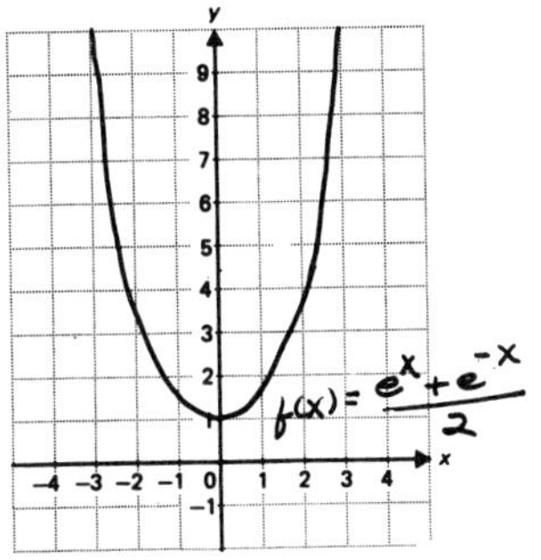

30.

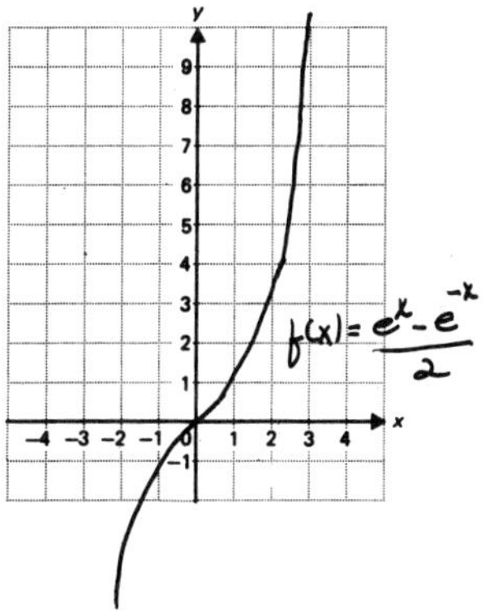

31. $N(t) = \frac{4800}{6 + 794e^{-0.4t}}$

a) $N(3) = \frac{4800}{6 + 794e^{-0.4(3)}} = \frac{4800}{6 + 794e^{-1.2}} \approx 20$

$N(5) = \frac{4800}{6 + 794e^{-0.4(5)}} = \frac{4800}{6 + 794e^{-2}} \approx 42$

$N(10) = \frac{4800}{6 + 794e^{-0.4(10)}} = \frac{4800}{6 + 794e^{-4}} \approx 234$

$N(15) = \frac{4800}{6 + 794e^{-0.4(15)}} = \frac{4800}{6 + 794e^{-6}} \approx 602$

b) Plot the points (0, 6), (3, 20), (5, 42), (10, 234), and (15, 602) and draw the graph.

31. (continued)

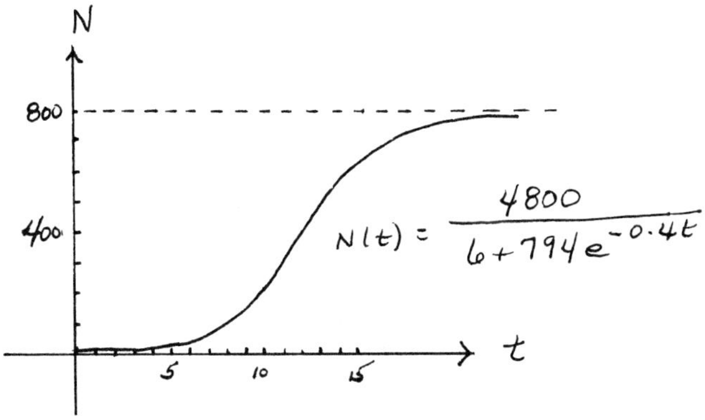

32. a) 33, 183, 758, 1339, 1952, 1998

b)

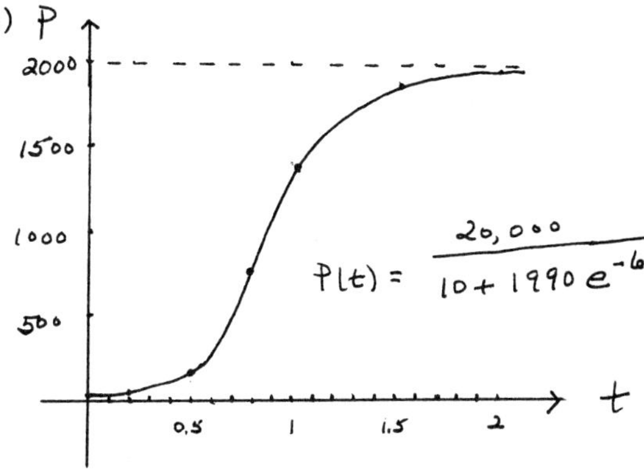

Exercise Set 6.6

1. $2^x = 32$

$2^x = 2^5 \quad (32 = 2^5)$

Since the base is the same, 2, on both sides, the exponents must be the same.

Thus, x = 5.

2. 4

3. $4^{2x} = 8^{3x-4}$

$(2^2)^{2x} = (2^3)^{3x-4}$

$2^{4x} = 2^{9x-12}$

$4x = 9x - 12$ (Since the bases are the same, the exponents must be the same.)

$12 = 5x$

$\frac{12}{5} = x$

4. -3, -1

5. $3^{5x} \cdot 9^{x^2} = 27$

$3^{5x} \cdot 3^{2x^2} = 3^3$

$3^{5x+2x^2} = 3^3$

$5x + 2x^2 = 3$ (Since the bases are the same, the exponents must be the same.)

$2x^2 + 5x - 3 = 0$

$(2x - 1)(x + 3) = 0$

$2x - 1 = 0$ or $x + 3 = 0$

$2x = 1$ or $x = -3$

$x = \frac{1}{2}$ or $x = -3$

The solutions are $\frac{1}{2}$ and -3.

6. 1.4037

7. $2^x = 3^{x-1}$

$\log 2^x = \log 3^{x-1}$ (Taking the log on both sides)

$x \log 2 = (x - 1) \log 3$ (Theorem 5)

$x \log 2 = x \log 3 - \log 3$

$x \log 2 - x \log 3 = -\log 3$

$x(\log 2 - \log 3) = -\log 3$

$x = \frac{-\log 3}{\log 2 - \log 3}$

$x = \frac{\log 3}{\log 3 - \log 2}$

$x \approx \frac{0.4771}{0.4771 - 0.3010}$

$\approx \frac{0.4771}{0.1761}$

≈ 2.7095

8. 7.4516

9. $\log x + \log (x - 9) = 1$

$\log x(x - 9) = 1$ (Theorem 4)

$x^2 - 9x = 10^1$ (Equivalent exponential equation)

$x^2 - 9x - 10 = 0$

$(x + 1)(x - 10) = 0$

$x + 1 = 0$ or $x - 10 = 0$

$x = -1$ or $x = 10$

Only 10 checks. The number -1 is not a solution because negative numbers do not have logarithms. The solution is 10.

10. $\frac{1}{3}$

11. $\log (x + 9) - \log x = 1$

$\log \frac{x + 9}{x} = 1$ (Theorem 6)

$\frac{x + 9}{x} = 10^1$ (Equivalent exponential equation)

$x + 9 = 10x$

$9 = 9x$

$1 = x$

The number 1 checks and is the solution.

12. $\frac{21}{8}$

13. $\log_4 (x + 3) + \log_4 (x - 3) = 2$

$\log_4 (x + 3)(x - 3) = 2$ (Theorem 4)

$(x + 3)(x - 3) = 4^2$ (Equivalent exponential equation)

$x^2 - 9 = 16$

$x^2 = 25$

$x = \pm 5$

Only 5 checks. The number -5 is not a solution because negative numbers do not have logarithms. The solution is 5.

14. $\frac{1}{3}$

15. $\log x^2 = (\log x)^2$

$2 \log x = (\log x)^2$ (Theorem 5)

$0 = (\log x)^2 - 2 \log x$

$0 = \log x (\log x - 2)$

$\log x = 0$ or $\log x - 2 = 0$

$x = 10^0$ or $\log x = 2$

$x = 1$ or $x = 10^2$

$x = 1$ or $x = 100$

Both values check. The solutions are 1 and 100.

16. 27, $\frac{1}{3}$

17. $\log_3 (\log_4 x) = 0$

$\log_4 x = 3^0$ (Equivalent exponential equation)

$\log_4 x = 1$

$x = 4^1$ (Equivalent exponential equation)

$x = 4$

The number 4 checks and is the solution.

18. 10^{100}

19. $e^t = 100$

$\ln e^t = \ln 100$ (Taking the natural logarithm on both sides)

$t \ln e = \ln 100$ (Theorem 5)

$t = \ln 100$ ($\ln e = 1$)

$t \approx 4.6052$

20. 6.9078

21. $e^x = 60$

$\ln e^x = \ln 60$ (Taking the natural logarithm on both sides)

$x \ln e = \ln 60$ (Theorem 5)

$x = \ln 60$ ($\ln e = 1$)

$x \approx 4.0943$

22. 4.4998

23. $e^{-t} = 0.1$

$\ln e^{-t} = \ln 0.1$ (Taking the natural logarithm on both sides)

$-t \ln e = \ln 0.1$ (Theorem 5)

$-t = \ln 0.1$ ($\ln e = 1$)

$t = -\ln 0.1$

$t \approx 2.3026$

24. 4.6052

25. $e^{-0.02k} = 0.06$

$\ln e^{-0.02k} = \ln 0.06$ (Taking the natural logarithm on both sides)

$-0.02k \ln e = \ln 0.06$ (Theorem 5)

$-0.02k = \ln 0.06$ ($\ln e = 1$)

$k = \dfrac{\ln 0.06}{-0.02}$

$k \approx 140.67$

26. 9.9021

27. $\dfrac{e^x - e^{-x}}{t} = 5$

$e^x - e^{-x} = 5t$

$e^x - \dfrac{1}{e^x} = 5t$

$(e^x)^2 - 1 = 5te^x$ (Multiplying by e^x)

$(e^x)^2 - 5te^x - 1 = 0$

Solve for e^x using the quadratic formula.

$e^x = \dfrac{-(-5t) + \sqrt{(-5t)^2 - 4(1)(-1)}}{2\cdot 1}$ [Taking the positive square root. ($e^x > 0$ for all x.)]

$e^x = \dfrac{5t + \sqrt{25t^2 + 4}}{2}$

Then take the natural logarithm on both sides.

27. (continued)

$\ln e^x = \ln \dfrac{5t + \sqrt{25t^2 + 4}}{2}$

$x \ln e = \ln \dfrac{5t + \sqrt{25t^2 + 4}}{2}$

$x = \ln \dfrac{5t + \sqrt{25t^2 + 4}}{2}$

28. 1.567, -1.567

29. $\dfrac{e^x + e^{-x}}{e^x - e^{-x}} = t$

$\dfrac{e^x + e^{-x}}{e^x - e^{-x}} \cdot \dfrac{e^x}{e^x} = t$ (Multiplying by 1)

$\dfrac{e^{2x} + 1}{e^{2x} - 1} = t$

$e^{2x} + 1 = t(e^{2x} - 1)$

$e^{2x} + 1 = te^{2x} - t$

$t + 1 = te^{2x} - e^{2x}$

$t + 1 = e^{2x}(t - 1)$

$\dfrac{t + 1}{t - 1} = e^{2x}$

$\ln \dfrac{t + 1}{t - 1} = \ln e^{2x}$ (Taking the natural logarithm on both sides)

$\ln \dfrac{t + 1}{t - 1} = 2x$

$\dfrac{1}{2} \ln \dfrac{t + 1}{t - 1} = x$

30. $\dfrac{1}{2} \log_5 \dfrac{t + 1}{1 - t}$

31. $A = P(1 + r)^t$

$2000 = 1000(1 + 0.06)^t$ (Substituting)

$2 = (1.06)^t$

$\log 2 = \log (1.06)^t$ (Taking the log on both sides)

$\log 2 = t \log 1.06$ (Theorem 5)

$\dfrac{\log 2}{\log 1.06} = t$

$\dfrac{0.3010}{0.0253} \approx t$

$11.9 \text{ yr} \approx t$

32. 22.5 yr

33. We first substitute and solve for k.

$$P = P_0e^{kt}$$

$$815{,}000 = 680{,}000e^{k \cdot 9} \quad (t = 1969 - 1960 = 9)$$

$$\frac{815}{680} = e^{9k}$$

$$\ln \frac{815}{680} = \ln e^{9k} \quad \text{(Taking the ln on both sides)}$$

$$\ln 815 - \ln 680 = 9k \quad \text{(Theorem 6)}$$

$$\frac{\ln 815 - \ln 680}{9} = k$$

$$\frac{6.7032 - 6.5220}{9} \approx k$$

$$0.020 \approx k$$

The population equation in this situation is now $P = 680{,}000e^{0.02t}$

In the year 1990, t will be 30. Substitute 30 for t and solve for P.

$$P = 680{,}000e^{0.02(30)}$$
$$= 680{,}000e^{0.6}$$
$$\approx 680{,}000(1.8221)$$
$$\approx 1{,}239{,}000$$

34. $k = 0.0065$, $P = 616{,}042$

35. We first substitute $\frac{1}{2} N_0$ for N and 3 for t and then solve for k.

$$N = N_0e^{-kt}$$

$$\frac{1}{2} N_0 = N_0e^{-k \cdot 3}$$

$$0.5 = e^{-3k}$$

$$\ln 0.5 = \ln e^{-3k} \quad \text{(Taking the ln on both sides)}$$

$$\ln 0.5 = -3k$$

$$\frac{\ln 0.5}{-3} = k$$

$$0.231 \approx k$$

Now we use the formula $N = 410e^{-0.231t}$

We substitute 30 for t and solve for N.

$$N = 410e^{-0.231(30)}$$
$$= 410e^{-6.93}$$
$$\approx 0.4 \text{ g}$$

36. 125 g

37. We first substitute and solve for k.

$$N = N_0e^{-kt}$$

$$6.5 = 66{,}560e^{-16k}$$

$$\frac{6.5}{66{,}560} = e^{-16k}$$

$$\ln \frac{6.5}{66{,}560} = \ln e^{-16k} \quad \text{(Taking the ln on both sides)}$$

$$\frac{\ln 6.5 - \ln 66{,}560}{-16} = k$$

$$\frac{1.8718 - 11.1059}{-16} \approx k$$

$$0.577 \approx k$$

Now we use the formula $N = N_0e^{-0.577t}$

We substitute $\frac{1}{2} N_0$ for N and solve for t.

$$\frac{1}{2} N_0 = N_0e^{-0.577t}$$

$$0.5 = e^{-0.577t}$$

$$\ln 0.5 = \ln e^{-0.577t} \quad \text{(Taking the ln on both sides)}$$

$$\ln 0.5 = -0.577t$$

$$\frac{\ln 0.5}{-0.577} = t$$

$$\frac{-0.6931}{-0.577} \approx t$$

$$1.2 \text{ days} \approx t$$

38. 248,000 yr

39. a) We first must determine the decay rate of carbon-14. The half-life is 5750 years. Substitute $\frac{1}{2} N_0$ for N and 5750 for t and solve for k.

$$N = N_0e^{-kt}$$

$$\frac{1}{2} N_0 = N_0 \, e^{-5750k}$$

$$0.5 = e^{-5750k}$$

$$\ln 0.5 = \ln e^{-5750k} \quad \text{(Taking the ln on both sides)}$$

$$\ln 0.5 = -5750k$$

$$\frac{\ln 0.5}{-5750} = k$$

$$\frac{-0.6931}{-5750} \approx k$$

$$0.00012 \approx k$$

Now we use the formula $N = N_0e^{-0.00012t}$

An animal bone that has lost 30% of its carbon-14 contains 70% of its carbon-14. Substitute $0.70N_0$ for N and solve for t.

$$N = N_0e^{-0.00012t}$$

$$0.70N_0 = N_0e^{-0.00012t}$$

$$0.7 = e^{-0.00012t}$$

$$\ln 0.7 = \ln e^{-0.00012t} \quad \text{(Taking the ln on both sides)}$$

39. (continued)

$$\ln 0.7 = -0.00012t$$

$$\frac{\ln 0.7}{-0.00012} = t$$

$$\frac{-0.3567}{-0.00012} \approx t$$

$$2972 \text{ yr} \approx t$$

b) An animal bone that has lost 46% of its carbon-14 contains 54% of its carbon-14. Substitute $0.54N_0$ for N and solve for t.

$$N = N_0e^{-0.00012t}$$

$$0.54N_0 = N_0e^{-0.00012t}$$

$$0.54 = e^{-0.00012t}$$

$$\ln 0.54 = \ln e^{-0.00012t}$$ (Taking the ln on both sides)

$$\ln 0.54 = -0.00012t$$

$$\frac{\ln 0.54}{-0.00012} = t$$

$$\frac{-0.6162}{-0.00012} \approx t$$

$$5135 \text{ yr} \approx t$$

40. a) 1860 yr, b) 3590 yr

41. $P = P_0e^{-kh}$

$$P = 1013e^{-3.85\times10^{-5}\times14,162}$$

$$\approx 1013e^{-0.545}$$

$$\approx 1013(0.5798)$$

$$\approx 587 \text{ millibars}$$

42. 72,559 ft

43. a) $S(t) = 82 - 18 \log (t + 1)$

$$S(0) = 82 - 18 \log (0 + 1)$$ (Substituting 0 for t)

$$= 82 - 18 \log 1$$

$$= 82 - 18\cdot0$$

$$= 82$$

b) $S(5) = 82 - 18 \log (5 + 1)$ (Substituting 5 for t)

$$= 82 = 18 \log 6$$

$$\approx 82 - 18(0.7782)$$

$$\approx 82 - 14.0076$$

$$\approx 68$$

44. a) 75, b) 58

45. $S(t) = 82 - 18 \log (t + 1)$

$$64 = 82 - 18 \log (t + 1)$$ (Substituting 64 for S(t))

$$-18 = -18 \log (t + 1)$$

$$1 = \log (t + 1)$$

$$10^1 = t + 1$$

$$9 = t$$

46. 4 months

47. $L = 10 \log \frac{I}{I_0}$

$$L = 10 \log \frac{3,100,000I_0}{I_0}$$ (Substituting $3,100,000I_0$ for I)

$$= 10 \log 3,100,000$$

$$\approx 10(6.4914)$$

$$\approx 65 \text{ decibels}$$

48. 64 db

49. $L = 10 \log \frac{I}{I_0}$

$$L = 10 \log \frac{10^{14}\cdot I_0}{I_0}$$ (Substituting $10^{14}\cdot I_0$ for I)

$$= 10 \log 10^{14}$$

$$= 10\cdot14$$

$$= 140 \text{ decibels}$$

50. 120 db

51. $R = \log \frac{I}{I_0}$

$$R = \log \frac{10^{6.7}\cdot I_0}{I_0}$$ (Substituting $10^{6.7}\cdot I_0$ for I)

$$= \log 10^{6.7}$$

$$= 6.7$$

52. 8.25

53. $R = \log \frac{I}{I_0}$

$$5 = \log \frac{I}{I_0}$$

$$10^5 = \frac{I}{I_0}$$

$$10^5\cdot I_0 = I$$

54. $10^7\cdot I_0$

55. a)
$$P = P_0e^{kt}$$
$$184.50 = 100e^{k\cdot 10} \quad \text{(Substituting)}$$
$$1.8450 = e^{10k}$$
$$\ln 1.845 = \ln e^{10k} \quad \text{(Taking the ln on both sides)}$$
$$\ln 1.845 = 10k$$
$$\frac{\ln 1.845}{10} = k$$
$$\frac{0.6125}{10} \approx k$$
$$0.061 \approx k$$

The equation is $P = 100e^{0.061t}$.

b)
$$P = 100e^{0.061t}$$
$$P = 100e^{0.061(20)} \quad (1987 - 1967 = 20)$$
$$= 100e^{1.22}$$
$$\approx 100(3.3872)$$
$$\approx \$338.72$$

c)
$$P = P_0e^{0.061t}$$
$$200 = 100e^{0.061t}$$
$$2 = e^{0.061t}$$
$$\ln 2 = \ln e^{0.061t}$$
$$\frac{\ln 2}{0.061} = t$$
$$\frac{0.6931}{0.061} \approx t$$
$$11.4 \text{ yr} \approx t$$

56. a) $k = 0.03$, $P = 0.52e^{0.03t}$, b) $0.84, c) 2001

57.
$$p^H = -\log [H^+]$$
$$p^H = -\log (6.3 \times 10^{-5})$$
$$= -(\log 6.3 + \log 10^{-5})$$
$$= -(\log 6.3 - 5)$$
$$= -\log 6.3 + 5$$
$$\approx -0.7993 + 5$$
$$\approx 4.2$$

58. 7.8

59.
$$\log \sqrt{x} = \sqrt{\log x}$$
$$\log x^{1/2} = \sqrt{\log x}$$
$$\frac{1}{2} \log x = \sqrt{\log x} \quad \text{(Theorem 5)}$$

Let $u = \sqrt{\log x}$
$$\frac{1}{2} u^2 = u$$
$$u^2 = 2u$$
$$u^2 - 2u = 0$$
$$u(u - 2) = 0$$

59. (continued)

$u = 0$ or $u - 2 = 0$

$u = 0$ or $u = 2$

Replacing u with $\sqrt{\log x}$, we get

$\sqrt{\log x} = 0$ or $\sqrt{\log x} = 2$

$\log x = 0$ or $\log x = 4$

$x = 10^0$ or $x = 10^4$

$x = 1$ or $x = 10{,}000$

Both values check. The solutions are 1 and 10,000.

60. $\pm 2\sqrt{6}$

61.
$$(\log_a x)^{-1} = \log_a x^{-1}$$
$$\frac{1}{\log_a x} = -1 \log_a x \quad \text{(Theorem 5)}$$
$$1 = -(\log_a x)^2 \quad \text{(Multiplying by } \log_a x)$$
$$-1 = (\log_a x)^2$$

Since $(\log_a x)^2$ is never negative, there is no solution. The solution set is $\emptyset$.

62. 25, $\frac{1}{25}$

63.
$$\log_3 |x| = 2$$
$$|x| = 3^2 \quad \text{(Equivalent exponential equation)}$$
$$|x| = 9$$
$$x = \pm 9$$

64. $-\frac{1}{2}$

65.
$$\frac{\sqrt{(e^{2x}\cdot e^{-5x})^{-4}}}{e^x \div e^{-x}} = e^7$$
$$\frac{\sqrt{(e^{-3x})^{-4}}}{e^{2x}} = e^7$$
$$\frac{\sqrt{e^{12x}}}{e^{2x}} = e^7$$
$$\frac{e^{6x}}{e^{2x}} = e^7$$
$$e^{4x} = e^7$$
$$4x = 7$$
$$x = \frac{7}{4}$$

66. 100, $\frac{1}{100}$

67.
$$y = ax^n$$
$$\frac{y}{a} = x^n$$
$$\log_x \frac{y}{a} = \log_x x^n \quad \text{(Taking the log on both sides)}$$
$$\log_x y - \log_x a = n \quad \text{(Theorem 6)}$$

68. $t = \dfrac{\log_e y - \log_e k}{a}$

69.
$$P = P_0e^{kt}$$
$$\frac{P}{P_0} = e^{kt}$$
$$\ln \frac{P}{P_0} = \ln e^{kt} \quad \text{(Taking the ln on both sides)}$$
$$\ln P - \ln P_0 = kt \quad \text{(Theorem 6)}$$
$$\frac{\ln P - \ln P_0}{k} = t$$

70. $t = \dfrac{\ln P_0 - \ln P}{k}$

71.
$$T = T_0 + (T_1 - T_0)10^{-kt}$$
$$T - T_0 = (T_1 - T_0)10^{-kt}$$
$$\frac{T - T_0}{T_1 - T_0} = 10^{-kt}$$
$$\log \frac{T - T_0}{T_1 - T_0} = \log 10^{-kt} \quad \text{(Taking the log on both sides)}$$
$$\log \frac{T - T_0}{T_1 - T_0} = -kt$$
$$-\frac{1}{k} \log \frac{T - T_0}{T_1 - T_0} = t$$

72. $n = \log_V c - \log_V P$, or $\log_V \frac{c}{P}$

73.
$$\log_a Q = \frac{1}{3} \log_a y + b$$
$$\log_a Q = \log_a \sqrt[3]{y} + b \quad \text{(Theorem 5)}$$
$$\log_a Q - \log_a \sqrt[3]{y} = b$$
$$\log_a \frac{Q}{\sqrt[3]{y}} = b \quad \text{(Theorem 6)}$$
$$\frac{Q}{\sqrt[3]{y}} = a^b \quad \text{(Equivalent exponential equation)}$$
$$Q = a^b \sqrt[3]{y}$$

74. $y = xa^{2x}$

75.
$$x^{\log x} = \frac{x^3}{100}$$
$$\log x^{\log x} = \log \frac{x^3}{100} \quad \text{(Taking the log on both sides)}$$
$$\log x \log x = \log x^3 - \log 100 \quad \text{(Theorems 5 and 6)}$$
$$(\log x)^2 = 3 \log x - 2 \quad \text{(Theorem 5)}$$
$$(\log x)^2 - 3 \log x + 2 = 0$$
$$u^2 - 3u + 2 = 0 \quad \text{(Letting } u = \log x\text{)}$$
$$(u - 1)(u - 2) = 0$$

$u - 1 = 0$ or $u - 2 = 0$
$u = 1$ or $u = 2$

If $u = 1$, then $\log x = 1$. Thus $x = 10^1 = 10$.
If $u = 2$, then $\log x = 2$. Thus $x = 10^2 = 100$.

Both values check. The solutions are 10 and 100.

76. 100, $\frac{1}{10}$

77. $f(x) = (1 + x)^{1/x}$
$f(1) = (1 + 1)^{1/1} = 2^1 = 2$
$f(0.5) = (1 + 0.5)^{1/0.5} = (1.5)^2 = 2.25$
$f(0.2) = (1 + 0.2)^{1/0.2} = (1.2)^5 = 2.48832$
$f(0.1) = (1 + 0.1)^{1/0.1} = (1.1)^{10} \approx 2.593742$
$f(0.01) = (1 + 0.01)^{1/0.01} = (1.01)^{100} \approx 2.704814$
$f(0.001) = (1 + 0.001)^{1/0.001} = (1.001)^{1000}$
≈ 2.716924

78. 4.0, 2.867972, 2.731999, 2.719642, 2.718418

Exercise Set 7.1

1. $\sin \theta = \frac{\text{opposite}}{\text{hypotenuse}} = \frac{8}{17}$ $\csc \theta = \frac{\text{hypotenuse}}{\text{opposite}} = \frac{17}{8}$

$\cos \theta = \frac{\text{adjacent}}{\text{hypotenuse}} = \frac{15}{17}$ $\sec \theta = \frac{\text{hypotenuse}}{\text{adjacent}} = \frac{17}{15}$

$\tan \theta = \frac{\text{opposite}}{\text{adjacent}} = \frac{8}{15}$ $\cot \theta = \frac{\text{adjacent}}{\text{opposite}} = \frac{15}{8}$

2. $\sin \phi = \frac{15}{17}$, $\cos \phi = \frac{8}{17}$, $\tan \phi = \frac{15}{8}$

$\csc \phi = \frac{17}{15}$, $\sec \phi = \frac{17}{8}$, $\cot \phi = \frac{8}{15}$

3. $\sin \theta = \frac{\text{opposite}}{\text{hypotenuse}} = \frac{3}{h}$ $\csc \theta = \frac{\text{hypotenuse}}{\text{opposite}} = \frac{h}{3}$

$\cos \theta = \frac{\text{adjacent}}{\text{hypotenuse}} = \frac{7}{h}$ $\sec \theta = \frac{\text{hypotenuse}}{\text{adjacent}} = \frac{h}{7}$

$\tan \theta = \frac{\text{opposite}}{\text{adjacent}} = \frac{3}{7}$ $\cot \theta = \frac{\text{adjacent}}{\text{opposite}} = \frac{7}{3}$

4. $\sin \phi = \frac{7}{h}$ $\cos \phi = \frac{4}{h}$, $\tan \phi = \frac{7}{4}$

$\csc \phi = \frac{h}{7}$, $\sec \phi = \frac{h}{4}$, $\cot \phi = \frac{4}{7}$

5. $\sin \theta = \frac{7.8023}{8.8781} = 0.8788$

$\cos \theta = \frac{4.2361}{8.8781} = 0.4771$

$\tan \theta = \frac{7.8023}{4.2361} = 1.8419$

$\cot \theta = \frac{4.2361}{7.8023} = 0.5429$

$\sec \theta = \frac{8.8781}{4.2361} = 2.0958$

$\csc \theta = \frac{8.8781}{7.8023} = 1.1379$

6. $\sin \phi = 0.4771$, $\cos \phi = 0.8788$, $\tan \phi = 0.5429$
$\csc \phi = 2.0958$, $\sec \phi = 1.1379$, $\cot \phi = 1.8419$

7. $\sin 37.5^\circ = \frac{\text{opp.}}{\text{hyp.}} = \frac{28}{\ell}$ $\csc 37.5^\circ = \frac{\text{hyp.}}{\text{opp.}} = \frac{\ell}{28}$

$\cos 37.5^\circ = \frac{\text{adj.}}{\text{hyp.}} = \frac{d}{\ell}$ $\sec 37.5^\circ = \frac{\text{hyp.}}{\text{adj.}} = \frac{\ell}{d}$

$\tan 37.5^\circ = \frac{\text{opp.}}{\text{adj.}} = \frac{28}{d}$ $\cot 37.5^\circ = \frac{\text{adj.}}{\text{opp.}} = \frac{d}{28}$

8. $\sin 36^\circ = \frac{36}{c}$, $\cos 36^\circ = \frac{a}{c}$, $\tan 36^\circ = \frac{36}{a}$

$\csc 36^\circ = \frac{c}{36}$, $\sec 36^\circ = \frac{c}{a}$, $\cot 36^\circ = \frac{a}{36}$

9.

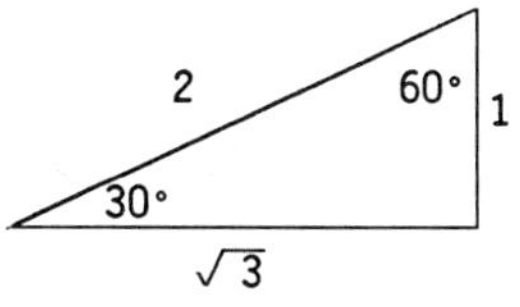

$\sin 30^\circ = \frac{\text{opp.}}{\text{hyp.}} = \frac{1}{2}$, or 0.5

9. (continued)

$\cos 30^\circ = \frac{\text{adj.}}{\text{hyp.}} = \frac{\sqrt{3}}{2} \approx \frac{1.732}{2} = 0.866$

$\tan 30^\circ = \frac{\text{opp.}}{\text{adj.}} = \frac{1}{\sqrt{3}} \approx \frac{1}{1.732} \approx 0.577$

$\cot 30^\circ = \frac{\text{adj.}}{\text{opp.}} = \frac{\sqrt{3}}{1} \approx 1.732$

$\sec 30^\circ = \frac{\text{hyp.}}{\text{adj.}} = \frac{2}{\sqrt{3}} \approx \frac{2}{1.732} \approx 1.155$

$\csc 30^\circ = \frac{\text{hyp.}}{\text{opp.}} = \frac{2}{1} = 2$

10. $\sin 60^\circ \approx 0.866$, $\cos 60^\circ = 0.5$, $\tan 60^\circ \approx 1.732$,
$\cot 60^\circ \approx 0.577$, $\sec 60^\circ = 2$, $\csc 60^\circ \approx 1.155$

11. Use the trigonometric function keys on a calculator to find the following:
$\sin 30^\circ = 0.5$, $\cos 30^\circ \approx 0.866$, $\tan 30^\circ \approx 0.577$,
$\cot 30^\circ \approx 1.732$, $\sec 30^\circ \approx 1.155$, $\csc 30^\circ = 2$

12. $\sin 60^\circ \approx 0.866$, $\cos 60^\circ = 0.5$, $\tan 60^\circ \approx 1.732$,
$\cot 60^\circ \approx 0.577$, $\sec 60^\circ = 2$, $\csc 60^\circ \approx 1.155$

13. $8.6^\circ = 8^\circ + 0.6 \times 1^\circ$

Now substituting 60' for 1°,
$0.6 \times 1^\circ = 0.6 \times 60' = 36'$.

So $8.6^\circ = 8^\circ\ 36'$.

14. 47° 48'

15. $72.25^\circ = 72^\circ + 0.25 \times 1^\circ$

Substituting 60' for 1°,
$0.25 \times 1^\circ = 0.25 \times 60' = 15'$.

So $72.25^\circ = 72^\circ\ 15'$.

16. 11° 45'

17. $46.38^\circ = 46^\circ + 0.38 \times 1^\circ$
$= 46^\circ + 0.38 \times 60'$ $(1^\circ = 60')$
$= 46^\circ + 22.8'$
$\approx 46^\circ\ 23'$

18. 85° 13'

19. $67.84^\circ = 67^\circ + 0.84 \times 1^\circ$
$= 67^\circ + 0.84 \times 60'$ $(1^\circ = 60')$
$= 67^\circ + 50.4'$
$\approx 67^\circ\ 50'$

20. 38° 29'

21. $9° \; 45' = 9° + \left(\frac{45}{60}\right)°$ $\left[1' = \frac{1°}{60}\right]$

$= 9° + 0.75°$

$= 9.75°$

22. 52.25°

23. $35° \; 50' = 35° + \left(\frac{50}{60}\right)°$ $\left[1' = \frac{1°}{60}\right]$

$\approx 35° + 0.83°$

$= 35.83°$

24. 64.67°

25. $80° \; 33' = 80° + \left(\frac{33}{60}\right)°$ $\left[1' = \frac{1°}{60}\right]$

$= 80° + 0.55°$

$= 80.55°$

26. 27.32°

27. $3° \; 2' = 3° + \left(\frac{2}{60}\right)°$ $\left[1' = \frac{1°}{60}\right]$

$\approx 3° + 0.03°$

$= 3.03°$

28. 10.13°

29. Use a calculator to find cos 18°.

Enter 18 and then press the [cos] key.

$\cos 18° \approx 0.9511$

30. 0.6018

31. Use a calculator to find tan 2.6°.

Enter 2.6 and then press the [tan] key.

$\tan 2.6° \approx 0.0454$

32. 0.8211

33. Use a calculator to find sin 62° 20'.

First convert 62° 20' to 62.33°. Then enter 62.33 and press the [sin] key.

$\sin 62° \; 20' \approx 0.8857$

34. 1.455

35. Use a calculator to find cos 15° 35'.

First convert 15° 35' to 15.58°. Then enter 15.58 and press the [cos] key.

$\cos 15° \; 35' \approx 0.9632$

36. 0.6550

37. Use a calculator to find csc 29°.

The cosecant function value can be found by taking the reciprocal of sine function value. Enter 29 and press the [sin] key.

$\sin 29° \approx 0.4848$

Then press the reciprocal key ([1/x]), or divide 1 by 0.4848.

$\csc 29° \approx 2.063$

38. 0.7265

39. Use a calculator to find sec 10° 30'.

The secant function value can be found by taking the reciprocal of the cosine function value. First convert 10° 30' to 10.5°. Then enter 10.5 and press the [cos] key.

$\cos 10.5° \approx 0.9833$

Now press the reciprocal key ([1/x]), or divide 1 by 0.9833.

$\sec 10° \; 30' \approx 1.017$

40. 1.408

41. Enter 0.5125 and press the [$\sin^{-1}$] key.

$\theta = \sin^{-1} 0.5125 \approx 30.83°$, or $30° \; 50'$.

42. 34.50°, or 34° 30'

43. Enter 0.6512 and press the [$\cos^{-1}$] key.

$\theta = \cos^{-1} 0.6512 \approx 49.37°$, or $49° \; 22'$

44. 55.29°, or 55° 17'

45. Enter 3.163 and press the [$\tan^{-1}$] key.

$\theta = \tan^{-1} 3.163 \approx 72.46°$, or $72° \; 27'$

46. 82.33°, or 82° 20'

47. $\csc \theta = \frac{1}{\sin \theta} = 6.277$

$\sin \theta = \frac{1}{6.277} \approx 0.1593$

Enter 0.1593 and press the [$\sin^{-1}$] key.

$\theta = \sin^{-1} 0.1593 \approx 9.17°$, or $9° \; 10'$

48. 31.67°, or 31° 40'

49. Since 25° and 65° are complements, we have

$\sin 25° = \cos 65° = 0.4226$

$\cos 25° = \sin 65° = 0.9063$

$\tan 25° = \cot 65° = 0.4663$

$\cot 25° = \tan 65° = 2.145$

$\sec 25° = \csc 65° = 1.103$

$\csc 25° = \sec 65° = 2.366$

50. $\sin 58° = 0.8480$ $\csc 58° = 1.179$

$\cos 58° = 0.5299$ $\sec 58° = 1.887$

$\tan 58° = 1.600$ $\cot 58° = 0.6249$

51. We know from the definition of the sine function that the ratio $\frac{24}{25}$ is $\frac{\text{side opposite } \theta}{\text{hypotenuse}}$. Let us consider a right triangle in which the hypotenuse has length 25 and the side opposite θ has length 24.

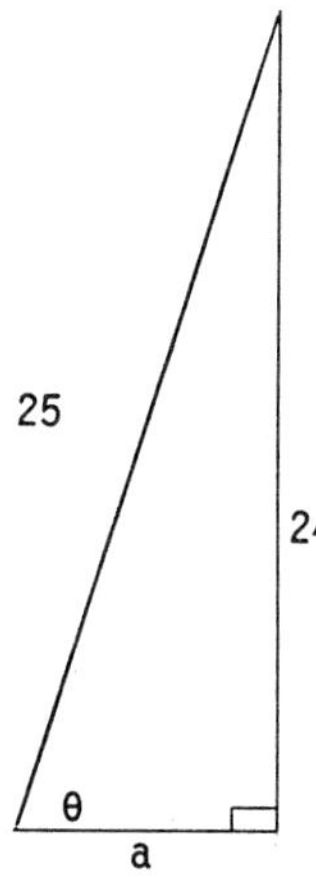

Use the Pythagorean theorem to find the length of the side adjacent to θ.

$a^2 + 24^2 = 25^2$

$a^2 = 625 - 576 = 49$

$a = 7$

Use a = 7, b = 24, and c = 25 to find the other five ratios in the triangle.

$\cos \theta = \frac{7}{25}$, $\tan \theta = \frac{24}{7}$, $\cot \theta = \frac{7}{24}$,

$\sec \theta = \frac{25}{7}$, $\csc \theta = \frac{25}{24}$

52. $\sin \theta = \frac{\sqrt{51}}{10}$; $\tan \theta = \frac{\sqrt{51}}{7}$; $\cot \theta = \frac{7}{\sqrt{51}}$, or $\frac{7\sqrt{51}}{51}$; $\sec \theta = \frac{10}{7}$; $\csc \theta = \frac{10}{\sqrt{51}}$, or $\frac{10\sqrt{51}}{51}$

53. We know from the definition of the tangent function that 2, or $\frac{2}{1}$, is the ratio $\frac{\text{side opposite } \phi}{\text{side adjacent } \phi}$. Let us consider a right triangle in which the side opposite φ has length 2 and the side adjacent to φ has length 1.

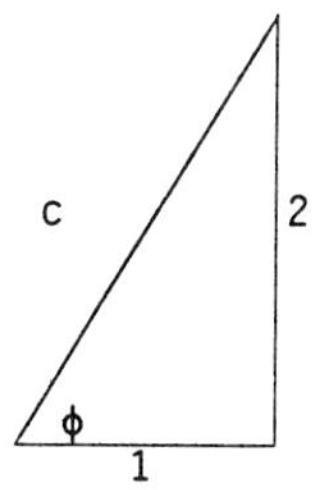

Use the Pythagorean theorem to find the length of the hypotenuse.

$1^2 + 2^2 = c^2$

$1 + 4 = c^2$

$5 = c^2$

$\sqrt{5} = c$

Use a = 1, b = 2, and c = $\sqrt{5}$ to find the other five ratios in the triangle.

$\sin \phi = \frac{2}{\sqrt{5}}$, or $\frac{2\sqrt{5}}{5}$; $\cos \phi = \frac{1}{\sqrt{5}}$, or $\frac{\sqrt{5}}{5}$;

$\cot \phi = \frac{1}{2}$; $\sec \phi = \sqrt{5}$; $\csc \phi = \frac{\sqrt{5}}{2}$

54. $\sin \phi = \frac{4}{\sqrt{17}}$, or $\frac{4\sqrt{17}}{17}$; $\cos \phi = \frac{1}{\sqrt{17}}$, or $\frac{\sqrt{17}}{17}$;

$\tan \phi = 4$; $\cot \phi = \frac{1}{4}$; $\csc \phi = \frac{\sqrt{17}}{4}$

55.

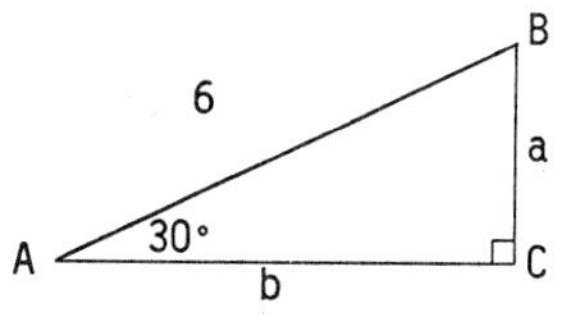

To solve this triangle find B, a, and b.

$B = 90° - 30° = 60°$

$\frac{a}{6} = \sin 30°$

$a = 6 \sin 30°$

$a = 6 \times \frac{1}{2}$

$a = 3$

$\frac{b}{6} = \cos 30°$

$b = 6 \cos 30°$

$b = 6 \times \frac{\sqrt{3}}{2}$

$b = 3\sqrt{3} \approx 5.20$

56. $B = 30°$, $b = \frac{27}{\sqrt{3}} \approx 15.59$, $C = \frac{54}{\sqrt{3}} \approx 31.18$

57.

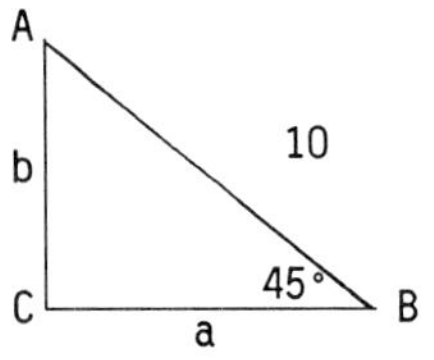

To solve this triangle find A, a, and b.

$A = 90° - 45° = 45°$

$\frac{a}{10} = \cos 45°$

$a = 10 \cos 45°$

$a = 10 \times \frac{\sqrt{2}}{2}$

$a = 5\sqrt{2} \approx 7.07$

$\frac{b}{10} = \sin 45°$

$b = 10 \sin 45°$

$b = 10 \times \frac{\sqrt{2}}{2}$

$b = 5\sqrt{2} \approx 7.07$

58. $B = 45°$, $a = 3$, $c = 3\sqrt{2} \approx 4.24$

59.

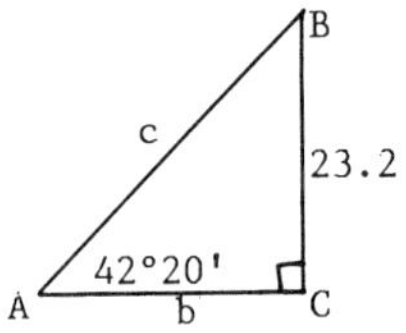

To solve this triangle find B, b, and c.

$B = 90° - 42°\ 20' = 47°\ 40'$

$\frac{b}{23.2} = \cot 42°\ 20'$

$b = 23.2 \cot 42°\ 20'$

$b = 23.2 \times 1.0977$

$b = 25.5$

$\frac{c}{23.2} = \csc 42°\ 20'$

$c = 23.2 \csc 42°\ 20'$

$c = 23.2 \times 1.4849$

$c = 34.4$

60. $B = 61°\ 30'$, $b = 31.9$, $c = 36.3$

61.

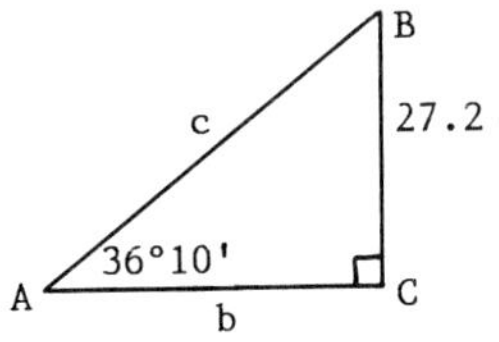

To solve this triangle find B, b, and c.

$B = 90° - 36°\ 10' = 53°\ 50'$

$\frac{b}{27.2} = \cot 36°\ 10'$

$b = 27.2 \cot 36°\ 10'$

$b = 27.2 \times 1.3680$

$b = 37.2$

$\frac{c}{27.2} = \csc 36°\ 10'$

$c = 27.2 \csc 36°\ 10'$

$c = 27.2 \times 1.6945$

$c = 46.1$

62. $B = 2°\ 20'$, $b = 0.40$, $c = 9.74$

63.

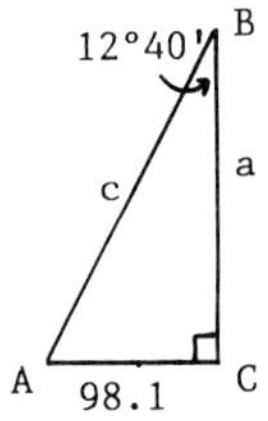

To solve this triangle find A, a, and c.

$A = 90° - 12°\ 40' = 77°\ 20'$

$\frac{a}{98.1} = \cot 12°\ 40'$

$a = 98.1 \cot 12°\ 40'$

$a = 98.1 \times 4.4494$

$a = 436.5$

$\frac{c}{98.1} = \csc 12°\ 40'$

$c = 98.1 \csc 12°\ 40'$

$c = 98.1 \times 4.5604$

$c = 447.4$

64. $A = 20°\ 10'$, $a = 46.6$, $c = 135$

65.

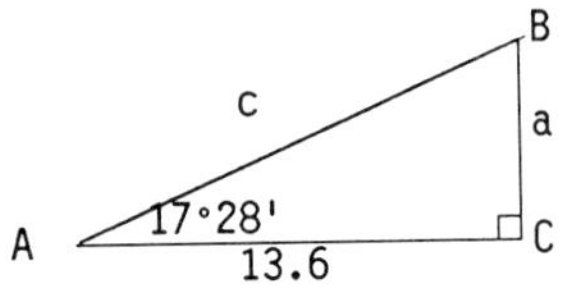

To solve this triangle find B, a, and c.

$B = 90° - 17°\ 28' = 72°\ 32'$

65. (continued)

$\frac{a}{13.6} = \tan 17° \ 28'$

$a = 13.6 \tan 17° \ 28'$

$a = 13.6 \times 0.3147$

$a = 4.3$

$\frac{c}{13.6} = \sec 17° \ 28'$

$c = 13.6 \sec 17° \ 28'$

$c = 13.6 \times 1.048$

$c = 14.3$

66. B = 11° 18', a = 6706, c = 6839

67.

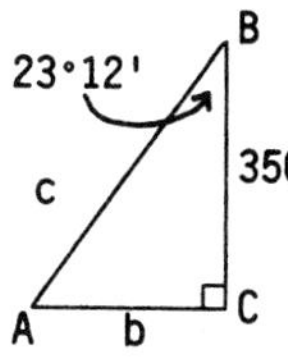

To solve this triangle find A, b, and c.

$A = 90° - 23° \ 12' = 66° \ 48'$

$\frac{b}{350} = \tan 23° \ 12'$

$b = 350 \tan 23° \ 12'$

$b = 350 \times 0.4286$

$b = 150$

$\frac{c}{350} = \sec 23° \ 12'$

$c = 350 \sec 23° \ 12'$

$c = 350 \times 1.088$

$c = 381$

68. A = 20° 38', b = 637, c = 681

69.

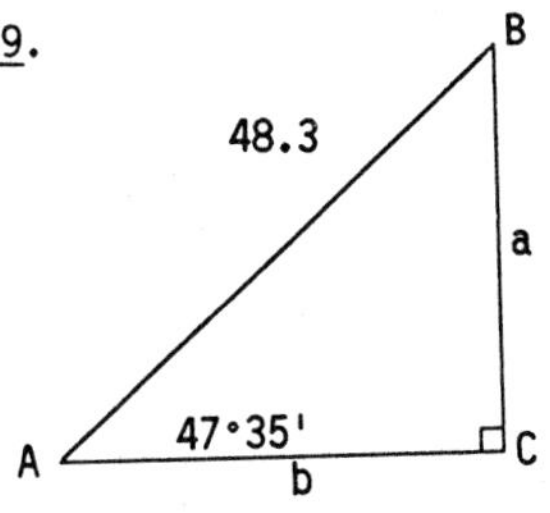

To solve this triangle find B, a, and b.

$B = 90° - 47° \ 35' = 42° \ 25'$

$\frac{a}{48.3} = \sin 47° \ 35'$

$a = 48.3 \sin 47° \ 35'$

$a = 48.3 \times 0.7383$

$a = 35.7$

69. (continued)

$\frac{b}{48.3} = \cos 47° \ 35'$

$b = 48.3 \cos 47° \ 35'$

$b = 48.3 \times 0.6745$

$b = 32.6$

70. B = 1° 5', a = 3949, b = 75

71.

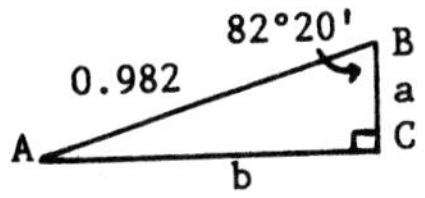

To solve this triangle find A, a, and b.

$A = 90° - 82° \ 20' = 7° \ 40'$

$\frac{a}{0.982} = \cos 82° \ 20'$

$a = 0.982 \cos 82° \ 20'$

$a = 0.982 \times 0.1334$

$a = 0.131$

$\frac{b}{0.982} = \sin 82° \ 20'$

$b = 0.982 \sin 82° \ 20'$

$b = 0.982 \times 0.9911$

$b = 0.973$

72. A = 33° 30', a = 0.0247, b = 0.0372

73.

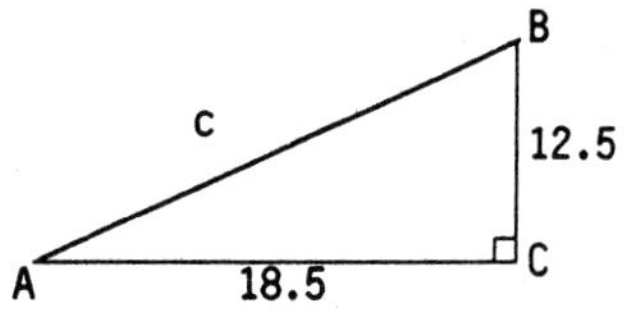

To solve this triangle find c, A, and B.

$c^2 = (12.5)^2 + (18.5)^2$

$c^2 = 156.25 + 342.25$

$c^2 = 498.5$

$c = 22.3$

$\tan A = \frac{12.5}{18.5} \approx 0.6757$

$A = 34.05°$, or $34° \ 3'$

$B = 90° - 34.05° = 55.95°$, or $55° \ 57'$

74. A = 26.56°, or 26° 34'; B = 63.44°, or 63° 26'; c = 22.8

75.

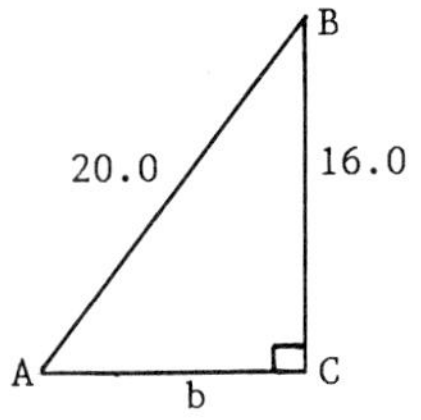

To solve this triangle find b, A, and B.

$b^2 = (20.0)^2 - (16.0)^2$

$b^2 = 400 - 256$

$b^2 = 144$

$b = 12$

$\sin A = \frac{16.0}{20.0} = 0.8000$

Thus $A = 53.13°$, or $53°\ 10'$

$B = 90° - 53.13° = 36.87°$, or $36°\ 50'$

76. $b = 42.4$, $A = 19.47°$, or $19°\ 28'$, $B = 70.53°$, or $70°\ 32'$

77.

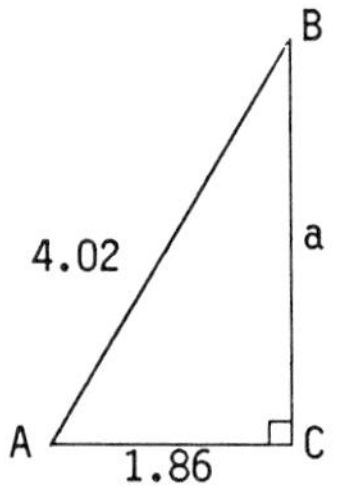

To solve this triangle find a, A, and B.

$a^2 + (1.86)^2 = (4.02)^2$

$a^2 + 3.4596 = 16.1604$

$a^2 = 12.7008$

$a = 3.56$

$\cos A = \frac{1.86}{4.02} \approx 0.4627$

$A = 62.44°$, or $62°\ 26'$

$B = 90° - 62.44° = 27.56°$ or $27°\ 32'$

78. $a = 439$, $A = 77.16°$, or $77°\ 10'$, $B = 12.84°$, or $12°\ 50'$

79.

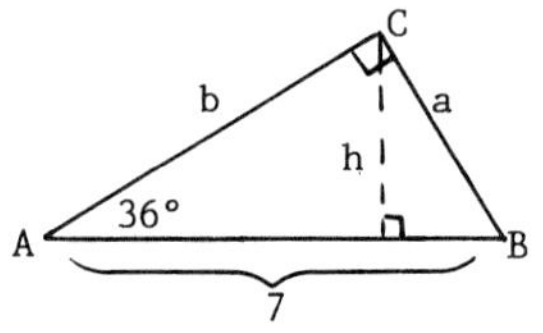

$\frac{b}{7} = \cos 36°$

$b = 7 \cos 36°$

$b = 7 \times 0.8090$

$b = 5.66$

79. (continued)

$\frac{h}{5.66} = \sin 36°$

$h = 5.66 \sin 36°$

$h = 5.66 \times 0.5878$

$h = 3.33$

80. 5.88

81. a) We know from the definition of the sine function that 0.45399, or $\frac{0.45399}{1}$ is the ratio $\frac{\text{side opposite } 27°}{\text{hypotenuse}}$. Let us consider a right triangle in which the side opposite 27° has length 0.45399 and the hypotenuse has length 1.

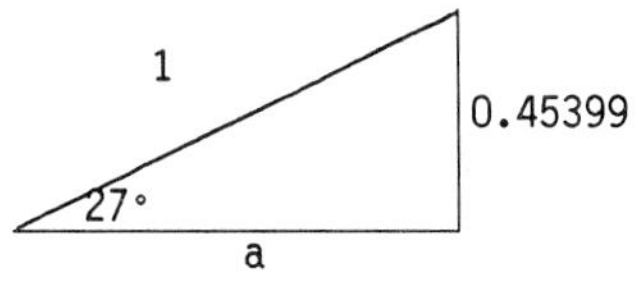

Use the Pythagorean theorem to find the length of the side adjacent to 27°.

$a^2 + 0.45399^2 = 1^2$

$a^2 = 1 - 0.45399^2$

$a \approx 0.89101$

Use $a = 0.89101$, $b = 0.45399$, and $c = 1$ to find the other five function values for 27°.

$\cos 27° = 0.89101$, $\tan 27° = \frac{0.45399}{0.89101} \approx 0.50952$

$\cot 27° = \frac{0.89101}{0.45399} \approx 1.9626$, $\sec 27° = 1.1223$,

$\csc 27° = 2.2027$

b) Since 63° and 27° are complements, we have

$\sin 63° = \cos 27° = 0.89101$

$\cos 63° = \sin 27° = 0.45399$

$\tan 63° = \cot 27° = 1.9626$

$\cot 63° = \tan 27° = 0.50952$

$\sec 63° = \csc 27° = 2.2027$

$\csc 63° = \sec 27° = 1.1223$

82. a) $\sin 54° = 0.80900$, $\cos 54° = 0.58777$, $\tan 54° = 1.3764$, $\sec 54° = 1.7014$, $\csc 54° = 1.2361$

b) $\sin 36° = 0.58777$, $\cos 36° = 0.80900$, $\tan 36° = 0.72654$, $\cot 36° = 1.3764$, $\sec 36° = 1.2361$, $\csc 36° = 1.7014$

Exercise Set 7.2

1.

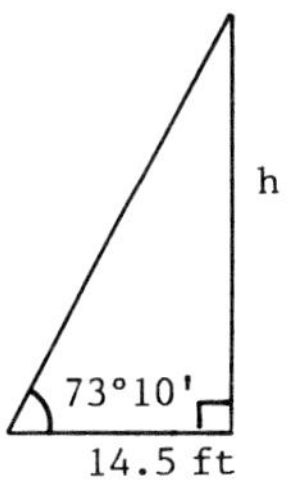

$$\frac{h}{14.5} = \tan 73^\circ\ 10'$$

$$h = 14.5 \tan 73^\circ\ 10'$$

$$h = 14.5 \times 3.3052$$

$$h = 47.9 \text{ ft}$$

2. 9.59 ft

3.

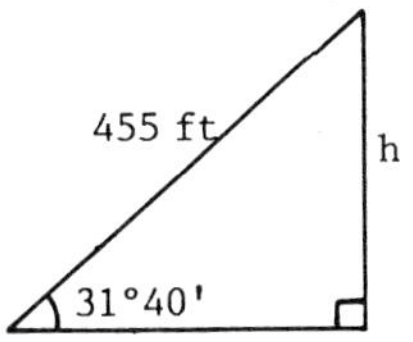

$$\frac{h}{455} = \sin 31^\circ\ 40'$$

$$h = 455 \sin 31^\circ\ 40'$$

$$h = 455 \times 0.5250$$

$$h = 239 \text{ ft}$$

4. 171 ft

5.

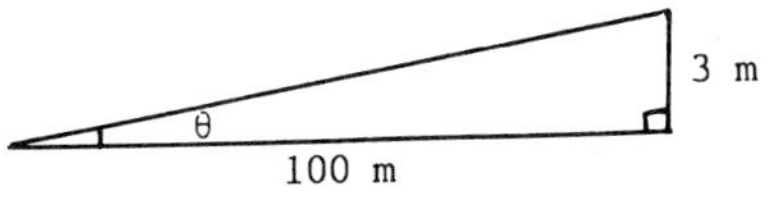

$$\tan\theta = \frac{3}{100} = 0.03$$

Thus $\theta = 1.72^\circ$, or $1^\circ\ 43'$.

6. 10.32°, or $10^\circ\ 19'$

7.

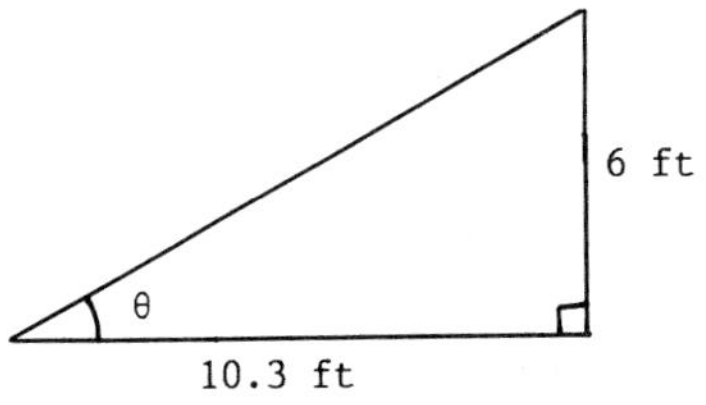

$$\tan\theta = \frac{6}{10.3} = 0.5825$$

Thus $\theta = 30.22^\circ$, or $30^\circ\ 13'$.

8. 60.26°, or $60^\circ\ 15'$

9.

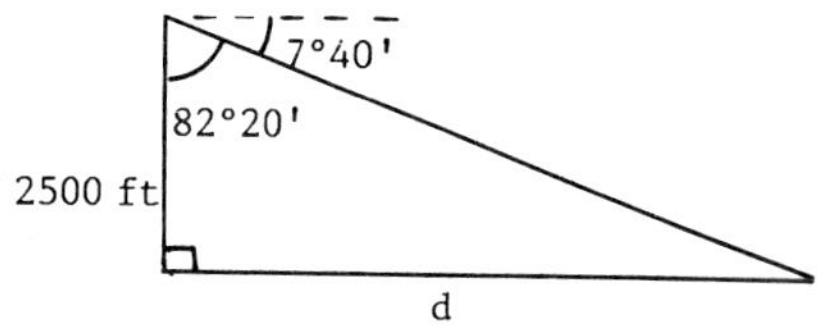

$$\frac{d}{2500} = \tan 82^\circ\ 20'$$

$$d = 2500 \tan 82^\circ\ 20'$$

$$d = 2500 \times 7.4287$$

$$d = 18{,}571 \text{ ft, or } 3.52 \text{ mi}$$

10. 274 ft

11.

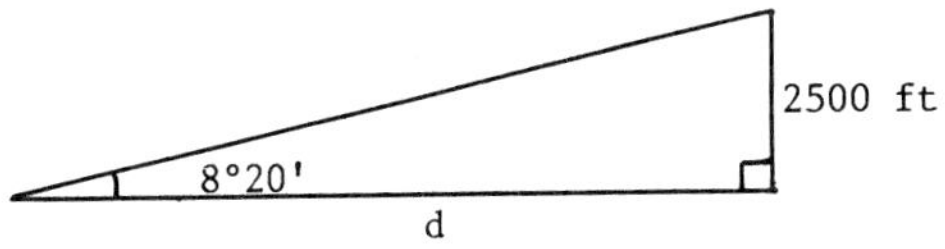

$$\frac{d}{2500} = \cot 8^\circ\ 20'$$

$$d = 2500 \cot 8^\circ\ 20'$$

$$d = 2500 \times 6.8269$$

$$d = 17{,}067 \text{ ft, Or } 3.23 \text{ mi}$$

12. 328 ft

13.

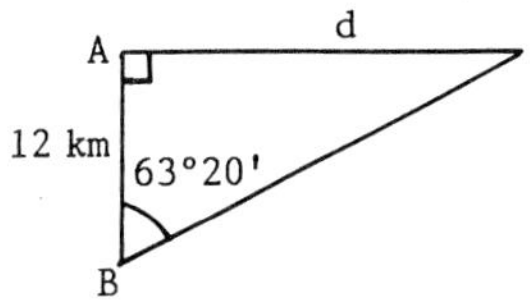

$$\frac{d}{12} = \tan 63^\circ\ 20'$$

$$d = 12 \tan 63^\circ\ 20'$$

$$d = 12 \times 1.9912$$

$$d = 23.9 \text{ km}$$

14. 19.3 km

15.

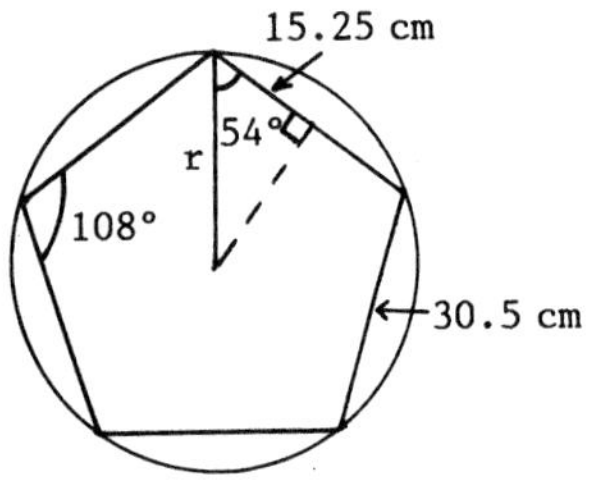

$$\frac{r}{15.25} = \sec 54°$$

$r = 15.25 \sec 54°$

$r = 15.25 \times 1.7013$

$r = 25.9$ cm

16. 36.4 cm

17.

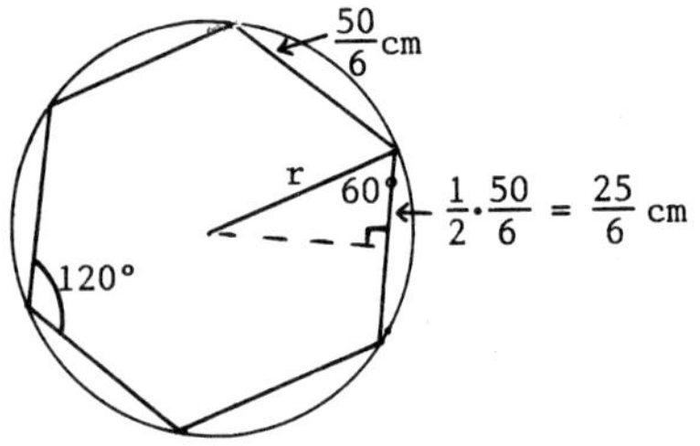

$$\frac{r}{\frac{25}{6}} = \sec 60°$$

$r = \frac{25}{6} \sec 60°$

$r = \frac{25}{6} \times 2$

$r = \frac{25}{3}$, or 8.33 cm

18. 96.7 cm

19.

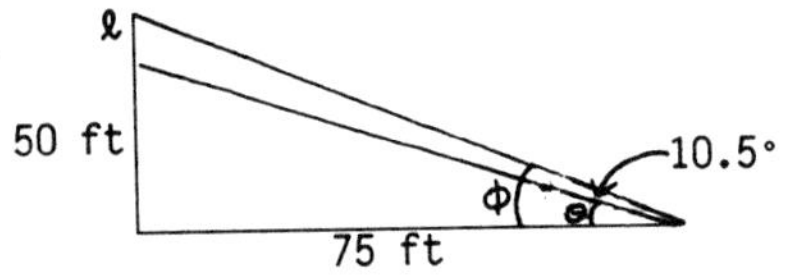

$\tan \theta = \frac{50}{75} \approx 0.6667$

$\theta = 33.7°$

$\phi = \theta + 10.5° = 33.7° + 10.5° = 44.2°$

$\tan 44.2° = \frac{50 + \ell}{75}$

$75 \tan 44.2° = 50 + \ell$

$75 \times 0.9721 = 50 + \ell$

$72.9 = 50 + \ell$

$22.9 = \ell$

The length of the antenna is 22.9 ft.

20. 44.9 ft

21.

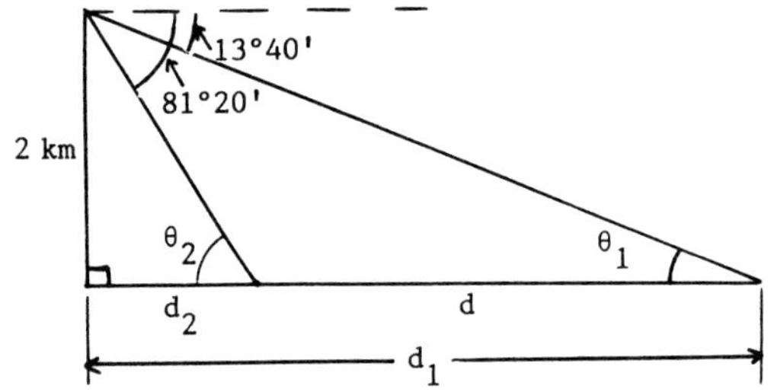

The distance to find is d, which is $d_1 - d_2$.

$\frac{d_1}{2} = \cot \theta_1$ $\qquad$ $\frac{d_2}{2} = \cot \theta_2$

$d_1 = 2 \cot 13°\ 40'$ $\qquad$ $d_2 = 2 \cot 81°\ 20'$

$d = d_1 - d_2 = 2 \cot 13°\ 40' - 2 \cot 81°\ 20'$

$= 2(\cot 13°\ 40' - \cot 81°\ 20')$

$= 2(4.1126 - 0.1524)$

$= 7.92$ km

22. 4671 m

23.

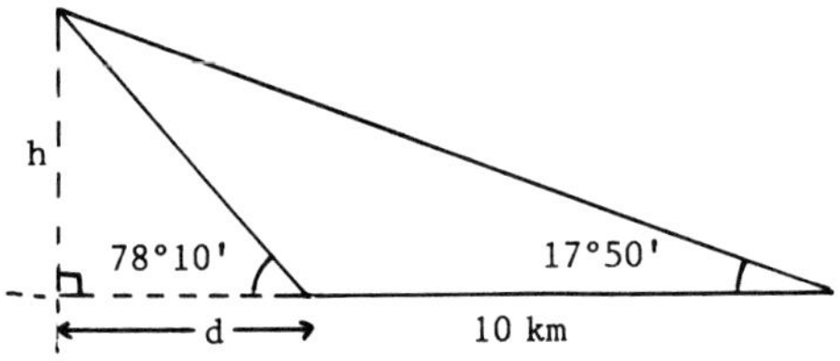

First find d. Then solve for h.

$\frac{h}{d} = \tan 78°\ 10'$ $\qquad$ $\frac{h}{d + 10} = \tan 17°\ 50'$

$h = d \tan 78°\ 10'$ $\qquad$ $h = (d + 10) \tan 17°\ 50'$

$d \tan 78°\ 10' = d \tan 17°\ 50' + 10 \tan 17°\ 50'$

$d(\tan 78°\ 10' - \tan 17°\ 50') = 10 \tan 17°\ 50'$

$$d = \frac{10 \tan 17°\ 50'}{\tan 78°\ 10' - \tan 17°\ 50'}$$

$$d = \frac{10 \times 0.3217}{4.7729 - 0.3217}$$

$$d = \frac{3.217}{4.4512}$$

$d = 0.7227$

$$\frac{h}{0.7227} = \tan 78°\ 10'$$

$h = 0.7227 \tan 78°\ 10'$

$h = 0.7227(4.7729)$

$h = 3.45$ km

24. 225 ft

25. 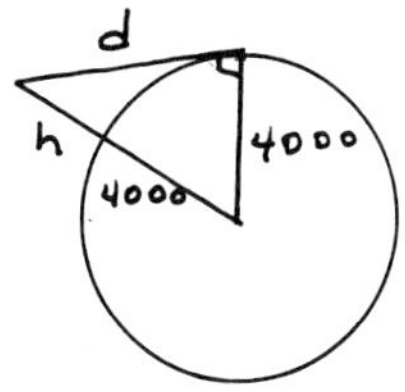

The earth's radius is approximately 4000 miles.
Use the Pythagorean theorem to find d (in miles):

$$d^2 + 4000^2 = (4000 + h)^2$$

$$d^2 + 4000^2 = 4000^2 + 8000h + h^2$$

$$d^2 = 8000h + h^2$$

$$d = \sqrt{8000h + h^2}$$

When h = 1000 ft:

$$1000 \text{ ft} \cdot \frac{1 \text{ mi}}{5280 \text{ ft}} = \frac{1000}{5280} \text{ mi} = \frac{25}{132} \text{ mi}$$

$$d = \sqrt{8000\left(\frac{25}{132}\right) + \left(\frac{25}{132}\right)^2}$$

$$\approx 38.9 \text{ mi}$$

26. 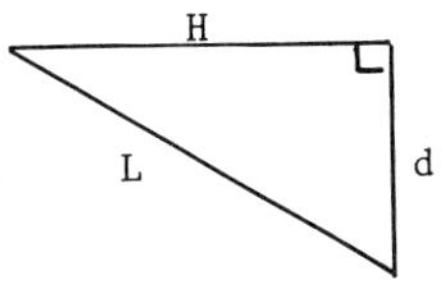

$$C = L - H$$

$$d^2 = L^2 - H^2 = (L + H)(L - H) = C(L + H)$$

$$C = \frac{d^2}{L + H} \approx \frac{d^2}{2L}$$

27. $\tan \theta = \frac{8}{1.5} \approx 5.3333$

$\theta = 79.38°$ or $79° \ 23'$

28. a) 1.96 cm
b) 4.00 cm
c) d = 1.02(VP)
d) $VP = \frac{d}{1.02}$, or VP = 0.98d

Exercise Set 7.3

1. 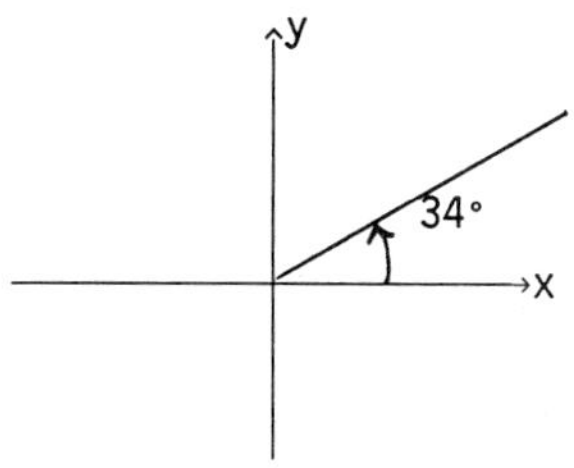

The terminal side lies in the first quadrant.

2. IV

3. 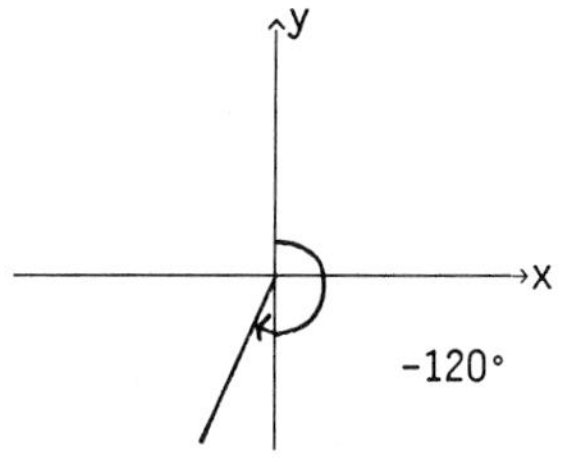

The terminal side lies in the third quadrant.

4. II

5. 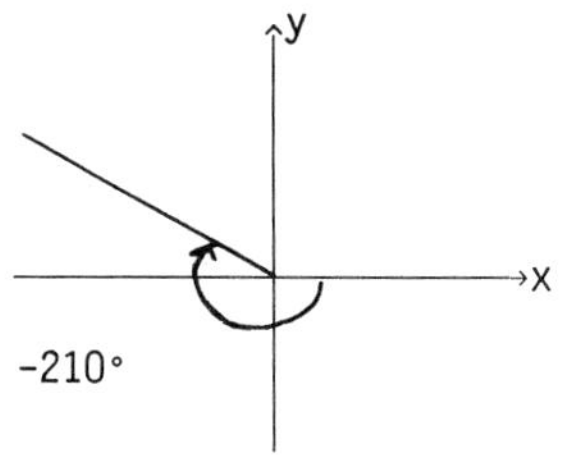

The terminal side lies in the second quadrant.

6. I

7. 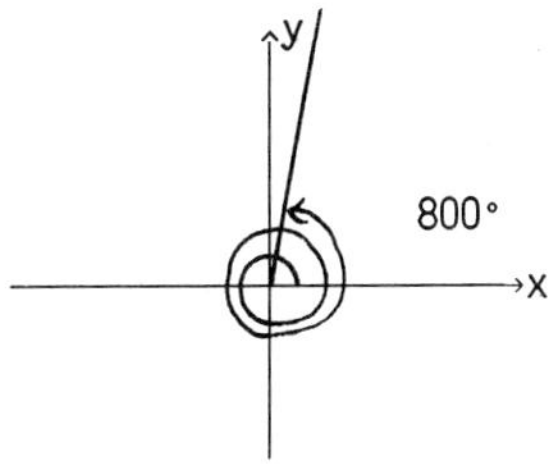

The terminal side lies in the first quadrant.

8. IV

9. 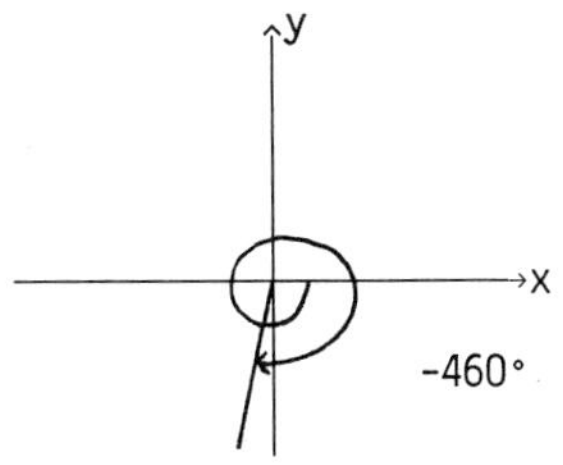

The terminal side lies in the third quadrant.

10. II

11.

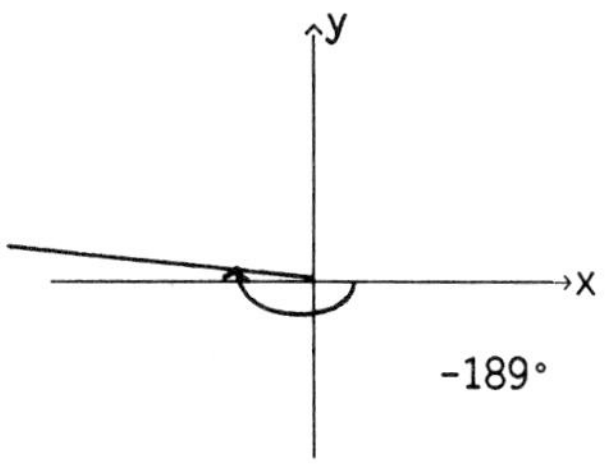

The terminal side lies in the <u>second</u> quadrant.

12. II

13.

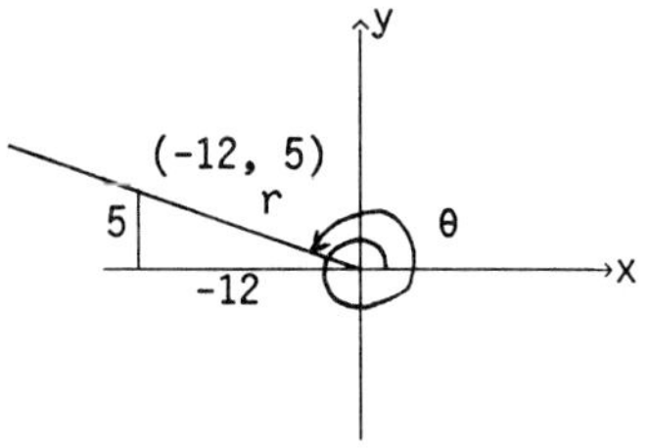

We first determine r.

$r = \sqrt{x^2 + y^2}$

$r = \sqrt{(-12)^2 + 5^2}$

$= \sqrt{144 + 25}$

$= \sqrt{169}$

$= 13$

Substituting -12 for x, 5 for y, and 13 for r, the trigonometric function values of θ are

$\sin\theta = \frac{y}{r} = \frac{5}{13}$

$\cos\theta = \frac{x}{r} = \frac{-12}{13} = -\frac{12}{13}$

$\tan\theta = \frac{y}{x} = \frac{5}{-12} = -\frac{5}{12}$

$\cot\theta = \frac{x}{y} = \frac{-12}{5} = -\frac{12}{5}$

$\sec\theta = \frac{r}{x} = \frac{13}{-12} = -\frac{13}{12}$

$\csc\theta = \frac{r}{y} = \frac{13}{5}$

14. $\sin\theta = -\frac{5}{13}$

$\cos\theta = -\frac{12}{13}$

$\tan\theta = \frac{5}{12}$

$\cot\theta = \frac{12}{5}$

$\sec\theta = -\frac{13}{12}$

$\csc\theta = -\frac{13}{5}$

15.

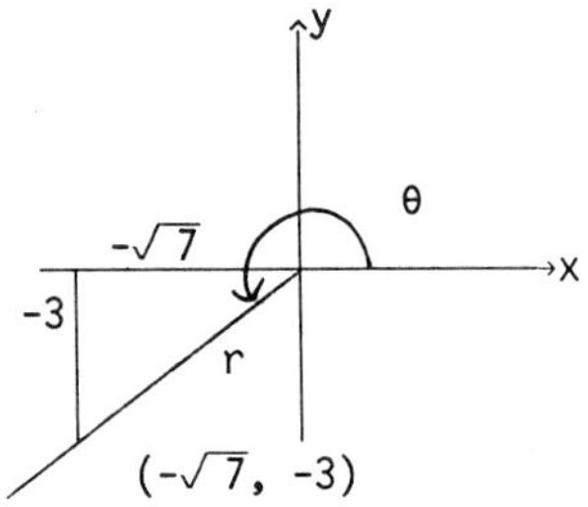

$r = \sqrt{x^2 + y^2}$

$= \sqrt{(-\sqrt{7})^2 + (-3)^2}$

$= \sqrt{7 + 9}$

$= \sqrt{16}$

$= 4$

$\sin\theta = \frac{y}{r} = \frac{-3}{4} = -\frac{3}{4}$

$\cos\theta = \frac{x}{r} = \frac{-\sqrt{7}}{4} = -\frac{\sqrt{7}}{4}$

$\tan\theta = \frac{y}{x} = \frac{-3}{-\sqrt{7}} = \frac{3}{\sqrt{7}}$, or $\frac{3\sqrt{7}}{7}$

$\cot\theta = \frac{x}{y} = \frac{-\sqrt{7}}{-3} = \frac{\sqrt{7}}{3}$

$\sec\theta = \frac{r}{x} = \frac{4}{-\sqrt{7}} = -\frac{4}{\sqrt{7}}$, or $-\frac{4\sqrt{7}}{7}$

$\csc\theta = \frac{r}{y} = \frac{4}{-3} = -\frac{4}{3}$

16. $\sin\theta = -\frac{3}{4}$

$\cos\theta = \frac{\sqrt{7}}{4}$

$\tan\theta = -\frac{3}{\sqrt{7}}$, or $-\frac{3\sqrt{7}}{7}$

$\cot\theta = -\frac{\sqrt{7}}{3}$

$\sec\theta = \frac{4}{\sqrt{7}}$, or $\frac{4\sqrt{7}}{7}$

$\csc\theta = -\frac{4}{3}$

17. 270° < 319° < 360°, so R(x, y) is in the fourth quadrant. The cosine and secant are positive, and the other four function values are negative.

18. The cosine and secant are positive, and the other four function values are negative.

19.

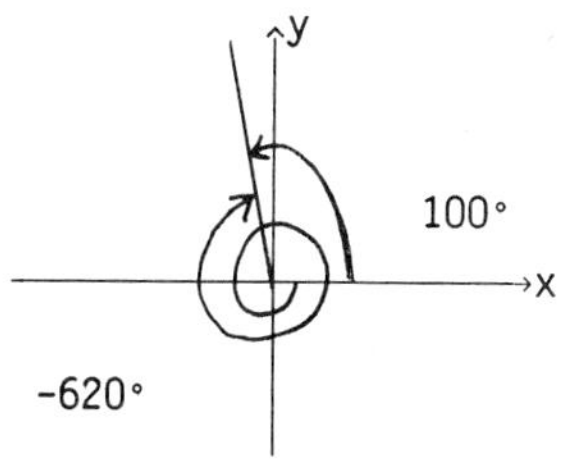

-620° has the same terminal side as an angle of 100°.

90° < 100° < 180°, so R(x, y) is in the second quadrant. The sine and cosecant are positive and the other four function values are negative.

20. The tangent and cotangent are positive, and the other four function values are negative.

21.

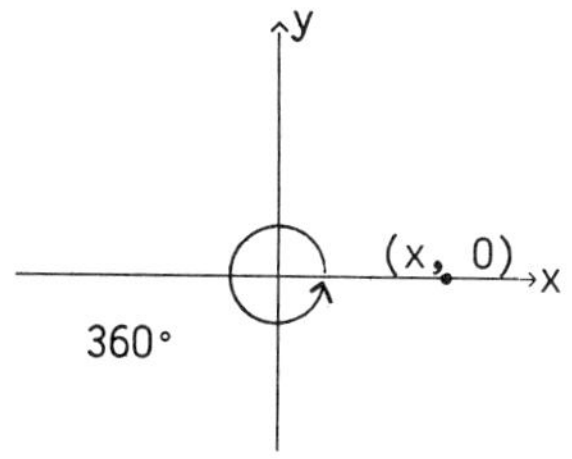

Note that the first coordinate is positive, the second coordinate is zero, and that x = r (r is always positive).

$\sin \theta = \frac{y}{r}$

$\sin 360° = \frac{y}{r} = \frac{0}{r} = 0$

22. -1

23.

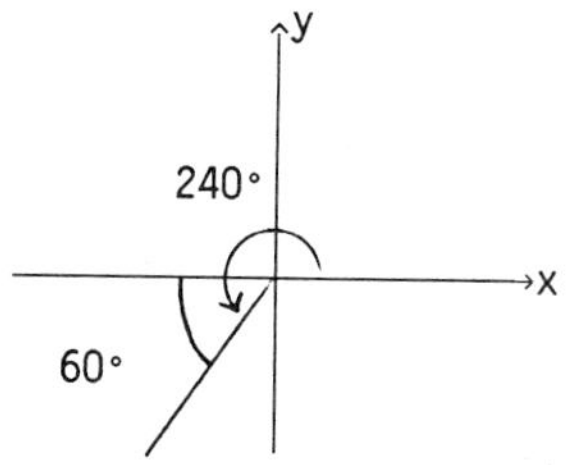

The terminal side makes a 60° angle with the x-axis. The reference angle is 60°. Find tan 60° and prefix the appropriate sign.

$\tan 60° = \sqrt{3}$

Since the tangent function is also positive in the third quadrant,

$\tan 240° = \sqrt{3}$.

24. $\frac{1}{\sqrt{3}}$, or $\frac{\sqrt{3}}{3}$

25.

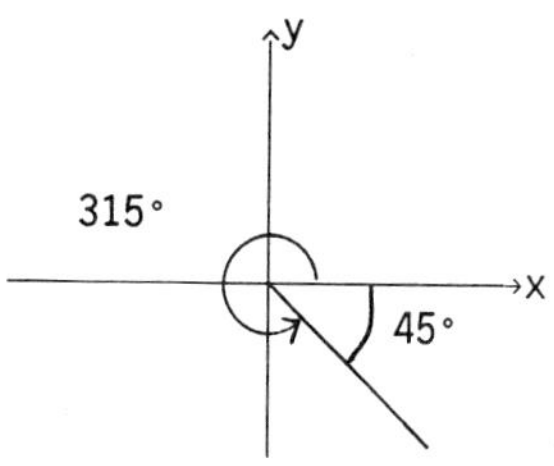

The terminal side makes a 45° angle with the x-axis. The reference angle is 45°. Find sec 45° and prefix the appropriate sign.

$\sec 45° = \frac{2}{\sqrt{2}}$, or $\sqrt{2}$

Since the secant function is also positive in the fourth quadrant,

$\sec 315° = \frac{2}{\sqrt{2}}$, or $\sqrt{2}$.

26. $-\frac{2}{\sqrt{2}}$, or $-\sqrt{2}$

27.

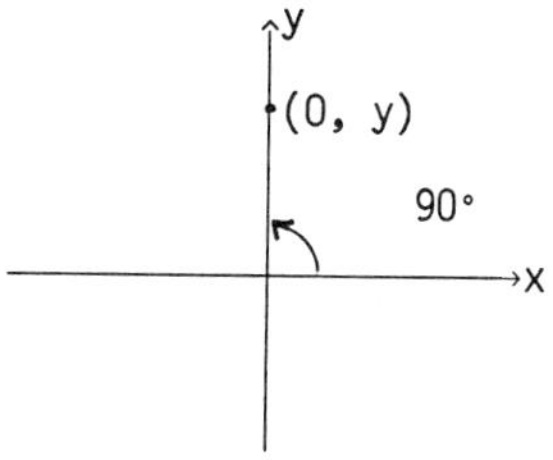

Note that the first coordinate is 0, and the second coordinate is positive.

$\tan \theta = \frac{y}{x}$

$\tan 90° = \frac{y}{x} = \frac{y}{0}$

Since $\frac{y}{0}$ is undefined, tan 90° is undefined.

28. 0

29.

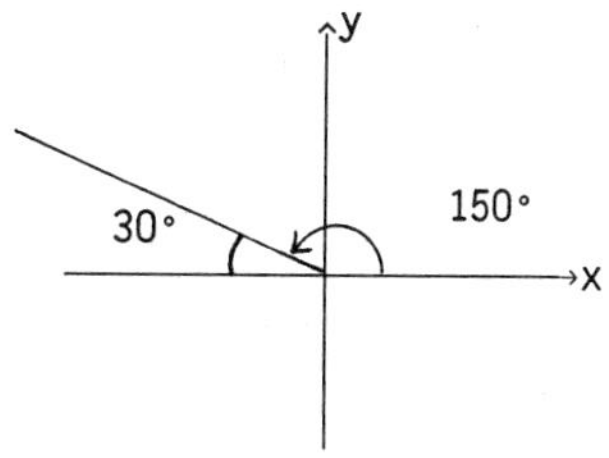

The terminal side makes a 30° angle with the x-axis. The reference angle is 30°. Find sin 30° and prefix the appropriate sign.

$\sin 30° = \frac{1}{2}$

Since the sine function is also positive in the second quadrant,

$\sin 150° = \frac{1}{2}$.

30. $-\frac{\sqrt{3}}{2}$

31.

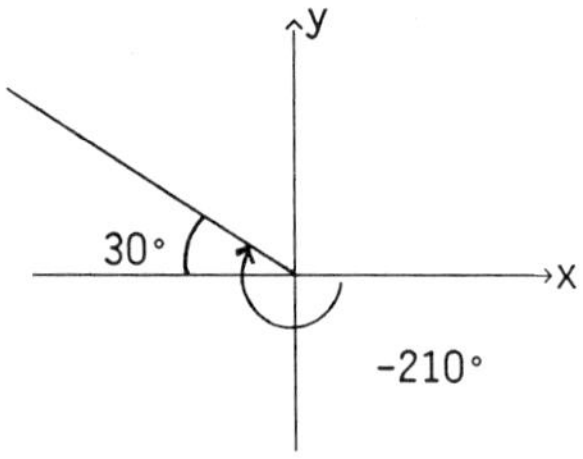

The terminal side makes a 30° angle with the x-axis. The reference angle is 30°. Find sec 30° and prefix the appropriate sign.

$\sec 30° = \frac{2}{\sqrt{3}}$, or $\frac{2\sqrt{3}}{3}$

Since the secant function is negative in the second quadrant,

$\sec (-210°) = -\frac{2}{\sqrt{3}}$, or $-\frac{2\sqrt{3}}{3}$

32. 2

33.

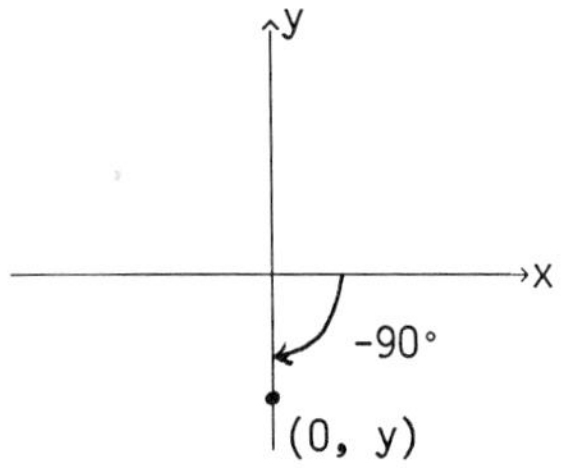

Note that the first coordinate is 0, the second coordinate is negative, and that $|y| = r$ (r is always positive).

$\sec \theta = \frac{r}{x}$

$\sec -90° = \frac{r}{x} = \frac{r}{0}$

Since $\frac{r}{0}$ is undefined, sec -90° is undefined.

34. Undefined

35.

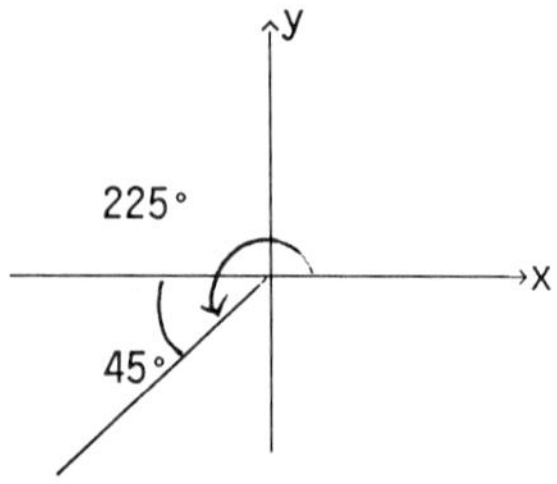

The terminal side makes a 45° angle with the x-axis. The reference angle is 45°. Find csc 45° and prefix the appropriate sign.

$\csc 45° = \frac{2}{\sqrt{2}}$, or $\sqrt{2}$

35. (continued)

Since the cosecant function is negative in the third quadrant,

$\csc 225° = -\frac{2}{\sqrt{2}}$, or $-\sqrt{2}$.

36. $-\frac{\sqrt{2}}{2}$

37.

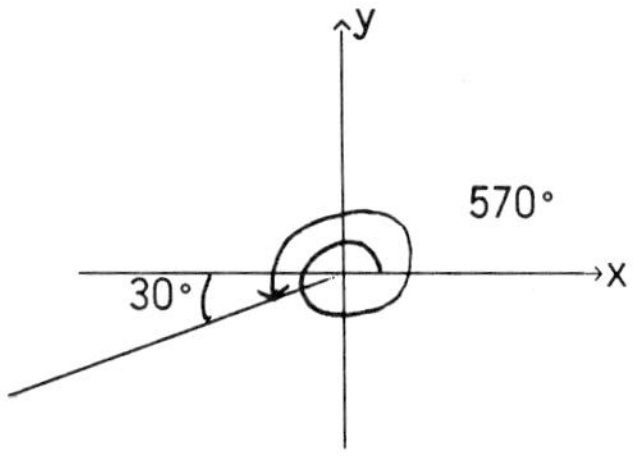

The terminal side makes a 30° angle with the x-axis. The reference angle is 30°. Find cot 30° and prefix the appropriate sign.

$\cot 30° = \frac{3}{\sqrt{3}}$, or $\sqrt{3}$

Since the cotangent function is also positive in the third quadrant,

$\cot 570° = \frac{3}{\sqrt{3}}$, or $\sqrt{3}$.

38. $\frac{1}{\sqrt{3}}$, or $\frac{\sqrt{3}}{3}$

39.

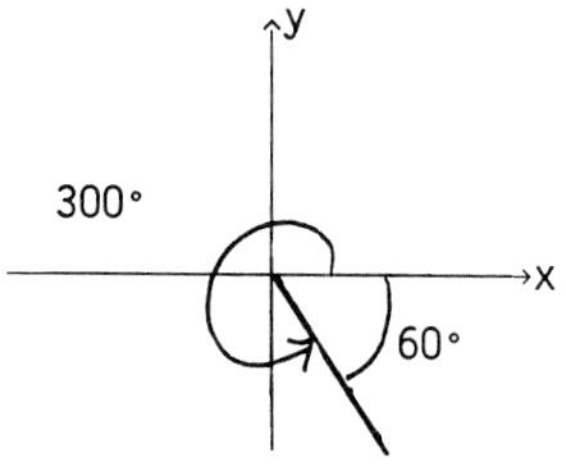

The terminal side makes a 60° angle with the x-axis. The reference angle is 60°. Find sec 60° and prefix the appropriate sign.

$\sec 60° = 2$

Since the cosine function is also positive in the fourth quadrant,

$\sec 300° = 2$.

40. $-\frac{2}{\sqrt{3}}$, or $-\frac{2\sqrt{3}}{3}$

41. 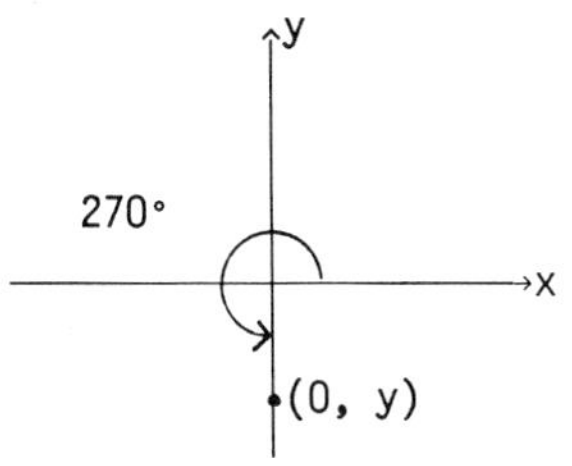

Note that the first coordinate is 0, the second coordinate is negative, and that $|y| = r$ (r is always positive).

$\csc \theta = \frac{r}{y}$

$\csc 270° = \frac{r}{y} = -1$ $(|y| = r$, y is negative)

42. -1

43. 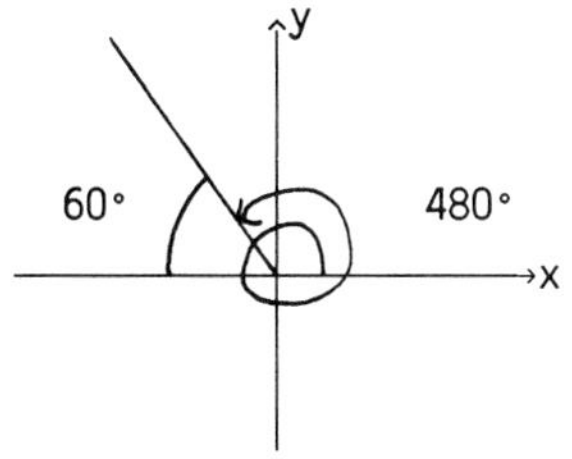

The terminal side makes a 60° angle with the x-axis. The reference angle is 60°. Find tan 60° and prefix the appropriate sign.

$\tan 60° = \sqrt{3}$

Since the tangent function is negative in the second quadrant,

$\tan 480° = -\sqrt{3}$.

44. -1

45. 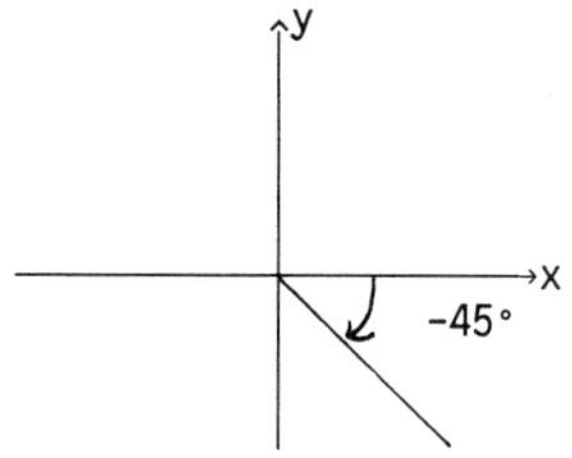

The terminal side makes a 45° angle with the x-axis. The reference angle is 45°. Find sin 45° and prefix the appropriate sign.

$\sin 45° = \frac{\sqrt{2}}{2}$

Since the sine function is negative in the fourth quadrant,

$\sin (-45°) = -\frac{\sqrt{2}}{2}$.

46. $-\frac{1}{2}$

47. 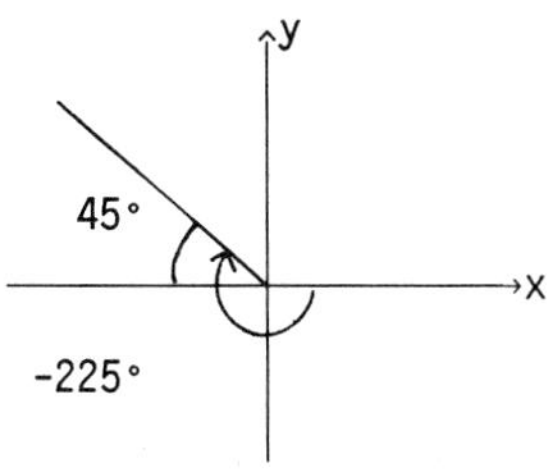

The terminal side makes a 45° angle with the x-axis. The reference angle is 45°. Find cot 45° and prefix the appropriate sign.

$\cot 45° = 1$

Since the cotangent function is negative in the second quadrant,

$\cot (-225°) = -1$.

48. $-\sqrt{2}$

49. 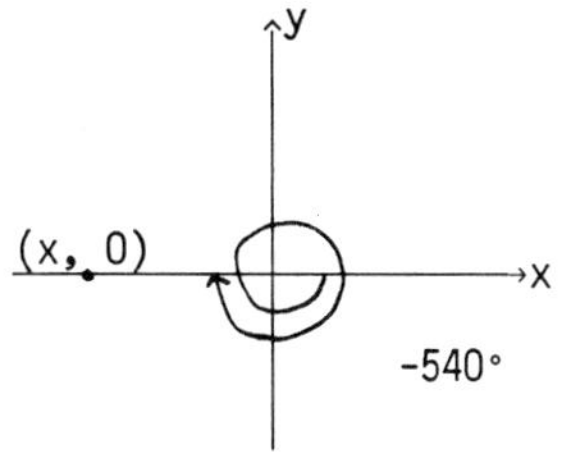

Note that the first coordinate is negative, the second coordinate is 0, and that $|x| = r$, (r is always positive).

$\cos \theta = \frac{x}{r}$

$\cos (-540°) = \frac{x}{r} = -1$ $(|x| = r$, x is negative)

50. -1

51. 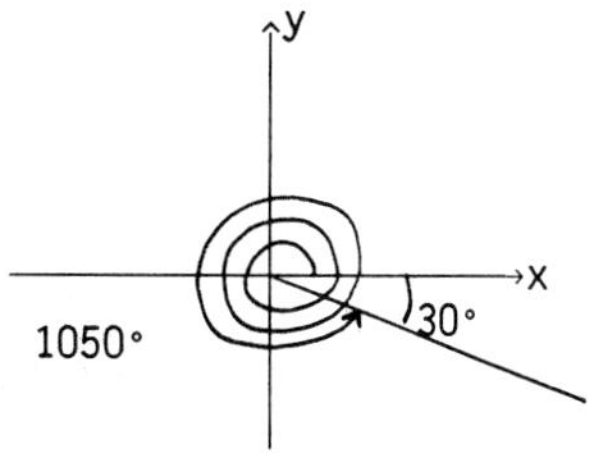

The terminal side makes a 30° angle with the x-axis. The reference angle is 30°. Find sin 30° and prefix the appropriate sign.

$\sin 30° = \frac{1}{2}$

Since the sine function is negative in the fourth quadrant,

$\sin 1050° = -\frac{1}{2}$.

52. $\frac{\sqrt{2}}{2}$

53.

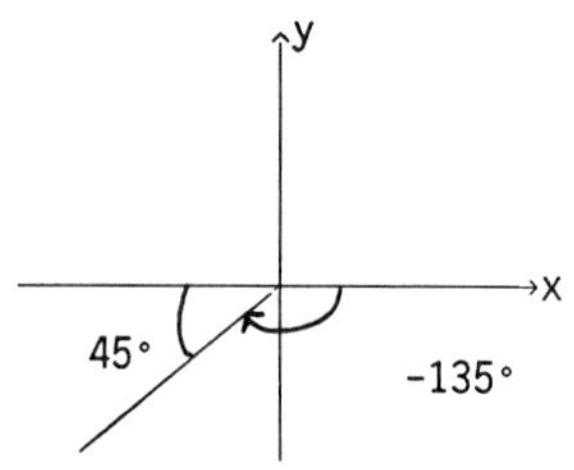

The terminal side makes a 45° angle with the x-axis. The reference angle is 45°. Find tan 45° and prefix the appropriate sign.

$\tan 45° = 1$

Since the tangent function is also positive in the third quadrant,

$\tan (-135°) = 1.$

54. $-\frac{\sqrt{2}}{2}$

55.

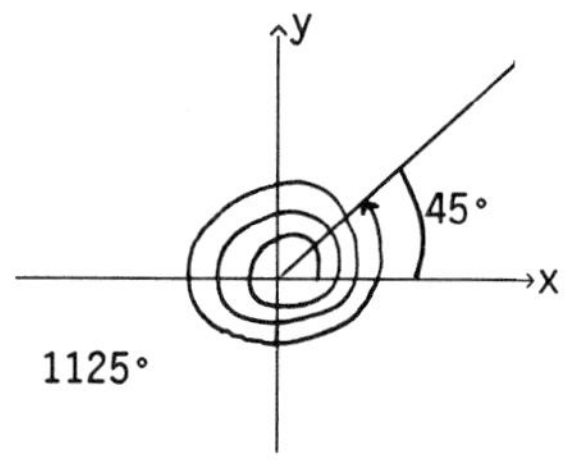

The terminal side makes a 45° angle with the x-axis. The reference angle is 45°. Find sec 45° and prefix the appropriate sign.

$\sec 45° = \frac{2}{\sqrt{2}}$, or $\sqrt{2}$

Note that the terminal side of 1125° also lies in the first quadrant. Thus,

$\sec 1125° = \frac{2}{\sqrt{2}}$, or $\sqrt{2}$.

56. $\sqrt{2}$

57.

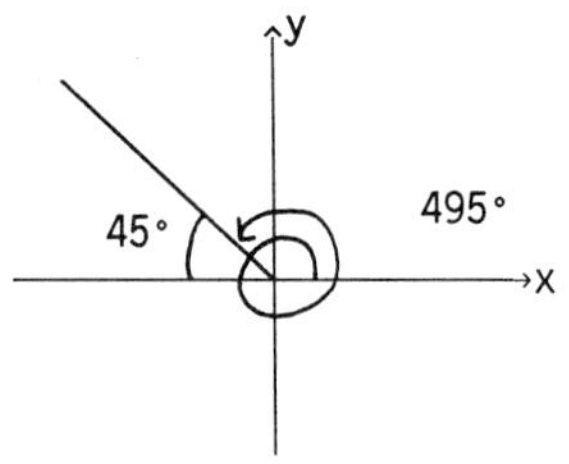

The terminal side makes a 45° angle with the x-axis. The reference angle is 45°. Find sin 45° and prefix the appropriate sign.

$\sin 45° = \frac{\sqrt{2}}{2}$

Since the sine function is also positive in the second quadrant,

$\sin 495° = \frac{\sqrt{2}}{2}$.

58. $-\frac{\sqrt{2}}{2}$

59.

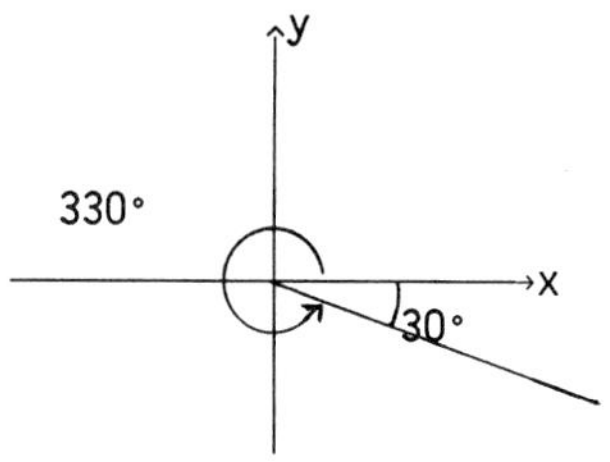

The terminal side makes a 30° angle with the x-axis. The reference angle is 30°. Find tan 30° and prefix the appropriate sign.

$\tan 30° = \frac{\sqrt{3}}{3}$

Since the tangent function is negative in the fourth quadrant,

$\tan 330° = -\frac{\sqrt{3}}{3}$.

60. $-\sqrt{3}$

61.

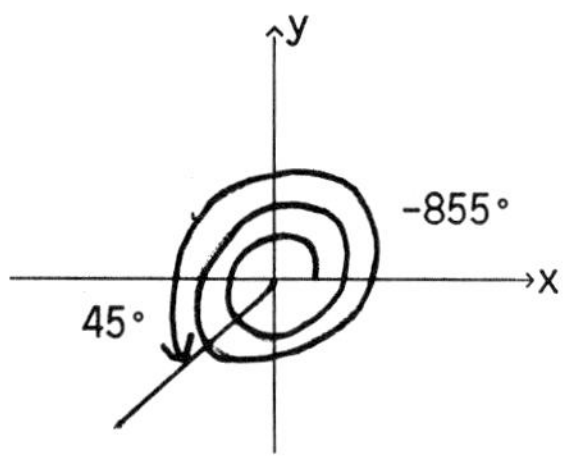

The terminal side makes a 45° angle with the x-axis. The reference angle is 45°. Find sec 45° and prefix the appropriate sign.

$\sec 45° = \frac{2}{\sqrt{2}}$, or $\sqrt{2}$

Since the secant function is negative in the third quadrant,

$\sec -855° = -\sqrt{2}$.

62. $-\frac{2}{\sqrt{3}}$, or $-\frac{2\sqrt{3}}{3}$

63. $\cos 5220° = \cos (5040° + 180°)$

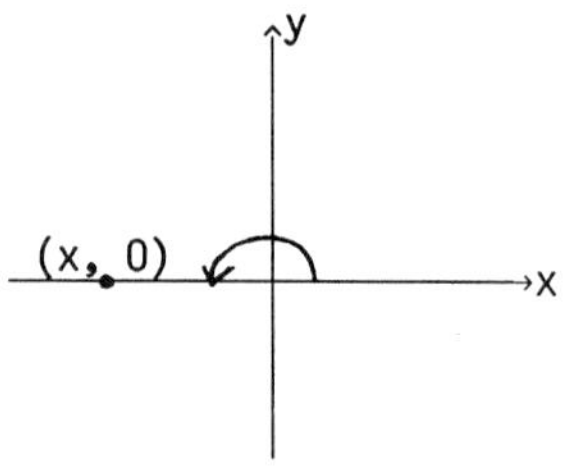

The first coordinate is negative, the second coordinate is 0, and $|x| = r$ (r is always positive).

63. (continued)

$\cos\theta = \frac{x}{r}$

$\cos 5220° = \frac{x}{r} = -1$ $(|x| = r,\ r \text{ is negative})$

64. 0

65.

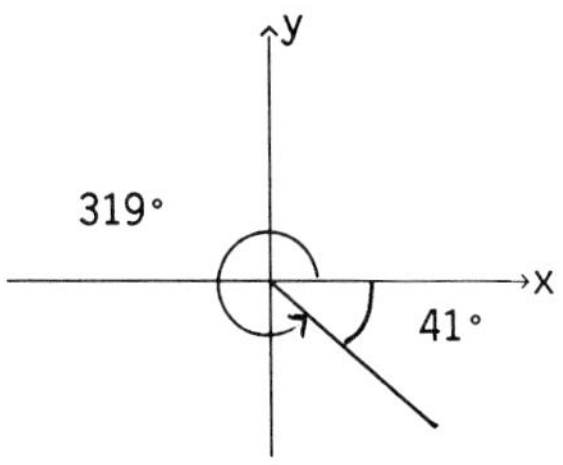

The terminal side makes a 41° angle with the x-axis. The reference angle is 41°. Find the trigonometric function values for 41° and prefix the appropriate signs.

sin 41° = 0.6561	csc 41° = 1.5242
cos 41° = 0.7547	sec 41° = 1.3250
tan 41° = 0.8693	cot 41° = 1.1504

In the fourth quadrant, the cosine and secant functions are positive and the other four are negative.

sin 319° = -0.6561	csc 319° = -1.5242
cos 319° = 0.7547	sec 319° = 1.3250
tan 319° = -0.8693	cot 319° = -1.1504

66.

sin 333° = -0.4540	csc 333° = -2.2026
cos 333° = 0.8910	sec 333° = 1.1223
tan 333° = -0.5095	cot 333° = -1.9627

67.

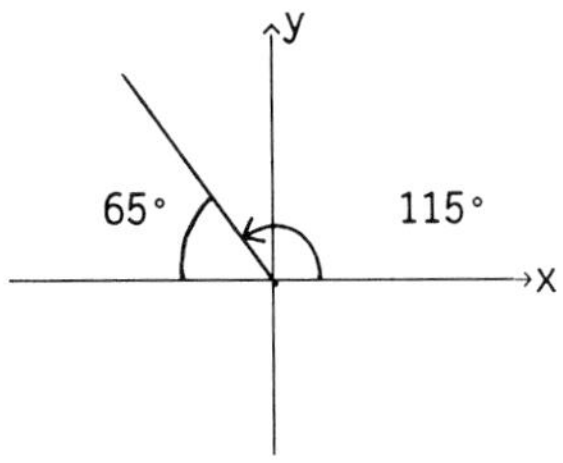

The terminal side makes a 65° angle with the x-axis. The reference angle is 65°. Find the trigonometric function values for 65° and prefix the appropriate signs.

sin 65° = 0.9063	csc 65° = 1.1034
cos 65° = 0.4226	sec 65° = 2.3663
tan 65° = 2.1445	cot 65° = 0.4663

In the second quadrant, the sine and cosecant functions are positive and the other four are negative.

sin 115° = 0.9063	csc 115° = 1.1034
cos 115° = -0.4226	sec 115° = -2.3663
tan 115° = -2.1445	cot 115° = -0.4663

68.

sin 215° = -0.5736	csc 215° = -1.7434
cos 215° = -0.8192	sec 215° = -1.2207
tan 215° = 0.7002	cot 215° = 1.4282

69.

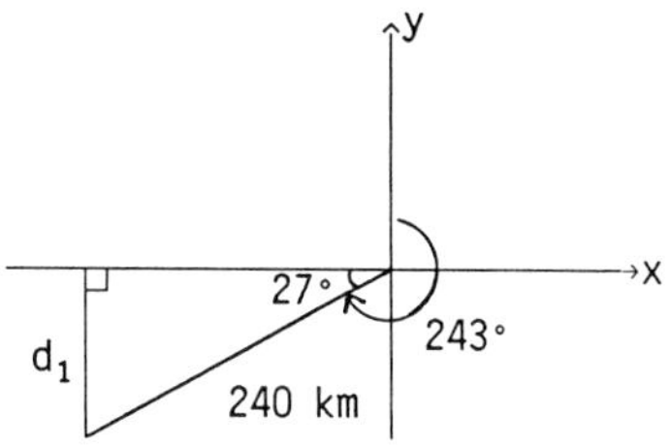

$d = rt$

$d = 120\ \frac{\text{km}}{\text{h}} \cdot 2\text{h} = 240\ \text{km}$

$\frac{d_1}{240} = \sin 27°$

$d_1 = 240 \sin 27°$

$d_1 = 240 \times 0.4540$

$d_1 = 109\ \text{km}$

71. Find tan 295° 14'. Use a calculator.

tan 295° 14' ≈ tan 295.23°

Enter 295.23 and press the [tan] key.

tan 295° 14' ≈ -2.122

72. -0.6309

73. Find sec 146.9°. Use a calculator. The secant function value is the reciprocal of the cosine function value.

Enter 146.9, press the [cos] key, and then press the reciprocal key.

$\sec 146.9° = \frac{1}{\cos 146.9°} \approx -1.194$

74. 0.9893

75. Find sin 756° 25'. Use a calculator.

756° 25' ≈ 756.42°

Enter 756.42 and press the [sin] key.

sin 756° 25' ≈ 0.5937

76. -0.1883

77. Find cos (-1000.85°). Use a calculator.

Enter -1000.85 and press the [cos] key.

cos (-1000.85°) ≈ 0.1882

78. -0.1078

79. Find cot (-16° 37'). Use a calculator.

-16° 37' ≈ -16.62°

The cotangent function value is the reciprocal of the tangent function value.

Enter -16.62, press the [tan] key, and then press the reciprocal key.

$$\cot(-16°\ 37') = \frac{1}{\tan(-16°\ 37')} \approx -3.351$$

80. -4.294

81. Find sin 3824°. Use a calculator.

Enter 3824 and press the [sin] key.

sin 3824° ≈ -0.6947

82. 0.9563

83. sin θ = -0.9956, 270° < θ < 360°

We find the reference angle, ignoring the fact that sin θ is negative. Enter 0.9956 and press the [$\sin^{-1}$] key. The reference angle is 84.62°, or 84° 37'.

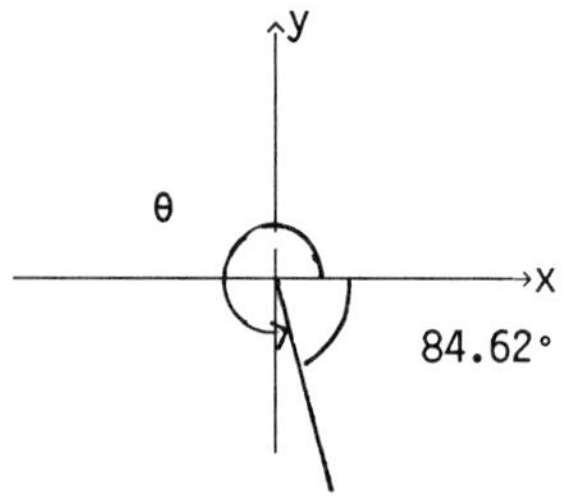

θ = 360° - 84.62° = 275.38°, or 275° 23'.

84. 154.45°, or 154° 27'

85. cos θ = -0.9388, 180° < θ < 270°

We find the reference angle, ignoring the fact that cos θ is negative. Enter 0.9388 and press the [$\cos^{-1}$] key. The reference angle is 20.15°, or 20° 9'.

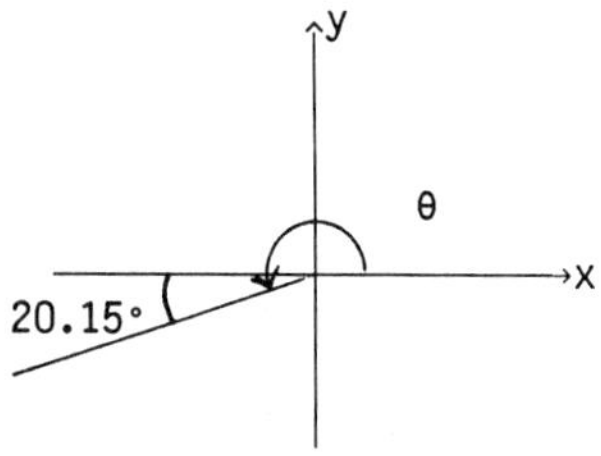

θ = 180° + 20.15° = 200.15°, or 200° 9'.

86. 95.68°, or 95° 41'

87. tan θ = 0.2460, 180° < θ < 270°

Enter 0.2460 and press the [$\tan^{-1}$] key. The reference angle is 13.82°, or 13° 49'.

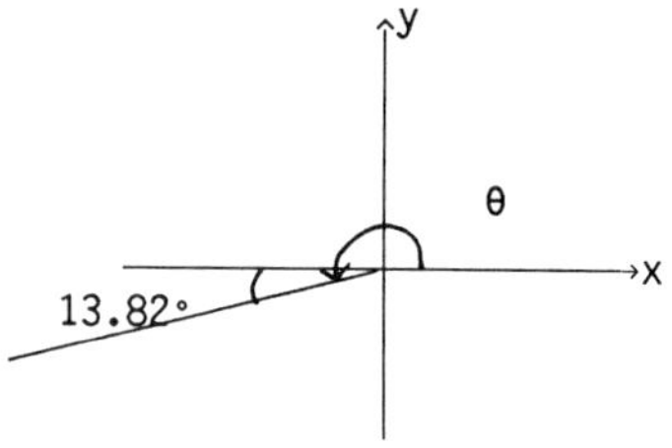

θ = 180° + 13.82° = 193.82°, or 193° 49'.

88. 288.13°, or 288° 8'

89. sec θ = -1.0485, 90° < θ < 180°

Find the reference angle, ignoring the fact that sec θ is negative. Enter 1.0485, press the reciprocal key, and then press the [$\cos^{-1}$] key. The reference angle is 17.49°, or 17° 30'.

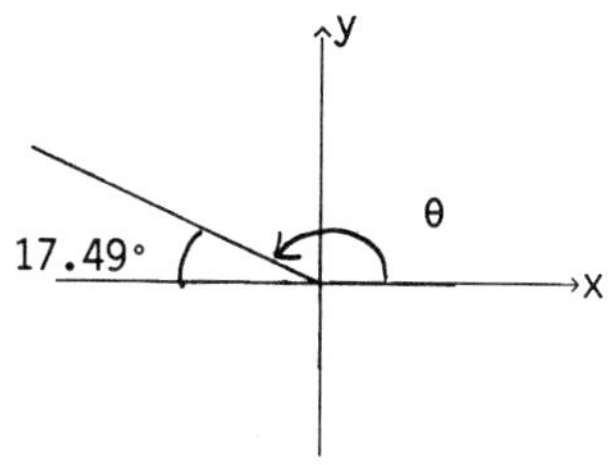

θ = 180° - 17.49° = 162.51°, or 162° 30'.

90. 72.59°, or 72° 35'

91. Sketch a third quadrant triangle. Since $\sin\theta = -\frac{1}{3}$, the length of the leg parallel to the y-axis is 1 and the length of the hypotenuse is 3. Using the Pythagorean Theorem, we know the other leg has length $\sqrt{8}$, or $2\sqrt{2}$.

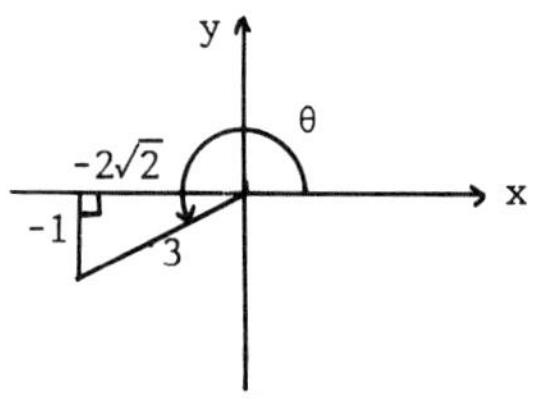

$\sin\theta = -\frac{1}{3}$ $\qquad$ $\csc\theta = -3$

$\cos\theta = -\frac{2\sqrt{2}}{3}$ $\qquad$ $\sec\theta = -\frac{3}{2\sqrt{2}}$, or $-\frac{3\sqrt{2}}{4}$

$\tan\theta = \frac{1}{2\sqrt{2}}$, or $\frac{\sqrt{2}}{4}$ $\qquad$ $\cot\theta = \frac{4}{\sqrt{2}}$, or $2\sqrt{2}$

92. $\sin\theta = -\frac{1}{5}$ $\csc\theta = -5$

$\cos\theta = \frac{2\sqrt{6}}{5}$ $\sec\theta = \frac{5\sqrt{6}}{12}$

$\tan\theta = -\frac{\sqrt{6}}{12}$ $\cot\theta = -2\sqrt{6}$

93. Sketch a fourth quadrant triangle. Since $\cos\theta = \frac{3}{5}$, label the leg on the x-axis 3 and the hypotenuse 5. Using the Pythagorean Theorem we know the other leg has length 4.

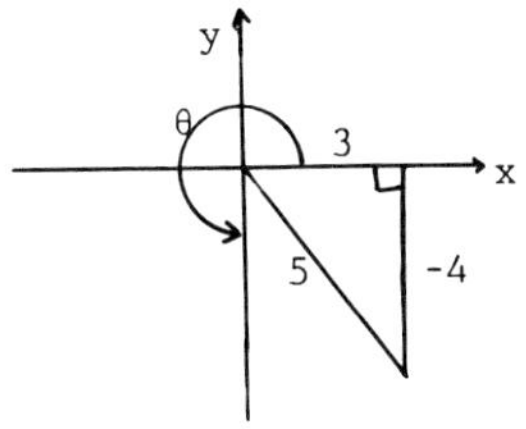

$\sin\theta = -\frac{4}{5}$ $\csc\theta = -\frac{5}{4}$

$\cos\theta = \frac{3}{5}$ $\sec\theta = \frac{5}{3}$

$\tan\theta = -\frac{4}{3}$ $\cot\theta = -\frac{3}{4}$

94. $\sin\theta = \frac{3}{5}$ $\csc\theta = \frac{5}{3}$

$\cos\theta = -\frac{4}{5}$ $\sec\theta = -\frac{5}{4}$

$\tan\theta = -\frac{3}{4}$ $\cot\theta = -\frac{4}{3}$

95. Sketch a fourth quadrant triangle. Since $\cot\theta = -2$, the length of the leg on the x-axis is 2 and the length of the leg parallel to the y-axis is 1. Using the Pythagorean Theorem, we know that the hypotenuse is $\sqrt{5}$.

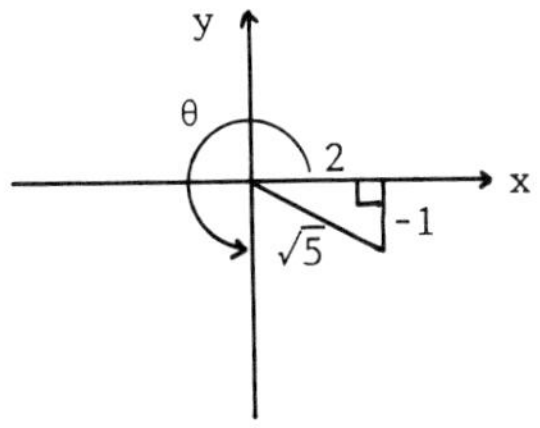

$\sin\theta = -\frac{1}{\sqrt{5}}$, or $-\frac{\sqrt{5}}{5}$ $\csc\theta = -\sqrt{5}$

$\cos\theta = \frac{2}{\sqrt{5}}$, or $\frac{2\sqrt{5}}{5}$ $\sec\theta = \frac{\sqrt{5}}{2}$

$\tan\theta = -\frac{1}{2}$ $\cot\theta = -2$

96. $\sin\theta = -\frac{5\sqrt{26}}{26}$ $\csc\theta = -\frac{\sqrt{26}}{5}$

$\cos\theta = -\frac{\sqrt{26}}{26}$ $\sec\theta = -\sqrt{26}$

$\tan\theta = 5$ $\cot\theta = \frac{1}{5}$

97. We draw the terminal side of an angle of 390°.

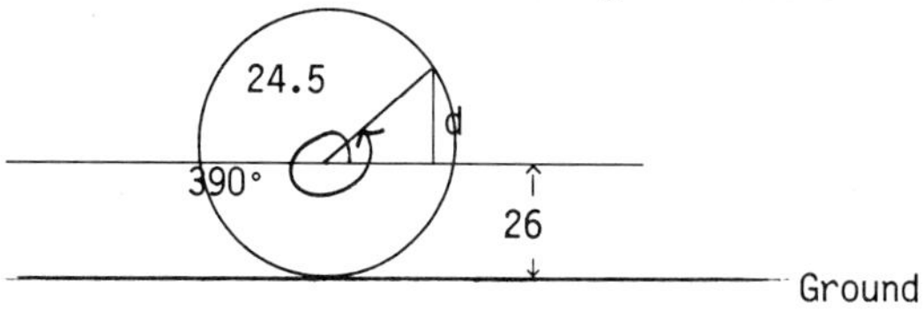

The reference angle is 30°.

Let d be the vertical distance of the valve cap above the center of the wheel.

$$\frac{d}{24.5} = \sin 30°$$

$$d = 24.5 \sin 30°$$

$$= 24.5 \times \frac{1}{2}$$

$$= 12.25$$

The distance above the ground is then

12.25 in. + 26 in. = 38.25 in.

98. 15.3 ft

Exercise Set 7.4

1.

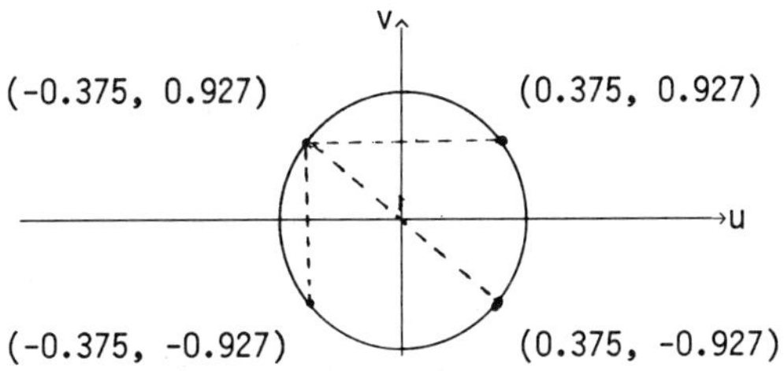

a) The reflection of (-0.375, 0.927) across the u-axis is (-0.375, -0.927). The first coordinates are the same, and the second coordinates are additive inverses of each other.

b) The reflection of (-0.375, 0.927) across the v-axis is (0.375, 0.927). The first coordinates are additive inverses of each other, and the second coordinates are the same.

c) The reflection of (-0.375, 0.927) across the origin is (0.375, -0.927). Both the first coordinates and the second coordinates are additive inverses of each other.

d) By symmetry, all these points are on the circle.

2. a) (0.625, 0.781) b) (-0.625, -0.781)

c) (-0.625, 0.781) d) Yes, by symmetry.

3.

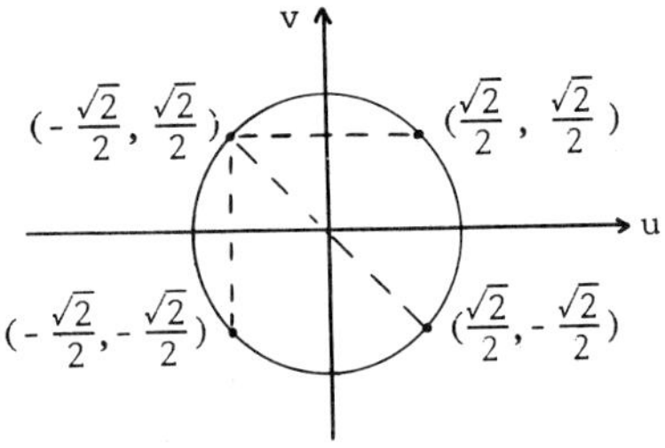

a) $\left(-\frac{\sqrt{2}}{2}, -\frac{\sqrt{2}}{2}\right)$, b) $\left(\frac{\sqrt{2}}{2}, \frac{\sqrt{2}}{2}\right)$,

c) $\left(\frac{\sqrt{2}}{2}, -\frac{\sqrt{2}}{2}\right)$

4. a) $\left(-\frac{3}{4}, -\frac{\sqrt{7}}{4}\right)$, b) $\left(\frac{3}{4}, \frac{\sqrt{7}}{4}\right)$, c) $\left(\frac{3}{4}, -\frac{\sqrt{7}}{4}\right)$

5.

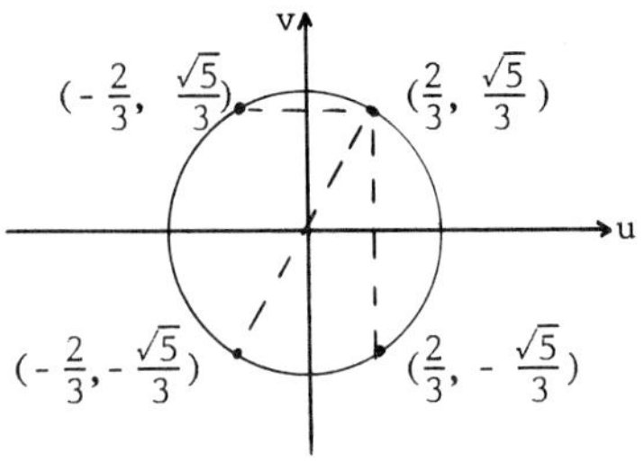

a) $\left(\frac{2}{3}, -\frac{\sqrt{5}}{3}\right)$, b) $\left(-\frac{2}{3}, \frac{\sqrt{5}}{3}\right)$, c) $\left(-\frac{2}{3}, -\frac{\sqrt{5}}{3}\right)$

6. a) $\left(-\frac{\sqrt{3}}{2}, \frac{1}{2}\right)$, b) $\left(\frac{\sqrt{3}}{2}, -\frac{1}{2}\right)$, c) $\left(\frac{\sqrt{3}}{2}, \frac{1}{2}\right)$

7. See figure in text.

8.

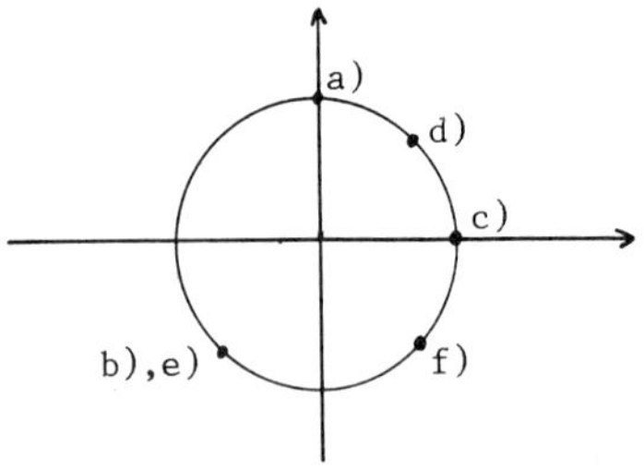

9. See figure in text.

10.

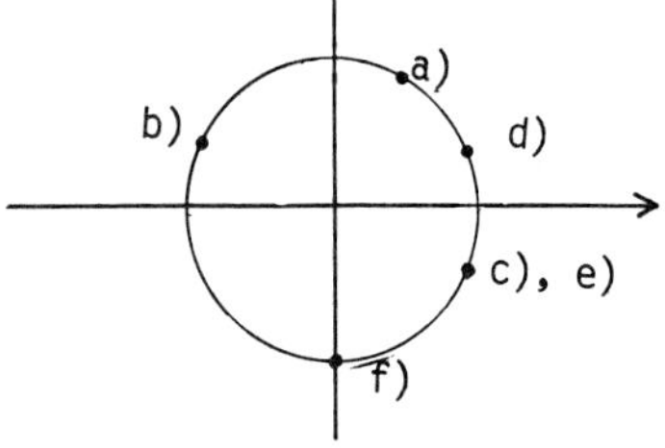

11.

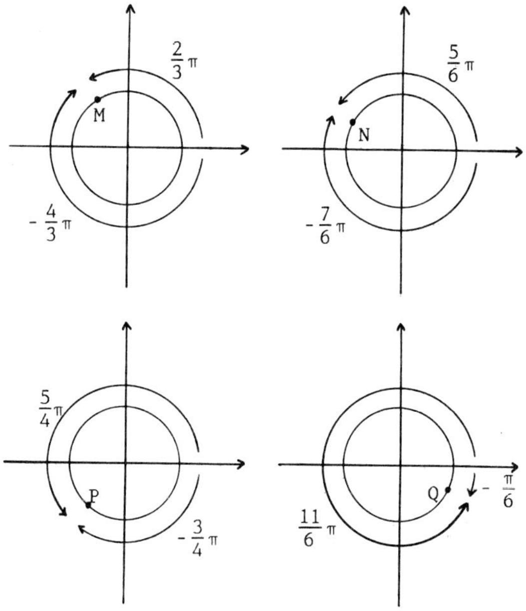

12. M: $\frac{\pi}{3}$, $-\frac{5}{3}\pi$; N: $\frac{7}{4}\pi$, $-\frac{\pi}{4}$

P: $\frac{4}{3}\pi$, $-\frac{2}{3}\pi$; Q: $\frac{7}{6}\pi$, $-\frac{5}{6}\pi$

13. See figure in text.

14.

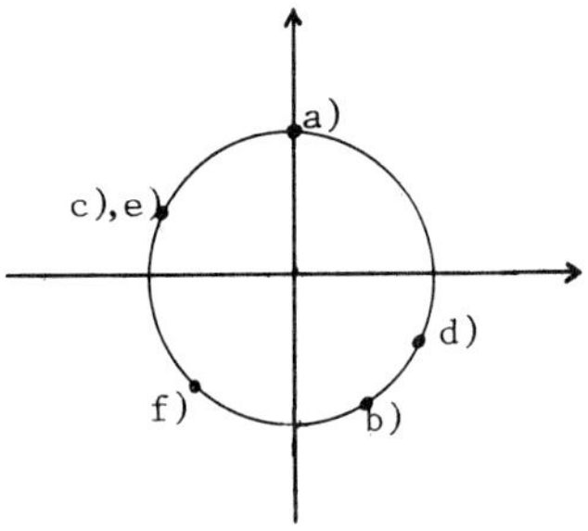

15. The point determined by $-\frac{\pi}{6}$ is a reflection across the u-axis of the point determined by $\frac{\pi}{6}$, $\left(\frac{\sqrt{3}}{2}, \frac{1}{2}\right)$. The first coordinates are the same, and the second coordinates are additive inverses of each other. The coordinates of the point determined by $-\frac{\pi}{6}$ are $\left(\frac{\sqrt{3}}{2}, -\frac{1}{2}\right)$.

16. $\left(\frac{1}{2}, -\frac{\sqrt{3}}{2}\right)$

17. The point determined by $-\alpha$ is a reflection across the u-axis of the point α, $\left(\frac{3}{4}, -\frac{\sqrt{7}}{4}\right)$. The first coordinates are the same, and the second coordinates are additive inverses of each other. The coordinates of the point determined by $-\alpha$ are $\left(\frac{3}{4}, \frac{\sqrt{7}}{4}\right)$.

18. $\left[-\frac{2}{3}, -\frac{\sqrt{5}}{3}\right]$

19. $30° = 30° \cdot \frac{\pi \text{ radians}}{180°} = \frac{\pi}{6}$ radians, or $\frac{\pi}{6}$.

20. $\frac{\pi}{12}$

21. $60° = 60° \cdot \frac{\pi \text{ radians}}{180°} = \frac{\pi}{3}$ radians, or $\frac{\pi}{3}$.

22. $\frac{10}{9}\pi$

23. $75° = 75° \cdot \frac{\pi \text{ radians}}{180°} = \frac{5}{12}\pi$ radians, or $\frac{5}{12}\pi$.

24. $\frac{5}{3}\pi$

25. $37.71° = 37.71° \cdot \frac{\pi \text{ radians}}{180°}$

$= 0.2095\pi$ radians, or 0.2095π

26. 0.0707π

27. $214.6° = 214.6° \cdot \frac{\pi \text{ radians}}{180°}$

$\approx 1.1922\pi$ radians, or 1.1922π

28. 0.41π

29. $120° = 120° \cdot \frac{\pi \text{ radians}}{180°} = \frac{2}{3}\pi \approx 2.0933$

30. 4.187

31. $320° = 320° \cdot \frac{\pi \text{ radians}}{180°} = \frac{16}{9}\pi \approx 5.5822$

32. 1.308

33. $200° = 200° \cdot \frac{\pi \text{ radians}}{180°} = \frac{10}{9}\pi \approx 3.4889$

34. 5.233

35. $117.8° = 117.8° \cdot \frac{\pi \text{ radians}}{180°} = \frac{117.8}{180}\pi \approx 2.0550$

36. 4.0332

37. $1.354° = 1.354° \cdot \frac{\pi \text{ radians}}{180°} = \frac{1.354}{180}\pi \approx 0.0236$

38. 5.7200

39. $1 \text{ radian} = 1 \text{ radian} \cdot \frac{180°}{\pi \text{ radians}}$

$\approx \left(\frac{180}{3.14}\right)° \approx 57.32°$

40. 114.6°

41. $8\pi = 8\pi \text{ radians} \cdot \frac{180°}{\pi \text{ radians}}$

$= 8 \cdot 180° = 1440°$

42. -2160°

43. $\frac{3}{4}\pi = \frac{3}{4}\pi \text{ radians} \cdot \frac{180°}{\pi \text{ radians}}$

$= \frac{3}{4} \cdot 180° = 135°$

44. 225°

45. $1.303 = 1.303 \text{ radians} \cdot \frac{180°}{\pi \text{ radians}}$

$\approx \frac{1.303 \cdot 180°}{3.14} \approx 74.69°$

46. 134.5°

47. $0.7532\pi = 0.7532\pi \text{ radians} \cdot \frac{180°}{\pi \text{ radians}}$

$= 0.7532 \cdot 180° = 135.576°$

48. -216.9°

49. $0° = 0$ radians

$30° = \frac{\pi}{6}$

$45° = \frac{\pi}{4}$

$60° = \frac{\pi}{3}$

$90° = \frac{\pi}{2}$

$135° = \frac{3}{4}\pi$

$180° = \pi$

$225° = \frac{5}{4}\pi$

$270° = \frac{3}{2}\pi$

$315° = \frac{7}{4}\pi$

$360° = 2\pi$

50. $-30° = -\frac{\pi}{6}$, $-60° = -\frac{\pi}{3}$, $-90° = -\frac{\pi}{2}$, $-135° = -\frac{3}{4}\pi$,

$-225° = -\frac{5}{4}\pi$, $-270° = -\frac{3}{2}\pi$, $-315° = -\frac{7}{4}\pi$

51. $\theta = \frac{s}{r}$

θ is radian measure of central angle, s is arc length, and r is radius length.

$\theta = \frac{132 \text{ cm}}{120 \text{ cm}}$ (Substituting 132 cm for s and 120 cm for r)

$\theta = \frac{11}{10}$, or 1.1 (The unit is understood to be radians)

$1.1 \text{ radians} = 1.1 \text{ radians} \cdot \frac{180°}{\pi \text{ radians}} \approx 63°$.

52. 0.325 radians, 19°

53. In 60 minutes the minute hand rotates 2π radians. In 50 minutes it rotates

$\frac{50}{60} \cdot 2\pi$, or 5.233 radians.

54. 4526 radians

55. $\theta = \frac{s}{r}$

$1.6 = \frac{s}{10\text{ m}}$ (Substituting 1.6 for θ and 10 m for r)

$1.6(10\text{ m}) = s$

$16\text{ m} = s$

56. 10.5 m

57. Since the linear speed must be in cm/min, the given angular speed, 7 radians/sec, must be changed to radians/min.

$\omega = \frac{7\text{ radians}}{1\text{ sec}} = \frac{7}{1\text{ sec}} \cdot \frac{60\text{ sec}}{1\text{ min}} = \frac{420}{1\text{ min}}$

$r = \frac{d}{2} = \frac{15\text{ cm}}{2} = 7.5\text{ cm}$

Using $v = r\omega$, we have

$v = 7.5\text{ cm} \cdot \frac{420}{1\text{ min}}$ (Substituting)

$= 3150\ \frac{\text{cm}}{\text{min}}$

58. 54 m/min

59. First change ω to radians per second.

$\omega = 33\frac{1}{3}\frac{\text{rev}}{\text{min}} = 33\frac{1}{3} \cdot \frac{2\pi}{\text{min}}$ (Substituting 2π for 1 rev)

$= 33\frac{1}{3} \cdot \frac{2\pi}{\text{min}} \cdot \frac{1\text{ min}}{60\text{ sec}}$

$\approx 3.4888\ \frac{\text{radians}}{\text{sec}}$

Using $v = r\omega$, we have

$v = 15\text{ cm} \cdot \frac{3.4888}{1\text{ sec}} \approx 52.33\ \frac{\text{cm}}{\text{sec}}$

60. 41.0 cm/sec

61. First find ω in radians per hour.

$\omega = \frac{1\text{ rev}}{24\text{ hr}} = \frac{2\pi}{24\text{ hr}} = \frac{\pi}{12}\frac{\text{radians}}{\text{hr}}$

Using $v = r\omega$, we have

$v = 4000\text{ mi} \cdot \frac{\pi}{12\text{ hr}} = 1047\text{ mph}$

62. 66,626 mph

63. The units of distance for v and r must be the same. Thus,

$v = 11\ \frac{\text{m}}{\text{s}} \cdot \frac{100\text{ cm}}{1\text{ m}} = 1100\ \frac{\text{cm}}{\text{s}}$

$r = \frac{d}{2} = \frac{32\text{ cm}}{2} = 16\text{ cm}$

Using $v = r\omega$, we have

$1100\ \frac{\text{cm}}{\text{s}} = 16 \cdot \text{cm} \cdot \omega$

$\frac{1100}{16} \cdot \frac{\text{radians}}{\text{s}} = \omega$

$68.75\ \frac{\text{radians}}{\text{sec}} \approx \omega$

64. 1429 radians/hr

65. First change ω to radians per hour.

$\omega = 14\ \frac{\text{rev}}{\text{min}} = 14 \cdot \frac{2\pi}{\text{min}} \cdot \frac{60\text{ min}}{1\text{ hr}} \approx 5275.2\ \frac{\text{radians}}{\text{hr}}$

Next change 10 ft to miles.

$r = 10\text{ ft} \cdot \frac{1\text{ mi}}{5280\text{ ft}} = \frac{1}{528}\text{ mi}$

Using $v = r\omega$, we have

$v = \frac{1}{528}\text{ mi} \cdot \frac{5275.2}{1\text{ hr}} \approx 10\text{ mph}$

66. 11.4 mph

67. $\omega = 12\ \frac{\text{rev}}{\text{min}} = 12 \cdot \frac{2\pi}{\text{min}}$

$= 24\pi\ \frac{\text{radians}}{\text{min}} \approx 75.36\ \frac{\text{radians}}{\text{min}}$

$r = \frac{d}{2} = \frac{24\text{ in}}{2} = 12\text{ in} = 1\text{ ft}$

Using $v = r\omega$, we have

$v = 1\text{ ft} \cdot \frac{75.36}{\text{min}} = 75.36\ \frac{\text{ft}}{\text{min}}$

The bike will travel 75.36 ft in 1 min.

68. 4710 ft

69. First find v in ft/sec.

$v = 30\ \frac{\text{mi}}{\text{hr}} = 30 \cdot \frac{5280\text{ ft}}{3600\text{ sec}} = 44\ \frac{\text{ft}}{\text{sec}}$

Next find r in ft.

$r = 14\text{ in} = 14\text{ in} \cdot \frac{1\text{ ft}}{12\text{ in}} = \frac{7}{6}\text{ ft.}$

Using $v = r\omega$, we have

$44\ \frac{\text{ft}}{\text{sec}} = \frac{7}{6}\text{ ft} \cdot \omega$

$\frac{6}{7} \cdot 44\ \frac{\text{radians}}{\text{sec}} = \omega$

$37.7\ \frac{\text{radians}}{\text{sec}} \approx \omega$

69. (continued)

If $\omega = 37.7 \frac{\text{radians}}{\text{sec}}$, then the angle through which a wheel rotates is $37.7 \frac{\text{radians}}{\text{sec}} \cdot 10$ sec, or 377 radians.

70. 563 radians

71. $90° = \frac{\pi}{2}$ radians = 100 grads

Note the following:

$1 = \frac{90°}{100 \text{ grads}}$; also $\frac{100 \text{ grads}}{90°} = 1$

$1 = \frac{\frac{\pi}{2} \text{ radians}}{100 \text{ grads}} = \frac{\pi \text{ radians}}{200 \text{ grads}}$; also $\frac{200 \text{ grads}}{\pi \text{ radians}} = 1$

a) $48° = 48° \cdot \frac{100 \text{ grads}}{90°}$

$= \frac{48}{90} \cdot 100 \text{ grads} = 53.3 \text{ grads}$

b) $153° = 153° \cdot \frac{100 \text{ grads}}{90°}$

$= \frac{153}{90} \cdot 100 \text{ grads} = 170 \text{ grads}$

c) $\frac{\pi}{8}$ radians $= \frac{\pi}{8}$ radians $\cdot \frac{200 \text{ grads}}{\pi \text{ radians}}$

$= \frac{200}{8}$ grads = 25 grads

d) $\frac{5\pi}{7}$ radians $= \frac{5\pi}{7}$ radians $\cdot \frac{200 \text{ grads}}{\pi \text{ radians}}$

$= \frac{5}{7} \cdot 200 \text{ grads} \approx 142.9 \text{ grads}$

72. a) 5°37'30", b) 19°41'15"

73. One degree of latitude is $\frac{1}{360}$ of the circumference of the earth.

$C = \pi d$, or $2\pi r$

When r = 6400 km, $C = 2\pi \cdot 6400 \approx 40{,}192$ km.
Thus 1° of latitude is $\frac{1}{360} \cdot 40{,}192$, or ≈ 112 km.

When r = 4000 mi, $C = 2\pi \cdot 4000 \approx 25{,}120$ mi.
Thus 1° of latitude is $\frac{1}{360} \cdot 25{,}120$, or ≈ 70 mi.

74. Circumference: 21,600 NM, Radius: 3439 NM

75.

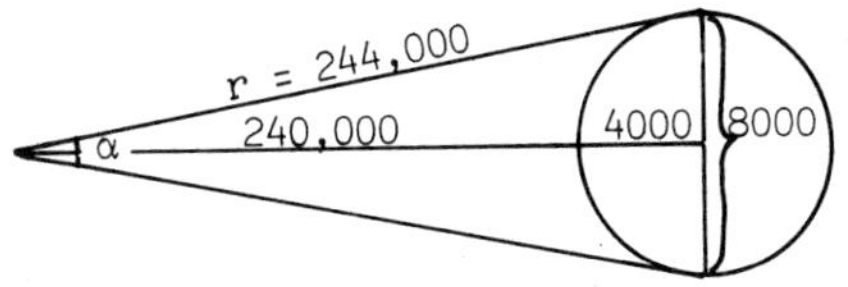

Use $\alpha = \frac{s}{r}$.

75. (continued)

The arc, s, is approximately 8000 miles, the approximate length of the earth's diameter. The radius, r, is the sum of the earth's distance away from the astronaut, 240,000 miles, and the earth's radius, 4000 miles. Thus r is approximately 244,000 miles.

$\alpha = \frac{s}{r}$

$\alpha = \frac{8{,}000}{244{,}000}$ (Substituting 8000 for s and 244,000 for r)

$\alpha \approx 0.03$ radian

76. 25,000 miles

77.

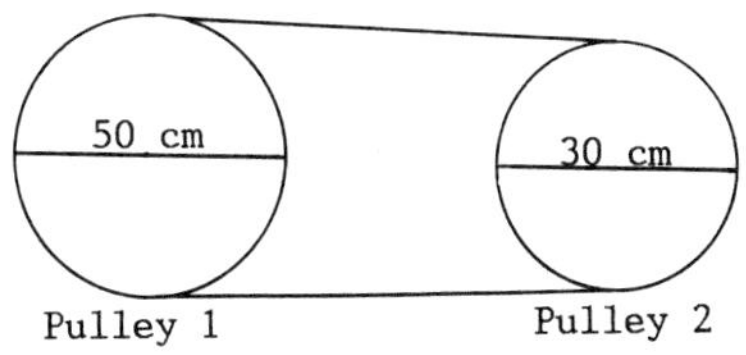

Since the pulleys are connected by the same belt, their linear speed v, will be the same.

$v = r_1\omega_1 = r_2\omega_2$

Then $\frac{r_1}{r_2} = \frac{\omega_2}{\omega_1}$

Find ω_1, the angular speed of the larger pulley, in radians/sec. The larger pulley makes 12 revolutions per minute (each revolution is 2π radians).

$\omega_1 = \frac{12 \cdot 2\pi \text{ radians}}{1 \text{ min}}$

$= \frac{24\pi \text{ radians}}{1 \text{ min}}$

$= \frac{24\pi \text{ radians}}{1 \text{ min}} \cdot \frac{1 \text{ min}}{60 \text{ sec}}$ (Multiplying by 1)

$= \frac{1.256 \text{ radians}}{1 \text{ sec}}$

$\frac{r_1}{r_2} = \frac{\omega_2}{\omega_1}$

$\frac{25}{15} = \frac{\omega_2}{1.256}$ (Substituting)

$1.256 \cdot \frac{25}{15} = \omega_2$

$2.093 = \omega_2$

The smaller pulley has an angular speed of 2.093 radians/sec.

78. 1.675 radians/sec

79. a) Angular acceleration $\approx \frac{(2500 - 800) \text{ rpm}}{4.3 \text{ sec}}$

≈ 395.35 rpm/sec

79. (continued)

b) $395.35\ \frac{\text{rpm}}{\text{sec}} = 395.35\ \frac{\text{rev}}{\text{min}} \cdot \frac{1}{\text{sec}}$

$= 395.35\ \frac{2\pi \text{ radians}}{\text{min}} \cdot \frac{1}{\text{sec}} \cdot \frac{1 \text{ min}}{60 \text{ sec}}$

$= \frac{395.35 \times 2}{60}\ \pi \cdot \frac{\text{radians}}{\text{sec}^2}$

≈ 41.38 radians/sec^2

80. a) 12.5 knots/sec

b) 1.01×10^{-6} radians/sec^2

Exercise Set 7.5

1. $\sin \pi = \sin 180° = 0$

2. 0

3. $\tan \frac{3}{4}\pi = \tan 135°$

The reference angle for $\frac{3}{4}\pi$ is $\frac{\pi}{4}$, or 45°. Recall that the tangent function value is negative in the second quadrant. Thus,

$\tan \frac{3}{4}\pi = -1.$

4. $-\frac{1}{\sqrt{3}}$, or $-\frac{\sqrt{3}}{3}$

5. $\cos \left(-\frac{\pi}{3}\right) = \cos (-60°)$

The reference angle for $-\frac{\pi}{3}$ is $\frac{\pi}{3}$, or 60°. The cosine function value is positive in the fourth quadrant. Thus,

$\cos \left(-\frac{\pi}{3}\right) = \frac{1}{2}.$

6. $-\frac{\sqrt{3}}{2}$

7. $\sin \frac{\pi}{2} = \sin 90° = 1$

8. $-\frac{1}{2}$

9. $\cos (-11\pi) = \cos (-11 \cdot 180°) = \cos (-1980°)$

The terminal side of the angle lies on the negative x-axis. Hence,

$\cos (-11\pi) = -1.$

10. 0

11. $\tan \frac{11}{4}\pi = \tan 495°$

The reference angle for $\frac{11}{4}\pi$ is $\frac{\pi}{4}$, or 45°. The tangent function value is negative in the second quadrant. Thus,

$\tan \frac{11}{4}\pi = -1.$

12. 1

13. $\sin \left(-\frac{9}{4}\pi\right) = \sin (-405°)$

The reference angle for $-\frac{9}{4}\pi$ is $\frac{\pi}{4}$, or 45°. The sine function value is negative in the fourth quadrant. Thus,

$\sin \left(-\frac{9}{4}\pi\right) = -\frac{1}{\sqrt{2}}$, or $-\frac{\sqrt{2}}{2}$.

14. 0

15. $\tan \frac{4}{3}\pi = \tan 240°$

The reference angle for $\frac{4}{3}\pi$ is $\frac{\pi}{3}$, or 60°. The tangent function value is positive in the third quadrant. Thus,

$\tan \frac{4}{3}\pi = \sqrt{3}.$

16. $\sqrt{3}$

17. Find sin 37 using a calculator.

With the calculator in radian mode, enter 37 and then press the [sin] key.

$\sin 37 \approx -0.6435$

18. 0.8623

19. Find cos (-10) using a calculator.

With the calculator in radian mode, enter -10 and then press the [cos] key.

$\cos (-10) \approx -0.8391$

20. 0.7568

21. Find tan 5π using a calculator.

Using the [π] key, we find $5\pi \approx 15.7080$. Then, with the calculator in radian mode, press the [tan] key.

$\tan 5\pi = 0$

22. Undefined

23. Find sec $\left(-\frac{\pi}{5}\right)$ using a calculator.

Using the [π] key we find $-\frac{\pi}{5} \approx -0.6283$.
The secant function value can be found by taking the reciprocal of the cosine function value. Using radian mode, press the [cos] key and then press the reciprocal key [1/x].

$\sec\left(-\frac{\pi}{5}\right) \approx 1.236$

24. 1.051

25. Find cot 1000 using a calculator.

The cotangent function value can be found by taking the reciprocal of the tangent function value. With the calculator in radian mode, enter 1000, press the [tan] key, and then press the reciprocal key [1/x].

cos 1000 ≈ 0.6801

26. 0.4630

27. Find sec $\frac{10\pi}{7}$ using a calculator.

Using the [π] key, we find $\frac{10\pi}{7} \approx 4.4880$.
With the calculator in radian mode, press the [cos] key and then press the reciprocal key [1/x].
$\left(\sec\theta = \frac{1}{\cos\theta}\right)$

$\sec\frac{10\pi}{7} \approx -4.494$

28. 0.7975

29. Find cos (-13π) using a calculator.

Using the [π] key, we find $-13\pi \approx -40.8407$. Then press the [cos] key.

$\cos(-13\pi) = -1$

30. 0

31. Find tan 2.5 using a calculator.

Enter 2.5 and then press the [tan] key.
tan 2.5 ≈ -0.7470

32. -0.5303

33. See graph in text.

34.

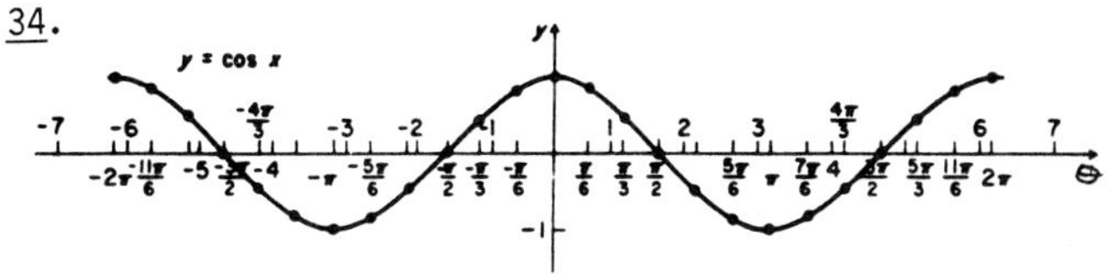

35. See graph in text.

36. a) Same as Exercise 34.
b) Same as a).
c)

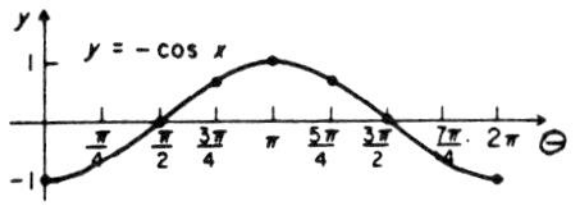

d) Same

37. See graph in text.

38. a) Same as Exercise 33.
b)

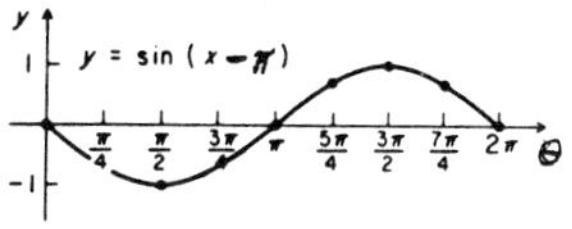

c) Same as b).
d) Same

39. See graph in text.

40. a) Same as Exercise 34.
b)

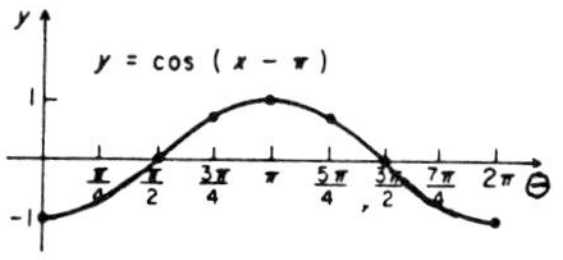

c) Same as b).
d) Same

41. $\cos(-\theta) \equiv \cos\theta$

42. $\sin(-\theta) \equiv -\sin\theta$

43. $\sin(\theta + \pi) \equiv -\sin\theta$

44. $\sin(\theta - \pi) \equiv -\sin\theta$

45. $\cos(\pi - \theta) \equiv -\cos\theta$

46. $\sin(\pi - \theta) \equiv \sin\theta$

47. $\cos(\theta + 2k\pi) \equiv \cos\theta$

48. $\sin(\theta + 2k\pi) \equiv \sin\theta$

49. $\cos(\theta - \pi) \equiv -\cos\theta$

50. $\cos(\theta + \pi) \equiv -\cos\theta$

51. $\cot\frac{\pi}{4} = \dfrac{\cos\frac{\pi}{4}}{\sin\frac{\pi}{4}} = \dfrac{\frac{\sqrt{2}}{2}}{\frac{\sqrt{2}}{2}} = 1$

52. -1

53. $\tan\frac{\pi}{6} = \dfrac{\sin\frac{\pi}{6}}{\cos\frac{\pi}{6}} = \dfrac{\frac{1}{2}}{\frac{\sqrt{3}}{2}} = \dfrac{1}{\sqrt{3}}$, or $\dfrac{\sqrt{3}}{3}$

54. $-\sqrt{3}$

55. $\sec\frac{\pi}{4} = \dfrac{1}{\cos\frac{\pi}{4}} = \dfrac{1}{\frac{\sqrt{2}}{2}} = \dfrac{2}{\sqrt{2}} = \sqrt{2}$

56. $\sqrt{2}$

57. $\tan\frac{3}{2}\pi = \dfrac{\sin\frac{3}{2}\pi}{\cos\frac{3}{2}\pi} = \dfrac{-1}{0}$

Division by 0 is undefined. Therefore, $\tan\frac{3}{2}\pi$ is undefined.

58. Undefined

59. $\tan\frac{2}{3}\pi = \dfrac{\sin\frac{2}{3}\pi}{\cos\frac{2}{3}\pi} = \dfrac{\frac{\sqrt{3}}{2}}{-\frac{1}{2}} = -\sqrt{3}$

60. $\dfrac{1}{\sqrt{3}}$, or $\dfrac{\sqrt{3}}{3}$

61. $\sec\left(-\frac{19\pi}{4}\right) = \dfrac{1}{\cos\left(-\frac{19\pi}{4}\right)} = \dfrac{1}{-\frac{\sqrt{2}}{2}} = -\dfrac{2}{\sqrt{2}}$

$= -\sqrt{2}$

62. $-\dfrac{2}{\sqrt{3}}$, or $-\dfrac{2\sqrt{3}}{3}$

63. $\csc 9\pi = \dfrac{1}{\sin 9\pi} = \dfrac{1}{0}$

Division by 0 is undefined. Therefore, $\csc 9\pi$ is undefined.

64. 1

65. $\cot\frac{11}{6}\pi = \dfrac{\cos\frac{11}{6}\pi}{\sin\frac{11}{6}\pi} = \dfrac{\frac{\sqrt{3}}{2}}{-\frac{1}{2}} = -\sqrt{3}$

66. -1

67. See table in text.

$\frac{\pi}{16}$: $\tan\frac{\pi}{16} = \dfrac{\sin(\pi/16)}{\cos(\pi/16)} \approx \dfrac{0.19509}{0.98079} \approx 0.19891$

$\cot\frac{\pi}{16} = \dfrac{\cos(\pi/16)}{\sin(\pi/16)} \approx \dfrac{0.98079}{0.19509} \approx 5.02737$

$\sec\frac{\pi}{16} = \dfrac{1}{\cos(\pi/16)} \approx \dfrac{1}{0.98079} \approx 1.01959$

$\csc\frac{\pi}{16} = \dfrac{1}{\sin(\pi/16)} = \dfrac{1}{0.19509} = 5.12584$

$\frac{\pi}{6}$: $\sin\frac{\pi}{6} = \frac{1}{2} = 0.50000$

$\cos\frac{\pi}{6} = \dfrac{\sqrt{3}}{2} \approx \dfrac{1.73205}{2} \approx 0.86603$

$\tan\frac{\pi}{6} = \dfrac{\sin(\pi/6)}{\cos(\pi/6)} \approx \dfrac{0.50000}{0.86603} \approx 0.57735$

$\cot\frac{\pi}{6} = \dfrac{\cos(\pi/6)}{\sin(\pi/6)} \approx \dfrac{0.86603}{0.50000} \approx 1.73206$

$\sec\frac{\pi}{6} = \dfrac{1}{\cos(\pi/6)} \approx \dfrac{1}{0.86603} \approx 1.15469$

$\csc\frac{\pi}{6} = \dfrac{1}{\sin(\pi/6)} = \dfrac{1}{0.50000} = 2.00000$

$\frac{\pi}{4}$: $\sin\frac{\pi}{4} = \dfrac{\sqrt{2}}{2} \approx \dfrac{1.41421}{2} \approx 0.70711$

$\cos\frac{\pi}{4} = \dfrac{\sqrt{2}}{2} \approx 0.70711$

$\tan\frac{\pi}{4} = \dfrac{\sin(\pi/4)}{\cos(\pi/4)} \approx \dfrac{0.70711}{0.70711} = 1.00000$

$\cot\frac{\pi}{4} = \dfrac{\cos(\pi/4)}{\sin(\pi/4)} \approx \dfrac{0.70711}{0.70711} = 1.00000$

$\sec\frac{\pi}{4} = \dfrac{1}{\cos(\pi/4)} \approx \dfrac{1}{0.70711} = 1.41421$

$\csc\frac{\pi}{4} = \dfrac{1}{\sin(\pi/4)} \approx \dfrac{1}{0.70711} = 1.41421$

68.

	$-\frac{\pi}{16}$	$-\frac{\pi}{8}$	$-\frac{\pi}{6}$	$-\frac{\pi}{4}$	$-\frac{\pi}{3}$
sin	-0.19509	-0.38268	-0.50000	-0.70711	-0.86603
cos	0.98079	0.92388	0.86603	0.70711	0.50000
tan	-0.19891	-0.41421	-0.57735	-1.00000	-1.73206
cot	-5.02737	-2.41424	-1.73206	-1.00000	-0.57735
sec	1.01959	1.08239	1.15469	1.41421	2.00000
csc	-5.12584	-2.61315	-2.00000	-1.41421	-1.15469

69. See graph in text.

70. $y = \cot \theta$

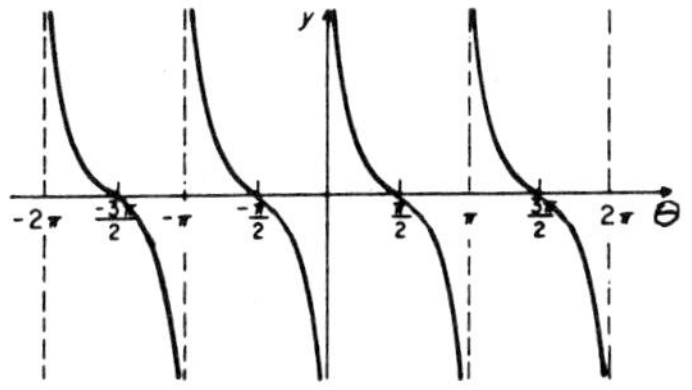

71. See graph in text.

72. $y = \csc \theta$

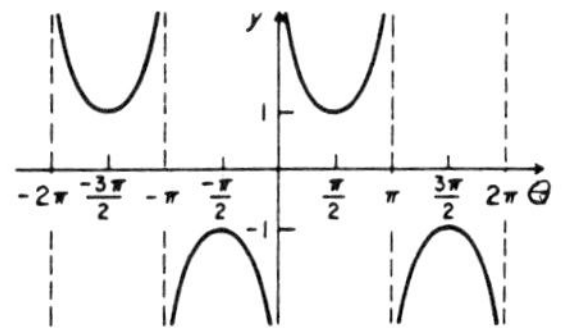

73. $\sin(-\theta) \equiv -\sin \theta$

The sine function is an odd function.

$\cos(-\theta) \equiv \cos \theta$

The cosine function is an even function.

$$\tan(-\theta) = \frac{\sin(-\theta)}{\cos(-\theta)} = \frac{-\sin \theta}{\cos \theta} = -\frac{\sin \theta}{\cos \theta} = -\tan \theta$$

The tangent function is an odd function.

$$\cot(-\theta) = \frac{\cos(-\theta)}{\sin(-\theta)} = \frac{\cos \theta}{-\sin \theta} = -\frac{\cos \theta}{\sin \theta} = -\cot \theta$$

The cotangent function is an odd function.

$$\sec(-\theta) = \frac{1}{\cos(-\theta)} = \frac{1}{\cos \theta} = \sec \theta$$

The secant function is an even function.

$$\csc(-\theta) = \frac{1}{\sin(-\theta)} = \frac{1}{-\sin \theta} = -\frac{1}{\sin \theta} = -\csc \theta$$

The cosecant function is an odd function.

74. See Exercise 73.

75. See the graph and the list of properties for each function in section 7.5 in the text. The sin, cos, sec, and csc functions are periodic, with period 2π.

76. tangent, cotangent

77.

Function	I	II	III	IV
$\sin \theta$	+	+	-	-
$\cos \theta$	+	-	-	+
$\tan \theta = \frac{\sin \theta}{\cos \theta}$	+	-	+	-

The tangent function is positive in the first and third quadrants and negative in the second and fourth quadrants.

78. Positive: I and III

Negative: II and IV

79.

	I	II	III	IV
$\cos \theta$	+	-	-	+
$\sec \theta = \frac{1}{\cos \theta}$	+	-	-	+

The secant function is positive in the first and fourth quadrants and negative in the second and third quadrants.

80. Positive: I and II

Negative: III and IV

81. See text answer section.

82. $\csc(-\theta) \equiv \frac{1}{\sin(-\theta)} \equiv \frac{1}{-\sin \theta} \equiv -\csc \theta$

83. See text answer section.

84. $\cot(\theta - \pi) \equiv \frac{\cos(\theta - \pi)}{\sin(\theta - \pi)} \equiv \frac{-\cos \theta}{-\sin \theta} \equiv \cot \theta$

85. See text answer section.

86. $\tan(\pi - \theta) \equiv \frac{\sin(\pi - \theta)}{\cos(\pi - \theta)} \equiv \frac{\sin \theta}{-\cos \theta} \equiv -\tan \theta$

87. a) $\sin \frac{\pi}{2} = 1$

$\sin\left(\frac{\pi}{2} + 2\pi\right) = 1$ $\quad$ $\sin\left(\frac{\pi}{2} - 2\pi\right) = 1$

$\sin\left(\frac{\pi}{2} + 2\cdot 2\pi\right) = 1$ $\quad$ $\sin\left(\frac{\pi}{2} - 2\cdot 2\pi\right) = 1$

$\sin\left(\frac{\pi}{2} + 3\cdot 2\pi\right) = 1$ $\quad$ $\sin\left(\frac{\pi}{2} - 3\cdot 2\pi\right) = 1$

$\sin\left(\frac{\pi}{2} + k\cdot 2\pi\right) = 1$, k any integer

Thus

$x = \frac{\pi}{2} + 2k\pi$, k any integer

87. (continued)

b) $\sin \frac{3}{2}\pi = -1$

$\sin\left(\frac{3}{2}\pi + 2\pi\right) = -1 \qquad \sin\left(\frac{3}{2}\pi - 2\pi\right) = -1$

$\sin\left(\frac{3}{2}\pi + 2\cdot 2\pi\right) = -1 \qquad \sin\left(\frac{3}{2}\pi - 2\cdot 2\pi\right) = -1$

$\sin\left(\frac{3}{2}\pi + 3\cdot 2\pi\right) = -1 \qquad \sin\left(\frac{3}{2}\pi - 3\cdot 2\pi\right) = -1$

$\sin\left(\frac{3}{2}\pi + k\cdot 2\pi\right) = -1$, k any integer

Thus

$x = \frac{3}{2}\pi + 2k\pi$, k any integer

88. a) $2k\pi$, k any integer
b) $(2k + 1)\pi$, k any integer

89. For which numbers, θ, is $\sin \theta = 0$?

$\sin 0 = 0$

$\sin \pi = 0 \qquad \sin(-\pi) = 0$

$\sin 2\pi = 0 \qquad \sin(-2\pi) = 0$

$\sin 3\pi = 0 \qquad \sin(-3\pi) = 0$

$\sin 4\pi = 0 \qquad \sin(-4\pi) = 0$

$\sin k\pi = 0$, k any integer

Thus

$\theta = k\pi$, k any integer

90. $\theta = \frac{\pi}{2} + k\pi$, k any integer

91. $f(\theta) = \theta^2 + 2\theta \qquad g(\theta) = \cos \theta$

$f \circ g(\theta) = f(g(\theta)) = f(\cos\theta)$
$= (\cos\theta)^2 + 2\cos\theta$
$= \cos^2\theta + 2\cos\theta$

$g \circ f(\theta) = g(f(\theta)) = g(\theta^2 + 2\theta)$
$= \cos(\theta^2 + 2\theta)$

92. For example,

$\sin\left(\frac{\pi}{4} + \frac{\pi}{4}\right) = \sin\frac{\pi}{2} = 1$, but

$\sin\frac{\pi}{4} + \sin\frac{\pi}{4} = \frac{\sqrt{2}}{2} + \frac{\sqrt{2}}{2} = \sqrt{2}$

93. See text answer section.

94.

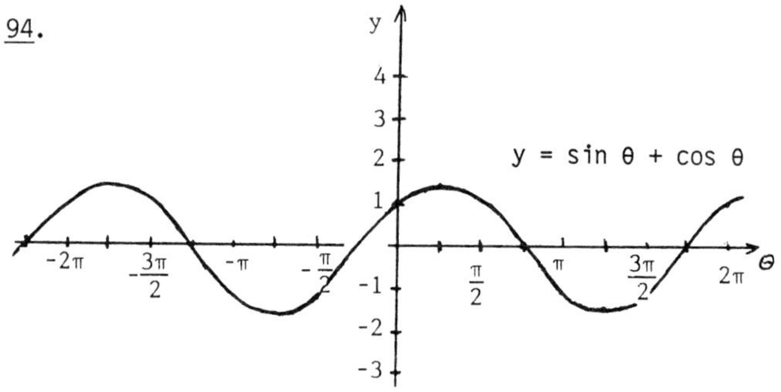

95. $y = \sin^2\theta = \sin\theta \cdot \sin\theta$

It is helpful to graph the function.

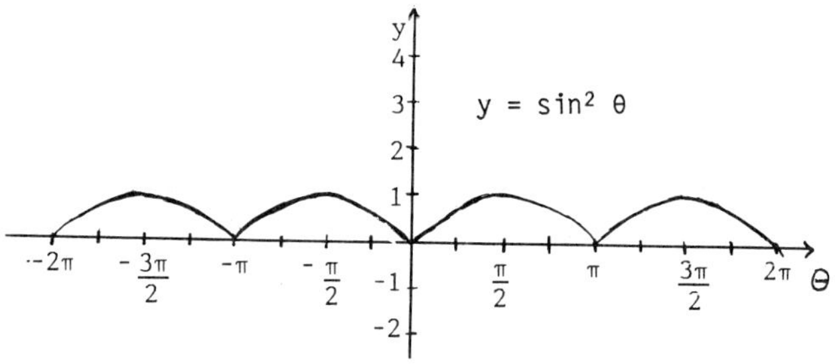

Domain: Set of all real numbers

Range: Set of all real numbers from 0 to 1 inclusive.

Period: π $[\sin^2\theta = \sin^2(\theta + k\pi)$, k any integer$]$

Amplitude: $\frac{1}{2}$ $\left(\frac{1}{2} \text{ of } 1 - 0 \text{ is } \frac{1}{2}\right)$

96. Domain: Set of all real numbers

Range: Set of all real numbers from 1 to 2 inclusive.

Period: π

Amplitude: $\frac{1}{2}$

97. $f(x) = \sqrt{\cos\theta}$

Solve $\cos\theta \geqslant 0$.

Since $\cos\theta \geqslant 0$ when θ is in the intervals $\left[-\frac{\pi}{2} + 2k\pi, \frac{\pi}{2} + 2k\pi\right]$, the domain consists of the intervals $\left[-\frac{\pi}{2} + 2k\pi, \frac{\pi}{2} + 2k\pi\right]$.

98. The domain is the set of all real numbers except $k\pi$ for any integer k.

99. $f(\theta) = \dfrac{\sin \theta}{\cos \theta}$

Note that $\cos \theta \neq 0$ and $\cos \theta = 0$ when $\theta = \frac{\pi}{2} + k\pi$ for any integer k. Thus the domain is the set of all real numbers except $\frac{\pi}{2} + k\pi$ for any integer k.

100. The domain consists of the intervals $(0 + 2k\pi, \pi + 2k\pi)$, k any integer.

101. $f(\theta) = \dfrac{\sin \theta}{\theta}$, when $0 < \theta < \frac{\pi}{2}$

$\dfrac{\sin \pi/2}{\pi/2} \approx 0.6369$

$\dfrac{\sin 3\pi/8}{3\pi/8} \approx 0.7846$

$\dfrac{\sin \pi/4}{\pi/4} \approx 0.9008$

$\dfrac{\sin \pi/8}{\pi/8} \approx 0.9750$

The limit of $\dfrac{\sin \theta}{\theta}$ as θ approaches 0 is 1.

102. $\left[-\frac{3}{4}\pi, \frac{\pi}{4}\right]$

103. See text answer section.

104. The graph of $\tan(\theta + \pi)$ is like that of $\tan \theta$, but moved π units to the left. This graph is identical to that of $\tan \theta$.

105. See text answer section.

106. If the graph of $\sec \theta$ were translated to the right $\frac{\pi}{2}$ units, the graph of $\csc \theta$ would be obtained. There are other descriptions.

107. Studying the graphs of the six circular functions we see that

$\sin \theta = 0$ when $\theta = 0, \pm\pi, \pm 2\pi, \pm 3\pi, \ldots$

$\cos \theta = 0$ when $\theta = \pm\frac{\pi}{2}, \pm\frac{3}{2}\pi, \pm\frac{5}{2}\pi, \ldots$

$\tan \theta = 0$ when $\theta = 0, \pm\pi, \pm 2\pi, \pm 3\pi, \ldots$

$\cot \theta = 0$ when $\theta = \pm\frac{\pi}{2}, \pm\frac{3}{2}\pi, \pm\frac{5}{2}\pi, \ldots$

$\sec \theta$ is never 0.

$\left[\text{Think: } \sec \theta \equiv \dfrac{1}{\cos \theta}; \dfrac{1}{\cos \theta} \text{ is never } 0\right]$

$\csc \theta$ is never 0.

$\left[\text{Think: } \csc \theta \equiv \dfrac{1}{\sin \theta}; \dfrac{1}{\sin \theta} \text{ is never } 0\right]$

In the sine and tangent functions, the same inputs give outputs of 0. Also in the cosine and cotangent functions, the same inputs give outputs of 0.

108. The tangent and secant functions have the same asymptotes. The cotangent and cosecant functions have the same asymptotes. The inputs producing zero outputs of the first two occur where the asymptotes of the latter occur, and conversely.

109. See text answer section.

110.

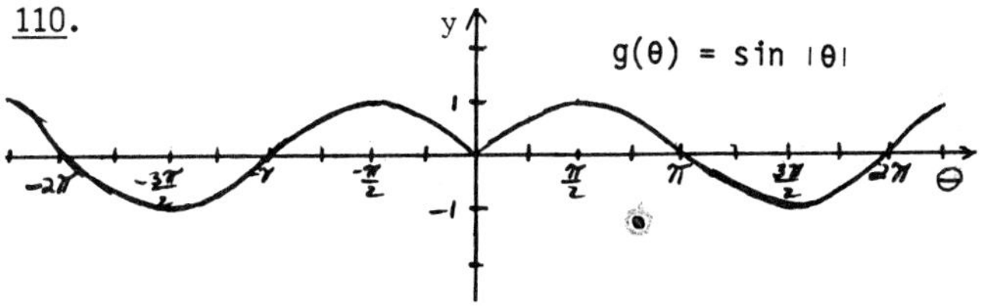

Exercise Set 7.6

1. $1 + \cot^2\theta \equiv \csc^2\theta$ (Pythagorean Identity)

$1 \equiv \csc^2\theta - \cot^2\theta$ (Adding $-\cot^2\theta$)

or

$\cot^2\theta \equiv \csc^2\theta - 1$ (Adding -1)

2. $\sec^2\theta - 1 \equiv \tan^2\theta$, $\sec^2\theta - \tan^2\theta \equiv 1$

3. a) $\csc^2\theta \equiv 1 + \cot^2\theta$ (Pythagorean Identity)

$\csc \theta \equiv \pm\sqrt{1 + \cot^2\theta}$

b) $1 + \cot^2\theta \equiv \csc^2\theta$ (Pythagorean Identity)

$\cot^2\theta \equiv \csc^2\theta - 1$

$\cot \theta \equiv \pm\sqrt{\csc^2\theta - 1}$

4. a) $\tan \theta \equiv \pm\sqrt{\sec^2\theta - 1}$

b) $\sec \theta \equiv \pm\sqrt{1 + \tan^2\theta}$

5. $\tan\left(\theta - \frac{\pi}{2}\right) \equiv \dfrac{\sin\left(\theta - \frac{\pi}{2}\right)}{\cos\left(\theta - \frac{\pi}{2}\right)}$ (Definition)

$\equiv \dfrac{-\cos \theta}{\sin \theta}$ (Cofunction identities)

$\equiv -\dfrac{\cos \theta}{\sin \theta}$

$\equiv -\cot \theta$

$\tan\left(\theta - \frac{\pi}{2}\right) \equiv -\cot \theta$

6. $\cot\left(\theta + \frac{\pi}{2}\right) \equiv \dfrac{\cos\left(\theta + \frac{\pi}{2}\right)}{\sin\left(\theta + \frac{\pi}{2}\right)}$

$\equiv \dfrac{-\sin \theta}{\cos \theta} \equiv -\dfrac{\sin \theta}{\cos \theta} \equiv -\tan \theta$

$\cot\left(\theta + \frac{\pi}{2}\right) \equiv -\tan \theta$

7. $\sec\left(\frac{\pi}{2} - \theta\right) \equiv \frac{1}{\cos\left(\frac{\pi}{2} - \theta\right)}$ (Definition)

$\equiv \frac{1}{\cos\left[-\left(\theta - \frac{\pi}{2}\right)\right]}$

$\equiv \frac{1}{\cos\left(\theta - \frac{\pi}{2}\right)}$ (Even function)

$\equiv \frac{1}{\sin\theta}$ (Cofunction identity)

$\equiv \csc\theta$

$\sec\left(\frac{\pi}{2} - \theta\right) \equiv \csc\theta$

8. $\csc\left(\frac{\pi}{2} - \theta\right) \equiv \frac{1}{\sin\left(\frac{\pi}{2} - \theta\right)}$

$\equiv \frac{1}{\sin\left[-\left(\theta - \frac{\pi}{2}\right)\right]}$

$\equiv \frac{1}{-\sin\left(\theta - \frac{\pi}{2}\right)} \equiv \frac{1}{\cos\theta} \equiv \sec\theta$

$\csc\left(\frac{\pi}{2} - \theta\right) \equiv \sec\theta$

9. $\sin\left(\theta \pm \frac{\pi}{2}\right) \equiv \pm\cos\theta$ (Developed in text)

$\cos\left(\theta \pm \frac{\pi}{2}\right) \equiv \mp\sin\theta$ (Developed in text)

$\tan\left(\theta \pm \frac{\pi}{2}\right) \equiv -\cot\theta$ (See Example 3, Section 7.6 and Exercise 5 above)

$\cot\left(\theta \pm \frac{\pi}{2}\right) \equiv -\tan\theta$ (Developed below)

$\sec\left(\theta \pm \frac{\pi}{2}\right) \equiv \mp\csc\theta$ (Developed below)

$\csc\left(\theta \pm \frac{\pi}{2}\right) \equiv \pm\sec\theta$ (Developed below)

$\cot\left(\theta + \frac{\pi}{2}\right) \equiv \frac{1}{\tan\left(\theta + \frac{\pi}{2}\right)} \equiv \frac{1}{-\cot\theta} \equiv -\tan\theta$

$\cot\left(\theta - \frac{\pi}{2}\right) \equiv \frac{1}{\tan\left(\theta - \frac{\pi}{2}\right)} \equiv \frac{1}{-\cot\theta} \equiv -\tan\theta$

$\sec\left(\theta + \frac{\pi}{2}\right) \equiv \frac{1}{\cos\left(\theta + \frac{\pi}{2}\right)} \equiv \frac{1}{-\sin\theta} \equiv -\csc\theta$

$\sec\left(\theta - \frac{\pi}{2}\right) \equiv \frac{1}{\cos\left(\theta - \frac{\pi}{2}\right)} \equiv \frac{1}{\sin\theta} \equiv \csc\theta$

$\csc\left(\theta + \frac{\pi}{2}\right) \equiv \frac{1}{\sin\left(\theta + \frac{\pi}{2}\right)} \equiv \frac{1}{\cos\theta} \equiv \sec\theta$

$\csc\left(\theta - \frac{\pi}{2}\right) \equiv \frac{1}{\sin\left(\theta - \frac{\pi}{2}\right)} \equiv \frac{1}{-\cos\theta} \equiv -\sec\theta$

10. Quadrants I & II

11. $\sin\theta = 0.1425$ (Quadrant II)

In quadrant II the sin and csc are positive; the cos, sec, tan, and cot are negative.

Since $\sin^2\theta + \cos^2\theta = 1$, or $\cos\theta = \pm\sqrt{1 - \sin^2\theta}$, and we are restricted to quadrant II, we get

$\cos\theta = -\sqrt{1 - \sin^2\theta}$

$\cos\theta = -\sqrt{1 - (0.1425)^2}$ (Substituting)

≈ -0.9898

The others follow from the definitions.

$\tan\theta = \frac{\sin\theta}{\cos\theta} = \frac{0.1425}{-0.9898} \approx -0.1440$

$\cot\theta = \frac{\cos\theta}{\sin\theta} = \frac{-0.9898}{0.1425} \approx -6.9460$

$\sec\theta = \frac{1}{\cos\theta} = \frac{1}{-0.9898} \approx -1.0103$

$\csc\theta = \frac{1}{\sin\theta} = \frac{1}{0.1425} \approx 7.0175$

12. $\sin\theta = -0.6253$

$\cos\theta = -0.7804$

$\cot\theta = 1.2481$

$\sec\theta = -1.2814$

$\csc\theta = -1.5993$

13. Given: $\cot\theta = 0.7534$

$\sin\theta > 0$

Since the sine function is positive in only quadrants I and II and the cotangent function is positive in only quadrants I and III, this problem concerns function values in only quadrant I.

Since $\cot\theta = 0.7534$,

then $\tan\theta = \frac{1}{0.7534} \approx 1.3273$.

Since $1 + \cot^2\theta \equiv \csc^2\theta$,

then $\pm\sqrt{1 + \cot^2\theta} \equiv \csc\theta$.

Since we are restricted to quadrant I, we find

$\csc\theta = \sqrt{1 + \cot^2\theta}$

$= \sqrt{1 + (0.7534)^2}$

≈ 1.2520

Since $\csc\theta = 1.2520$,

then $\sin\theta = \frac{1}{1.2520} \approx 0.7987$.

Since $1 + \tan^2\theta \equiv \sec^2\theta$,

then $\pm\sqrt{1 + \tan^2\theta} \equiv \sec\theta$.

Since we are restricted to quadrant I, we find

$\sec\theta = \sqrt{1 + \tan^2\theta}$

$= \sqrt{1 + (1.3273)^2}$

≈ 1.6618

Since $\sec\theta = 1.6618$,

then $\cos\theta = \frac{1}{1.6618} \approx 0.6018$.

14. a) 0.92388, b) -0.38268, c) 0.92388
d) -0.92388, e) 0.38268

15.

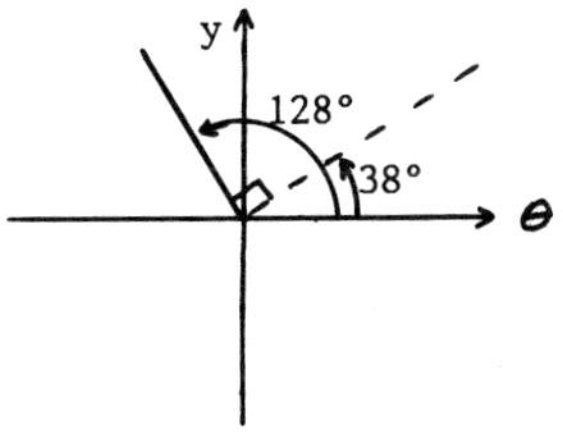

$\sin 128° = \sin (90° + 38°) = \sin \left[\frac{\pi}{2} + 38°\right]$
$= \cos 38°$
$= 0.78801$

Using $\cos^2\theta + \sin^2\theta = 1$, find cos 128°.
$(\cos 128°)^2 + (\sin 128°)^2 = 1$
$(\cos 128°)^2 + (0.78801)^2 = 1$
$(\cos 128°)^2 \approx 1 - 0.62096$
$\cos 128° \approx -0.61566$
(cosine is negative in quadrant II)

$\tan 128° = \frac{0.78801}{-0.61566} \approx -1.27994$

$\cot 128° = \frac{1}{-1.27994} \approx -0.78129$

$\sec 128° = \frac{1}{-0.61566} \approx -1.62427$

$\csc 128° = \frac{1}{0.78801} \approx 1.26902$

16. sin 343° = -0.29237 csc 343° = -3.42030
cos 343° = 0.95631 sec 343° = 1.04569
tan 343° = -0.30573 cot 343° = -3.27085

17. A circle of radius r is $x^2 + y^2 = r^2$. Compare with circle of radius 1. By similar triangles, $x = r \cos \theta$ and $y = r \sin \theta$ so the coordinates of P are $(r \cos \theta, r \sin \theta)$.

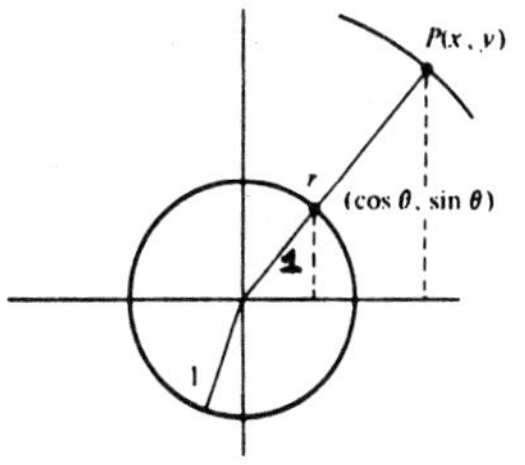

Exercise Set 7.7

1. $y = 2 + \sin x$, or $y - 2 = \sin x$
(A = 2, B = C = 1, D = 0)

The graph of $y = 2 + \sin x$ is a translation of $y = \sin x$ upward 2 units. See graph in text.

2.

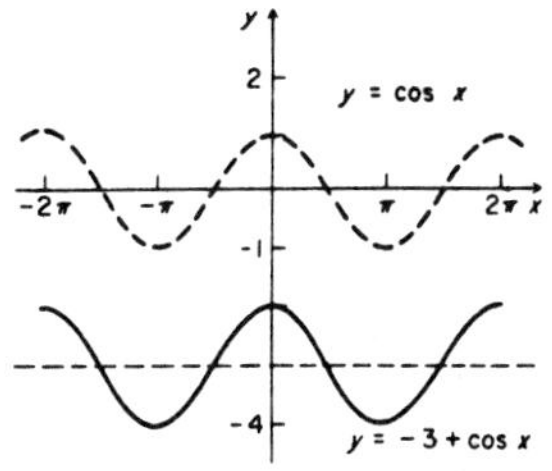

3. $y = \frac{1}{2} \sin x$ $\left[A = 0, B = \frac{1}{2}, C = 1, D = 0\right]$

The graph of $y = \frac{1}{2} \sin x$ is a vertical shrinking of the graph of $y = \sin x$. The amplitude of this function is $\frac{1}{2}$. See graph in text.

4.

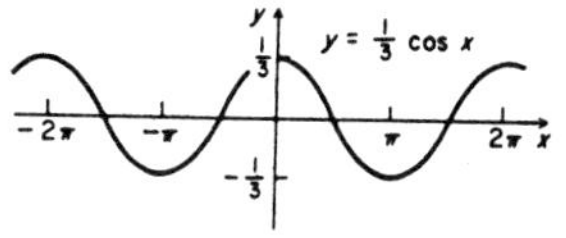

5. $y = 2 \cos x$ (A = 0, B = 2, C = 1, D = 0)

The graph of $y = 2 \cos x$ is a vertical stretching of the graph of $y = \cos x$. The amplitude of this function is 2. See graph in text.

6.

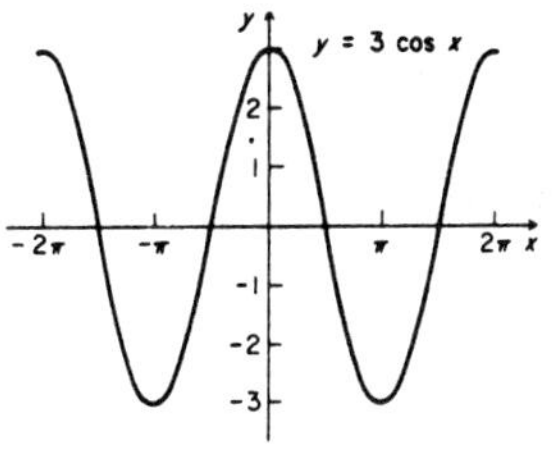

7. $y = -\frac{1}{2} \cos x$ $\left[A = 0, B = -\frac{1}{2}, C = 1, D = 0\right]$

The graph of $y = -\frac{1}{2} \cos x$ is a vertical shrinking of the graph of $y = \cos x$. The amplitude of this function is $\left|-\frac{1}{2}\right|$, or $\frac{1}{2}$. There is also a reflection across the x-axis since $B < 0$. See graph in text.

8.

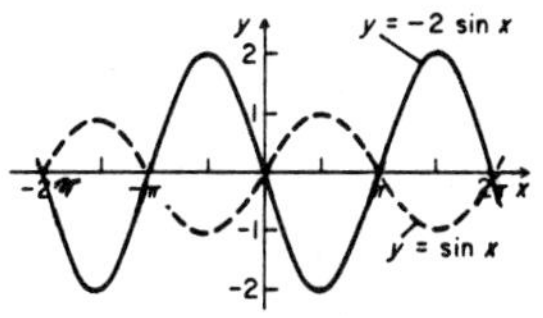

9. $y = \cos 2x$ (A = 0, B = 1, C = 2, D = 0)

The graph of $y = \cos 2x$ is a horizontal shrinking of the graph of $y = \cos x$. The period of this function is π ($2\pi \div 2 = \pi$). See graph in text.

10.

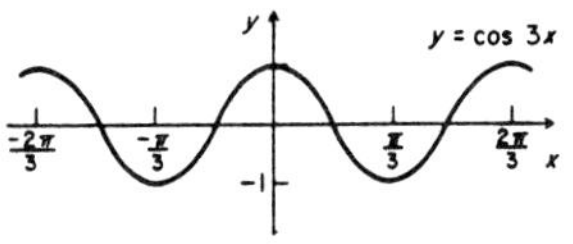

11. y = cos (-2x) (A = 0, B = 1, C = -2, D = 0)

The graph of y = cos (-2x) is a horizontal shrinking of the graph of y = cos x. The period of this function is π, $(2\pi \div 2 = \pi)$. There is also a reflection across the y-axis since C < 0. Note that the graphs of y = cos 2x and y = cos (-2x) are identical. The cosine function is an even function, cos (-2x) = cos 2x. See graph in text.

12. The graph of y = cos (-3x) is the same as the graph of y = cos 3x. See the graph in Exercise 10 above.

13. $y = \sin \frac{1}{2} x$ $\left[A = 0,\ B = 1,\ C = \frac{1}{2},\ D = 0\right]$

The graph of $y = \sin \frac{1}{2} x$ is a horizontal stretching of the graph of y = sin x. The period of this function is 4π, $\left[2\pi \div \frac{1}{2} = 4\pi\right]$. See graph in text.

14.

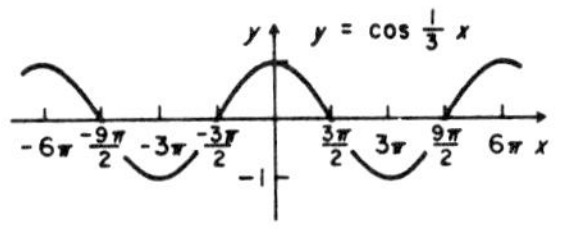

15. $y = \sin\left[-\frac{1}{2} x\right]$ $\left[A = 0,\ B = 1,\ C = -\frac{1}{2},\ D = 0\right]$

The graph of $y = \sin\left[-\frac{1}{2} x\right]$ is a horizontal stretching of the graph of y = sin x. The period of the function is 4π, $\left[2\pi \div \frac{1}{2} = 4\pi\right]$. There is also a reflection across the y-axis since C < 0. See graph in text.

16.

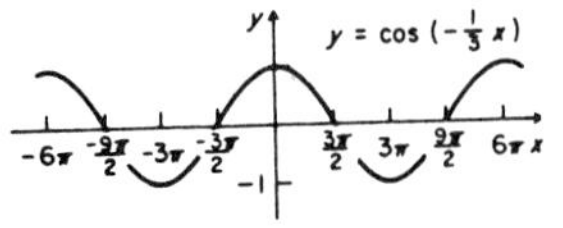

17. $y = \cos (2x - \pi)$, or $y = \cos 2\left[x - \frac{\pi}{2}\right]$

$(A = 0,\ B = 1,\ C = 2,\ D = \pi)$

The $\frac{\pi}{2}$ translates the graph of y = cos 2x (see Exercise 9) a distance of $\frac{\pi}{2}$ to the right. The period of the function is π $(2\pi \div 2 = \pi)$. See graph in text.

18.

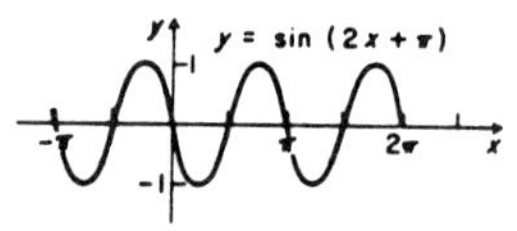

19. $y = 2 \cos \left[\frac{1}{2} x - \frac{\pi}{2}\right]$, or $y = 2 \cos \frac{1}{2} (x - \pi)$

$\left[A = 0,\ B = 2,\ C = \frac{1}{2},\ D = \frac{\pi}{2}\right]$

The graph of $y = 2 \cos \frac{1}{2} (x - \pi)$ has an amplitude of 2, a period of 4π $\left[2\pi \div \frac{1}{2} = 4\pi\right]$, and a phase shift of π.

First graph y = cos x.

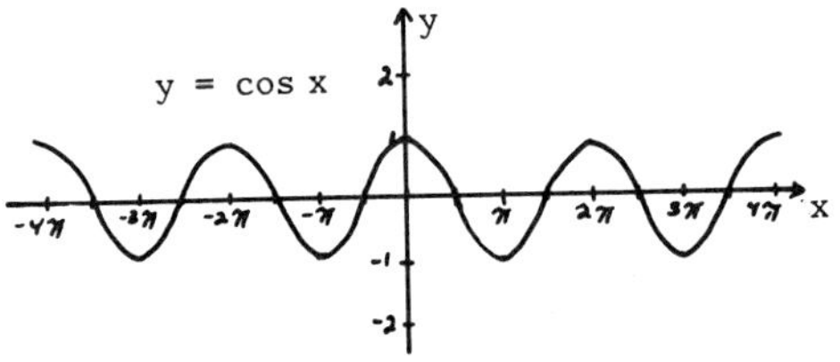

The graph of $y = \cos \frac{1}{2} x$ is a horizontal stretching of the graph of y = cos x. The period of y = cos x is 2π. The period of $y = \cos \frac{1}{2} x$ is 4π.

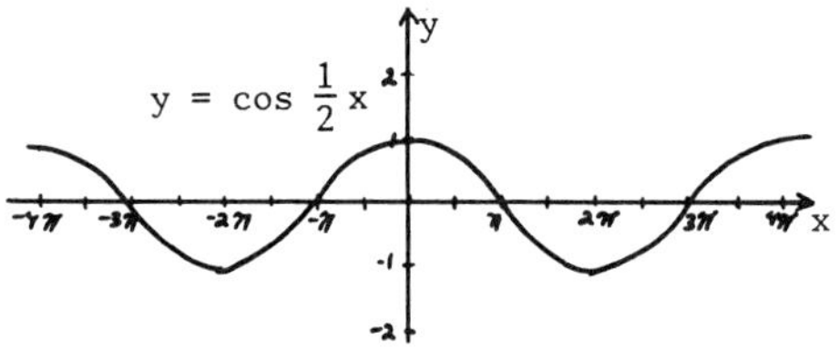

The graph of $y = 2 \cos \frac{1}{2} x$ is a vertical stretching of the graph of $y = \cos \frac{1}{2} x$. The amplitude of $y = \cos \frac{1}{2} x$ is 1. The amplitude of $y = 2 \cos \frac{1}{2} x$ is 2.

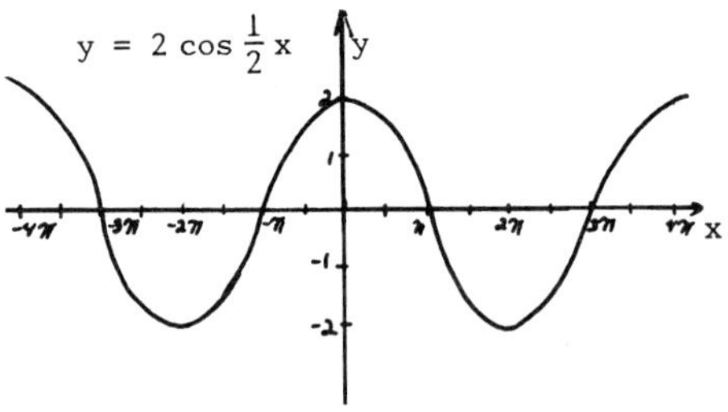

The graph of $y = 2 \cos \frac{1}{2} (x - \pi)$ is a translation of the graph of $y = 2 \cos \frac{1}{2} x$ a distance of π to the right.

19. (continued)

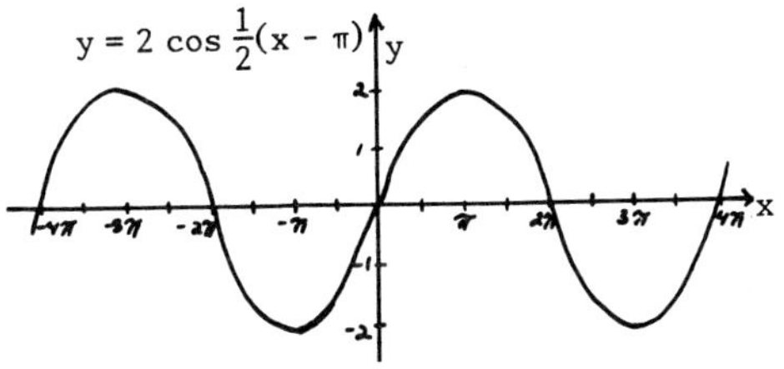

20.

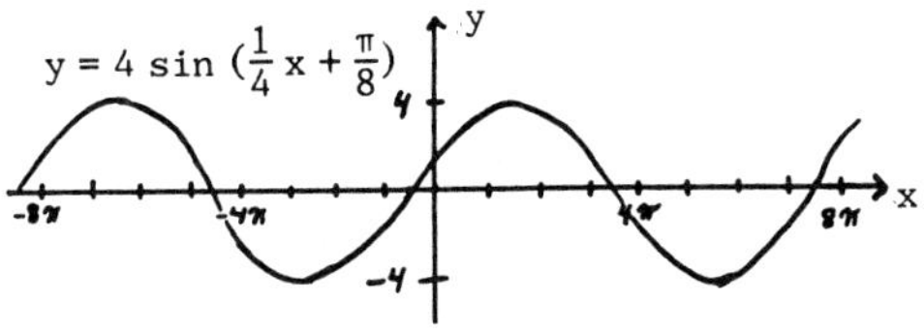

21. $y = -3\cos(4x - \pi)$, or $y = -3\cos 4\left[x - \frac{\pi}{4}\right]$

$(A = 0,\ B = -3,\ C = 4,\ D = \pi)$

The graph of $y = -3\cos 4\left[x - \frac{\pi}{4}\right]$ has an amplitude of 3, a period of $\frac{\pi}{2}$ $\left[2\pi \div 4 = \frac{\pi}{2}\right]$, and a phase shift of $\frac{\pi}{4}$.

First graph $y = \cos x$.

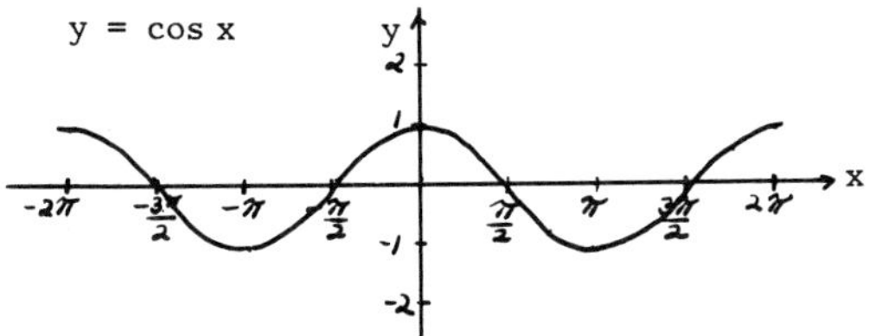

The graph of $y = \cos 4x$ is a horizontal shrinking of the graph of $y = \cos x$. The period of $y = \cos x$ is 2π. The period of $y = \cos 4x$ is $\frac{\pi}{2}$.

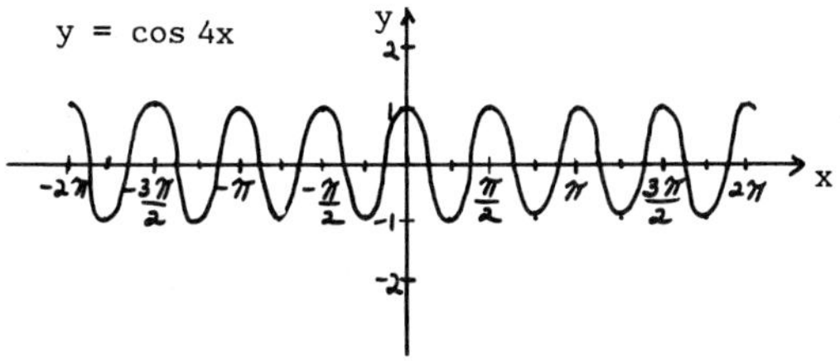

The graph of $y = -3\cos 4x$ is a vertical stretching of the graph of $y = \cos 4x$. The amplitude of $y = \cos 4x$ is 1. The amplitude of $y = -3\cos 4x$ is 3. The graph is also a reflection across the x-axis since B is negative.

21. (continued)

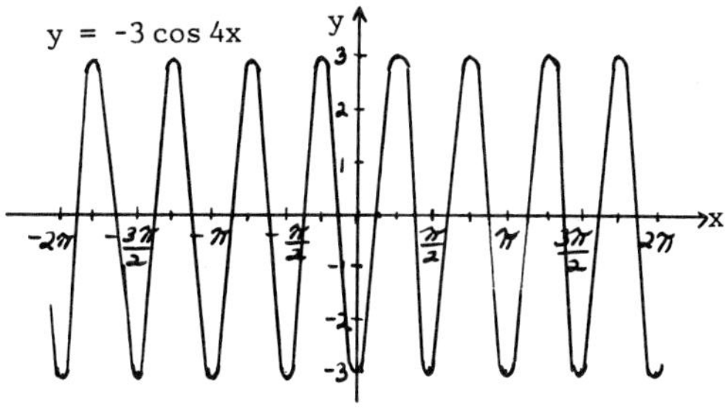

The graph of $y = -3\cos 4\left[x - \frac{\pi}{4}\right]$ is a translation of the graph of $y = -3\cos 4x$ a distance of $\frac{\pi}{4}$ to the right.

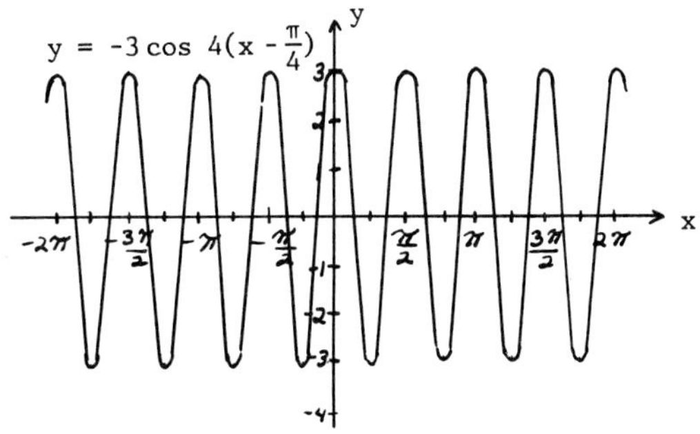

22.

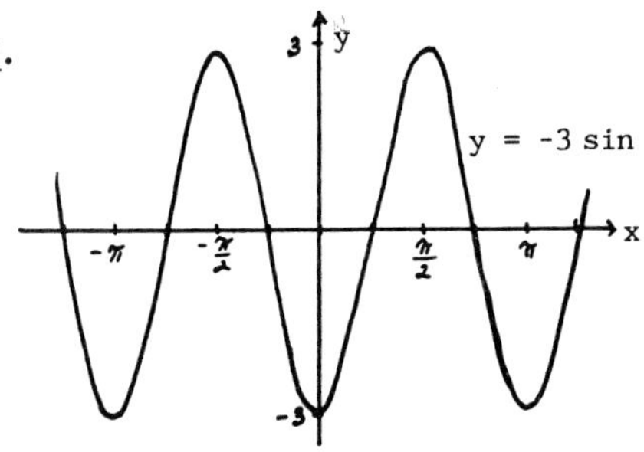

23. $y = 2\cos x + \cos 2x$

Graph $y = 2\cos x$ and $y = \cos 2x$ on the same set of axes. Then graphically add some ordinates to obtain points on the graph of $y = 2\cos x + \cos 2x$. See graph in text.

24.

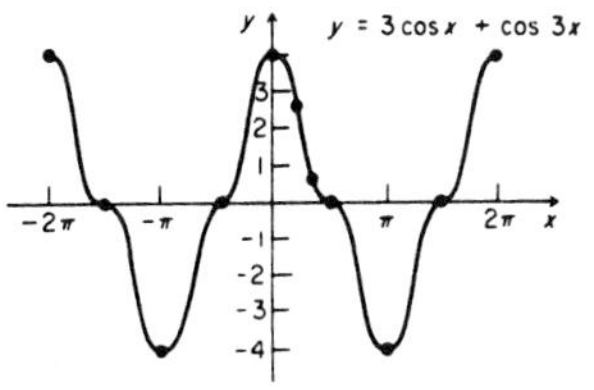

25. $y = \sin x + \cos 2x$

Graph $y = \sin x$ and $y = \cos 2x$ on the same set of axes. Then graphically add some ordinates to obtain points on the graph of $y = \sin x + \cos 2x$. See graph in text.

26.

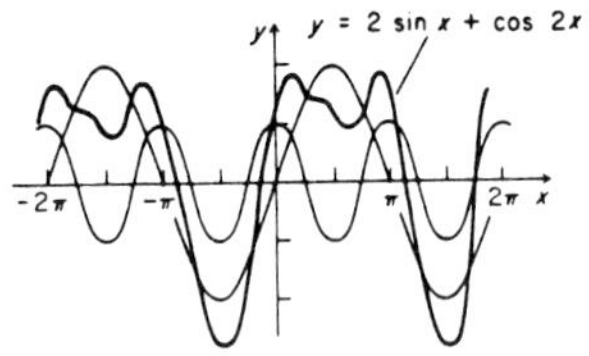

27. $y = 3 \cos x - \sin 2x$

Graph $y = 3 \cos x$ and $y = \sin 2x$ on the same set of axes. Then graphically subtract some ordinates to obtain points on the graph of $y = 3 \cos x - \sin 2x$. See graph in text.

28.

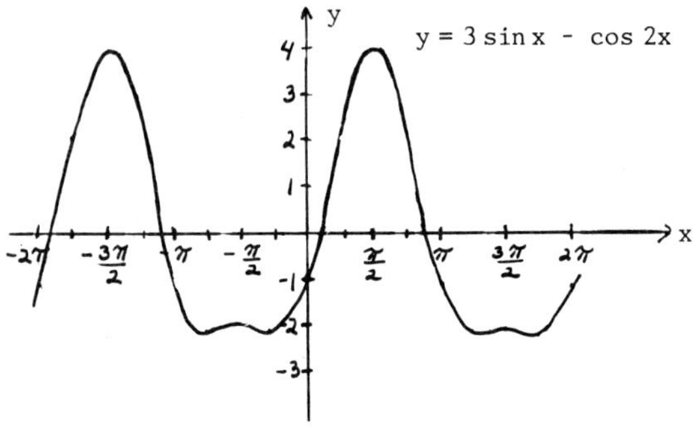

29. $y = 3 \cos \left(3x - \frac{\pi}{2}\right)$ $\left(A = 0,\ B = 3,\ C = 3,\ D = \frac{\pi}{2}\right)$

$y = 3 \cos 3\left(x - \frac{\pi}{6}\right)$

The amplitude is 3.

The period is $2\pi \div 3$, or $\frac{2}{3}\pi$.

The phase shift is $\frac{\pi}{6}$.

30. Amplitude: 4, Period: $\frac{\pi}{2}$, Phase shift: $\frac{\pi}{12}$

31. $y = -5 \cos \left(4x + \frac{\pi}{3}\right)$

$y = -5 \cos \left[4x - \left(-\frac{\pi}{3}\right)\right]$ $\left(A = 0,\ B = -5,\ C = 4\ D = -\frac{\pi}{3}\right)$

$y = -5 \cos 4\left[x - \left(-\frac{\pi}{12}\right)\right]$

The amplitude is $|-5|$, or 5.

The period is $2\pi \div 4$, or $\frac{\pi}{2}$.

The phase shift is $-\frac{\pi}{12}$.

32. Amplitude: 4, Period: $\frac{2}{5}\pi$, Phase shift: $-\frac{\pi}{10}$

33. $y = \frac{1}{2} \sin (2\pi x + \pi)$

$y = \frac{1}{2} \sin [2\pi x - (-\pi)]$

$\left(A = 0,\ B = \frac{1}{2},\ C = 2\pi,\ D = -\pi\right)$

$y = \frac{1}{2} \sin 2\pi\left[x - \left(-\frac{1}{2}\right)\right]$

The amplitude is $\frac{1}{2}$.

The period is $2\pi \div 2\pi$, or 1.

The phase shift is $-\frac{1}{2}$.

34. Amplitude: $\frac{1}{4}$, Period: 2, Phase shift: $\frac{4}{\pi}$

35. $y = \sec^2 x = \frac{1}{\cos^2 x}$

First make a table of values. Since the cosine and secant functions are even, we know

$\cos(-\theta) \equiv \cos \theta$

$\sec(-\theta) \equiv \sec \theta$

x	cos x	$\frac{1}{\cos x} \equiv \sec x$	$\sec^2 x = y$
0	1	1	1
$\pm\frac{\pi}{6}$	$\frac{\sqrt{3}}{2}$	$\frac{2}{\sqrt{3}}$	$\frac{4}{3}$
$\pm\frac{\pi}{4}$	$\frac{\sqrt{2}}{2}$	$\frac{2}{\sqrt{2}}$	$\frac{4}{2} = 2$
$\pm\frac{\pi}{3}$	$\frac{1}{2}$	2	4
$\pm\frac{\pi}{2}$	0	Undefined	
$\pm\frac{2}{3}\pi$	$-\frac{1}{2}$	-2	4
$\pm\frac{3}{4}\pi$	$-\frac{\sqrt{2}}{2}$	$-\frac{2}{\sqrt{2}}$	$\frac{4}{2} = 2$
$\pm\frac{5}{6}\pi$	$-\frac{\sqrt{3}}{2}$	$-\frac{2}{\sqrt{3}}$	$\frac{4}{3}$
$\pm\pi$	-1	-1	1
$\pm\frac{3}{2}\pi$	0	Undefined	
$\pm 2\pi$	1	1	1

Plot these values and complete the graph. See graph in text.

36. Graph is the same as the graph of $y = \sec x$.

37. $y = 3\cos 2\left[x + \frac{\pi}{4}\right]$

$= 3\cos\left[2x + \frac{\pi}{2}\right]$

$= 3(-\sin 2x)$

$= -3\sin 2x$

See graph in text.

38.

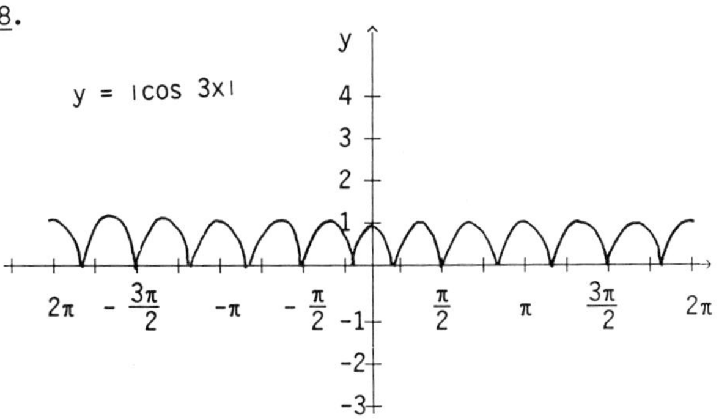

Exercise Set 7.8A

1. $(\sin x - \cos x)(\sin x + \cos x)$
$= \sin^2 x - \cos^2 x$

2. $\tan^2 y - \cot^2 y$

3. $\tan x(\cos x - \csc x)$
$= \frac{\sin x}{\cos x}\left[\cos x - \frac{1}{\sin x}\right]$
$= \sin x - \frac{1}{\cos x}$
$= \sin x - \sec x$

4. $\cos x + \csc x$

5. $\cos y \sin y(\sec y + \csc y)$
$= \cos y \sin y\left[\frac{1}{\cos y} + \frac{1}{\sin y}\right]$
$= \sin y + \cos y$

6. $\sin y - \tan y$

7. $(\sin x + \cos x)(\csc x - \sec x)$
$= \sin x \csc x - \sin x \sec x + \cos x \csc x - \cos x \sec x$
$= \sin x \cdot \frac{1}{\sin x} - \sin x \cdot \frac{1}{\cos x} + \cos x \cdot \frac{1}{\sin x} - \cos x \cdot \frac{1}{\cos x}$
$= 1 - \frac{\sin x}{\cos x} + \frac{\cos x}{\sin x} - 1$
$= \cot x - \tan x$

8. $2 + \tan x + \cot x$

9. $(\sin y - \cos y)^2$
$= \sin^2 y - 2\sin y\cos y + \cos^2 y$
$= 1 - 2\sin y\cos y$ $\quad(\sin^2 y + \cos^2 y = 1)$

10. $1 + 2\sin y\cos y$

11. $(1 + \tan x)^2$
$= 1 + 2\tan x + \tan^2 x$
$= \sec^2 x + 2\tan x$ $\quad(1 + \tan^2 x = \sec^2 x)$

12. $2\cot x + \csc^2 x$

13. $(\sin y - \csc y)^2$
$= \sin^2 y - 2\sin y\csc y + \csc^2 y$
$= \sin^2 y - 2\sin y \cdot \frac{1}{\sin y} + \csc^2 y$
$= \sin^2 y + \csc^2 y - 2$

14. $\cos^2 y + \sec^2 y + 2$

15. $(\cos x - \sec x)(\cos^2 x + \sec^2 x + 1)$
$= \cos^3 x + \cos x\sec^2 x + \cos x - \sec x\cos^2 x - \sec^3 x - \sec x$
$= \cos^3 x + \cos x \cdot \frac{1}{\cos^2 x} + \cos x - \frac{1}{\cos x} \cdot \cos^2 x - \sec^3 x - \frac{1}{\cos x}$
$= \cos^3 x + \frac{1}{\cos x} + \cos x - \cos x - \sec^3 x - \frac{1}{\cos x}$
$= \cos^3 - \sec^3 x$

16. $\sin^3 x + \csc^3 x$

17. $(\cot x - \tan x)(\cot^2 x + 1 + \tan^2 x)$
$= \cot^3 x + \cot x + \cot x\tan^2 x - \tan x\cot^2 x - \tan x - \tan^3 x$
$= \cot^3 x + \cot x + \cot x \cdot \frac{1}{\cot^2 x} - \frac{1}{\cot x} \cdot \cot^2 x - \frac{1}{\cot x} - \tan^3 x$
$= \cot^3 x + \cot x + \frac{1}{\cot x} - \cot x - \frac{1}{\cot x} - \tan^3 x$
$= \cot^3 x - \tan^3 x$

18. $\cot^3 y + \tan^3 y$

19. $\sin\left(\frac{\pi}{2} - x\right)[\sec x - \cos x]$

$= -\sin\left(x - \frac{\pi}{2}\right)\left[\frac{1}{\cos x} - \cos x\right]$

$\left[\sin\left(\frac{\pi}{2} - x\right) = \sin\left[-\left(x - \frac{\pi}{2}\right)\right] = -\sin\left(x - \frac{\pi}{2}\right)\right]$

$= -(-\cos x)\left(\frac{1 - \cos^2x}{\cos x}\right)$ $\left[\sin\left(x - \frac{\pi}{2}\right) = -\cos x\right]$

$= 1 - \cos^2x$

$= \sin^2x$

20. $\cos^2x$

21. $\sin x \cos x + \cos^2x$

$= \cos x (\sin x + \cos x)$

22. $\csc x (\sec x - \csc x)$

23. $\sin^2y - \cos^2y$

$= (\sin y + \cos y)(\sin y - \cos y)$

24. $(\tan y - \cot y)(\tan y + \cot y)$

25. $\tan x + \sin (\pi - x)$

$= \frac{\sin x}{\cos x} + \sin x$ $[\sin (\pi - x) = \sin x]$

$= \sin x \left(\frac{1}{\cos x} + 1\right)$

$= \sin x (\sec x + 1)$

26. $\cos x (\csc x + 1)$

27. $\sin^4x - \cos^4x$

$= (\sin^2x + \cos^2x)(\sin^2x - \cos^2x)$

$= \sin^2x - \cos^2x$

$= (\sin x + \cos x)(\sin x - \cos x)$

28. $-1(\tan^2x + \sec^2x)$

29. $3 \cot^2y + 6 \cot y + 3$

$= 3(\cot^2y + 2 \cot y + 1)$

$= 3(\cot y + 1)^2$

30. $4(\sin y + 1)^2$

31. $\csc^4x + 4 \csc^2x - 5$

$= (\csc^2x + 5)(\csc^2x - 1)$

$= (\csc^2x + 5)(\csc x + 1)(\csc x - 1)$

32. $(\tan^2x + 2)(\tan x - 2)(\tan x + 2)$

33. $\sin^3y + 27$

$= (\sin y)^3 + 3^3$

$= (\sin y + 3)(\sin^2y - 3 \sin y + 9)$

34. $(1 - 5 \tan y)(1 + 5 \tan y + 25 \tan^2y)$

35. $\sin^3y - \csc^3y$

$= (\sin y - \csc y)(\sin^2y + \sin y \csc y + \csc^2y)$

$= (\sin y - \csc y)(\sin^2y + 1 + \csc^2y)$

$\left[\sin y \csc y = \sin y \cdot \frac{1}{\sin y} = 1\right]$

36. $(\cos x - \sec x)(\cos^2x + 1 + \sec^2x)$

37. $\sin x \cos y - \cos\left(x + \frac{\pi}{2}\right) \tan y$

$= \sin x \cos y - (-\sin x) \tan y$

$\left[\cos\left(x + \frac{\pi}{2}\right) = -\sin x\right]$

$= \sin x \cos y + \sin x \tan y$

$= \sin x (\cos y + \tan y)$

38. $\sin x (\tan y + \cot y)$

39. $\cos (\pi - x) + \cot x \sin\left(x - \frac{\pi}{2}\right)$

$= -\cos x + \cot x (-\cos x)$ $\left[\sin\left(x - \frac{\pi}{2}\right) = -\cos x\right]$

40. $\sin x (1 - \tan x)$

Exercise Set 7.8B

1. $\frac{\sin^2x \cos x}{\cos^2x \sin x}$

$= \frac{\sin x}{\cos x} \cdot \frac{\sin x \cos x}{\sin x \cos x}$

$= \frac{\sin x}{\cos x}$

$= \tan x$

2. $\cot x$

3. $\frac{4 \sin x \cos^3x}{18 \sin^2x \cos x}$

$= \frac{2}{9} \cdot \frac{\cos^2x}{\sin x} \cdot \frac{\sin x \cos x}{\sin x \cos x}$

$= \frac{2 \cos^2x}{9 \sin x}$

4. $\frac{5 \sin^2x}{\cos x}$

5. $\dfrac{\cos^2 x - 2\cos x + 1}{\cos x - 1}$

$= \dfrac{(\cos x - 1)^2}{\cos x - 1}$

$= \cos x - 1$

6. $\sin x + 1$

7. $\dfrac{\cos^2 x - 1}{\cos x - 1}$

$= \dfrac{(\cos x + 1)(\cos x - 1)}{\cos x - 1}$

$= \cos x + 1$

8. $\sin x - 1$

9. $\dfrac{4\tan x \sec x + 2\sec x}{6\sin x \sec x + 2\sec x}$

$= \dfrac{2\sec x\,(2\tan x + 1)}{2\sec x\,(3\sin x + 1)}$

$= \dfrac{2\tan x + 1}{3\sin x + 1}$

10. $\dfrac{2\tan x - 1}{3\sin x + 1}$

11. $\dfrac{\cos^2 x - 1}{\sin\left(\frac{\pi}{2} - x\right) - 1}$

$= \dfrac{(\cos x + 1)(\cos x - 1)}{\cos x - 1}$

$\left[\sin\left(\frac{\pi}{2} - x\right) = -\sin\left(x - \frac{\pi}{2}\right) = \cos x\right]$

$= \cos x + 1$

12. $\sin x - 1$

13. $\dfrac{\sin x - \cos\left(x - \frac{\pi}{2}\right)\cos x}{-\sin x - \cos\left(x - \frac{\pi}{2}\right)\tan x}$

$= \dfrac{\sin x - \sin x\cos x}{-\sin x - \sin x\tan x}$ $\left[\cos\left(x - \frac{\pi}{2}\right) = \sin x\right]$

$= \dfrac{\sin x\,(1 - \cos x)}{-\sin x\,(1 + \tan x)}$

$= \dfrac{\cos x - 1}{1 + \tan x}$ $[-1(1 - \cos x) = \cos x - 1]$

14. $\dfrac{1 - \sin x}{1 + \tan x}$

15. $\dfrac{\sin^4 x - \cos^4 x}{\sin^2 x - \cos^2 x}$

$= \dfrac{(\sin^2 x + \cos^2 x)(\sin^2 x - \cos^2 x)}{\sin^2 x - \cos^2 x}$

$= 1$ $(\sin^2 x + \cos^2 x = 1)$

16. 1

17. $\dfrac{\cos^2 x + 2\sin\left(x - \frac{\pi}{2}\right) + 1}{\sin\left(\frac{\pi}{2} - x\right) - 1}$

$= \dfrac{\cos^2 x + 2(-\cos x) + 1}{-\sin\left(x - \frac{\pi}{2}\right) - 1}$ $\left[\sin\left(x - \frac{\pi}{2}\right) = -\cos x\right]$

$= \dfrac{\cos^2 x - 2\cos x + 1}{\cos x - 1}$

$= \dfrac{(\cos x - 1)^2}{\cos x - 1}$

$= \cos x - 1$

18. $\sin x - 1$

19. $\dfrac{\sin^2 y\cos\left(y + \frac{\pi}{2}\right)}{\cos^2 y\cos\left(\frac{\pi}{2} - y\right)}$

$= \dfrac{\sin^2 y\,(-\sin y)}{\cos^2 y\,(\sin y)}$ $\left[\cos\left(\frac{\pi}{2} - y\right) = \cos\left(y - \frac{\pi}{2}\right) = \sin y\right]$

$= -\tan^2 y$

20. $\cot^2 y$

21. $\dfrac{2\sin^2 x}{\cos^3 x} \cdot \left(\dfrac{\cos x}{2\sin x}\right)^2$

$= \dfrac{2\sin^2 x}{\cos^3 x} \cdot \dfrac{\cos^2 x}{4\sin^2 x}$

$= \dfrac{1}{2} \cdot \dfrac{1}{\cos x} \cdot \dfrac{\cos^2 x\sin^2 x}{\cos^2 x\sin^2 x}$

$= \dfrac{1}{2\cos x}$, or $\dfrac{1}{2}\sec x$

22. $\dfrac{\cos x}{4}$

23. $\dfrac{3\sin x}{\cos^2 x} \cdot \dfrac{\cos^2 x + \cos x\sin x}{\cos^2 x - \sin^2 x}$

$= \dfrac{3\sin x}{\cos^2 x} \cdot \dfrac{\cos x\,(\cos x + \sin x)}{(\cos x - \sin x)(\cos x + \sin x)}$

$= 3 \cdot \dfrac{\sin x}{\cos x} \cdot \dfrac{1}{\cos x - \sin x} \cdot \dfrac{\cos x(\cos x + \sin x)}{\cos x(\cos x + \sin x)}$

$= \dfrac{3\tan x}{\cos x - \sin x}$

24. $\dfrac{5\cot x}{\sin x + \cos x}$

25. $\dfrac{\tan^2 y}{\sec y} \div \dfrac{3\tan^3 y}{\sec y}$

$= \dfrac{\tan^2 y}{\sec y} \cdot \dfrac{\sec y}{3\tan^3 y}$

$= \dfrac{1}{3\tan y} \cdot \dfrac{\sec y\tan^2 y}{\sec y\tan^2 y}$

$= \dfrac{1}{3}\cot y$

26. $\dfrac{1}{4}\cot y$

27. $\dfrac{1}{\sin^2 y - \cos^2 y} - \dfrac{2}{\cos y + \sin y}$

$= \dfrac{1}{\sin^2 y - \cos^2 y} - \dfrac{2}{\sin y + \cos y} \cdot \dfrac{\sin y - \cos y}{\sin y - \cos y}$

$= \dfrac{1 - 2 \sin y + 2 \cos y}{\sin^2 y - \cos^2 y}$

28. $\dfrac{3 \cos y + 3 \sin y + 2}{\cos^2 y - \sin^2 y}$ or $\dfrac{-3 \cos y - 3 \sin y - 2}{\sin^2 y - \cos^2 y}$

29. $\left(\dfrac{\sin x}{\cos x}\right)^2 - \dfrac{1}{\cos^2 x}$

$= \tan^2 x - \sec^2 x$

$= -1$ $(1 + \tan^2 x \equiv \sec^2 x)$

30. 1

31. $\dfrac{\sin^2 x - 9}{2 \cos x + 1} \cdot \dfrac{10 \cos x + 5}{3 \sin x + 9}$

$= \dfrac{(\sin x + 3)(\sin x - 3)}{2 \cos x + 1} \cdot \dfrac{5(2 \cos x + 1)}{3(\sin x + 3)}$

$= \dfrac{5(\sin x - 3)}{3}$

32. $\dfrac{(3 \cos x + 5)(\cos x + 1)}{4}$

Exercise Set 7.8C

1. $\sqrt{\sin^2 x \cos x} \cdot \sqrt{\cos x}$

$= \sqrt{\sin^2 x \cos^2 x}$

$= \sin x \cos x$

2. $\cos x \sin x$

3. $\sqrt{\sin^3 y} + \sqrt{\sin y \cos^2 y}$

$= \sin y \sqrt{\sin y} + \cos y \sqrt{\sin y}$

$= \sqrt{\sin y}\,(\sin y + \cos y)$

4. $\sqrt{\cos y}\,(\sin y - \cos y)$

5. $\sqrt{\sin^2 x + 2 \cos x \sin x + \cos^2 x}$

$= \sqrt{(\sin x + \cos x)^2}$

$= \sin x + \cos x$

6. $\tan x - \sin x$

7. $(1 - \sqrt{\sin y})(\sqrt{\sin y} + 1)$

$= (1 - \sqrt{\sin y})(1 + \sqrt{\sin y})$

$= 1 - \sin y$ $[(\sqrt{\sin y})^2 = \sin y]$

8. $4 - \tan y$

9. $\sqrt{\sin x}\,(\sqrt{2 \sin x} + \sqrt{\sin x \cos x})$

$= \sqrt{2 \sin^2 x} + \sqrt{\sin^2 x \cos x}$

$= \sin x \cdot \sqrt{2} + \sin x \cdot \sqrt{\cos x}$

$= \sin x\,(\sqrt{2} + \sqrt{\cos x})$

10. $\cos x\,(\sqrt{3} - \sqrt{\sin x})$

11. $\sqrt{\dfrac{\sin x}{\cos x}}$

$= \sqrt{\dfrac{\sin x}{\cos x} \cdot \dfrac{\cos x}{\cos x}}$

$= \sqrt{\dfrac{\sin x \cos x}{\cos^2 x}}$

$= \dfrac{\sqrt{\sin x \cos x}}{\cos x}$

12. $\dfrac{\sqrt{\sin x \cos x}}{\sin x}$ or $\sqrt{\cot x}$

13. $\sqrt{\dfrac{\sin x}{\cot x}}$

$= \sqrt{\dfrac{\sin x}{\cot x} \cdot \dfrac{\cot x}{\cot x}}$

$= \sqrt{\dfrac{\sin x \cot x}{\cot^2 x}}$

$= \dfrac{\sqrt{\cos x}}{\cot x}$ $\left[\sqrt{\sin x \cot x} = \sqrt{\sin x \cdot \dfrac{\cos x}{\sin x}} = \sqrt{\cos x}\right]$

14. $\dfrac{\sqrt{\sin x}}{\tan x}$

15. $\sqrt{\dfrac{\cos^2 x}{2 \sin^2 x}}$

$= \sqrt{\dfrac{\cot^2 x}{2} \cdot \dfrac{2}{2}}$

$= \dfrac{\sqrt{2} \cot x}{2}$

16. $\dfrac{\sqrt{3} \tan x}{3}$

17. $\sqrt{\dfrac{1 + \sin x}{1 - \sin x}}$

$= \sqrt{\dfrac{1 + \sin x}{1 - \sin x} \cdot \dfrac{1 - \sin x}{1 - \sin x}}$

$= \sqrt{\dfrac{1 - \sin^2 x}{(1 - \sin x)^2}}$

$= \dfrac{\sqrt{\cos^2 x}}{1 - \sin x}$ $(1 - \sin^2 x = \cos^2 x)$

$= \dfrac{\cos x}{1 - \sin x}$

18. $\dfrac{\sin x}{1 + \cos x}$

19. $\tan^2 x + 4 \tan x = 21$

$\tan^2 x + 4 \tan x - 21 = 0$

$(\tan x - 3)(\tan x + 7) = 0$

$\tan x - 3 = 0$ or $\tan x + 7 = 0$

$\tan x = 3$ or $\tan x = -7$

20. $\sec x = 2, 5$

21. $8 \sin^2 x - 2 \sin x = 3$

$8 \sin^2 x - 2 \sin x - 3 = 0$

$(4 \sin x - 3)(2 \sin x + 1) = 0$

$4 \sin x - 3 = 0$ or $2 \sin x + 1 = 0$

$4 \sin x = 3$ or $2 \sin x = -1$

$\sin x = \frac{3}{4}$ or $\sin x = -\frac{1}{2}$

22. $\cos x = -\frac{1}{3}, -\frac{5}{2}$

23. $\cot^2 x + 9 \cot x - 10 = 0$

$(\cot x + 10)(\cot x - 1) = 0$

$\cot x + 10 = 0$ or $\cot x - 1 = 0$

$\cot x = -10$ or $\cot x = 1$

24. $\csc x = -5, 2$

25. $\sin^2 x - \cos\left(x + \frac{\pi}{2}\right) = 6$

$\sin^2 x - (-\sin x) = 6$

$\sin^2 x + \sin x - 6 = 0$

$(\sin x - 2)(\sin x + 3) = 0$

$\sin x - 2 = 0$ or $\sin x + 3 = 0$

$\sin x = 2$ or $\sin x = -3$

26. $\sin x = \frac{1}{2}, -2$

27. $\tan^2 x - 6 \tan x = 4$

$\tan^2 x - 6 \tan x - 4 = 0$

$$\tan x = \frac{-(-6) \pm \sqrt{(-6)^2 - 4(1)(-4)}}{2 \cdot 1}$$

$$= \frac{6 \pm \sqrt{52}}{2} = \frac{6 \pm 2\sqrt{13}}{2}$$

$$= 3 \pm \sqrt{13}$$

28. $\cot x = -4 \pm \sqrt{21}$

29. $6 \sec^2 x - 5 \sec x - 2 = 0$

$$\sec x = \frac{-(-5) \pm \sqrt{(-5)^2 - 4(6)(-2)}}{2 \cdot 6}$$

$$= \frac{5 \pm \sqrt{73}}{12}$$

30. $\csc x = \dfrac{3 \pm \sqrt{41}}{4}$

Exercise Set 8.1

1. $\sin 75° = \sin(45° + 30°)$
$= \sin 45° \cos 30° + \cos 45° \sin 30°$
$= \frac{\sqrt{2}}{2} \cdot \frac{\sqrt{3}}{2} + \frac{\sqrt{2}}{2} \cdot \frac{1}{2}$
$= \frac{\sqrt{6}}{4} + \frac{\sqrt{2}}{4} = \frac{\sqrt{6} + \sqrt{2}}{4} \approx 0.9659$

2. $\frac{\sqrt{6} - \sqrt{2}}{4} \approx 0.2588$

3. $\sin 15° = \sin(45° - 30°)$
$= \sin 45° \cos 30° - \cos 45° \sin 30°$
$= \frac{\sqrt{2}}{2} \cdot \frac{\sqrt{3}}{2} - \frac{\sqrt{2}}{2} \cdot \frac{1}{2}$
$= \frac{\sqrt{6}}{4} - \frac{\sqrt{2}}{4} = \frac{\sqrt{6} - \sqrt{2}}{4} \approx 0.2588$

4. $\frac{\sqrt{6} + \sqrt{2}}{4} \approx 0.9659$

5. $\sin 105° = \sin(75° + 30°)$
$= \sin 75° \cos 30° + \cos 75° \sin 30°$
$= \frac{\sqrt{6} + \sqrt{2}}{4} \cdot \frac{\sqrt{3}}{2} + \frac{\sqrt{6} - \sqrt{2}}{4} \cdot \frac{1}{2}$
$= \frac{3\sqrt{2} + \sqrt{6}}{8} + \frac{\sqrt{6} - \sqrt{2}}{8}$
$= \frac{2\sqrt{6} + 2\sqrt{2}}{8} = \frac{\sqrt{6} + \sqrt{2}}{4} \approx 0.9659$

6. $\frac{\sqrt{2} - \sqrt{6}}{4} \approx -0.2588$

7. $\tan 75° = \tan(45° + 30°)$
$= \frac{\tan 45° + \tan 30°}{1 - \tan 45° \tan 30°}$
$= \frac{1 + \frac{\sqrt{3}}{3}}{1 - 1 \cdot \frac{\sqrt{3}}{3}} = \frac{3 + \sqrt{3}}{3 - \sqrt{3}} \approx 3.7321$

8. $\frac{3 + \sqrt{3}}{3 - \sqrt{3}} \approx 3.7321$

9. - 14.

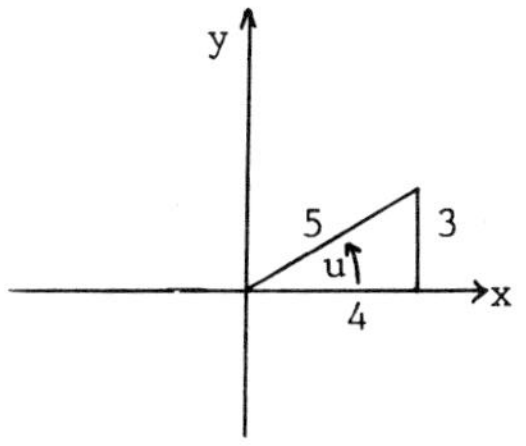

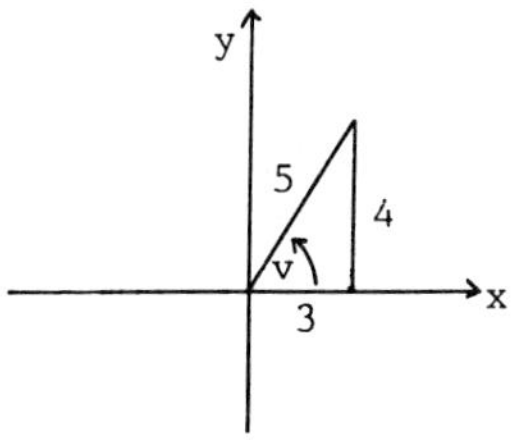

9. - 14. (continued)

$\sin u = \frac{3}{5}$ $\sin v = \frac{4}{5}$

$\cos u = \frac{4}{5}$ $\cos v = \frac{3}{5}$

$\tan u = \frac{3}{4}$ $\tan v = \frac{4}{3}$

9. $\sin(u + v) = \sin u \cos v + \cos u \sin v$
$= \frac{3}{5} \cdot \frac{3}{5} + \frac{4}{5} \cdot \frac{4}{5}$
$= \frac{9}{25} + \frac{16}{25} = \frac{25}{25} = 1$

10. $-\frac{7}{25}$

11. Use drawings above Exercise 9.
$\cos(u + v) = \cos u \cos v - \sin u \sin v$
$= \frac{4}{5} \cdot \frac{3}{5} - \frac{3}{5} \cdot \frac{4}{5}$
$= \frac{12}{25} - \frac{12}{25} = 0$

12. $\frac{24}{25}$

13. Use drawings above Exercise 9.
$\tan(u + v) = \frac{\tan u + \tan v}{1 - \tan u \tan v}$
$= \frac{\frac{3}{4} + \frac{4}{3}}{1 - \frac{3}{4} \cdot \frac{4}{3}} = \frac{\frac{3}{4} + \frac{4}{3}}{0}$

Division by 0 is undefined. Thus $\tan(u + v)$ is undefined.

14. $-\frac{7}{24}$

15. Given: $\sin\theta = 0.6249$ and $\cos\phi = 0.1102$.

Find $\cos\theta$ and $\sin\phi$ using the following identity:
$\sin^2 x + \cos^2 x \equiv 1$

If $\sin\theta = 0.6249$, then
$(0.6249)^2 + \cos^2\theta = 1$
$\cos^2\theta = 1 - 0.3905 = 0.6095$
$\cos\theta = 0.7807$

If $\cos\phi = 0.1102$, then
$\sin^2\phi + (0.1102)^2 = 1$
$\sin^2\phi = 1 - 0.0121 = 0.9879$
$\sin\phi = 0.9939$

15. (continued)

$$\sin(\theta + \phi) = \sin\theta\cos\phi + \cos\theta\sin\phi$$
$$= (0.6249)(0.1102) + (0.7807)(0.9939)$$
$$= 0.0689 + 0.7759$$
$$= 0.8448$$

16. -0.5350

17. $\sin 37° \cos 22° + \cos 37° \sin 22° = \sin(37° + 22°) = \sin 59°$, or 0.8572

18. $\cos 59°$, or 0.5150

19. $\dfrac{\tan 20° + \tan 32°}{1 - \tan 20° \tan 32°} = \tan(20° + 32°) = \tan 52°$, or 1.2799

20. $\tan 23°$, or 0.4245

21.
$$\sin(\alpha + \beta) + \sin(\alpha - \beta)$$
$$= (\sin\alpha\cos\beta + \cos\alpha\sin\beta) + (\sin\alpha\cos\beta - \cos\alpha\sin\beta)$$
$$= 2\sin\alpha\cos\beta$$

22. $2\cos\alpha\sin\beta$

23.
$$\cos(\alpha + \beta) + \cos(\alpha - \beta)$$
$$= (\cos\alpha\cos\beta - \sin\alpha\sin\beta) + (\cos\alpha\cos\beta + \sin\alpha\sin\beta)$$
$$= 2\cos\alpha\cos\beta$$

24. $-2\sin\alpha\sin\beta$

25.
$$\cos(u + v)\cos v + \sin(u + v)\sin v$$
$$= (\cos u\cos v - \sin u\sin v)\cos v + (\sin u\cos v + \cos u\sin v)\sin v$$
$$= \cos u\cos^2 v - \sin u\sin v\cos v + \sin u\cos v\sin v + \cos u\sin^2 v$$
$$= \cos u(\cos^2 v + \sin^2 v)$$
$$= \cos u$$

26. $\sin u$

27. Solve each equation for y; then determine the slope of each line.

ℓ_1: $3y = \sqrt{3}\,x + 2$, $y = \frac{\sqrt{3}}{3}x + \frac{2}{3}$. Thus $m_1 = \frac{\sqrt{3}}{3}$.

ℓ_2: $3y + \sqrt{3}\,x = -3$, $y = -\frac{\sqrt{3}}{3}x - 1$. Thus $m_2 = -\frac{\sqrt{3}}{3}$.

Let ϕ be the smallest angle from ℓ_1 to ℓ_2.

$$\tan\phi = \frac{-\frac{\sqrt{3}}{3} - \frac{\sqrt{3}}{3}}{1 + \left(-\frac{\sqrt{3}}{3}\right)\left(\frac{\sqrt{3}}{3}\right)} \qquad \left(\tan\phi = \frac{m_2 - m_1}{1 + m_2 m_1}\right)$$

$$= \frac{-\frac{2\sqrt{3}}{3}}{\frac{2}{3}} = -\sqrt{3}$$

Since $\tan\phi$ is negative, ϕ is obtuse. Hence ϕ is $\frac{2\pi}{3}$.

28. $\frac{\pi}{6}$

29. Solve each equation for y; then determine the slope of each line.

ℓ_1: $2x = 3 - 2y$, $y = -x + \frac{3}{2}$. Thus $m_1 = -1$ and the y-intercept is $\frac{3}{2}$.

ℓ_2: $x + y = 5$, $y = -x + 5$. Thus $m_2 = -1$ and the y-intercept is 5.

The lines do not form an angle. The lines are parallel. When the formula is used, the result is 0°.

30. The lines do not form an angle. The lines are parallel. When the formula is used, the result is 0°.

31. Solve each equation for y; then determine the slope of each line.

ℓ_1: $2x - 5y + 1 = 0$
$$y = \frac{2}{5}x + \frac{1}{5}$$

Thus $m_1 = \frac{2}{5}$.

ℓ_2: $3x + y - 7 = 0$
$$y = -3x + 7$$

Thus $m_2 = -3$.

Let ϕ be the smallest angle from ℓ_1 to ℓ_2.

31. (continued)

$$\tan\phi = \frac{-3 - \frac{2}{5}}{1 + (-3)\left(\frac{2}{5}\right)} \qquad \left[\tan\phi = \frac{m_2 - m_1}{1 + m_2 m_1}\right]$$

$$= \frac{-\frac{17}{5}}{-\frac{1}{5}} = 17$$

Since tan ϕ is positive we know that ϕ is acute. Using a calculator we find $\phi \approx 86.63°$, or $86°\ 38'$.

32. $126.87°$, or $126°\ 52'$

33. Find the slope of each line.

ℓ_1: $y = 3$ $\qquad (y = 0x + 3)$

Thus $m = 0$.

ℓ_2: $x + y = 5$

$y = -x + 5$

Thus $m = -1$.

Let ϕ be the smallest angle from ℓ_1 to ℓ_2.

$$\tan\phi = \frac{-1 - 0}{1 + (-1)(0)} \qquad \left[\tan\phi = \frac{m_2 - m_1}{1 + m_2 m_1}\right]$$

$$= \frac{-1}{1} = -1$$

Since tan ϕ is negative, we know that ϕ is obtuse. Thus $\phi = \frac{3\pi}{4}$, or $135°$.

34. $45°$

35. $\sin 2\theta = \sin(\theta + \theta)$

$= \sin\theta\cos\theta + \cos\theta\sin\theta$

$= 2\sin\theta\cos\theta$

36. $\cos^2\theta - \sin^2\theta$, or $1 - 2\sin^2\theta$, or $2\cos^2\theta - 1$

37. See text answer section.

38. $\cot(\alpha - \beta) \equiv \dfrac{\cot\alpha\cot\beta + 1}{\cot\beta - \cot\alpha}$

39. $\sin\left(\frac{\pi}{2} - x\right) = \sin\frac{\pi}{2}\cos x - \cos\frac{\pi}{2}\sin x$

$= 1 \cdot \cos x - 0 \cdot \sin x$

$= \cos x$

Thus $\sin\left(\frac{\pi}{2} - x\right) \equiv \cos x$.

40. $\cos\left(\frac{\pi}{2} - x\right) \equiv \sin x$

41. See text answer section.

42. $\cos\left(x - \frac{3\pi}{2}\right) = \cos x\cos\frac{3\pi}{2} + \sin x\sin\frac{3\pi}{2}$

$= (\cos x)(0) + (\sin x)(-1)$

$= -\sin x$

43. See text answer section.

44. $\sin\left(x - \frac{3\pi}{2}\right) = \sin x\cos\frac{3\pi}{2} - \cos x\sin\frac{3\pi}{2}$

$= (\sin x)(0) - (\cos x)(-1)$

$= \cos x$

45. See text answer section.

46. $$\tan\left(x - \frac{\pi}{4}\right) = \frac{\tan x - \tan\frac{\pi}{4}}{1 + \tan x\tan\frac{\pi}{4}}$$

$$= \frac{\tan x - 1}{\tan x + 1}$$

47. See text answer section.

48. $$\frac{\sin(\alpha + \beta)}{\sin(\alpha - \beta)} = \frac{\sin\alpha\cos\beta + \cos\alpha\sin\beta}{\sin\alpha\cos\beta - \cos\alpha\sin\beta}$$

$$= \frac{\frac{\sin\alpha\cos\beta + \cos\alpha\sin\beta}{\cos\alpha\cos\beta}}{\frac{\sin\alpha\cos\beta - \cos\alpha\sin\beta}{\cos\alpha\cos\beta}}$$

$$= \frac{\frac{\sin\alpha}{\cos\alpha} + \frac{\sin\beta}{\cos\beta}}{\frac{\sin\alpha}{\cos\alpha} - \frac{\sin\beta}{\cos\beta}} = \frac{\tan\alpha + \tan\beta}{\tan\alpha - \tan\beta}$$

49. See text answer section.

50. $\cos(\alpha + \beta) \cdot \cos(\alpha - \beta)$

$= (\cos\alpha\cos\beta - \sin\alpha\sin\beta)(\cos\alpha\cos\beta + \sin\alpha\sin\beta)$

$= \cos^2\alpha\cos^2\beta - \sin^2\alpha\sin^2\beta$

$= (1 - \sin^2\alpha)(1 - \sin^2\beta) - \sin^2\alpha\sin^2\beta$

$= 1 - \sin^2\beta - \sin^2\alpha + \sin^2\alpha\sin^2\beta - \sin^2\alpha\sin^2\beta$

$= 1 - \sin^2\beta - \sin^2\alpha$

$= \cos^2\alpha - \sin^2\beta$

51. $$\tan\phi = \frac{m_2 - m_1}{1 + m_2 m_1}$$

$$\tan 45° = \frac{\frac{4}{3} - m_1}{1 + \frac{4}{3} \cdot m_1}$$

$$1 = \frac{4 - 3m_1}{3 + 4m_1}$$

$3 + 4m_1 = 4 - 3m_1$

$7m_1 = 1$

$m_1 = \frac{1}{7}$

52. $-\frac{11}{16}$

53. Find the slope of ℓ_1.

$$m_1 = \frac{-1 - 4}{5 - (-2)} = \frac{-5}{7} = -\frac{5}{7}$$

Use the following formula to determine the slope of ℓ_2.

$$\tan \phi = \frac{m_2 - m_1}{1 + m_2 m_1}$$

$$\tan 45° = \frac{m_2 - \left(-\frac{5}{7}\right)}{1 + m_2\left(-\frac{5}{7}\right)}$$

$$1 = \frac{7m_2 + 5}{7 - 5m_2}$$

$$7 - 5m_2 = 7m_2 + 5$$

$$2 = 12m_2$$

$$\frac{1}{6} = m_2$$

54. $-\frac{5}{2}$

55. Find the slope of each line.

$$m_1 = \frac{3.899 - (-0.9012)}{-2.123 - (-4.892)} = \frac{4.8002}{2.769} = 1.734$$

$$m_2 = \frac{4.013 - (-3.814)}{5.925 - 0} = \frac{7.827}{5.925} = 1.321$$

Let ϕ be the smallest angle from ℓ_1 to ℓ_2.

$$\tan \phi = \frac{1.321 - 1.734}{1 + (1.321)(1.734)} = \frac{-0.413}{3.291} = -0.126$$

Since $\tan \phi$ is negative, we know that ϕ is obtuse. Thus $\phi = 172° \ 50'$.

56. 168.7°

57.

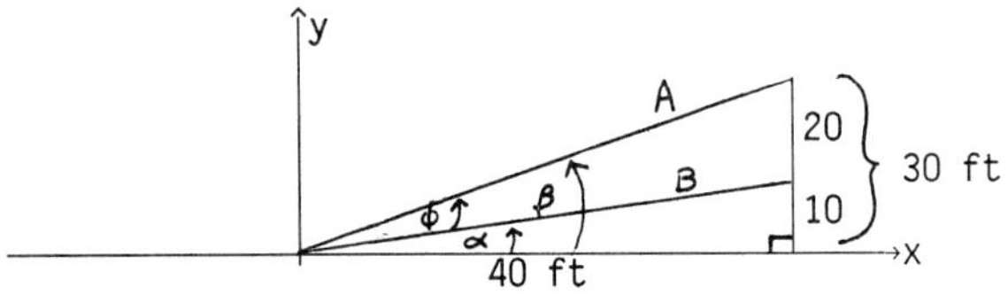

Position a set of coordinate axes as shown above. Let ℓ_1 be the line containing the points (0, 0) and (40, 10). Find its slope.

$$m_1 = \tan \alpha = \frac{10}{40} = \frac{1}{4}$$

Let ℓ_2 be the line containing the points (0, 0) and (40, 30). Find its slope.

$$m_2 = \tan \beta = \frac{30}{40} = \frac{3}{4}$$

Now ϕ is the smallest positive angle from ℓ_1 to ℓ_2.

57. (continued)

$$\tan \phi = \frac{\frac{3}{4} - \frac{1}{4}}{1 + \frac{3}{4} \cdot \frac{1}{4}} = \frac{\frac{1}{2}}{\frac{19}{16}} = \frac{8}{19} \approx 0.4211$$

$\phi \approx 22.83°$, or $22° \ 50'$

Exercise Set 8.2

1. Sketch a first quadrant triangle. Since $\sin \theta = \frac{4}{5}$, label the leg parallel to the y-axis 4 and the hypotenuse 5. Using the Pythagorean theorem we know the other leg has length 3.

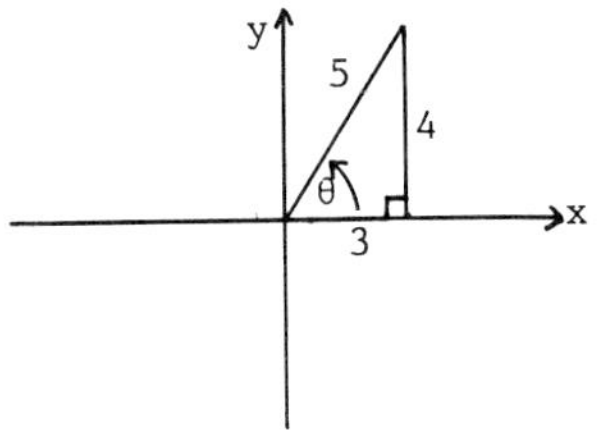

From this diagram we find that $\cos \theta = \frac{3}{5}$ and $\tan \theta = \frac{4}{3}$.

$$\sin 2\theta = 2 \sin \theta \cos \theta = 2 \cdot \frac{4}{5} \cdot \frac{3}{5} = \frac{24}{25}$$

$$\cos 2\theta = \cos^2\theta - \sin^2\theta = \left(\frac{3}{5}\right)^2 - \left(\frac{4}{5}\right)^2 = \frac{9}{25} - \frac{16}{25} = -\frac{7}{25}$$

$$\tan 2\theta = \frac{2 \tan \theta}{1 - \tan^2\theta} = \frac{2 \cdot \frac{4}{3}}{1 - \left(\frac{4}{3}\right)^2} = \frac{\frac{8}{3}}{-\frac{7}{9}} = -\frac{24}{7}$$

Since $\sin 2\theta$ is positive and $\cos 2\theta$ is negative, we know that 2θ is in quadrant II.

2. $\sin 2\theta = \frac{120}{169}$, $\cos 2\theta = \frac{119}{169}$, $\tan 2\theta = \frac{120}{119}$; 2θ is in quadrant I.

3. Sketch a third quadrant triangle. Since $\cos \theta = -\frac{4}{5}$, label the leg on the x-axis -4 and the hypotenuse 5. Using the Pythagorean theorem we know the other leg has length 3.

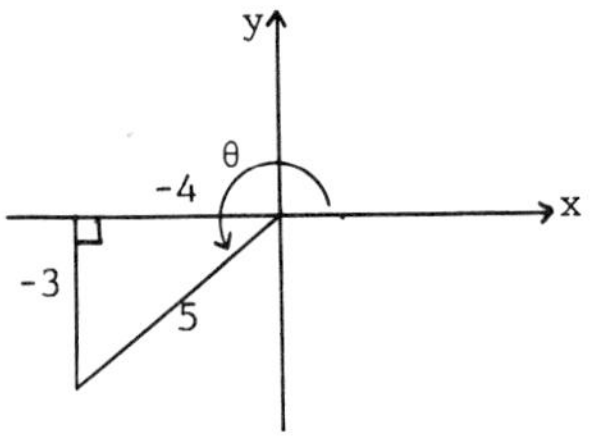

From this diagram we find that $\sin \theta = -\frac{3}{5}$ and $\tan \theta = \frac{3}{4}$.

3. (continued)

$$\sin 2\theta = 2 \sin\theta \cos\theta = 2\left(-\frac{3}{5}\right)\left(-\frac{4}{5}\right) = \frac{24}{25}$$

$$\cos 2\theta = \cos^2\theta - \sin^2\theta = \left(-\frac{4}{5}\right)^2 - \left(-\frac{3}{5}\right)^2 = \frac{16}{25} - \frac{9}{25} = \frac{7}{25}$$

$$\tan 2\theta = \frac{2\tan\theta}{1 - \tan^2\theta} = \frac{2 \cdot \frac{3}{4}}{1 - \left(\frac{3}{4}\right)^2} = \frac{\frac{3}{2}}{\frac{7}{16}} = \frac{24}{7}$$

Since both sin 2θ and cos 2θ are positive, we know that 2θ is in quadrant I.

4. $\sin 2\theta = \frac{24}{25}$, $\cos 2\theta = -\frac{7}{25}$, $\tan 2\theta = -\frac{24}{7}$; 2θ is in quadrant II.

5. Sketch a third quadrant triangle. Since $\tan\theta = \frac{4}{3}$, label the leg on the x-axis -3 and the leg parallel to the y-axis -4. Using the Pythagorean theorem we know the hypotenuse has length 5.

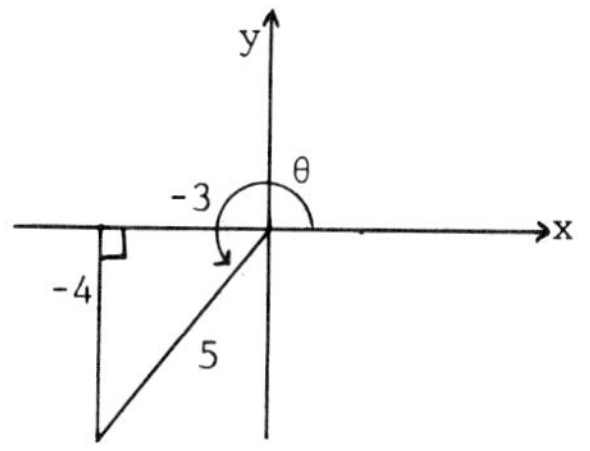

From this diagram we find that $\sin\theta = -\frac{4}{5}$ and $\cos\theta = -\frac{3}{5}$.

$$\sin 2\theta = 2\sin\theta\cos\theta = 2\left(-\frac{4}{5}\right)\left(-\frac{3}{5}\right) = \frac{24}{25}$$

$$\cos 2\theta = \cos^2\theta - \sin^2\theta = \left(-\frac{3}{5}\right)^2 - \left(-\frac{4}{5}\right)^2 = \frac{9}{25} - \frac{16}{25} = -\frac{7}{25}$$

$$\tan 2\theta = \frac{2\tan\theta}{1 - \tan^2\theta} = \frac{2 \cdot \frac{4}{3}}{1 - \left(\frac{4}{3}\right)^2} = \frac{\frac{8}{3}}{-\frac{7}{9}} = -\frac{24}{7}$$

Since sin 2θ is positive and cos 2θ is negative, 2θ is in quadrant II.

6. Same as Exercise 3.

7.
$$\begin{aligned}\sin 4\theta &= \sin 2(2\theta)\\ &= 2\sin 2\theta\cos 2\theta\\ &= 2(2\sin\theta\cos\theta)(\cos^2\theta - \sin^2\theta)\\ &= 4\sin\theta\cos^3\theta - 4\sin^3\theta\cos\theta\end{aligned}$$

8. $\cos^4\theta - 6\cos^2\theta\sin^2\theta + \sin^4\theta$, or $8\cos^4\theta - 8\cos^2\theta + 1$

9.
$$\begin{aligned}\sin^4\theta &= \sin^2\theta \cdot \sin^2\theta\\ &= \frac{1-\cos 2\theta}{2} \cdot \frac{1-\cos 2\theta}{2}\\ &= \frac{1 - 2\cos 2\theta + \cos^2 2\theta}{4}\\ &= \frac{1 - 2\cos 2\theta + \frac{1+\cos 4\theta}{2}}{4}\\ &= \frac{2 - 4\cos 2\theta + 1 + \cos 4\theta}{8}\\ &= \frac{3 - 4\cos 2\theta + \cos 4\theta}{8}\end{aligned}$$

10. $\frac{3 + 4\cos 2\theta + \cos 4\theta}{8}$

11.
$$\begin{aligned}\sin 75^\circ &= \sin\frac{150^\circ}{2}\\ &= \sqrt{\frac{1 - \cos 150^\circ}{2}} = \sqrt{\frac{1 - (-\sqrt{3}/2)}{2}}\\ &= \sqrt{\frac{2+\sqrt{3}}{4}} = \frac{\sqrt{2+\sqrt{3}}}{2}\end{aligned}$$

The expression is positive because 75° is in the first quadrant.

12. $\frac{\sqrt{2-\sqrt{3}}}{2}$

13.
$$\begin{aligned}\tan 75^\circ &= \tan\frac{150^\circ}{2}\\ &= \frac{\sin 150^\circ}{1 + \cos 150^\circ} = \frac{\frac{1}{2}}{1 + \left(-\frac{\sqrt{3}}{2}\right)}\\ &= \frac{1}{2-\sqrt{3}}, \text{ or } 2 + \sqrt{3}\end{aligned}$$

14. $1 + \sqrt{2}$

15.
$$\begin{aligned}\sin\frac{5\pi}{8} &= \sin\frac{\frac{5\pi}{4}}{2}\\ &= \sqrt{\frac{1 - \cos(5\pi/4)}{2}} = \sqrt{\frac{1 - (-\sqrt{2}/2)}{2}}\\ &= \sqrt{\frac{2+\sqrt{2}}{4}} = \frac{\sqrt{2+\sqrt{2}}}{2}\end{aligned}$$

The expression is positive because $\frac{5\pi}{8}$ is in the second quadrant.

16. $\dfrac{-\sqrt{2 - \sqrt{2}}}{2}$

17. First find $\cos\theta$ using the following identity:

$$\sin^2\theta + \cos^2\theta = 1$$
$$(0.3416)^2 + \cos^2\theta = 1$$
$$\cos^2\theta = 1 - 0.11669056,$$
$$\cos\theta = 0.9398 \quad (\theta \text{ is in quadrant I})$$

$$\sin 2\theta = 2\sin\theta\cos\theta$$
$$= 2(0.3416)(0.9398)$$
$$= 0.6421$$

18. 0.7666

19. $\sin 4\theta = 4\sin\theta\cos^3\theta - 4\sin^3\theta\cos\theta$ (See Exercise 7)

$$= 4(0.3416)(0.9398)^3 - 4(0.3416)^3(0.9398)$$
$$= 1.1342 - 0.1498$$
$$= 0.9844$$

20. 0.1754

21. $\sin\frac{\theta}{2} = \sqrt{\dfrac{1 - \cos\theta}{2}} \quad \left(\frac{\theta}{2} \text{ is in quadrant I}\right)$

$$= \sqrt{\frac{1 - 0.9398}{2}}$$
$$= 0.1735$$

22. 0.9848

23. $\dfrac{\sin 2x}{2\sin x} = \dfrac{2\sin x\cos x}{2\sin x} = \cos x$

24. $\sin x$

25. $1 - 2\sin^2\frac{x}{2} = 1 - 2\left(\frac{1 - \cos x}{2}\right)$

$$= 1 - 1 + \cos x$$
$$= \cos x$$

26. $\cos x$

27. $2\sin\frac{x}{2}\cos\frac{x}{2} = \sin 2\left(\frac{x}{2}\right) = \sin x$

28. $\sin 4x$

29. $\cos^2\frac{x}{2} - \sin^2\frac{x}{2} = \cos 2\left(\frac{x}{2}\right) = \cos x$

30. $\cos 2x$

31. $(\sin x + \cos x)^2 - \sin 2x$

$$= \sin^2 x + 2\sin x\cos x + \cos^2 x - \sin 2x$$
$$= \sin^2 x + \sin 2x + \cos^2 x - \sin 2x$$
$$= 1$$

32. 1

33. $2\sin^2\frac{x}{2} + \cos x$

$$= 2\left(\frac{1 - \cos x}{2}\right) + \cos x$$
$$= 1 - \cos x + \cos x$$
$$= 1$$

34. 1

35. $(-4\cos x\sin x + 2\cos 2x)^2 + (2\cos 2x + 4\sin x\cos x)^2$

$$= (4\cos^2 2x - 16\sin x\cos x\cos 2x + 16\cos^2 x\sin^2 x) + (4\cos^2 2x + 16\sin x\cos x\cos 2x + 16\sin^2 x\cos^2 x)$$
$$= 8\cos^2 2x + 32\cos^2 x\sin^2 x$$
$$= 8(\cos^2 x - \sin^2 x)^2 + 32\cos^2 x\sin^2 x$$
$$= 8(\cos^4 x - 2\cos^2 x\sin^2 x + \sin^4 x) + 32\cos^2 x\sin^2 x$$
$$= 8\cos^4 x + 16\cos^2 x\sin^2 x + 8\sin^4 x$$
$$= 8(\cos^4 x + 2\cos^2 x\sin^2 x + \sin^4 x)$$
$$= 8(\cos^2 x + \sin^2 x)$$
$$= 8$$

36. 32

37. $2\sin x\cos^3 x + 2\sin^3 x\cos x$

$$= 2\sin x\cos x(\cos^2 x + \sin^2 x)$$
$$= 2\sin x\cos x$$
$$= \sin 2x$$

38. $\sin 2x\cos 2x$

39. $(\sin x + \cos x)^2$

$$= \sin^2 x + 2\sin x\cos x + \cos^2 x$$
$$= 1 + \sin 2x$$

Thus $(\sin x + \cos x)^2 \equiv 1 + \sin 2x$.

40. $\cos^4 x - \sin^4 x \equiv \cos 2x$

41. $\dfrac{2\cot x}{\cot^2 x - 1} = \dfrac{2\,\dfrac{\cos x}{\sin x}}{\dfrac{\cos^2 x}{\sin^2 x} - 1}$

$= \dfrac{2\cos x}{\sin x} \cdot \dfrac{\sin^2 x}{\cos^2 x - \sin^2 x}$

$= \dfrac{2\cos x}{\sin x} \cdot \dfrac{\sin^2 x}{\cos 2x}$

$= \dfrac{2\cos x \sin x}{\cos 2x}$

$= \dfrac{\sin 2x}{\cos 2x}$

$= \tan 2x$

Thus $\dfrac{2\cot x}{\cot^2 x - 1} \equiv \tan 2x$.

42. $\dfrac{2 - \sec^2 x}{\sec^2 x} \equiv \cos 2x$

43. $2\sin^2 2x + \cos 4x$

$= 2\sin^2 2x + \cos 2(2x)$

$= 2\sin^2 2x + (\cos^2 2x - \sin^2 2x)$

$= \sin^2 2x + \cos^2 2x$

$= 1$

Thus $2\sin^2 2x + \cos 4x \equiv 1$.

44. $\dfrac{1 + \sin 2x + \cos 2x}{1 + \sin 2x - \cos 2x} \equiv \cot x$

45. See text answer section.

46. By Exercise 45,

$2\sin\dfrac{x+y}{2}\cos\dfrac{x-y}{2}$

$= 2 \cdot \dfrac{1}{2}\left[\sin\left(\dfrac{x+y}{2} + \dfrac{x-y}{2}\right) + \sin\left(\dfrac{x+y}{2} - \dfrac{x-y}{2}\right)\right]$

$= \sin x + \sin y$

Other formulas follow similarly.

Exercise Set 8.3A

Note: Answers for the odd-numbered exercises 1 - 31 are in the answer section in the text.

2. $\sec x - \sin x \tan x \equiv \cos x$

$\dfrac{1}{\cos x} - \sin x \cdot \dfrac{\sin x}{\cos x}$	$\cos x$
$\dfrac{1 - \sin^2 x}{\cos x}$	
$\dfrac{\cos^2 x}{\cos x}$	
$\cos x$	

4. $\dfrac{1}{\sin\theta\cos\theta} - \dfrac{\cos\theta}{\sin\theta} \equiv \dfrac{\sin\theta\cos\theta}{1 - \sin^2\theta}$

$\dfrac{1 - \cos^2\theta}{\sin\theta\cos\theta}$	$\dfrac{\sin\theta\cos\theta}{\cos^2\theta}$
$\dfrac{\sin^2\theta}{\sin\theta\cos\theta}$	$\dfrac{\sin\theta}{\cos\theta}$
$\dfrac{\sin\theta}{\cos\theta}$	

6. $\dfrac{1 - \cos x}{\sin x} \equiv \dfrac{\sin x}{1 + \cos x}$

$\dfrac{1 - \cos x}{\sin x}$	$\dfrac{\sin x}{1 + \cos x} \cdot \dfrac{1 - \cos x}{1 - \cos x}$
	$\dfrac{\sin x\,(1 - \cos x)}{1 - \cos^2 x}$
	$\dfrac{\sin x\,(1 - \cos x)}{\sin^2 x}$
	$\dfrac{1 - \cos x}{\sin x}$

8.
$$\begin{array}{c|c}
\multicolumn{2}{c}{\dfrac{\cot\theta - 1}{1 - \tan\theta} \equiv \dfrac{\csc\theta}{\sec\theta}} \\ \hline
\dfrac{\dfrac{\cos\theta}{\sin\theta} - 1}{1 - \dfrac{\sin\theta}{\cos\theta}} & \dfrac{\dfrac{1}{\sin\theta}}{\dfrac{1}{\cos\theta}} \\
\dfrac{\dfrac{\cos\theta - \sin\theta}{\sin\theta}}{\dfrac{\cos\theta - \sin\theta}{\cos\theta}} & \dfrac{\cos\theta}{\sin\theta} \\
\dfrac{\cos\theta}{\sin\theta} &
\end{array}$$

10.
$$\begin{array}{c|c}
\multicolumn{2}{c}{\dfrac{\sin x - \cos x}{\sec x - \csc x} \equiv \dfrac{\cos x}{\csc x}} \\ \hline
\dfrac{\sin x - \cos x}{\dfrac{1}{\cos x} - \dfrac{1}{\sin x}} & 2\sin x\cos x \\
\dfrac{\sin x - \cos x}{\dfrac{\sin x - \cos x}{\sin x\cos x}} & \\
\sin x\cos x &
\end{array}$$

12.
$$\begin{array}{c|c}
\multicolumn{2}{c}{\dfrac{\cos^2\theta + \cot\theta}{\cos^2\theta - \cot\theta} \equiv \dfrac{\cos^2\theta\tan\theta + 1}{\cos^2\theta\tan\theta - 1}} \\ \hline
\dfrac{\cos^2\theta + \dfrac{\cos\theta}{\sin\theta}}{\cos^2\theta - \dfrac{\cos\theta}{\sin\theta}} & \dfrac{\cos^2\theta\dfrac{\sin\theta}{\cos\theta} + 1}{\cos^2\theta\dfrac{\sin\theta}{\cos\theta} - 1} \\
\dfrac{\cos\theta\left(\cos\theta + \dfrac{1}{\sin\theta}\right)}{\cos\theta\left(\cos\theta - \dfrac{1}{\sin\theta}\right)} & \dfrac{\sin\theta\cos\theta + 1}{\sin\theta\cos\theta - 1} \\
\dfrac{\cos\theta + \dfrac{1}{\sin\theta}}{\cos\theta - \dfrac{1}{\sin\theta}} & \\
\dfrac{\dfrac{\sin\theta\cos\theta + 1}{\sin\theta}}{\dfrac{\sin\theta\cos\theta - 1}{\sin\theta}} & \\
\dfrac{\sin\theta\cos\theta + 1}{\sin\theta\cos\theta - 1} &
\end{array}$$

14.
$$\begin{array}{c|c}
\multicolumn{2}{c}{\dfrac{2\tan\theta}{1 + \tan^2\theta} \equiv \sin 2\theta} \\ \hline
\dfrac{2\tan\theta}{\sec^2\theta} & 2\sin\theta\cos\theta \\
\dfrac{2\sin\theta}{\cos\theta}\cdot\dfrac{\cos^2\theta}{1} & \\
2\sin\theta\cos\theta &
\end{array}$$

16.
$$\begin{array}{c|c}
\multicolumn{2}{c}{\cot 2\theta \equiv \dfrac{\cot^2\theta - 1}{2\cot\theta}} \\ \hline
\dfrac{\cos 2\theta}{\sin 2\theta} & \dfrac{\dfrac{\cos^2\theta}{\sin^2\theta} - 1}{2\dfrac{\cos\theta}{\sin\theta}} \\
\dfrac{\cos^2\theta - \sin^2\theta}{2\sin\theta\cos\theta} & \dfrac{1}{2}\left[\dfrac{\cos^2\theta - \sin^2\theta}{\sin^2\theta}\cdot\dfrac{\sin\theta}{\cos\theta}\right] \\
 & \dfrac{\cos^2\theta - \sin^2\theta}{2\sin\theta\cos\theta}
\end{array}$$

18.
$$\begin{array}{c|c}
\multicolumn{2}{c}{\dfrac{\cos(\alpha - \beta)}{\cos\alpha\sin\beta} \equiv \tan\alpha + \cot\beta} \\ \hline
\dfrac{\cos\alpha\cos\beta + \sin\alpha\sin\beta}{\cos\alpha\sin\beta} & \tan\alpha + \cot\beta \\
\dfrac{\cos\alpha\cos\beta}{\cos\alpha\sin\beta} + \dfrac{\sin\alpha\sin\beta}{\cos\alpha\sin\beta} & \\
\dfrac{\cos\beta}{\sin\beta} + \dfrac{\sin\alpha}{\cos\alpha} & \\
\cot\beta + \tan\alpha &
\end{array}$$

20.
$$\begin{array}{c|c}
\multicolumn{2}{c}{2\sin\theta\cos^3\theta + 2\sin^3\theta\cos\theta \equiv \sin 2\theta} \\ \hline
2\sin\theta\cos\theta(\cos^2\theta + \sin^2\theta) & 2\sin\theta\cos\theta \\
2\sin\theta\cos\theta &
\end{array}$$

22.
$$\begin{array}{c|c}
\multicolumn{2}{c}{\dfrac{\tan\theta - \sin\theta}{2\tan\theta} \equiv \sin^2\dfrac{\theta}{2}} \\ \hline
\dfrac{1}{2}\left[\dfrac{\dfrac{\sin\theta}{\cos\theta} - \sin\theta}{\dfrac{\sin\theta}{\cos\theta}}\right] & \dfrac{1 - \cos\theta}{2} \\
\dfrac{1}{2}\,\dfrac{\sin\theta - \sin\theta\cos\theta}{\cos\theta}\cdot\dfrac{\cos\theta}{\sin\theta} & \\
\dfrac{1 - \cos\theta}{2} &
\end{array}$$

24.
$$\begin{array}{c|c}
\multicolumn{2}{c}{\dfrac{\cos^4 x - \sin^4 x}{1 - \tan^4 x} \equiv \cos^4 x} \\ \hline
\dfrac{\cos^4 x - \sin^4 x}{1 - \dfrac{\sin^4 x}{\cos^4 x}} & \cos^4 x \\
\dfrac{\cos^4 x - \sin^4 x}{\dfrac{\cos^4 x - \sin^4 x}{\cos^4 x}} & \\
\cos^4 x &
\end{array}$$

26.
$$\begin{array}{c|c}
\multicolumn{2}{c}{\left(\dfrac{1 + \tan\theta}{1 - \tan\theta}\right)^2 \equiv \dfrac{1 + \sin 2\theta}{1 - \sin 2\theta}} \\ \hline
\left[\dfrac{\dfrac{\cos\theta + \sin\theta}{\cos\theta}}{\dfrac{\cos\theta - \sin\theta}{\cos\theta}}\right]^2 & \dfrac{1 + 2\sin\theta\cos\theta}{1 - 2\sin\theta\cos\theta} \\
\dfrac{\cos^2\theta + 2\sin\theta\cos\theta + \sin^2\theta}{\cos^2\theta - 2\sin\theta\cos\theta + \sin^2\theta} & \\
\dfrac{1 + 2\sin\theta\cos\theta}{1 - 2\sin\theta\cos\theta} &
\end{array}$$

28.
$$\frac{\sin^3 t + \cos^3 t}{\sin t + \cos t} \equiv \frac{2 - \sin 2t}{2}$$

$\frac{(\sin t + \cos t)(\sin^2 t - \sin t \cos t + \cos^2 t)}{\sin t + \cos t}$	$\frac{2 - 2 \sin t \cos t}{2}$
$1 - \sin t \cos t$	$1 - \sin t \cos t$

30.
$$\cos(\alpha + \beta)\cos(\alpha - \beta) \equiv \cos^2\alpha - \sin^2\beta$$

$(\cos\alpha\cos\beta - \sin\alpha\sin\beta)(\cos\alpha\cos\beta + \sin\alpha\sin\beta)$	$\cos^2\alpha - \sin^2\beta$
$\cos^2\alpha\cos^2\beta - \sin^2\alpha\sin^2\beta$	
$(1 - \sin^2\alpha)(1 - \sin^2\beta) - \sin^2\alpha\sin^2\beta$	
$1 - \sin^2\alpha - \sin^2\beta + \sin^2\alpha\sin^2\beta - \sin^2\alpha\sin^2\beta$	
$1 - \sin^2\alpha - \sin^2\beta$	
$\cos^2\alpha - \sin^2\beta$	

32.
$$\sin(\alpha + \beta) + \sin(\alpha - \beta) \equiv 2\sin\alpha\cos\beta$$

$\sin\alpha\cos\beta + \cos\alpha\sin\beta + \sin\alpha\cos\beta - \cos\alpha\sin\beta$	$2\sin\alpha\cos\beta$
$2\sin\alpha\cos\beta$	

33. Theorem 2: For any numbers a, c, and d, $c \sin ax + d \cos ax \equiv A \sin(ax + b)$, where $A = \sqrt{c^2 + d^2}$ and b is a number whose cosine is $\frac{c}{A}$ and whose sine is $\frac{d}{A}$.

$\sin 2x + \sqrt{3} \cos 2x$

$a = 2, \quad c = 1, \quad$ and $d = \sqrt{3}$

Find A:

$A = \sqrt{c^2 + d^2} = \sqrt{1^2 + (\sqrt{3})^2} = \sqrt{4} = 2$

Find b:

We know that b is a number such that

$\cos b = \frac{1}{2}$ and $\sin b = \frac{\sqrt{3}}{2}$.

Therefore, we can use 60° or $\frac{\pi}{3}$ for b.

Thus, $\sin 2x + \sqrt{3} \cos 2x \equiv 2 \sin\left(2x + \frac{\pi}{3}\right)$.

34. $2 \sin\left(3x + \frac{11\pi}{6}\right)$, or $2 \sin\left(3x - \frac{\pi}{6}\right)$

35. $4 \sin x + 3 \cos x$

$a = 1, \quad c = 4, \quad$ and $d = 3$

Find A:

$A = \sqrt{c^2 + d^2} = \sqrt{4^2 + 3^2} = \sqrt{25} = 5$

Find b:

We know that b is a number such that

$\cos b = \frac{4}{5}$ and $\sin b = \frac{3}{5}$.

Thus, $4 \sin x + 3 \cos x \equiv 5 \sin(x + b)$, where b is a number whose cosine is $\frac{4}{5}$ and whose sine is $\frac{3}{5}$.

36. $5 \sin(2x + b)$, where $\cos b = \frac{4}{5}$, $\sin b = -\frac{3}{5}$

37. $6.75 \sin 0.374x + 4.08 \cos 0.374x$

$a = 0.374$, $c = 6.75$, and $d = 4.08$

Find A:

$A = \sqrt{(6.75)^2 + (4.08)^2} = \sqrt{45.5625 + 16.6464} = \sqrt{62.2089} \approx 7.89$

Find b:

We know that b is a number such that

$\cos b = \frac{6.75}{7.89} \approx 0.85$ and $\sin b = \frac{4.08}{7.89} \approx 0.517$.

Therefore, we can use 31.2° for b.

$6.75 \sin 0.374x + 4.08 \cos 0.374x$

$\equiv 7.89 \sin (0.374x + 31.2°)$

38. $97.93 \sin (0.8081x - 2.85°)$

39. See text answer section.

40. $2 \sin \left(x + \frac{\pi}{2}\right)$

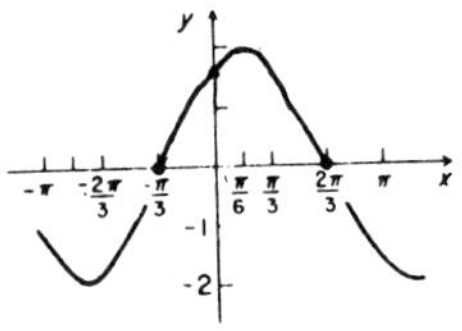

41. See text answer section.

42. $\sin \theta = \cos \phi$

43. See text answer section.

Exercise Set 8.3B

Note: Answers for the odd-numbered exercises 1 - 55 are in the answer section in the text.

2. $\cos^2 x (1 - \sec^2 x) \equiv -\sin^2 x$

$\cos^2 x (-\tan^2 x)$	$-\sin^2 x$
$\cos^2 x \left(-\frac{\sin^2 x}{\cos^2 x}\right)$	
$-\sin^2 x$	

4. $\tan \theta (\tan \theta + \cot \theta) \equiv \sec^2 \theta$

$\tan^2 \theta + \tan \theta \cot \theta$	$\sec^2 \theta$
$\tan^2 \theta + 1$	
$\sec^2 \theta$	

6. $\frac{\tan x + \sin x}{1 + \sec x} \equiv \sin x$

$\dfrac{\frac{\sin x}{\cos x} + \frac{\sin x}{1}}{1 + \frac{1}{\cos x}}$	$\sin x$
$\dfrac{\frac{\sin x + \sin x \cos x}{\cos x}}{\frac{\cos x + 1}{\cos x}}$	
$\frac{\sin x (1 + \cos x)}{\cos x} \cdot \frac{\cos x}{\cos x + 1}$	
$\sin x$	

8. $\frac{1 + \tan^2 \theta}{\csc^2 \theta} \equiv \tan^2 \theta$

$\frac{\sec^2 \theta}{\csc^2 \theta}$	$\frac{\sin^2 \theta}{\cos^2 \theta}$
$\frac{1}{\cos^2 \theta} \cdot \frac{\sin^2 \theta}{1}$	
$\frac{\sin^2 \theta}{\cos^2 \theta}$	

10. $\frac{1 + \cos^2 x}{\sin^2 x} \equiv 2 \csc^2 x - 1$

$\frac{1}{\sin^2 x} + \frac{\cos^2 x}{\sin^2 x}$	$2 \csc^2 x - 1$
$\csc^2 x + \cot^2 x$	
$\csc^2 x + \csc^2 x - 1$	
$2 \csc^2 x - 1$	

12. $\frac{\csc x - \sin x}{\cot x} \equiv \cos x$

$\dfrac{\frac{1}{\sin x} - \sin x}{\frac{\cos x}{\sin x}}$	$\cos x$
$\frac{1 - \sin^2 x}{\sin x} \cdot \frac{\sin x}{\cos x}$	
$\frac{1 - \sin^2 x}{\cos x}$	
$\frac{\cos^2 x}{\cos x}$	
$\cos x$	

14. $\frac{\csc\theta - \sin\theta}{\cos^2\theta} \equiv \csc\theta$

$\frac{\frac{1}{\sin\theta}-\sin\theta}{\cos^2\theta}$	$\frac{1}{\sin\theta}$
$\frac{1-\sin^2\theta}{\sin\theta}\cdot\frac{1}{\cos^2\theta}$	
$\frac{\cos^2\theta}{\sin\theta}\cdot\frac{1}{\cos^2\theta}$	
$\frac{1}{\sin\theta}$	

16. $\frac{1+\sin x}{1-\sin x}+\frac{\sin x - 1}{1+\sin x} \equiv 4\sec x\tan x$

$\frac{(1+\sin x)^2-(1-\sin x)^2}{1-\sin^2 x}$	$4\cdot\frac{1}{\cos x}\cdot\frac{\sin x}{\cos x}$
$\frac{(1+2\sin x+\sin^2 x)-(1-2\sin x+\sin^2 x)}{\cos^2 x}$	$\frac{4\sin x}{\cos^2 x}$
$\frac{4\sin x}{\cos^2 x}$	

18. $\cos^2\theta\cot^2\theta \equiv \cot^2\theta-\cos^2\theta$

$(1-\sin^2\theta)\cot^2\theta$	$\cot^2\theta-\cos^2\theta$
$\cot^2\theta - \sin^2\theta\cdot\frac{\cos^2\theta}{\sin^2\theta}$	
$\cot^2\theta-\cos^2\theta$	

20. $\frac{\cos^2 x-1}{1-\sec^2 x}\equiv\frac{1}{\tan^2 x+1}$

$\frac{-\sin^2 x}{-\tan^2 x}$	$\frac{1}{\sec^2 x}$
$\sin^2 x\cdot\frac{\cos^2 x}{\sin^2 x}$	$\cos^2 x$
$\cos^2 x$	

22. $\tan\theta-\cot\theta\equiv(\sec\theta-\csc\theta)(\sin\theta+\cos\theta)$

$\frac{\sin\theta}{\cos\theta}-\frac{\cos\theta}{\sin\theta}$	$\left(\frac{1}{\cos\theta}-\frac{1}{\sin\theta}\right)(\sin\theta+\cos\theta)$
$\frac{\sin^2\theta-\cos^2\theta}{\cos\theta\sin\theta}$	$\left(\frac{\sin\theta-\cos\theta}{\sin\theta\cos\theta}\right)\left(\frac{\sin\theta+\cos\theta}{1}\right)$
	$\frac{\sin^2\theta-\cos^2\theta}{\sin\theta\cos\theta}$

24. $\csc x-\cot x\equiv\frac{1}{\csc x+\cot x}$

$\frac{1}{\sin x}-\frac{\cos x}{\sin x}$	$\frac{1}{\frac{1}{\sin x}+\frac{\cos x}{\sin x}}$
$\frac{1-\cos x}{\sin x}\cdot\frac{1+\cos x}{1+\cos x}$	$\frac{1}{\frac{1+\cos x}{\sin x}}$
$\frac{1-\cos^2 x}{\sin x\,(1+\cos x)}$	$\frac{\sin x}{1+\cos x}$
$\frac{\sin^2 x}{\sin x\,(1+\cos x)}$	
$\frac{\sin x}{1+\cos x}$	

26. $2\sin^2\theta\cos^2\theta+\cos^4\theta\equiv 1-\sin^4\theta$

$\cos^2\theta(2\sin^2\theta+\cos^2\theta)$	$(1+\sin^2\theta)(1-\sin^2\theta)$
$\cos^2\theta(\sin^2\theta+\sin^2\theta+\cos^2\theta)$	$(1+\sin^2\theta)(\cos^2\theta)$
$\cos^2\theta(\sin^2\theta+1)$	

28. $\frac{\cot\theta}{\csc\theta-1}\equiv\frac{\csc\theta+1}{\cot\theta}$

$\frac{\cot\theta}{\csc\theta-1}\cdot\frac{\csc\theta+1}{\csc\theta+1}$	$\frac{\csc\theta+1}{\cot\theta}$
$\frac{\cot\theta(\csc\theta+1)}{\csc^2\theta-1}$	
$\frac{\cot\theta(\csc\theta+1)}{\cot^2\theta}$	
$\frac{\csc\theta+1}{\cot\theta}$	

30. $\frac{1+\sin x}{1-\sin x}\equiv(\sec x+\tan x)^2$

$\frac{1+\sin x}{1-\sin x}\cdot\frac{1+\sin x}{1+\sin x}$	$\left(\frac{1}{\cos x}+\frac{\sin x}{\cos x}\right)^2$
$\frac{(1+\sin x)^2}{1-\sin^2 x}$	$\frac{(1+\sin x)^2}{\cos^2 x}$
$\frac{(1+\sin x)^2}{\cos^2 x}$	

32. $\sec^4\theta-\tan^2\theta\equiv\tan^4\theta+\sec^2\theta$

$\sec^4\theta-(\sec^2\theta-1)$	$(\tan^2\theta)^2+\sec^2\theta$
$\sec^4\theta-\sec^2\theta+1$	$(\sec^2\theta-1)^2+\sec^2\theta$
	$\sec^4\theta-2\sec^2\theta+1+\sec^2\theta$
	$\sec^4\theta-\sec^2\theta+1$

34. $1+\sec x\equiv\csc x(\sin x+\tan x)$

$1+\frac{1}{\cos x}$	$\frac{1}{\sin x}\left(\frac{\sin x}{1}+\frac{\sin x}{\cos x}\right)$
	$1+\frac{1}{\cos x}$

36. $\frac{\sin^3\theta-\cos^3\theta}{\sin\theta-\cos\theta}\equiv\sin\theta\cos\theta+1$

$\frac{(\sin\theta-\cos\theta)(\sin^2\theta+\sin\theta\cos\theta+\cos^2\theta)}{\sin\theta-\cos\theta}$	$\sin\theta\cos\theta+1$
$\sin^2\theta+\sin\theta\cos\theta+\cos^2\theta$	
$\sin\theta\cos\theta+1$	

38. $2\sec^2 x + \frac{\csc x}{1-\csc x} \equiv \frac{\csc x}{\csc x + 1}$

$2\cdot\frac{1}{\cos^2 x} + \frac{\frac{1}{\sin x}}{1-\frac{1}{\sin x}}$	$\frac{\frac{1}{\sin x}}{\frac{1}{\sin x}+1}$
$\frac{2}{\cos^2 x} + \frac{\frac{1}{\sin x}}{\frac{\sin x - 1}{\sin x}}$	$\frac{\frac{1}{\sin x}}{\frac{1+\sin x}{\sin x}}$
$\frac{2}{\cos^2 x} + \frac{1}{\sin x - 1}$	$\frac{1}{1+\sin x}\cdot\frac{1-\sin x}{1-\sin x}$
$\frac{2}{\cos^2 x} + \frac{1}{\sin x - 1}\cdot\frac{\sin x + 1}{\sin x + 1}$	$\frac{1-\sin x}{1-\sin^2 x}$
$\frac{2}{\cos^2 x} + \frac{\sin x + 1}{\sin^2 x - 1}$	$\frac{1-\sin x}{\cos^2 x}$
$\frac{2}{\cos^2 x} + \frac{\sin x + 1}{-\cos^2 x}$	
$\frac{1-\sin x}{\cos^2 x}$	

40. $\cos\theta + \frac{\sin\theta}{\cot\theta - 1} \equiv \frac{\cos\theta}{1-\tan\theta} - \sin\theta$

$\cos\theta + \frac{\sin\theta}{\frac{\cos\theta}{\sin\theta}-1}$	$\frac{\cos\theta}{1-\frac{\sin\theta}{\cos\theta}} - \sin\theta$
$\cos\theta + \frac{\sin\theta}{1}\cdot\frac{\sin\theta}{\cos\theta-\sin\theta}$	$\frac{\cos\theta}{1}\cdot\frac{\cos\theta}{\cos\theta-\sin\theta} - \sin\theta$
$\cos\theta + \frac{\sin^2\theta}{\cos\theta-\sin\theta}$	$\frac{\cos^2\theta}{\cos\theta-\sin\theta} - \sin\theta$
$\frac{\cos^2\theta - \cos\theta\sin\theta + \sin^2\theta}{\cos\theta - \sin\theta}$	$\frac{\cos^2\theta - \sin\theta\cos\theta + \sin^2\theta}{\cos\theta - \sin\theta}$

42. $\frac{\cos x + \cot x}{1+\csc x} \equiv \cos x$

$\frac{\frac{\cos x}{1} + \frac{\cos x}{\sin x}}{1+\frac{1}{\sin x}}$	$\cos x$
$\frac{\sin x\cos x + \cos x}{\sin x}\cdot\frac{\sin x}{\sin x + 1}$	
$\frac{\cos x(\sin x + 1)}{\sin x + 1}$	
$\cos x$	

44. $\sec^2\theta - \csc^2\theta \equiv \frac{\tan\theta - \cot\theta}{\sin\theta\cos\theta}$

$\frac{1}{\cos^2\theta} - \frac{1}{\sin^2\theta}$	$\frac{\frac{\sin\theta}{\cos\theta} - \frac{\cos\theta}{\sin\theta}}{\sin\theta\cos\theta}$
$\frac{\sin^2\theta - \cos^2\theta}{\cos^2\theta\sin^2\theta}$	$\frac{\sin^2\theta - \cos^2\theta}{\sin\theta\cos\theta}\cdot\frac{1}{\sin\theta\cos\theta}$
	$\frac{\sin^2\theta - \cos^2\theta}{\sin^2\theta\cos^2\theta}$

46. $\sec^4 x - 4\tan^2 x \equiv (1-\tan^2 x)^2$

$(\sec^2 x)^2 - 4\tan^2 x$	$1 - 2\tan^2 x + \tan^4 x$
$(1+\tan^2 x)^2 - 4\tan^2 x$	
$1 + 2\tan^2 x + \tan^4 x - 4\tan^2 x$	
$1 - 2\tan^2 x + \tan^4 x$	

48. $\frac{\tan^2 x + \sec^2 x}{\sec^4 x} \equiv 1 - \sin^4 x$

$\frac{\frac{\sin^2 x}{\cos^2 x} + \frac{1}{\cos^2 x}}{\frac{1}{\cos^4 x}}$	$(1-\sin^2 x)(1+\sin^2 x)$
$\frac{\frac{\sin^2 x + 1}{\cos^2 x}}{\frac{1}{\cos^4 x}}$	$\cos^2 x\ (\sin^2 x + 1)$
$\cos^2 x(\sin^2 x + 1)$	

50. $\tan 2x(\cos x + \cos 3x) \equiv \sin x + \sin 3x$

$\frac{\sin 2x}{\cos 2x}\left[2\cos\frac{4x}{2}\cos\frac{-2x}{2}\right]$	$2\sin\frac{4x}{2}\cos\frac{-2x}{2}$
$\frac{\sin 2x}{\cos 2x}(2\cos 2x\cos x)$	$2\cos 2x\cos x$
$2\sin 2x\cos x$	

52. $\tan\frac{x+y}{2} \equiv \frac{\sin x + \sin y}{\cos x + \cos y}$

$\frac{\sin\frac{x+y}{2}}{\cos\frac{x+y}{2}}$	$\frac{2\sin\frac{x+y}{2}\cos\frac{x-y}{2}}{2\cos\frac{x+y}{2}\cos\frac{x-y}{2}}$
	$\frac{\sin\frac{x+y}{2}}{\cos\frac{x+y}{2}}$

54. $\tan\frac{\theta+\phi}{2}\tan\frac{\phi-\theta}{2} \equiv \frac{\cos\theta - \cos\phi}{\cos\theta + \cos\phi}$

$\tan\frac{\theta+\phi}{2}\tan\frac{\phi-\theta}{2}$	$\frac{2\sin\frac{\phi+\theta}{2}\sin\frac{\phi-\theta}{2}}{2\cos\frac{\phi+\theta}{2}\cos\frac{\phi-\theta}{2}}$
	$\tan\frac{\theta+\phi}{2}\tan\frac{\phi-\theta}{2}$

56. $\sin 2\theta + \sin 4\theta + \sin 6\theta \equiv 4\cos\theta\cos 2\theta\sin 3\theta$

$2\sin\frac{6\theta}{2}\cos\frac{-2\theta}{2} + \sin 6\theta$	$4\cos\theta\cos 2\theta\sin 3\theta$
$2\sin 3\theta\cos\theta + 2\sin 3\theta\cos 3\theta$	
$2\sin 3\theta(\cos\theta + \cos 3\theta)$	
$2\sin 3\theta\left[2\cos\frac{4\theta}{2}\cos\frac{2\theta}{2}\right]$	
$2\sin 3\theta(2\cos 2\theta\cos\theta)$	
$4\cos\theta\cos 2\theta\sin 3\theta$	

Exercise Set 8.4

1. $\arcsin\frac{\sqrt{2}}{2}$

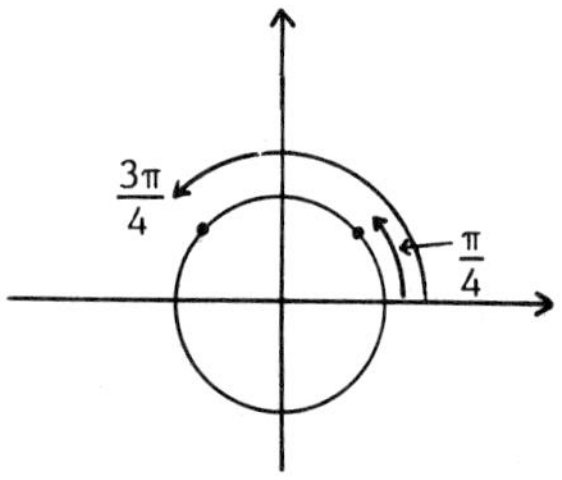

1. (continued)

On the unit circle there are two points at which the sine is $\frac{\sqrt{2}}{2}$. The arc length for the point in the first quadrant is $\frac{\pi}{4}$ plus any multiple of 2π. The arc length for the point in the second quadrant is $\frac{3\pi}{4}$ plus any multiple of 2π.

All values of arcsin $\frac{\sqrt{2}}{2}$ are

$\frac{\pi}{4} + 2k\pi$ or $\frac{3\pi}{4} + 2k\pi$, where k is any integer.

2. $\frac{\pi}{3} + 2k\pi$, $\frac{2\pi}{3} + 2k\pi$

3. $\cos^{-1} \frac{\sqrt{2}}{2}$

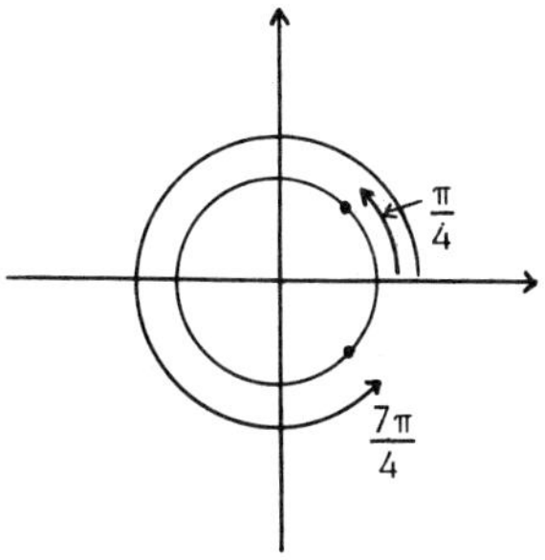

On the unit circle there are two points at which the cosine is $\frac{\sqrt{2}}{2}$. The arc length for the point in the first quadrant is $\frac{\pi}{4}$ plus any multiple of 2π. The arc length for the point in the fourth quadrant is $\frac{7\pi}{4}$ plus any multiple of 2π.

All values of $\cos^{-1} \frac{\sqrt{2}}{2}$ are

$\frac{\pi}{4} + 2k\pi$ or $\frac{7\pi}{4} + 2k\pi$, where k is any integer.

4. $\frac{\pi}{6} + 2k\pi$, $\frac{11\pi}{6} + 2k\pi$

5. $\sin^{-1} \left(-\frac{\sqrt{2}}{2}\right)$

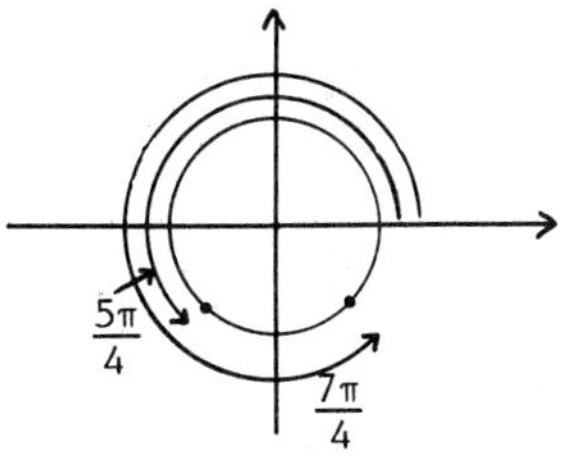

On the unit circle there are two points at which the sine is $-\frac{\sqrt{2}}{2}$. The arc length for the point in the third quadrant is $\frac{5\pi}{4}$ plus any multiple of 2π. The arc length for the point in the fourth quadrant is $\frac{7\pi}{4}$ plus any multiple of 2π.

5. (continued)

All values of $\sin^{-1} \left(-\frac{\sqrt{2}}{2}\right)$ are

$\frac{5\pi}{4} + 2k\pi$ or $\frac{7\pi}{4} + 2k\pi$, where k is any integer.

6. $\frac{4\pi}{3} + 2k\pi$, $\frac{5\pi}{3} + 2k\pi$

7. arccos $\left(-\frac{\sqrt{2}}{2}\right)$

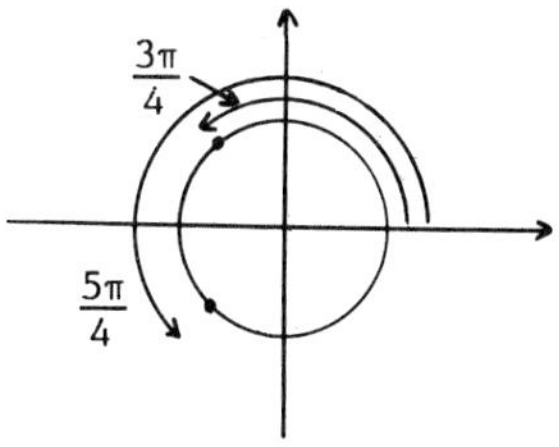

On the unit circle there are two points at which the cosine is $-\frac{\sqrt{2}}{2}$. The arc length for the point in the second quadrant is $\frac{3\pi}{4}$ plus any multiple of 2π. The arc length for the point in the third quadrant is $\frac{5\pi}{4}$ plus any multiple of 2π.

All values of arccos $\left(-\frac{\sqrt{2}}{2}\right)$ are

$\frac{3\pi}{4} + 2k\pi$ or $\frac{5\pi}{4} + 2k\pi$, where k is any integer.

8. $\frac{5\pi}{6} + 2k\pi$, $\frac{7\pi}{6} + 2k\pi$

9. arctan $\sqrt{3}$

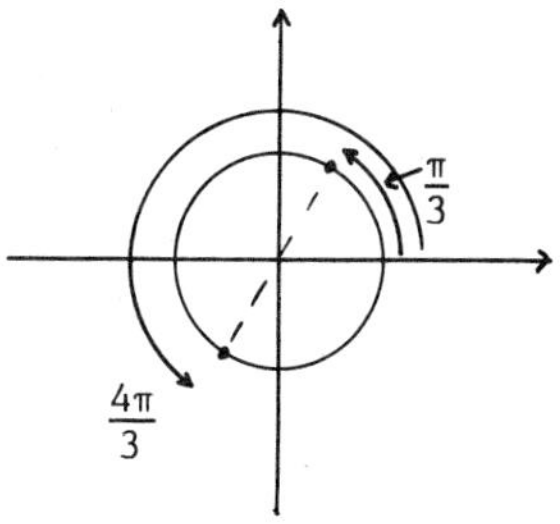

On the unit circle there are two points at which the tangent is $\sqrt{3}$. Since these points are opposite ends of a diameter, the arc lengths differ by π. The arc length for the point in the first quadrant is $\frac{\pi}{3}$.

All values of arctan $\sqrt{3}$ are

$\frac{\pi}{3} + k\pi$, where k is any integer.

10. $\frac{\pi}{6} + k\pi$

11. $\cot^{-1} 1$

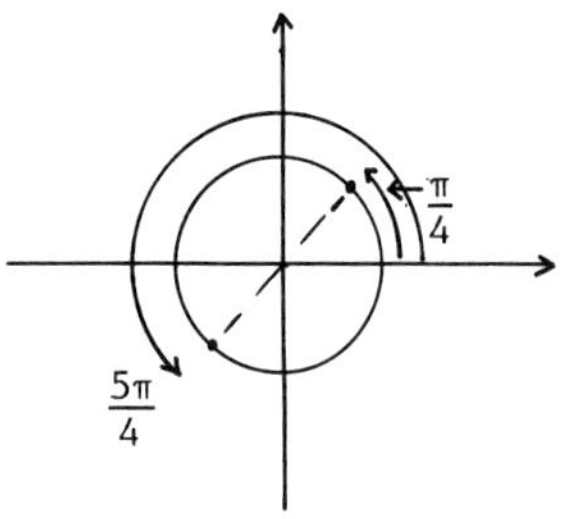

On the unit circle there are two points at which the cotangent is 1. Since these points are opposite ends of a diameter, the arc lengths differ by π. The arc length for the point in the first quadrant is $\frac{\pi}{4}$.

All values of $\cot^{-1} 1$ are

$\frac{\pi}{4} + k\pi$, where k is any integer.

12. $\frac{\pi}{6} + k\pi$

13. $\arctan\left(-\frac{\sqrt{3}}{3}\right)$

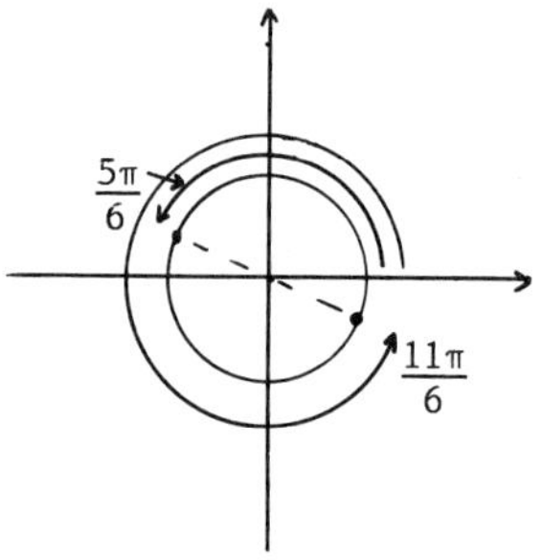

On the unit circle there are two points at which the tangent is $-\frac{\sqrt{3}}{3}$. Since these points are opposite ends of a diameter, the arc lengths differ by π. The arc length for the point in the second quadrant is $\frac{5\pi}{6}$.

All values of $\arctan\left(-\frac{\sqrt{3}}{3}\right)$ are

$\frac{5\pi}{6} + k\pi$, where k is any integer.

14. $\frac{2\pi}{3} + k\pi$

15. arccot (-1)

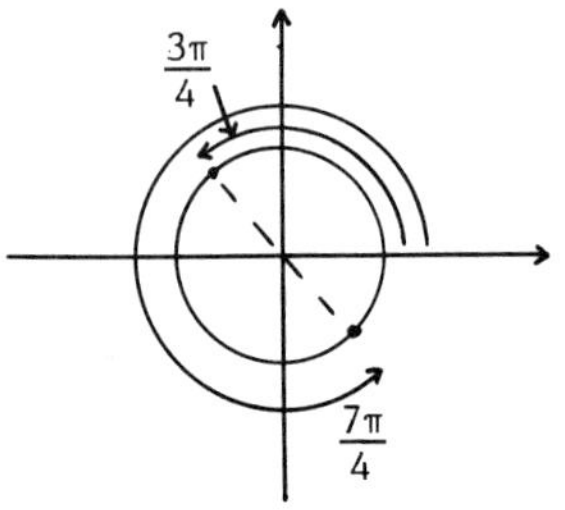

15. (continued)

On the unit circle there are two points at which the cotangent is -1. Since these points are opposite ends of a diameter, the arc lengths differ by π. The arc length for the point in the second quadrant is $\frac{3\pi}{4}$.

All values of arccot (-1) are
$\frac{3\pi}{4} + k\pi$, where k is any integer.

16. $\frac{5\pi}{6} + k\pi$

17. arcsec 1

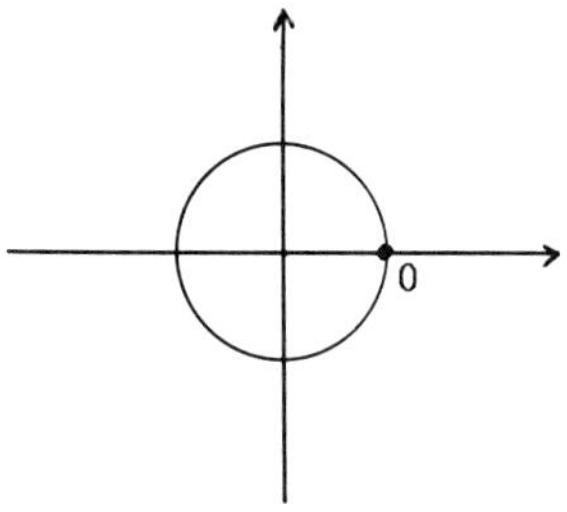

On the unit circle there is only one point at which the secant is 1. The arc length for this point is 0 plus any multiple of 2π.

All values of arcsec 1 are
$0 + 2k\pi$, where k is any integer.

18. $\frac{\pi}{3} + 2k\pi$, $\frac{5\pi}{3} + 2k\pi$

19. $\csc^{-1} 1$

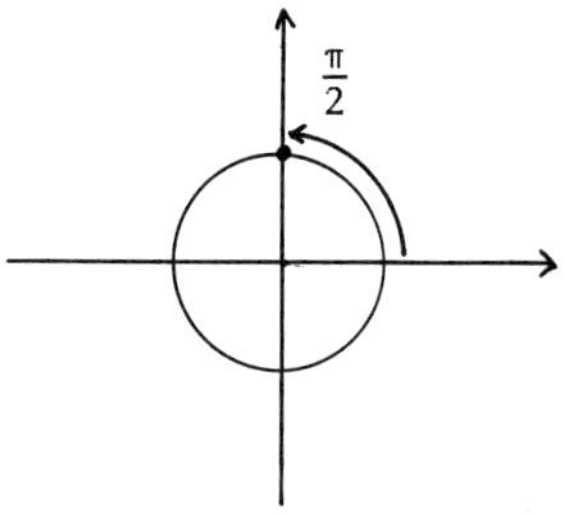

On the unit circle there is only one point at which the cosecant is 1. The arc length for this point is $\frac{\pi}{2}$ plus any multiple of 2π.

All values of $\csc^{-1} 1$ are
$\frac{\pi}{2} + 2k\pi$, where k is any integer.

20. $\frac{\pi}{6} + 2k\pi$, $\frac{5\pi}{6} + 2k\pi$

21. arcsin 0.3907

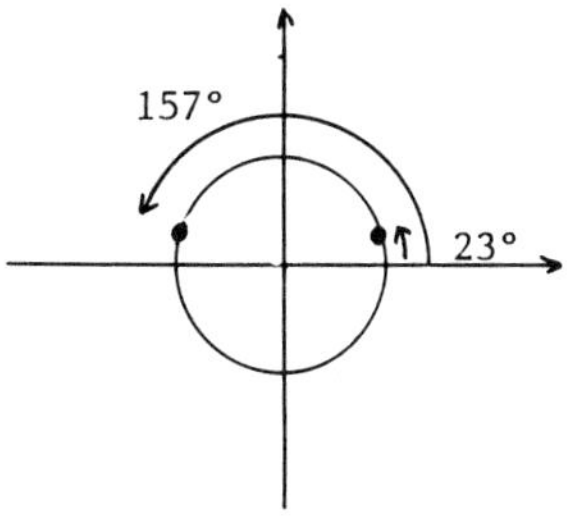

Using a calculator we find that the angle whose sine is 0.3907 is 23°. This is a reference angle. Sketch this on a unit circle to find the two points where the sine is 0.3907. The two angles are 23° and 157°, plus any multiple of 360°.

Thus all values of arcsin 0.3907 are

23° + k·360° or 157° + k·360°,

where k is any integer.

22. 74° + k·360°, 106° + k·360°

23. $\sin^{-1}$ 0.6293

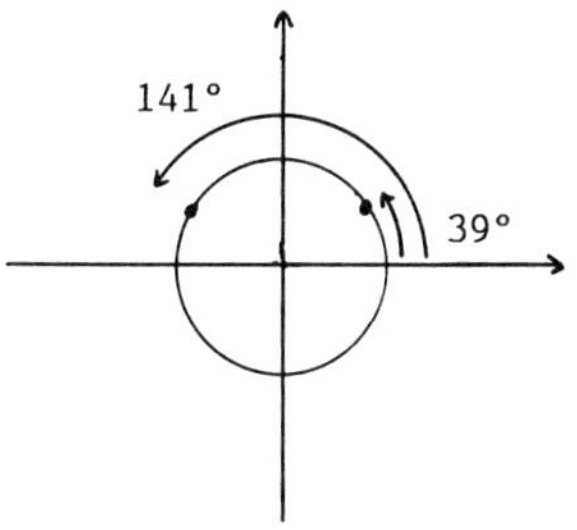

Using a calculator we find that the angle whose sine is 0.6293 is 39°. This is a reference angle. Sketch this on a unit circle to find the two points where the sine is 0.6293. The two angles are 39° and 141°, plus any multiple of 360°.

Thus all values of $\sin^{-1}$ 0.6293 are

39° + k·360° or 141° + k·360°,

where k is any integer.

24. 61° + k·360°, 119° + k·360°

25. arccos 0.7990

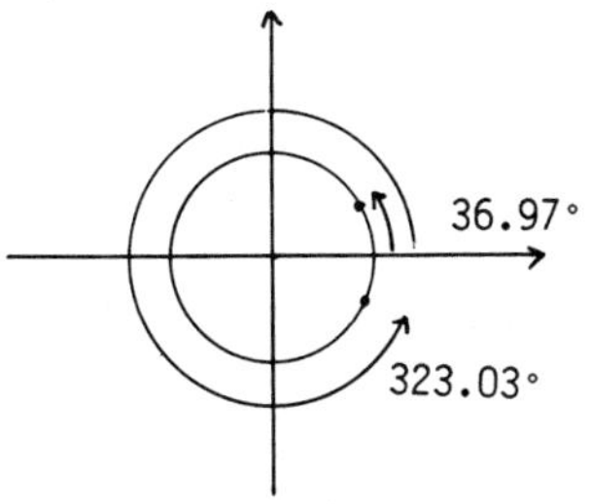

Using a calculator we find that the angle whose cosine is 0.7990 is 36.97°, or 36° 58'. This is a reference angle. Sketch this on a unit circle to find the two points where the cosine is 0.7990. The two angles are 36.97° and 323.03°, plus any multiple of 360°.

25. (continued)

Thus all values of arccos 0.7990 are

36.97° (or 36° 58') + k·360° or

323.03° (or 323° 02') + k·360°,

where k is any integer.

26. 22.10° (or 22° 06') + k·360°,

337.90° (or 337° 54') + k·360°

27. $\cos^{-1}$ 0.9310

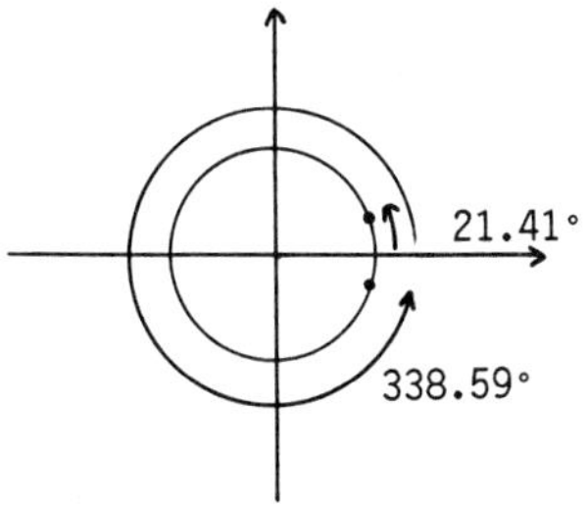

Using a calculator we find that the angle whose cosine is 0.9310 is 21.41°, or 21° 25'. This is a reference angle. On a unit circle we find the two points where the cosine is 0.9310. The two angles are 21.41° and 338.59°, plus any multiple of 360°.

Thus all values of $\cos^{-1}$ 0.9310 are

21.41° (or 21° 25') + k·360° or

338.59° (or 338° 35') + k·360°,

where k is any integer.

28. 74.13° (or 74° 08') + k·360°,

285.87° (or 285° 52') + k·360°

29. $\tan^{-1}$ 0.3673

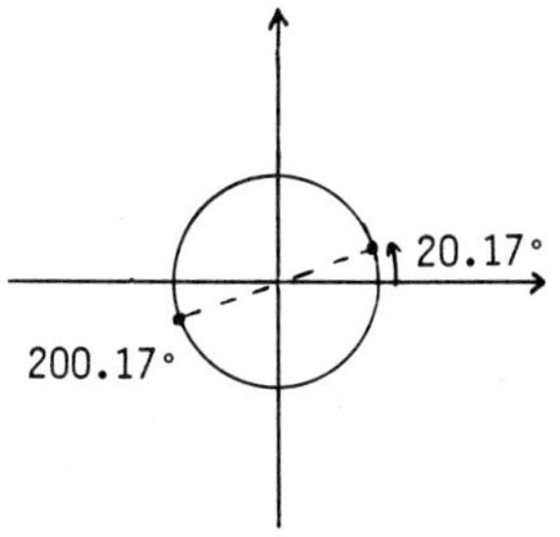

Using a calculator we find the angle whose tangent is 0.3673 is 20.17°, or 20° 10'. This is a reference angle. On a unit circle we find the two points where the tangent is 0.3673. The two points are opposite ends of a diameter. Hence the angles differ by 180°.

Thus all values of $\tan^{-1}$ 0.3673 are

20.17° (or 20° 10') + k·180°, where k is any integer.

30. 47.49° (or 47° 29') + k·180°

31. $\cot^{-1} 1.265$

[Note: $\cot \theta = \frac{1}{\tan \theta}$]

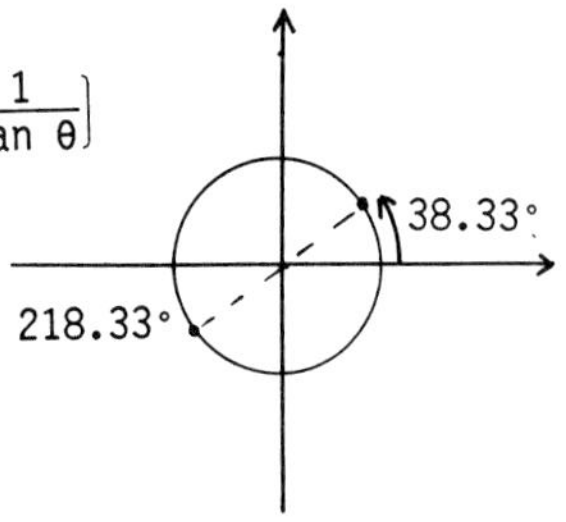

Using a calculator we find the angle whose cotangent is 1.265 is 38.33°, or 38° 20'. This is a reference angle. On a unit circle we find the two points where the cotangent is 1.265. The two points are opposite ends of a diameter. Hence the angles differ by 180°.

Thus all values of $\cot^{-1} 1.265$ are

38.33° (or 38° 20') + k·180°, where k is any integer.

32. 64.50° (or 64° 30') + k·180°

33. $\sec^{-1} 1.167$

[Note: $\sec \theta = \frac{1}{\cos \theta}$]

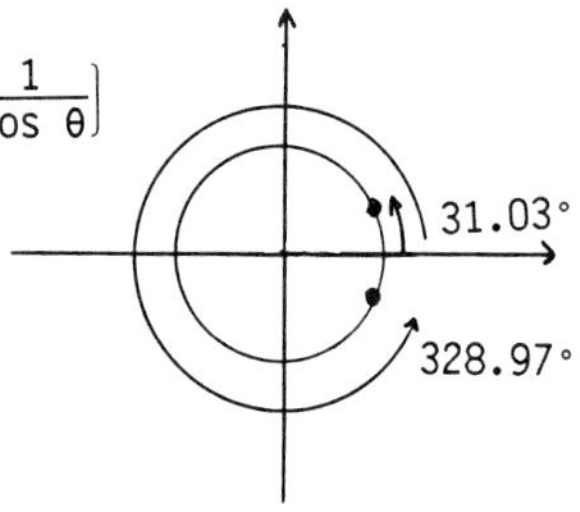

Using a calculator we find that the angle whose secant is 1.167 is 31.03°, or 31° 02'. This is a reference angle. On a unit circle we find the two points where the secant is 1.167. The two angles are 31.03° and 328.97°, plus any multiple of 360°.

Thus all values of $\sec^{-1} 1.167$ are

31.03° (or 31° 02') + k·360° or

328.97° (or 328° 58') + k·360°,

where k is any integer.

34. 46.02° (or 46° 1') + k·360°,

313.98° (or 313° 59') + k·360°

35. arccsc 6.277

[Note: $\csc \theta = \frac{1}{\sin \theta}$]

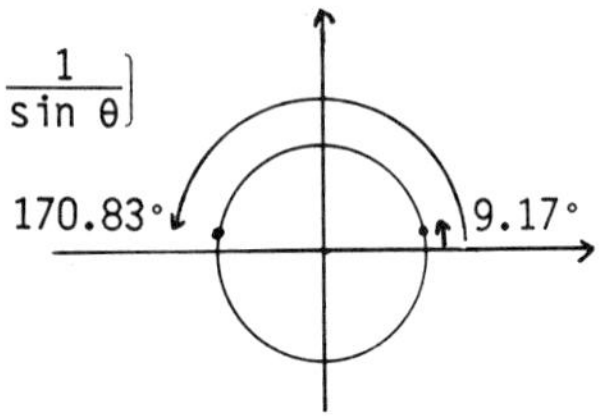

Using a calculator we find that the angle whose cosecant is 6.277 is 9.17°, or 9° 10'. This is a reference angle. On a unit circle we find two points where the cosecant is 6.277. The two angles are 9.17° and 170.83°, plus any multiple of 360°.

35. (continued)

Thus all values of arccsc 6.277 are

9.17° (or 9° 10') + k·360° or

170.83° (or 170° 50') + k·360°,

where k is any integer.

36. 64.17° (or 64° 10') + k·360°, or

115.83° (or 115° 50') + k·360°

37. Find Arcsin $\frac{\sqrt{2}}{2}$.

Note: Range $\left[-\frac{\pi}{2}, \frac{\pi}{2}\right]$

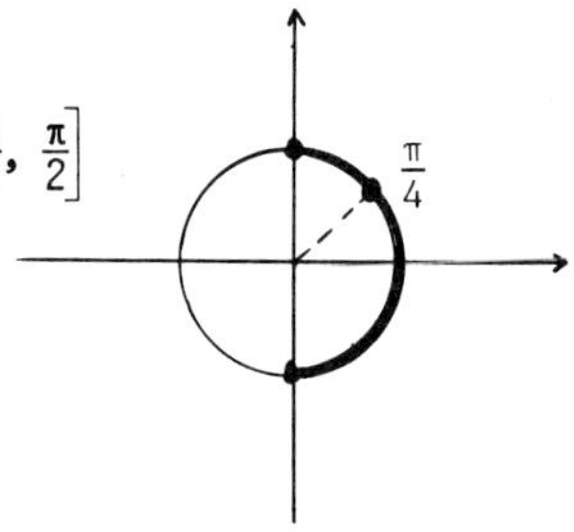

In the restricted range, the only number whose sine is $\frac{\sqrt{2}}{2}$ is $\frac{\pi}{4}$. Hence Arcsin $\frac{\sqrt{2}}{2} = \frac{\pi}{4}$.

38. $\frac{\pi}{6}$

39. Find $\mathrm{Cos}^{-1} \frac{1}{2}$.

Note: Range $[0, \pi]$

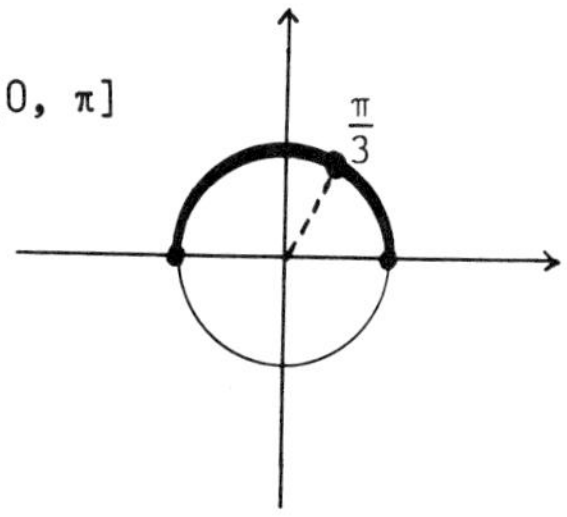

In the restricted range, the only number whose cosine is $\frac{1}{2}$ is $\frac{\pi}{3}$. Hence $\mathrm{Cos}^{-1} \frac{1}{2} = \frac{\pi}{3}$.

40. $\frac{\pi}{4}$

41. Find $\mathrm{Sin}^{-1} \left(-\frac{\sqrt{3}}{2}\right)$.

Note: Range $\left[-\frac{\pi}{2}, \frac{\pi}{2}\right]$

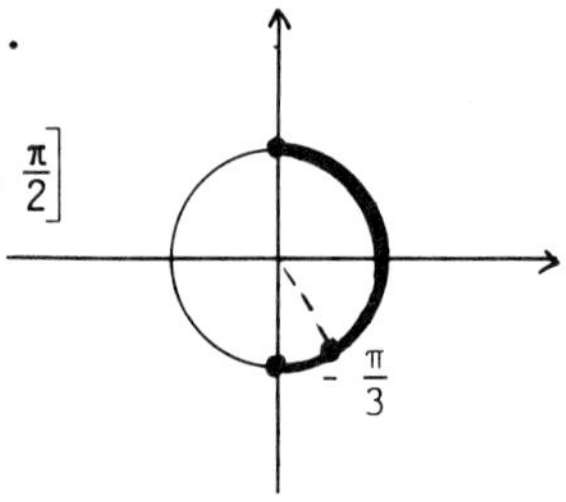

In the restricted range, the only number whose sine is $-\frac{\sqrt{3}}{2}$ is $-\frac{\pi}{3}$. Hence $\mathrm{Sin}^{-1} \left(-\frac{\sqrt{3}}{2}\right) = -\frac{\pi}{3}$.

42. $-\frac{\pi}{6}$

43. Find Arccos $\left(-\frac{\sqrt{2}}{2}\right)$.

Note: Range $[0, \pi]$

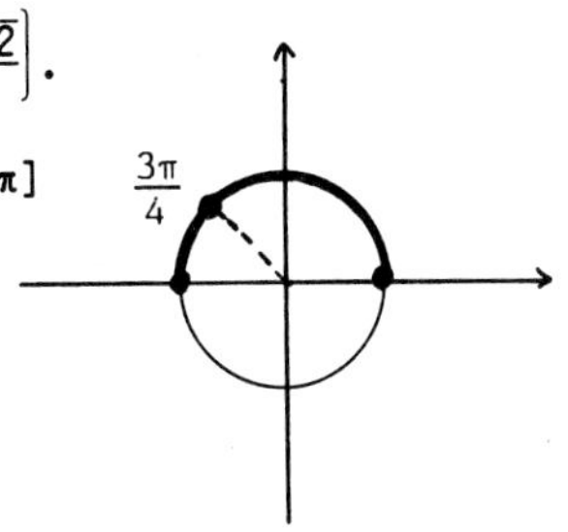

In the restricted range, the only number whose cosine is $-\frac{\sqrt{2}}{2}$ is $\frac{3\pi}{4}$. Hence Arccos $\left(-\frac{\sqrt{2}}{2}\right) = \frac{3\pi}{4}$.

44. $\frac{5\pi}{6}$

45. Find $\text{Tan}^{-1}\left(-\frac{\sqrt{3}}{3}\right)$.

Note: Range $\left(-\frac{\pi}{2}, \frac{\pi}{2}\right)$

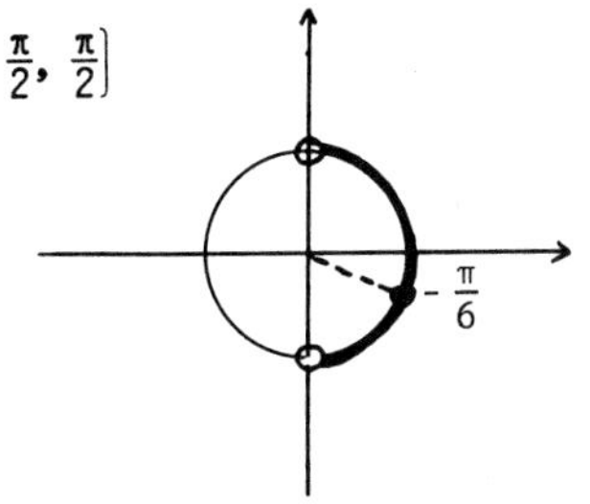

In the restricted range, the only number whose tangent is $-\frac{\sqrt{3}}{3}$ is $-\frac{\pi}{6}$. Hence $\text{Tan}^{-1}\left(-\frac{\sqrt{3}}{3}\right) = -\frac{\pi}{6}$.

46. $-\frac{\pi}{3}$

47. Find Arccot $\left(-\frac{\sqrt{3}}{3}\right)$.

Note: Range $(0, \pi)$

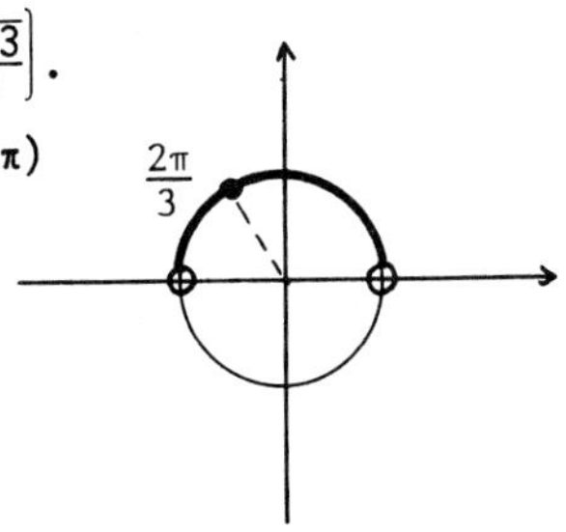

In the restricted range, the only number whose cotangent is $-\frac{\sqrt{3}}{3}$ is $\frac{2\pi}{3}$. Hence Arccot $\left(-\frac{\sqrt{3}}{3}\right)$ is $\frac{2\pi}{3}$.

48. $\frac{5\pi}{6}$

49. Arcsin 0.2334

Note: Range [-90°, 90°]

Using a calculator we find that the only angle in the restricted range whose sine is 0.2334 is 13.50°, or 13° 30'. Hence Arcsin 0.2334 is 13.50°, or 13° 30'.

50. 26.83°, or 26° 50'

51. Find Sin^{-1} (-0.6361).

Note: Range [-90°, 90°]

Using a calculator we find that the only angle in the restricted range whose sine is -0.6361 is -39.50°, or -39° 30'. Hence Sin^{-1} (-0.6361) = -39.50°, or -39° 30'.

52. -55.00°

53. Find Arccos (-0.8886).

Note: Range [0°, 180°]

Using a calculator we find that the only angle in the restricted range whose cosine is -0.8886 is 152.70°, or 152° 42'. Hence Arccos (-0.8886) = 152.70°, or 152° 42'.

54. 107.07°, or 107° 4'

55. Find Tan^{-1} (-0.4087).

Note: Range (-90°, 90°)

Using a calculator we find that the only angle in the restricted range whose tangent is -0.4087 is -22.23°, or -22° 14'. Hence Tan^{-1} (-0.4087) = -22.23°, or -22° 14'.

56. -13.55°, or -13° 33'

57. Find Cot^{-1} (-5.396).

Note: Range (0, 180°) and $\cot\theta = \frac{1}{\tan\theta}$.

Using a calculator we find that the only angle in the restricted range whose cotangent is -5.396 is 169.50°, or 169° 30'. Hence Cot^{-1} (-5.396) = 169.50°, or 169° 30'.

58. 142.83°, or 142° 50'

Exercise Set 8.5

1. sin Arcsin 0.3 = 0.3 (sin Arcsin $x \equiv x$)

2. 0.2

3. tan Tan^{-1} (-4.2) = -4.2 (tan Arctan $x \equiv x$)

4. -1.5

5. Arcsin $\sin \frac{2\pi}{3}$

$= \text{Arcsin } \frac{\sqrt{3}}{2}$ $\left[\sin \frac{2\pi}{3} = \frac{\sqrt{3}}{2}\right]$

$= \frac{\pi}{3}$ [The restricted range of the Arcsin function is $[-\frac{\pi}{2}, \frac{\pi}{2}]$]

Note: $\text{Arcsin } \sin \frac{2\pi}{3} \neq \frac{2\pi}{3}$ because $\frac{2\pi}{3}$ is not in the range of the Arcsine function.

6. $\frac{\pi}{2}$

7. $\text{Sin}^{-1} \sin \left(-\frac{3\pi}{4}\right)$

$= \text{Sin}^{-1} \left(-\frac{\sqrt{2}}{2}\right)$ $\left[\sin \left(-\frac{3\pi}{4}\right) = -\frac{\sqrt{2}}{2}\right]$

$= -\frac{\pi}{4}$ [The restricted range of the Arcsin function is $[-\frac{\pi}{2}, \frac{\pi}{2}]$]

Note: $\text{Sin}^{-1} \sin \left(-\frac{3\pi}{4}\right) \neq -\frac{3\pi}{4}$ because $-\frac{3\pi}{4}$ is not in the range of the Arcsine function.

8. $\frac{\pi}{4}$

9. $\text{Sin}^{-1} \sin \frac{\pi}{5} = \frac{\pi}{5}$ because $\frac{\pi}{5}$ is in the range of the Arcsine function.

10. $\frac{\pi}{7}$

11. $\text{Tan}^{-1} \tan \frac{2\pi}{3}$

$= \text{Tan}^{-1} (-\sqrt{3})$ $\left[\tan \frac{2\pi}{3} = -\sqrt{3}\right]$

$= -\frac{\pi}{3}$ [The restricted range of the Arctangent function is $\left(-\frac{\pi}{2}, \frac{\pi}{2}\right)$.]

Note: $\text{Tan}^{-1} \tan \frac{2\pi}{3} \neq \frac{2\pi}{3}$ because $\frac{2\pi}{3}$ is not in the range of the Arctangent function.

12. $\frac{2\pi}{3}$

13. $\sin \text{Arctan } \sqrt{3}$

$= \sin \frac{\pi}{3}$

$= \frac{\sqrt{3}}{2}$

14. $\frac{1}{2}$

15. $\cos \text{Arcsin } \frac{\sqrt{3}}{2}$

$= \cos \frac{\pi}{3}$

$= \frac{1}{2}$

16. $\frac{\sqrt{2}}{2}$

17. $\tan \text{Cos}^{-1} \frac{\sqrt{2}}{2}$

$= \tan \frac{\pi}{4}$

$= 1$

18. $\frac{\sqrt{3}}{3}$

19. $\text{Cos}^{-1} \sin \frac{\pi}{3}$

$= \text{Cos}^{-1} \frac{\sqrt{3}}{2}$

$= \frac{\pi}{6}$

20. $\frac{\pi}{2}$

21. $\text{Arcsin } \cos \frac{\pi}{6}$

$= \text{Arcsin } \frac{\sqrt{3}}{2}$

$= \frac{\pi}{3}$

22. $\frac{\pi}{4}$

23. $\text{Sin}^{-1} \tan \frac{\pi}{4}$

$= \text{Sin}^{-1} 1$

$= \frac{\pi}{2}$

24. $-\frac{\pi}{2}$

25. $\sin \text{Arctan } \frac{x}{2}$

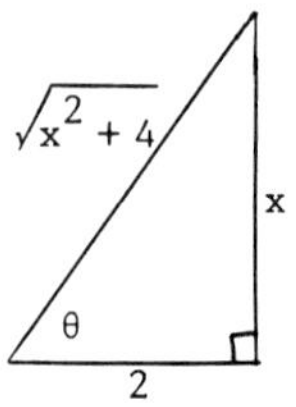

We wish to find the sine of an angle whose tangent is $\frac{x}{2}$.

The hypotenuse has length $\sqrt{x^2 + 4}$.

Thus, $\sin \text{Arctan } \frac{x}{2} = \frac{x}{\sqrt{x^2 + 4}}$.

26. $\frac{a}{\sqrt{a^2 + 9}}$

27. $\tan \text{Cos}^{-1} \frac{3}{x}$

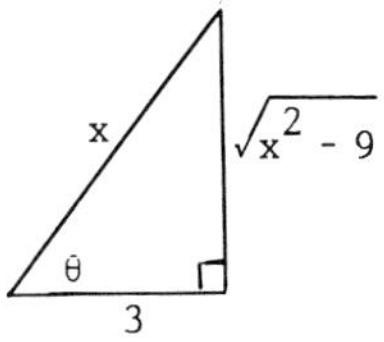

We wish to find the sine of an angle whose cosine is $\frac{3}{x}$.

The other leg has length $\sqrt{x^2 - 9}$.

Thus, $\tan \text{Cos}^{-1} \frac{3}{x} = \frac{\sqrt{x^2 - 9}}{3}$.

28. $\frac{\sqrt{y^2 - 5^2}}{5}$

29. $\cot \text{Sin}^{-1} \frac{a}{b}$

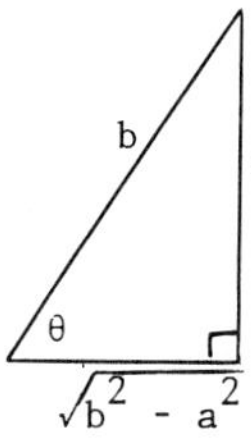

We wish to find the cotangent of an angle whose sine is $\frac{a}{b}$.

The length of the other leg is $\sqrt{b^2 - a^2}$.

Thus, $\cot \text{Sin}^{-1} \frac{a}{b} = \frac{\sqrt{b^2 - a^2}}{a}$.

30. $\frac{\sqrt{q^2 - p^2}}{p}$

31. $\cos \text{Tan}^{-1} \frac{\sqrt{2}}{3}$

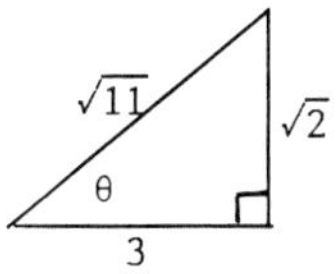

We wish to find the cosine of an angle whose tangent is $\frac{\sqrt{2}}{3}$.

The length of the hypotenuse is $\sqrt{11}$.

Thus, $\cos \text{Tan}^{-1} \frac{\sqrt{2}}{3} = \frac{3}{\sqrt{11}}$.

32. $\frac{4}{\sqrt{19}}$

33. tan Arcsin 0.1

We wish to find the tangent of an angle whose sine is 0.1, or $\frac{1}{10}$.

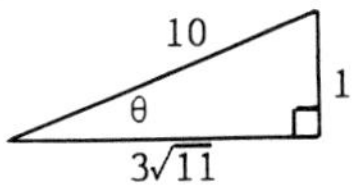

The length of the other leg is $3\sqrt{11}$.

Thus, $\tan \text{Arcsin } 0.1 = \frac{1}{3\sqrt{11}}$.

34. $\frac{1}{2\sqrt{6}}$

35. $\cot \text{Cos}^{-1} (-0.2)$

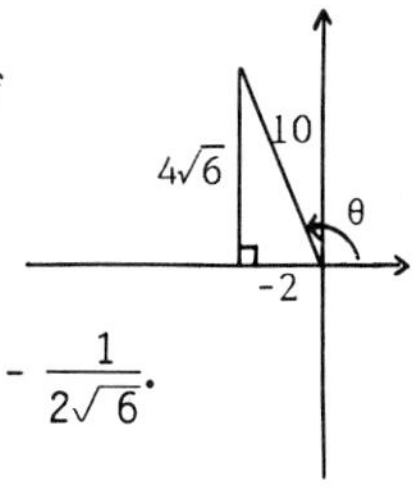

We wish to find the cotangent of an angle whose cosine is -0.2.

The length of the other leg is $4\sqrt{6}$.

Thus, $\cot \text{Cos}^{-1} (-0.2) = \frac{-2}{4\sqrt{6}} = -\frac{1}{2\sqrt{6}}$.

36. $-\frac{3}{\sqrt{91}}$

37. sin Arccot y

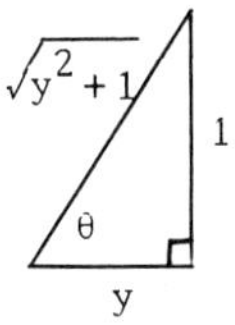

We wish to find the sine of an angle whose cotangent is y.

The length of the hypotenuse is $\sqrt{y^2 + 1}$.

Thus, $\sin \text{Arccot } y = \frac{1}{\sqrt{y^2 + 1}}$.

38. $\frac{1}{\sqrt{1 + x^2}}$

39. cos Arctan t

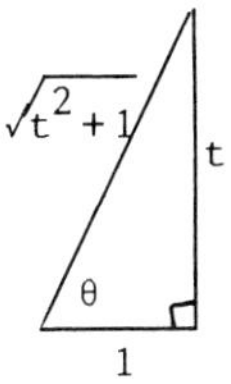

We wish to find the cosine of an angle whose tangent is t.

The length of the hypotenuse is $\sqrt{t^2 + 1}$.

Thus, $\cos \text{Arctan } t = \frac{1}{\sqrt{t^2 + 1}}$.

40. $\frac{t}{\sqrt{1 + t^2}}$

41. $\cot \text{Sin}^{-1} y$

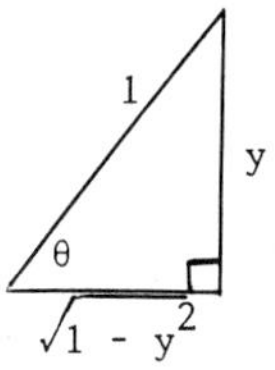

We wish to find the cotangent of an angle whose sine is y.

The length of the other leg is $\sqrt{1 - y^2}$.

Thus, $\cot \text{Sin}^{-1} y = \frac{\sqrt{1 - y^2}}{y}$.

42. $\frac{\sqrt{1 - y^2}}{y}$

43. $\sin \text{Cos}^{-1} x$

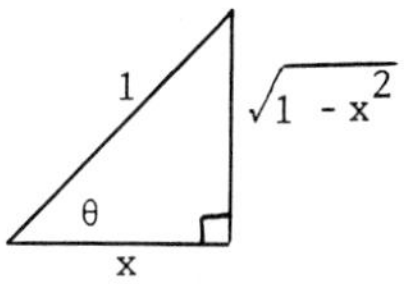

We wish to find the sine of an angle whose cosine is x.

The length of the other leg is $\sqrt{1 - x^2}$.

Thus, $\sin \text{Cos}^{-1} x = \sqrt{1 - x^2}$.

44. $\sqrt{1 - x^2}$

45. $\tan\left[\frac{1}{2}\text{ Arcsin }\frac{4}{5}\right]$

$= \tan\frac{\theta}{2}$ $\left[\text{Letting } \theta = \text{Arcsin }\frac{4}{5}\right]$

$= \frac{\sin\theta}{1 + \cos\theta}$ (Half-angle identity)

$= \frac{\frac{4}{5}}{1 + \frac{3}{5}}$

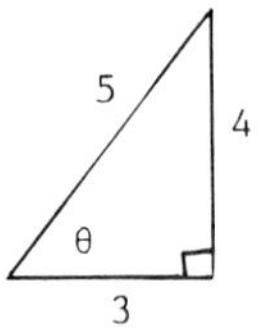

$= \frac{\frac{4}{5}}{\frac{8}{5}}$

$= \frac{1}{2}$

46. $2 - \sqrt{3}$

47. $\cos\left[\frac{1}{2}\text{ Arcsin }\frac{1}{2}\right]$

$= \cos\frac{\theta}{2}$ $\left[\text{Letting } \theta = \text{Arcsin }\frac{1}{2}\right]$

$= \sqrt{\frac{1 + \cos\theta}{2}}$ (Half-angle identity) (θ is a 1st quadrant angle)

$= \sqrt{\frac{1 + \frac{\sqrt{3}}{2}}{2}}$

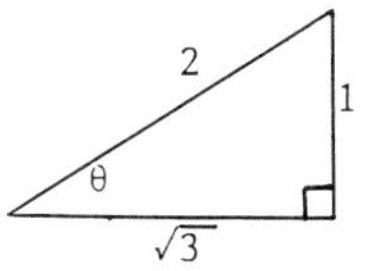

$= \sqrt{\frac{2 + \sqrt{3}}{4}}$

$= \frac{\sqrt{2 + \sqrt{3}}}{2}$

48. $\frac{\sqrt{3}}{2}$

49. $\sin\left[2\text{ Cos}^{-1}\frac{3}{5}\right]$

$= \sin 2\theta$ $\left[\text{Letting } \theta = \text{Cos}^{-1}\frac{3}{5}\right]$

$= 2\sin\theta\cos\theta$ (Double-angle identity)

$= 2 \cdot \frac{4}{5} \cdot \frac{3}{5}$

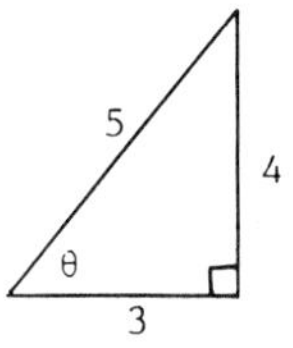

$= \frac{24}{25}$

50. $\frac{\sqrt{3}}{2}$

51. $\cos\left[2\text{ Sin}^{-1}\frac{5}{13}\right]$

$= \cos 2\theta$ $\left[\text{Letting } \theta = \text{Sin}^{-1}\frac{5}{13}\right]$

$= 1 - 2\sin^2\theta$ (Double-angle identity)

51. (continued)

$= 1 - 2 \cdot \left[\frac{5}{13}\right]^2$

$= 1 - \frac{50}{169}$

$= \frac{119}{169}$

52. $\frac{7}{25}$

53. $\sin\left[\text{Sin}^{-1}\frac{1}{2} + \text{Cos}^{-1}\frac{3}{5}\right]$

$= \sin(u + v)$ $\left[\text{Letting } u = \text{Sin}^{-1}\frac{1}{2} \text{ and } v = \text{Cos}^{-1}\frac{3}{5}\right]$

$= \sin u\cos v + \cos u\sin v$ (Sum identity)

$= \sin\text{Sin}^{-1}\frac{1}{2} \cdot \cos\text{Cos}^{-1}\frac{3}{5} +$

$\cos\text{Sin}^{-1}\frac{1}{2} \cdot \sin\text{Cos}^{-1}\frac{3}{5}$

$= \frac{1}{2} \cdot \frac{3}{5} + \frac{\sqrt{3}}{2} \cdot \frac{4}{5}$

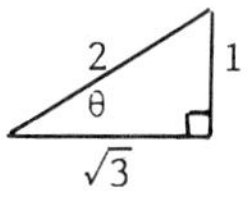

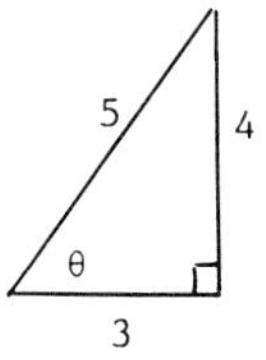

$= \frac{3 + 4\sqrt{3}}{10}$

54. $\frac{4 - 3\sqrt{3}}{10}$

55. $\cos\left[\text{Sin}^{-1}\frac{\sqrt{2}}{2} + \text{Cos}^{-1}\frac{3}{5}\right]$

$= \cos(u + v)$ $\left[\text{Letting } u = \text{Sin}^{-1}\frac{\sqrt{2}}{2} \text{ and } v = \text{Cos}^{-1}\frac{3}{5}\right]$

$= \cos u\cos v - \sin u\sin v$ (Sum identity)

$= \cos\text{Sin}^{-1}\frac{\sqrt{2}}{2} \cdot \cos\text{Cos}^{-1}\frac{3}{5} -$

$\sin\text{Sin}^{-1}\frac{\sqrt{2}}{2} \cdot \sin\text{Cos}^{-1}\frac{3}{5}$

$= \frac{\sqrt{2}}{2} \cdot \frac{3}{5} - \frac{\sqrt{2}}{2} \cdot \frac{4}{5}$

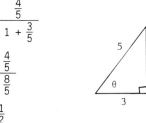

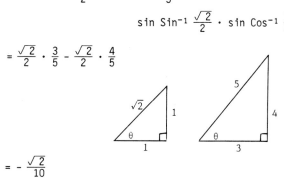

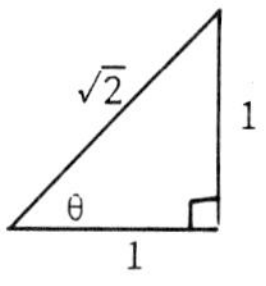

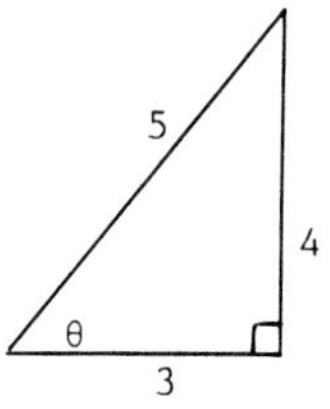

$= -\frac{\sqrt{2}}{10}$

56. $\frac{3 + 4\sqrt{3}}{10}$

57. $\sin(\text{Sin}^{-1} x + \text{Cos}^{-1} y)$

$= \sin(u + v)$ (Letting $u = \text{Sin}^{-1} x$ and $v = \text{Cos}^{-1} y$)

$= \sin u \cos v + \cos u \sin v$ (Sum identity)

$= \sin \text{Sin}^{-1} x \cdot \cos \text{Cos}^{-1} y + \cos \text{Sin}^{-1} x \cdot \sin \text{Cos}^{-1} y$

$= x \cdot y + \sqrt{1 - x^2} \cdot \sqrt{1 - y^2}$

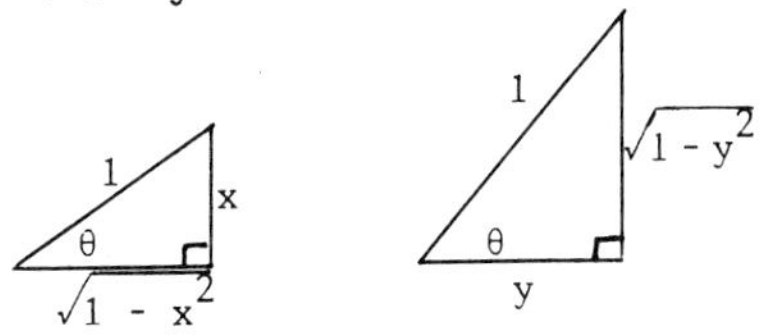

$= xy + \sqrt{(1 - x^2)(1 - y^2)}$

58. $xy - \sqrt{(1 - x^2)(1 - y^2)}$

59. $\cos(\text{Sin}^{-1} x + \text{Cos}^{-1} y)$

$= \cos(u + v)$ (Letting $u = \text{Sin}^{-1} x$ and $v = \text{Cos}^{-1} y$)

$= \cos u \cos v - \sin u \sin v$ (Sum identity)

$= \cos \text{Sin}^{-1} x \cdot \cos \text{Cos}^{-1} y - \sin \text{Sin}^{-1} x \cdot \sin \text{Cos}^{-1} y$

$= \sqrt{1 - x^2} \cdot y - x \cdot \sqrt{1 - y^2}$

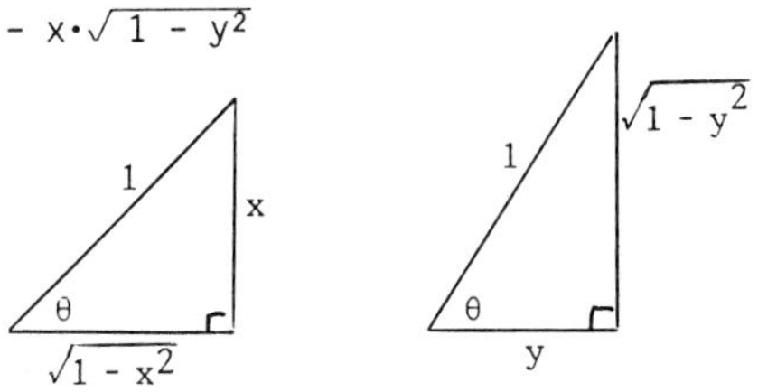

$= y\sqrt{1 - x^2} - x\sqrt{1 - y^2}$

60. $y\sqrt{1 - x^2} + x\sqrt{1 - y^2}$

61. $\sin(\text{Sin}^{-1} 0.6032 + \text{Cos}^{-1} 0.4621)$

$= \sin \text{Sin}^{-1} 0.6032 \cdot \cos \text{Cos}^{-1} 0.4621 + \cos \text{Sin}^{-1} 0.6032 \cdot \sin \text{Cos}^{-1} 0.4621$

$= 0.6032 \cdot 0.4621 + 0.7976 \cdot 0.8868$

$= 0.2787 + 0.7073$

$= 0.9860$

62. 0.9704

63.

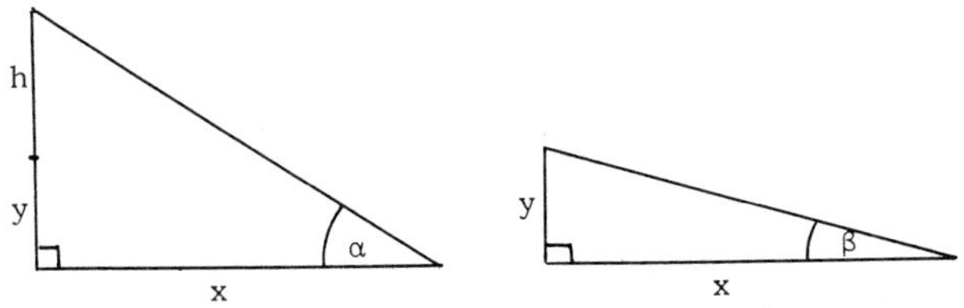

63. (continued)

Let $\theta = \alpha - \beta$

$\tan \alpha = \frac{h + y}{x}$, $\alpha = \text{Arctan}\, \frac{h + y}{x}$

$\tan \beta = \frac{y}{x}$, $\beta = \text{Arctan}\, \frac{y}{x}$

Thus, $\theta = \text{Arctan}\, \frac{h + y}{x} - \text{Arctan}\, \frac{y}{x}$

64. 38.7°

Exercise Set 8.6

1. $\sin x = \frac{\sqrt{3}}{2}$

Since sin x is positive, the solutions are to be found in the first and second quadrants.

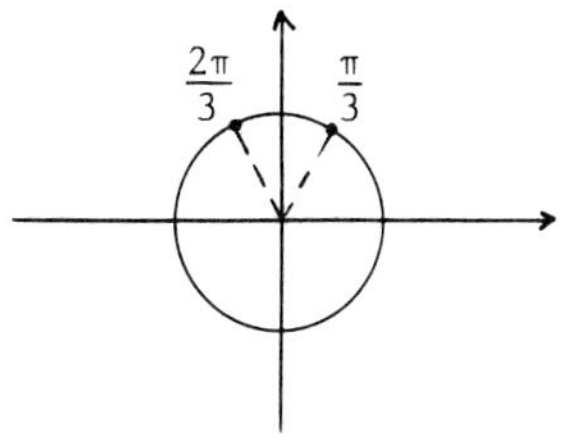

The solutions are $\frac{\pi}{3} + 2k\pi$ or $\frac{2\pi}{3} + 2k\pi$, where k is any integer.

2. $\frac{\pi}{6} + 2k\pi$, $\frac{11\pi}{6} + 2k\pi$

3. $\cos x = \frac{1}{\sqrt{2}}$, or $\frac{\sqrt{2}}{2}$

Since cos x is positive, the solutions are to be found in the first and fourth quadrants.

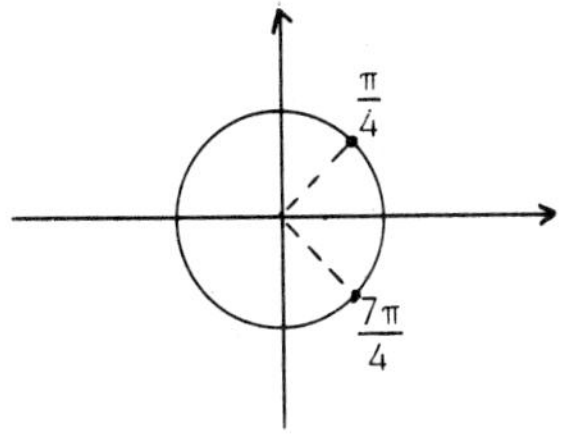

The solutions are $\frac{\pi}{4} + 2k\pi$ or $\frac{7\pi}{4} + 2k\pi$, where k is any integer.

4. $\frac{\pi}{3} + k\pi$

5. $\sin x = 0.3448$

Using a calculator, we find the reference angle, $x = 20.17°$, or $20° \ 10'$. Since $\sin x$ is positive, the solutions are to be found in the first and second quadrants.

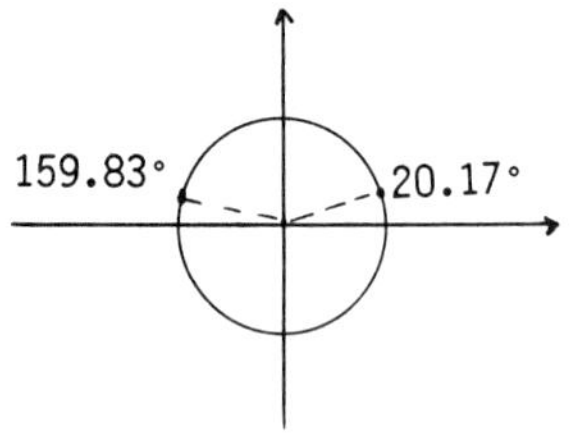

$(180° - 20.17° = 159.83°)$

The solutions are $20.17°$ (or $20° \ 10'$) $+ k \cdot 360°$ or $159.83°$ (or $159° \ 50'$) $+ k \cdot 360°$, where k is any integer.

6. $50.16°$ (or $50° \ 10'$) $+ k \cdot 360°$,
$309.83°$ (or $309° \ 50'$) $+ k \cdot 360°$

7. $\cos x = -0.5495$

Using a calculator, we find the reference angle, $x = 56.67°$, or $56° \ 40'$. Since $\cos x$ is negative, the solutions are in the second and third quadrants.

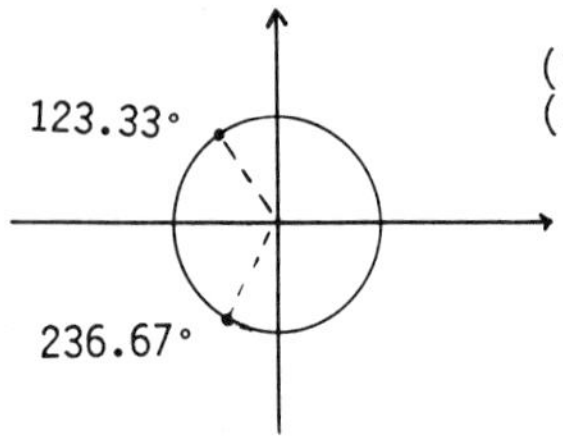

$(180° - 56.67° = 123.33°)$
$(180° + 56.67° = 236.67°)$

The solutions in $[0°, 360°)$ are $123.33°$ (or $123° \ 20'$) and $236.67°$ (or $236° \ 40'$).

8. $205.33°$ (or $205° \ 20'$), $334.67°$ (or $334° \ 40'$)

9. $2 \sin x + \sqrt{3} = 0$

$$2 \sin x = -\sqrt{3}$$

$$\sin x = -\frac{\sqrt{3}}{2}$$

There are two points on the unit circle for which the sine is $-\frac{\sqrt{3}}{2}$.

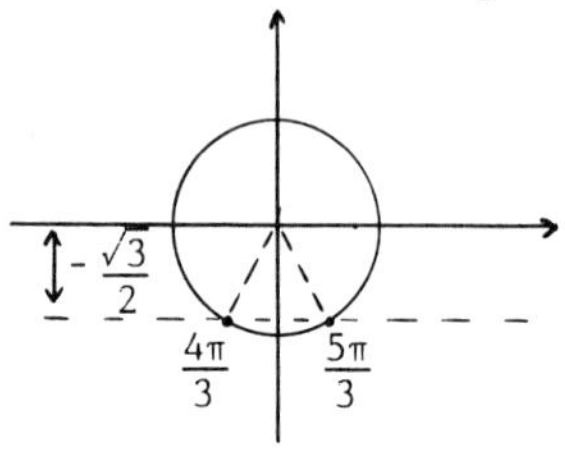

The solutions in $[0, 2\pi)$ are $\frac{4\pi}{3}$ and $\frac{5\pi}{3}$.

10. $\frac{5\pi}{6}, \frac{11\pi}{6}$

11. $2 \tan x + 3 = 0$

$$2 \tan x = -3$$

$$\tan x = -1.5$$

Using a calculator, we find the reference angle, $x = 56.31°$, or $56° \ 19'$. Since $\tan x$ is negative, the solutions are in the second and fourth quadrants.

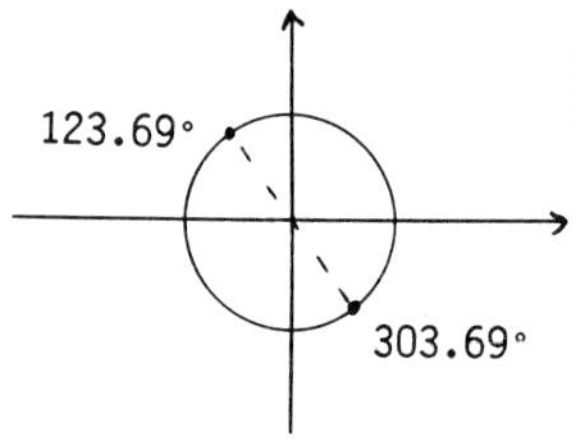

$(180° - 56.31° = 123.69°)$
$(360° - 56.31° = 303.69°)$

The solutions in $[0°, 360°)$ are $123.69°$ (or $123° \ 41'$) and $303.69°$ (or $303° \ 41'$).

12. $14.48°$ (or $14° \ 29'$), $165.52°$ (or $165° \ 31'$)

13. $4 \sin^2 x - 1 = 0$

$$4 \sin^2 x = 1$$

$$\sin^2 x = \frac{1}{4}$$

$$|\sin x| = \frac{1}{2}$$

$$\sin x = \pm \frac{1}{2}$$

Use the unit circle to find those numbers having a sine of $\frac{1}{2}$ or $-\frac{1}{2}$.

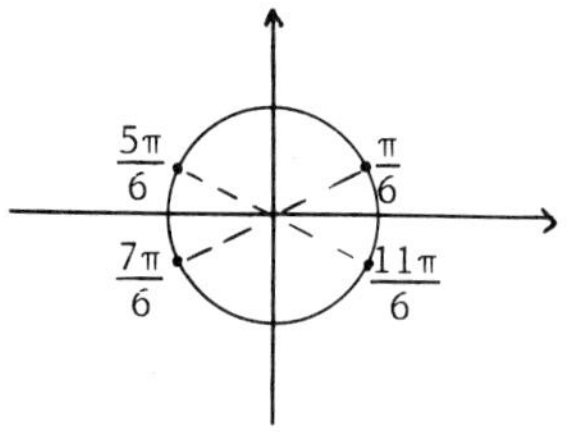

The solutions in $[0, 2\pi)$ are $\frac{\pi}{6}, \frac{5\pi}{6}, \frac{7\pi}{6}$, and $\frac{11\pi}{6}$.

14. $\frac{\pi}{4}, \frac{3\pi}{4}, \frac{5\pi}{4}, \frac{7\pi}{4}$

15. $\cot^2 x - 3 = 0$

$$\cot^2 x = 3$$

$$|\cot x| = \sqrt{3}$$

$$\cot x = \pm \sqrt{3}$$

Use the unit circle to find those numbers having a cotangent of $\sqrt{3}$ or $-\sqrt{3}$.

15. (continued)

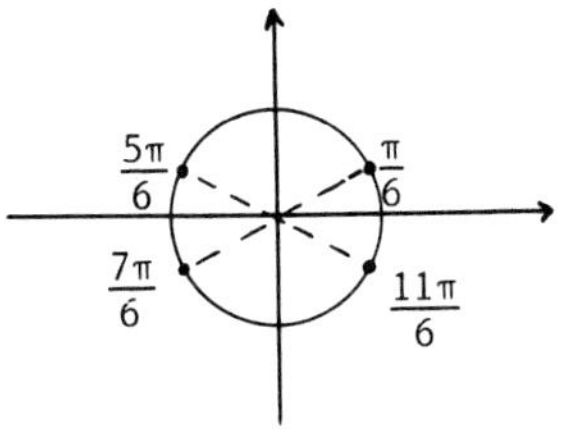

The solutions in $[0, 2\pi)$ are $\frac{\pi}{6}, \frac{5\pi}{6}, \frac{7\pi}{6}$, and $\frac{11\pi}{6}$.

16. $\frac{\pi}{6}, \frac{5\pi}{6}, \frac{7\pi}{6}, \frac{11\pi}{6}$

17.
$$2 \sin^2 x + \sin x = 1$$
$$2 \sin^2 x + \sin x - 1 = 0$$
$$(2 \sin x - 1)(\sin x + 1) = 0$$

$2 \sin x - 1 = 0$ or $\sin x + 1 = 0$

$2 \sin x = 1$ or $\sin x = -1$

$\sin x = \frac{1}{2}$ or $\sin x = -1$

Use a unit circle to find those numbers having a sine of $\frac{1}{2}$ or -1.

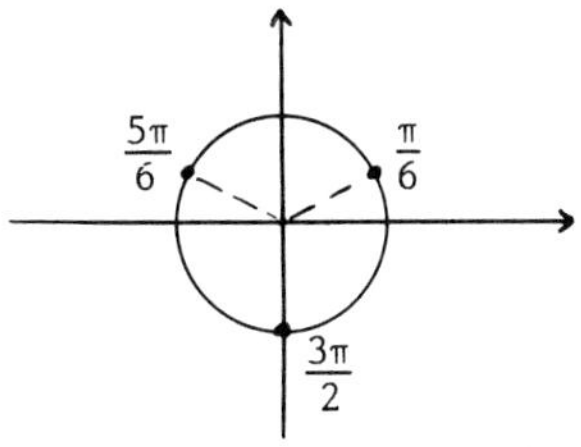

The solutions in $[0, 2\pi)$ are $\frac{\pi}{6}, \frac{5\pi}{6}$, and $\frac{3\pi}{2}$.

18. $\frac{2\pi}{3}, \frac{4\pi}{3}, \pi$

19.
$$\cos^2 x + 2 \cos x = 3$$
$$\cos^2 x + 2 \cos x - 3 = 0$$
$$(\cos x + 3)(\cos x - 1) = 0$$

$\cos x + 3 = 0$ or $\cos x - 1 = 0$

$\cos x = -3$ or $\cos x = 1$

Since cosines are never less than -1, $\cos x = -3$ has no solution. There is one point on the unit circle for which the cosine is 1. Thus the solution in $[0, 2\pi)$ is 0.

20. $\frac{3\pi}{2}$

21.
$$4 \sin^3 x - \sin x = 0$$
$$\sin x(4 \sin^2 x - 1) = 0$$
$$\sin x(2 \sin x + 1)(2 \sin x - 1) = 0$$

$\sin x = 0$ or $2 \sin x + 1 = 0$

$\sin x = 0$ or $\sin x = -\frac{1}{2}$

$x = 0, \pi$ or $x = \frac{7\pi}{6}, \frac{11\pi}{6}$

or $2 \sin x - 1 = 0$

or $\sin x = \frac{1}{2}$

or $x = \frac{\pi}{6}, \frac{5\pi}{6}$

The solutions in $[0, 2\pi)$ are $0, \frac{\pi}{6}, \frac{5\pi}{6}, \pi, \frac{7\pi}{6}$, and $\frac{11\pi}{6}$.

22. $\frac{\pi}{2}, \frac{3\pi}{2}, \frac{\pi}{6}, \frac{11\pi}{6}$

23.
$$2 \sin^2\theta + 7 \sin \theta = 4$$
$$2 \sin^2\theta + 7 \sin \theta - 4 = 0$$
$$(2 \sin \theta - 1)(\sin \theta + 4) = 0$$

$2 \sin \theta - 1 = 0$ or $\sin \theta + 4 = 0$

$\sin \theta = \frac{1}{2}$ or $\sin \theta = -4$

$\theta = \frac{\pi}{6}, \frac{5\pi}{6}$

Since sines are never less than -1, $\sin \theta = -4$ has no solution.

The solutions in $[0, 2\pi)$ are $\frac{\pi}{6}$ and $\frac{5\pi}{6}$.

24. $\frac{\pi}{6}, \frac{5\pi}{6}$

25.
$$6 \cos^2\phi + 5 \cos \phi + 1 = 0$$
$$(3 \cos \phi + 1)(2 \cos \phi + 1) = 0$$

$3 \cos \phi + 1 = 0$ or $2 \cos \phi + 1 = 0$

$\cos \phi = -\frac{1}{3}$ or $\cos \phi = -\frac{1}{2}$

$\phi = 109.47°$ (or $109°28'$), $250.53°$ (or $250°32'$) or $\phi = 120°, 240°$

The solutions in $[0, 360°)$ are $109.47°$ (or $109°\ 28'$), $120°$, $240°$, $250.53°$ (or $250°\ 32'$).

26. $\frac{\pi}{6}, \frac{5\pi}{6}, \frac{3\pi}{2}$

27. $2 \sin t \cos t + 2 \sin t - \cos t - 1 = 0$
$$2 \sin t(\cos t + 1) - (\cot t + 1) = 0$$
$$(2 \sin t - 1)(\cos t + 1) = 0$$
$2 \sin t - 1 = 0$ or $\cos t + 1 = 0$
$\sin t = \frac{1}{2}$ or $\cos t = -1$
$t = \frac{\pi}{6}, \frac{5\pi}{6}$ or $t = \pi$
The solutions in $[0, 2\pi)$ are $\frac{\pi}{6}, \frac{5\pi}{6}$, and π.

28. $\frac{\pi}{4}, \frac{5\pi}{4}, \frac{7\pi}{6}, \frac{11\pi}{6}$

29. $\cos 2x \sin x + \sin x = 0$
$$(1 - 2 \sin^2 x)\sin x + \sin x = 0$$
$$\sin x - 2 \sin^3 x + \sin x = 0$$
$$-2 \sin^3 x + 2 \sin x = 0$$
$$2 \sin x(1 - \sin^2 x) = 0$$
$2 \sin x = 0$ or $1 - \sin^2 x = 0$
$\sin x = 0$ or $\sin^2 x = 1$
$x = 0, \pi$ or $\sin x = \pm 1$
$x = 0, \pi$ or $x = \frac{\pi}{2}, \frac{3\pi}{2}$
The solutions in $[0, 2\pi)$ are $0, \frac{\pi}{2}, \pi$, and $\frac{3\pi}{2}$.

30. $\frac{\pi}{4}, \frac{5\pi}{4}, \frac{\pi}{2}, \frac{3\pi}{2}$

31. $|\sin x| = \frac{\sqrt{3}}{2}$
$\sin x = \pm \frac{\sqrt{3}}{2}$
$x = \frac{\pi}{3}, \frac{2\pi}{3}, \frac{4\pi}{3}, \frac{5\pi}{3}$
or 60°, 120°, 240°, 300°

32. 60°, 120°, 240°, 300°

33. $\sqrt{\tan x} = \sqrt[4]{3}$
$(\tan x)^{1/2} = 3^{1/4}$
$\tan^2 x = 3$
$|\tan x| = \sqrt{3}$
$\tan x = \pm \sqrt{3}$
$x = \frac{\pi}{3}, \frac{2\pi}{3}, \frac{4\pi}{3}, \frac{5\pi}{3}$
When $\tan x$ is negative $\sqrt{\tan x}$ is undefined. Thus the only solutions are $\frac{\pi}{3}$ and $\frac{4\pi}{3}$ (or 60° and 240°).

34. 3.58°, 6.38°, 173.62°, 176.42°

35. $16 \cos^4 x - 16 \cos^2 x + 3 = 0$
$(4 \cos^2 x - 1)(4 \cos^2 x - 3) = 0$
$4 \cos^2 x - 1 = 0$ or $4 \cos^2 x - 3 = 0$
$\cos^2 x = \frac{1}{4}$ or $\cos^2 x = \frac{3}{4}$
$\cos x = \pm \frac{1}{2}$ or $\cos x = \pm \frac{\sqrt{3}}{2}$
$x = \frac{\pi}{3}, \frac{2\pi}{3}, \frac{4\pi}{3}, \frac{5\pi}{3}$ or $x = \frac{\pi}{6}, \frac{5\pi}{6}, \frac{7\pi}{6}, \frac{11\pi}{6}$
The solutions are
$\frac{\pi}{6}, \frac{\pi}{3}, \frac{2\pi}{3}, \frac{5\pi}{6}, \frac{7\pi}{6}, \frac{4\pi}{3}, \frac{5\pi}{3}$, and $\frac{11\pi}{6}$.
or 30°, 60°, 120°, 150°, 210°, 240°, 300°, and 330°.

36. 0

37. $e^{\sin x} = 1$
$\ln e^{\sin x} = \ln 1$
$\sin x = 0$
$x = 0, \pi$

38. $e^{(3\pi)/2}$

39. $e^{\ln \sin x} = 1$
$\sin x = 1$
$x = \frac{\pi}{2}$

40. 0, ± 1.56, ± 4.72, etc.

Exercise Set 8.7

1. $\tan x \sin x - \tan x = 0$
$\tan x(\sin x - 1) = 0$
$\tan x = 0$ or $\sin x - 1 = 0$
$x = 0, \pi$ or $\sin x = 1$
$x = \frac{\pi}{2}$
The value $\frac{\pi}{2}$ does not check, but the other values do. Thus the solutions in $[0, 2\pi)$ are 0 and π.

2. $0, \pi, \frac{2\pi}{3}, \frac{4\pi}{3}$

3. $2 \sec x \tan x + 2 \sec x + \tan x + 1 = 0$

$$2 \sec x(\tan x + 1) + (\tan x + 1) = 0$$
$$(2 \sec x + 1)(\tan x + 1) = 0$$

$2 \sec x + 1 = 0$ or $\tan x + 1 = 0$

$\sec x = -\frac{1}{2}$ or $\tan x = -1$

$\cos x = -2$ $x = \frac{3\pi}{4}, \frac{7\pi}{4}$

No solution

Both values check. The solutions in $[0, 2\pi)$ are $\frac{3\pi}{4}$ and $\frac{7\pi}{4}$.

4. $\frac{\pi}{6}, \frac{5\pi}{6}, \frac{\pi}{3}, \frac{5\pi}{3}$

5. $\sin 2x - \cos x = 0$

$$2 \sin x \cos x - \cos x = 0$$
$$\cos x(2 \sin x - 1) = 0$$

$\cos x = 0$ or $2 \sin x - 1 = 0$

$x = \frac{\pi}{2}, \frac{3\pi}{2}$ or $\sin x = \frac{1}{2}$

$x = \frac{\pi}{6}, \frac{5\pi}{6}$

All values check. The solutions in $[0, 2\pi)$ are $\frac{\pi}{6}, \frac{\pi}{2}, \frac{5\pi}{6}$, and $\frac{3\pi}{2}$.

6. $0, \pi, \frac{7\pi}{6}, \frac{11\pi}{6}$

7. $\sin 2x \sin x - \cos x = 0$

$$2 \sin x \cos x \cdot \sin x - \cos x = 0$$
$$\cos x(2 \sin^2 x - 1) = 0$$

$\cos x = 0$ or $2 \sin^2 x - 1 = 0$

$x = \frac{\pi}{2}, \frac{3\pi}{2}$ $\sin^2 x = \frac{1}{2}$

$\sin x = \pm \frac{\sqrt{2}}{2}$

$x = \frac{\pi}{4}, \frac{3\pi}{4}, \frac{5\pi}{4}, \frac{7\pi}{4}$

All values check. The solutions in $[0, 2\pi)$ are $\frac{\pi}{4}, \frac{\pi}{2}, \frac{3\pi}{4}, \frac{5\pi}{4}, \frac{3\pi}{2}, \frac{7\pi}{4}$.

8. $0, \frac{\pi}{4}, \frac{3\pi}{4}, \pi, \frac{5\pi}{4}, \frac{7\pi}{4}$

9. $\sin 2x + 2 \sin x \cos x = 0$

$$\sin 2x + \sin 2x = 0$$
$$2 \sin 2x = 0$$
$$\sin 2x = 0$$
$$2x = 0, \pi, 2\pi, 3\pi$$
$$x = 0, \frac{\pi}{2}, \pi, \frac{3\pi}{2}$$

9. (continued)

All values check. The solutions in $[0, 2\pi)$ are $0, \frac{\pi}{2}, \pi$, and $\frac{3\pi}{2}$.

10. $0, \frac{\pi}{2}, \frac{3\pi}{2}, \pi$

11. $\cos 2x \cos x + \sin 2x \sin x = 1$

$$(1 - 2 \sin^2 x)\cos x + 2 \sin x \cos x \cdot \sin x = 1$$
$$(1 - 2 \sin^2 x)\cos x + 2 \sin^2 x \cos x = 1$$
$$\cos x = 1$$
$$x = 0$$

The value 0 checks. The only solution in $[0, 2\pi)$ is 0.

12. $0, \frac{\pi}{2}, \pi, \frac{3\pi}{2}$

13. $\sin 4x - 2 \sin 2x = 0$

$$\sin 2(2x) - 2 \sin 2x = 0$$
$$2 \sin 2x \cos 2x - 2 \sin 2x = 0$$
$$\sin 2x(\cos 2x - 1) = 0$$

$\sin 2x = 0$ or $\cos 2x - 1 = 0$

$\sin 2x = 0$ or $\cos 2x = 1$

$2x = 0, \pi, 2\pi, 3\pi$ or $2x = 0, 2\pi$

$x = 0, \frac{\pi}{2}, \pi, \frac{3\pi}{2}$ $x = 0, \pi$

All values check. The solutions in $[0, 2\pi)$ are $0, \frac{\pi}{2}, \pi$, and $\frac{3\pi}{2}$.

14. $0, \frac{\pi}{2}, \pi, \frac{3\pi}{2}$

15. $\sin 2x + 2 \sin x - \cos x - 1 = 0$

$$2 \sin x \cos x + 2 \sin x - \cos x - 1 = 0$$
$$2 \sin x(\cos x + 1) - (\cos x + 1) = 0$$
$$(2 \sin x - 1)(\cos x + 1) = 0$$

$2 \sin x - 1 = 0$ or $\cos x + 1 = 0$

$\sin x = \frac{1}{2}$ or $\cos x = -1$

$x = \frac{\pi}{6}, \frac{5\pi}{6}$ or $x = \pi$

All values check. The solutions in $[0, 2\pi)$ are $\frac{\pi}{6}, \frac{5\pi}{6}$, and π.

16. $\frac{2\pi}{3}, \frac{4\pi}{3}, \frac{3\pi}{2}$

17. $\sec^2 x = 4\tan^2 x$

$\sec^2 x = 4(\sec^2 x - 1)$

$\sec^2 x = 4\sec^2 x - 4$

$4 = 3\sec^2 x$

$\frac{4}{3} = \sec^2 x$

$\pm\frac{2}{\sqrt{3}} = \sec x$

$\pm\frac{\sqrt{3}}{2} = \cos x$

$x = \frac{\pi}{6}, \frac{5\pi}{6}, \frac{7\pi}{6}, \frac{11\pi}{6}$

All values check. The solutions in $[0, 2\pi)$ are $\frac{\pi}{6}, \frac{5\pi}{6}, \frac{7\pi}{6}$, and $\frac{11\pi}{6}$.

18. $\frac{\pi}{4}, \frac{3\pi}{4}, \frac{5\pi}{4}, \frac{7\pi}{4}$

19. $\sec^2 x + 3\tan x - 11 = 0$

$(1 + \tan^2 x) + 3\tan x - 11 = 0$

$\tan^2 x + 3\tan x - 10 = 0$

$(\tan x + 5)(\tan x - 2) = 0$

$\tan x + 5 = 0$ or $\tan x - 2 = 0$

$\tan x = -5$ or $\tan x = 2$

$x = 101.31°$ (or $101° 19'$), $281.31°$ (or $281° 19'$) or $x = 63.43°$ (or $63° 26'$), $243.43°$ (or $243° 26'$)

All values check. The solutions in $[0, 360°)$ are $63.43°$ (or $63° 26'$), $101.31°$ (or $101° 19'$), $243.43°$ (or $243° 26'$), and $281.31°$ (or $281° 19'$).

20. $45°$, $225°$, $116.57°$ (or $116° 34'$), $296.57°$ (or $296° 34'$)

21. $\cot x = \tan(2x - 3\pi)$

$\cot x = \frac{\tan 2x - \tan 3\pi}{1 + \tan 2x \tan 3\pi}$

$\cot x = \frac{\tan 2x - 0}{1 + \tan 2x \cdot 0}$

$\cot x = \tan 2x$

$\frac{1}{\tan x} = \frac{2\tan x}{1 - \tan^2 x}$

(In this form of the equation, x cannot be $\pi/2$ and $3\pi/2$. Therefore $\pi/2$ and $3\pi/2$ must also be checked in the original equation.)

$1 - \tan^2 x = 2\tan^2 x$

$1 = 3\tan^2 x$

$\frac{1}{3} = \tan^2 x$

$\frac{1}{\sqrt{3}} = |\tan x|$

21. (continued)

$\pm\frac{\sqrt{3}}{3} = \tan x$

$\frac{\pi}{6}, \frac{5\pi}{6}, \frac{7\pi}{6}, \frac{11\pi}{6} = x$

These values and also $\frac{\pi}{2}$ and $\frac{3\pi}{2}$ check. The solutions in $[0, 2\pi)$ are $\frac{\pi}{6}, \frac{\pi}{2}, \frac{5\pi}{6}, \frac{7\pi}{6}, \frac{3\pi}{2}$, and $\frac{11\pi}{6}$.

22. $\frac{\pi}{6}, \frac{5\pi}{6}, \frac{7\pi}{6}, \frac{11\pi}{6}$

23. $\cos(\pi - x) + \sin\left(x - \frac{\pi}{2}\right) = 1$

$(-\cos x) + (-\cos x) = 1$

$-2\cos x = 1$

$\cos x = -\frac{1}{2}$

$x = \frac{2\pi}{3}, \frac{4\pi}{3}$

Both values check. The solutions in $[0, 2\pi)$ are $\frac{2\pi}{3}$ and $\frac{4\pi}{3}$.

24. $\frac{\pi}{6}, \frac{5\pi}{6}$

25. $\frac{\cos^2 x - 1}{\sin\left(\frac{\pi}{2} - x\right) - 1} = \frac{\sqrt{2}}{2} + 1$

$\frac{(\cos x + 1)(\cos x - 1)}{\cos x - 1} = \frac{\sqrt{2}}{2} + 1$

$\cos x + 1 = \frac{\sqrt{2}}{2} + 1$

$\cos x = \frac{\sqrt{2}}{2}$

$x = \frac{\pi}{4}, \frac{7\pi}{4}$

Both values check. The solutions in $[0, 2\pi)$ are $\frac{\pi}{4}$ and $\frac{7\pi}{4}$.

26. $\frac{\pi}{4}, \frac{3\pi}{4}$

27. $2\cos x + 2\sin x = \sqrt{6}$

$\cos x + \sin x = \frac{\sqrt{6}}{2}$

$\cos^2 x + 2\sin x\cos x + \sin^2 x = \frac{6}{4}$

$\sin 2x + 1 = \frac{3}{2}$

$\sin 2x = \frac{1}{2}$

$2x = \frac{\pi}{6}, \frac{5\pi}{6}, \frac{13\pi}{6}, \frac{17\pi}{6}$

27. (continued)

$$x = \frac{\pi}{12}, \frac{5\pi}{12}, \frac{13\pi}{12}, \frac{17\pi}{12}$$

The values $\frac{13\pi}{12}$ and $\frac{17\pi}{12}$ do not check, but the other values do. The solutions in $[0, 2\pi)$ are $\frac{\pi}{12}$ and $\frac{5\pi}{12}$.

28. $\frac{7\pi}{12}, \frac{23\pi}{12}$

29. $\sqrt{3}\cos x - \sin x = 1$

$$\begin{aligned}
\sqrt{3}\cos x &= 1 + \sin x \\
3\cos^2 x &= 1 + 2\sin x + \sin^2 x \\
3(1 - \sin^2 x) &= 1 + 2\sin x + \sin^2 x \\
3 - 3\sin^2 x &= 1 + 2\sin x + \sin^2 x \\
0 &= 4\sin^2 x + 2\sin x - 2 \\
0 &= 2\sin^2 x + \sin x - 1 \\
0 &= (2\sin x - 1)(\sin x + 1)
\end{aligned}$$

$2\sin x - 1 = 0$ or $\sin x + 1 = 0$

$\sin x = \frac{1}{2}$ or $\sin x = -1$

$x = \frac{\pi}{6}, \frac{5\pi}{6}$ or $x = \frac{3\pi}{2}$

The value $\frac{5\pi}{6}$ does not check, but the other values do check. The solutions in $[0, 2\pi)$ are $\frac{\pi}{6}$ and $\frac{3\pi}{2}$.

30. $\frac{7\pi}{4}$

31. Sketch a triangle having an angle θ whose cosine is $\frac{3}{5}$. Using the Pythagorean theorem we know the other leg has length 4. From the diagram we find that $\sin\theta = \frac{4}{5}$. Thus, $\text{Arccos}\,\frac{3}{5} = \text{Arcsin}\,\frac{4}{5}$.

$$\begin{aligned}
\text{Arccos}\,x &= \text{Arccos}\,\frac{3}{5} - \text{Arcsin}\,\frac{4}{5} \\
\text{Arccos}\,x &= 0 \\
\cos 0 &= x \\
1 &= x
\end{aligned}$$

32. $\frac{\sqrt{2}}{2}$

33. First graph $y = \sin x - \cos x$. Then graph $y = \cot x$. The solutions are 1.15, 5.65, -0.63, -0.513, etc.

Exercise Set 9.1

1\.

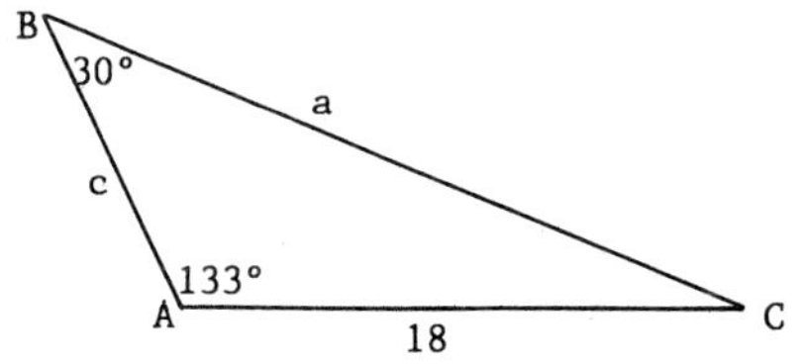

To solve this triangle find C, a, and c.

C = 180° - (133° + 30°) = 17°

Use the law of sines to find a and c.

Find a:

$$\frac{a}{\sin A} = \frac{b}{\sin B}$$

$$a = \frac{b \sin A}{\sin B}$$

$$a = \frac{18 \sin 133^\circ}{\sin 30^\circ} = \frac{18 \times 0.7314}{0.5} = 26.3$$

Find c:

$$\frac{c}{\sin C} = \frac{b}{\sin B}$$

$$c = \frac{b \sin C}{\sin B}$$

$$c = \frac{18 \sin 17^\circ}{\sin 30^\circ} = \frac{18 \times 0.2924}{0.5} = 10.5$$

2\. A = 30°, b = 27.7, c = 16.0

3\.

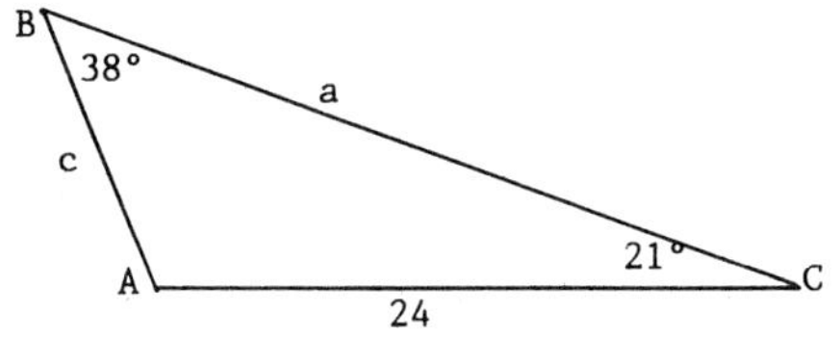

To solve this triangle find A, a, and c.

A = 180° - (38° + 21°) = 121°

Use the law of sines to find a and c.

Find a:

$$\frac{a}{\sin A} = \frac{b}{\sin B}$$

$$a = \frac{b \sin A}{\sin B}$$

$$a = \frac{24 \sin 121^\circ}{\sin 38^\circ} = \frac{24 \times 0.8572}{0.6157} = 33.4$$

Find c:

$$\frac{c}{\sin C} = \frac{b}{\sin B}$$

$$c = \frac{b \sin C}{\sin B}$$

$$c = \frac{24 \sin 21^\circ}{\sin 38^\circ} = \frac{24 \times 0.3584}{0.6157} = 14.0$$

4\. B = 26°, a = 17.2, c = 8.91

5\.

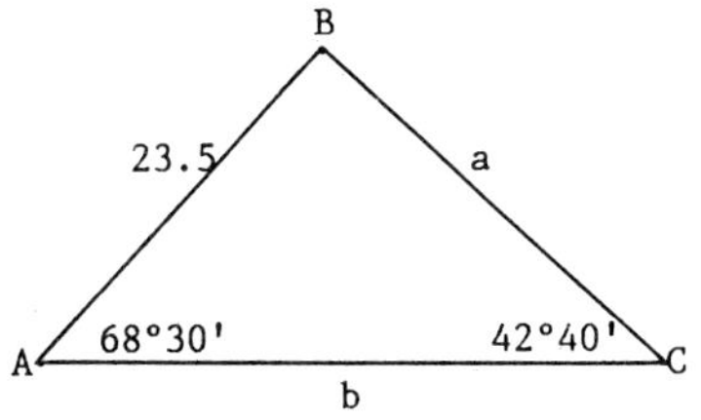

To solve this triangle find B, a, and b.

B = 180° - (68° 30' + 42° 40')

= 180° - 111° 10'

= 68° 50'

Use the law of sines to find a and b.

$$\frac{a}{\sin A} = \frac{c}{\sin C}$$

$$a = \frac{c \sin A}{\sin C}$$

$$a = \frac{23.5 \sin 68^\circ\ 30'}{\sin 42^\circ\ 40'} = \frac{23.5 \times 0.9304}{0.6777} = 32.3$$

$$\frac{b}{\sin B} = \frac{c}{\sin C}$$

$$b = \frac{c \sin B}{\sin C}$$

$$b = \frac{23.5 \sin 68^\circ\ 50'}{\sin 42^\circ\ 40'} = \frac{23.5 \times 0.9325}{0.6777} = 32.3$$

6\. A = 16°, a = 13.2, c = 34.2

7\. To solve this triangle we find c, B, and C. There will be two solutions. An arc of radius 24 meets the base at two points.

Solution I

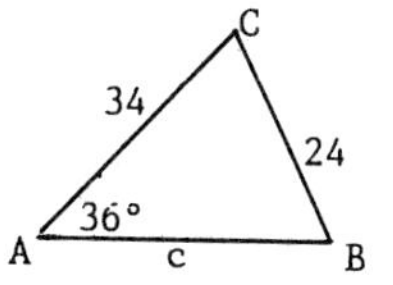

Solution II

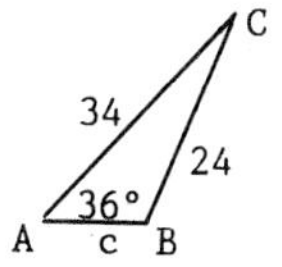

Find B:

$$\frac{b}{\sin B} = \frac{a}{\sin A}$$

$$\sin B = \frac{b \sin A}{a}$$

$$\sin B = \frac{34 \sin 36^\circ}{24} = \frac{34 \times 0.5878}{24} = 0.8327$$

There are two angles less than 180° having a sine of 0.8327. They are 56° 23' and 123° 37'. This gives us two possible solutions.

Solution I

If B = 56° 23'

then C = 180° - (36° + 56° 23') = 87° 37'.

7. (continued)

Find c:

$$\frac{c}{\sin C} = \frac{a}{\sin A}$$

$$c = \frac{a \sin C}{\sin A}$$

$$c = \frac{24 \sin 87° \; 37'}{\sin 36°} = \frac{24 \times 0.9991}{0.5878} = 40.8$$

Solution II

If B = 123° 37',
then C = 180° - (36° + 123° 37') = 20° 23'.

Find c:

$$\frac{c}{\sin C} = \frac{a}{\sin A}$$

$$c = \frac{a \sin C}{\sin A}$$

$$c = \frac{24 \sin 20° \; 23'}{\sin 36°} = \frac{24 \times 0.3483}{0.5878} = 14.2$$

8. A = 95° 53', B = 41° 7', a = 40.8

9.

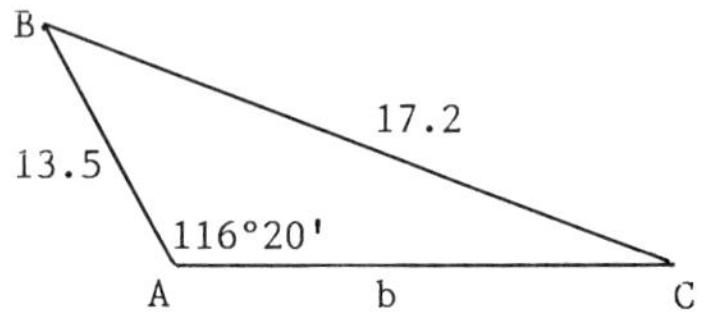

To solve this triangle find B, C, and b.

Find C:

$$\frac{c}{\sin C} = \frac{a}{\sin A}$$

$$\sin C = \frac{c \sin A}{a}$$

$$\sin C = \frac{13.5 \sin 116° \; 20'}{17.2} = \frac{13.5 \times 0.8962}{17.2} = 0.7034$$

Then C = 44° 42' or C = 135° 18'. An angle of 135° 18' cannot be an angle of this triangle because it already has an angle of 116° 20' and these two would total more than 180°. Thus C = 44° 42'.

Find B:

B = 180° - (116° 20' + 44° 42') = 18° 58'

Find b:

$$\frac{b}{\sin B} = \frac{a}{\sin A}$$

$$b = \frac{a \sin B}{\sin A}$$

$$b = \frac{17.2 \sin 18° \; 58'}{\sin 116° \; 20'} = \frac{17.2 \times 0.3250}{0.8962} = 6.24$$

10. B = 28° 28', C = 103° 42', c = 37.1

11.

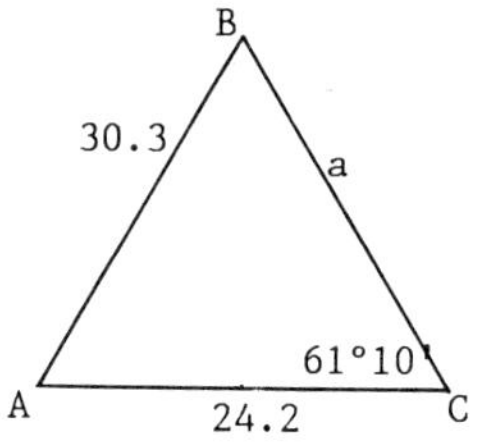

To solve this triangle find A, B, and a.

Find B:

$$\frac{b}{\sin B} = \frac{c}{\sin C}$$

$$\sin B = \frac{b \sin C}{c}$$

$$\sin B = \frac{24.2 \sin 61° \; 10'}{30.3} = \frac{24.2 \times 0.8760}{30.3} = 0.6996$$

Then B = 44° 24' or B = 135° 36'. An angle of 135° 36' cannot be an angle of this triangle because it already has an angle of 61° 10' and the two would total more than 180°. Thus B = 44° 24'.

Find A:

A = 180° - (61° 10' + 44° 24') = 74° 26'

Find a:

$$\frac{a}{\sin A} = \frac{c}{\sin C}$$

$$a = \frac{c \sin A}{\sin C}$$

$$a = \frac{30.3 \sin 74° \; 26'}{\sin 61° \; 10'} = \frac{30.3 \times 0.9633}{0.8760} = 33.3$$

12. A = 41° 7', C = 80° 13', c = 37.6

13.

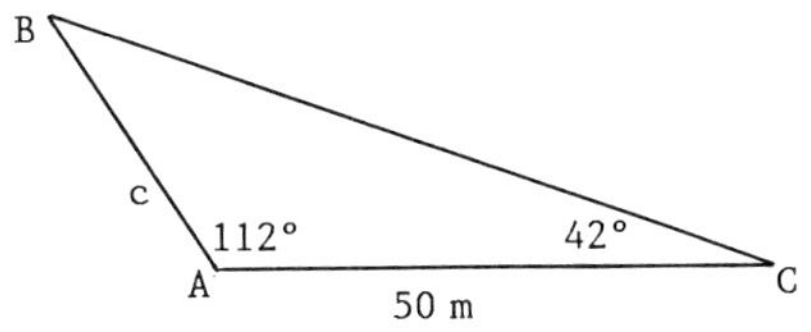

Find B:

B = 180° - (112° + 42°) = 26°

Find c:

$$\frac{c}{\sin C} = \frac{b}{\sin B}$$

$$c = \frac{b \sin C}{\sin B}$$

$$c = \frac{50 \sin 42°}{\sin 26°} = \frac{50 \times 0.6991}{0.4384} = 76.3$$

The width of the crater is 76.3 m.

14. 26.9 ft

15.

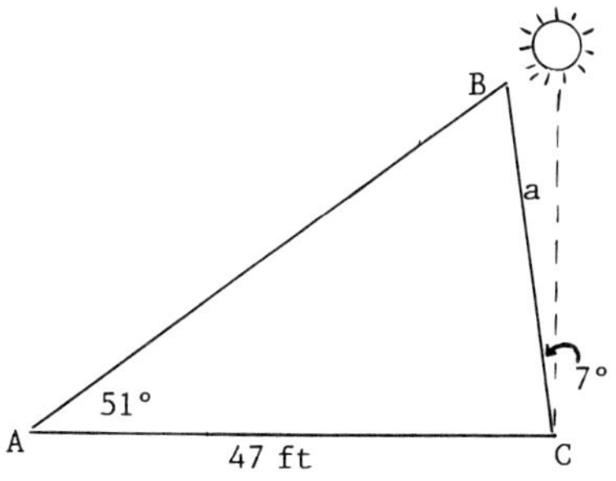

Find C: $C = 90° - 7° = 83°$

Find B: $B = 180° - (51° + 83°) = 46°$

Find a: (a is the length of the pole.)

$$\frac{a}{\sin A} = \frac{b}{\sin B}$$

$$a = \frac{b \sin A}{\sin B}$$

$$a = \frac{47 \sin 51°}{\sin 46°} = \frac{47 \times 0.7771}{0.7193} = 50.8$$

The pole is 50.8 ft long.

16. 9.29 ft

17.

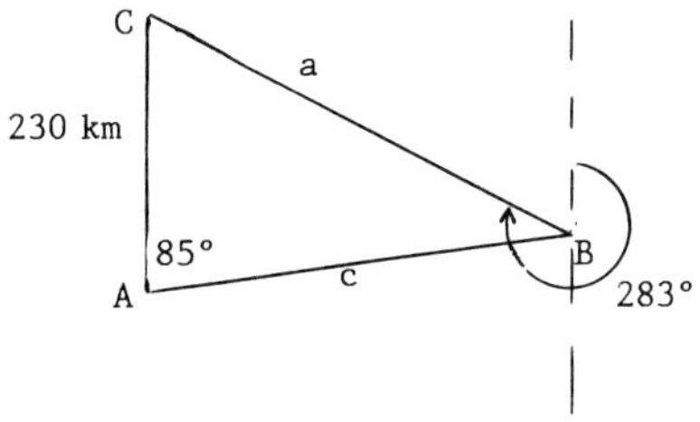

Find B: $B = 283° - 180° - 85° = 18°$

Find C: $C = 180° - (85° + 18°) = 77°$

Find a:

$$\frac{a}{\sin A} = \frac{b}{\sin B}$$

$$a = \frac{b \sin A}{\sin B}$$

$$a = \frac{230 \sin 85°}{\sin 18°} = \frac{230 \times 0.9962}{0.3090} = 742$$

Find c:

$$\frac{c}{\sin C} = \frac{b}{\sin B}$$

$$c = \frac{b \sin C}{\sin B}$$

$$c = \frac{230 \sin 77°}{\sin 18°} = \frac{230 \times 0.9744}{0.3090} = 725$$

The plane flew a total of
742 + 725, or 1467 km.

18. 24.7 km, 28.4 km

19.

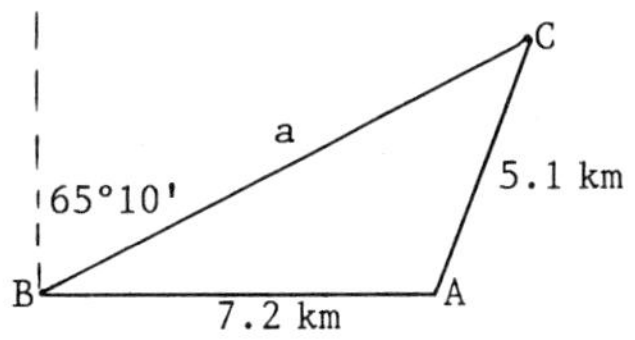

Find B: $B = 90° - 65°\ 10' = 24°\ 50'$

Find C:

$$\frac{c}{\sin C} = \frac{b}{\sin B}$$

$$\sin C = \frac{c \sin B}{b}$$

$$\sin C = \frac{7.2 \sin 24°\ 50'}{5.1} = \frac{7.2 \times 0.4200}{5.1} = 0.5929$$

Then $C = 36°\ 20'$ or $C = 143°\ 40'$.

If $C = 143°\ 40'$,
then $A = 180° - (24°\ 50' + 143°\ 40') = 11°\ 30'$.

Find a:

$$\frac{a}{\sin A} = \frac{b}{\sin B}$$

$$a = \frac{b \sin A}{\sin B}$$

$$a = \frac{5.1 \sin 11°\ 30'}{\sin 24°\ 50'} = \frac{5.1 \times 0.1994}{0.4200} = 2.42$$

If $C = 36°\ 20'$,
then $A = 180° - (24°\ 50' + 36°\ 20') = 118°\ 50'$.

Find a:

$$\frac{a}{\sin A} = \frac{b}{\sin B}$$

$$a = \frac{b \sin A}{\sin B}$$

$$a = \frac{5.1 \sin 118°\ 50'}{\sin 24°\ 50'} = \frac{5.1 \times 0.8760}{0.4200} = 10.6$$

The boat is either 2.42 km or 10.6 km from lighthouse B.

20. 373 km

21.

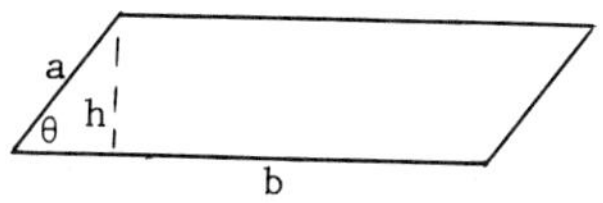

$A = bh$, $h = a \sin \theta$, so $A = ab \sin \theta$

22.

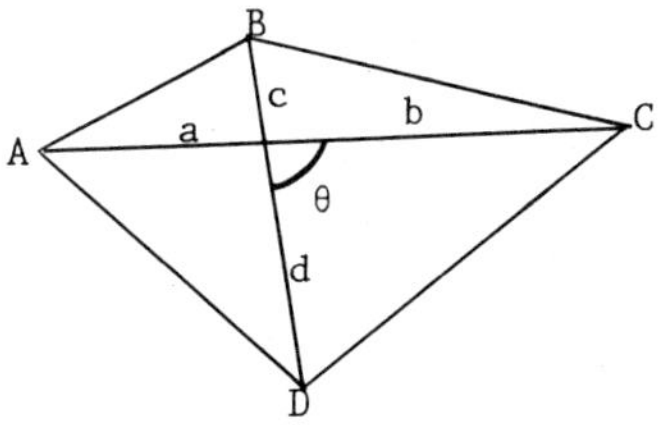

22. (continued)

$$A = \frac{1}{2}\,bd \sin\theta + \frac{1}{2}\,ac \sin\theta + \frac{1}{2}\,ad \sin(180° - \theta) + \frac{1}{2}\,bc \sin(180° - \theta)$$

$$= \frac{1}{2}\,(bd + ac + ad + bc) \sin\theta$$

$$= \frac{1}{2}\,(a + b)(c + d) \sin\theta$$

Exercise Set 9.2

1.

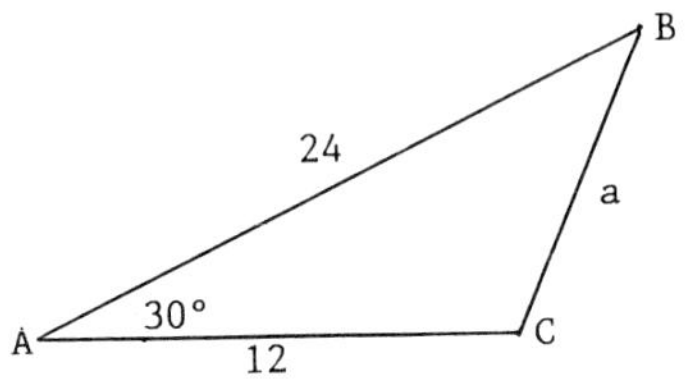

To solve this triangle find a, B, and C.

From the law of cosines,

$a^2 = b^2 + c^2 - 2bc \cos A$

$a^2 = 12^2 + 24^2 - 2 \cdot 12 \cdot 24 \cos 30°$

$= 144 + 576 - 576(0.8660)$

$= 221$

Then $a = \sqrt{221} = 14.9$.

Next we use the law of sines to find a second angle.

$$\frac{b}{\sin B} = \frac{a}{\sin A}$$

$$\sin B = \frac{b \sin A}{a}$$

$$\sin B = \frac{12 \sin 30°}{14.9} = \frac{12 \times 0.5}{14.9} = 0.4027$$

Thus $B = 23°\ 45'$.

The third angle is easy to find.

$C = 180° - (30° + 23°\ 45') = 126°\ 15'$

2. $c = 13.7$, $A = 71°\ 29'$, $B = 48°\ 31'$

3.

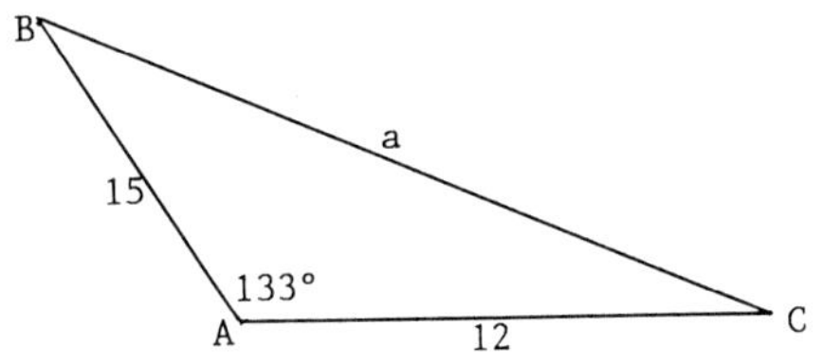

To solve this triangle find a, B, and C.

3. (continued)

From the law of cosines,

$a^2 = b^2 + c^2 - 2bc \cos A$

$a^2 = 12^2 + 15^2 - 2 \cdot 12 \cdot 15 \cos 133°$

$= 144 + 225 - 360(-0.6820)$

$= 615$

Then $a = \sqrt{615} = 24.8$.

Next we use the law of sines to find a second angle.

$$\frac{b}{\sin B} = \frac{a}{\sin A}$$

$$\sin B = \frac{b \sin A}{a}$$

$$\sin B = \frac{12 \sin 133°}{24.8} = \frac{12 \times 0.7314}{24.8} = 0.3539$$

Thus $B = 20°\ 43'$.

The third angle is easy to find.

$C = 180° - (133° + 20°\ 43') = 26°\ 17'$.

4. $b = 47.6$, $A = 35°\ 50'$, $C = 28°\ 10'$

5.

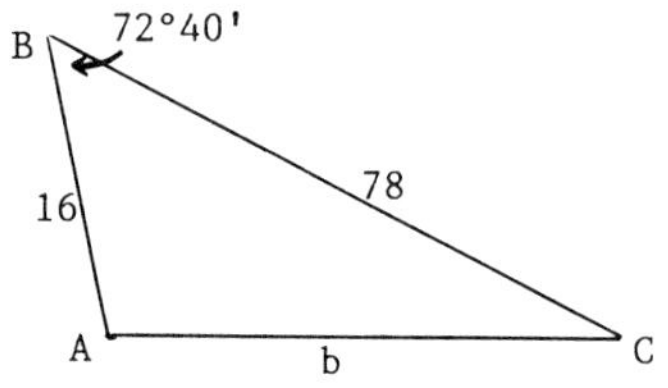

To solve this triangle find b, A, and C.

From the law of cosines,

$b^2 = a^2 + c^2 - 2ac \cos B$

$b^2 = 78^2 + 16^2 - 2 \cdot 78 \cdot 16 \cos 72°\ 40'$

$= 6084 + 256 - 2496(0.2979)$

$= 5596$

Then $b = \sqrt{5596} = 74.8$

Next we use the law of sines to find a second angle.

$$\frac{c}{\sin C} = \frac{b}{\sin B}$$

$$\sin C = \frac{c \sin B}{b}$$

$$\sin C = \frac{16 \sin 72°\ 40'}{74.8} = \frac{16 \times 0.9546}{74.8} = 0.2042$$

Thus $C = 11°\ 47'$.

The third angle is easy to find.

$A = 180° - (72°\ 40' + 11°\ 47') = 95°\ 33'$

6. $a = 55.6$, $B = 149°\ 30'$, $C = 6°$

7.

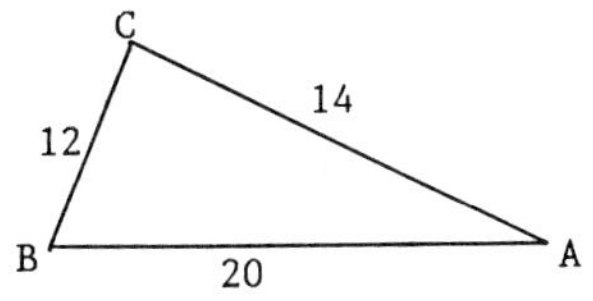

To solve this triangle find A, B, and C.

Find B:

$b^2 = a^2 + c^2 - 2ac \cos B$

$14^2 = 12^2 + 20^2 - 2 \cdot 12 \cdot 20 \cos B$

$196 = 144 + 400 - 480 \cos B$

$\cos B = \frac{348}{480} = 0.7250$

Thus B = 43° 32'.

Find A:

$a^2 = b^2 + c^2 - 2bc \cos A$

$12^2 = 14^2 + 20^2 - 2 \cdot 14 \cdot 20 \cos A$

$144 = 196 + 400 - 560 \cos A$

$\cos A = \frac{452}{560} = 0.8071$

Thus A = 36° 11'.

Then C = 180° - (43° 30' + 36° 11') = 100° 17'.

8. A = 37° 18', B = 37° 18', C = 105° 24'

9.

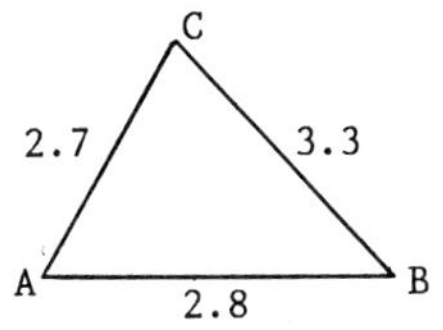

To solve this triangle find A, B, and C.

Find A:

$a^2 = b^2 + c^2 - 2bc \cos A$

$(3.3)^2 = (2.7)^2 + (2.8)^2 - 2(2.7)(2.8) \cos A$

$10.89 = 7.29 + 7.84 - 15.12 \cos A$

$\cos A = \frac{4.24}{15.12} = 0.2804$

Thus A = 73° 43'.

Find B:

$b^2 = a^2 + c^2 - 2ac \cos B$

$(2.7)^2 = (3.3)^2 + (2.8)^2 - 2(3.3)(2.8) \cos B$

$7.29 = 10.89 + 7.84 - 18.48 \cos B$

$\cos B = \frac{11.44}{18.48} = 0.6190$

Thus B = 51° 45'

Then C = 180° - (73° 43' + 51° 45') = 54° 32'

10. A = 24° 9', B = 30° 45', C = 125° 6'

11.

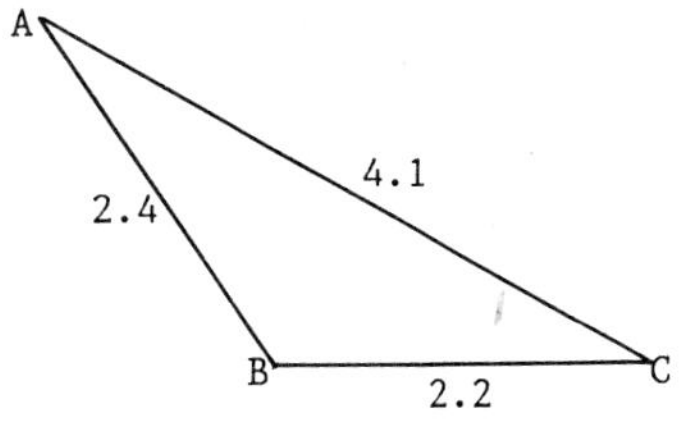

To solve this triangle find A, B, and C.

Find A:

$a^2 = b^2 + c^2 - 2bc \cos A$

$(2.2)^2 = (4.1)^2 + (2.4)^2 - 2(4.1)(2.4) \cos A$

$4.84 = 16.81 + 5.76 - 19.68 \cos A$

$\cos A = \frac{17.73}{19.68} = 0.9009$

Thus A = 25° 43'.

Find C:

$c^2 = a^2 + b^2 - 2ab \cos C$

$(2.4)^2 = (2.2)^2 + (4.1)^2 - 2(2.2)(4.1) \cos C$

$5.76 = 4.84 + 16.81 - 18.04 \cos C$

$\cos C = \frac{15.89}{18.04} = 0.8808$

Thus C = 28° 16'.

Then B = 180° - (25° 43' + 28° 16') = 126° 1'

12. A = 33° 43', B = 107° 5', C = 39° 12'

13.

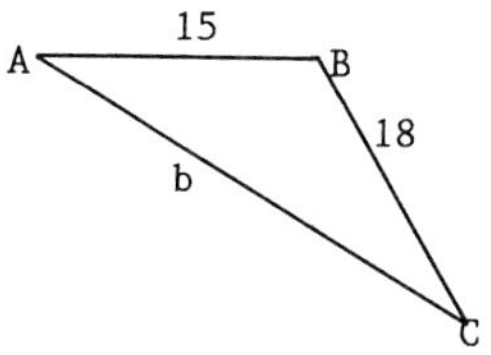

Find B: B = 90° + 27° = 117°

Find b:

From the law of cosines,

$b^2 = a^2 + c^2 - 2ac \cos B$

$b^2 = 18^2 + 15^2 - 2 \cdot 18 \cdot 15 \cos 117°$

$= 324 + 225 - 540(-0.4540)$

$= 794.16$

Then $b = \sqrt{794.16} = 28.2$.

Find A:

From the law of sines

$\frac{a}{\sin A} = \frac{b}{\sin B}$

$\sin A = \frac{a \sin B}{b}$

$\sin A = \frac{18 \sin 117°}{28.2} = \frac{18 \times 0.8910}{28.2} = 0.5687$

Thus A = 34° 40' and 90° - 34° 40' = 55° 20'.

The ship is 28.2 nautical mi from the harbor in a direction of S55° 20'E.

14. 315 km, 241°

15.

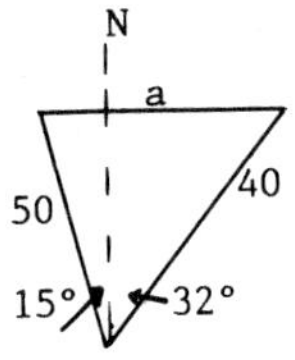

2 hr × 25 knots = 50 nautical mi

2 hr × 20 knots = 40 nautical mi

Find A: A = 15° + 32° = 47°

Find a:

$a^2 = b^2 + c^2 - 2bc \cos A$

$a^2 = 50^2 + 40^2 - 2\cdot 50\cdot 40 \cos 47°$

$= 2500 + 1600 - 4000(0.6820)$

$= 1372$

Then $a = \sqrt{1372} = 37.0$.

The ships are 37 nautical mi apart.

16. 912 km

17.

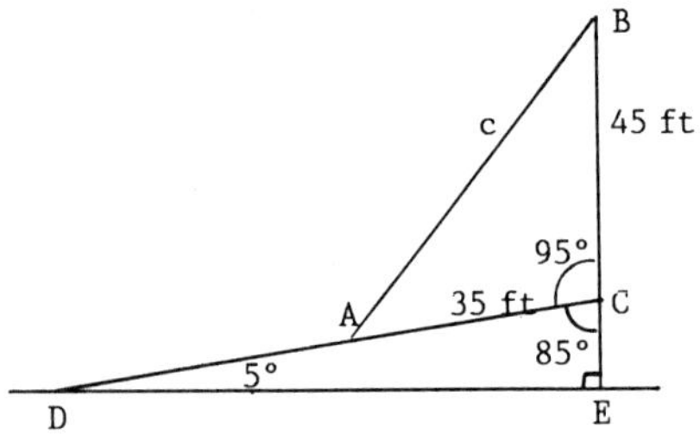

Since D = 5° and E = 90°, ∠DCE = 85° and ∠ACB = 95°.

Using the law of cosines,

$c^2 = 45^2 + 35^2 - 2\cdot 45\cdot 35 \cos 95°$

$= 2025 + 1225 - 3150(-0.0872)$

$= 3525$

Then $c = \sqrt{3525} = 59.4$.

The length of the rope must be 59.4 ft.

18. 87.4 ft

19. The perimeter is 5.5 m. The length of the third side is 5.5 - (1.5 + 2), or 2 m.

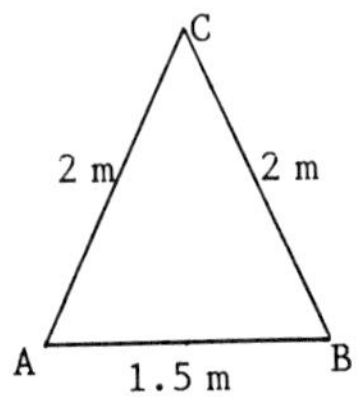

19. (continued)

Find A:

$a^2 = b^2 + c^2 - 2bc \cos A$

$2^2 = (1.5)^2 + 2^2 - 2(1.5)(2) \cos A$

$4 = 2.25 + 4 - 6 \cos A$

$\cos A = \frac{2.25}{6} = 0.375$

Thus A = 68°. Since △ABC is isosceles B is also 68°.

Find C:

C = 180° - (68° + 68°) = 44°

The angles are 68°, 68°, and 44°.

20. 52° 50', 85° 30', 41° 40'

21.

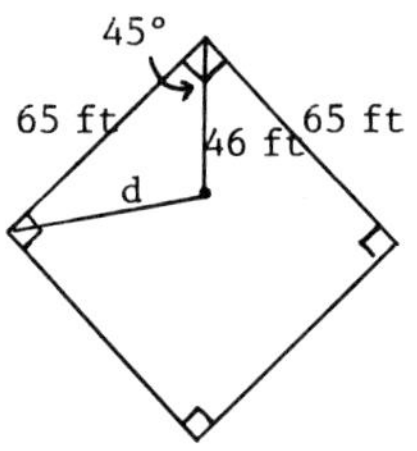

Using the law of cosines,

$d^2 = 65^2 + 46^2 - 2\cdot 65\cdot 46\cdot \cos 45°$

$= 4225 + 2116 - 5980(0.7071)$

$= 2112.5$

Then $d = \sqrt{2112.5} = 45.96$ ft

The distance from the pitcher's mound to first base is 45.96 ft.

22. 63.7 ft

23.

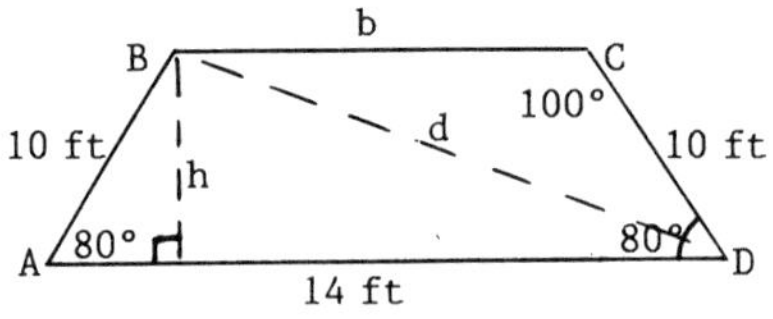

Let b represent the other base, d represent the diagonal, and h the height.

a) Find d:

Using the law of cosines,

$d^2 = 10^2 + 14^2 - 2\cdot 10\cdot 14 \cos 80°$

$= 100 + 196 - 280(0.1736)$

$= 247.4$

Then $d = \sqrt{247.4} = 15.7$.

The length of the diagonal is 15.7 ft.

23. (continued)

b) Find h:

$\frac{h}{10} = \sin 80°$

$h = 10 \sin 80°$

$h = 10(0.9848)$

$h = 9.85$

Find ∠CBD:

Using the law of sines,

$\frac{10}{\sin \angle CBD} = \frac{15.7}{\sin 100°}$

$\sin \angle CBD = \frac{10 \sin 100°}{15.7} = \frac{10 \times 0.9848}{15.7} = 0.6273$

Thus ∠CBD = 38° 50'.

Then ∠CDB = 180° - (100° + 38° 50') = 41° 10'.

Find b:

Using the law of sines,

$\frac{b}{\sin 41° 10'} = \frac{15.7}{\sin 100°}$

$b = \frac{15.7 \sin 41° 10'}{\sin 100°} = \frac{15.7 \times 0.6583}{0.9848}$

$= 10.5$

$\text{Area} = \frac{1}{2} h(b_1 + b_2)$

$\text{Area} = \frac{1}{2} (9.85)(10.5 + 14)$

$= 121 \text{ ft}^2$

24. 123.1

25.

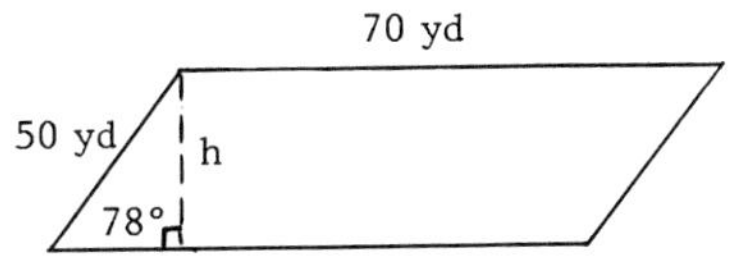

Let b = 70 yd; then find h.

$\frac{h}{50} = \sin 78°$

$h = 50 \sin 78°$

$h = 50(0.9781)$

$h = 48.9$

$\text{Area} = b \cdot h$

$\text{Area} = 70 \cdot 48.9$

$= 3423 \text{ yd}^2$

26. 13°

27.

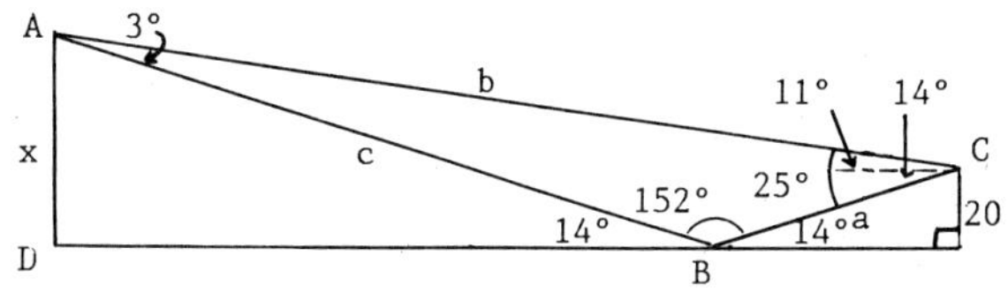

Find ∠ABC, ∠CAB, and ∠ACB:

The angle of incidence equals the angle of reflection. Thus ∠ABC = 180° - (14° + 14°), or 152°. Then ∠CAB = 180° - (25° + 152°), or 3°, ∠ACB = 11° + 14°, or 25°.

Find a:

$\frac{a}{20} = \csc 14°$

$a = 20 \csc 14° = 20 \times 4.1336 = 82.7$

Find c:

Using the law of sines,

$\frac{c}{\sin 25°} = \frac{82.7}{\sin 3°}$

$c = \frac{82.7 \sin 25°}{\sin 3°} = \frac{82.7 \times 0.4226}{0.0523} = 668$

Find x:

$\frac{x}{668} = \sin 14°$

$x = 668 \sin 14° = 668 \times 0.2419 = 162 \text{ ft}$

The height of the tree is 162 ft.

28. 9386 ft

29.

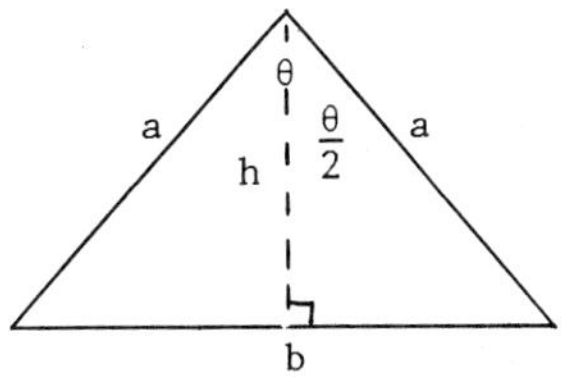

Let b represent the base, h the height, and θ the included angle.

Find h:

$\frac{h}{a} = \cos \frac{\theta}{2}$

$h = a \cos \frac{\theta}{2}$

Find b:

$b^2 = a^2 + a^2 - 2 \cdot a \cdot a \cos \theta$

$b^2 = 2a^2 - 2a^2 \cos \theta$

$b^2 = 2a^2(1 - \cos \theta)$

$b = \sqrt{2}\, a\sqrt{1 - \cos \theta}$

29. (continued)

Find A (area):

$A = \frac{1}{2}bh$

$A = \frac{1}{2}\left[\sqrt{2}\,a\sqrt{1 - \cos\theta}\right]\left[a\cos\frac{\theta}{2}\right]$

$= \left[a\sqrt{\frac{1 - \cos\theta}{2}}\right]\left[a\sqrt{\frac{1 + \cos\theta}{2}}\right]$

$= \frac{1}{2}a^2\sqrt{1 - \cos^2\theta} = \frac{1}{2}a^2\sqrt{\sin^2\theta}$

$= \frac{1}{2}a^2 \sin\theta$

The maximum value of $\sin\theta$ is 1 when θ is 90°.

30. a) S 87° 36' E, b) 504.9 ft N 70° 17' E

Exercise Set 9.3

1.

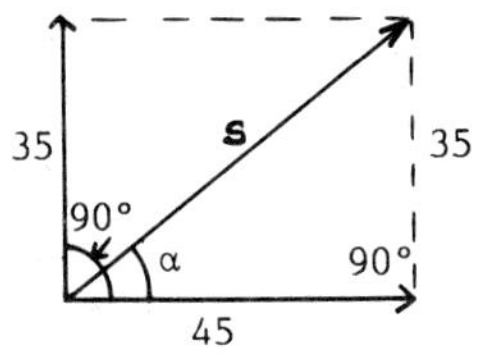

Find the resultant, **s**, and the angle, α, it makes with vector **a**. Consider the following triangle:

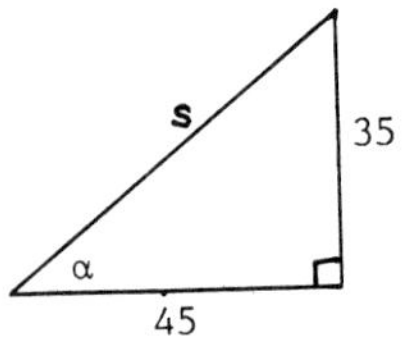

Find $|s|$:

Using the Pythagorean theorem,

$|s|^2 = 35^2 + 45^2$ ($|s|$ denotes the length of s)

$= 1225 + 2025$

$= 3250$

Then $|s| = \sqrt{3250} = 57.0$

Find α:

$\tan\alpha = \frac{35}{45} = 0.7778$

Thus α = 38°.

2. 69, 39°

3.

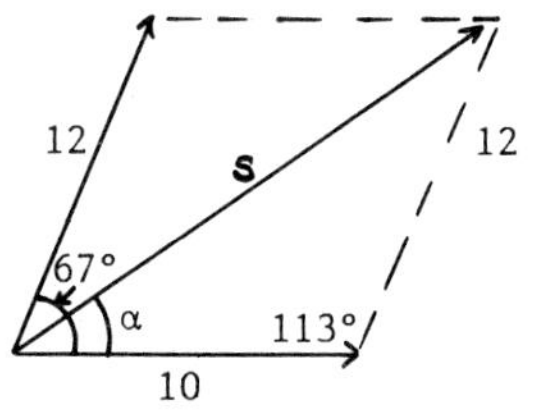

Find the resultant, **s**, and the angle, α, it makes with vector **a**. Consider the following triangle:

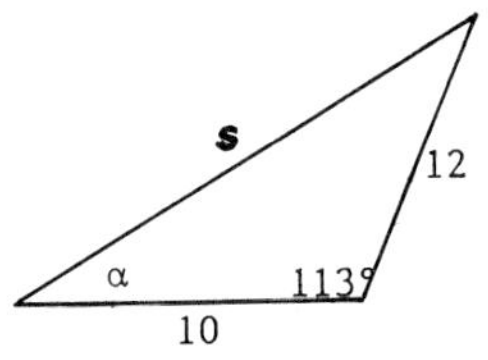

Find $|s|$:

Using the law of cosines,

$|s|^2 = 10^2 + 12^2 - 2 \cdot 10 \cdot 12 \cos 113°$

$= 100 + 144 - 240(-0.3907)$

$= 338$

Then $|s| = \sqrt{338} = 18.4$.

Find α:

Using the law of sines,

$\frac{12}{\sin\alpha} = \frac{18.4}{\sin 113°}$

$\sin\alpha = \frac{12 \sin 113°}{18.4} = \frac{12 \times 0.9205}{18.4} = 0.6003$

Thus α = 37°.

4. 43.7, 42°

5.

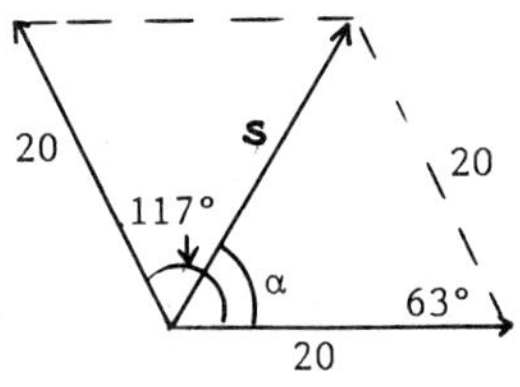

Find the resultant, s, and the angle, α, it makes with vector **a**. Consider the following triangle:

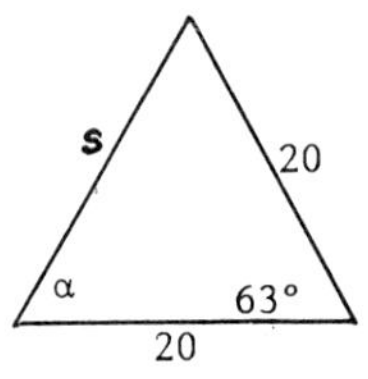

5. (continued)

Find $|s|$:

Using the law of cosines,

$|s|^2 = 20^2 + 20^2 - 2 \cdot 20 \cdot 20 \cos 63°$

$= 400 + 400 - 800(0.4540)$

$= 437$

Then $|s| = \sqrt{437} = 20.9$.

Find α:

Since the triangle is isosceles and the angle between the congruent sides is 63°, we know that

$\alpha = \frac{1}{2}(180° - 63°) = \frac{1}{2}(117°) = 58° \ 30'$.

6. 28.6, 62°

7.

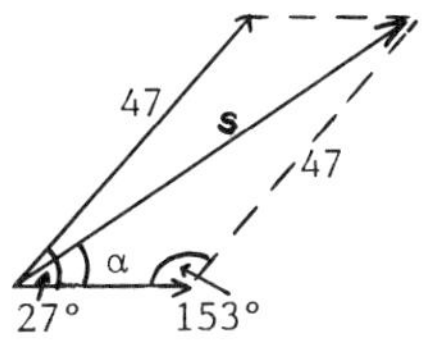

Find the resultant, s, and the angle, α, it makes with vector **a**. Consider the following triangle:

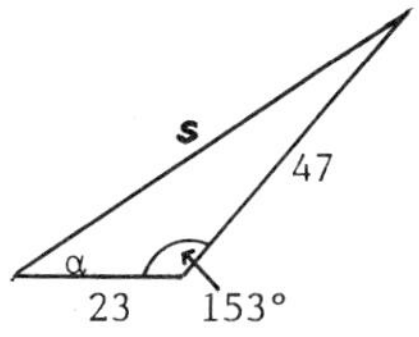

Find $|s|$:

Using the law of cosines,

$|s|^2 = 23^2 + 47^2 - 2 \cdot 23 \cdot 47 \cos 153°$

$= 529 + 2209 - 2162(-0.8910)$

$= 4664$

Then $|s| = \sqrt{4664} = 68.3$.

Find α:

Using the law of sines,

$\frac{47}{\sin \alpha} = \frac{68.3}{\sin 153°}$

$\sin \alpha = \frac{47 \sin 153°}{68.3} = \frac{47 \times 0.4540}{68.3} = 0.3124$

Thus $\alpha = 18°$.

8. 89.2, 52°

9.

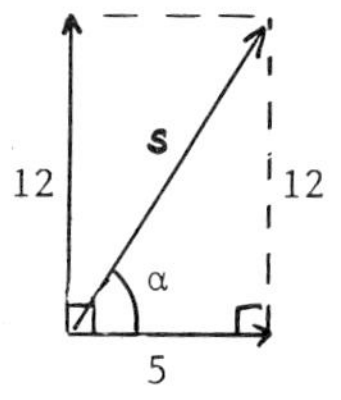

Find $|s|$:

Using the Pythagorean theorem,

$|s|^2 = 12^2 + 5^2$

$= 144 + 25$

$= 169$

Then $|s| = \sqrt{169} = 13$ kg.

Find α:

$\tan \alpha = \frac{12}{5} = 2.4$

Then $\alpha = 67°$.

10. 50 kg, 53°

11.

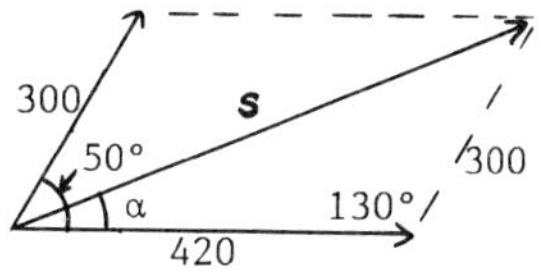

Find $|s|$:

Using the law of cosines,

$|s|^2 = 420^2 + 300^2 - 2 \cdot 420 \cdot 300 \cos 130°$

$= 176{,}400 + 90{,}000 - 252{,}000(-0.6428)$

$= 428{,}386$

Then $|s| = \sqrt{428{,}386} = 655$ kg

Find α:

Using the law of sines,

$\frac{300}{\sin \alpha} = \frac{655}{\sin 130°}$

$\sin \alpha = \frac{300 \sin 130°}{655} = \frac{300 \times 0.7660}{655} = 0.3508$

Thus $\alpha = 21°$.

12. 929 kg, 28°

13.

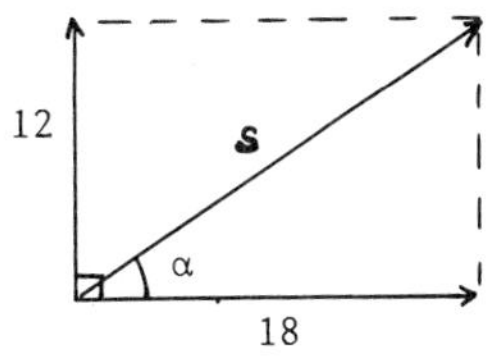

13. (continued)

Find $|s|$, the speed of the balloon, and α, the angle it makes with the horizontal. Consider the following triangle:

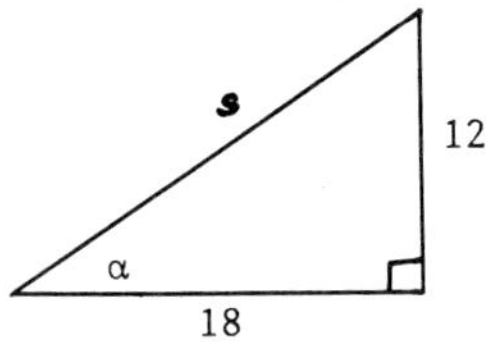

Find $|s|$:

$|s|^2 = 18^2 + 12^2 = 324 + 144 = 468$

Then $|s| = \sqrt{468} = 21.6$.

Find α:

$\tan \alpha = \frac{12}{18} = 0.6667$

Thus $\alpha = 34°$.

The speed of the balloon is 21.6 ft/sec, and it makes a 34° angle with the horizontal.

14. $11.2 \frac{\text{ft}}{\text{sec}}$, 63°

15.

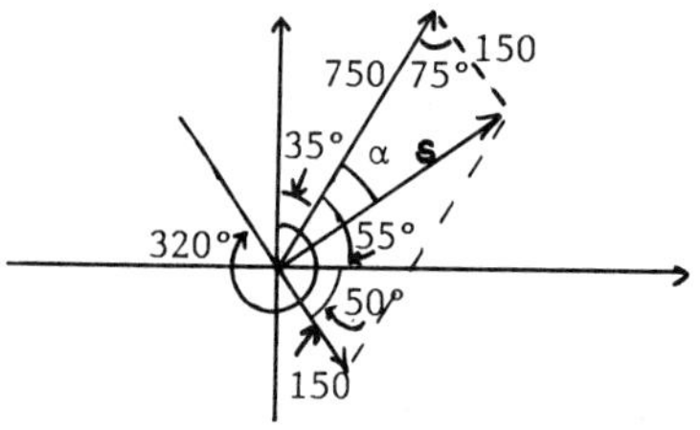

Find $|s|$:

Using the law of cosines,

$|s|^2 = 750^2 + 150^2 - 2\cdot750\cdot150 \cos 75°$

$= 562{,}500 + 22{,}500 - 225{,}000(0.2588)$

$= 526{,}770$

Then $|s| = \sqrt{526{,}770} = 726$.

Find α:

Using the law of sines,

$\frac{150}{\sin \alpha} = \frac{726}{\sin 75°}$

$\sin \alpha = \frac{150 \sin 75°}{726} = \frac{150 \times 0.9659}{726} = 0.1996$

Thus $\alpha = 12°$.

The resultant force is 726 lb, and the boat is moving in the direction 35° + 12°, or 47°.

16. 729 lb, 225°

17.

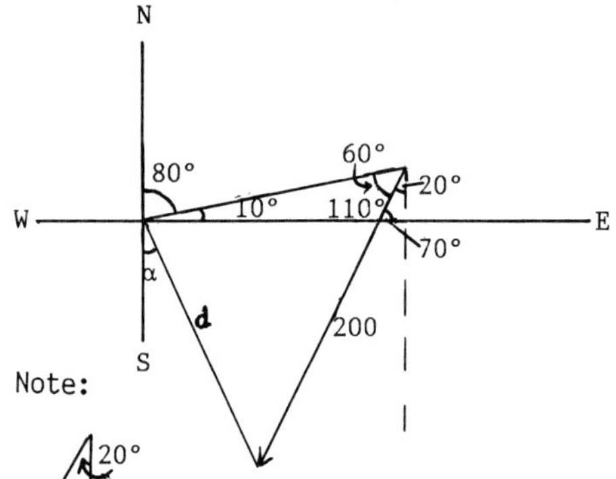

Note:

$180° - (20° + 90°) = 70°$

$180° - 70° = 110°$

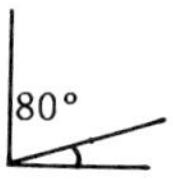

$90° - 80° = 10°$

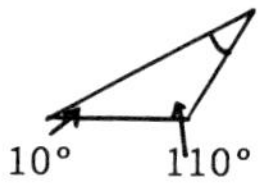

$180° - (10° + 110°) = 60°$

Find $|d|$:

Using the law of cosines,

$|d|^2 = 120^2 + 200^2 - 2\cdot120\cdot200 \cos 60°$

$= 14{,}400 + 40{,}000 - 48{,}000(0.5)$

$= 30{,}400$

Then $|d| = \sqrt{30{,}400} = 174$.

Use the law of sines to find θ in this triangle.

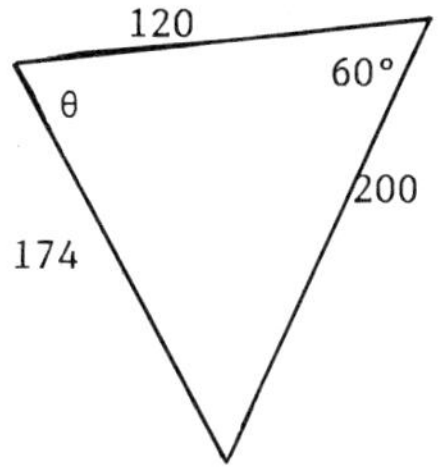

$\frac{200}{\sin \theta} = \frac{174}{\sin 60°}$

$\sin \theta = \frac{200 \sin 60°}{174} = \frac{200(0.8660)}{174} = 0.9954$

Thus $\theta = 85°$.

Now we can find α.

$\alpha = 180° - (80° + 85°) = 15°$

The ship is 174 nautical miles from the starting point in the direction S 15° 30' E.

18. 215 km, 345°

19. 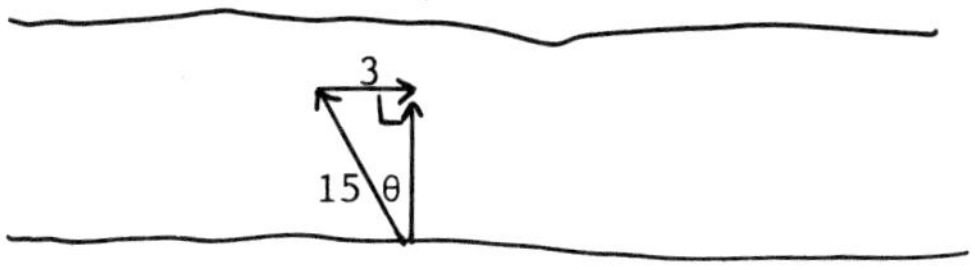

Find θ:

$\sin\theta = \frac{3}{15} = 0.2000$

Thus θ = 12°.

The boat should be pointed at an angle of 12° upstream.

20. 070°

Exercise Set 9.4

1. <3, 7> + <2, 9>
   = <3 + 2, 7 + 9>
   = <5, 16>

2. <2, -5>

3. <17, 7.6> + <-12.2, 6.1>
   = <17 + (-12.2), 7.6 + 6.1>
   = <4.8, 13.7>

4. <-7.3, 19.3>

5. <-650, -750> + <-12, 324>
   = <-650 + (-12), -750 + 324>
   = <-662, -426>

6. <-429, -717>

7. Let **v** = <3, 4>.

   $\tan\theta = \frac{4}{3} = 1.333$

   θ = 53° 8'

   $|\mathbf{v}| = \sqrt{3^2 + 4^2} = \sqrt{25} = 5$

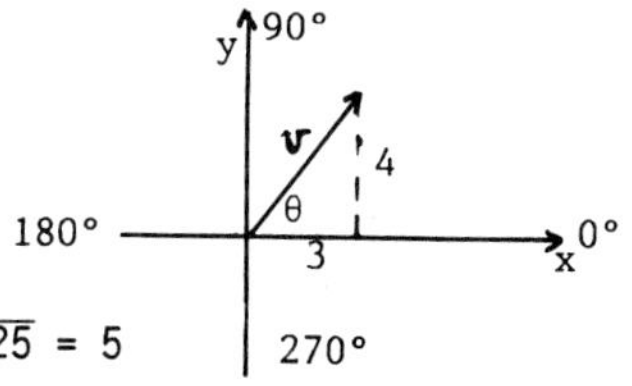

Thus polar notation for **v** is (5, 53° 8').

8. (5, 36° 52')

9. Let **v** = <10, -15>.

   The reference angle is ϕ.

   $\tan\phi = \frac{-15}{10} = -1.5$

   ϕ = -56° 19'

   θ = 360° - 56° 19' = 303° 41'

   $|\mathbf{v}| = \sqrt{10^2 + (-15)^2} = \sqrt{325} \approx 18.0$

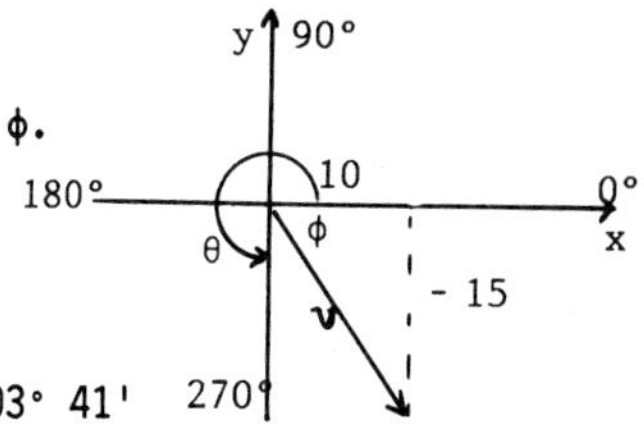

Thus polar notation for **v** is (18.0, 303° 41') or (18.0, -56° 19').

10. (19.7, 329° 32')

11. Let **v** = <-3, -4>.

    The reference angle is ϕ.

    $\tan\phi = \frac{-4}{-3} = 1.3333$

    ϕ = 53° 8'

    θ = 180° + 53° 8' = 233° 8'

    $|\mathbf{v}| = \sqrt{(-3)^2 + (-4)^2} = \sqrt{25} = 5$

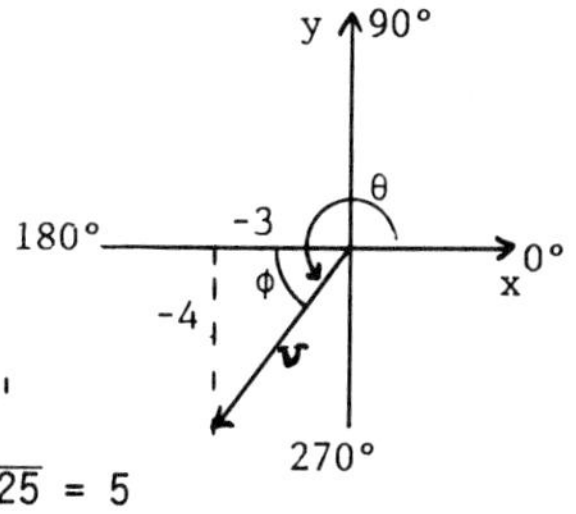

Thus polar notation for **v** is (5, 233° 8').

12. (5, 216° 52')

13. Let **v** = <-10, 15>.

    The reference angle is ϕ.

    $\tan\phi = \frac{15}{-10} = -1.5$

    ϕ = -56° 19'

    θ = 180° - 56° 19' = 123° 41'

    $|\mathbf{v}| = \sqrt{(-10)^2 + 15^2} = \sqrt{325} \approx 18.0$

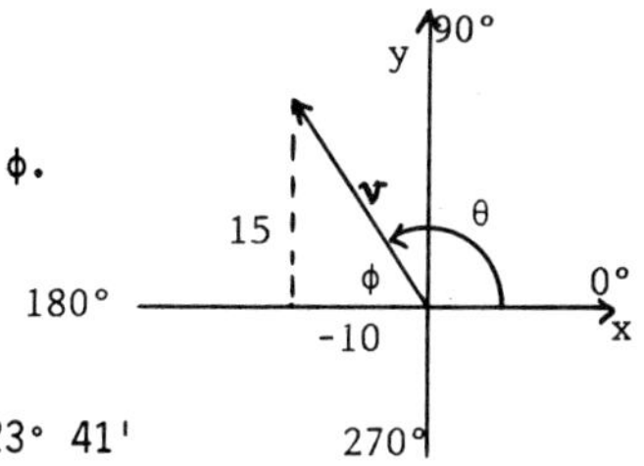

Thus polar notation for **v** is (18.0, 123° 41').

14. (19.7, 149° 32')

15. **v** = (4, 30°)

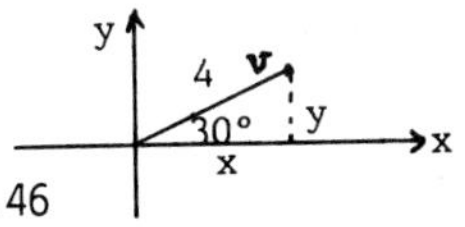

From the drawing we have

x = 4 cos 30° = 4(0.8660) = 3.46

y = 4 sin 30° = 4(0.5) = 2

Thus rectangular notation for **v** is <3.46, 2>.

16. <4, 6.93>

17. v = (10, 235°)

From the drawing we have

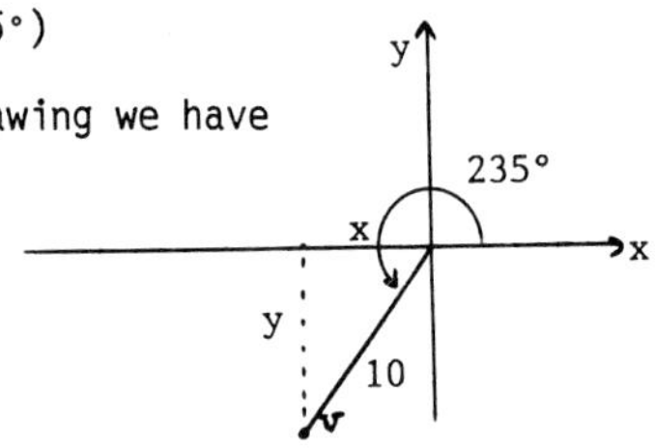

$x = 10 \cos 235° = 10(-0.5736) = -5.74$
$y = 10 \sin 235° = 10(-0.8192) = -8.19$

Thus rectangular notation for **v** is <-5.74, -8.19>.

18. <-13.0, -7.50>

19. v = (20, 330°)

From the drawing we have

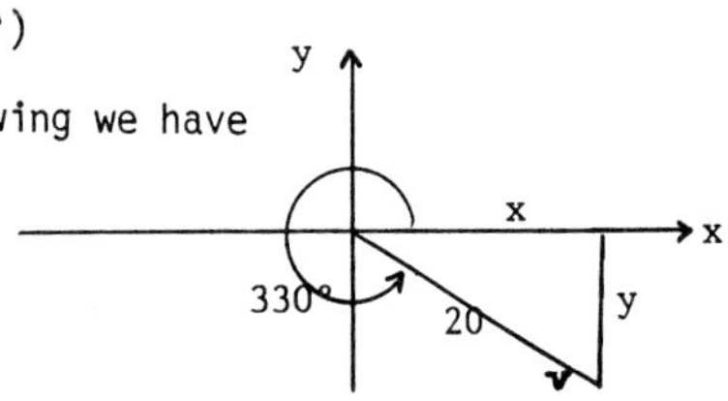

$x = 20 \cos 330° = 20(0.8660) = 17.3$
$y = 20 \sin 330° = 20(-0.5) = -10$

Thus rectangular notation for **v** is <17.3, -10>.

20. <-18.8, -6.84>

21. v = (100, -45°)

From the drawing we have

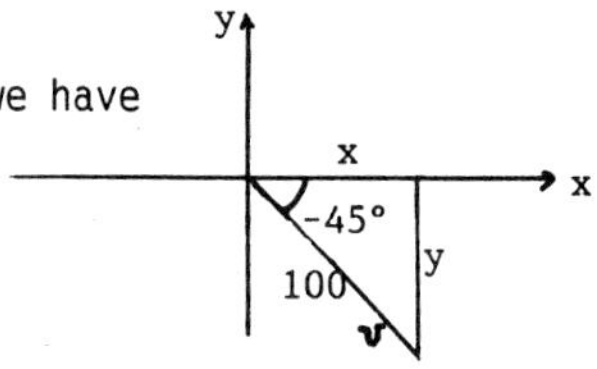

$x = 100 \cos (-45°) = 100(0.7071) = 70.7$
$y = 100 \sin (-45°) = 100(-0.7071) = -70.7$

Thus rectangular notation for **v** is <70.7, -70.7>.

22. <75, -130>

23.

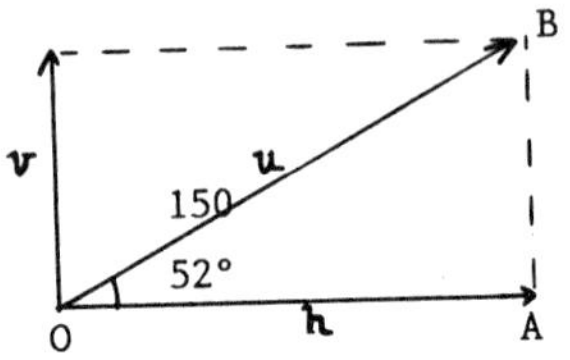

From Δ OAB, we find

$|\mathbf{h}| = 150 \cos 52° = 150(0.6157) = 92.3$
$|\mathbf{v}| = 150 \sin 52° = 150(0.7880) = 118.2$

24. Vert: 151.5 downward,
Hort: 77.2 to the right

25.

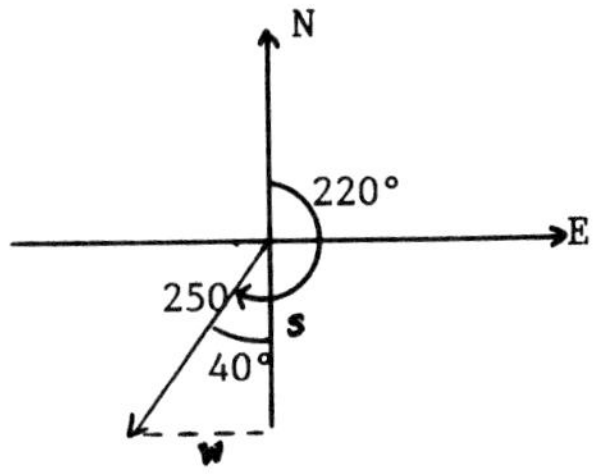

From the drawing we find

$|\mathbf{s}| = 250 \cos 40° = 250(0.7660) = 192$ km/h
$|\mathbf{w}| = 250 \sin 40° = 250(0.6428) = 161$ km/h

26. S: 16.1 mph, E: 19.2 mph

27.

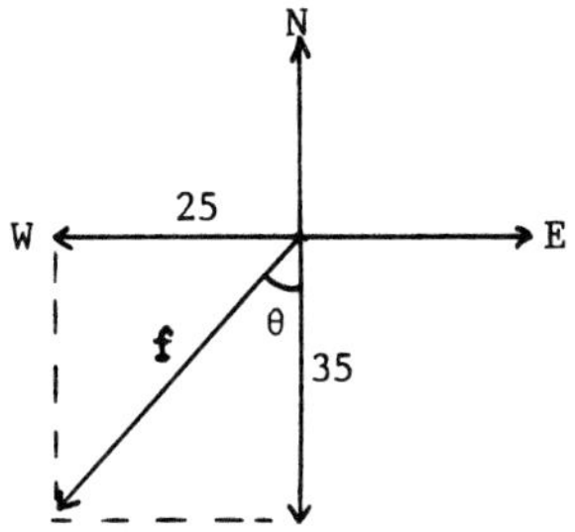

Find $|\mathbf{f}|$:

$|\mathbf{f}| = \sqrt{25^2 + 35^2} = \sqrt{1850} = 43.0$

Find θ:

$\tan \theta = \frac{25}{35} = 0.7143$

$\theta = 35° 32'$

The vector **f** has a magnitude of 43 and a direction of S 35° 32' W.

28. 111 kg, 35° 50' with horizontal

29.

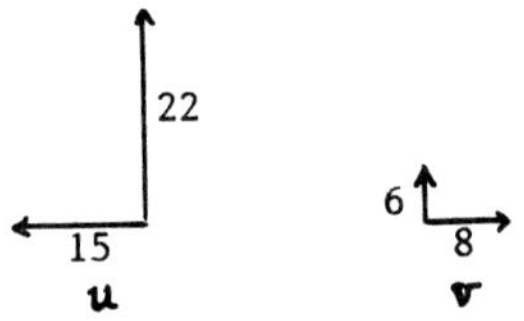

a) Adding the east-west components, we obtain 7 west. Adding the north-south components, we obtain 28 north. The components of **u** + **v** are 7 west and 28 north.

b)

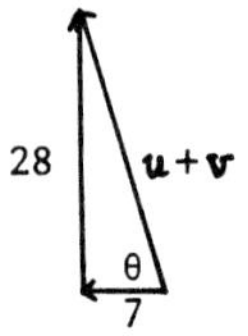

29. (continued)

Find θ:

$\tan\theta = \frac{28}{7} = 4$

$\theta = 75°\ 58'$

Thus $90° - 75°\ 58' = 14°\ 2'$.

Find $|\mathbf{u} + \mathbf{v}|$:

$|\mathbf{u} + \mathbf{v}| = 28 \csc 75°\ 58' = 28(1.0308) = 28.9$

The vector **u** + **v** has a magnitude of 28.9 and a direction of N 14° 2' W.

30. a) Up: 13, Right: 23
b) 26.4, 29° 29' with horizontal

31. - 40.
Given: $\mathbf{u} = \langle 3, 4\rangle$, $\mathbf{v} = \langle 5, 12\rangle$, and $\mathbf{w} = \langle -6, 8\rangle$

31. $3\mathbf{u} + 2\mathbf{v} = 3\langle 3, 4\rangle + 2\langle 5, 12\rangle$
$= \langle 9, 12\rangle + \langle 10, 24\rangle$
$= \langle 9 + 10, 12 + 24\rangle$
$= \langle 19, 36\rangle$

32. $\langle 27, 20\rangle$

33. $(\mathbf{u} + \mathbf{v}) - \mathbf{w} = (\langle 3, 4\rangle + \langle 5, 12\rangle) - \langle -6, 8\rangle$
$= \langle 8, 16\rangle + \langle 6, -8\rangle$
$= \langle 14, 8\rangle$

34. $\langle 4, -16\rangle$

35. $|\mathbf{u}| = |\langle 3, 4\rangle| = \sqrt{3^2 + 4^2} = \sqrt{25} = 5$

$|\mathbf{v}| = |\langle 5, 12\rangle| = \sqrt{5^2 + 12^2} = \sqrt{169} = 13$

$|\mathbf{u}| + |\mathbf{v}| = 5 + 13 = 18$

36. -8

37. $|\mathbf{u} + \mathbf{v}| = |\langle 3, 4\rangle + \langle 5, 12\rangle|$
$= |\langle 8, 16\rangle|$
$= \sqrt{8^2 + 16^2} = \sqrt{64 + 256}$
$= \sqrt{320}$
$= 17.9$

38. 8.246

39. $2|\mathbf{u} + \mathbf{v}| = 2(17.9)$ (See Exercise 37.)
$= 35.8$

40. 36

41. Let $\mathbf{PQ} = \langle x, y\rangle$. Then $\mathbf{QP} = \langle -x, -y\rangle$ and $\mathbf{PQ} + \mathbf{QP} = \langle x, y\rangle + \langle -x, -y\rangle = \langle 0, 0\rangle = \mathbf{0}$

42. If then $\theta = 90°$, so $\mathbf{u}\cdot\mathbf{v} = 0$. If $\mathbf{u}\cdot\mathbf{v} = 0$, then since, $|\mathbf{u}| \neq 0$ and $|\mathbf{v}| \neq 0$, $\cos\theta = 0$, and $\theta = 90°$.

Exercise Set 9.5

See text answer section for odd exercise answers 1. - 23.

Even exercises 2. - 24.

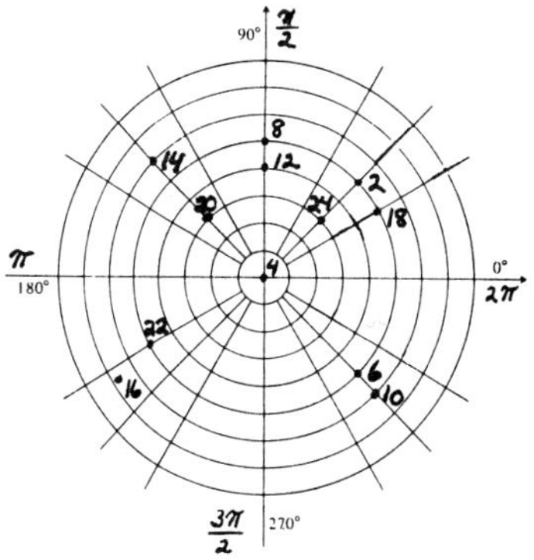

25. Find $|\mathbf{v}|$:

$|\mathbf{v}| = \sqrt{4^2 + 4^2} = \sqrt{32} = 4\sqrt{2}$

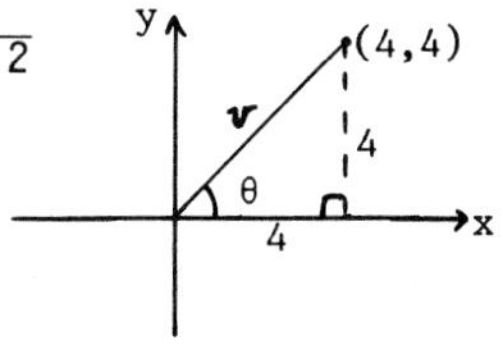

Find θ:

$\tan\theta = \frac{4}{4} = 1$

$\theta = 45°$

Polar coordinates of the point (4, 4) are $(4\sqrt{2}, 45°)$.

26. $(5\sqrt{2}, 45°)$

27. $|\mathbf{v}| = 5$

$\theta = 90°$

Polar coordinates of the point (0, 5) are (5, 90°)

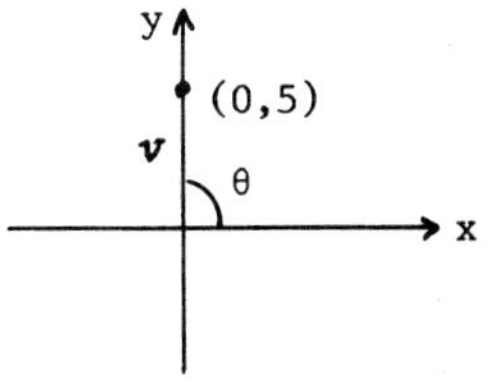

28. (3, 270°)

29. $|\mathbf{v}| = 4$

$\theta = 0°$

Polar coordinates of the point (4, 0) are (4, 0°)

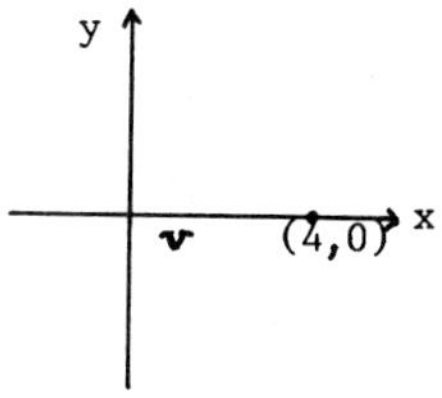

30. (5, 180°)

31. Find $|\mathbf{v}|$:

$|\mathbf{v}| = \sqrt{3^2 + (3\sqrt{3})^2} = \sqrt{36} = 6$

Find θ:

$\tan\theta = \frac{3\sqrt{3}}{3} = \sqrt{3}$

$\theta = 60°$

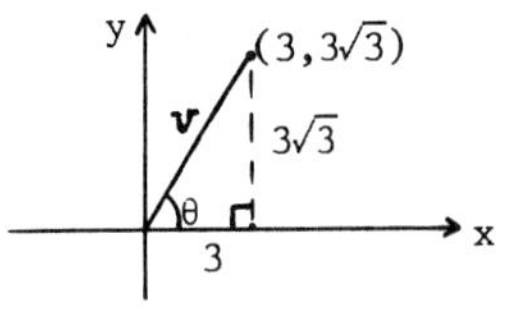

Polar coordinates of the point $(3, 3\sqrt{3})$ are $(6, 60°)$

32. $(6, 240°)$

33. Find $|\mathbf{v}|$:

$|\mathbf{v}| = \sqrt{(\sqrt{3})^2 + 1^2} = \sqrt{4} = 2$

Find θ:

$\tan\theta = \frac{1}{\sqrt{3}}$

$\theta = 30°$

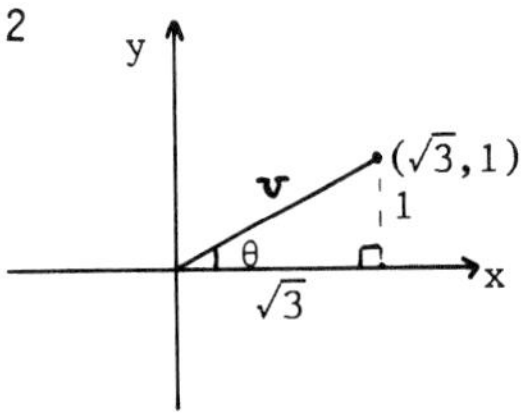

Polar coordinates of the point $(\sqrt{3}, 1)$ are $(2, 30°)$.

34. $(2, 150°)$

35. Find $|\mathbf{v}|$:

$|\mathbf{v}| = \sqrt{(3\sqrt{3})^2 + 3^2} = \sqrt{36} = 6$

Find θ:

$\tan\theta = \frac{3}{3\sqrt{3}} = \frac{1}{\sqrt{3}}$

$\theta = 30°$

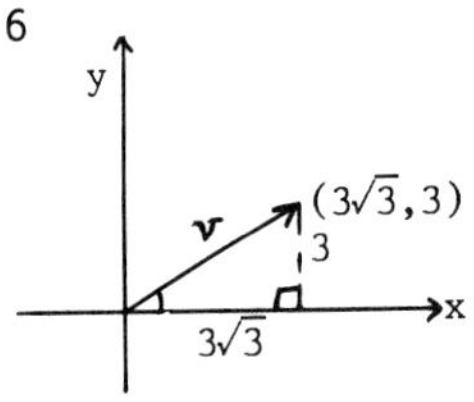

Polar coordinates of the point $(3\sqrt{3}, 3)$ are $(6, 30°)$.

36. $(8, 330°)$

37. $x = 4\cos 45° = 4 \cdot \frac{\sqrt{2}}{2} = 2\sqrt{2}$

$y = 4\sin 45° = 4 \cdot \frac{\sqrt{2}}{2} = 2\sqrt{2}$

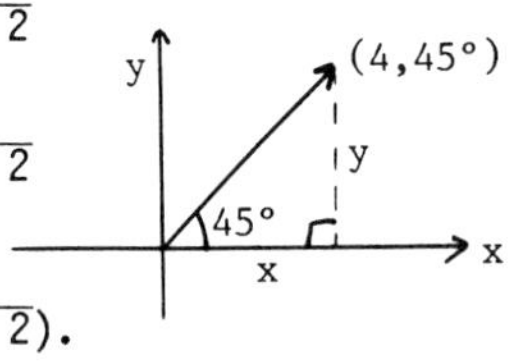

Cartesian coordinates of the point $(4, 45°)$ are $(2\sqrt{2}, 2\sqrt{2})$.

38. $\left(\frac{5}{\sqrt{2}}, \frac{5\sqrt{3}}{2}\right)$

39. $x = 0\cos 23° = 0$

$y = 0\sin 23° = 0$

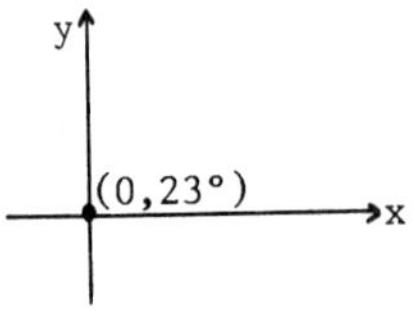

Cartesian coordinates of the point $(0, 23°)$ are $(0, 0)$.

40. $(0, 0)$

41. $x = -3\cos 45° = -3 \cdot \frac{\sqrt{2}}{2} = -\frac{3\sqrt{2}}{2}$

$y = -3\sin 45° = -3 \cdot \frac{\sqrt{2}}{2} = -\frac{3\sqrt{2}}{2}$

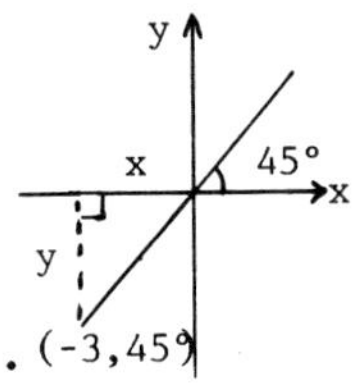

Cartesian coordinates of the point $(-3, 45°)$ are $\left(-\frac{3\sqrt{2}}{2}, -\frac{3\sqrt{2}}{2}\right)$.

42. $\left(-\frac{5\sqrt{3}}{2}, -\frac{5}{2}\right)$

43. $x = 6\cos(-60°) = 6 \cdot \frac{1}{2} = 3$

$y = 6\sin(-60°) = 6\left(-\frac{\sqrt{3}}{2}\right) = -3\sqrt{3}$

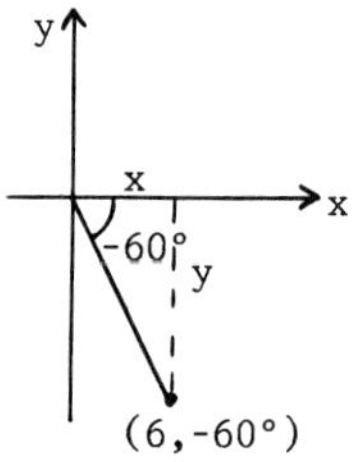

Cartesian coordinates of the point $(6, -60°)$ are $(3, -3\sqrt{3})$.

44. $\left(-\frac{3}{2}, -\frac{3\sqrt{3}}{2}\right)$

45. $x = 10\cos\frac{\pi}{6} = 10 \cdot \frac{\sqrt{3}}{2} = 5\sqrt{3}$

$y = 10\sin\frac{\pi}{6} = 10 \cdot \frac{1}{2} = 5$

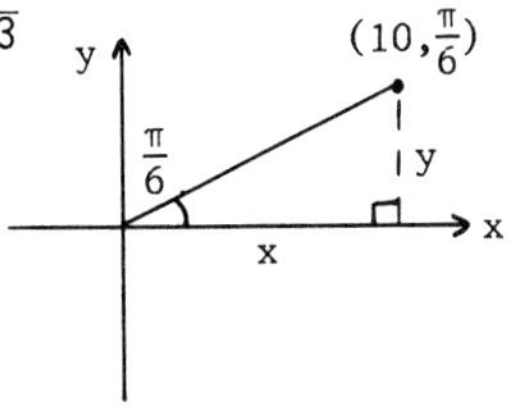

Cartesian coordinates of the point $\left(10, \frac{\pi}{6}\right)$ are $(5\sqrt{3}, 5)$.

46. $(-6\sqrt{2}, 6\sqrt{2})$

47. $x = -5\cos\frac{5\pi}{6} = -5\left(-\frac{\sqrt{3}}{2}\right) = \frac{5\sqrt{3}}{2}$

$y = -5\sin\frac{5\pi}{6} = -5 \cdot \frac{1}{2} = -\frac{5}{2}$

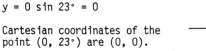

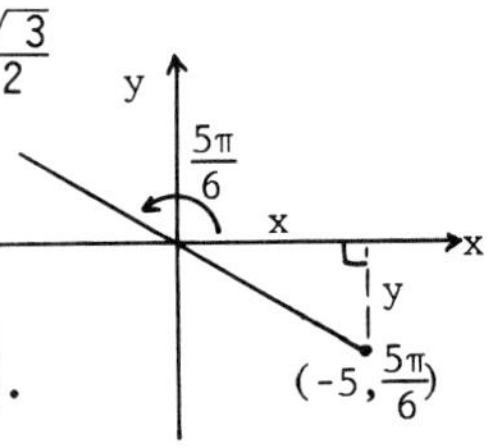

Cartesian coordinates of the point $\left(-5, \frac{5\pi}{6}\right)$ are $\left(\frac{5\sqrt{3}}{2}, -\frac{5}{2}\right)$.

48. $(3\sqrt{2}, -3\sqrt{2})$

49. $3x + 4y = 5$

$3r\cos\theta + 4r\sin\theta = 5$ $\quad (x = r\cos\theta,\ y = r\sin\theta)$

50. $5r\cos\theta + 3r\sin\theta = 4$

51. $x = 5$

$r\cos\theta = 5$ $\quad (x = r\cos\theta)$

52. $r\sin\theta = 4$

53. $$x^2 + y^2 = 36$$
$$(r\cos\theta)^2 + (r\sin\theta)^2 = 36 \qquad (x = r\cos\theta,\ y = r\sin\theta)$$
$$r^2\cos^2\theta + r^2\sin^2\theta = 36$$
$$r^2(\cos^2\theta + \sin^2\theta) = 36$$
$$r^2 = 36 \qquad (\cos^2\theta + \sin^2\theta = 1)$$

54. $r^2 = 16$

55. $$x^2 - 4y^2 = 4$$
$$(r\cos\theta)^2 - 4(r\sin\theta)^2 = 4$$
$$r^2\cos^2\theta - 4r^2\sin^2\theta = 4$$
$$r^2(\cos^2\theta - 4\sin^2\theta) = 4$$

56. $r^2(\cos^2\theta - 5\sin^2\theta) = 5$

57. $$r = 5$$
$$+\sqrt{x^2 + y^2} = 5 \qquad \text{(Substituting for } r\text{)}$$
$$x^2 + y^2 = 25 \qquad \text{(Squaring)}$$

58. $x^2 + y^2 = 64$

59. $$\tan\theta = \frac{y}{x}$$
$$\tan\frac{\pi}{4} = \frac{y}{x} \qquad \left[\theta = \frac{\pi}{4}\right]$$
$$1 = \frac{y}{x}$$
$$x = y$$

60. $y = -x$

61. $$r\sin\theta = 2$$
$$y = 2 \qquad (y = r\sin\theta)$$

62. $x = 5$

63. $$r = 4\cos\theta$$
$$r^2 = 4r\cos\theta \qquad \text{(Multiplying by } r\text{)}$$
$$x^2 + y^2 = 4x \qquad (x^2 + y^2 = r^2,\ x = r\cos\theta)$$

64. $x^2 + y^2 = -3y$

65. $$r - r\sin\theta = 2$$
$$r - y = 2 \qquad (y = r\sin\theta)$$
$$r = 2 + y$$
$$r^2 = 4 + 4y + y^2 \qquad \text{(Squaring)}$$
$$x^2 + y^2 = 4 + 4y + y^2 \qquad (x^2 + y^2 = r^2)$$
$$x^2 - 4y = 4$$

66. $y^2 + 6x = 9$

67. $$r - 2\cos\theta = 3\sin\theta$$
$$r^2 - 2r\cos\theta = 3r\sin\theta \qquad \text{(Multiplying by } r\text{)}$$
$$x^2 + y^2 - 2x = 3y \qquad (x^2 + y^2 = r^2,\ x = r\cos\theta,\ y = r\sin\theta)$$
$$x^2 + y^2 - 2x - 3y = 0$$

68. $x^2 - 7x + y^2 + 5y = 0$

69. Graph: $r = 4\cos\theta$

First make a table of values.

θ	$\cos\theta$	$r = 4\cos\theta$
0°	1	4
30°	0.866	3.46
45°	0.707	2.83
60°	0.5	2
90°	0	0
120°	-0.5	-2
135°	-0.707	-2.83
150°	-0.866	-3.46
180°	-1	-4
210°	-0.866	-3.46
225°	-0.707	-2.83

Note that points are beginning to repeat. Plot these points and draw the graph.

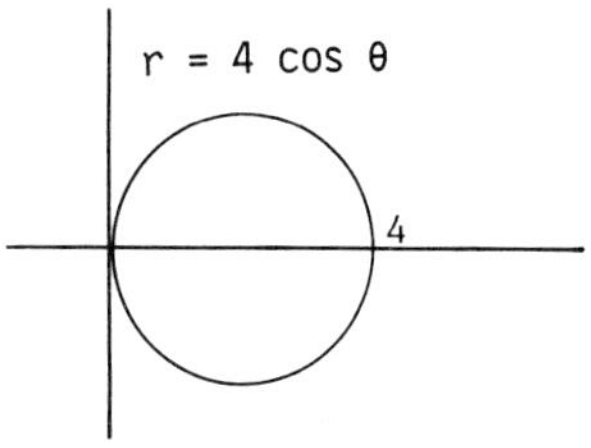

70.

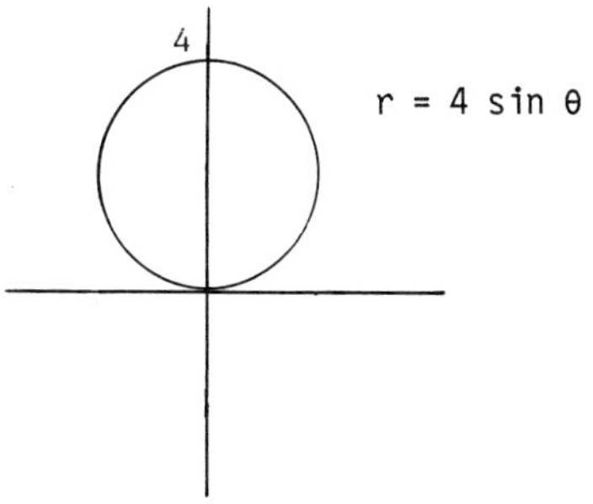

71. Graph: $r = 1 - \cos\theta$

First make a table of values.

θ	cos θ	r = 1 - cos θ
0°	1	0
30°	0.866	0.134
45°	0.707	0.293
60°	0.5	0.5
90°	0	1
120°	-0.5	1.5
135°	-0.707	1.707
150°	-0.866	1.866
180°	-1	2
210°	-0.866	1.866
225°	-0.707	1.707
240°	-0.5	1.5
270°	0	1
300°	0.5	0.5
315°	0.707	0.293
330°	0.866	0.134
360°	1	0
390°	0.866	0.134

Note that points are beginning to repeat. Plot these points and draw the graph.

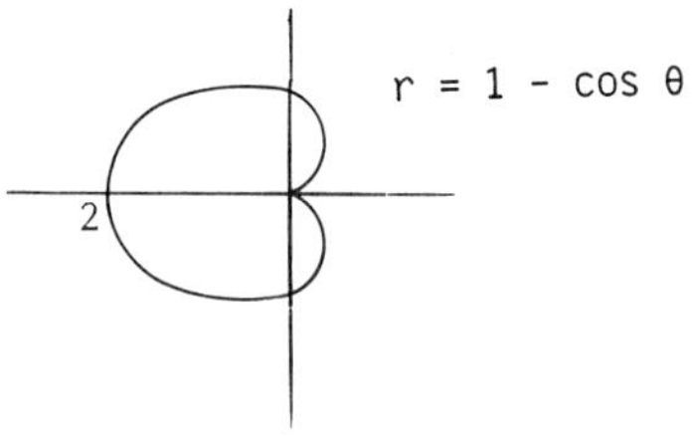

72.

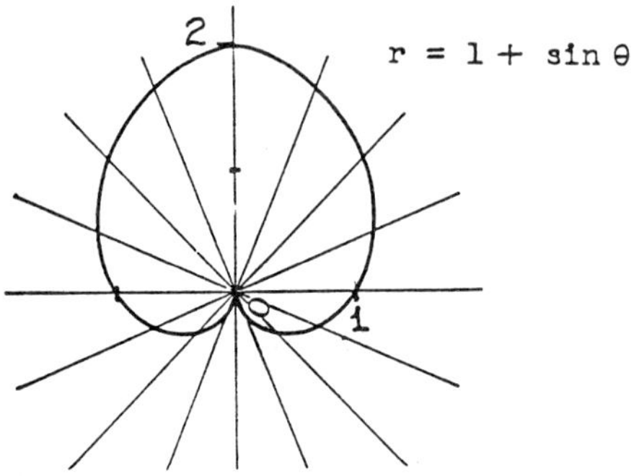

73. Graph: $r = \sin 2\theta$

First make a table of values.

θ	2θ	r = sin 2θ
0°	0°	0
30°	60°	0.866
45°	90°	1
60°	120°	0.866
90°	180°	0
120°	240°	-0.866
135°	270°	-1
150°	300°	-0.866
180°	360°	0
210°	420°	0.866
225°	450°	1
240°	480°	0.866
270°	540°	0
300°	600°	-0.866
315°	630°	-1
330°	660°	-0.866
360°	720°	0
390°	780°	0.866

Note that points are beginning to repeat. Plot these points and draw the graph.

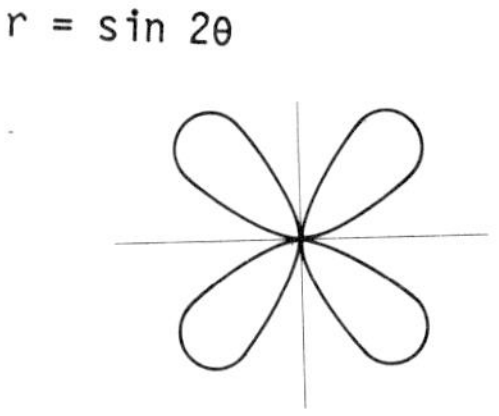

74. $r = 3\cos 2\theta$

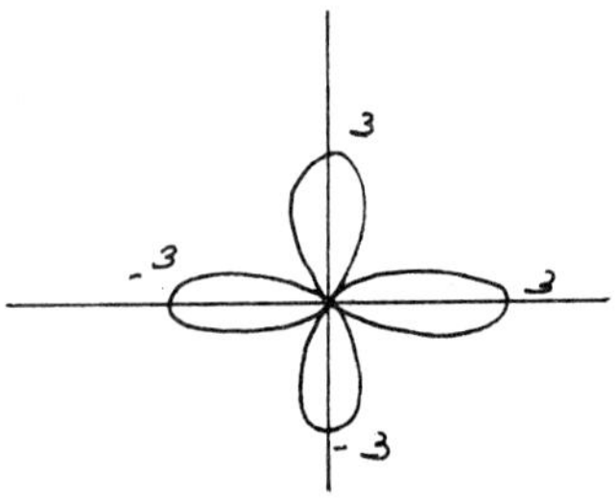

75. Graph: $r = 2\cos 3\theta$

First make a table of values.

θ	3θ	$r = 2\cos 3\theta$
0°	0°	2
30°	90°	0
45°	135°	-1.414
60°	180°	-2
90°	270°	0
120°	360°	2
135°	405°	1.414
150°	450°	0
180°	540°	-2
210°	630°	0
225°	675°	1.414
240°	720°	2
270°	810°	0
300°	900°	-2
315°	945°	-1.414
330°	990°	0
360°	1080°	2
390°	1170°	0

Note that points are beginning to repeat. Plot these points and draw the graph.

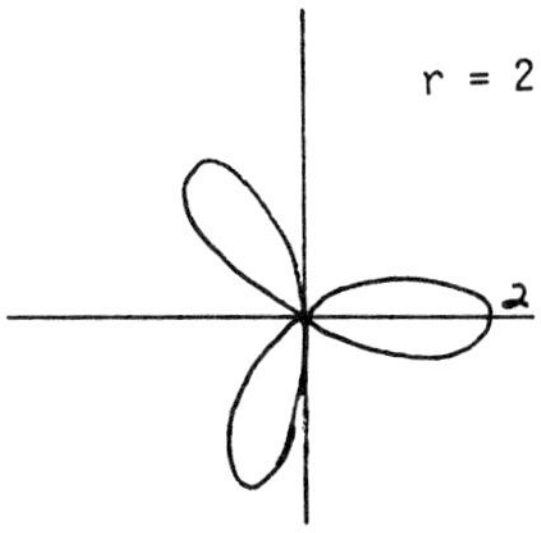

76.

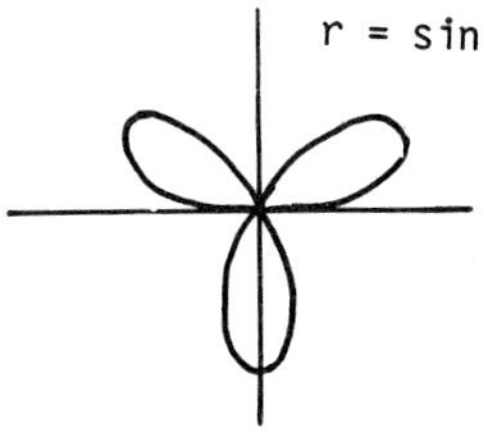

77. Graph: $r\cos\theta = 4$

Express the function as $r = \frac{4}{\cos\theta}$. Then make a table of values.

θ	$\cos\theta$	$r = \frac{4}{\cos\theta}$
0°	1	4
30°	0.866	4.62
45°	0.707	5.66
60°	0.5	8
90°	0	Undefined
120°	-0.5	-8
135°	-0.707	-5.66
150°	-0.866	-4.62
180°	-1	-4
210°	-0.866	-4.62
225°	-0.707	-5.66

Note that points are begining to repeat. Plot these points and draw the graph.

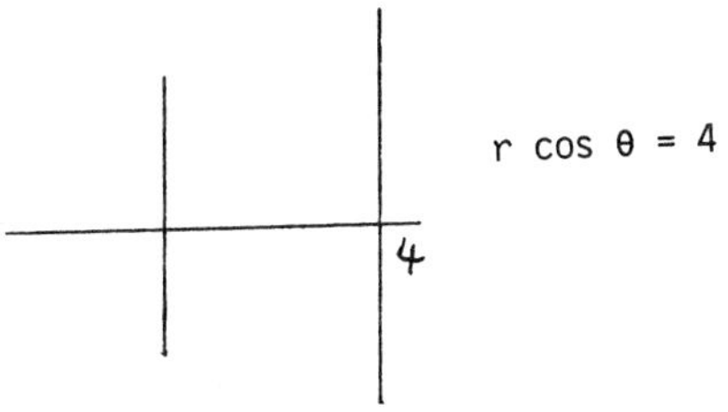

78. $r\sin\theta = 6$

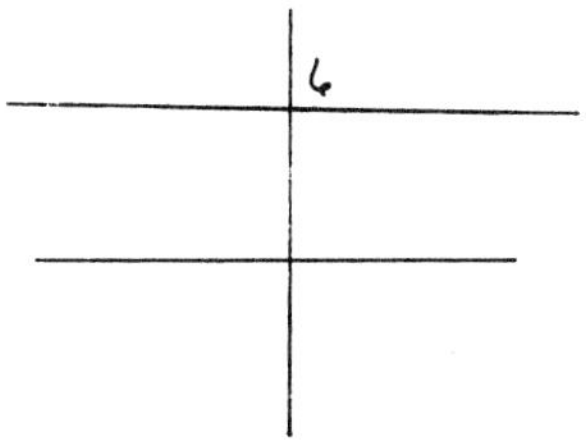

79. Graph: $r = \dfrac{5}{1 + \cos \theta}$

Make a table of values.

θ	$\cos \theta$	$r = \dfrac{5}{1 + \cos \theta}$
0°	1	2.5
30°	0.866	2.68
45°	0.707	2.93
60°	0.5	3.33
90°	0	5
120°	-0.5	10
135°	-0.707	17.07
150°	-0.866	37.32
180°	-1	Undefined
210°	-0.866	37.32
225°	-0.707	17.07
240°	-0.5	10
270°	0	5
300°	0.5	3.33
315°	0.707	2.93
330°	0.866	2.68
360°	1	2.5
390°	0.866	2.68

Note that points are beginning to repeat. Plot these points and draw the graph.

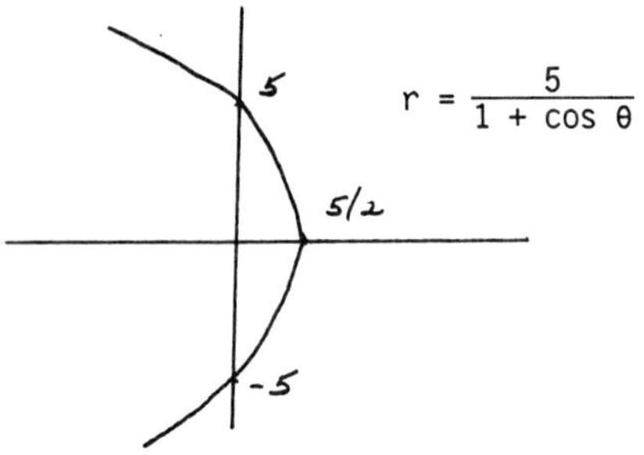

80.

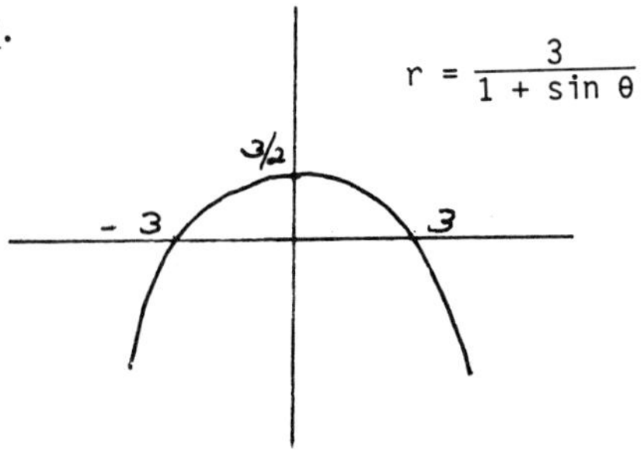

Exercise Set 9.6

1. At the end of the boom there are three forces acting at a point:

   1) The 150-lb weight of the sign acting down.

   2) The cable pulling up and to the left.

   3) The boom pushing to the right.

   It helps to draw a force diagram.

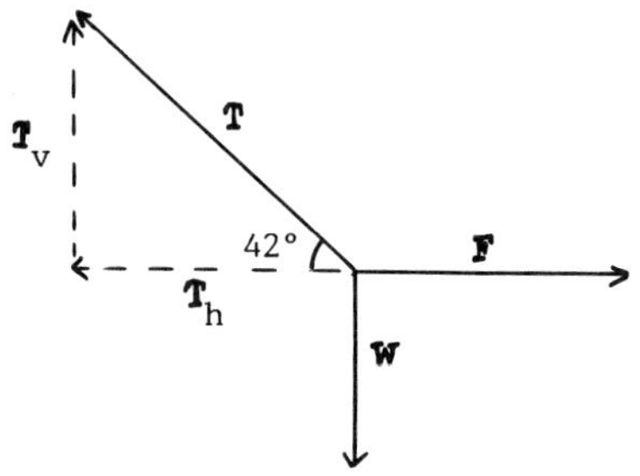

We let T represent the tension in the cable, F the compression in the boom, and W the weight. We resolve the tension in the cable into horizontal (T_h) and vertical (T_v) components. The sum of the horizontal components must be 0, and the sum of the vertical components must also be 0.

Thus, $|T_v| = 150$ lb and $|F| = |T_h|$.

$$\sin 42° = \frac{|T_v|}{|T|} = \frac{150}{|T|}$$

$$|T| = \frac{150}{\sin 42°} = \frac{150}{0.6691} = 224 \text{ lb}$$

$$\cos 42° = \frac{|T_h|}{|T|} = \frac{|T_h|}{224}$$

$$|T_h| = 224 \cos 42° = 224(0.7431) = 166$$

$$|F| = 166 \text{ lb}$$

Tension in the cable: 224 lb
Compression in the boom: 166 lb

2. Tension in the cable: 386 lb
Compression in the boom: 243 lb

3. There are three forces acting at point C:

   1) The 200 kg weight acting down.

   2) Rod AC pushing up and to the right.

   3) Rod BC pulling to the left.

   Draw a force diagram.

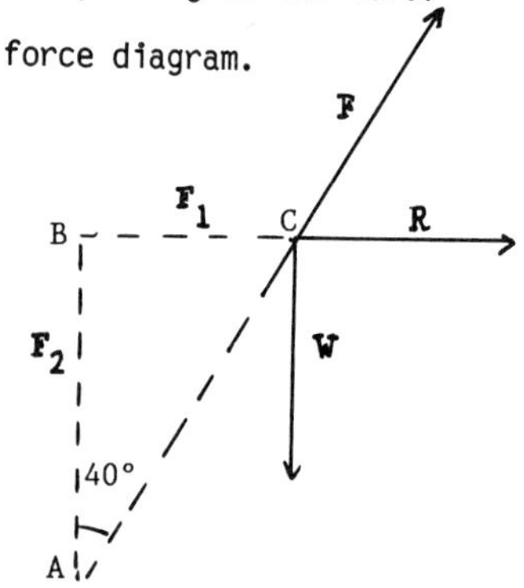

3. (continued)

F is the force exerted by rod AC.
R is the force exerted by rod BC.

Resolve the force in rod AC into horizontal and vertical components, F_1 and F_2. There is a balance so $F + R + W = 0$. The sum of the horizontal components $(F_1 + R)$ is 0, and the sum of the vertical components $(F_2 + W)$ must also be 0.

Thus, $|F_2| = 200\text{kg}$ and $|R| = |F_1|$.

$$\cos 40° = \frac{|F_2|}{|F|} = \frac{200}{|F|}$$

$$|F| = \frac{200}{\cos 40°} = \frac{200}{0.7660} = 261 \text{ kg}$$

$$\sin 40° = \frac{|F_1|}{|F|} = \frac{|F_1|}{261}$$

$$|F_1| = 261 \sin 40° = 261(0.6428) = 168$$

$$|R| = 168 \text{ kg}$$

The force exerted by rod AC is 261 kg.
The force exerted by rod BC is 168 kg.

4. The force exerted by rod AC is 467 kg.
The force exerted by rod BC is 358 kg.

5.

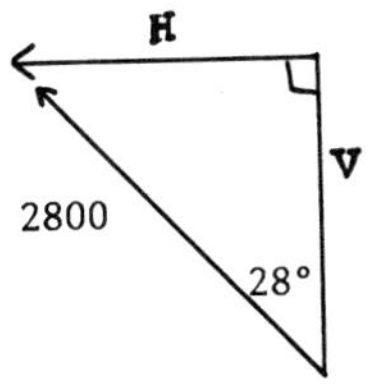

Let **V** represent the vertical component and **H** represent the horizontal component.

$$\sin 28° = \frac{|H|}{2800}$$

$$|H| = 2800 \sin 28° = 2800(0.4695) = 1315 \text{ lb}$$

$$\cos 28° = \frac{|V|}{2800}$$

$$|V| = 2800 \cos 28° = 2800(0.8829) = 2472 \text{ lb}$$

The lift is 2472 lb. The drag is 1315 lb.

6. Life: 2968 lb, Drag: 1855 lb

7.

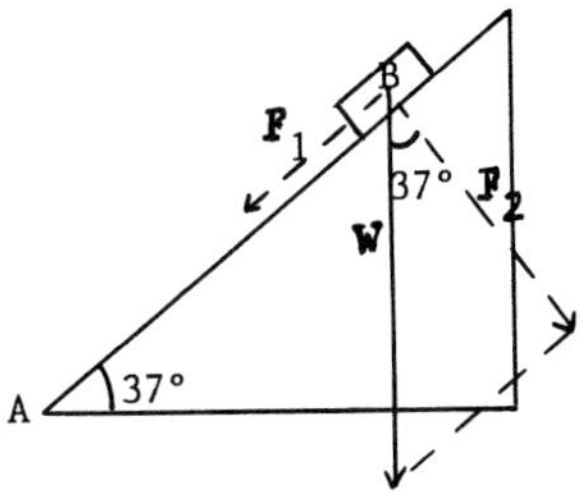

7. (continued)

The weight **W** is acting downward. F_1 and F_2 are the components. The angle at B has the same measure as the angle at A because their sides are respectively perpendicular.

Find $|F_1|$.

$$\frac{|F_1|}{W} = \sin 37°$$

$$|F_1| = W \sin 37°$$

$$|F_1| = 100(0.6018) = 60$$

A force of 60 kg is needed to keep the block from sliding down.

8. 792 kg

9.

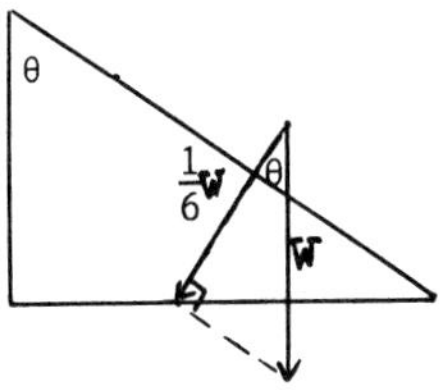

$$\cos \theta = \frac{\frac{1}{6} W}{W} = \frac{1}{6} = 0.16667$$

$$\theta \approx 80° \ 24'$$

10. 67° 59'

11. First draw a force diagram.

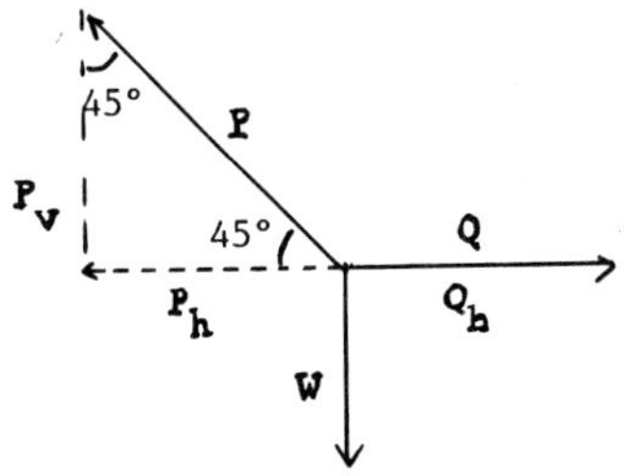

The forces exerted by the ropes are **P** and **Q**. The horizontal components of **P** and **Q** have magnitudes $|P| \cos 45°$ and $|Q| \cos 0°$. Since there is a balance, these must be the same.

$$|P| \cos 45° = |Q| \cos 0°$$

The vertical components of **P** and **Q** have magnitudes $|P| \sin 45°$ and $|Q| \sin 0°$. Since there is a balance, these must total 400 kg, the weight of the block.

$$|P| \sin 45° + |Q| \sin 0° = 400$$

We now have two equations with unknowns $|P|$ and $|Q|$.

$$|P| \cos 45° - |Q| \cos 0° = 0 \quad (1)$$

$$|P| \sin 45° + |Q| \sin 0° = 400 \quad (2)$$

11. (continued)

Solve Equation (1) for $|P|$.

$$|P| = |Q| \frac{\cos 0°}{\cos 45°} \qquad (3)$$

Substitute in Equation (2) and solve for $|Q|$.

$$|Q| \frac{\cos 0° \sin 45°}{\cos 45°} + |Q| \sin 0° = 400$$

$$|Q| \frac{\cos 0° \sin 45°}{\cos 45°} + |Q| \frac{\sin 0° \cos 45°}{\cos 45°} = 400$$

$$|Q|(\cos 0° \sin 45° + \sin 0° \cos 45°) = 400 \cos 45°$$

$$|Q|[\sin (45° + 0°)] = 400 \cos 45°$$

(Using an identity)

$$|Q|(\sin 45°) = 400 \cos 45°$$

$$|Q| = \frac{400 \cos 45°}{\sin 45°}$$

$$|Q| = 400 \tan 45°$$

$$|Q| = 400$$

Substituting in Equation (3), we obtain

$$|P| = 400 \frac{\cos 0°}{\cos 45°} = 400 \cdot \frac{1}{0.7071} = 566.$$

The tension in rope P is 566 kg, and the tension in rope Q is 400 kg.

12. Horizontal rope: 1730 lb, Other rope: 2000 lb

13. First draw force diagram.

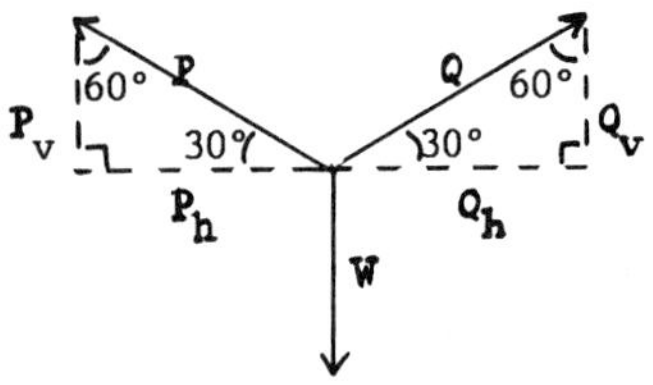

The forces exerted by the ropes are P and Q. The horizontal components of P and Q have magnitudes $|P| \cos 30°$ and $|Q| \cos 30°$. Since there is a balance, these must be the same.

$|P| \cos 30° = |Q| \cos 30°$, or $|P| = |Q|$.

The vertical components of P and Q have magnitudes $|P| \sin 30°$ and $|Q| \sin 30°$. Since there is a balance, these must total 2000 kg, the weight of the block.

$$|P| \sin 30° + |Q| \sin 30° = 2000$$

Substitute $|Q|$ for $|P|$ and solve for $|Q|$.

$$|Q| \sin 30° + |Q| \sin 30° = 2000$$

$$2|Q| \sin 30° = 2000$$

$$2|Q| \cdot \frac{1}{2} = 2000$$

$$|Q| = 2000$$

Since $|Q| = |P|$, the tension in each rope is 2000 kg.

14. 1060 lb in each

15. First draw a force diagram.

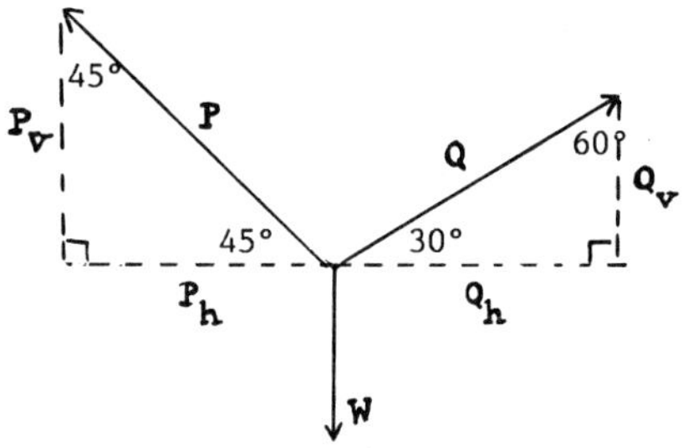

The forces exerted by the ropes are P and Q. The horizontal components of P and Q have magnitudes $|P| \cos 45°$ and $|Q| \cos 30°$. Since there is a balance, these must be the same.

$$|P| \cos 45° = |Q| \cos 30°$$

The vertical components of P and Q have magnitudes $|P| \sin 45°$ and $|Q| \sin 30°$. Since there is a balance, these must total 2500 kg, the weight of the block.

$$|P| \sin 45° + |Q| \sin 30° = 2500$$

We now have two equations with unknowns $|P|$ and $|Q|$.

$$|P| \cos 45° - |Q| \cos 30° = 0 \qquad (1)$$

$$|P| \sin 45° + |Q| \sin 30° = 2500 \qquad (2)$$

Solve Equation (1) for $|P|$.

$$|P| = \frac{|Q| \cos 30°}{\cos 45°}$$

Substitute in Equation (2) and solve for $|Q|$.

$$\frac{|Q| \cos 30° \sin 45°}{\cos 45°} + |Q| \sin 30° = 2500$$

$$\frac{|Q| \cos 30° \sin 45°}{\cos 45°} + \frac{|Q| \sin 30° \cos 45°}{\cos 45°} = 2500$$

$$|Q|(\cos 30° \sin 45° + \sin 30° \cos 45°) = 2500 \cos 45°$$

$$|Q|[\sin (30° + 45°)] = 2500 \cos 45°$$

$$|Q| = \frac{2500 \cos 45°}{\sin 75°}$$

$$|Q| = \frac{2500(0.7071)}{0.9659}$$

$$|Q| = 1830 \text{ kg.}$$

Substituting in Equation (3), we obtain

$$|P| = \frac{1830 \cos 30°}{\cos 45°} = \frac{1830(0.8660)}{0.7071} = 2241 \text{ kg}$$

The tension in rope P is 2241 kg.
The tension in rope Q is 1830 kg.

16. Left rope: 1035 kg, Right rope: 1464 kg

Exercise Set 9.7

1.

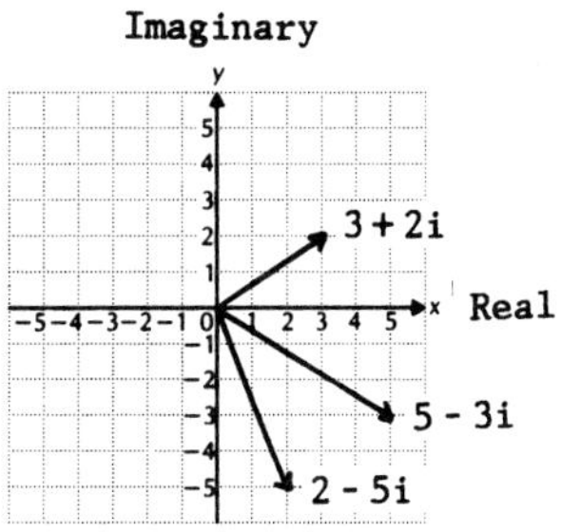

$(3 + 2i) + (2 - 5i)$
$= (3 + 2) + (2 - 5)i$
$= 5 - 3i$

2.

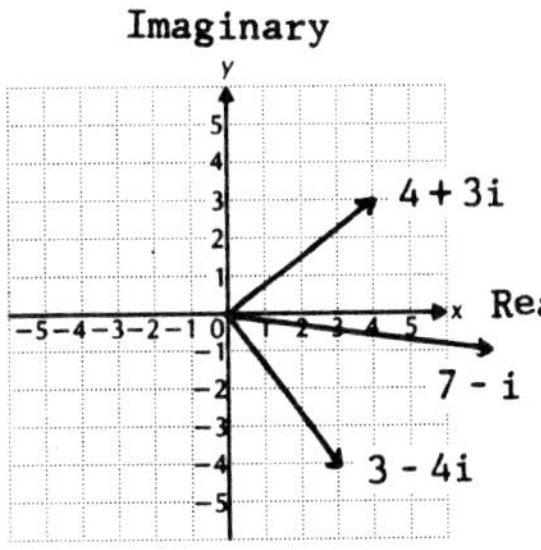

3.

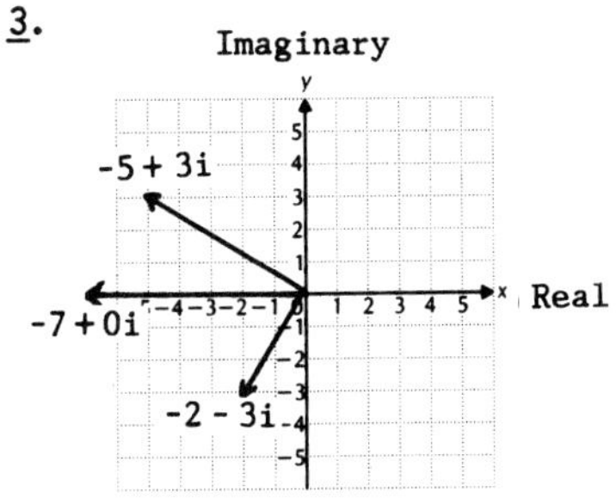

$(-5 + 3i) + (-2 - 3i)$
$= (-5 - 2) + (3 - 3)i$
$= -7 + 0i = -7$

4.

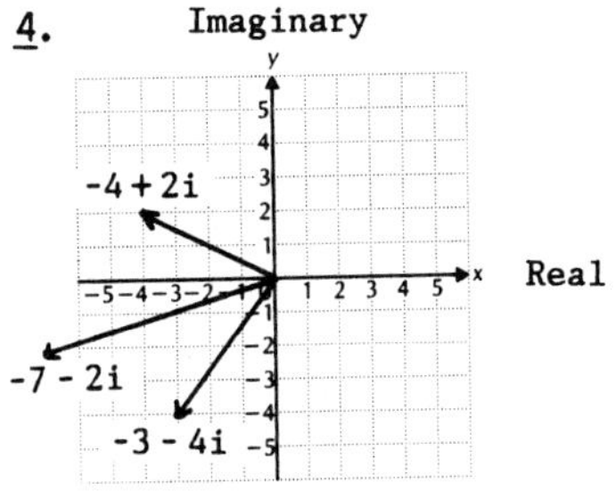

5.

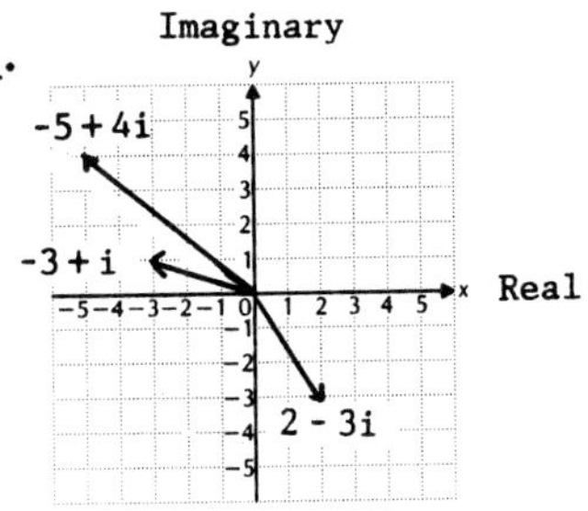

$(2 - 3i) + (-5 + 4i)$
$= (2 - 5) + (-3 + 4)i$
$= -3 + i$

6.

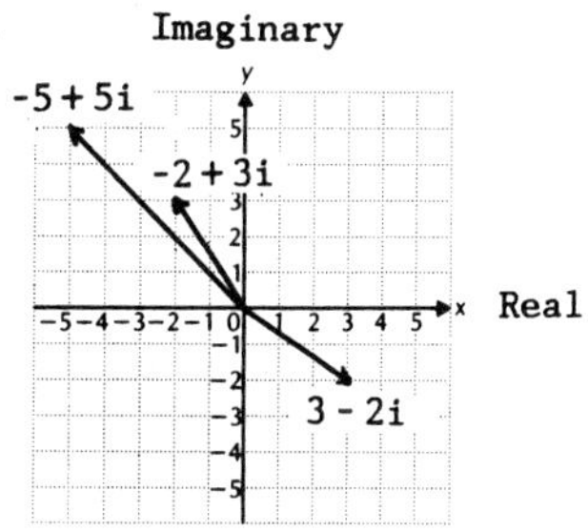

7.

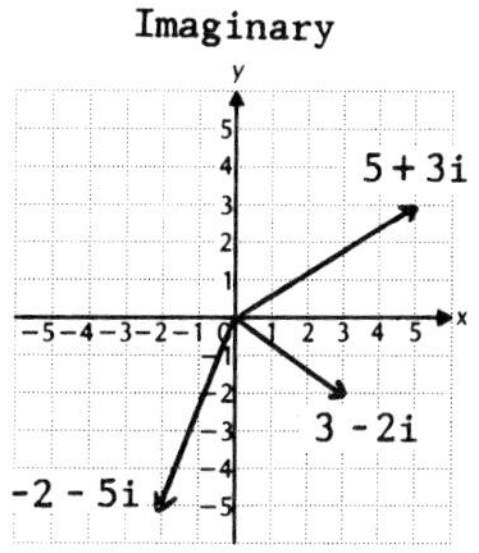

$(-2 - 5i) + (5 + 3i)$
$= (-2 + 5) + (-5 + 3)i$
$= 3 - 2i$

8.

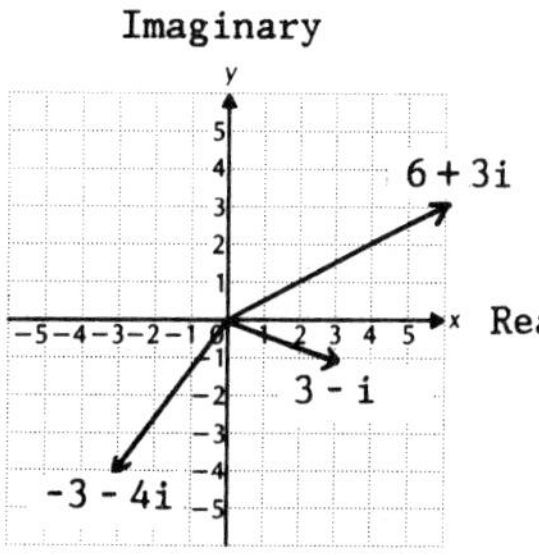

9. $3(\cos 30° + i \sin 30°)$

$a = 3 \cos 30° = 3 \cdot \frac{\sqrt{3}}{2} = \frac{3\sqrt{3}}{2}$

$b = 3 \sin 30° = 3 \cdot \frac{1}{2} = \frac{3}{2}$

Thus $3(\cos 30° + i \sin 30°) = \frac{3\sqrt{3}}{2} + \frac{3}{2}i$.

10. $-3\sqrt{3} + 3i$

11. $10 \text{ cis } 270° = 10(\cos 270° + i \sin 270°)$

$a = 10 \cos 270° = 10 \cdot 0 = 0$
$b = 10 \sin 270° = 10(-1) = -10$

Thus $10 \text{ cis } 270° = 0 + (-10)i = -10i$.

12. $\frac{5}{2} - \frac{5\sqrt{3}}{2}i$

13. $\sqrt{8}\left[\cos \frac{\pi}{4} + i \sin \frac{\pi}{4}\right]$

$a = \sqrt{8} \cos \frac{\pi}{4} = \sqrt{8} \cdot \frac{\sqrt{2}}{2} = 2$

$b = \sqrt{8} \sin \frac{\pi}{4} = \sqrt{8} \cdot \frac{\sqrt{2}}{2} = 2$

Thus $\sqrt{8}\left[\cos \frac{\pi}{4} + i \sin \frac{\pi}{4}\right] = 2 + 2i$

14. $\frac{5}{2} + \frac{5\sqrt{3}}{2}i$

15. $\sqrt{8}\text{ cis }\frac{5\pi}{4} = \sqrt{8}\left[\cos\frac{5\pi}{4} + i\sin\frac{5\pi}{4}\right]$

$a = \sqrt{8}\cos\frac{5\pi}{4} = \sqrt{8}\left[-\frac{\sqrt{2}}{2}\right] = -2$

$b = \sqrt{8}\sin\frac{5\pi}{4} = \sqrt{8}\left[-\frac{\sqrt{2}}{2}\right] = -2$

Thus $\sqrt{8}\text{ cis }\frac{5\pi}{4} = -2 - 2i$.

16. $2 - 2i$

17. $1 - i$

$a = 1$ and $b = -1$

$r = \sqrt{a^2 + b^2} = \sqrt{1^2 + (-1)^2} = \sqrt{2}$

$\sin\theta = \frac{b}{r} = \frac{-1}{\sqrt{2}} = -\frac{\sqrt{2}}{2}$

$\cos\theta = \frac{a}{r} = \frac{1}{\sqrt{2}} = \frac{\sqrt{2}}{2}$

Thus $\theta = \frac{7\pi}{4}$, or 315°, and we have

$1 - i = \sqrt{2}\text{ cis }\frac{7\pi}{4}$, or $\sqrt{2}\text{ cis }315°$.

18. 2 cis 30°

19. $10\sqrt{3} - 10i$

$a = 10\sqrt{3}$ and $b = -10$

$r = \sqrt{a^2 + b^2} = \sqrt{(10\sqrt{3})^2 + (-10)^2} = \sqrt{400} = 20$

$\sin\theta = \frac{b}{r} = \frac{-10}{20} = -\frac{1}{2}$

$\cos\theta = \frac{a}{r} = \frac{10\sqrt{3}}{20} = \frac{\sqrt{3}}{2}$

Thus $\theta = \frac{11\pi}{6}$, or 330°, and we have

$10\sqrt{3} - 10i = 20\text{ cis }\frac{11\pi}{6}$, or 20 cis 330°.

20. 20 cis 150°

21. -5 (Think $-5 + 0i$)

$a = -5$ and $b = 0$

$r = \sqrt{a^2 + b^2} = \sqrt{(-5)^2 + 0^2} = \sqrt{25} = 5$

$\sin\theta = \frac{b}{r} = \frac{0}{5} = 0$

$\cos\theta = \frac{-5}{5} = -1$

Thus $\theta = \pi$, or 180°, and we have

$-5 = 5\text{ cis }\pi$, or 5 cis 180°.

22. 5 cis 270°

23. $(1 - i)(2 + 2i)$

Find polar notation for $1 - i$.

$a = 1$, $b = -1$, and $r = \sqrt{1^2 + (-1)^2} = \sqrt{2}$

$\sin\theta = \frac{-1}{\sqrt{2}} = -\frac{\sqrt{2}}{2}$, $\cos\theta = \frac{1}{\sqrt{2}} = \frac{\sqrt{2}}{2}$

Thus $\theta = \frac{7\pi}{4}$, or 315°, and $1 - i = \sqrt{2}\text{ cis }315°$.

Find polar notation for $2 + 2i$.

$a = 2$, $b = 2$, and $r = \sqrt{2^2 + 2^2} = \sqrt{8} = 2\sqrt{2}$

$\sin\theta = \frac{2}{2\sqrt{2}} = \frac{\sqrt{2}}{2}$, $\cos\theta = \frac{2}{2\sqrt{2}} = \frac{\sqrt{2}}{2}$

Thus $\theta = \frac{\pi}{4}$, or 45°, and $2 + 2i = 2\sqrt{2}\text{ cis }45°$

$(1 - i)(2 + 2i)$

$= \sqrt{2}\text{ cis }315° \cdot 2\sqrt{2}\text{ cis }45°$

$= \sqrt{2} \cdot 2\sqrt{2}\text{ cis }(315° + 45°)$ (Theorem 3)

$= 4\text{ cis }0°$, or 4

24. $2\sqrt{2}\text{ cis }105°$

25. $(2\sqrt{3} + 2i)(2i)$

Find polar notation for $2\sqrt{3} + 2i$.

$a = 2\sqrt{3}$, $b = 2$, and $r = \sqrt{(2\sqrt{3})^2 + 2^2}$

$= \sqrt{16} = 4$

$\sin\theta = \frac{2}{4} = \frac{1}{2}$, $\cos\theta = \frac{2\sqrt{3}}{4} = \frac{\sqrt{3}}{2}$

Thus $\theta = \frac{\pi}{6}$, or 30°, and $2\sqrt{3} + 2i = 4\text{ cis }30°$.

Find polar notation for $2i$.

$a = 0$, $b = 2$, and $r = \sqrt{0^2 + 2^2} = 2$

$\sin\theta = \frac{2}{2} = 1$, $\cos\theta = \frac{0}{2} = 0$

Thus $\theta = \frac{\pi}{2}$, or 90°, and $2i = 2\text{ cis }90°$.

$(2\sqrt{3} + 2i)(2i)$

$= 4\text{ cis }30° \cdot 2\text{ cis }90°$

$= 4 \cdot 2\text{ cis }(30° + 90°)$ (Theorem 3)

$= 8\text{ cis }120°$

26. 12 cis 60°

27. $\frac{1 - i}{1 + i}$

Find polar notation for $1 - i$.

$a = 1$, $b = -1$, and $r = \sqrt{2}$

$\sin \theta = \frac{-1}{\sqrt{2}} = -\frac{\sqrt{2}}{2}$, $\cos \theta = \frac{1}{\sqrt{2}} = \frac{\sqrt{2}}{2}$

Thus $\theta = \frac{7\pi}{4}$, or 315°, and $1 - i = \sqrt{2}$ cis 315°.

Find polar notation for $1 + i$.

$a = 1$, $b = 1$, and $r = \sqrt{2}$

$\sin \theta = \frac{1}{\sqrt{2}} = \frac{\sqrt{2}}{2}$, $\cos \theta = \frac{1}{\sqrt{2}} = \frac{\sqrt{2}}{2}$

Thus $\theta = \frac{\pi}{4}$, or 45°, and $1 + i = \sqrt{2}$ cis 45°.

$$\frac{1 - i}{1 + i} = \frac{\sqrt{2} \text{ cis } 315°}{\sqrt{2} \text{ cis } 45°}$$

$$= \frac{\sqrt{2}}{\sqrt{2}} \text{ cis } (315° - 45°) \quad \text{(Theorem 4)}$$

$$= \text{cis } 270°, \text{ or } -i$$

28. $\frac{\sqrt{2}}{2}$ cis 345°

29. $\frac{2\sqrt{3} - 2i}{1 + \sqrt{3}\, i}$

Find polar notation for $2\sqrt{3} - 2i$.

$a = 2\sqrt{3}$, $b = -2$, and $r = \sqrt{(2\sqrt{3})^2 + (-2)^2}$

$= \sqrt{16} = 4$

$\sin \theta = \frac{-2}{4} = -\frac{1}{2}$, $\cos \theta = \frac{2\sqrt{3}}{4} = \frac{\sqrt{3}}{2}$

Thus $\theta = \frac{11\pi}{6}$, or 330°, and $2\sqrt{3} - 2i = 4$ cis 330°.

Find polar notation for $1 + \sqrt{3}\, i$.

$a = 1$, $b = \sqrt{3}$, and $r = \sqrt{1^2 + (\sqrt{3})^2} = \sqrt{4} = 2$

$\sin \theta = \frac{\sqrt{3}}{2}$, $\cos \theta = \frac{1}{2}$

Thus $\theta = \frac{\pi}{3}$, or 60°, and $1 + \sqrt{3}\, i = 2$ cis 60°.

$$\frac{2\sqrt{3} - 2i}{1 + \sqrt{3}\, i} = \frac{4 \text{ cis } 330°}{2 \text{ cis } 60°}$$

$$= \frac{4}{2} \text{ cis } (330° - 60°)$$

$$= 2 \text{ cis } 270°, \text{ or } -2i$$

30. 3 cis 330°

31. See answer section in text.

32. $z = a + bi$, $|z| = \sqrt{a^2 + b^2}$

$\bar{z} = a - bi$, $|\bar{z}| = \sqrt{a^2 + (-b)^2} = \sqrt{a^2 + b^2}$

$|z| = |\bar{z}|$

33. See answer section in text.

34. $|(a + bi)^2| = |a^2 - b^2 + 2abi|$

$= \sqrt{(a^2 - b^2)^2 + 4a^2b^2}$

$= \sqrt{a^4 + 2a^2b^2 + b^4}$

$= a^2 + b^2$

$|a + bi|^2 = (\sqrt{a^2 + b^2})^2 = a^2 + b^2$

35. See answer section in text.

36. $\frac{z}{w} = \frac{r_1 \text{cis } \theta_1}{r_2 \text{cis } \theta_2} = \frac{r_1}{r_2} \text{ cis } (\theta_1 - \theta_2)$

$$\left|\frac{z}{w}\right| = \sqrt{\left[\frac{r_1}{r_2} \cos (\theta_1 - \theta_2)\right]^2 + \left[\frac{r_1}{r_2} \sin (\theta_1 - \theta_2)\right]^2}$$

$$= \sqrt{\frac{r_1^2}{r_2^2}} = \frac{|r_1|}{|r_2|}$$

$|z| = \sqrt{(r_1 \cos \theta_1)^2 + (r_1 \sin \theta_1)^2}$

$= \sqrt{r_1^2} = |r_1|$

$|w| = \sqrt{(r_2 \cos \theta_2)^2 + (r_2 \sin \theta_2)^2}$

$= \sqrt{r_2^2} = |r_2|$

Then $\left|\frac{z}{w}\right| = \frac{|r_1|}{|r_2|} = \frac{|z|}{|w|}$.

37. Graph: $|z| = 1$

Let $z = a + bi$

Then $|a + bi| = 1$

$\sqrt{a^2 + b^2} = 1$ (Definition)

$a^2 + b^2 = 1$

The graph of $a^2 + b^2 = 1$ is a circle whose radius is 1 and whose center is (0, 0).

38.

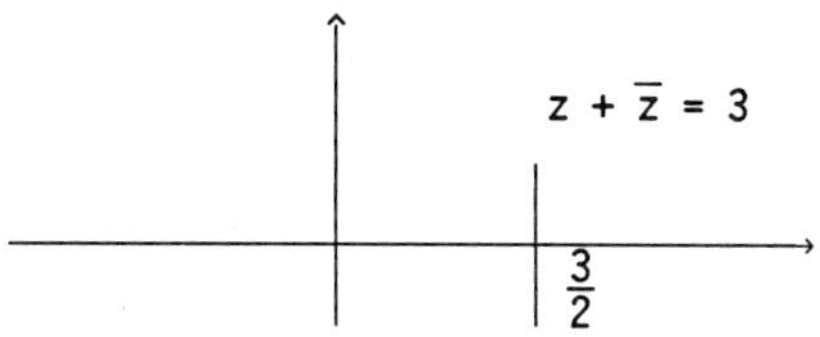

Exercise Set 9.8

1. $\left[2 \text{ cis } \frac{\pi}{3}\right]^3 = 2^3 \text{ cis } 3 \cdot \frac{\pi}{3}$ (Theorem 5)

$= 8 \text{ cis } \pi$

2. $81 \text{ cis } 2\pi$

3. $\left[2 \text{ cis } \frac{\pi}{6}\right]^6 = 2^6 \text{ cis } 6 \cdot \frac{\pi}{6}$ (Theorem 5)

$= 64 \text{ cis } \pi$

4. $32 \text{ cis } \pi$

5. $(1 + i)^6$

We first find polar notation for $1 + i$. See Exercise 27 in Exercise Set 9.7.

$1 + i = \sqrt{2} \text{ cis } 45°$

$(1 + i)^6 = (\sqrt{2} \text{ cis } 45°)^6$

$= (\sqrt{2})^6 \text{ cis } 6 \cdot 45°$ (Theorem 5)

$= 8 \text{ cis } 270°$

6. $8 \text{ cis } 90°$

7. $(2 \text{ cis } 240°)^4 = 2^4 \text{ cis } 4 \cdot 240°$ (Theorem 5)

$= 16 \text{ cis } 960°$

$= 16 \text{ cis } 240°$

$= 16(\cos 240° + i \sin 240°)$

$= 16\left[-\frac{1}{2} - \frac{\sqrt{3}}{2} i\right]$

$= -8 - 8\sqrt{3}\, i$

8. $-8 + 8\sqrt{3}\, i$

9. $(1 + \sqrt{3}\, i)^4$

We first find polar notation for $1 + \sqrt{3}\, i$. See Exercise 29 in Exercise Set 9.7.

$1 + \sqrt{3}\, i = 2 \text{ cis } 60°$

$(1 + \sqrt{3}\, i)^4 = (2 \text{ cis } 60°)^4$

$= 2^4 \text{ cis } 4 \cdot 60°$ (Theorem 5)

$= 16 \text{ cis } 240°$

$= 16(\cos 240° + i \sin 240°)$

$= 16\left[-\frac{1}{2} - \frac{\sqrt{3}}{2} i\right]$

$= -8 - 8\sqrt{3}\, i$

10. -64

11. $\left[\frac{1}{\sqrt{2}} + \frac{1}{\sqrt{2}} i\right]^{10}$

First find polar notation for $\frac{1}{\sqrt{2}} + \frac{1}{\sqrt{2}} i$.

$a = \frac{1}{\sqrt{2}}$, $b = \frac{1}{\sqrt{2}}$, and $r = \sqrt{\left[\frac{1}{\sqrt{2}}\right]^2 + \left[\frac{1}{\sqrt{2}}\right]^2}$

$= \sqrt{1} = 1$

$\sin \theta = \frac{\frac{1}{\sqrt{2}}}{1} = \frac{1}{\sqrt{2}}$, or $\frac{\sqrt{2}}{2}$

$\cos \theta = \frac{\frac{1}{\sqrt{2}}}{1} = \frac{1}{\sqrt{2}}$, or $\frac{\sqrt{2}}{2}$

Thus $\theta = \frac{\pi}{4}$, or $45°$ and $\frac{1}{\sqrt{2}} + \frac{1}{\sqrt{2}} i = 1 \cdot \text{cis } 45°$.

$\left[\frac{1}{\sqrt{2}} + \frac{1}{\sqrt{2}} i\right]^{10} = (\text{cis } 45°)^{10}$

$= \text{cis } 450°$ (Theorem 5)

$= \text{cis } 90°$

$= \cos 90° + i \sin 90°$

$= 0 + i$, or i

12. -1

13. $\left[\frac{\sqrt{3}}{2} + \frac{1}{2} i\right]^{12}$

First find polar notation for $\frac{\sqrt{3}}{2} + \frac{1}{2} i$.

$a = \frac{\sqrt{3}}{2}$, $b = \frac{1}{2}$, and $r = \sqrt{\left[\frac{\sqrt{3}}{2}\right]^2 + \left[\frac{1}{2}\right]^2} = \sqrt{1} = 1$

$\sin \theta = \frac{\frac{1}{2}}{1} = \frac{1}{2}$, $\cos \theta = \frac{\frac{\sqrt{3}}{2}}{1} = \frac{\sqrt{3}}{2}$

Thus $\theta = \frac{\pi}{6}$, or $30°$, and $\frac{\sqrt{3}}{2} + \frac{1}{2} i = \text{cis } 30°$.

$\left[\frac{\sqrt{3}}{2} + \frac{1}{2} i\right]^{12} = (\text{cis } 30°)^{12}$

$= \text{cis } 12 \cdot 30°$ (Theorem 5)

$= \text{cis } 360°$, or $\text{cis } 0°$

$= \cos 0° + i \sin 0°$

$= 1 + 0i$, or 1

14. $\frac{1}{2} - \frac{\sqrt{3}}{2} i$

15. $x^2 = i$

Find the square roots of i.

Polar notation for i is $\text{cis } 90°$.

Using Theorem 6, we know

$(\text{cis } 90°)^{1/2} = \text{cis } \left[\frac{90°}{2} + k \cdot \frac{360°}{2}\right]$, $k = 0, 1$

$= \text{cis } (45° + k \cdot 180°)$, $k = 0, 1$

15. (continued)

Thus the roots are

cis 45° when k = 0 and cis 225° when k = 1.

Rectangular notation for these roots is as follows:

cis 45° = cos 45° + i sin 45°

$= \frac{\sqrt{2}}{2} + \frac{\sqrt{2}}{2}i$

cis 225° = cos 225° + i sin 225°

$= -\frac{\sqrt{2}}{2} - \frac{\sqrt{2}}{2}i$

The roots are $\frac{\sqrt{2}}{2} + \frac{\sqrt{2}}{2}i$ and $-\frac{\sqrt{2}}{2} - \frac{\sqrt{2}}{2}i$.

16. cis 135°, cis 315°, $-\frac{\sqrt{2}}{2} + \frac{\sqrt{2}}{2}i$, $\frac{\sqrt{2}}{2} - \frac{\sqrt{2}}{2}i$

17. $x^2 = 2\sqrt{2} - 2\sqrt{2}\,i$

Find the square roots of $2\sqrt{2} - 2\sqrt{2}\,i$.

Polar notation for $2\sqrt{2} - 2\sqrt{2}\,i$ is 4 cis 315°.

Using Theorem 6, we know

$(4 \text{ cis } 315°)^{1/2} = 4^{1/2} \text{ cis } \left(\frac{315°}{2} + \frac{k \cdot 360°}{2}\right)$, k = 0, 1

$= 2 \text{ cis } (157.5° + k \cdot 180°)$, k = 0, 1

Thus the roots are

2 cis 157.5° when k = 0 and 2 cis 337.5° when k = 1.

18. $\sqrt[4]{2}$ cis 22.5°, $\sqrt[4]{2}$ cis 202.5°

19. $x^2 = -1 + \sqrt{3}\,i$

Find the square roots of $-1 + \sqrt{3}\,i$.

Polar notation for $-1 + \sqrt{3}\,i$ is 2 cis 120°.

Using Theorem 6, we know

$(2 \text{ cis } 120°)^{1/2} = 2^{1/2} \text{ cis } \left(\frac{120°}{2} + k \cdot \frac{360°}{2}\right)$, k = 0, 1

$= \sqrt{2} \text{ cis } (60° + k \cdot 180°)$, k = 0, 1

Thus the roots are

$\sqrt{2}$ cis 60° when k = 0 and $\sqrt{2}$ cis 240° when k = 1.

Rectangular notation for these roots is as follows:

$\sqrt{2} \text{ cis } 60° = \sqrt{2}(\cos 60° + i \sin 60°)$

$= \sqrt{2}\left(\frac{1}{2} + \frac{\sqrt{3}}{2}i\right) = \frac{\sqrt{2}}{2} + \frac{\sqrt{6}}{2}i$

19. (continued)

$\sqrt{2} \text{ cis } 240° = \sqrt{2}(\cos 240° + i \sin 240°)$

$= \sqrt{2}\left(-\frac{1}{2} - \frac{\sqrt{3}}{2}i\right) = -\frac{\sqrt{2}}{2} - \frac{\sqrt{6}}{2}i$

The roots are $\frac{\sqrt{2}}{2} + \frac{\sqrt{6}}{2}i$ and $-\frac{\sqrt{2}}{2} - \frac{\sqrt{6}}{2}i$.

20. $\sqrt{2}$ cis 105°, $\sqrt{2}$ cis 285°

21. $x^3 = i$

Find the cube roots of i.

Polar notation for i is cis 90°.

Using Theorem 6, we know

$(\text{cis } 90°)^{1/3} = 1^{1/3} \text{ cis } \left(\frac{90°}{3} + k \cdot \frac{360°}{3}\right)$, k = 0, 1, 2

$= \text{cis } (30° + k \cdot 120°)$, k = 0, 1, 2

Thus the roots are

cis 30° when k = 0, cis 150° when k = 1, and cis 270° when k = 2.

Rectangular notation for these roots is as follows:

$\text{cis } 30° = \cos 30° + i \sin 30° = \frac{\sqrt{3}}{2} + \frac{1}{2}i$

$\text{cis } 150° = \cos 150° + i \sin 150° = -\frac{\sqrt{3}}{2} + \frac{1}{2}i$

$\text{cis } 270° = \cos 270° + i \sin 270° = -i$

22. 4.09, -2.045 + 3.542i, -2.045 - 3.542i

23. Solve $x^4 = 16$.

Find the fourth roots of 16.

Polar notation for 16 is 16 cis 0°.

Using Theorem 6, we know

$(16 \text{ cis } 0°)^{1/4} = 16^{1/4} \text{ cis } \left(\frac{0°}{4} + k \cdot \frac{360°}{4}\right)$, k = 0, 1, 2, 3

$= 2 \text{ cis } k \cdot 90°$, k = 0, 1, 2, 3

Thus the roots are

2 cis 0° = 2(cos 0° + i sin 0°) when k = 0
= 2(1 + 0i) = 2

2 cis 90° = 2(cos 90° + i sin 90°) when k = 1
= 2(0 + i) = 2i

2 cis 180° = 2(cos 180° + i sin 180°) when k = 2
= 2(-1 + 0i) = -2

23. (continued)

$2 \text{ cis } 270° = 2(\cos 270° + i \sin 270°)$ when $k = 3$

$= 2(0 - i) = -2i$

The graph of ± 2 and $\pm 2i$ is as follows:

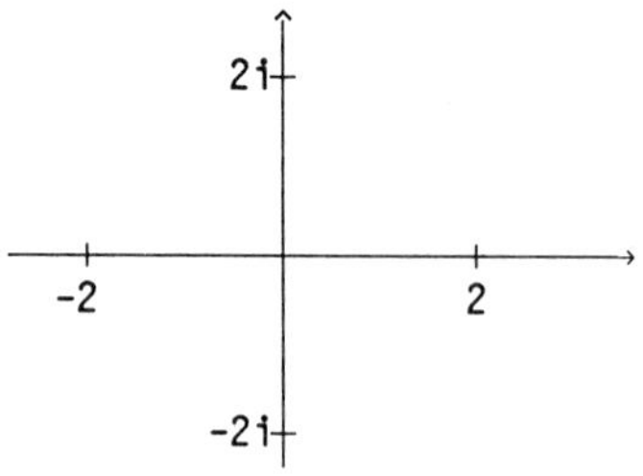

24. $\text{cis } 22\frac{1}{2}°$, $\text{cis } 112\frac{1}{2}°$, $\text{cis } 202\frac{1}{2}°$, $\text{cis } 292\frac{1}{2}°$

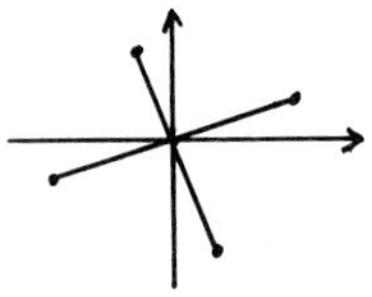

25. $x^2 + (1 - i)x + i = 0$

$a = 1, \quad b = 1 - i, \quad c = i$

$$x = \frac{-(1 - i) \pm \sqrt{(1 - i)^2 - 4 \cdot 1 \cdot i}}{2 \cdot 1}$$

$$= \frac{-1 + i \pm \sqrt{1 - 2i + i^2 - 4i}}{2}$$

$$= \frac{-1 + i \pm \sqrt{-6i}}{2}$$

Let $z = \sqrt{-6i}$. then $z^2 = -6i$.

Let $z = a + bi = \sqrt{-6i}$.

Then $z^2 = a^2 - b^2 + 2abi = -6i$ and $a^2 - b^2 = 0$ and $2ab = -6$.

Solve the second equation for b and substitute in the first to solve for a.

$2ab = -6$

$b = \frac{-6}{2a}$

$= -\frac{3}{a}$

$a^2 - b^2 = 0$

$a^2 - \left(-\frac{3}{a}\right)^2 = 0$

$a^2 - \frac{9}{a^2} = 0$

$a^4 - 9 = 0$

$(a^2 + 3)(a^2 - 3) = 0$

$a^2 + 3 = 0$ or $a^2 - 3 = 0$

No real solution $\quad a^2 = 3$

$a = \pm\sqrt{3}$

If $a = \sqrt{3}$, then $b = -\frac{3}{\sqrt{3}} = -\sqrt{3}$.

If $a = -\sqrt{3}$, then $b = -\frac{3}{-\sqrt{3}} = \sqrt{3}$.

25. (continued)

The solutions of $z = \sqrt{-6i}$ are $\sqrt{3} - \sqrt{3}\,i$ and $-\sqrt{3} + \sqrt{3}\,i$.

Thus, substituting for $\sqrt{-6i}$, we get

$$x = \frac{-1 + i + \sqrt{3} - \sqrt{3}\,i}{2} \approx 0.366 - 0.366i$$

$$x = \frac{-1 + i - \sqrt{3} + \sqrt{3}\,i}{2} \approx -1.366 + 1.366i$$

26. $0.1437 + 0.38281i, \quad -0.477 - 1.049i$

Exercise Set 10.1

1. We replace x by $\frac{1}{2}$ and y by 1.

$3x + y = \frac{5}{2}$	
$3 \cdot \frac{1}{2} + 1$	$\frac{5}{2}$
$\frac{3}{2} + \frac{2}{2}$	
$\frac{5}{2}$	

$2x - y = \frac{1}{4}$	
$2 \cdot \frac{1}{2} - 1$	$\frac{1}{4}$
$1 - 1$	
0	

The ordered pair $\left(\frac{1}{2}, 1\right)$ is not a solution of $2x - y = \frac{1}{4}$. Therefore it is not a solution of the system of equations.

2. Yes

3. Graph both lines on the same set of axes. The x and y-intercepts of the line x + y = 2 are (2, 0) and (0, 2). Plot these points and draw the line they determine. Next graph the line 3x + y = 0. Three of the points on the line are (0, 0), (-1, 3), and (2, -6). Plot these points and draw the line they determine.

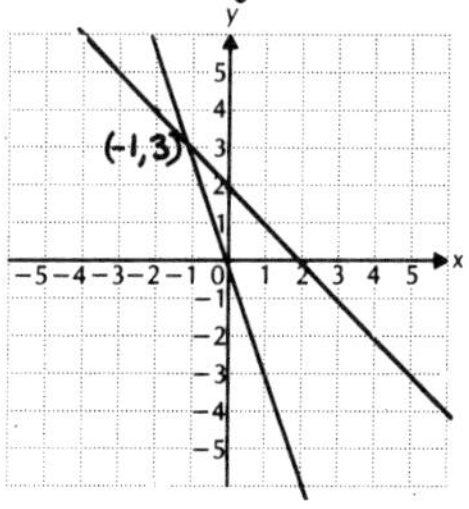

The solution (point of intersection) seems to be the point (-1, 3).

Check:

$x + y = 2$	
$-1 + 3$	2
2	

$3x + y = 0$	
$3(-1) + 3$	0
$-3 + 3$	
0	

The solution is (-1, 3).

4. (3, -2)

5. $x - 5y = 4$ (1)

$2x + y = 7$ (2)

Solve equation (1) for x.

$x - 5y = 4$

$x = 5y + 4$

5. (continued)

Substitute 5y + 4 for x in equation (2) and solve for y.

$$2x + y = 7$$
$$2(5y + 4) + y = 7$$
$$10y + 8 + y = 7$$
$$11y + 8 = 7$$
$$11y = -1$$
$$y = -\frac{1}{11}$$

Now substitute $-\frac{1}{11}$ for y in either equation (1) or (2) and solve for x. It's easier to use equation (1).

$$x - 5y = 4$$
$$x - 5\left(-\frac{1}{11}\right) = 4$$
$$x + \frac{5}{11} = \frac{44}{11}$$
$$x = \frac{39}{11}$$

Check:

$x - 5y = 4$	
$\frac{39}{11} - 5\left(-\frac{1}{11}\right)$	4
$\frac{39}{11} + \frac{5}{11}$	
$\frac{44}{11}$	
4	

$2x + y = 7$	
$2\left(\frac{39}{11}\right) + \left(-\frac{1}{11}\right)$	7
$\frac{78}{11} - \frac{1}{11}$	
$\frac{77}{11}$	
7	

The solution is $\left(\frac{39}{11}, -\frac{1}{11}\right)$.

6. $\left(\frac{11}{8}, -\frac{7}{8}\right)$

7. $x - 3y = 2$ (1)

$6x + 5y = -34$ (2)

Multiply equation (1) by -6 and add the result to equation (2).

$$-6x + 18y = -12$$
$$6x + 5y = -34$$
$$23y = -46$$
$$y = -2 \quad \text{(Solving for y)}$$

Now substitute -2 for y in either equation (1) or (2) and solve for x. It's easier to use equation (1).

$$x - 3y = 2$$
$$x - 3(-2) = 2$$
$$x + 6 = 2$$
$$x = -4$$

7. (continued)

Check:

$x - 3y = 2$	
$-4 - 3(-2)$	2
$-4 + 6$	
2	

$6x + 5y = -34$	
$6(-4) + 5(-2)$	-34
$-24 - 10$	
-34	

The solution is (-4, -2).

8. (3, -1)

9. $0.3x + 0.2y = -0.9$

$0.2x - 0.3y = -0.6$

Multiply each equation by 10 to clear of decimal points.

$3x + 2y = -9$ (1)

$2x - 3y = -6$ (2)

Multiply equation (2) by 3 to make the x-coefficient a multiple of 3.

$3x + 2y = -9$

$6x - 9y = -18$

Then multiply equation (1) by -2 and add the result to equation (2).

$$\begin{array}{r} -6x - 4y = 18 \\ \underline{6x - 9y = -18} \\ -13y = 0 \\ y = 0 \end{array}$$ (Solving for y)

Now substitute 0 for y in either equation (1) or (2) and solve for x. Here we use equation (2).

$2x - 3y = -6$

$2x - 3 \cdot 0 = -6$

$2x = -6$

$x = -3$

The ordered pair (-3, 0) checks. It is the solution.

10. $\left(\frac{1}{2}, -\frac{2}{3}\right)$

11. $\frac{1}{5}x + \frac{1}{2}y = 6$

$\frac{3}{5}x - \frac{1}{2}y = 2$

Multiply each equation by 10 to clear of fractions and then add.

$2x + 5y = 60$ (1)

$\underline{6x - 5y = 20}$ (2)

$8x = 80$

$x = 10$ (Solving for x)

11. (continued)

Substitute 10 for x in either equation (1) or (2) and solve for y. Here we use equation(1).

$2x + 5y = 60$

$2 \cdot 10 + 5y = 60$

$20 + 5y = 60$

$5y = 40$

$y = 8$

The ordered pair (10, 8) checks and is the solution.

12. (-12, -15)

13. $2a = 5 - 3b$

$4a = 11 - 7b$

First write each equation in the form Ax + By = C.

$2a + 3b = 5$ (1)

$4a + 7b = 11$ (2)

Multiply equation (1) by -2 and add the result to equation (2).

$$\begin{array}{r} -4a - 6b = -10 \\ \underline{4a + 7b = 11} \\ b = 1 \end{array}$$

Substitute 1 for b in either equation (1) or (2) and solve for a. Here we use equation (1).

$2a + 3b = 5$

$2a + 3 \cdot 1 = 5$

$2a + 3 = 5$

$2a = 2$

$a = 1$

The ordered pair (1, 1) checks. It is the solution.

14. (3, 1)

15. Translate:

Let x and y represent the numbers. A system of equations results directly from the statements in the problem.

$x + y = -10$ (The sum is -10.)

$x - y = 1$ (The difference is 1.)

Carry out: We solve the system.

Multiply the first equation by -1 and add the result to the second equation.

$$\begin{array}{r} -x - y = 10 \\ \underline{x - y = 1} \\ -2y = 11 \\ y = -\frac{11}{2} \end{array}$$ (Solving for y)

15. (continued)

Substitute $-\frac{11}{2}$ for y in either equation of the system and solve for x.

$$x + y = -10$$

$$x + \left(-\frac{11}{2}\right) = -10$$

$$x - \frac{11}{2} = -\frac{20}{2}$$

$$x = -\frac{9}{2}$$

Check:

The sum of $-\frac{9}{2}$ and $-\frac{11}{2}$ is $-\frac{20}{2}$, or -10. The difference is $-\frac{9}{2} - \left(-\frac{11}{2}\right)$, or $\frac{2}{2}$ which is 1.

Thus the numbers are $-\frac{9}{2}$ and $-\frac{11}{2}$.

16. $\frac{9}{2}$, $-\frac{11}{2}$

17. Familiarize:

It helps to make a drawing. Then organize the information in a table. Let b represent the speed of the boat and s represent the speed of the stream. The speed upstream is b - s. The speed downstream is b + s.

$d_1 = 46$ $t_1 = 2$ $r_1 = b + s$

Downstream

$d_2 = 51$ $t_2 = 3$ $r_2 = b - s$

Upstream

	Distance	Speed	Time
Downstream	46	b + s	2
Upstream	51	b - s	3

Translate:

Using d = rt in each row of the table, we get a system of equations.

$46 = (b + s)2 = 2b + 2s$ or $b + s = 23$

$51 = (b - s)3 = 3b - 3s$ $b - s = 17$

Carry out:

Multiply the first equation by -1 and add the result to the second equation.

$$\begin{aligned} -b - s &= -23 \\ \underline{b - s} &= \underline{17} \\ -2s &= -6 \\ s &= 3 \quad \text{(Solving for s)} \end{aligned}$$

Substitute 3 for s in either equation of the system and solve for b.

$$b + s = 23$$

$$b + 3 = 23$$

$$b = 20$$

17. (continued)

Check:

The speed downstream is 20 + 3, or 23. The distance downstream is 23·2, or 46. The speed upstream is 20 - 3, or 17. The distance upstream is 17·3, or 51. The values check.

The speed of the boat is 20 km/h. The speed of the stream is 3 km/h.

18. Plane: 875 km/h, Wind: 125 km/h

19. Familiarize:

We organize the information in a table. We let x represent the number of liters of antifreeze A and y represent the number of liters of antifreeze B.

	Amount of antifreeze	Percent of alcohol	Amount of alcohol in antifreeze
A	x liters	18%	18%x
B	y liters	10%	10%y
Mixture	20 liters	15%	0.15 × 20, or 3

Translate:

If we add x and y in the first column, we get 20.

$$x + y = 20$$

If we add the amounts of alcohol in the third column, we get 3.

$$18\%x + 10\%y = 3$$

After changing percents to decimals and clearing, we have the following system:

$$x + y = 20$$

$$18x + 10y = 300$$

Carry out:

Multiply -18 times the first equation and add the result to the second equation.

$$\begin{aligned} -18x - 18y &= -360 \\ \underline{18x + 10y} &= \underline{300} \\ -8y &= -60 \\ y &= 7.5 \quad \text{(Solving for y)} \end{aligned}$$

Substitute 7.5 for y in one of the original equations and solve for x.

$$x + y = 20$$

$$x + 7.5 = 20$$

$$x = 12.5$$

Thus 12.5 L of A and 7.5 L of B is a possible solution.

19. (continued)

Check:

We add the amounts of antifreeze: 12.5 L + 7.5 L = 20 L. Thus the amount of antifreeze checks. Next, we check the amount of alcohol: 18%(12.5) + 10%(7.5) = 2.25 + 0.75, or 3 L. The amount of alcohol also checks.

The solution of the problem is 12.5 L of A and 7.5 L of B.

20. Beer A: 15 L, Beer B: 35 L

21. Familiarize:

We first make a drawing.

t hours 80 km/h 96 km/h t hours

d_1 kilometers d_2 kilometers

528 km

Then we organize the information in a table. Each car travels the same amount of time. Let t represent the time and d_1 and d_2 represent the distances of the slow car and the fast car respectively.

	Distance	Speed	Time
Slow car	d_1	80	t
Fast car	d_2	96	t
Total	528		

Translate:

The sum of the distances is 528 km.

$d_1 + d_2 = 528$

Using d = rt in each row of the table, we get

$d_1 = 80t$ and $d_2 = 96t$.

Using substitution we get the following equation:

$80t + 96t = 528$

Carry out:

$176t = 528$

$t = 3$

Check:

If t = 3, then $d_1 = 80\cdot 3$, or 240 and $d_2 = 96\cdot 3$, or 288. The total distance is 240 + 288, or 528 km. The value checks.

In 3 hours the cars will be 528 km apart.

22. 375 km

23. Familiarize:

We first make a drawing.

t hours 190 km/h 200 km/h t hours

d_1 kilometers d_2 kilometers

780 km

Then we organize the information in a table. The time is the same for each plane. Let t represent the time and d_1 and d_2 represent the distances of the planes.

	Distance	Speed	Time
Slow plane	d_1	190	t
Fast plane	d_2	200	t
Total	780		

Translate:

The sum of the distances is 780 km.

$d_1 + d_2 = 780$

Using d = rt in each row of the table, we get

$d_1 = 190t$ and $d_2 = 200t$.

Using substitution we get the following equation:

$190t + 200t = 780$

Carry out:

$390t = 780$

$t = 2$

Check:

If t = 2, then $d_1 = 190\cdot 2$, or 380 and $d_2 = 200\cdot 2$, or 400. The total distance is 380 + 400, or 780 km. The value checks.

In 2 hours the planes will meet.

24. $1\frac{3}{4}$ hours

25. Familiarize and Translate:

Let x represent the number of white scarves and y represent the number of printed scarves. Thus, \$4.95x worth of white ones and \$7.95y worth of printed ones were sold. The total number of scarves was 40.

$x + y = 40$

The total sales were \$282.

$4.95x + 7.95y = 282$, or $495x + 795y = 28,200$

We solve the following system of equations:

$$x + y = 40$$
$$495x + 795y = 28,200$$

25. (continued)

Carry out:

Multiply -495 times the first equation and add the result to the second equation.

$$\begin{aligned} -495x - 495y &= -19{,}800 \\ \underline{495x + 795y} &= \underline{28{,}200} \\ 300y &= 8400 \\ y &= 28 \end{aligned}$$

Substitute 28 for y in one of the equations of the system and solve for x.

$$\begin{aligned} x + y &= 40 \\ x + 28 &= 40 \\ x &= 12 \end{aligned}$$

Check:

The total number of scarves was 12 + 28, or 40. Total sales were 4.95(12) + 7.95(28), or 59.40 + 222.60, or $282. The values check.

Thus, 12 white ones and 28 printed ones were sold.

26. 8 white, 22 yellow

27. Familiarize and Translate:

Let x represent Paula's age now and y represent Bob's age now. It helps to organize the information in a table.

	Age now	Age four years from now
Paula	x	x + 4
Bob	y	y + 4

Paula is 12 years older than Bob.

$x = y + 12$, or $x - y = 12$

Four years from now, Bob will be $\frac{2}{3}$ as old as Paula.

$$y + 4 = \frac{2}{3}(x + 4)$$

or

$$\begin{aligned} 3(y + 4) &= 2(x + 4) \\ 3y + 12 &= 2x + 8 \\ 4 &= 2x - 3y \end{aligned}$$

We now have a system of equations.

$$\begin{aligned} x - y &= 12 \\ 2x - 3y &= 4 \end{aligned}$$

Carry out:

Multiply the first equation by -2 and add the result to the second equation.

$$\begin{aligned} -2x + 2y &= -24 \\ \underline{2x - 3y} &= \underline{4} \\ -y &= -20 \\ y &= 20 \end{aligned}$$

27. (continued)

Substitute 20 for y in either of the original equations and solve for x.

$$\begin{aligned} x - y &= 12 \\ x - 20 &= 12 \\ x &= 32 \end{aligned}$$

Check:

If Paula is 32 and Bob 20, Paula is 12 years older than Bob. In four years, Paula will be 36 and Bob 24. Thus Bob will be $\frac{2}{3}$ as old as Paula $\left(24 = \frac{2}{3} \cdot 36\right)$. The ages check.

Paula is 32, and Bob is 20.

28. Maria is 20, and Carlos is 28.

29. Familiarize:

If helps to make a drawing. We let ℓ represent the length and w represent the width.

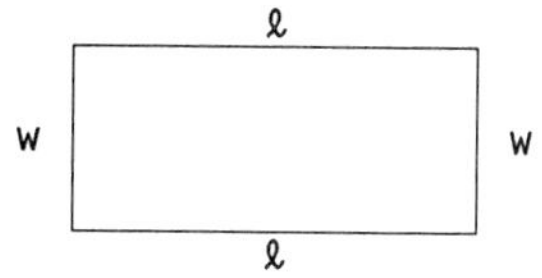

The formula for the perimeter is $P = 2\ell + 2w$.

Translate:

The perimeter is 190 m.

$2\ell + 2w = 190$, or $\ell + w = 95$

The width is one-fourth the length.

$$w = \frac{1}{4}\ell$$

The resulting system is

$$\begin{aligned} \ell + w &= 95 \\ w &= \frac{1}{4}\ell \end{aligned}$$

Carry out:

Substitute $\frac{1}{4}\ell$ for w in the first equation and solve for ℓ.

$$\begin{aligned} \ell + w &= 95 \\ \ell + \frac{1}{4}\ell &= 95 \\ \frac{5}{4}\ell &= 95 \\ \frac{4}{5} \cdot \frac{5}{4}\ell &= \frac{4}{5} \cdot 95 \\ \ell &= 76 \end{aligned}$$

Then substitute 76 for ℓ in one of the equations of the system and solve for w.

$$\begin{aligned} \ell + w &= 95 \\ 76 + w &= 95 \\ w &= 19 \end{aligned}$$

29. (continued)

Check:

If $\ell = 76$ m and $w = 19$ m, then the width is one-fourth the length. The perimeter is $2 \cdot 76 + 2 \cdot 19$, or 190 m. The values check.

The length is 76 m; the width is 19 m.

30. The length is 160 m; the width is 154 m.

31. Familiarize:

We first make a drawing.

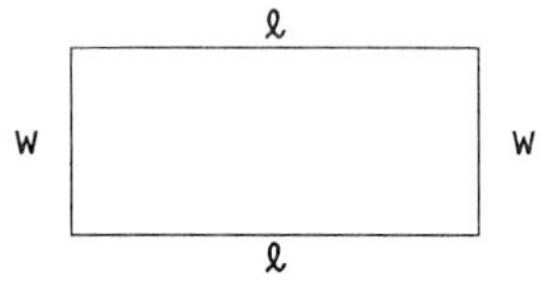

We let ℓ represent the length and w represent the width. The formula for the perimeter is $P = 2\ell + 2w$.

Translate:

The perimeter is 384 m.

$2\ell + 2w = 384$, or $\ell + w = 192$

The length is 82 m greater than the width.

$\ell = w + 82$

The resulting system is

$\ell + w = 192$

$\ell = w + 82$

Carry out:

Substitute w + 82 for ℓ in the first equation and solve for w.

$$\begin{aligned} \ell + w &= 192 \\ w + 82 + w &= 192 \\ 2w + 82 &= 192 \\ 2w &= 110 \\ w &= 55 \end{aligned}$$

Then substitute 55 for w in one of the original equations and solve for ℓ.

$\ell = w + 82$

$\ell = 55 + 82$

$\ell = 137$

Check:

The length is 82 greater than the width ($137 = 55 + 82$). The perimeter is $2 \cdot 137 + 2 \cdot 55$, or 384 m. The values check.

The length is 137 m; the width is 55 m.

32. 372 cm^2

33. $\frac{x + y}{4} - \frac{x - y}{3} = 1$

$\frac{x - y}{2} + \frac{x + y}{4} = -9$

First multiply each equation by the LCM to clear fractions.

$3(x + y) - 4(x - y) = 12$ (Multiplying by 12)

$2(x - y) + (x + y) = -36$ (Multiplying by 4)

After simplifying, the resulting system is

$-x + 7y = 12$

$3x - y = -36$.

Multiply the first equation by 3 and add the result to the second equation.

$$\begin{aligned} -3x + 21y &= 36 \\ \underline{3x - \quad y} &\underline{= -36} \\ 20y &= 0 \\ y &= 0 \end{aligned}$$

Substitute 0 for y in one of the original equations and solve for x.

$$\begin{aligned} 3x - y &= -36 \\ 3x - 0 &= -36 \\ 3x &= -36 \\ x &= -12 \end{aligned}$$

The solution is (-12, 0).

34. (0, 0)

35. $2.35x - 3.18y = 4.82$

$1.92x + 6.77y = -3.87$

It is helpful to multiply each equation by 100 to clear of decimal points.

$235x - 318y = 482$

$192x + 677y = -387$

Multiply the first equation by 677 and the second equation by 318. Then add the results.

$$\begin{aligned} 159{,}095x - 215{,}286y &= 326{,}314 \\ \underline{61{,}056x + 215{,}286y} &\underline{= -123{,}066} \\ 220{,}151x \qquad\qquad &= 203{,}248 \\ x &\approx 0.923 \end{aligned}$$

Multiply the first equation by 192 and the second equation by -235. Then add the results and solve for y, $y \approx -0.833$.

The solution is (0.923, -0.833).

36. (0.13122, -1.10518)

37. Familiarize:

We first make a drawing.

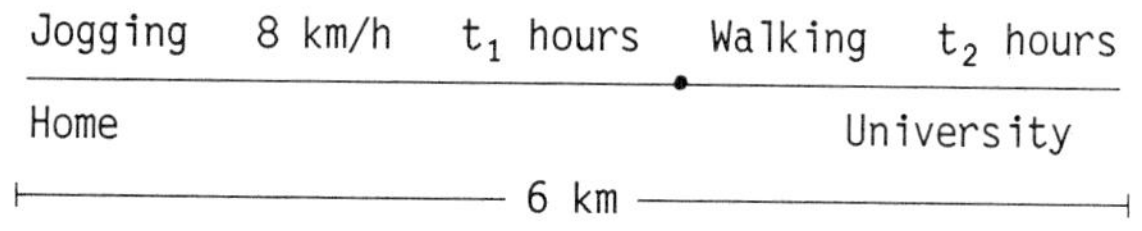

It helps to organize the information in a table. Using $d = rt$, we let $8t_1$ represent the distance jogging and $4t_2$ represent the distance walking.

	Distance	Speed	Time
Jogging	$8t_1$	8	t_1
Walking	$4t_2$	4	t_2
Total	6		1

Translate:

The total distance is 6 km. The total time is 1 hour. A system of equations results from the columns in the table.

$$t_1 + t_2 = 1$$
$$8t_1 + 4t_2 = 6$$

Carry out:

Multiply the first equation by -8 and add the result to the second equation.

$$-8t_1 - 8t_2 = -8$$
$$\underline{8t_1 + 4t_2 = 6}$$
$$-4t_2 = -2$$
$$t_2 = \frac{1}{2}$$

Substitute $\frac{1}{2}$ for t_2 in $t_1 + t_2 = 1$ and solve for t_1. Thus $t_1 = \frac{1}{2}$.

The jogging distance, $8t_1$, is $8 \cdot \frac{1}{2}$, or 4 km.

The walking distance, $4t_2$, is $4 \cdot \frac{1}{2}$, or 2 km.

Check:

The total distance is 4 + 2, or 6 km. The total time is $\frac{1}{2} + \frac{1}{2}$, or 1 hour. The values check.

She jogs 4 km on each trip.

38. Joan: 14 yr; James: 32 yr

39. Familiarize:

Let x represent the number of one-book orders. Let y represent the number of two-book orders. Then 12x represents the amount taken in from the one-book orders and 20y represents the amount taken in from the two-book orders. It helps to organize the information in a table.

39. (continued)

	Number of orders	Number of books	Amount taken in
One-book orders	x	x	12x
Two-book orders	y	2y	20y
Total		880	9840

Translate:

The total number of books sold was 880.

$$x + 2y = 880$$

The total amount of money taken in was \$9840.

$$12x + 20y = 9840, \quad \text{or} \quad 3x + 5y = 2460$$

The resulting system of equations is

$$x + 2y = 880$$
$$3x + 5y = 2460$$

Multiply the first equation by -3 and add the result to the second equation.

$$-3x - 6y = -2640$$
$$\underline{3x + 5y = 2460}$$
$$-y = -180$$
$$y = 180$$

Substitute 180 for y in one of the equations of the system and solve for x.

$$x + 2y = 880$$
$$x + 2 \cdot 180 = 880$$
$$x + 360 = 880$$
$$x = 520$$

Check:

The total number of books sold was $520 + 2 \cdot 180$, or 880. The total amount taken in was $12 \cdot 520 + 20 \cdot 180$, or \$9840. The values check.

Thus 180 members ordered two books.

40. 82

41. Familiarize and Translate:

Let x represent the numerator and y the denominator.

The numerator is 12 more than the denominator.

$$x = y + 12, \quad \text{or} \quad x - y = 12$$

The sum of the numerator and the denominator is 5 more than three times the denominator.

$$x + y = 3y + 5, \quad \text{or} \quad x - 2y = 5$$

The resulting system is

$$x - y = 12$$
$$x - 2y = 5$$

41. (continued)

Carry out:

Multiply the first equation by -1 and add the result to the second equation.

$$\begin{array}{r} -x + y = -12 \\ x - 2y = 5 \\ \hline -y = -7 \\ y = 7 \end{array}$$

Substitute 7 for y in one of the equations of the system and solve for x.

$x - y = 12$

$x - 7 = 12$

$x = 19$

Check:

If $x = 19$ and $y = 7$, the fraction is $\frac{19}{7}$. The numerator is 12 more than the denominator. The sum of the numerator and the denominator, 19 + 7 (or 26), is 5 more than three times the denominator ($26 = 5 + 3\cdot 7$). The numbers check.

The reciprocal of $\frac{19}{7}$ is $\frac{7}{19}$.

42. 137°

43. Familiarize and Translate:

First situation:

3 hours r_1 km/h — r_2 km/h 2 hours

Union — Central

216 km

Second situation:

1.5 hours r_1 km/h — r_2 km/h 3 hours

Union — Central

216 km

We let r_1 and r_2 represent the speeds of the trains and list the information in a table. Using $d = rt$, we let $3r_1$, $2r_2$, $1.5r_1$, and $3r_2$ represent the distances the trains travel.

	Distance traveled in first situation	Distance traveled in second situation
Train_1 (from Union to Central)	$3r_1$	$1.5r_1$
Train_2 (from Central to Union)	$2r_2$	$3r_2$
Total	216	216

The total distance in each situation is 216 km. Thus, we have a system of equations.

$3\ r_1 + 2r_2 = 216$

$1.5r_1 + 3r_2 = 216$

43. (continued)

Carry out:

Multiply the first equation by -1 and the second equation by 2. Then add the results.

$$\begin{array}{r} -3r_1 - 2r_2 = -216 \\ 3r_1 + 6r_2 = 432 \\ \hline 4r_2 = 216 \\ r_2 = 54 \end{array}$$

Substitute 54 for r_2 in either of the original equations and solve for r_1.

$3r_1 + 2r_2 = 216$

$3r_1 + 2\cdot 54 = 216$

$3r_1 + 108 = 216$

$3r_1 = 108$

$r_1 = 36$

Check:

If $r_1 = 36$ and $r_2 = 54$, the total distance the trains travel in the first situation is $3\cdot 36 + 2\cdot 54$, or 216 km. The total distance they travel in the second situation is $1.5\cdot 36 + 3\cdot 54$, or 216 km. The values check.

The speed of the first train is 36 km/h.
The speed of the second train is 54 km/h.

44. $4\frac{4}{7}$ L

45. Let x represent the value of the horse. If the stablehand had worked the entire year, he would have received 240 + x. After seven months, he received 100 + x. This amount is $\frac{7}{12}$ of what he would have received if he had worked all year. This gives us an equation to solve.

$100 + x = \frac{7}{12}(240 + x)$

$100 + x = 140 + \frac{7}{12}x$

$\frac{5}{12}x = 40$

$x = 96$

The value of the horse is $96.

46. 5

47. Familiarize and Translate:

Let x represent the number of girls in the family and y represent the number of boys. Then Phil has x sisters and y - 1 brothers, and Phyllis has x - 1 sisters and y brothers.

47. (continued)

	Brothers	Sisters
Phil	$y - 1$	x
Phyllis	y	$x - 1$

Phil has the same number of brothers and sisters.

$y - 1 = x$, or $x - y = -1$

Phyllis has twice as many brothers as sisters.

$y = 2(x - 1)$, or $2x - y = 2$

We now have a system of equations.

$x - y = -1$

$2x - y = 2$

Carry out:

Multiply the first equation by -2 and add the result to the second equation.

$-2x + 2y = 2$

$\underline{2x - y = 2}$

$y = 4$

Substitute 4 for y in either of the equations of the original system and solve for x.

$x - y = -1$

$x - 4 = -1$

$x = 3$

Check:

If there are 3 girls and 4 boys in the family, Phil has 3 brothers and 3 sisters and Phyllis has 2 sisters and 4 brothers. Thus, Phil has the same number of brothers and sisters, and Phyllis has twice as many brothers as sisters.

There are 3 girls and 4 boys in the family.

48. City: 261 mi, Highway: 204 mi

49. $\frac{1}{x} - \frac{3}{y} = 2$

$\frac{6}{x} + \frac{5}{y} = -34$

Substitute u for $\frac{1}{x}$ and v for $\frac{1}{y}$ to obtain a linear system.

$u - 3v = 2$

$6u + 5v = -34$

Multiply the first equation by -6 and add the result to the second equation.

$-6u + 18v = -12$

$\underline{6u + 5v = -34}$

$23v = -46$

$v = -2$

49. (continued)

Substitute -2 for v in either of the equations of the linear system and solve for u.

$u - 3v = 2$

$u - 3(-2) = 2$

$u + 6 = 2$

$u = -4$

If $v = -2$, then $\frac{1}{y} = -2$, or $y = -\frac{1}{2}$.

If $u = -4$, then $\frac{1}{x} = -4$, or $x = -\frac{1}{4}$.

Both values check. The solution of the original system is $\left(-\frac{1}{4}, -\frac{1}{2}\right)$.

50. (-125, 100)

51. $3|x| + 5|y| = 30$

$5|x| + 3|y| = 34$

Substitute u for $|x|$ and v for $|y|$ to obtain a linear system.

$3u + 5v = 30$

$5u + 3v = 34$

Multiply the first equation by -5 and the second equation by 3. Then add the results.

$-15u - 25v = -150$

$\underline{15u + 9v = 102}$

$-16v = -48$

$v = 3$

Substitute 3 for v in one of the equations of the linear system and solve for u.

$3u + 5v = 30$

$3u + 5 \cdot 3 = 30$

$3u + 15 = 30$

$3u = 15$

$u = 5$

If $v = 3$, then $|y| = 3$ and $y = 3$ or -3.

If $u = 5$, then $|x| = 5$ and $x = 5$ or -5.

Thus, the solution set for the original system is $\{(5, 3), (5, -3), (-5, 3), (-5, -3)\}$.

52. $(0, \sqrt[3]{3})$

53. $x \cos\theta - y \sin\theta = 0$ (1)

$x \sin\theta + y \cos\theta = 1$ (2)

Multiply equation (2) by $-\cos\theta$ to make the x-coefficient a multiple of $\cos\theta$.

$x \cos\theta - y \sin\theta = 0$

$-x \sin\theta \cos\theta - y \cos^2\theta = -\cos\theta$

Then multiply equation (1) by $\sin\theta$ and add the result to equation (2).

53. (continued)

$$x \sin\theta \cos\theta - y \sin^2\theta = 0$$
$$-x \sin\theta \cos\theta - y \cos^2\theta = -\cos\theta$$
$$-y \sin^2\theta - y \cos^2\theta = -\cos\theta$$
$$-y(\sin^2\theta + \cos^2\theta) = -\cos\theta$$
$$-y = -\cos\theta$$
$$(\sin^2\theta + \cos^2\theta = 1)$$
$$y = \cos\theta$$

Substitute $\cos\theta$ for y in either equation (1) or (2) and solve for x. We use equation (1).

$$x \cos\theta - \cos\theta \sin\theta = 0$$
$$\cos\theta(x - \sin\theta) = 0$$
$$x - \sin\theta = 0$$
$$x = \sin\theta$$

The solution is $x = \sin\theta$, $y = \cos\theta$.

54. First graph $y = 2 \sin \frac{x}{2}$. Then graph $y = \mathrm{Tan}^{-1}\, x$. The solutions are 0, $\pm$ 4.8, $\pm$ 14.3, $\pm$ 17.1, etc.

Exercise Set 10.2

1. We substitute $\left(\frac{1}{2}, \frac{1}{3}, \frac{1}{5}\right)$ into the equations, using alphabetical order.

$2x + 3y - 5z = 1$	
$2 \cdot \frac{1}{2} + 3 \cdot \frac{1}{3} - 5 \cdot \frac{1}{5}$	1
$1 + 1 - 1$	
1	

$6x - 6y + 10z = 3$	
$6 \cdot \frac{1}{2} - 6 \cdot \frac{1}{3} + 10 \cdot \frac{1}{5}$	3
$3 - 2 + 2$	
3	

$4x - 9y + 5z = 0$	
$4 \cdot \frac{1}{2} - 9 \cdot \frac{1}{3} + 5 \cdot \frac{1}{5}$	0
$2 - 3 + 1$	
0	

The triple $\left(\frac{1}{2}, \frac{1}{3}, \frac{1}{5}\right)$ makes all three equations true, so it is the solution.

2. No

3.
$$x + y + z = 2 \qquad (1)$$
$$6x - 4y + 5z = 31 \qquad (2)$$
$$5x + 2y + 2z = 13 \qquad (3)$$

Multiply (1) by -6 and add the result to (2). We also multiply (1) by -5 and add the result to (3).

3. (continued)

$$x + y + z = 2$$
$$-10y - z = 19$$
$$-3y - 3z = 3$$

Multiply (3) by 10 to make the y-coefficient a multiple of the y-coefficient in (2).

$$x + y + z = 2$$
$$-10y - z = 19$$
$$-30y - 30z = 30$$

Multiply (2) by -3 and add the result to (3).

$$x + y + z = 2$$
$$-10y - z = 19$$
$$-27z = -27$$

Solve (3) for z.

$$-27z = -27$$
$$z = 1$$

Substitute 1 for z in (2) and solve for y.

$$-10y - z = 19$$
$$-10y - 1 = 19$$
$$-10y = 20$$
$$y = -2$$

Substitute 1 for z and -2 for y in (1) and solve for x.

$$x + y + z = 2$$
$$x + (-2) + 1 = 2$$
$$x - 1 = 2$$
$$x = 3$$

The solution is (3, -2, 1).

4. (-2, -1, 4)

5.
$$x - y + 2z = -3 \qquad (1)$$
$$x + 2y + 3z = 4 \qquad (2)$$
$$2x + y + z = -3 \qquad (3)$$

Multiply (1) by -1 and add the result to (2). Also multiply (1) by -2 and add the result to (3).

$$x - y + 2z = -3$$
$$3y + z = 7$$
$$3y - 3z = 3$$

Multiply (2) by -1 and add the result to (3).

$$x - y + 2z = -3$$
$$3y + z = 7$$
$$-4z = -4$$

Solve (3) for z.

$$-4z = -4$$
$$z = 1$$

5. (continued)

Substitute 1 for z in (2) and solve for y.

$3y + z = 7$

$3y + 1 = 7$

$3y = 6$

$y = 2$

Substitute 1 for z and 2 for y in (1) and solve for x.

$x - y + 2z = -3$

$x - 2 + 2\cdot 1 = -3$

$x = -3$

The solution is (-3, 2, 1).

6. (1, 2, 3)

7.
$$\begin{aligned} 4a + 9b \quad\quad &= 8 \quad (1)\\ 8a \quad + 6c &= -1 \quad (2)\\ 6b + 6c &= -1 \quad (3)\end{aligned}$$

Multiply (1) by -2 and add the result to (2).

$$\begin{aligned} 4a + 9b \quad\quad &= 8\\ -18b + 6c &= -17\\ 6b + 6c &= -1\end{aligned}$$

Multiply (3) by 3 to make the b-coefficient a multiple of the b-coefficient in (2).

$$\begin{aligned} 4a + 9b \quad\quad &= 8\\ -18b + 6c &= -17\\ 18b + 18c &= -3\end{aligned}$$

Add (2) to (3).

$$\begin{aligned} 4a + 9b \quad\quad &= 8\\ -18b + 6c &= -17\\ 24c &= -20\end{aligned}$$

Solve (3) for c.

$24c = -20$

$c = -\frac{20}{24} = -\frac{5}{6}$

Substitute $-\frac{5}{6}$ for c in (2) and solve for b.

$-18b + 6c = -17$

$-18b + 6\left(-\frac{5}{6}\right) = -17$

$-18b - 5 = -17$

$-18b = -12$

$b = \frac{12}{18} = \frac{2}{3}$

7. (continued)

Substitute $\frac{2}{3}$ for b in (1) and solve for a.

$4a + 9b = 8$

$4a + 9 \cdot \frac{2}{3} = 8$

$4a + 6 = 8$

$4a = 2$

$a = \frac{1}{2}$

The solution is $\left(\frac{1}{2}, \frac{2}{3}, -\frac{5}{6}\right)$.

8. $\left(4, \frac{1}{2}, -\frac{1}{2}\right)$

9.
$$\begin{aligned} w + x + y + z &= 2 \quad (1)\\ w + 2x + 2y + 4x &= 1 \quad (2)\\ -w + x - y - z &= -6 \quad (3)\\ -w + 3x + y - z &= -2 \quad (4)\end{aligned}$$

Multiply (1) by - 1 and add to (2). Add (1) to (3) and to (4).

$$\begin{aligned} w + x + y + z &= 2\\ x + y + 3z &= -1\\ 2x \quad\quad &= -4\\ 4x + 2y \quad &= 0\end{aligned}$$

Solve (3) for x.

$2x = -4$

$x = -2$

Substitute -2 for x in (4) and solve for y.

$4(-2) + 2y = 0$

$-8 + 2y = 0$

$2y = 8$

$y = 4$

Substitute -2 for x and 4 for y in (2) and solve for z.

$-2 + 4 + 3z = -1$

$3z = -3$

$z = -1$

Substitute -2 for x, 4 for y, and -1 for z in (1) and solve for w.

$w - 2 + 4 - 1 = 2$

$w = 1$

The solution is (1, -2, 4, -1).

10. (-3, -1, 0, 4)

11. Familiarize and Translate:

We let x, y, z represent the first, second, and third numbers respectively.

The sum of the three numbers is 26.

$x + y + z = 26$

Twice the first minus the second is 2 less than the third.

$2x - y = z - 2$, or $2x - y - z = -2$

The third is the second minus three times the first.

$z = y - 3x$, or $3x - y + z = 0$

We now have a system of three equations.

$x + y + z = 26$

$2x - y - z = -2$

$3x - y + z = 0$

Carry out:

We solve the system. The solution is (8, 21, -3).

Check:

The sum of the numbers is $8 + 21 + (-3)$, or 26. Twice the first minus the second $(2 \cdot 8 - 21 = -5)$ is two less than the third $(-3 - 2 = -5)$. The third (-3) is the second minus three times the first $(21 - 3 \cdot 8 = -3)$. The numbers check.

The numbers are 8, 21, and -3.

12. 4, 2, -1

13. Familiarize:

We make a drawing and use x, y, and z for the measures of the angles.

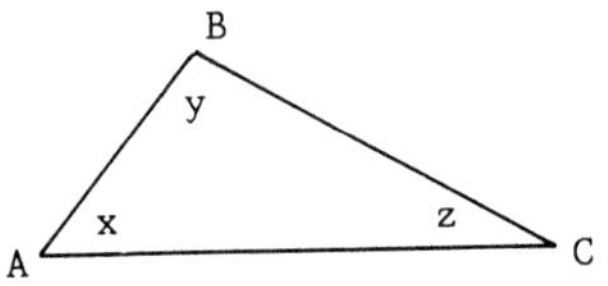

We must use the fact that the measures of the angles of a triangle add up to 180°.

Translate:

The sum of the angle measures is 180°.

$x + y + z = 180$

The measure of angle B is three times the measure of angle A.

$y = 3x$

The measure of angle C is 30° greater than the measure of angle A.

$z = x + 30$

We now have a system of equations.

$x + y + z = 180$

$y = 3x$

$z = x + 30$

13. (continued)

Carry out:

Substitute 3x for y and x + 30 for z in the first equation and solve for x.

$$x + y + z = 180$$
$$x + 3x + (x + 30) = 180$$
$$5x + 30 = 180$$
$$5x = 150$$
$$x = 30$$

If $x = 30$, then $y = 3x = 3 \cdot 30$, or 90, and $z = x + 30 = 30 + 30$, or 60.

Check:

The sum of the angle measures is $30 + 90 + 60$, or 180°. The measure of angle B, 90°, is three times 30°, the measure of angle A. The measure of angle C, 60°, is 30° greater than the measure of angle A, 30°. The values check.

The measures of angles A, B, and C are 30°, 90°, and 60°, respectively.

14. A = 34°, B = 104°, C = 42°

15. Familiarize and Translate:

Let x, y, and z represent the number of quarts Pat picked on Monday, Tuesday, and Wednesday, respectively.

He picked a total of 87 quarts.

$x + y + z = 87$

On Tuesday he picked 15 quarts more than on Monday.

$y = x + 15$, or $x - y = -15$

On Wednesday he picked 3 quarts fewer than on Tuesday.

$z = y - 3$, or $y - z = 3$

We now have a system of equations.

$x + y + z = 87$

$x - y = -15$

$y - z = 3$

Carry out:

We solve the system. The solution is (20, 35, 32).

Check:

The total quarts picked in three days was $20 + 35 + 32$, or 87 quarts. On Tuesday Pat picked 35 quarts which is 15 more quarts than on Monday $(20 + 15 = 35)$. On Wednesday he picked 32 quarts which is 3 quarts less than on Tuesday $(35 - 3 = 32)$. The amounts check.

Pat picked 20 quarts on Monday, 35 quarts on Tuesday, and 32 quarts on Wednesday.

16. \$21 on Thurs., \$18 on Fri., \$27 on Sat.

17. Familiarize and Translate:

Let x, y, and z represent the number of board-feet of lumber produced per day by sawmills A, B, and C, respectively.

All three together can produce 7400 board-feet in a day.

$x + y + z = 7400$

A and B together can produce 4700 board-feet in a day.

$x + y = 4700$

B and C together can produce 5200 board-feet in a day.

$y + z = 5200$

We now have a system of equations.

$$x + y + z = 7400$$
$$x + y \quad = 4700$$
$$y + z = 5200$$

Carry out:

It is easy to solve the system using substitution. For example, substitute 4700 for x + y in the first equation and solve for z. The solution is (2200, 2500, 2700).

Check:

All three can produce 2200 + 2500 + 2700, or 7400 board-feet per day. A and B together can produce 2200 + 2500, or 4700 board-feet. B and C together can produce 2500 + 2700, or 5200 board-feet. All values check.

In a day, sawmill A can produce 2200 board-feet, sawmill B can produce 2500 board-feet, and sawmill C can produce 2700 board-feet.

18. A: 1500, B: 1900, C: 2300

19. Familiarize and Translate:

Let x, y, and z represent the number of linear feet per hour welders A, B, and C can weld, respectively.

Together A, B, and C can weld 37 linear feet.

$x + y + z = 37$

A and B together can weld 22 linear feet.

$x + y = 22$

A and C together can weld 25 linear feet.

$x + z = 25$

We now have a system of equations.

$$x + y + z = 37$$
$$x + y \quad = 22$$
$$x \quad + z = 25$$

Carry out:

We solve the system. Using the substitution method is easy. For example, substitute 22 for x + y in the first equation and solve for z. The solution is (10, 12, 15).

19. (continued)

Check:

Together all three can weld 10 + 12 + 15, or 37 linear feet per hour. A and B together can weld 10 + 12, or 22 linear feet. A and C together can weld 10 + 15, or 25 linear feet. The values check.

Welder A can weld 10 linear feet per hour.
Welder B can weld 12 linear feet per hour.
Welder C can weld 15 linear feet per hour.

20. A: 900 gal/hr, B: 1300 gal/hr, C: 1500 gal/hr

21. We wish to find a quadratic function

$$f(x) = ax^2 + bx + c$$

containing the three given points.

When we substitute, we get

for (1, 4) $\quad 4 = a\cdot 1^2 + b\cdot 1 + c$

for (-1, -2) $\quad -2 = a\cdot(-1)^2 + b\cdot(-1) + c$

for (2, 13) $\quad 13 = a\cdot 2^2 + b\cdot 2 + c$

We now have a system of equations in three unknowns, a, b, and c:

$$a + b + c = 4$$
$$a - b + c = -2$$
$$4a + 2b + c = 13$$

We solve this system, obtaining (2, 3, -1). Thus the function we are looking for is

$f(x) = 2x^2 + 3x - 1$.

22. $f(x) = 3x^2 - x + 2$

23. a) We wish to find a quadratic function

$$E(t) = at^2 + bt + c$$

containing the three given points.

When we substitute, we get

for (1, 38) $\quad 38 = a\cdot 1^2 + b\cdot 1 + c$

for (2, 66) $\quad 66 = a\cdot 2^2 + b\cdot 2 + c$

for (3, 86) $\quad 86 = a\cdot 3^2 + b\cdot 3 + c$

We now have a system of equations in three unknowns, a, b, and c.

$$a + b + c = 38$$
$$4a + 2b + c = 66$$
$$9a + 3b + c = 86$$

We solve the system, obtaining (-4, 40, 2). Thus, the function we are looking for is $E(t) = -4t^2 + 40t + 2$.

b) $E(t) = -4t^2 + 40t + 2$

$$E(4) = -4\cdot 4^2 + 40\cdot 4 + 2$$
$$= -64 + 160 + 2$$
$$= \$98$$

24. a) $E(t) = 2500t^2 - 6500t + 5000$, b) \$19,000

25. $\dfrac{-4x^2 - 2x + 10}{(3x + 5)(x + 1)^2}$

The decomposition looks like

$\dfrac{A}{3x + 5} + \dfrac{B}{x + 1} + \dfrac{C}{(x + 1)^2}$.

Add and equate the numerators.

$-4x^2 - 2x + 10 = A(x + 1)^2 + B(3x + 5)(x + 1) + C(3x + 5)$

$= A(x^2 + 2x + 1) + B(3x^2 + 8x + 5) + C(3x + 5)$

or

$-4x^2 - 2x + 10 = (A + 3B)x^2 + (2A + 8B + 3C)x + (A + 5B + 5C)$

Then equate corresponding coefficients.

$-4 = A + 3B$ (Coefficients of x^2-terms)

$-2 = 2A + 8B + 3C$ (Coefficients of x-terms)

$10 = A + 5B + 5C$ (Constant terms)

We solve this system of three equations and find $A = 5$, $B = -3$, $C = 4$.

The decomposition is

$\dfrac{5}{3x + 5} - \dfrac{3}{x + 1} + \dfrac{4}{(x + 1)^2}$.

26. $\dfrac{6}{x - 4} + \dfrac{1}{2x - 1} - \dfrac{3}{(2x - 1)^2}$

27. $\dfrac{36x + 1}{12x^2 - 7x - 10} = \dfrac{36x + 1}{(4x - 5)(3x + 2)}$

The decomposition looks like

$\dfrac{A}{4x - 5} + \dfrac{B}{3x + 2}$.

Add and equate the numerators.

$36x + 1 = A(3x + 2) + B(4x - 5)$

or $36x + 1 = (3A + 4B)x + (2A - 5B)$

Then equate corresponding coefficients.

$36 = 3A + 4B$ (Coefficients of x-terms)

$1 = 2A - 5B$ (Constant terms)

We solve this system of equations and find $A = 8$ and $B = 3$.

The decomposition is

$\dfrac{8}{4x - 5} + \dfrac{3}{3x + 2}$.

28. $\dfrac{7}{6x - 3} - \dfrac{4}{x + 7}$

29. $\dfrac{-4x^2 - 9x + 8}{(3x^2 + 1)(x - 2)}$

The decomposition looks like

$\dfrac{Ax + B}{3x^2 + 1} + \dfrac{C}{x - 2}$.

Add and equate the numerators.

$-4x^2 - 9x + 8 = (Ax + B)(x - 2) + C(3x^2 + 1)$

$= Ax^2 - 2Ax + Bx - 2B + 3Cx^2 + C$

or $-4x^2 - 9x + 8 = (A + 3C)x^2 + (-2A + B)x + (-2B + C)$

Then equate corresponding coefficients.

$-4 = A + 3C$ (Coefficients of x^2-terms)

$-9 = -2A + B$ (Coefficients of x-terms)

$8 = -2B + C$ (Constant terms)

We solve this system of equations and find $A = 2$, $B = -5$, $C = -2$.

The decomposition is

$\dfrac{2x - 5}{3x^2 + 1} - \dfrac{2}{x - 2}$.

30. $\dfrac{5x + 1}{x^2 + 4} + \dfrac{6}{x - 8}$

31. $\dfrac{2}{x} - \dfrac{1}{y} - \dfrac{3}{z} = -1$

$\dfrac{2}{x} - \dfrac{1}{y} + \dfrac{1}{z} = -9$

$\dfrac{1}{x} + \dfrac{2}{y} - \dfrac{4}{z} = 17$

First substitute u for $\dfrac{1}{x}$, v for $\dfrac{1}{y}$, and w for $\dfrac{1}{z}$ and solve for u, v, and w.

$2u - v - 3w = -1$

$2u - v + w = -9$

$u + 2v - 4w = 17$

Solving this system we get $(-1, 5, -2)$.

If $u = -1$ and $u = \dfrac{1}{x}$, then $\dfrac{1}{x} = -1$, or $x = -1$.

If $v = 5$ and $v = \dfrac{1}{y}$, then $\dfrac{1}{y} = 5$, or $y = \dfrac{1}{5}$.

If $w = -2$ and $w = \dfrac{1}{z}$, then $\dfrac{1}{z} = -2$, or $z = -\dfrac{1}{2}$.

The solution of the original system is

$\left(-1, \dfrac{1}{5}, -\dfrac{1}{2}\right)$.

32. $\left(-\dfrac{1}{2}, -1, -\dfrac{1}{3}\right)$

33. If A, working alone, can do the job in x hours, then A does $\frac{1}{x}$ of it in 1 hour. If B, working alone, can do the job in y hours, then B does $\frac{1}{y}$ of it in 1 hour. If C, working alone, can do the job in z hours, then C does $\frac{1}{z}$ of it in 1 hour.

If A, B, and C, working together, can do the job in 2 hr, then they can do $\frac{1}{2}$ of it in 1 hour. If $\frac{1}{x} + \frac{1}{y} + \frac{1}{z}$ represents the total amount of work done in 1 hour by A, B, and C working together, then

$\frac{1}{x} + \frac{1}{y} + \frac{1}{z} = \frac{1}{2}$

If B and C, working together, can do the job in 4 hr, then they can do $\frac{1}{4}$ of it in 1 hour. If $\frac{1}{y} + \frac{1}{z}$ represents the total amount of work done in 1 hr by B and C, working together, then

$\frac{1}{y} + \frac{1}{z} = \frac{1}{4}$

If A and B, working together, can do the job in $\frac{12}{5}$ hr, then they can do $\frac{5}{12}$ of it in 1 hour. If $\frac{1}{x} + \frac{1}{y}$ represents the total amount of work done in 1 hr by A and B, working together, then

$\frac{1}{x} + \frac{1}{y} = \frac{5}{12}$

We get the following system:

$$\frac{1}{x} + \frac{1}{y} + \frac{1}{z} = \frac{1}{2}$$
$$\frac{1}{y} + \frac{1}{z} = \frac{1}{4}$$
$$\frac{1}{x} + \frac{1}{y} = \frac{5}{12}$$

Substitute u for $\frac{1}{x}$, v for $\frac{1}{y}$, and w for $\frac{1}{z}$ and solve for u, v, and w.

$$u + v + w = \frac{1}{2}$$
$$v + w = \frac{1}{4}$$
$$u + v = \frac{5}{12}$$

Solving this system, we get $\left(\frac{1}{4}, \frac{1}{6}, \frac{1}{12}\right)$.

If $u = \frac{1}{4}$ and $u = \frac{1}{x}$, then $\frac{1}{x} = \frac{1}{4}$, or $x = 4$.

If $v = \frac{1}{6}$ and $v = \frac{1}{y}$, then $\frac{1}{y} = \frac{1}{6}$, or $y = 6$.

If $w = \frac{1}{12}$ and $w = \frac{1}{z}$, then $\frac{1}{z} = \frac{1}{12}$, or $z = 12$.

The solution of the original system is (4, 6, 12).

To do the job alone, it would take A: 4 hours, B: 6 hours, C: 12 hours.

34. A: 24 hr, B: 12 hr, C: $4\frac{4}{5}$ hr

35. Label the angle measures at the tips of the stars a, b, c, d, and e. Also label the angles of the pentagon 1, 2, 3, 4, and 5.

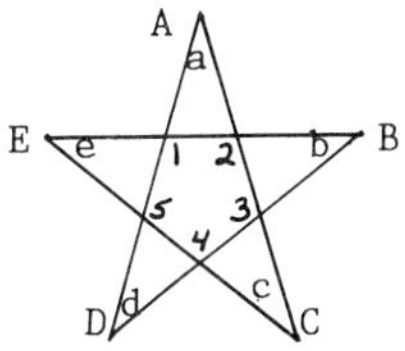

Using the geometric fact that the sum of the angle measures of a triangle is 180°, we get 5 equations.

$1 + b + d = 180$

$2 + c + e = 180$

$3 + a + d = 180$

$4 + b + e = 180$

$5 + a + c = 180$

Adding these equations, we get

$$(1 + 2 + 3 + 4 + 5) + 2a + 2b + 2c + 2d + 2e = 5(180)$$

The sum of the angle measures of any convex polygon with n sides is given by the formula $S = (n - 2)180$. Thus $1 + 2 + 3 + 4 + 5 = (5 - 2)180$, or 540. We substitute and solve for $a + b + c + d + e$.

$$540 + 2(a + b + c + d + e) = 900$$
$$2(a + b + c + d + e) = 360$$
$$a + b + c + d + e = 180$$

The sum of the angle measures at the tips of the star is 180°.

36. 1869

Exercise Set 10.3

Note: The answers to Exercises 19 and 20 are included in the answers to Exercises 1 - 18.

1. $9x - 3y = 15$ (1)
$6x - 2y = 10$ (2)

Multiply equation (2) by 3 to make the x-coefficient a multiple of 9.

$9x - 3y = 15$

$18x - 6y = 30$

Multiply equation (1) by -2 and add the result to equation (2).

$$-18x + 6y = -30$$
$$\underline{18x - 6y = 30}$$
$$0 = 0$$

1. (continued)

Now we have

$9x - 3y = 15$

$0 = 0$

which illustrates that the original system of equations is a dependent system of equations. For this particular system, there is an infinite number of solutions. The system is consistent.

These solutions can be described by expressing one variable in terms of the other. Since 0 = 0 contributes nothing, we solve 9x - 3y = 15 for x and obtain

$x = \frac{3y + 15}{9}$, or $\frac{y + 5}{3}$

The ordered pairs in the solution set can be described in terms of y only.

$\left(\frac{y + 5}{3}, y\right)$

Any value chosen for y gives a value for x, and thus an ordered pair in the solution set.

We could have solved 9x - 3y = 15 for y obtaining $y = \frac{9x - 15}{3}$ or 3x - 5 and described the solution set in terms of x only.

(x, 3x - 5)

Then any value chosen for x would give a value for y, and thus an ordered pair in the solution set.

A few of the solutions are (0, -5), (1, -2), and (-1, -8).

2. The solution set is Ø. The system is inconsistent and independent.

3. $5c + 2d = 24$ (1)

$30c + 12d = 10$ (2)

Multiply equation (1) by -6 and add the result to equation (2).

$$\begin{array}{r} -30c - 12d = -144 \\ \underline{30c + 12d = 10} \\ 0 = -134 \end{array}$$

This gives us

$5c + 2d = 24$

$0 = -134$

The second equation says that $0 \cdot x + 0 \cdot y = -134$. There are no numbers x and y for which this is true. Thus, the system has no solutions. The solution set is Ø. The system is inconsistent. This system is not equivalent to a system of fewer than 2 linear equations. The system is independent.

4. The solution set is Ø. The system is inconsistent and independent.

5. $3x + 2y = 5$

$4y = 10 - 6x$

Rewrite each equation in the form Ax + By = C.

$3x + 2y = 5$ (1)

$6x + 4y = 10$ (2)

If we multiply equation (1) by 2, we get equation (2). Thus the two equations are equivalent, and the system is dependent. For this particular system there is an infinite number of solutions. The system is consistent. The solutions can be described by expressing one variable in terms of the other.

If we solve 3x + 2y = 5 for x obtaining

$x = \frac{5 - 2y}{3}$,

the ordered pairs in the solution set can be described in terms of y only.

$\left(\frac{5 - 2y}{3}, y\right)$

If we solve 3x + 2y = 5 for y obtaining

$y = \frac{5 - 3x}{2}$,

the ordered pairs in the solution set can be described in terms of x only.

$\left(x, \frac{5 - 3x}{2}\right)$

A few of the solutions are $\left(0, \frac{5}{2}\right)$, (3, -2), and (-1, 4).

6. $\left(\frac{7y - 2}{5}, y\right)$ or $\left(x, \frac{5x + 2}{7}\right)$

$\left(0, \frac{2}{7}\right)$, (1, 1), $\left(-1, -\frac{3}{7}\right)$, etc.

The system is consistent and dependent.

7. $12y - 8x = 6$

$4x + 3 = 6y$

Rewrite each equation in the form Ax + By = C.

$-8x + 12y = 6$ (1)

$4x - 6y = -3$ (2)

If we multiply equation (2) by -2, we get equation (1). Thus the two equations are equivalent, and the system is dependent. For this particular system there is an infinite number of solutions. The system is consistent. The solutions can be described by expressing one variable in terms of the other.

If we solve 4x - 6y = -3 for x obtaining

$x = \frac{6y - 3}{4}$,

the ordered pairs in the solution set can be described in terms of y only.

$\left(\frac{6y - 3}{4}, y\right)$

7. (continued)

If we solve $4x - 6y = -3$ for y obtaining

$$y = \frac{4x + 3}{6},$$

the ordered pairs in the solution set can be described in terms of x only.

$$\left(x, \frac{4x + 3}{6}\right)$$

A few of the solutions are $\left(-\frac{3}{4}, 0\right)$, $\left(\frac{9}{4}, 2\right)$ and $\left(-\frac{9}{4}, -1\right)$.

8. $\left(\frac{6y + 5}{8}, y\right)$ or $\left(x, \frac{8x - 5}{6}\right)$

$\left(0, -\frac{5}{6}\right)$, $\left(1, \frac{1}{2}\right)$, $\left(-1, -\frac{13}{6}\right)$, etc.

The system is consistent and dependent.

9.
$$\begin{aligned} x + 2y - z &= -8 \quad (1)\\ 2x - y + z &= 4 \quad (2)\\ 8x + y + z &= 2 \quad (3)\end{aligned}$$

Multiply equation (1) by -2 and add the result to equation (2). Multiply equation (1) by -8 and add the result to equation (3).

$$\begin{aligned} x + 2y - z &= -8\\ -5y + 3z &= 20\\ -15y + 9z &= 66\end{aligned}$$

Multiply equation (2) by -3 and add the result to equation (3).

$$\begin{aligned} x + 2y - z &= -8\\ -5y + 3z &= 20\\ 0 &= 6\end{aligned}$$

Since in (3) we obtain a false equation $0 = 6$, the system has no solution. The solution set is $\emptyset$. The system is <u>inconsistent</u>. This system is not equivalent to a system of fewer than 3 equations. The system is <u>independent</u>.

10. $\left(\frac{28 + z}{11}, \frac{8 + 5z}{11}, z\right)$

$\left(\frac{28}{11}, \frac{8}{11}, 0\right)$, $\left(\frac{29}{11}, \frac{13}{11}, 1\right)$, $\left(\frac{27}{11}, \frac{3}{11}, -1\right)$, etc.

The system is consistent and dependent.

11.
$$\begin{aligned} 2x + y - 3z &= 1\\ x - 4y + z &= 6\\ 4x - 16y + 4z &= 24\end{aligned}$$

First interchange the first two equations so all the x-coefficients are multiples of the first.

$$\begin{aligned} x - 4y + z &= 6 \quad (1)\\ 2x + y - 3z &= 1 \quad (2)\\ 4x - 16y + 4z &= 24 \quad (3)\end{aligned}$$

11. (continued)

Multiply equation (1) by -2 and add the result to equation (2). Multiply equation (1) by -4 and add the result to equation (3).

$$\begin{aligned} x - 4y + z &= 6\\ 9y - 5z &= -11\\ 0 &= 0\end{aligned}$$

Now we know the system is <u>dependent</u>. It may or may not have an infinite set of solutions. We need further analysis.

Solve equation (2) for y.

$$\begin{aligned} 9y - 5z &= -11\\ 9y &= 5z - 11\\ y &= \frac{5z - 11}{9}\end{aligned}$$

Substitute $\frac{5z - 11}{9}$ for y in equation (1) and solve for x.

$$\begin{aligned} x - 4y + z &= 6\\ x - 4\left(\frac{5z - 11}{9}\right) + z &= 6\\ 9x - 4(5z - 11) + 9z &= 54\\ 9x - 20z + 44 + 9z &= 54\\ 9x &= 11z + 10\\ x &= \frac{11z + 10}{9}\end{aligned}$$

The ordered triples in the solution set can be described as follows:

$$\left(\frac{11z + 10}{9}, \frac{5z - 11}{9}, z\right)$$

There is an infinite number of solutions. The system is <u>consistent</u>.

A few of the solutions are $\left(\frac{10}{9}, -\frac{11}{9}, 0\right)$, $\left(\frac{7}{3}, -\frac{2}{3}, 1\right)$, and $\left(\frac{32}{9}, -\frac{1}{9}, 2\right)$.

12. $\left(\frac{5 + z}{5}, -\frac{7}{5}z, z\right)$

$(1, 0, 0)$, $\left(\frac{6}{5}, -\frac{7}{5}, 1\right)$, $(2, -7, 5)$, etc.

The system is consistent and dependent.

13.
$$\begin{aligned} 2x + y - 3z &= 0\\ x - 4y + z &= 0\\ 4x - 16y + 4z &= 0\end{aligned}$$

Note that this is a system of homogeneous equations. The trivial solution is $(0, 0, 0)$. There may or may not be other solutions.

First we interchange the first two equations so all the x-coefficients are multiples of the first:

$$\begin{aligned} x - 4y + z &= 0 \quad (1)\\ 2x + y - 3z &= 0 \quad (2)\\ 4x - 16y + 4z &= 0 \quad (3)\end{aligned}$$

13. (continued)

Multiply (1) by -2 and add to (2).
Multiply (1) by -4 and add to (3).

$x - 4y + z = 0$
$9y - 5z = 0$
$0 = 0$

Now we know that the system is dependent. It may or may not have an infinite set of solutions. We need further analysis.

Solve (2) for y.

$9y - 5z = 0$
$9y = 5z$
$y = \frac{5}{9} z$

Substitute $\frac{5}{9} z$ for y in (1) and solve for x.

$x - 4y + z = 0$
$x - 4\left(\frac{5}{9} z\right) + z = 0$
$x - \frac{20}{9} z + \frac{9}{9} z = 0$
$x - \frac{11}{9} z = 0$
$x = \frac{11}{9} z$

The ordered triples in the solution set can be described as follows:

$\left(\frac{11}{9} z, \frac{5}{9} z, z\right)$

There is an infinite number of solutions. The system is consistent.

A few of the solutions are (0, 0, 0), $\left(\frac{11}{8}, \frac{5}{18}, \frac{1}{2}\right)$ and $\left(-\frac{11}{9}, -\frac{5}{9}, -1\right)$.

14. $\left(\frac{1}{5} z, -\frac{7}{5} z, z\right)$

(0, 0, 0), $\left(\frac{1}{5}, -\frac{7}{5}, 1\right)$, (1, -7, 5), etc.

The system is consistent and dependent.

15. $x + y - z = -3$ (1)
$x + 2y + 2z = -1$ (2)

Multiply (1) by -1 and add to (2).

$x + y - z = -3$
$y + 3z = 2$

Solve (2) for y.

$y + 3z = 2$
$y = -3z + 2$

15. (continued)

Substitute -3z + 2 for y in (1) and solve for x.

$x + y - z = -3$
$x + (-3z + 2) - z = -3$
$x - 4z + 2 = -3$
$x = 4z - 5$

The system has an infinite set of solutions. The system is consistent. The ordered triples in the solution set can be described as follows:

(4z - 5, -3z + 2, z)

A few of the solutions are (-5, 2, 0), (-9, 5, -1), and (3, -4, 2).

The system is independent.

16. $\left(-\frac{7}{2} z, -\frac{19}{2} z, z\right)$

(0, 0, 0), $\left(-\frac{7}{2}, -\frac{19}{2}, 1\right)$, (-7, -19, 2), etc.

The system is consistent and independent.

17. $2x + y + z = 0$
$x + y - z = 0$
$x + 2y + 2z = 0$

Any homogeneous system like this always has a solution, because (0, 0, 0) is a solution. There may or may not be other solutions.

Interchange the first two equations.

$x + y - z = 0$ (1)
$2x + y + z = 0$ (2)
$x + 2y + 2z = 0$ (3)

Multiply (1) by -2 and add to (2). Multiply (1) by -1 and add to (3).

$x + y - z = 0$
$-y + 3z = 0$
$y + 3z = 0$

Add (2) to (3)

$x + y - z = 0$
$-y + 3z = 0$
$6z = 0$

Solve (3) for z.

$6z = 0$
$z = 0$

Substitute 0 for z in (2) and solve for y.

$-y + 3z = 0$
$-y + 3 \cdot 0 = 0$
$-y = 0$
$y = 0$

17. (continued)

Substitute 0 for z and 0 for y in (1) and solve for x.

$x + y - z = 0$

$x + 0 - 0 = 0$

$x = 0$

The only solution is (0, 0, 0). The system is consistent. The system is not equivalent to a system of fewer than 3 equations. Thus, the system is independent.

18. (0, 0, 0) The system is consistent and independent.

19. - 20. Answers in Exercises 1 - 18.

21. $4.026x - 1.448y = 18.32$

$0.724y = -9.16 + 2.013x$

Multiply each equation by 1000 and rewrite in the form Ax + By = C.

$4026x - 1448y = 18{,}320 \quad (1)$

$2013x - 724y = 9160 \quad (2)$

If we multiply equation (2) by 2, we get (1). The equations are equivalent. The system is dependent. The solutions can be described by expressing one variable in terms of the other.

If we solve $2013x - 724y = 9160$ for x obtaining

$x = \frac{9160 + 724y}{2013}$,

the ordered pairs in the solution set can be described in terms of y only.

$\left(\frac{9160 + 724y}{2013}, y\right)$

If we solve $2013x - 724y = 9160$ for y obtaining

$y = \frac{2013x - 9160}{724}$,

the ordered pairs in the solution set can be described in terms of x only.

$\left(x, \frac{2013x - 9160}{724}\right)$

22. $\left(\frac{5260 - 142y}{40{,}570}, y\right)$ or $\left(x, \frac{5260 - 40{,}570x}{142}\right)$

23. $w + x + y + z = 4 \quad (1)$

$w + x + y + z = 3 \quad (2)$

$w + x + y + z = 3 \quad (3)$

Multiply (1) by -1 and add the result to (2) and to (3).

$w + x + y + z = 4$

$0 = -1$

$0 = -1$

23. (continued)

Since we obtain the false equation 0 = -1, the system is inconsistent. The solution set is ∅. The system is equivalent to a system of fewer than 3 equations. Thus the system is dependent.

24. a) (2x + 5y, x, y, 3x - 4y)

b) Consistent

c) Dependent

Exercise Set 10.4

1. $4x + 2y = 11$

$3x - y = 2$

Write a matrix using only the constants.

$$\begin{bmatrix} 4 & 2 & 11 \\ 3 & -1 & 2 \end{bmatrix}$$

Multiply row 2 by 4 to make the first number in row 2 a multiple of 4.

$$\begin{bmatrix} 4 & 2 & 11 \\ 12 & -4 & 8 \end{bmatrix}$$

Multiply row 1 by -3 and add it to row 2.

$$\begin{bmatrix} 4 & 2 & 11 \\ 0 & -10 & -25 \end{bmatrix}$$

Putting the variables back in, we have

$4x + 2y = 11 \quad (1)$

$-10y = -25 \quad (2)$

Solve (2) for y.

$-10y = -25$

$y = \frac{5}{2}$

Substitute $\frac{5}{2}$ for y in (1) and solve for x.

$4x + 2y = 11$

$4x + 2 \cdot \frac{5}{2} = 11$

$4x + 5 = 11$

$4x = 6$

$x = \frac{3}{2}$

The solution is $\left(\frac{3}{2}, \frac{5}{2}\right)$.

2. $\left(-\frac{1}{3}, -4\right)$

3. $x + 2y - 3z = 9$
$2x - y + 2z = -8$
$3x - y - 4z = 3$

Write a matrix using only the constants.

$$\begin{bmatrix} 1 & 2 & -3 & 9 \\ 2 & -1 & 2 & -8 \\ 3 & -1 & -4 & 3 \end{bmatrix}$$

Multiply row 1 by -2 and add it to row 2.
Multiply row 1 by -3 and add it to row 3.

$$\begin{bmatrix} 1 & 2 & -3 & 9 \\ 0 & -5 & 8 & -26 \\ 0 & -7 & 5 & -24 \end{bmatrix}$$

Multiply row 3 by 5.

$$\begin{bmatrix} 1 & 2 & -3 & 9 \\ 0 & -5 & 8 & -26 \\ 0 & -35 & 25 & -120 \end{bmatrix}$$

Multiply row 2 by -7 and add to row 3.

$$\begin{bmatrix} 1 & 2 & -3 & 9 \\ 0 & -5 & 8 & -26 \\ 0 & 0 & -31 & 62 \end{bmatrix}$$

Putting the variables back in, we have

$x + 2y - 3z = 9$ (1)
$-5y + 8z = -26$ (2)
$-31z = 62$ (3)

Solve (3) for z.

$-31z = 62$
$z = -2$

Substitute -2 for z in (2) and solve for y.

$-5y + 8z = -26$
$-5y + 8(-2) = -26$
$-5y - 16 = -26$
$-5y = -10$
$y = 2$

Substitute -2 for z and 2 for y in (1) and solve for x.

$x + 2y - 3z = 9$
$x + 2(2) - 3(-2) = 9$
$x + 10 = 9$
$x = -1$

The solution is (-1, 2, -2).

4. (0, 2, 1)

5. $5x - 3y = -2$
$4x + 2y = 5$

Write a matrix using only the constants.

$$\begin{bmatrix} 5 & -3 & -2 \\ 4 & 2 & 5 \end{bmatrix}$$

Multiply row 2 by 5.

$$\begin{bmatrix} 5 & -3 & -2 \\ 20 & 10 & 25 \end{bmatrix}$$

Multiply row 1 by -4 and add it to row 2.

$$\begin{bmatrix} 5 & -3 & -2 \\ 0 & 22 & 33 \end{bmatrix}$$

Putting the variables back in, we have

$5x - 3y = -2$ (1)
$22y = 33$ (2)

Solve (2) for y.

$22y = 33$
$y = \frac{3}{2}$

Substitute $\frac{3}{2}$ for y in (1) and solve for x.

$5x - 3y = -2$
$5x - 3 \cdot \frac{3}{2} = -2$
$5x - \frac{9}{2} = -\frac{4}{2}$
$5x = \frac{5}{2}$
$x = \frac{1}{2}$

The solution is $\left(\frac{1}{2}, \frac{3}{2}\right)$

6. $\left(-1, \frac{5}{2}\right)$

7. $4x - y - 3z = 1$
$8x + y - z = 5$
$2x + y + 2z = 5$

Write a matrix using only the constants.

$$\begin{bmatrix} 4 & -1 & -3 & 1 \\ 8 & 1 & -1 & 5 \\ 2 & 1 & 2 & 5 \end{bmatrix}$$

First interchange rows 1 and 3 so that each number below the first number in the first row is a multiple of that number.

$$\begin{bmatrix} 2 & 1 & 2 & 5 \\ 8 & 1 & -1 & 5 \\ 4 & -1 & -3 & 1 \end{bmatrix}$$

7. (continued)

Multiply row 1 by -4 and add it to row 2.
Multiply row 1 by -2 and add it to row 3.

$$\begin{bmatrix} 2 & 1 & 2 & 5 \\ 0 & -3 & -9 & -15 \\ 0 & -3 & -7 & -9 \end{bmatrix}$$

Multiply row 2 by -1 and add it to row 3.

$$\begin{bmatrix} 2 & 1 & 2 & 5 \\ 0 & -3 & -9 & -15 \\ 0 & 0 & 2 & 6 \end{bmatrix}$$

Putting the variables back in, we have

$$\begin{aligned} 2x + y + 2z &= 5 \quad (1) \\ -3y - 9z &= -15 \quad (2) \\ 2z &= 6 \quad (3) \end{aligned}$$

Solve (3) for z.

$2z = 6$

$z = 3$

Substitute 3 for z in (2) and solve for y.

$$\begin{aligned} -3y - 9z &= -15 \\ -3y - 9(3) &= -15 \\ -3y - 27 &= -15 \\ -3y &= 12 \\ y &= -4 \end{aligned}$$

Substitute 3 for z and -4 for y in (1) and solve for x.

$$\begin{aligned} 2x + y + 2z &= 5 \\ 2x + (-4) + 2(3) &= 5 \\ 2x - 4 + 6 &= 5 \\ 2x &= 3 \\ x &= \frac{3}{2} \end{aligned}$$

The solution is $\left(\frac{3}{2}, -4, 3\right)$.

8. $\left(2, \frac{1}{2}, -2\right)$

9.
$$\begin{aligned} p + q + r &= 1 \\ p + 2q + 3r &= 4 \\ 4p + 5q + 6r &= 7 \end{aligned}$$

Write a matrix using only the constants.

$$\begin{bmatrix} 1 & 1 & 1 & 1 \\ 1 & 2 & 3 & 4 \\ 4 & 5 & 6 & 7 \end{bmatrix}$$

Multiply row 1 by -1 and add it to row 2.
Multiply row 1 by -4 and add it to row 3.

9. (continued)

$$\begin{bmatrix} 1 & 1 & 1 & 1 \\ 0 & 1 & 2 & 3 \\ 0 & 1 & 2 & 3 \end{bmatrix}$$

Multiply row 2 by -1 and add it to row 3.

$$\begin{bmatrix} 1 & 1 & 1 & 1 \\ 0 & 1 & 2 & 3 \\ 0 & 0 & 0 & 0 \end{bmatrix}$$

Putting the variables back in, we have

$$\begin{aligned} p + q + r &= 1 \\ q + 2r &= 3 \\ 0 &= 0 \end{aligned}$$

This system is dependent. In this case it has an infinite set of solutions.

Solve $q + 2r = 3$ for q.

$q = -2r + 3$

Substitute $-2r + 3$ for q in the first equation and solve for p.

$$\begin{aligned} p + q + r &= 1 \\ p + (-2r + 3) + r &= 1 \\ p - r + 3 &= 1 \\ p - r &= -2 \\ p &= r - 2 \end{aligned}$$

The solutions can be described as follows:

$(r - 2, -2r + 3, r)$

10. No solution

11.
$$\begin{aligned} -2w + 2x + 2y - 2z &= -10 \\ w + x + y + z &= -5 \\ 3w + x - y + 4z &= -2 \\ w + 3x - 2y + 2z &= -6 \end{aligned}$$

Write a matrix using only the constants.

$$\begin{bmatrix} -2 & 2 & 2 & -2 & -10 \\ 1 & 1 & 1 & 1 & -5 \\ 3 & 1 & -1 & 4 & -2 \\ 1 & 3 & -2 & 2 & -6 \end{bmatrix}$$

Interchange rows 1 and 2.

$$\begin{bmatrix} 1 & 1 & 1 & 1 & -5 \\ -2 & 2 & 2 & -2 & -10 \\ 3 & 1 & -1 & 4 & -2 \\ 1 & 3 & -2 & 2 & -6 \end{bmatrix}$$

Multiply row 1 by 2 and add it to row 2.
Multiply row 1 by -3 and add it to row 3.
Multiply row 1 by -1 and add it to row 4.

11\. (continued)

$$\begin{bmatrix} 1 & 1 & 1 & 1 & -5 \\ 0 & 4 & 4 & 0 & -20 \\ 0 & -2 & -4 & 1 & 13 \\ 0 & 2 & -3 & 1 & -1 \end{bmatrix}$$

Interchange rows 2 and 3.

$$\begin{bmatrix} 1 & 1 & 1 & 1 & -5 \\ 0 & -2 & -4 & 1 & 13 \\ 0 & 4 & 4 & 0 & -20 \\ 0 & 2 & -3 & 1 & -1 \end{bmatrix}$$

Multiply row 2 by 2 and add it to row 3.
Add row 2 to row 4.

$$\begin{bmatrix} 1 & 1 & 1 & 1 & -5 \\ 0 & -2 & -4 & 1 & 13 \\ 0 & 0 & -4 & 2 & 6 \\ 0 & 0 & -7 & 2 & 12 \end{bmatrix}$$

Multiply row 4 by 4.

$$\begin{bmatrix} 1 & 1 & 1 & 1 & -5 \\ 0 & -2 & -4 & 1 & 13 \\ 0 & 0 & -4 & 2 & 6 \\ 0 & 0 & -28 & 8 & 48 \end{bmatrix}$$

Multiply row 3 by -7 and add it to row 4.

$$\begin{bmatrix} 1 & 1 & 1 & 1 & -5 \\ 0 & -2 & -4 & 1 & 13 \\ 0 & 0 & -4 & 2 & 6 \\ 0 & 0 & 0 & -6 & 6 \end{bmatrix}$$

Putting the variables back in, we get

$$\begin{aligned} w + x + y + z &= -5 \quad (1) \\ -2x - 4y + z &= 13 \quad (2) \\ -4y + 2z &= 6 \quad (3) \\ -6z &= 6 \quad (4) \end{aligned}$$

Solve (4) for z.

$-6z = 6$

$z = -1$

Substitute -1 for z in (3) and solve for y.

$-4y + 2(-1) = 6$

$-4y = 8$

$y = -2$

Substitute -2 for y and -1 for z in (2) and solve for x.

$-2x - 4(-2) + (-1) = 13$

$-2x + 8 - 1 = 13$

$-2x = 6$

$x = -3$

11\. (continued)

Substitute -3 for x, -2 for y, and -1 for z in (1) and solve for w.

$w + (-3) + (-2) + (-1) = -5$

$w - 6 = -5$

$w = 1$

The solution is (1, -3, -2, -1).

12\. (7, 4, 5, 6)

13\. Let d represent the number of dimes and n represent the number of nickels. Translate to a system of equations.

$d + n = 34$

$0.10d + 0.05n = 1.90$

Multiply the second equation by 100 to eliminate the decimal points.

$d + n = 34$

$10d + 5n = 190$

Use matrices to solve this system.

$$\begin{bmatrix} 1 & 1 & 34 \\ 10 & 5 & 190 \end{bmatrix}$$

Multiply row 1 by -10 and add it to row 2.

$$\begin{bmatrix} 1 & 1 & 34 \\ 0 & -5 & -150 \end{bmatrix}$$

Putting the variables back in, we have

$$\begin{aligned} d + n &= 34 \quad (1) \\ -5n &= -150 \quad (2) \end{aligned}$$

Solve (2) for n.

$-5n = -150$

$n = 30$

Substitute 30 for n in (1) and solve for d.

$d + n = 34$

$d + 30 = 34$

$d = 4$

The solution is (4, 30). Thus there are 4 dimes and 30 nickels.

14\. 21 dimes, 22 quarters

15\. Let x represent the number of nickels, y represent the number of dimes, and z represent the number of quarters. Translate to a system of equations.

$x + y + z = 22$

$x + 2y + 5z = 58$ ($2.90 = 290¢)
(5x + 10y + 25z = 290)

$x - y = 6$ (x = y + 6)

Use matrices to solve this system.

15. (continued)

$$\begin{bmatrix} 1 & 1 & 1 & 22 \\ 1 & 2 & 5 & 58 \\ 1 & -1 & 0 & 6 \end{bmatrix}$$

Multiply row 1 by -1 and add it to rows 2 and 3.

$$\begin{bmatrix} 1 & 1 & 1 & 22 \\ 0 & 1 & 4 & 36 \\ 0 & -2 & -1 & -16 \end{bmatrix}$$

Multiply row 2 by 2 and add it to row 3.

$$\begin{bmatrix} 1 & 1 & 1 & 22 \\ 0 & 1 & 4 & 36 \\ 0 & 0 & 7 & 56 \end{bmatrix}$$

Putting the variables back in, we have

$x + y + z = 22$ (1)

$y + 4z = 36$ (2)

$7z = 56$ (3)

Solve (3) for z.

$7z = 56$

$z = 8$

Substitute 8 for z in (2) and solve for y.

$y + 4z = 36$

$y + 4\cdot 8 = 36$

$y = 4$

Substitute 8 for z and 4 for y in (1) and solve for x.

$x + y + z = 22$

$x + 4 + 8 = 22$

$x = 10$

The solution is (10, 4, 8). Thus there are 10 nickels, 4 dimes, and 8 quarters.

16. 6 nickels, 5 dimes, 7 quarters

17. Let x represent the number of pounds of tobacco that sells for $4.05 per pound. Let y represent the number of pounds of tobacco that sells for $2.70 per pound.

Translate to a system of equations.

$x + y = 15$

$4.05x + 2.70y = 15(3.15)$

Multiply the second equation by 100 to eliminate the decimal points.

$x + y = 15$

$405x + 270y = 4725$

Use matrices to solve this system:

$$\begin{bmatrix} 1 & 1 & 15 \\ 405 & 270 & 4725 \end{bmatrix}$$

17. (continued)

Multiply row 1 by -405 and add to row 2.

$$\begin{bmatrix} 1 & 1 & 15 \\ 0 & -135 & -1350 \end{bmatrix}$$

Putting the variables back in, we have

$x + y = 15$ (1)

$-135y = -1350$ (2)

Solve (2) for y.

$-135y = -1350$

$y = 10$

Substitute 10 for y in (1) and solve for x.

$x + y = 15$

$x + 10 = 15$

$x = 5$

The solution is (5, 10). Thus, 5 lb of the $4.05 per lb tobacco and 10 lb of the $2.70 per lb tobacco should be used.

18. Candy: 14 lb, Nuts: 6 lb

19. Let x represent the amount invested at 12½% and y represent the amount invested at 13%. Translate to a system of equations.

$-x + y = 10,000$ $(y = x + 10,000)$

$125x + 130y = 8,950,000$ $(12\frac{1}{2}\%x + 13\%y = 8950)$

Use matrices to solve this system.

$$\begin{bmatrix} -1 & 1 & 10,000 \\ 125 & 130 & 8,950,000 \end{bmatrix}$$

Multiply row 1 by 125 and add to row 2.

$$\begin{bmatrix} -1 & 1 & 10,000 \\ 0 & 255 & 10,200,000 \end{bmatrix}$$

Putting the variables back in, we have

$-x + y = 10,000$ (1)

$225y = 10,200,000$ (2)

Solve (2) for y.

$225y = 10,200,000$

$y = 40,000$

Substitute 40,000 for y in (1) and solve for x.

$-x + y = 10,000$

$-x + 40,000 = 10,000$

$-x = -30,000$

$x = 30,000$

The solution is (30,000, 40,000). Thus, $30,000 was invested at 12½%, and $40,000 was invested at 13%.

20. $10,000 at $13\frac{1}{2}\%$, $14,000 at $13\frac{3}{4}\%$

21. $4.83x + 9.06y = -39.42$
$-1.35x + 6.67y = -33.99$

First multiply each equation by 100 to clear of decimal points.

$483x + 906y = -3942$
$-135x + 667y = -3399$

Use matrices to solve this system.

$$\begin{bmatrix} 483 & 906 & -3942 \\ -135 & 667 & -3399 \end{bmatrix}$$

Multiply row 2 by 483.

$$\begin{bmatrix} 483 & 906 & -3942 \\ -65{,}205 & 322{,}161 & -1{,}641{,}717 \end{bmatrix}$$

Multiply row 1 by 135 and add it to row 2.

$$\begin{bmatrix} 483 & 906 & -3942 \\ 0 & 444{,}471 & -2{,}173{,}887 \end{bmatrix}$$

Putting the variables back in, we have

$483x + 906y = -3942$ (1)
$444{,}471y = -2{,}173{,}887$ (2)

Solve (2) for y.

$444{,}471y = -2{,}173{,}887$
$y = -4.89095$

Substitute -4.89095 for y in (1) and solve for x.

$483x + 906y = -3942$
$483x + 906(-4.89095) = -3942$
$483x - 4431.2007 = -3942$
$483x = 489.2007$
$x = 1.01284$

The solution is (1.01284, -4.89095).

22. (-6.235, 2.451)

23. $3.55x - 1.35y + 1.03z = 9.16$
$-2.14x + 4.12y + 3.61z = -4.50$
$5.48x - 2.44y - 5.86z = 0.813$

Multiply each equation by 100.

$355x - 135y + 103z = 916$
$-214x + 412y + 361z = -450$
$548x - 244y - 586z = 81.3$

Use matrices to solve the system.

$$\begin{bmatrix} 355 & -135 & 103 & 916 \\ -214 & 412 & 361 & -450 \\ 548 & -244 & -586 & 81.3 \end{bmatrix}$$

Multiply rows 2 and 3 by 355.

23. (continued)

$$\begin{bmatrix} 355 & -135 & 103 & 916 \\ -75{,}970 & 146{,}260 & 128{,}155 & -159{,}750 \\ 194{,}540 & -86{,}620 & -208{,}030 & 28{,}861.5 \end{bmatrix}$$

Multiply row 1 by 214 and add to row 2.
Multiply row 1 by -548 and add to row 3.

$$\begin{bmatrix} 355 & -135 & 103 & 916 \\ 0 & 117{,}370 & 150{,}197 & 36{,}274 \\ 0 & -12{,}640 & -264{,}474 & -473{,}106.5 \end{bmatrix}$$

Multiply row 3 by 11,737.

$$\begin{bmatrix} 355 & -135 & 103 & 916 \\ 0 & 117{,}370 & 150{,}197 & 36{,}274 \\ 0 & -148{,}355{,}680 & -3{,}104{,}131{,}338 & -5{,}552{,}850{,}991 \end{bmatrix}$$

Multiply row 2 by 1264 and add to row 3.

$$\begin{bmatrix} 355 & -135 & 103 & 916 \\ 0 & 117{,}370 & 150{,}197 & 36{,}274 \\ 0 & 0 & -2{,}914{,}282{,}330 & -5{,}507{,}000{,}655 \end{bmatrix}$$

Putting the variables back in, we have

$355x - 135y + 103z = 916$ (1)
$117{,}370y + 150{,}197z = 36{,}274$ (2)
$-2{,}914{,}282{,}330z = -5{,}507{,}000{,}655$ (3)

Solve (3) for z.

$-2{,}914{,}282{,}330z = -5{,}507{,}000{,}655$
$z \approx 1.8897$

Substitute 1.8897 for z in (2) and solve for y.

$117{,}370y + 150{,}197z = 36{,}274$
$117{,}370y + 150{,}197(1.8897) = 36{,}274$
$117{,}370y + 283{,}827.2709 = 36{,}274$
$117{,}370y = -247{,}553.2709$
$y \approx -2.1092$

Substitute 1.8897 for z and -2.1092 for y in (1) and solve for x.

$355x - 135y + 103z = 916$
$355x - 135(-2.1092) + 103(1.8897) = 916$
$355x + 284.742 + 194.6391 = 916$
$355x + 479.3811 = 916$
$355x = 436.6189$
$x \approx 1.2299$

The solution is (1.2299, -2.1092, 1.8897).

24. (81.2, 39.7, -17.6)

Exercise Set 10.5

1. $A + B = \begin{bmatrix} 1 & 2 \\ 4 & 3 \end{bmatrix} + \begin{bmatrix} -3 & 5 \\ 2 & -1 \end{bmatrix} = \begin{bmatrix} 1 + (-3) & 2 + 5 \\ 4 + 2 & 3 + (-1) \end{bmatrix} = \begin{bmatrix} -2 & 7 \\ 6 & 2 \end{bmatrix}$

2. $\begin{bmatrix} -2 & 7 \\ 6 & 2 \end{bmatrix}$

3. $E + 0 = \begin{bmatrix} 1 & 3 \\ 2 & 6 \end{bmatrix} + \begin{bmatrix} 0 & 0 \\ 0 & 0 \end{bmatrix} = \begin{bmatrix} 1 + 0 & 3 + 0 \\ 2 + 0 & 6 + 0 \end{bmatrix} = \begin{bmatrix} 1 & 3 \\ 2 & 6 \end{bmatrix}$

4. $\begin{bmatrix} 2 & 4 \\ 8 & 6 \end{bmatrix}$

5. $3F = 3\begin{bmatrix} 3 & 3 \\ -1 & -1 \end{bmatrix} = \begin{bmatrix} 3\cdot 3 & 3\cdot 3 \\ 3\cdot(-1) & 3\cdot(-1) \end{bmatrix} = \begin{bmatrix} 9 & 9 \\ -3 & -3 \end{bmatrix}$

6. $\begin{bmatrix} -1 & -1 \\ -1 & -1 \end{bmatrix}$

7. $3F = 3\begin{bmatrix} 3 & 3 \\ -1 & -1 \end{bmatrix} = \begin{bmatrix} 9 & 9 \\ -3 & -3 \end{bmatrix}$, $2A = 2\begin{bmatrix} 1 & 2 \\ 4 & 3 \end{bmatrix} = \begin{bmatrix} 2 & 4 \\ 8 & 6 \end{bmatrix}$

   Then $3F + 2A = \begin{bmatrix} 9 & 9 \\ -3 & -3 \end{bmatrix} + \begin{bmatrix} 2 & 4 \\ 8 & 6 \end{bmatrix} = \begin{bmatrix} 9 + 2 & 9 + 4 \\ -3 + 8 & -3 + 6 \end{bmatrix} = \begin{bmatrix} 11 & 13 \\ 5 & 3 \end{bmatrix}$

8. $\begin{bmatrix} 4 & -3 \\ 2 & 4 \end{bmatrix}$

9. $B - A = \begin{bmatrix} -3 & 5 \\ 2 & -1 \end{bmatrix} - \begin{bmatrix} 1 & 2 \\ 4 & 3 \end{bmatrix} = \begin{bmatrix} -3 & 5 \\ 2 & -1 \end{bmatrix} + \begin{bmatrix} -1 & -2 \\ -4 & -3 \end{bmatrix}$ $[B - A = B + (-A)]$

   $= \begin{bmatrix} -3 + (-1) & 5 + (-2) \\ 2 + (-4) & -1 + (-3) \end{bmatrix} = \begin{bmatrix} -4 & 3 \\ -2 & -4 \end{bmatrix}$

10. $\begin{bmatrix} 1 & 3 \\ -6 & 17 \end{bmatrix}$

11. $BA = \begin{bmatrix} -3 & 5 \\ 2 & -1 \end{bmatrix}\begin{bmatrix} 1 & 2 \\ 4 & 3 \end{bmatrix} = \begin{bmatrix} -3\cdot 1 + 5\cdot 4 & -3\cdot 2 + 5\cdot 3 \\ 2\cdot 1 + (-1)4 & 2\cdot 2 + (-1)3 \end{bmatrix} = \begin{bmatrix} 17 & 9 \\ -2 & 1 \end{bmatrix}$

12. $\begin{bmatrix} 0 & 0 \\ 0 & 0 \end{bmatrix}$

13. $$CD = \begin{bmatrix} 1 & -1 \\ -1 & 1 \end{bmatrix}\begin{bmatrix} 1 & 1 \\ 1 & 1 \end{bmatrix} = \begin{bmatrix} 1\cdot1 + (-1)\cdot1 & 1\cdot1 + (-1)\cdot1 \\ -1\cdot1 + 1\cdot1 & -1\cdot1 + 1\cdot1 \end{bmatrix} = \begin{bmatrix} 0 & 0 \\ 0 & 0 \end{bmatrix}$$

14. $\begin{bmatrix} 0 & 0 \\ 0 & 0 \end{bmatrix}$

15. $$AI = \begin{bmatrix} 1 & 2 \\ 4 & 3 \end{bmatrix}\begin{bmatrix} 1 & 0 \\ 0 & 1 \end{bmatrix} = \begin{bmatrix} 1\cdot1 + 2\cdot0 & 1\cdot0 + 2\cdot1 \\ 4\cdot1 + 3\cdot0 & 4\cdot0 + 3\cdot1 \end{bmatrix} = \begin{bmatrix} 1 & 2 \\ 4 & 3 \end{bmatrix}$$ (Note: AI = A)

16. $\begin{bmatrix} 1 & 2 \\ 4 & 3 \end{bmatrix}$ or A

17. $$AB = \begin{bmatrix} 1 & 0 & -2 \\ 0 & -1 & 3 \\ 3 & 2 & 4 \end{bmatrix}\begin{bmatrix} -1 & -2 & 5 \\ 1 & 0 & -1 \\ 2 & -3 & 1 \end{bmatrix}$$

$$= \begin{bmatrix} 1(-1) + 0\cdot1 + (-2)2 & 1(-2) + 0\cdot0 + (-2)(-3) & 1\cdot5 + 0(-1) + (-2)1 \\ 0(-1) + (-1)1 + 3\cdot2 & 0(-2) + (-1)0 + 3(-3) & 0\cdot5 + (-1)(-1) + 3\cdot1 \\ 3(-1) + 2\cdot1 + 4\cdot2 & 3(-2) + 2\cdot0 + 4(-3) & 3\cdot5 + 2(-1) + 4\cdot1 \end{bmatrix}$$

$$= \begin{bmatrix} -5 & 4 & 3 \\ 5 & -9 & 4 \\ 7 & -18 & 17 \end{bmatrix}$$

18. $\begin{bmatrix} 14 & 12 & 16 \\ -2 & -2 & -6 \\ 5 & 5 & -9 \end{bmatrix}$

19. $$CI = \begin{bmatrix} -2 & 9 & 6 \\ -3 & 3 & 4 \\ 2 & -2 & 1 \end{bmatrix}\begin{bmatrix} 1 & 0 & 0 \\ 0 & 1 & 0 \\ 0 & 0 & 1 \end{bmatrix}$$

$$= \begin{bmatrix} -2\cdot1 + 9\cdot0 + 6\cdot0 & -2\cdot0 + 9\cdot1 + 6\cdot0 & -2\cdot0 + 9\cdot0 + 6\cdot1 \\ -3\cdot1 + 3\cdot0 + 4\cdot0 & -3\cdot0 + 3\cdot1 + 4\cdot0 & -3\cdot0 + 3\cdot0 + 4\cdot1 \\ 2\cdot1 + (-2)\cdot0 + 1\cdot0 & 2\cdot0 + (-2)1 + 1\cdot0 & 2\cdot0 + (-2)\cdot0 + 1\cdot1 \end{bmatrix}$$

$$= \begin{bmatrix} -2 & 9 & 6 \\ -3 & 3 & 4 \\ 2 & -2 & 1 \end{bmatrix}$$ (Note: CI = C)

20. $\begin{bmatrix} -2 & 9 & 6 \\ -3 & 3 & 4 \\ 2 & -2 & 1 \end{bmatrix}$ or C

21. $\begin{bmatrix} -3 & 2 \end{bmatrix}\begin{bmatrix} 4 \\ -2 \end{bmatrix} = \begin{bmatrix} -3\cdot 4 + 2(-2) \end{bmatrix} = \begin{bmatrix} -16 \end{bmatrix}$

22. $\begin{bmatrix} -14 \end{bmatrix}$

23. $\begin{bmatrix} -5 & 1 & 2 \end{bmatrix}\begin{bmatrix} 1 & 3 \\ -1 & 0 \\ 4 & -2 \end{bmatrix} = \begin{bmatrix} -5\cdot 1 + 1(-1) + 2\cdot 4 & -5\cdot 3 + 1\cdot 0 + 2(-2) \end{bmatrix} = \begin{bmatrix} 2 & -19 \end{bmatrix}$

24. $\begin{bmatrix} -8 \\ 2 \\ -6 \end{bmatrix}$

25. $\begin{bmatrix} 3 & -2 & 4 \\ 2 & 1 & -5 \end{bmatrix}\begin{bmatrix} x \\ y \\ z \end{bmatrix} = \begin{bmatrix} 17 \\ 13 \end{bmatrix}$

26. $\begin{bmatrix} 3 & 2 & 5 \\ 4 & -3 & 2 \end{bmatrix}\begin{bmatrix} x \\ y \\ z \end{bmatrix} = \begin{bmatrix} 9 \\ 10 \end{bmatrix}$

27. $\begin{bmatrix} 1 & -1 & 2 & -4 \\ 2 & -1 & -1 & 1 \\ 1 & 4 & -3 & -1 \\ 3 & 5 & -7 & 2 \end{bmatrix}\begin{bmatrix} x \\ y \\ z \\ w \end{bmatrix} = \begin{bmatrix} 12 \\ 0 \\ 1 \\ 9 \end{bmatrix}$

28. $\begin{bmatrix} 2 & 4 & -5 & 12 \\ 4 & -1 & 12 & -1 \\ -1 & 4 & 0 & 2 \\ 2 & 10 & 1 & 0 \end{bmatrix}\begin{bmatrix} x \\ y \\ z \\ w \end{bmatrix} = \begin{bmatrix} 2 \\ 5 \\ 13 \\ 5 \end{bmatrix}$

29. Let $\begin{bmatrix} 3.61 & -2.14 & 16.7 \\ -4.33 & 7.03 & 12.9 \\ 5.82 & -6.95 & 2.34 \end{bmatrix}\begin{bmatrix} 3.05 & 0.402 & -1.34 \\ 1.84 & -1.13 & 0.024 \\ -2.83 & 2.04 & 8.81 \end{bmatrix} = \begin{bmatrix} a & b & c \\ d & e & f \\ g & h & i \end{bmatrix}$

29. (continued)

Each matrix is a 3 × 3 matrix. The product is also a 3 × 3 matrix. We calculate each element in the product.

a) $3.61(3.05) + (-2.14)(1.84) + 16.7(-2.83) = -40.1881 \approx -40.19$

b) $3.61(0.402) + (-2.14)(-1.13) + 16.7(2.04) = 37.93742 \approx 37.94$

c) $3.61(-1.34) + (-2.14)(0.024) + 16.7(8.81) = 142.23824 \approx 142.24$

d) $-4.33(3.05) + 7.03(1.84) + 12.9(-2.83) = -36.7783 \approx -36.78$

e) $-4.33(0.402) + 7.03(-1.13) + 12.9(2.04) = 16.63144 \approx 16.63$

f) $-4.33(-1.34) + 7.03(0.024) + 12.9(8.81) = 119.61992 \approx 119.62$

g) $5.82(3.05) + (-6.95)(1.84) + 2.34(-2.83) = -1.6592 \approx -1.66$

h) $5.82(0.402) + (-6.95)(-1.13) + 2.34(2.04) = 14.96674 \approx 14.97$

i) $5.82(-1.34) + (-6.95)(0.024) + 2.34(8.81) = 12.6498 \approx 12.65$

The product is $\begin{bmatrix} -40.19 & 37.94 & 142.24 \\ -36.78 & 16.63 & 119.62 \\ -1.66 & 14.97 & 12.65 \end{bmatrix}$

30. $\begin{bmatrix} 142.1 & -66.62 & -136.5 \\ 257.2 & 1038.8 & 2694.1 \\ 182.4 & 169.6 & 452.8 \end{bmatrix}$

31. $A = \begin{bmatrix} -1 & 0 \\ 2 & 1 \end{bmatrix}, \quad B = \begin{bmatrix} 1 & -1 \\ 0 & 2 \end{bmatrix}$

$(A + B)(A - B) = \begin{bmatrix} 0 & -1 \\ 2 & 3 \end{bmatrix} \begin{bmatrix} -2 & 1 \\ 2 & -1 \end{bmatrix} = \begin{bmatrix} -2 & 1 \\ 2 & -1 \end{bmatrix}$

$A^2 - B^2 = \begin{bmatrix} -1 & 0 \\ 2 & 1 \end{bmatrix}\begin{bmatrix} -1 & 0 \\ 2 & 1 \end{bmatrix} - \begin{bmatrix} 1 & -1 \\ 0 & 2 \end{bmatrix}\begin{bmatrix} 1 & -1 \\ 0 & 2 \end{bmatrix}$

$= \begin{bmatrix} 1 & 0 \\ 0 & 1 \end{bmatrix} - \begin{bmatrix} 1 & -3 \\ 0 & 4 \end{bmatrix} = \begin{bmatrix} 0 & 3 \\ 0 & -3 \end{bmatrix}$

32. $(A + B)(A + B) = \begin{bmatrix} -2 & -3 \\ 6 & 7 \end{bmatrix}, \quad A^2 + 2AB + B^2 = \begin{bmatrix} 0 & -1 \\ 4 & 5 \end{bmatrix}$

33. $(A + B)(A - B) = \begin{bmatrix} -2 & 1 \\ 2 & -1 \end{bmatrix}$ (See Exercise 31.)

$A^2 = \begin{bmatrix} -1 & 0 \\ 2 & 1 \end{bmatrix}\begin{bmatrix} -1 & 0 \\ 2 & 1 \end{bmatrix} = \begin{bmatrix} 1 & 0 \\ 0 & 1 \end{bmatrix}$

$BA = \begin{bmatrix} 1 & -1 \\ 0 & 2 \end{bmatrix}\begin{bmatrix} -1 & 0 \\ 2 & 1 \end{bmatrix} = \begin{bmatrix} -3 & -1 \\ 4 & 2 \end{bmatrix}$

$AB = \begin{bmatrix} -1 & 0 \\ 2 & 1 \end{bmatrix}\begin{bmatrix} 1 & -1 \\ 0 & 2 \end{bmatrix} = \begin{bmatrix} -1 & 1 \\ 2 & 0 \end{bmatrix}$

33. (continued)

$$B^2 = \begin{bmatrix} 1 & -1 \\ 0 & 2 \end{bmatrix}\begin{bmatrix} 1 & -1 \\ 0 & 2 \end{bmatrix} = \begin{bmatrix} 1 & -3 \\ 0 & 4 \end{bmatrix}$$

$$A^2 + BA - AB - B^2 = \begin{bmatrix} 1 & 0 \\ 0 & 1 \end{bmatrix} + \begin{bmatrix} -3 & -1 \\ 4 & 2 \end{bmatrix} - \begin{bmatrix} -1 & 1 \\ 2 & 0 \end{bmatrix} - \begin{bmatrix} 1 & -3 \\ 0 & 4 \end{bmatrix}$$

$$= \begin{bmatrix} -2 & 1 \\ 2 & -1 \end{bmatrix}$$

Therefore, $(A + B)(A - B) = A^2 + BA - AB - B^2$.

34. $(A + B)(A + B) = \begin{bmatrix} -2 & -3 \\ 6 & 7 \end{bmatrix} = A^2 + BA + AB + B^2$

35. $\begin{bmatrix} \cos x & \sin x \\ -\sin x & \cos x \end{bmatrix}\begin{bmatrix} \cos y & \sin y \\ -\sin y & \cos y \end{bmatrix}$

$$= \begin{bmatrix} \cos x \cos y - \sin x \sin y & \cos x \sin y + \sin x \cos y \\ -\sin x \cos y - \cos x \sin y & -\sin x \sin y + \cos x \cos y \end{bmatrix}$$

$$= \begin{bmatrix} \cos (x + y) & \sin (x + y) \\ -\sin (x + y) & \cos (x + y) \end{bmatrix}$$ (Using sum and difference identities)

Exercise Set 10.6

1. $\begin{vmatrix} -2 & -\sqrt{5} \\ -\sqrt{5} & 3 \end{vmatrix}$

$= -2\cdot3 - (-\sqrt{5})(-\sqrt{5}) = -6 - 5 = -11$

2. $2\sqrt{5} + 12$

3. $\begin{vmatrix} x & 4 \\ x & x^2 \end{vmatrix} = x\cdot x^2 - x\cdot4 = x^3 - 4x$

4. $3y^2 + 2y$

5. $\begin{vmatrix} 3 & 1 & 2 \\ -2 & 3 & 1 \\ 3 & 4 & -6 \end{vmatrix}$

$= 3\begin{vmatrix} 3 & 1 \\ 4 & -6 \end{vmatrix} - (-2)\begin{vmatrix} 1 & 2 \\ 4 & -6 \end{vmatrix} + 3\begin{vmatrix} 1 & 2 \\ 3 & 1 \end{vmatrix}$

$= 3(-22) + 2(-14) + 3(-5)$

$= -66 - 28 - 15$

$= -109$

6. -9

7. $\begin{vmatrix} x & 0 & -1 \\ 2 & x & x^2 \\ -3 & x & 1 \end{vmatrix}$

$= x\begin{vmatrix} x & x^2 \\ x & 1 \end{vmatrix} - 2\begin{vmatrix} 0 & -1 \\ x & 1 \end{vmatrix} + (-3)\begin{vmatrix} 0 & -1 \\ x & x^2 \end{vmatrix}$

$= x(x - x^3) - 2(x) - 3(x)$

$= x^2 - x^4 - 2x - 3x$

$= -x^4 + x^2 - 5x$

8. $-2x^3$

9. $-2x + 4y = 3$

$3x - 7y = 1$

$$x = \frac{\begin{vmatrix} 3 & 4 \\ 1 & -7 \end{vmatrix}}{\begin{vmatrix} -2 & 4 \\ 3 & -7 \end{vmatrix}} = \frac{3(-7) - 1(4)}{-2(-7) - 3(4)} = \frac{-25}{2}$$

9. (continued)

$$y = \frac{\begin{vmatrix} -2 & 3 \\ 3 & 1 \end{vmatrix}}{\begin{vmatrix} -2 & 4 \\ 3 & -7 \end{vmatrix}} = \frac{-2(1) - 3(3)}{-2(-7) - 3(4)} = \frac{-11}{2}$$

The solution is $\left(-\frac{25}{2}, -\frac{11}{2}\right)$.

10. $\left(\frac{9}{19}, \frac{51}{38}\right)$

11. $\sqrt{3}x + \pi y = -5$

$\pi x - \sqrt{3}y = 4$

$$x = \frac{\begin{vmatrix} -5 & \pi \\ 4 & -\sqrt{3} \end{vmatrix}}{\begin{vmatrix} \sqrt{3} & \pi \\ \pi & -\sqrt{3} \end{vmatrix}} = \frac{5\sqrt{3} - 4\pi}{-3 - \pi^2} = \frac{-5\sqrt{3} + 4\pi}{3 + \pi^2}$$

$$y = \frac{\begin{vmatrix} \sqrt{3} & -5 \\ \pi & 4 \end{vmatrix}}{\begin{vmatrix} \sqrt{3} & \pi \\ \pi & -\sqrt{3} \end{vmatrix}} = \frac{4\sqrt{3} + 5\pi}{-3 - \pi^2} = \frac{-4\sqrt{3} - 5\pi}{3 + \pi^2}$$

The solution is $\left(\frac{-5\sqrt{3} + 4\pi}{3 + \pi^2}, \frac{-4\sqrt{3} - 5\pi}{3 + \pi^2}\right)$.

12. $\left(\frac{2\pi - 3\sqrt{5}}{\pi^2 + 5}, \frac{-3\pi - 2\sqrt{5}}{\pi^2 + 5}\right)$

13. $3x + 2y - z = 4$

$3x - 2y + z = 5$

$4x - 5y - z = -1$

$$x = \frac{\begin{vmatrix} 4 & 2 & -1 \\ 5 & -2 & 1 \\ -1 & -5 & -1 \end{vmatrix}}{\begin{vmatrix} 3 & 2 & -1 \\ 3 & -2 & 1 \\ 4 & -5 & -1 \end{vmatrix}}$$

$$= \frac{4\begin{vmatrix} -2 & 1 \\ -5 & -1 \end{vmatrix} - 5\begin{vmatrix} 2 & -1 \\ -5 & -1 \end{vmatrix} + (-1)\begin{vmatrix} 2 & -1 \\ -2 & 1 \end{vmatrix}}{3\begin{vmatrix} -2 & 1 \\ -5 & -1 \end{vmatrix} - 3\begin{vmatrix} 2 & -1 \\ -5 & -1 \end{vmatrix} + 4\begin{vmatrix} 2 & -1 \\ -2 & 1 \end{vmatrix}}$$

$$= \frac{4(7) - 5(-7) - 1(0)}{3(7) - 3(-7) + 4(0)}$$

$$= \frac{28 + 35}{21 + 21} = \frac{63}{42} = \frac{3}{2}$$

13. (continued)

$$y = \frac{\begin{vmatrix} 3 & 4 & -1 \\ 3 & 5 & 1 \\ 4 & -1 & -1 \end{vmatrix}}{42}$$

$$= \frac{3\begin{vmatrix} 5 & 1 \\ -1 & -1 \end{vmatrix} - 3\begin{vmatrix} 4 & -1 \\ -1 & -1 \end{vmatrix} + 4\begin{vmatrix} 4 & -1 \\ 5 & 1 \end{vmatrix}}{42}$$

$$= \frac{3(-4) - 3(-5) + 4(9)}{42}$$

$$= \frac{-12 + 15 + 36}{42} = \frac{39}{42} = \frac{13}{14}$$

$$z = \frac{\begin{vmatrix} 3 & 2 & 4 \\ 3 & -2 & 5 \\ 4 & -5 & -1 \end{vmatrix}}{42}$$

$$= \frac{3\begin{vmatrix} -2 & 5 \\ -5 & -1 \end{vmatrix} - 3\begin{vmatrix} 2 & 4 \\ -5 & -1 \end{vmatrix} + 4\begin{vmatrix} 2 & 4 \\ -2 & 5 \end{vmatrix}}{42}$$

$$= \frac{3(27) - 3(18) + 4(18)}{42}$$

$$= \frac{81 - 54 + 72}{42} = \frac{99}{42} = \frac{33}{14}$$

The solution is $\left(\frac{3}{2}, \frac{13}{14}, \frac{33}{14}\right)$.

14. $\left(-1, -\frac{6}{7}, \frac{11}{7}\right)$

15. $0x + 6y + 6z = -1$

$8x + 0y + 6z = -1$

$4x + 9y + 0z = 8$

$$x = \frac{\begin{vmatrix} -1 & 6 & 6 \\ -1 & 0 & 6 \\ 8 & 9 & 0 \end{vmatrix}}{\begin{vmatrix} 0 & 6 & 6 \\ 8 & 0 & 6 \\ 4 & 9 & 0 \end{vmatrix}}$$

$$= \frac{-1\begin{vmatrix} 0 & 6 \\ 9 & 0 \end{vmatrix} - (-1)\begin{vmatrix} 6 & 6 \\ 9 & 0 \end{vmatrix} + 8\begin{vmatrix} 6 & 6 \\ 0 & 6 \end{vmatrix}}{0\begin{vmatrix} 0 & 6 \\ 9 & 0 \end{vmatrix} - 8\begin{vmatrix} 6 & 6 \\ 9 & 0 \end{vmatrix} + 4\begin{vmatrix} 6 & 6 \\ 0 & 6 \end{vmatrix}}$$

$$= \frac{-1(-54) + 1(-54) + 8(36)}{0(-54) - 8(-54) + 4(36)} = \frac{54 - 54 + 288}{0 + 432 + 144}$$

$$= \frac{288}{576} = \frac{1}{2}$$

15. (continued)

$$y = \frac{\begin{vmatrix} 0 & -1 & 6 \\ 8 & -1 & 6 \\ 4 & 8 & 0 \end{vmatrix}}{576} = \frac{384}{576} = \frac{2}{3}$$

$$z = \frac{\begin{vmatrix} 0 & 6 & -1 \\ 8 & 0 & -1 \\ 4 & 9 & 8 \end{vmatrix}}{576} = \frac{-480}{576} = -\frac{5}{6}$$

The solution is $\left(\frac{1}{2}, \frac{2}{3}, -\frac{5}{6}\right)$.

16. $\left(-\frac{31}{16}, \frac{25}{16}, -\frac{58}{16}\right)$

17. $\begin{vmatrix} x & 5 \\ -4 & x \end{vmatrix} = 24$

$x^2 + 20 = 24$

$x^2 = 4$

$x = \pm 2$

18. 3, -2

19. $\begin{vmatrix} x & -3 \\ -1 & x \end{vmatrix} \geqslant 0$

$x^2 - 3 \geqslant 0$

First solve the equality portion of $\geqslant$.

$x^2 - 3 = 0$

$x^2 = 3$

$x = \pm \sqrt{3}$

Thus $\sqrt{3}$ and $-\sqrt{3}$ are in the solution set; they divide the real number line as pictured below.

Note that

$x^2 - 3 \geqslant 0$

is equivalent to

$(x + \sqrt{3})(x - \sqrt{3}) \geqslant 0.$

The value of $x^2 - 3$ is positive or negative depending on the signs of the factors $x + \sqrt{3}$ and $x - \sqrt{3}$.

Tabulate signs in these intervals.

Interval	$x + \sqrt{3}$	$x - \sqrt{3}$	Product
$x < -\sqrt{3}$	–	–	+
$-\sqrt{3} < x < \sqrt{3}$	+	–	–
$x > \sqrt{3}$	+	+	+

19. (continued)

From the table we can see that for $x^2 - 3$ to be positive x must be less than $-\sqrt{3}$ or greater than $\sqrt{3}$. Thus the solution set for the inequality $x^2 - 3 \geqslant 0$ is $\{x \mid x \leqslant -\sqrt{3} \text{ or } x \geqslant \sqrt{3}\}$.

Note that the inequality $x^2 - 3 \geqslant 0$ could also have been solved by graphing.

20. $\{y \mid -\sqrt{10} < y < \sqrt{10}\}$

21. $\begin{vmatrix} x+3 & 4 \\ x-3 & 5 \end{vmatrix} = -7$

$(x + 3)(5) - (x - 3)4 = -7$

$5x + 15 - 4x + 12 = -7$

$x + 27 = -7$

$x = -34$

22. 3

23. $\begin{vmatrix} 2 & x & 1 \\ 1 & 2 & -1 \\ 3 & 4 & -2 \end{vmatrix} = -6$

$2\begin{vmatrix} 2 & -1 \\ 4 & -2 \end{vmatrix} - 1\begin{vmatrix} x & 1 \\ 4 & -2 \end{vmatrix} + 3\begin{vmatrix} x & 1 \\ 2 & -1 \end{vmatrix} = -6$

$2(-4 + 4) - 1(-2x - 4) + 3(-x - 2) = -6$

$2x + 4 - 3x - 6 = -6$

$-x - 2 = -6$

$-x = -4$

$x = 4$

24. 0

25. - 30. Answers may vary

25. $2L + 2W = \begin{vmatrix} L & -W \\ 2 & 2 \end{vmatrix}$

26. $\begin{vmatrix} \pi & -h \\ \pi & r \end{vmatrix}$

27. $a^2 + b^2 = \begin{vmatrix} a & b \\ -b & a \end{vmatrix}$

28. $\begin{vmatrix} \frac{1}{2}h & -b \\ \frac{1}{2}h & a \end{vmatrix}$

29. $2\pi r^2 + 2\pi rh = \begin{vmatrix} 2\pi r & 2\pi r \\ -h & r \end{vmatrix}$

30. $\begin{vmatrix} x^2 & 1 \\ Q^2 & y^2 \end{vmatrix}$

31. $\begin{vmatrix} \cos x & \sin x \\ -\sin x & \cos x \end{vmatrix} = \cos^2 x - (-\sin^2 x)$

$= \cos^2 x + \sin^2 x$

$= 1$

31. (continued)

$\begin{vmatrix} \cos x & -\sin x \\ \sin x & \cos x \end{vmatrix} = \cos^2 x - (-\sin^2 x)$

$= \cos^2 x + \sin^2 x$

$= 1$

Therefore, $\begin{vmatrix} \cos x & \sin x \\ -\sin x & \cos x \end{vmatrix} \equiv \begin{vmatrix} \cos x & -\sin x \\ \sin x & \cos x \end{vmatrix}$

$\equiv 1.$

Exercise Set 10.7

1. a_{11} is the element in the 1st row and 1st column. $a_{11} = 7$
   a_{32} is the element in the 3rd row and 2nd column. $a_{32} = 2$
   a_{22} is the element in the 2nd row and 2nd column. $a_{22} = 0$

2. $a_{13} = -6$, $a_{31} = 1$, $a_{23} = -3$

3. To find M_{11} we delete the 1st row and the 1st column.

$$A = \begin{bmatrix} 7 & -4 & -6 \\ 2 & 0 & -3 \\ 1 & 2 & -5 \end{bmatrix}$$

We calculate the determinant of the matrix formed by the remaining elements.

$$M_{11} = \begin{vmatrix} 0 & -3 \\ 2 & -5 \end{vmatrix} = 0(-5) - 2(-3) = 0 + 6 = 6$$

To find M_{32} we delete the 3rd row and the 2nd column.

$$A = \begin{bmatrix} 7 & -4 & -6 \\ 2 & 0 & -3 \\ 1 & 2 & -5 \end{bmatrix}$$

We calculate the determinant of the matrix formed by the remaining elements.

$$M_{32} = \begin{vmatrix} 7 & -6 \\ 2 & -3 \end{vmatrix} = 7(-3) - 2(-6) = -21 + 12 = -9$$

To find M_{22} we delete the 2nd row and the 2nd column.

$$A = \begin{bmatrix} 7 & -4 & -6 \\ 2 & 0 & -3 \\ 1 & 2 & -5 \end{bmatrix}$$

To calculate the determinant of the matrix formed by the remaining elements.

$$M_{22} = \begin{vmatrix} 7 & -6 \\ 1 & -5 \end{vmatrix} = 7(-5) - 1(-6) = -35 + 6 = -29.$$

4. $M_{13} = 4$, $M_{31} = 12$, $M_{23} = 18$

5. $A_{11} = (-1)^{1+1}\,M_{11} = (-1)^2 \cdot \begin{vmatrix} 0 & -3 \\ 2 & -5 \end{vmatrix} = 1 \cdot 6 = 6$

$A_{32} = (-1)^{3+2}\,M_{32} = (-1)^5 \cdot \begin{vmatrix} 7 & -6 \\ 2 & -3 \end{vmatrix} = -1 \cdot (-9) = 9$

$A_{22} = (-1)^{2+2}\,M_{22} = (-1)^4 \cdot \begin{vmatrix} 7 & -6 \\ 1 & -5 \end{vmatrix} = 1 \cdot (-29) = -29$

6. $A_{13} = 4, \quad A_{31} = 12, \quad A_{23} = -18$

7.

$$A = \begin{bmatrix} 7 & -4 & -6 \\ \boxed{2 \quad 0 \quad -3} \\ 1 & 2 & -5 \end{bmatrix}$$

Find $|A|$ by expanding about the second row.

$$|A| = (2)(-1)^{2+1} \cdot \begin{vmatrix} -4 & -6 \\ 2 & -5 \end{vmatrix} + (0)(-1)^{2+2} \cdot \begin{vmatrix} 7 & -6 \\ 1 & -5 \end{vmatrix} + (-3)(-1)^{2+3} \cdot \begin{vmatrix} 7 & -4 \\ 1 & 2 \end{vmatrix}$$

$= 2(-1)(32) + 0 + (-3)(-1)(18)$

$= -64 + 54 = -10$

8. $|A| = -10$

9.

$$A = \begin{bmatrix} 7 & -4 & \boxed{-6} \\ 2 & 0 & -3 \\ 1 & 2 & -5 \end{bmatrix}$$

Find $|A|$ by expanding about the third column.

$$|A| = (-6)(-1)^{1+3} \cdot \begin{vmatrix} 2 & 0 \\ 1 & 2 \end{vmatrix} + (-3)(-1)^{2+3} \cdot \begin{vmatrix} 7 & -4 \\ 1 & 2 \end{vmatrix} + (-5)(-1)^{3+3} \cdot \begin{vmatrix} 7 & -4 \\ 2 & 0 \end{vmatrix}$$

$= (-6)(1)(2 \cdot 2 - 1 \cdot 0) + (-3)(-1)[7 \cdot 2 - 1(-4)] + (-5)(1)[7 \cdot 0 - 2(-4)]$

$= (-6)(4) + (3)(18) + (-5)(8)$

$= -24 + 54 - 40 = -10$

10. $|A| = -10$

11. To find M_{41} we delete the 4th row and the 1st column.

$$A = \begin{bmatrix} 1 & 0 & 0 & -2 \\ 4 & 1 & 0 & 0 \\ 5 & 6 & 7 & 8 \\ -2 & -3 & -1 & 0 \end{bmatrix}$$

We calculate the determinant of the matrix formed by the remaining elements.

$$M_{41} = \begin{vmatrix} 0 & 0 & -2 \\ 1 & 0 & 0 \\ 6 & 7 & 8 \end{vmatrix} = 0 + 0 + (-2)(-1)^{1+3} \cdot \begin{vmatrix} 1 & 0 \\ 6 & 7 \end{vmatrix} = -2 \cdot 1 \cdot 7 = -14$$

11. (continued)

To find M_{33} we delete the 3rd row and the 3rd column.

$$A = \begin{bmatrix} 1 & 0 & 0 & -2 \\ 4 & 1 & 0 & 0 \\ 5 & 6 & 7 & 8 \\ -2 & -3 & -2 & 0 \end{bmatrix}$$

We calculate the determinant of the matrix formed by the remaining elements.

$$M_{33} = \begin{vmatrix} 1 & 0 & \boxed{-2} \\ 4 & 1 & \boxed{0} \\ -2 & -3 & \boxed{0} \end{vmatrix} = -2(-1)^{1+3} \cdot \begin{vmatrix} 4 & 1 \\ -2 & -3 \end{vmatrix} + 0 + 0 = -2 \cdot 1 \cdot (-10) = 20$$

12. $M_{12} = 32, \quad M_{44} = 7$

13.

$$A = \begin{bmatrix} 1 & 0 & 0 & -2 \\ 4 & 1 & 0 & 0 \\ 5 & 6 & 7 & 8 \\ -2 & -3 & -1 & 0 \end{bmatrix} \qquad (a_{24} \text{ is } 0; \quad a_{43} \text{ is } -1)$$

A_{24} is the cofactor of the element a_{24}.

$A_{24} = (-1)^{2+4} M_{24}$ (M_{24} is the minor of a_{24})

$$= (-1)^6 \begin{vmatrix} 1 & 0 & 0 \\ 5 & 6 & 7 \\ -2 & -3 & -1 \end{vmatrix} \qquad \text{Think: } A = \begin{bmatrix} 1 & 0 & 0 & -2 \\ 4 & 1 & 0 & 0 \\ 5 & 6 & 7 & 8 \\ -2 & -3 & -1 & 0 \end{bmatrix}$$

$$= (1)\left[1(-1)^{1+1} \cdot \begin{vmatrix} 6 & 7 \\ -3 & -1 \end{vmatrix} + 0(-1)^{1+2} \cdot \begin{vmatrix} 5 & 7 \\ -2 & -1 \end{vmatrix} + 0(-1)^{1+3} \cdot \begin{vmatrix} 5 & 6 \\ -2 & -3 \end{vmatrix}\right]$$

(Here we expanded about the first row.)

$= 1(1)[6(-1) - (-3)7] + 0 + 0 = -6 + 21 = 15$

A_{43} is the cofactor of the element a_{43}.

$A_{43} = (-1)^{4+3} M_{43}$ (M_{43} is the minor of a_{43})

$$= (-1)^7 \begin{vmatrix} 1 & 0 & -2 \\ 4 & 1 & 0 \\ 5 & 6 & 8 \end{vmatrix} \qquad \text{Think: } A = \begin{bmatrix} 1 & 0 & 0 & -2 \\ 4 & 1 & 0 & 0 \\ 5 & 6 & 7 & 8 \\ -2 & -3 & -1 & 0 \end{bmatrix}$$

$$= (-1)\left[4(-1)^{2+1} \cdot \begin{vmatrix} 0 & -2 \\ 6 & 8 \end{vmatrix} + 1(-1)^{2+2} \cdot \begin{vmatrix} 1 & -2 \\ 5 & 8 \end{vmatrix} + 0(-1)^{2+3} \cdot \begin{vmatrix} 1 & 0 \\ 5 & 6 \end{vmatrix}\right]$$

(Here we expanded about the second row.)

$= (-1)[4(-1)(0 \cdot 8 - 6(-2)) + 1(1 \cdot 8 - 5(-2)) + 0]$

$= (-1)[-4(12) + 18] = (-1)(-30) = 30$

14. $A_{22} = -10, \quad A_{34} = 1$

15.

$$A = \begin{bmatrix} \boxed{1 \quad 0 \quad 0 \quad -2} \\ 4 & 1 & 0 & 0 \\ 5 & 6 & 7 & 8 \\ -2 & -3 & -1 & 0 \end{bmatrix}$$

Find $|A|$ by expanding about the first row.

$$|A| = (1)(-1)^{1+1} \cdot \begin{vmatrix} \boxed{1 \quad 0 \quad 0} \\ 6 & 7 & 8 \\ -3 & -1 & 0 \end{vmatrix} + 0 + 0 + (-2)(-1)^{1+4} \cdot \begin{vmatrix} \boxed{4 \quad 1 \quad 0} \\ 5 & 6 & 7 \\ -2 & -3 & -1 \end{vmatrix}$$

$$= 1 \cdot [1(-1)^{1+1} \cdot \begin{vmatrix} 7 & 8 \\ -1 & 0 \end{vmatrix} + 0 + 0] + 2 \cdot [4(-1)^{1+1} \cdot \begin{vmatrix} 6 & 7 \\ -3 & -1 \end{vmatrix} + 1(-1)^{1+2} \cdot \begin{vmatrix} 5 & 7 \\ -2 & -1 \end{vmatrix} + 0]$$

$$= 1[1(0 - (-8))] + 2[4(-6 - (-21)) + (-1)(-5 - (-14))]$$

$$= 1 \cdot 8 + 2 \cdot (60 - 9) = 8 + 102 = 110$$

16. $|A| = 110$

17.

$$\begin{vmatrix} \boxed{5} & -4 & 2 & -2 \\ \boxed{3} & -3 & -4 & 7 \\ \boxed{-2} & 3 & 2 & 4 \\ \boxed{-8} & 9 & 5 & -5 \end{vmatrix}$$

(Here we will expand about the first column.)

$$= 5(-1)^{1+1} \cdot \begin{vmatrix} -3 & -4 & 7 \\ 3 & 2 & 4 \\ 9 & 5 & -5 \end{vmatrix} + 3(-1)^{2+1} \cdot \begin{vmatrix} -4 & 2 & -2 \\ 3 & 2 & 4 \\ 9 & 5 & -5 \end{vmatrix} +$$

$$(-2)(-1)^{3+1} \cdot \begin{vmatrix} -4 & 2 & -2 \\ -3 & -4 & 7 \\ 9 & 5 & -5 \end{vmatrix} + (-8)(-1)^{4+1} \cdot \begin{vmatrix} -4 & 2 & -2 \\ -3 & -4 & 7 \\ 3 & 2 & 4 \end{vmatrix}$$

$$= 5\left[-3(-1)^{1+1} \cdot \begin{vmatrix} 2 & 4 \\ 5 & -5 \end{vmatrix} + 3(-1)^{2+1} \cdot \begin{vmatrix} -4 & 7 \\ 5 & -5 \end{vmatrix} + 9(-1)^{3+1} \cdot \begin{vmatrix} -4 & 7 \\ 2 & 4 \end{vmatrix}\right] +$$

$$(-3)\left[-4(-1)^{1+1} \cdot \begin{vmatrix} 2 & 4 \\ 5 & -5 \end{vmatrix} + 3(-1)^{2+1} \cdot \begin{vmatrix} 2 & -2 \\ 5 & -5 \end{vmatrix} + 9(-1)^{3+1} \cdot \begin{vmatrix} 2 & -2 \\ 2 & 4 \end{vmatrix}\right] +$$

$$(-2)\left[-4(-1)^{1+1} \cdot \begin{vmatrix} -4 & 7 \\ 5 & -5 \end{vmatrix} + (-3)(-1)^{2+1} \cdot \begin{vmatrix} 2 & -2 \\ 5 & -5 \end{vmatrix} + 9(-1)^{3+1} \cdot \begin{vmatrix} 2 & -2 \\ -4 & 7 \end{vmatrix}\right] +$$

$$8\left[-4(-1)^{1+1} \cdot \begin{vmatrix} -4 & 7 \\ 2 & 4 \end{vmatrix} + (-3)(-1)^{2+1} \cdot \begin{vmatrix} 2 & -2 \\ 2 & 4 \end{vmatrix} + 3(-1)^{3+1} \cdot \begin{vmatrix} 2 & -2 \\ -4 & 7 \end{vmatrix}\right]$$

$$= 5[-3(-30) + (-3)(-15) + 9(-30)] + (-3)[-4(-30) + (-3)(0) + 9(12)] +$$
$$(-2)[-4(-15) + 3(0) + 9(6)] + 8[-4(-30) + 3(12) + 3(6)]$$

$$= 5(90 + 45 - 270) + (-3)(120 + 0 + 108) + (-2)(60 + 0 + 54) + 8(120 + 36 + 18)$$

$$= 5(-135) + (-3)(228) + (-2)(114) + 8(174) = -675 - 684 - 228 + 1392 = -195$$

18. xyzw

19. $\begin{vmatrix} x & y & 1 \\ x_1 & y_1 & 1 \\ x_2 & y_2 & 1 \end{vmatrix} = 0$

We first evaluate the determinant. Here we evaluate by expanding about the 1st row.

$$x(-1)^{1+1} \cdot \begin{vmatrix} y_1 & 1 \\ y_2 & 1 \end{vmatrix} + y(-1)^{1+2} \cdot \begin{vmatrix} x_1 & 1 \\ x_2 & 1 \end{vmatrix} + 1(-1)^{1+3} \cdot \begin{vmatrix} x_1 & y_1 \\ x_2 & y_2 \end{vmatrix} = 0$$

$$x(y_1 - y_2) - y(x_1 - x_2) + 1(x_1y_2 - x_2y_1) = 0$$

$$xy_1 - xy_2 - yx_1 + yx_2 + x_1y_2 - x_2y_1 = 0$$

The two-point equation of the line that contains the points (x_1, y_1) and (x_2, y_2) is

$$y - y_1 = \frac{y_2 - y_1}{x_2 - x_1}(x - x_1)$$

$(y - y_1)(x_2 - x_1) = (y_2 - y_1)(x - x_1)$ (Multiplying by $x_2 - x_1$)

$yx_2 - yx_1 - y_1x_2 + y_1x_1 = y_2x - y_2x_1 - y_1x + y_1x_1$ (Using FOIL)

$yx_2 - yx_1 - y_1x_2 + y_1x_1 - y_2x + y_2x_1 + y_1x - y_1x_1 = 0$

$xy_1 - xy_2 - yx_1 + yx_2 + x_1y_2 - x_2y_1 = 0$

Thus $\begin{vmatrix} x & y & 1 \\ x_1 & y_1 & 1 \\ x_2 & y_2 & 1 \end{vmatrix} = 0$ and $y - y_1 = \dfrac{y_2 - y_1}{x_2 - x_1}(x - x_1)$ are equivalent.

Exercise Set 10.8

1. $\begin{vmatrix} -4 & 5 \\ 6 & 10 \end{vmatrix}$

$= 5\begin{vmatrix} -4 & 1 \\ 6 & 2 \end{vmatrix}$ (Theorem 6, factoring 5 out of the second column)

$= 5\cdot 2\begin{vmatrix} -4 & 1 \\ 3 & 1 \end{vmatrix}$ (Theorem 6, factoring 2 out of the second row)

$= 5\cdot 2\begin{vmatrix} -4 & 1 \\ 7 & 0 \end{vmatrix}$ (Theorem 7, multiplying each element of the first row by -1 and adding the products to the corresponding elements of the second row)

$= 5\cdot 2(0 - 7)$

$= 10(-7) = -70$

2. -6

3. $$\begin{vmatrix} 2 & 1 & 1 \\ 2 & -3 & -1 \\ -4 & 5 & 2 \end{vmatrix}$$

$$= \begin{vmatrix} 2 & 1 & 1 \\ 4 & -2 & 0 \\ -8 & 3 & 0 \end{vmatrix}$$ (Theorem 7, adding the first row to the second row; also multiplying each element of the first row by -2 and adding the products to the corresponding elements of the third row)

$$= 1(-1)^{1+3}\begin{vmatrix} 4 & -2 \\ -8 & 3 \end{vmatrix} + 0 + 0$$ (Expanding the determinant about the third column)

$$= 1(12 - 16) = -4$$

4. -9

5. $$\begin{vmatrix} 11 & -15 & 20 \\ 16 & 24 & -8 \\ 6 & 9 & 15 \end{vmatrix}$$

$$= 8\cdot\begin{vmatrix} 11 & -15 & 20 \\ 2 & 3 & -1 \\ 6 & 9 & 15 \end{vmatrix} = 8\cdot 3\begin{vmatrix} 11 & -15 & 20 \\ 2 & 3 & -1 \\ 2 & 3 & 5 \end{vmatrix}$$ (Theorem 6, factoring 8 out of the second row and 3 out of the third row)

$$= 8\cdot 3\cdot 3\begin{vmatrix} 11 & -5 & 20 \\ 2 & 1 & -1 \\ 2 & 1 & 5 \end{vmatrix}$$ (Theorem 6, factoring 3 out of the second column)

$$= 8\cdot 3\cdot 3\begin{vmatrix} 11 & -5 & 20 \\ 2 & 1 & -1 \\ 0 & 0 & 6 \end{vmatrix}$$ (Theorem 7, multiplying each element of the second row by -1 and adding the products to the corresponding elements of the third row)

$$= 8\cdot 3\cdot 3\cdot 6\begin{vmatrix} 11 & -5 & 20 \\ 2 & 1 & -1 \\ 0 & 0 & 1 \end{vmatrix}$$ (Theorem 6, factoring 6 out of the third row)

$$= 8\cdot 3\cdot 3\cdot 6\left(0 + 0 + 1(-1)^{3+3}\begin{vmatrix} 11 & -5 \\ 2 & 1 \end{vmatrix}\right)$$ (Expanding the determinant about the third row)

$$= 8\cdot 3\cdot 3\cdot 6\cdot 21 = 9072$$

6. -546

7. $$\begin{vmatrix} -3 & 0 & 2 & 6 \\ 2 & 4 & 0 & -1 \\ -1 & 0 & -5 & 2 \\ 0 & -1 & -2 & -3 \end{vmatrix}$$

7. (continued)

$$= \begin{vmatrix} 0 & 0 & 17 & 0 \\ 2 & 4 & 0 & -1 \\ -1 & 0 & -5 & 2 \\ 0 & -1 & -2 & -3 \end{vmatrix}$$

(Theorem 7, multiplying each element of the third row by -3 and adding the products to the corresponding elements of the first row)

$$= 17\begin{vmatrix} 0 & 0 & 1 & 0 \\ 2 & 4 & 0 & -1 \\ -1 & 0 & -5 & 2 \\ 0 & -1 & -2 & -3 \end{vmatrix}$$

(Theorem 6, factoring 17 out of the first row)

$$= 17\left[0 + 0 + 1(-1)^{1+3}\begin{vmatrix} 2 & 4 & -1 \\ -1 & 0 & 2 \\ 0 & -1 & -3 \end{vmatrix} + 0\right]$$

(Expanding the determinant about the first row)

$$= -17\begin{vmatrix} 2 & 4 & -1 \\ 1 & 0 & -2 \\ 0 & -1 & -3 \end{vmatrix}$$

(Theorem 6, factoring -1 out of the second row)

$$= -17\begin{vmatrix} 0 & 4 & 3 \\ 1 & 0 & -2 \\ 0 & -1 & -3 \end{vmatrix}$$

(Theorem 7, multiplying each element of the second row by -2 and adding the products to the corresponding elements in the first row)

$$= -17\left[0 + 1(-1)^{2+1}\begin{vmatrix} 4 & 3 \\ -1 & -3 \end{vmatrix} + 0\right]$$

(Expanding the determinant about the first column)

$$= -17\cdot(-1)\cdot(-12 + 3) = 17(-9) = -153$$

8. 746

9. $$\begin{vmatrix} x & y & z \\ 0 & 0 & 0 \\ p & q & r \end{vmatrix} = 0$$

(Theorem 3)

10. 0

11. $$\begin{vmatrix} 2a & t & -7a \\ 2b & u & -7b \\ 2c & v & -7c \end{vmatrix}$$

$$= 2(-7)\begin{vmatrix} a & t & a \\ b & u & b \\ c & v & c \end{vmatrix}$$

(Theorem 6, factoring 2 out of the first column and -7 out of the third column)

$$= 2\cdot(-7)\cdot 0 = 0$$

(Theorem 5; columns 1 and 3 are the same)

12. 0

13. $$\begin{vmatrix} x^2 & x & 1 \\ y^2 & y & 1 \\ z^2 & z & 1 \end{vmatrix}$$

$$= \begin{vmatrix} x^2 - y^2 & x - y & 0 \\ y^2 & y & 1 \\ z^2 - y^2 & z - y & 0 \end{vmatrix}$$ (Theorem 7, multiplying each element of the second row by -1 and adding the products to the corresponding elements of the first and third rows)

$$= (x - y)(z - y)\begin{vmatrix} x + y & 1 & 0 \\ y^2 & y & 1 \\ z + y & 1 & 0 \end{vmatrix}$$ (Theorem 6, factoring x - y out of the first row and z - y out of the third row)

$$= (x - y)(z - y)\begin{vmatrix} x - z & 0 & 0 \\ y^2 & y & 1 \\ z + y & 1 & 0 \end{vmatrix}$$ (Theorem 7, multiplying each element of the third row by -1 and adding the products to the corresponding elements of the first row)

$$= (x - y)(z - y)(x - z)(-1)$$ (Expanding the determinant about the first row)

$$= (x - y)(z - y)(z - x)$$

14. $(a - b)(b - c)(c - a)$

15. $$\begin{vmatrix} x & x^2 & x^3 \\ y & y^2 & y^3 \\ z & z^2 & z^3 \end{vmatrix}$$

$$= xyz \cdot \begin{vmatrix} 1 & x & x^2 \\ 1 & y & y^2 \\ 1 & z & z^2 \end{vmatrix}$$ (Theorem 6, factoring x out of the first row, y out of the second row, and z out of the third row)

$$= xyz \cdot \begin{vmatrix} 0 & x - z & x^2 - z^2 \\ 0 & y - z & y^2 - z^2 \\ 1 & z & z^2 \end{vmatrix}$$ (Theorem 7, multiplying each element of the third row by -1 and adding the products to the corresponding elements of the first and second rows)

$$= xyz(x - z)(y - z) \cdot \begin{vmatrix} 0 & 1 & x + z \\ 0 & 1 & y + z \\ 1 & z & z^2 \end{vmatrix}$$ (Theorem 6, factoring x - z out of the first row and y - z out of the second row)

$$= xyz(x - z)(y - z) \cdot \begin{vmatrix} 0 & 0 & x - y \\ 0 & 1 & y + z \\ 1 & z & z^2 \end{vmatrix}$$ (Theorem 7, multiplying each element of the second row by -1 and adding the products to the corresponding elements of the first row)

$$= xyz(x - z)(y - z)(x - y) \cdot \begin{vmatrix} 0 & 0 & 1 \\ 0 & 1 & y + z \\ 1 & z & z^2 \end{vmatrix}$$ (Theorem 6, factoring x - y out of the first row)

$$= xyz(x - z)(y - z)(x - y) \cdot \left(0 + 0 + 1(-1)^{1+3}\begin{vmatrix} 0 & 1 \\ 1 & z \end{vmatrix}\right)$$ (Expanding the determinant about the first row)

$$= xyz(x - z)(y - z)(x - y)(-1) \qquad \left(\begin{vmatrix} 0 & 1 \\ 1 & z \end{vmatrix} = 0 - 1 = -1\right)$$

15. (continued)

$= -xyz(x - z)(y - z)(x - y)$

or $xyz(z - x)(y - z)(x - y)$ $[-1(x - y) = y - x]$

or $xyz(x - z)(z - y)(x - y)$ $[-1(y - z) = z - y]$

or $xyz(x - z)(y - z)(y - x)$ $[-1(x - y) = y - x]$

16. $(a - b)(b - c)(c - a)(a + b + c)$

Exercise Set 10.9

1. $A = \begin{bmatrix} 3 & 2 \\ 5 & 3 \end{bmatrix}$

a) Find the cofactor of each element.

$A_{11} = (-1)^{1+1}(3) = 3$ $A_{12} = (-1)^{1+2}(5) = -5$

$A_{21} = (-1)^{2+1}(2) = -2$ $A_{22} = (-1)^{2+2}(3) = 3$

b) Replace each element by its cofactor.

$$\begin{bmatrix} A_{11} & A_{12} \\ A_{21} & A_{22} \end{bmatrix} = \begin{bmatrix} 3 & -5 \\ -2 & 3 \end{bmatrix}$$

c) Find the transpose of the matrix found in b).

The transpose of $\begin{bmatrix} 3 & -5 \\ -2 & 3 \end{bmatrix}$ is $\begin{bmatrix} 3 & -2 \\ -5 & 3 \end{bmatrix}$.

d) Multiply the matrix in c) by $\frac{1}{|A|}$.

$|A| = 3\cdot3 - 5\cdot2 = 9 - 10 = -1$ $\frac{1}{|A|} = \frac{1}{-1} = -1$

$$A^{-1} = -1\cdot\begin{bmatrix} 3 & -2 \\ -5 & 3 \end{bmatrix} = \begin{bmatrix} -3 & 2 \\ 5 & -3 \end{bmatrix}$$

2. $A^{-1} = \begin{bmatrix} 2 & -5 \\ -1 & 3 \end{bmatrix}$

3. $A = \begin{bmatrix} 11 & 3 \\ 7 & 2 \end{bmatrix}$

a) Find the cofactor of each element.

$A_{11} = (-1)^{1+1}(2) = 2$ $A_{12} = (-1)^{1+2}(7) = -7$

$A_{21} = (-1)^{2+1}(3) = -3$ $A_{22} = (-1)^{2+2}(11) = 11$

b) Replace each element by its cofator.

$$\begin{bmatrix} A_{11} & A_{12} \\ A_{21} & A_{22} \end{bmatrix} = \begin{bmatrix} 2 & -7 \\ -3 & 11 \end{bmatrix}$$

c) Find the transpose of the matrix found in b).

3. (continued)

The transpose of $\begin{bmatrix} 2 & -7 \\ -3 & 11 \end{bmatrix}$ is $\begin{bmatrix} 2 & -3 \\ -7 & 11 \end{bmatrix}$.

d) Multiply the matrix in c) by $\frac{1}{|A|}$.

$|A| = 11 \cdot 2 - 7 \cdot 3 = 22 - 21 = 1$ $\qquad \frac{1}{|A|} = \frac{1}{1} = 1$

$$A^{-1} = 1 \cdot \begin{bmatrix} 2 & -3 \\ -7 & 11 \end{bmatrix} = \begin{bmatrix} 2 & -3 \\ -7 & 11 \end{bmatrix}$$

4.

$$A^{-1} = \begin{bmatrix} -3 & 5 \\ 5 & -8 \end{bmatrix}$$

5.

$$A = \begin{bmatrix} 4 & -3 \\ 1 & 2 \end{bmatrix}$$

a) Find the cofactor of each element.

$A_{11} = (-1)^{1+1}(2) = 2$ $\qquad A_{12} = (-1)^{1+2}(1) = -1$

$A_{21} = (-1)^{2+1}(-3) = 3$ $\qquad A_{22} = (-1)^{2+2}(4) = 4$

b) Replace each element by its cofactor.

$$\begin{bmatrix} A_{11} & A_{12} \\ A_{21} & A_{22} \end{bmatrix} = \begin{bmatrix} 2 & -1 \\ 3 & 4 \end{bmatrix}$$

c) Find the transpose of the matrix found in b).

The transpose of $\begin{bmatrix} 2 & -1 \\ 3 & 4 \end{bmatrix}$ is $\begin{bmatrix} 2 & 3 \\ -1 & 4 \end{bmatrix}$.

d) Multiply the matrix in c) by $\frac{1}{|A|}$.

$|A| = 4 \cdot 2 - 1(-3) = 11$ $\qquad \frac{1}{|A|} = \frac{1}{11}$

$$A^{-1} = \frac{1}{11}\begin{bmatrix} 2 & 3 \\ -1 & 4 \end{bmatrix} = \begin{bmatrix} \frac{2}{11} & \frac{3}{11} \\ -\frac{1}{11} & \frac{4}{11} \end{bmatrix}$$

6.

$$A^{-1} = \begin{bmatrix} 0 & 1 \\ -1 & 0 \end{bmatrix}$$

7.

$$A = \begin{bmatrix} 3 & 1 & 0 \\ 1 & 1 & 1 \\ 1 & -1 & 2 \end{bmatrix}$$

a) Find the cofactor of each element.

$A_{11} = (-1)^{1+1}(3) = 3$ $\qquad A_{12} = (-1)^{1+2}(1) = -1$ $\qquad A_{13} = (-1)^{1+3}(-2) = -2$

7. (continued)

$A_{21} = (-1)^{2+1}(2) = -2$ $A_{22} = (-1)^{2+2}(6) = 6$ $A_{23} = (-1)^{2+3}(-4) = 4$

$A_{31} = (-1)^{3+1}(1) = 1$ $A_{32} = (-1)^{3+2}(3) = -3$ $A_{33} = (-1)^{3+3}(2) = 2$

b) Replace each element by its cofactor.

$$\begin{bmatrix} A_{11} & A_{12} & A_{13} \\ A_{21} & A_{22} & A_{23} \\ A_{31} & A_{32} & A_{33} \end{bmatrix} = \begin{bmatrix} 3 & -1 & -2 \\ -2 & 6 & 4 \\ 1 & -3 & 2 \end{bmatrix}$$

c) Find the transpose of the matrix found in b).

The transpose of $\begin{bmatrix} 3 & -1 & -2 \\ -2 & 6 & 4 \\ 1 & -3 & 2 \end{bmatrix}$ is $\begin{bmatrix} 3 & -2 & 1 \\ -1 & 6 & -3 \\ -2 & 4 & 2 \end{bmatrix}$.

d) Multiply the matrix in c) by $\frac{1}{|A|}$.

$$|A| = 3(-1)^{1+1} \cdot \begin{vmatrix} 1 & 1 \\ -1 & 2 \end{vmatrix} + 1(-1)^{1+2} \cdot \begin{vmatrix} 1 & 1 \\ 1 & 2 \end{vmatrix} + 0$$

(Expanding the determinant about the first row)

$$= 3(3) + (-1)(1) = 8$$

$$\frac{1}{|A|} = \frac{1}{8}$$

$$A^{-1} = \frac{1}{8}\begin{bmatrix} 3 & -2 & 1 \\ -1 & 6 & -3 \\ -2 & 4 & 2 \end{bmatrix} = \begin{bmatrix} \frac{3}{8} & -\frac{1}{4} & \frac{1}{8} \\ -\frac{1}{8} & \frac{3}{4} & -\frac{3}{8} \\ -\frac{1}{4} & \frac{1}{2} & \frac{1}{4} \end{bmatrix}$$

8.

$$A^{-1} = \begin{bmatrix} -\frac{1}{2} & \frac{1}{2} & \frac{1}{2} \\ 1 & 0 & -1 \\ \frac{3}{2} & -\frac{1}{2} & -\frac{1}{2} \end{bmatrix}$$

9.

$$A = \begin{bmatrix} 1 & -1 & 2 \\ 0 & 1 & 3 \\ 2 & 1 & -2 \end{bmatrix}$$

a) Find the cofactor of each element.

$A_{11} = (-1)^{1+1}(-5) = -5$ $A_{12} = (-1)^{1+2}(-6) = 6$ $A_{13} = (-1)^{1+3}(-2) = -2$

$A_{21} = (-1)^{2+1}(0) = 0$ $A_{22} = (-1)^{2+2}(-6) = -6$ $A_{23} = (-1)^{2+3}(3) = -3$

$A_{31} = (-1)^{3+1}(-5) = -5$ $A_{32} = (-1)^{3+2}(3) = -3$ $A_{33} = (-1)^{3+3}(1) = 1$

b) Replace each element by its cofactor.

$$\begin{bmatrix} A_{11} & A_{12} & A_{13} \\ A_{21} & A_{22} & A_{23} \\ A_{31} & A_{32} & A_{33} \end{bmatrix} = \begin{bmatrix} -5 & 6 & -2 \\ 0 & -6 & -3 \\ -5 & -3 & 1 \end{bmatrix}$$

c) Find the transpose of the matrix found in b).

9. (continued)

The transpose of $\begin{bmatrix} -5 & 6 & -2 \\ 0 & -6 & -3 \\ -5 & -3 & 1 \end{bmatrix}$ is $\begin{bmatrix} -5 & 0 & -5 \\ 6 & -6 & -3 \\ -2 & -3 & 1 \end{bmatrix}$.

d) Multiply the matrix in c) by $\frac{1}{|A|}$.

$$|A| = 1(-1)^{1+1}\begin{vmatrix} 1 & 3 \\ 1 & -2 \end{vmatrix} + 0 + 2(-1)^{3+1}\begin{vmatrix} -1 & 2 \\ 1 & 3 \end{vmatrix}$$ (Expanding the determinant about the first column)

$$= 1(-5) + 2(-5) = -15$$

$$\frac{1}{|A|} = -\frac{1}{15}$$

$$A^{-1} = -\frac{1}{15}\begin{bmatrix} -5 & 0 & -5 \\ 6 & -6 & -3 \\ -2 & -3 & 1 \end{bmatrix} = \begin{bmatrix} \frac{1}{3} & 0 & \frac{1}{3} \\ -\frac{2}{5} & \frac{2}{5} & \frac{1}{5} \\ \frac{2}{15} & \frac{1}{5} & -\frac{1}{15} \end{bmatrix}$$

10.

$$A^{-1} = \begin{bmatrix} -1 & 5 & 2 \\ -1 & 3 & 1 \\ \frac{1}{2} & -1 & -\frac{1}{2} \end{bmatrix}$$

11.

$$A = \begin{bmatrix} 1 & -4 & 8 \\ 1 & -3 & 2 \\ 2 & -7 & 10 \end{bmatrix}$$

$$|A| = \begin{vmatrix} 1 & -4 & 8 \\ 1 & -3 & 2 \\ 2 & -7 & 10 \end{vmatrix} = \begin{vmatrix} 1 & -4 & 8 \\ 0 & 1 & -6 \\ 0 & 1 & -6 \end{vmatrix}$$ (Theorem 6; adding -1 times the first row to the second row and adding -2 times the first row to the third row)

$= 0$ (Theorem 4; the second and third rows are the same)

If $|A| = 0$, then $\frac{1}{|A|}$ is not defined and A^{-1} does not exist.

12. A^{-1} does not exist.

13.

$$A = \begin{bmatrix} 1 & 2 & 3 & 4 \\ 0 & 1 & 3 & -5 \\ 0 & 0 & 1 & -2 \\ 0 & 0 & 0 & -1 \end{bmatrix}$$

a) Find the cofactor of each element.

$$A_{11} = (-1)^{1+1} \cdot \begin{vmatrix} 1 & 3 & -5 \\ 0 & 1 & -2 \\ 0 & 0 & -1 \end{vmatrix} = 1[1(-1)^{1+1} \cdot \begin{vmatrix} 1 & -2 \\ 0 & -1 \end{vmatrix} + 0 + 0] = -1$$

13. (continued)

$A_{12} = 0$ $A_{13} = 0$ $A_{14} = 0$ $A_{21} = 2$ $A_{22} = -1$ $A_{23} = 0$
$A_{24} = 0$ $A_{31} = -3$ $A_{32} = 3$ $A_{33} = -1$ $A_{34} = 0$ $A_{41} = -8$
$A_{42} = -1$ $A_{43} = 2$ $A_{44} = 1$

b) Replace each element by its cofactor.

$$\begin{bmatrix} A_{11} & A_{12} & A_{13} & A_{14} \\ A_{21} & A_{22} & A_{23} & A_{24} \\ A_{31} & A_{32} & A_{33} & A_{34} \\ A_{41} & A_{42} & A_{43} & A_{44} \end{bmatrix} = \begin{bmatrix} -1 & 0 & 0 & 0 \\ 2 & -1 & 0 & 0 \\ -3 & 3 & -1 & 0 \\ -8 & -1 & 2 & 1 \end{bmatrix}$$

c) Find the transpose of the matrix found in b).

The transpose of $\begin{bmatrix} -1 & 0 & 0 & 0 \\ 2 & -1 & 0 & 0 \\ -3 & 3 & -1 & 0 \\ -8 & -1 & 2 & 1 \end{bmatrix}$ is $\begin{bmatrix} -1 & 2 & -3 & -8 \\ 0 & -1 & 3 & -1 \\ 0 & 0 & -1 & 2 \\ 0 & 0 & 0 & 1 \end{bmatrix}$.

d) Multiply the matrix in c) by $\frac{1}{|A|}$.

$$|A| = 1\cdot(-1)^{1+1}\cdot \begin{vmatrix} 1 & 3 & -5 \\ 0 & 1 & -2 \\ 0 & 0 & -1 \end{vmatrix} = 1(-1)^{1+1}\begin{vmatrix} 1 & -2 \\ 0 & -1 \end{vmatrix} = -1 \qquad \frac{1}{|A|} = \frac{1}{-1} = -1$$

$$A^{-1} = -1\cdot\begin{bmatrix} -1 & 2 & -3 & -8 \\ 0 & -1 & 3 & -1 \\ 0 & 0 & -1 & 2 \\ 0 & 0 & 0 & 1 \end{bmatrix} = \begin{bmatrix} 1 & -2 & 3 & 8 \\ 0 & 1 & -3 & 1 \\ 0 & 0 & 1 & -2 \\ 0 & 0 & 0 & -1 \end{bmatrix}$$

14. A^{-1} does not exist.

15.

$$A = \begin{bmatrix} 3 & 2 \\ 5 & 3 \end{bmatrix}$$

$$\begin{bmatrix} 3 & 2 & 1 & 0 \\ 5 & 3 & 0 & 1 \end{bmatrix} = \begin{bmatrix} 3 & 2 & 1 & 0 \\ 15 & 9 & 0 & 3 \end{bmatrix} = \begin{bmatrix} 3 & 2 & 1 & 0 \\ 0 & -1 & -5 & 3 \end{bmatrix} =$$

$$\begin{bmatrix} 3 & 2 & 1 & 0 \\ 0 & 1 & 5 & -3 \end{bmatrix} = \begin{bmatrix} 3 & 0 & -9 & 6 \\ 0 & 1 & 5 & -3 \end{bmatrix} = \begin{bmatrix} 1 & 0 & -3 & 2 \\ 0 & 1 & 5 & -3 \end{bmatrix}$$

$$A^{-1} = \begin{bmatrix} -3 & 2 \\ 5 & -3 \end{bmatrix}$$

16.

$$A^{-1} = \begin{bmatrix} 2 & -5 \\ -1 & 3 \end{bmatrix}$$

17.
$$A = \begin{bmatrix} 11 & 3 \\ 7 & 2 \end{bmatrix}$$

$$\begin{bmatrix} 11 & 3 & 1 & 0 \\ 7 & 2 & 0 & 1 \end{bmatrix} = \begin{bmatrix} 7 & 2 & 0 & 1 \\ 11 & 3 & 1 & 0 \end{bmatrix} = \begin{bmatrix} 7 & 2 & 0 & 1 \\ 77 & 21 & 7 & 0 \end{bmatrix} =$$

$$\begin{bmatrix} 7 & 2 & 0 & 1 \\ 0 & -1 & 7 & -11 \end{bmatrix} = \begin{bmatrix} 7 & 2 & 0 & 1 \\ 0 & 1 & -7 & 11 \end{bmatrix} = \begin{bmatrix} 7 & 0 & 14 & -21 \\ 0 & 1 & -7 & 11 \end{bmatrix} =$$

$$\begin{bmatrix} 1 & 0 & 2 & -3 \\ 0 & 1 & -7 & 11 \end{bmatrix}$$

$$A^{-1} = \begin{bmatrix} 2 & -3 \\ -7 & 11 \end{bmatrix}$$

18.
$$A^{-1} = \begin{bmatrix} -3 & 5 \\ 5 & -8 \end{bmatrix}$$

19.
$$A = \begin{bmatrix} 4 & -3 \\ 1 & 2 \end{bmatrix}$$

$$\begin{bmatrix} 4 & -3 & 1 & 0 \\ 1 & 2 & 0 & 1 \end{bmatrix} = \begin{bmatrix} 1 & 2 & 0 & 1 \\ 4 & -3 & 1 & 0 \end{bmatrix} = \begin{bmatrix} 1 & 2 & 0 & 1 \\ 0 & -11 & 1 & -4 \end{bmatrix} =$$

$$\begin{bmatrix} 1 & 2 & 0 & 1 \\ 0 & 11 & -1 & 4 \end{bmatrix} = \begin{bmatrix} 11 & 22 & 0 & 11 \\ 0 & 11 & -1 & 4 \end{bmatrix} = \begin{bmatrix} 11 & 0 & 2 & 3 \\ 0 & 11 & -1 & 4 \end{bmatrix} =$$

$$\begin{bmatrix} 1 & 0 & \frac{2}{11} & \frac{3}{11} \\ 0 & 1 & -\frac{1}{11} & \frac{4}{11} \end{bmatrix}$$

$$A^{-1} = \begin{bmatrix} \frac{2}{11} & \frac{3}{11} \\ -\frac{1}{11} & \frac{4}{11} \end{bmatrix}$$

20.
$$A^{-1} = \begin{bmatrix} 0 & 1 \\ -1 & 0 \end{bmatrix}$$

21.

$$A = \begin{bmatrix} 3 & 1 & 0 \\ 1 & 1 & 1 \\ 1 & -1 & 2 \end{bmatrix}$$

$$\left[\begin{array}{ccc|ccc} 3 & 1 & 0 & 1 & 0 & 0 \\ 1 & 1 & 1 & 0 & 1 & 0 \\ 1 & -1 & 2 & 0 & 0 & 1 \end{array}\right] = \left[\begin{array}{ccc|ccc} 1 & 1 & 1 & 0 & 1 & 0 \\ 3 & 1 & 0 & 1 & 0 & 0 \\ 1 & -1 & 2 & 0 & 0 & 1 \end{array}\right] =$$

$$\left[\begin{array}{ccc|ccc} 1 & 1 & 1 & 0 & 1 & 0 \\ 0 & -2 & -3 & 1 & -3 & 0 \\ 0 & -2 & 1 & 0 & -1 & 1 \end{array}\right] = \left[\begin{array}{ccc|ccc} 1 & 1 & 1 & 0 & 1 & 0 \\ 0 & 2 & 3 & -1 & 3 & 0 \\ 0 & -2 & 1 & 0 & -1 & 1 \end{array}\right] =$$

$$\left[\begin{array}{ccc|ccc} 1 & 1 & 1 & 0 & 1 & 0 \\ 0 & 2 & 3 & -1 & 3 & 0 \\ 0 & 0 & 4 & -1 & 2 & 1 \end{array}\right] = \left[\begin{array}{ccc|ccc} 4 & 4 & 4 & 0 & 4 & 0 \\ 0 & 8 & 12 & -4 & 12 & 0 \\ 0 & 0 & 4 & -1 & 2 & 1 \end{array}\right] =$$

$$\left[\begin{array}{ccc|ccc} 4 & 4 & 0 & 1 & 2 & -1 \\ 0 & 8 & 0 & -1 & 6 & -3 \\ 0 & 0 & 4 & -1 & 2 & 1 \end{array}\right] = \left[\begin{array}{ccc|ccc} 8 & 8 & 0 & 2 & 4 & -2 \\ 0 & 8 & 0 & -1 & 6 & -3 \\ 0 & 0 & 4 & -1 & 2 & 1 \end{array}\right] =$$

$$\left[\begin{array}{ccc|ccc} 8 & 0 & 0 & 3 & -2 & 1 \\ 0 & 8 & 0 & -1 & 6 & -3 \\ 0 & 0 & 4 & -1 & 2 & 1 \end{array}\right] = \left[\begin{array}{ccc|ccc} 1 & 0 & 0 & \frac{3}{8} & -\frac{1}{4} & \frac{1}{8} \\ 0 & 1 & 0 & -\frac{1}{8} & \frac{3}{4} & -\frac{3}{8} \\ 0 & 0 & 1 & -\frac{1}{4} & \frac{1}{2} & \frac{1}{4} \end{array}\right]$$

$$A^{-1} = \begin{bmatrix} \frac{3}{8} & -\frac{1}{4} & \frac{1}{8} \\ -\frac{1}{8} & \frac{3}{4} & -\frac{3}{8} \\ -\frac{1}{4} & \frac{1}{2} & \frac{1}{4} \end{bmatrix}$$

22.

$$A^{-1} = \begin{bmatrix} -\frac{1}{2} & \frac{1}{2} & \frac{1}{2} \\ 1 & 0 & -1 \\ \frac{3}{2} & -\frac{1}{2} & -\frac{1}{2} \end{bmatrix}$$

23.

$$A = \begin{bmatrix} 1 & -1 & 2 \\ 0 & 1 & 3 \\ 2 & 1 & -2 \end{bmatrix}$$

$$\left[\begin{array}{ccc|ccc} 1 & -1 & 2 & 1 & 0 & 0 \\ 0 & 1 & 3 & 0 & 1 & 0 \\ 2 & 1 & -2 & 0 & 0 & 1 \end{array}\right] = \left[\begin{array}{ccc|ccc} 1 & -1 & 2 & 1 & 0 & 0 \\ 0 & 1 & 3 & 0 & 1 & 0 \\ 0 & 3 & -6 & -2 & 0 & 1 \end{array}\right] =$$

23. (continued)

$$\begin{bmatrix} 1 & -1 & 2 & 1 & 0 & 0 \\ 0 & 1 & 3 & 0 & 1 & 0 \\ 0 & 0 & -15 & -2 & -3 & 1 \end{bmatrix} = \begin{bmatrix} 15 & -15 & 30 & 15 & 0 & 0 \\ 0 & 5 & 15 & 0 & 5 & 0 \\ 0 & 0 & 15 & 2 & 3 & -1 \end{bmatrix} =$$

$$\begin{bmatrix} 15 & -15 & 0 & 11 & -6 & 2 \\ 0 & 5 & 0 & -2 & 2 & 1 \\ 0 & 0 & 15 & 2 & 3 & -1 \end{bmatrix} = \begin{bmatrix} 15 & 0 & 0 & 5 & 0 & 5 \\ 0 & 5 & 0 & -2 & 2 & 1 \\ 0 & 0 & 15 & 2 & 3 & -1 \end{bmatrix} =$$

$$\begin{bmatrix} 1 & 0 & 0 & \frac{1}{3} & 0 & \frac{1}{3} \\ 0 & 1 & 0 & -\frac{2}{5} & \frac{2}{5} & \frac{1}{5} \\ 0 & 0 & 1 & \frac{2}{15} & \frac{3}{15} & -\frac{1}{15} \end{bmatrix}$$

$$A^{-1} = \begin{bmatrix} \frac{1}{3} & 0 & \frac{1}{3} \\ -\frac{2}{5} & \frac{2}{5} & \frac{1}{5} \\ \frac{2}{15} & \frac{1}{5} & -\frac{1}{15} \end{bmatrix}$$

24. $$A^{-1} = \begin{bmatrix} -1 & 5 & 2 \\ -1 & 3 & 1 \\ \frac{1}{2} & -1 & -\frac{1}{2} \end{bmatrix}$$

25. We cannot obtain the identity matrix on the left using the Gauss-Jordan reduction method. Thus, A^{-1} does not exist. Also see Exercise 11.

26. A^{-1} does not exist.

27. $$A = \begin{bmatrix} 1 & 2 & 3 & 4 \\ 0 & 1 & 3 & -5 \\ 0 & 0 & 1 & -2 \\ 0 & 0 & 0 & -1 \end{bmatrix}$$

$$\begin{bmatrix} 1 & 2 & 3 & 4 & 1 & 0 & 0 & 0 \\ 0 & 1 & 3 & -5 & 0 & 1 & 0 & 0 \\ 0 & 0 & 1 & -2 & 0 & 0 & 1 & 0 \\ 0 & 0 & 0 & -1 & 0 & 0 & 0 & 1 \end{bmatrix} = \begin{bmatrix} 1 & 2 & 3 & 4 & 1 & 0 & 0 & 0 \\ 0 & 1 & 3 & -5 & 0 & 1 & 0 & 0 \\ 0 & 0 & 1 & -2 & 0 & 0 & 1 & 0 \\ 0 & 0 & 0 & 1 & 0 & 0 & 0 & -1 \end{bmatrix} =$$

$$\begin{bmatrix} 1 & 2 & 3 & 0 & 1 & 0 & 0 & 4 \\ 0 & 1 & 3 & 0 & 0 & 1 & 0 & -5 \\ 0 & 0 & 1 & 0 & 0 & 0 & 1 & -2 \\ 0 & 0 & 0 & 1 & 0 & 0 & 0 & -1 \end{bmatrix} = \begin{bmatrix} 1 & 2 & 0 & 0 & 1 & 0 & -3 & 10 \\ 0 & 1 & 0 & 0 & 0 & 1 & -3 & 1 \\ 0 & 0 & 1 & 0 & 0 & 0 & 1 & -2 \\ 0 & 0 & 0 & 1 & 0 & 0 & 0 & -1 \end{bmatrix} =$$

27. (continued)

$$\begin{bmatrix} 1 & 0 & 0 & 0 & 1 & -2 & 3 & 8 \\ 0 & 1 & 0 & 0 & 0 & 1 & -3 & 1 \\ 0 & 0 & 1 & 0 & 0 & 0 & 1 & -2 \\ 0 & 0 & 0 & 1 & 0 & 0 & 0 & -1 \end{bmatrix}$$

$$A^{-1} = \begin{bmatrix} 1 & -2 & 3 & 8 \\ 0 & 1 & -3 & 1 \\ 0 & 0 & 1 & -2 \\ 0 & 0 & 0 & -1 \end{bmatrix}$$

28. A^{-1} does not exist.

29. $7x - 2y = -3$
$9x + 3y = 4$

$$\begin{bmatrix} 7 & -2 \\ 9 & 3 \end{bmatrix} \begin{bmatrix} x \\ y \end{bmatrix} = \begin{bmatrix} -3 \\ 4 \end{bmatrix}$$

The inverse of the coefficient matrix, $\begin{bmatrix} 7 & -2 \\ 9 & 3 \end{bmatrix}$, is $\begin{bmatrix} \frac{1}{13} & \frac{2}{39} \\ -\frac{3}{13} & \frac{7}{39} \end{bmatrix}$

Multiply both sides of the matrix equation by the inverse.

$$\begin{bmatrix} \frac{1}{13} & \frac{2}{39} \\ -\frac{3}{13} & \frac{7}{39} \end{bmatrix} \begin{bmatrix} 7 & -2 \\ 9 & 3 \end{bmatrix} \begin{bmatrix} x \\ y \end{bmatrix} = \begin{bmatrix} \frac{1}{13} & \frac{2}{39} \\ -\frac{3}{13} & \frac{7}{39} \end{bmatrix} \begin{bmatrix} -3 \\ 4 \end{bmatrix}$$

$$\begin{bmatrix} 1 & 0 \\ 0 & 1 \end{bmatrix} \begin{bmatrix} x \\ y \end{bmatrix} = \begin{bmatrix} -\frac{1}{39} \\ \frac{55}{39} \end{bmatrix}$$

$$\begin{bmatrix} x \\ y \end{bmatrix} = \begin{bmatrix} -\frac{1}{39} \\ \frac{55}{39} \end{bmatrix}$$

The solution of the system of equations is $x = -\frac{1}{39}$ and $y = \frac{55}{39}$.

30. $\left(\frac{1}{17}, -\frac{13}{17}\right)$

31. $x_1 + x_3 = 1$
$2x_1 + x_2 = 3$
$x_1 - x_2 + x_3 = 4$

$$\begin{bmatrix} 1 & 0 & 1 \\ 2 & 1 & 0 \\ 1 & -1 & 1 \end{bmatrix} \begin{bmatrix} x_1 \\ x_2 \\ x_3 \end{bmatrix} = \begin{bmatrix} 1 \\ 3 \\ 4 \end{bmatrix}$$

31. (continued)

The inverse of the coefficient matrix $\begin{bmatrix} 1 & 0 & 1 \\ 2 & 1 & 0 \\ 1 & -1 & 1 \end{bmatrix}$ is $\begin{bmatrix} -\frac{1}{2} & \frac{1}{2} & \frac{1}{2} \\ 1 & 0 & -1 \\ \frac{3}{2} & -\frac{1}{2} & -\frac{1}{2} \end{bmatrix}$

Multiply both sides of the matrix equation by the inverse.

$$\begin{bmatrix} -\frac{1}{2} & \frac{1}{2} & \frac{1}{2} \\ 1 & 0 & -1 \\ \frac{3}{2} & -\frac{1}{2} & -\frac{1}{2} \end{bmatrix}\begin{bmatrix} 1 & 0 & 1 \\ 2 & 1 & 0 \\ 1 & -1 & 1 \end{bmatrix}\begin{bmatrix} x_1 \\ x_2 \\ x_3 \end{bmatrix} = \begin{bmatrix} -\frac{1}{2} & \frac{1}{2} & \frac{1}{2} \\ 1 & 0 & -1 \\ \frac{3}{2} & -\frac{1}{2} & -\frac{1}{2} \end{bmatrix}\begin{bmatrix} 1 \\ 3 \\ 4 \end{bmatrix}$$

$$\begin{bmatrix} 1 & 0 & 0 \\ 0 & 1 & 0 \\ 0 & 0 & 1 \end{bmatrix}\begin{bmatrix} x_1 \\ x_2 \\ x_3 \end{bmatrix} = \begin{bmatrix} 3 \\ -3 \\ -2 \end{bmatrix}$$

$$\begin{bmatrix} x_1 \\ x_2 \\ x_3 \end{bmatrix} = \begin{bmatrix} 3 \\ -3 \\ -2 \end{bmatrix}$$

The solution of the system of equations is $x_1 = 3$, $x_2 = -3$, and $x_3 = -2$.

32. (124, 14, -51)

33. $$\begin{bmatrix} a & b & c \\ d & e & f \\ g & h & i \end{bmatrix}\begin{bmatrix} 1 & 0 & 0 \\ 0 & 1 & 0 \\ 0 & 0 & 1 \end{bmatrix} = \begin{bmatrix} a & b & c \\ d & e & f \\ g & h & i \end{bmatrix}, \quad AI = A$$

$$\begin{bmatrix} 1 & 0 & 0 \\ 0 & 1 & 0 \\ 0 & 0 & 1 \end{bmatrix}\begin{bmatrix} a & b & c \\ d & e & f \\ g & h & i \end{bmatrix} = \begin{bmatrix} a & b & c \\ d & e & f \\ g & h & i \end{bmatrix} \quad IA = A$$

34. A^{-1} exists if and only if $x \neq 0$.

$A^{-1} = \begin{bmatrix} \frac{1}{x} \end{bmatrix}$

35. $A = \begin{bmatrix} x & 0 \\ 0 & y \end{bmatrix}$

a), b) Find the cofactor of each element and replace each element by its cofactor.

$$\begin{bmatrix} y & 0 \\ 0 & x \end{bmatrix}$$

35. (continued)

c) Find the transpose of the matrix found in a) - b).

The transpose of $\begin{bmatrix} y & 0 \\ 0 & x \end{bmatrix}$ is $\begin{bmatrix} y & 0 \\ 0 & x \end{bmatrix}$.

d) Multiply the matrix in c) by $\frac{1}{|A|}$.

$|A| = x \cdot y - 0 \cdot 0 = xy, \ \frac{1}{|A|} = \frac{1}{xy}$

$$A^{-1} = \frac{1}{xy} \cdot \begin{bmatrix} y & 0 \\ 0 & x \end{bmatrix} \begin{bmatrix} \frac{1}{x} & 0 \\ 0 & \frac{1}{y} \end{bmatrix}$$

A^{-1} exists if and only if $xy \neq 0$.

36. A^{-1} exists if and only if $xyz \neq 0$.

$$A^{-1} = \begin{bmatrix} 0 & 0 & \frac{1}{z} \\ 0 & \frac{1}{y} & 0 \\ \frac{1}{x} & 0 & 0 \end{bmatrix}$$

37.

$$A = \begin{bmatrix} x & 1 & 1 & 1 \\ 0 & y & 0 & 0 \\ 0 & 0 & z & 0 \\ 0 & 0 & 0 & w \end{bmatrix}$$

a), b) Find the cofactor of each element and replace each element by its cofactor.

$$\begin{bmatrix} yzw & 0 & 0 & 0 \\ -zw & xzw & 0 & 0 \\ -yw & 0 & xyw & 0 \\ -yz & 0 & 0 & xyz \end{bmatrix}$$

c) Find the transpose of the matrix in a) - b).

$$\begin{bmatrix} yzw & -zw & -yw & -yz \\ 0 & xzw & 0 & 0 \\ 0 & 0 & xyw & 0 \\ 0 & 0 & 0 & xyz \end{bmatrix}$$

d) Multiply the matrix in c) by $\frac{1}{|A|}$.

$|A| = x \cdot y \cdot z \cdot w = xyzw$ (Expanding about the 1st column)

37. (continued)

$$A^{-1} = \frac{1}{xyzw} \cdot \begin{bmatrix} yzw & -zw & -yw & -yz \\ 0 & xzw & 0 & 0 \\ 0 & 0 & xyw & 0 \\ 0 & 0 & 0 & xyz \end{bmatrix} = \begin{bmatrix} \frac{1}{x} & -\frac{1}{xy} & -\frac{1}{xz} & -\frac{1}{xw} \\ 0 & \frac{1}{y} & 0 & 0 \\ 0 & 0 & \frac{1}{z} & 0 \\ 0 & 0 & 0 & \frac{1}{w} \end{bmatrix}$$

A^{-1} exists if and only if $xyzw \neq 0$.

Exercise Set 10.10

1. Graph: $y < x$

First graph the equation $y = x$. A few solutions of this equation are (-4, -4), (0, 0), and (2, 2). Plot these points and draw the line dashed since the inequality is < and not ⩽.

Next determine which half-plane to shade by trying some point off the line. Here we use (-3, 4) as a check.

y < x	
4	-3

Since 4 < -3 is false, the point is not on the graph. We shade the half-plane which does not contain (-3, 4).

$y < x$

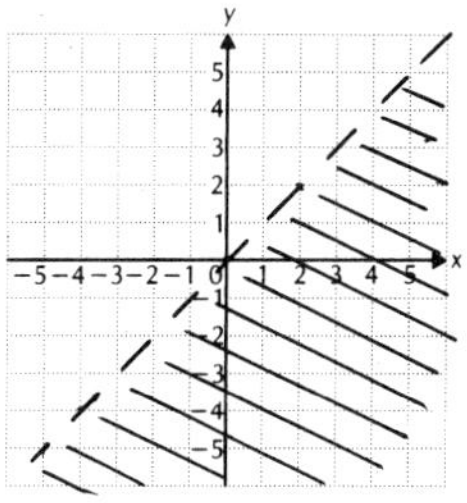

2. $y \geqslant x$

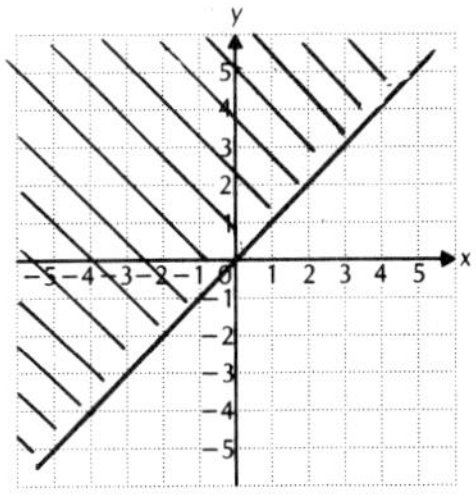

3. Graph: $y + x \geqslant 0$

First graph the equation $y + x = 0$. A few solutions of this equation are (-2, 2), (0, 0), and (3, -3). Plot these points and draw the line solid since the inequality is ⩾.

Next determine which half-plane to shade by trying some point off the line. Here we use (2, 2) as a check.

3. (continued)

y + x ⩾ 0	
2 + 2	0
4	

Since 4 ⩾ 0 is true, (2, 2) is a solution. Thus shade the half-plane containing (2, 2).

$y + x \geqslant 0$

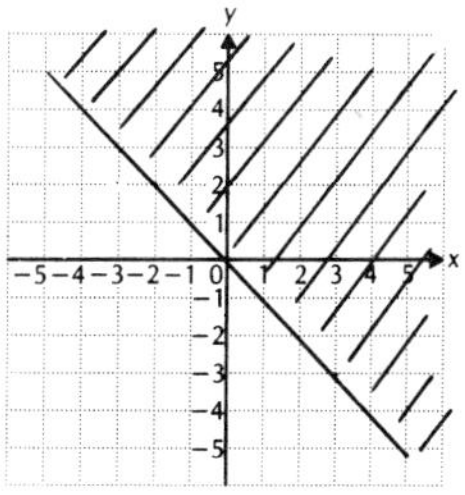

4. $y + x < 0$

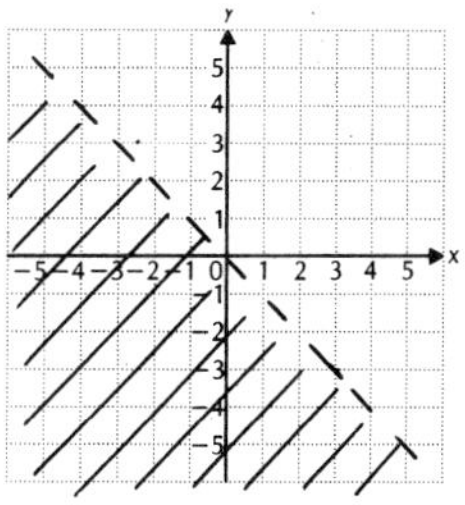

5. Graph: $3x - 2y < 6$

First graph the equation $3x - 2y = 6$. The y-intercept is (0, -3), and the x-intercept is (2, 0). Plot these points and draw the line dashed since the inequality is < and not ⩽.

Next complete the graph by shading the correct half-plane. Since the line does not contain the origin, check the point (0, 0) in the inequality to see if we get a true sentence.

3x - 2y < 6	
3(0) - 2(0)	6
0	6

Since 0 < 6 is a true sentence, (0, 0) is a solution. Thus shade the half-plane containing (0, 0).

5. (continued)

$3x - 2y < 6$

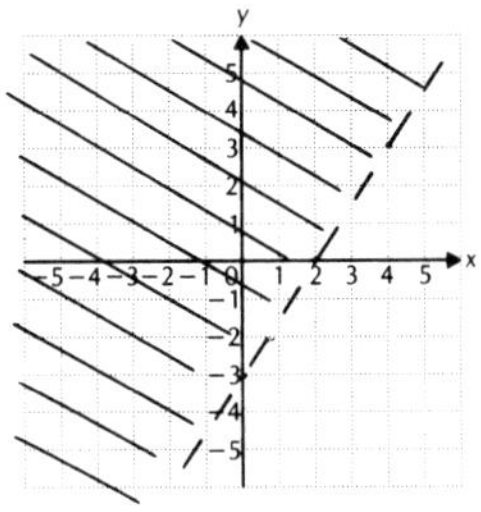

6. $2x - 5y > 10$

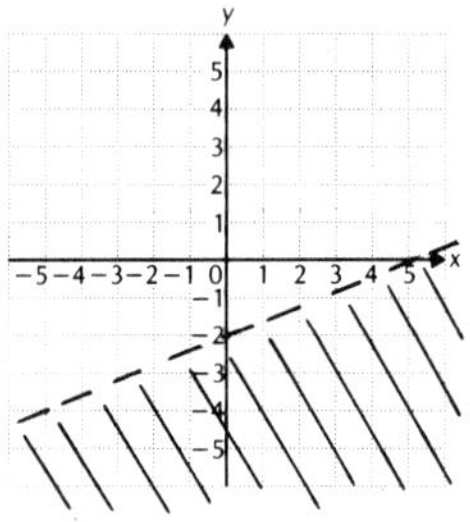

7. Graph: $2x + 3y \geqslant 6$

First graph the equation $2x + 3y = 6$. The y-intercept is (0, 2), and the x-intercept is (3, 0). Plot these points and draw the line solid since the inequality is $\geqslant$.

Next complete the graph by shading the correct half-plane. Since the line does not contain the origin, check the point (0, 0) in the inequality to see if we get a true sentence.

$2x + 3y \geqslant 6$	
$2 \cdot 0 + 3 \cdot 0$	6
0	

Since $0 \geqslant 6$ is false, (0, 0) is not a solution. We shade the half-plane which does not contain (0, 0).

$2x + 3y \geqslant 6$

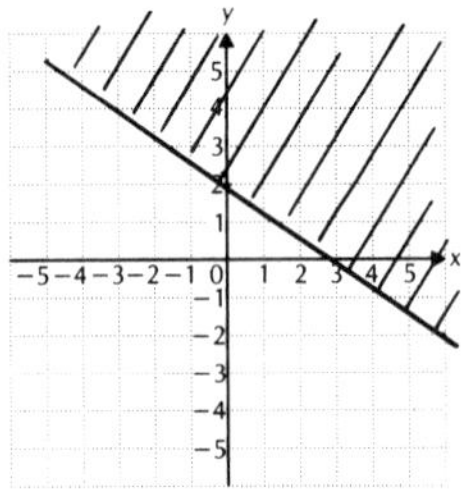

8. $x + 2y \leqslant 4$

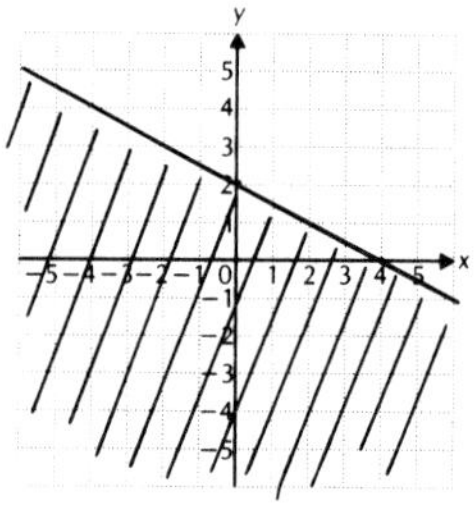

9. Graph: $3x - 2 \leqslant 5x + y$

$-2 \leqslant 2x + y$ (Adding $-3x$)

First graph $2x + y = -2$. The y-intercept is (0, -2), and the x-intercept is (-1, 0). Plot these points and draw the line solid since the inequality is $\leqslant$.

Next complete the graph by shading the correct half-plane. Since the line does not contain the origin, check the point (0, 0).

$2x + y \geqslant -2$	
$2(0) + 0$	-2
0	-2

Since $0 \geqslant -2$ is a true sentence, (0, 0) is a solution. Thus shade the half-plane containing the origin.

$3x - 2 \leqslant 5x + y$, or $2x + y \geqslant -2$

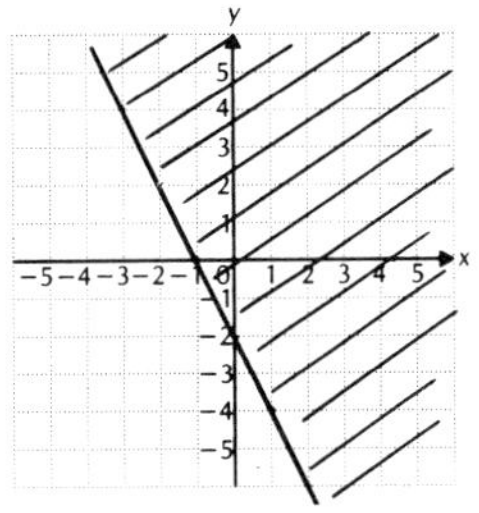

10. $2x - 6y \geqslant 8 + 4y$, or $2x - 10y \geqslant 8$

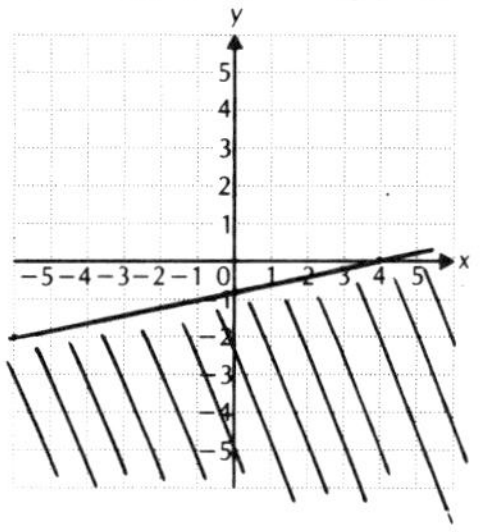

11. Graph: $x < -4$

We first graph the equation $x = -4$. The line is parallel to the y-axis with x-intercept (-4, 0). We draw the line dashed since the inequality is $<$ and not $\leqslant$.

Next complete the graph by shading the correct half-plane. Since the line does not contain the origin, check the point (0, 0).

$x < -4$	
0	-4

(Substituting 0 for x)

Since $0 < -4$ is false, (0, 0) is not a solution. Thus, we shade the half-plane which does not contain the origin.

$x < -4$

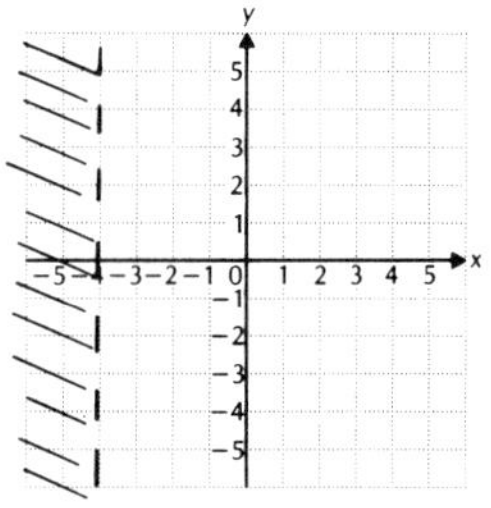

12. $y \geqslant 5$

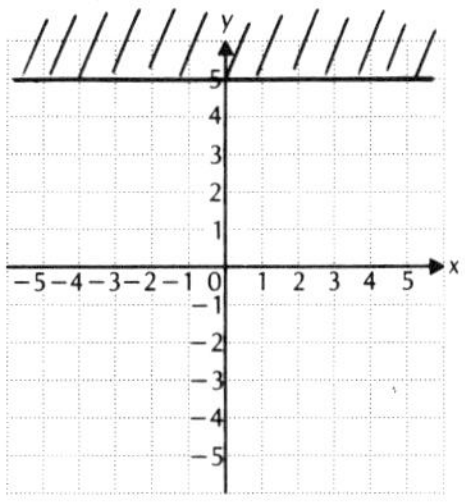

13. Graph: $0 \leqslant x < 5\frac{1}{2}$

This is the conjunction of two inequalities:

$0 \leqslant x$ and $x < 5\frac{1}{2}$

Graph $0 \leqslant x$ and $x < 5\frac{1}{2}$ separately and then graph the intersection.

$0 \leqslant x$

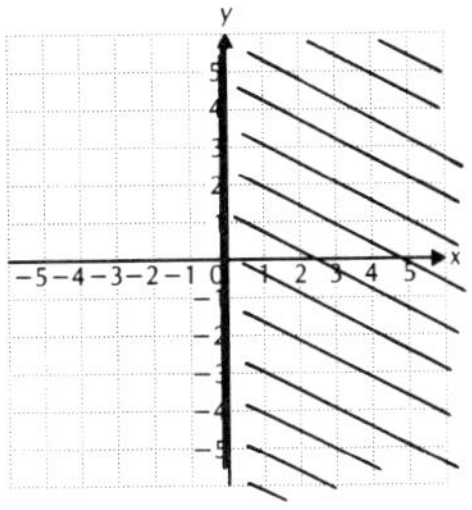

$x < 5\frac{1}{2}$

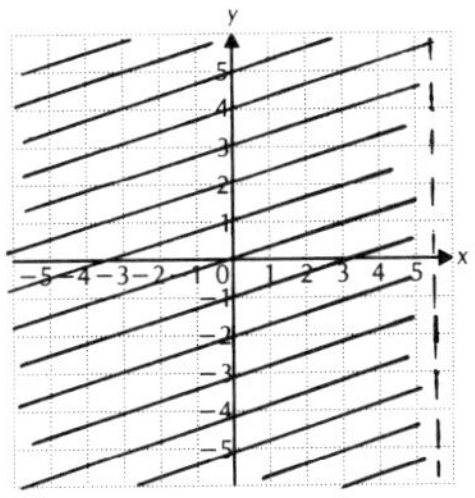

The graph of the intersection is as follows:

$0 \leqslant x < 5\frac{1}{2}$

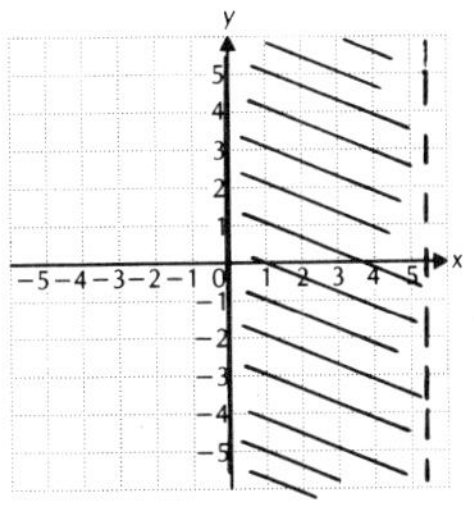

14. $-4 < y < -1$

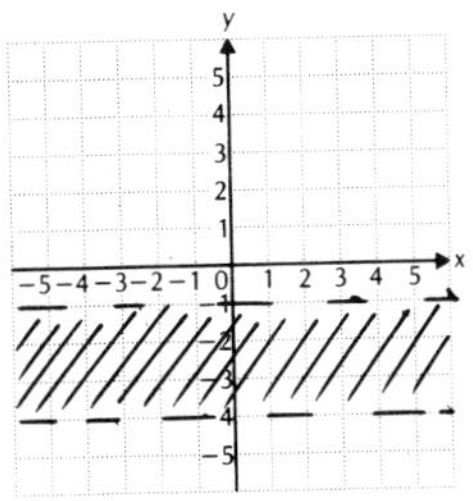

15. Graph: $y \geqslant |x|$

First graph $y = |x|$. A few solutions of this equation are (0, 0), (1, 1), (-1, 1), (4, 4), and (-4, 4). Plot these points and draw the graph solid since the inequality is $\geqslant$.

15. (continued)

$y = |x|$

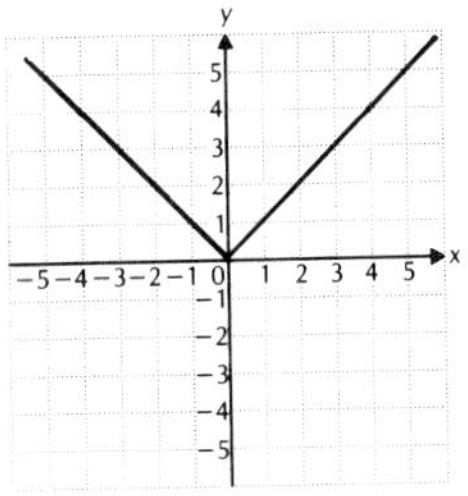

Note that in $y \geqslant |x|$, y is by itself. We interpret the graph of the inequality as the set of all ordered pairs (x, y) where the second coordinate y is greater than or equal to the absolute value of the first. Decide below which pairs satisfy the inequality and which do not.

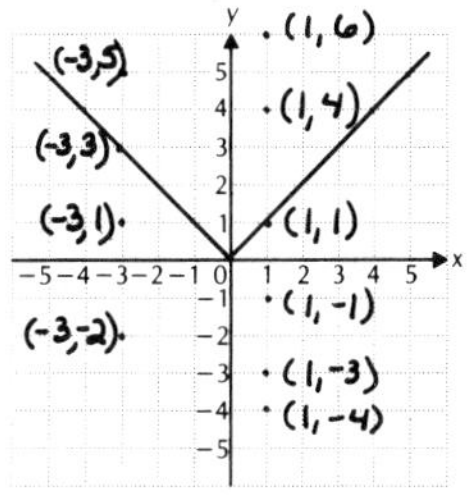

We see that any pair above the graph of $y = |x|$ is a solution as well as those in the graph of $y = |x|$. The graph is as follows.

$y \geqslant |x|$

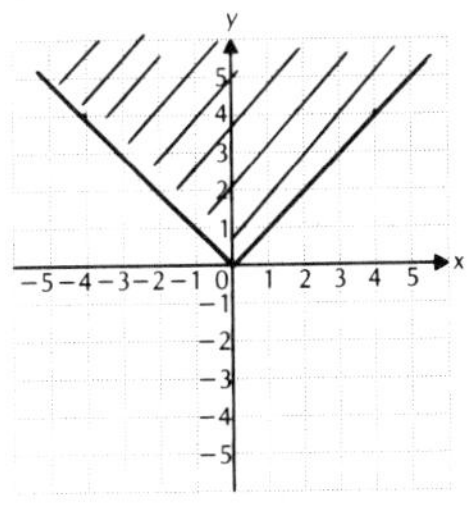

16. $y < |x|$

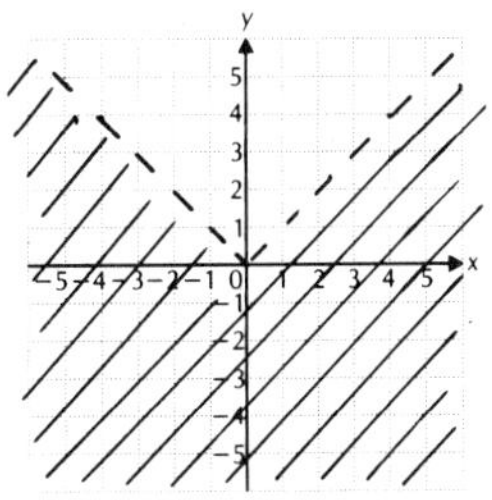

17. Graph this system: $x + y \leqslant 1$

$x - y \leqslant 2$

Graph the line $x + y = 1$. The y-intercept is (0, 1); the x-intercept is (1, 0). Plot these points and draw the line solid since the inequality is $\leqslant$. Since the origin is a solution of $x + y \leqslant 1$, shade the half-plane containing (0, 0).

17. (continued)

$x + y \leqslant 1$

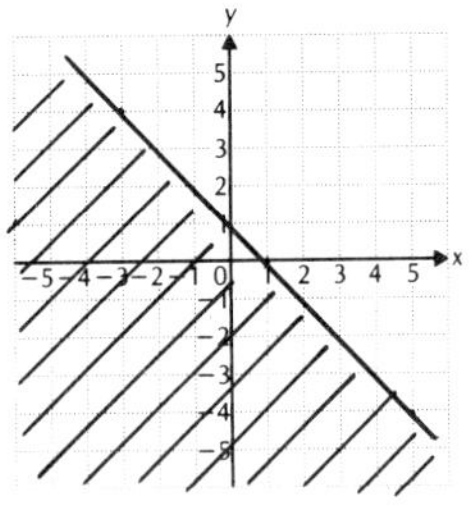

Graph the line $x - y = 2$. The y-intercept is $(0, -2)$; the x-intercept is $(2, 0)$. Plot these points and draw the line solid since the inequality is $\leqslant$. Since the origin is a solution of $x - y \leqslant 2$, shade the half-plane containing $(0, 0)$.

$x - y \leqslant 2$

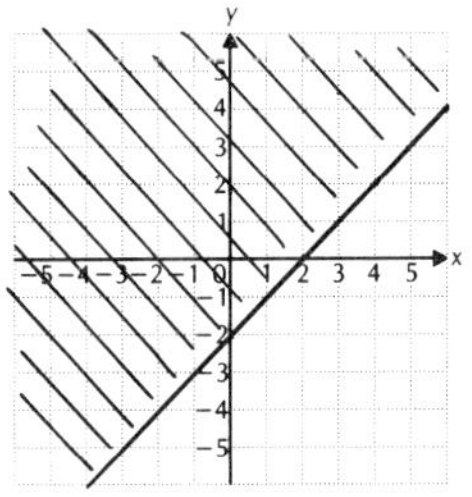

Now graph the intersection of the graphs.

$x + y \leqslant 1$
$x - y \leqslant 2$

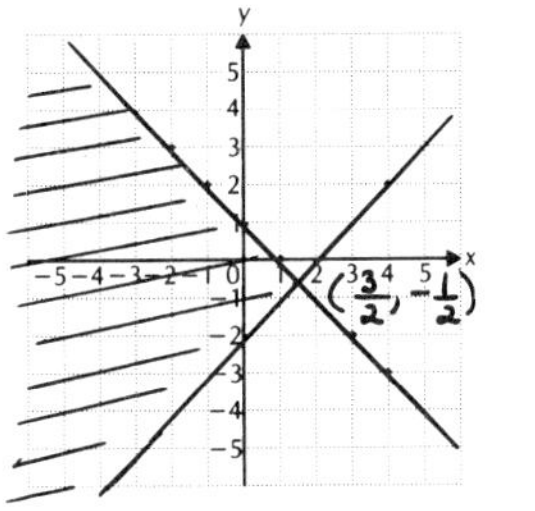

We find the vertex $\left(\frac{3}{2}, -\frac{1}{2}\right)$ by solving the system

$x + y = 1$

$x - y = 2.$

18. $x + y \leqslant 3$
$x - y \leqslant 4$

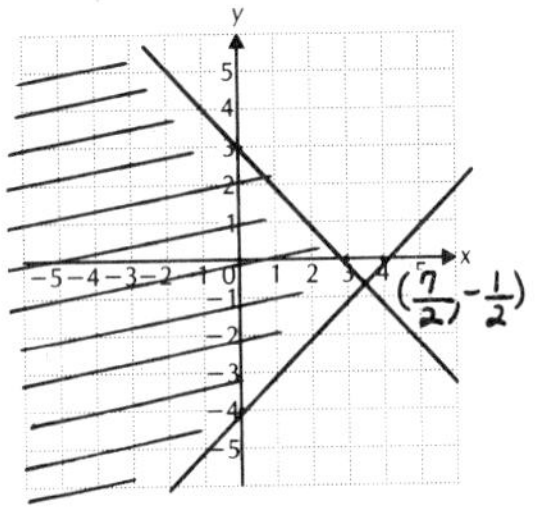

19. Graph this system: $y - 2x > 1$
$y - 2x < 3$

Graph the line $y - 2x = 1$. The intercepts are $(0, 1)$ and $\left(-\frac{1}{2}, 0\right)$. Plot these points and draw the line dashed since the inequality is $>$ and not $\geqslant$. Since the origin is not a solution of $y - 2x > 1$, we shade the half-plane which does not contain $(0, 0)$.

$y - 2x > 1$

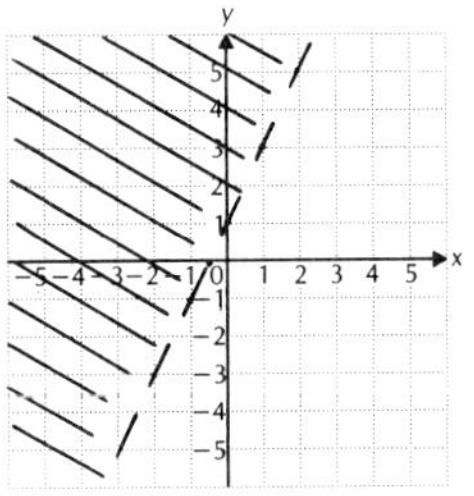

Graph the line $y - 2x = 3$. The intercepts are $(0, 3)$ and $\left(-\frac{3}{2}, 0\right)$. Plot these points and draw the line dashed since the inequality is $<$ and not $\leqslant$. Since the origin is a solution of $y - 2x < 3$, we shade the half-plane containing $(0, 0)$.

$y - 2x < 3$

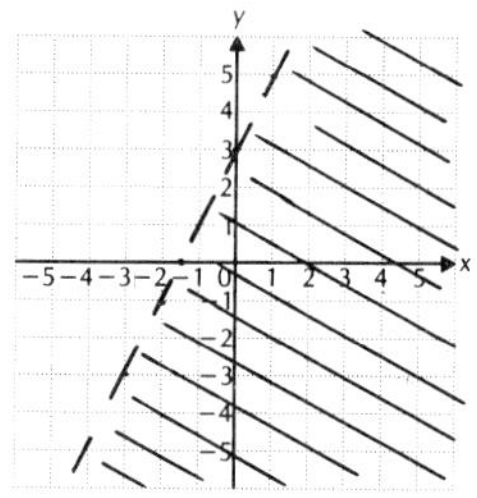

Now graph the intersection of the graphs.

$y - 2x > 1$
$y - 2x < 3$

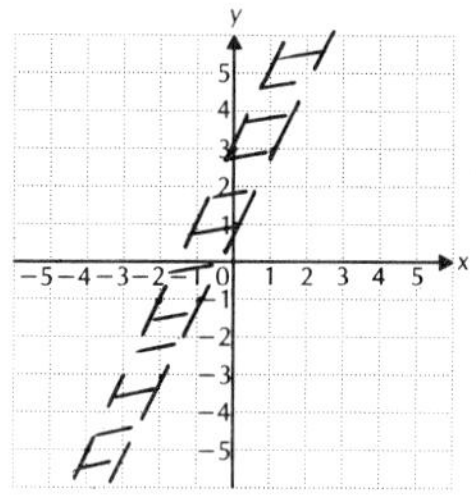

20. $y + x > 0$
$y + x < 2$

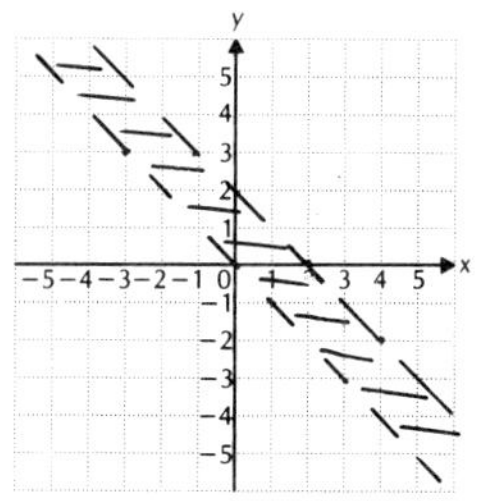

21. Graph this system: $2y - x \leqslant 2$
$y - 3x \geqslant -1$

Graph the line $2y - x = 2$. The intercepts are $(0, 1)$ and $(-2, 0)$. Plot these points and draw the line solid since the inequality is $\leqslant$. Since the origin is a solution of $2y - x \leqslant 2$, we shade the half-plane containing $(0, 0)$.

$2y - x \leqslant 2$

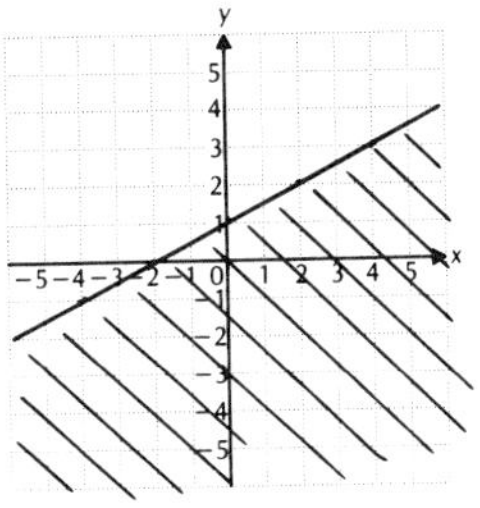

Graph the line $y - 3x = -1$. The intercepts are $(0, -1)$ and $\left(\frac{1}{3}, 0\right)$. Plot these points and draw the line solid since the inequality is $\geqslant$. Since the origin is a solution of $y - 3x \geqslant -1$, we shade the half-plane containing $(0, 0)$.

$y - 3x \geqslant -1$

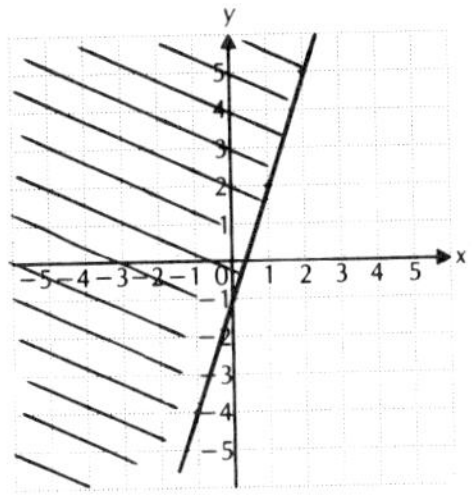

Now graph the intersection of the graphs.

$2y - x \leqslant 2$
$y - 3x \geqslant -1$

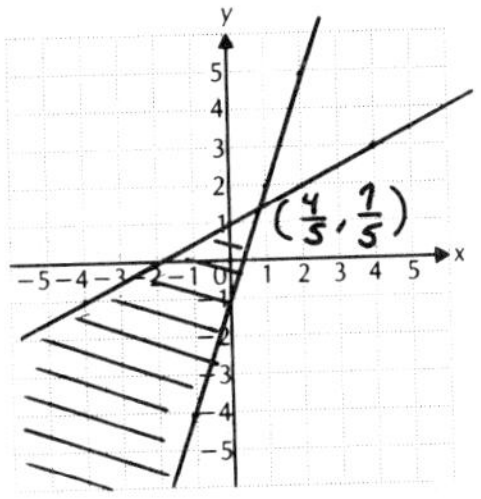

We find the vertex $\left(\frac{4}{5}, \frac{7}{5}\right)$ by solving the system

$2y - x = 2$
$y - 3x = -1.$

22. $x + 3y \geqslant 9$
$3x - 2y \leqslant 5$

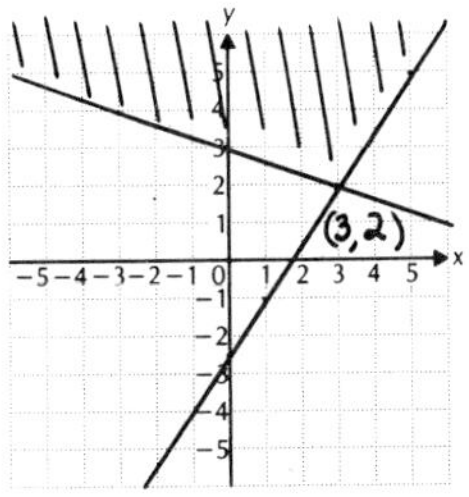

23. Graph this system: $y \leqslant 2x + 1$
$y \geqslant -2x + 1$
$x \leqslant 2$

Graph the line $y = 2x + 1$. The intercepts are $(0, 1)$ and $\left(-\frac{1}{2}, 0\right)$. Plot these points and draw the line solid since the inequality is $\leqslant$. Since the origin is a solution of $y \leqslant 2x + 1$, we shade the half-plane containing $(0, 0)$.

$y \leqslant 2x + 1$

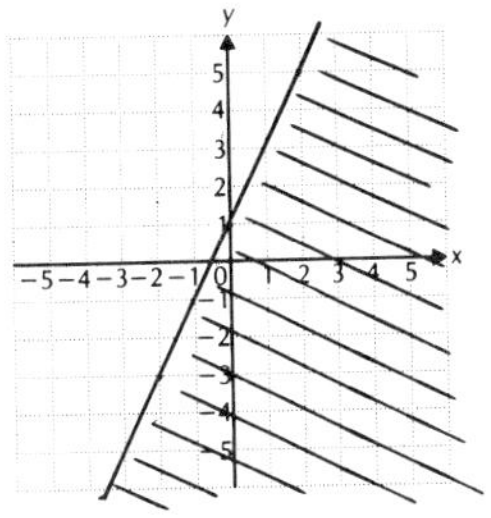

Graph the line $y = -2x + 1$. The intercepts are $(0, 1)$ and $\left(\frac{1}{2}, 0\right)$. Plot these points and draw the line solid since the inequality is $\geqslant$. Since the origin is not a solution of $y \geqslant -2x + 1$, we shade the half-plane not containing $(0, 0)$.

$y \geqslant -2x + 1$

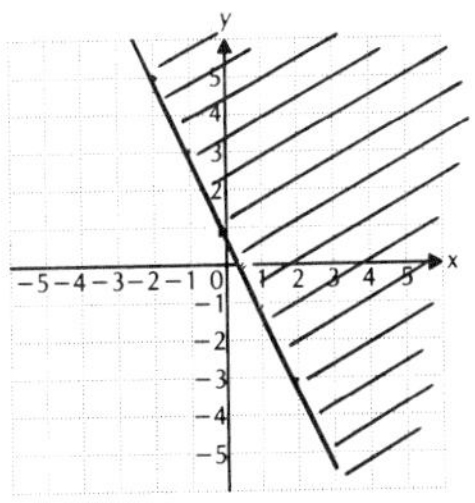

Graph the line $x = 2$. The line is parallel to the y-axis with x-intercept $(2, 0)$. We draw the line solid since the inequality is $\leqslant$. Since the origin is a solution of $x \leqslant 2$, we shade the half-plane containing $(0, 0)$.

$x \leqslant 2$

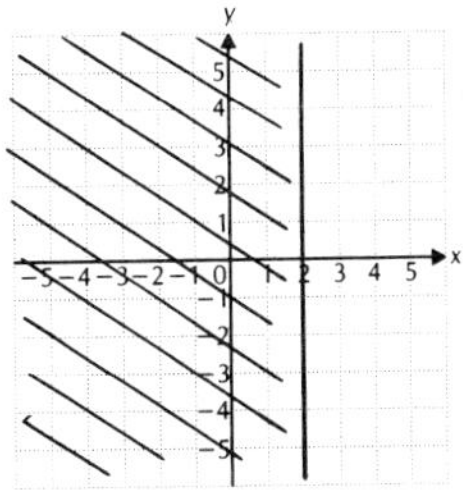

Now graph the intersection of the graphs.

$y \leqslant 2x + 1$
$y \geqslant -2x + 1$
$x \leqslant 2$

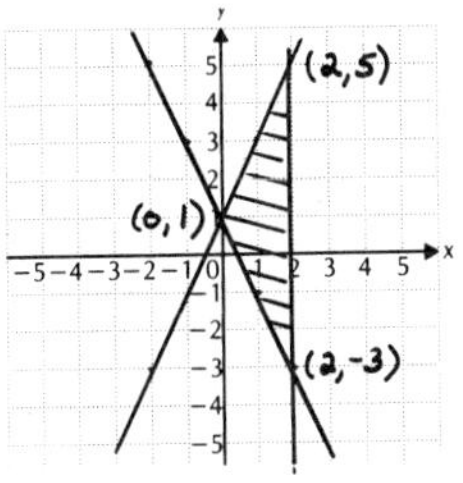

23. (continued)

We find the vertex (0, 1) by solving the system

$y = 2x + 1$

$y = -2x + 1.$

We find the vertex (2, 5) by solving the system

$y = 2x + 1$

$x = 2.$

We find the vertex (2, -3) by solving the system

$y = -2x + 1$

$x = 2.$

24. $x - y \leqslant 2$

$x + 2y \geqslant 8$

$y \leqslant 4$

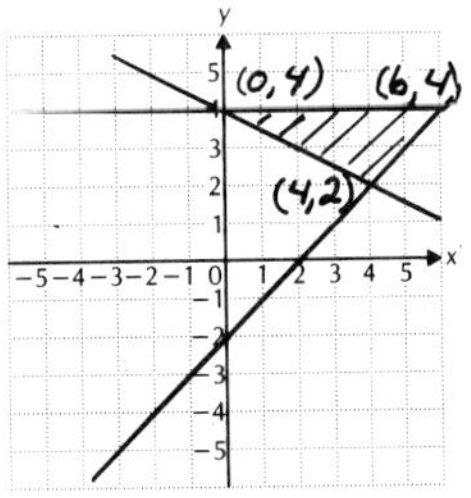

25. Graph this system: $x + 2y \leqslant 12$

$2x + y \leqslant 12$

$x \geqslant 0$

$y \geqslant 0$

Graph the line x + 2y = 12. The intercepts are (0, 6) and (12, 0). Plot these points and draw the line solid since the inequality is $\leqslant$. Since the origin is a solution of $x + 2y \leqslant 12$, we shade the half-plane containing (0, 0).

$x + 2y \leqslant 12$

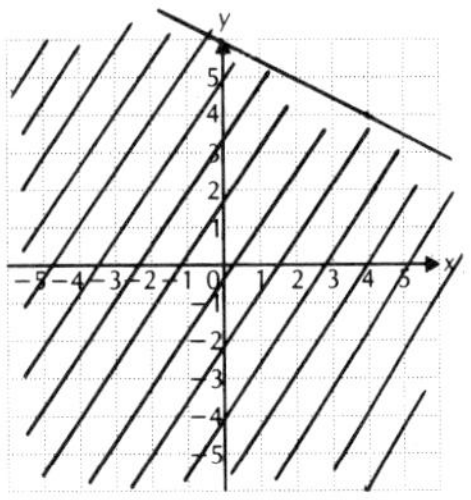

Graph the line 2x + y = 12. The intercepts are (0, 12) and (6, 0). Plot these points and draw the line solid since the inequality is $\leqslant$. Since the origin is a solution of $2x + y \leqslant 12$, we shade the half-plane containing (0, 0).

$2x + y \leqslant 12$

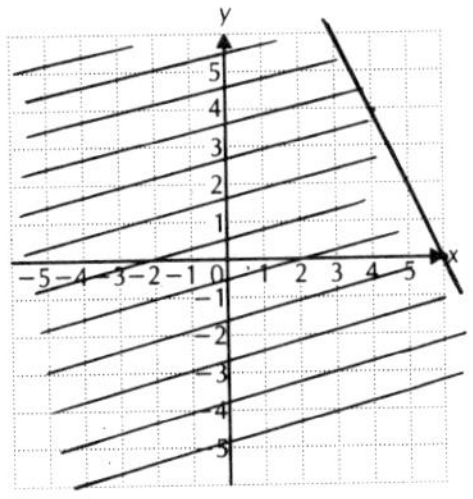

25. (continued)

Graph the lines x = 0 (the y-axis) and y = 0 (the x-axis), drawing them solid. Then shade the appropriate half-planes.

$x \geqslant 0$

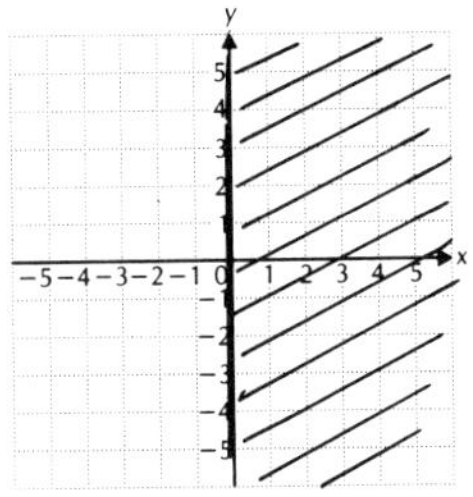

$y \geqslant 0$

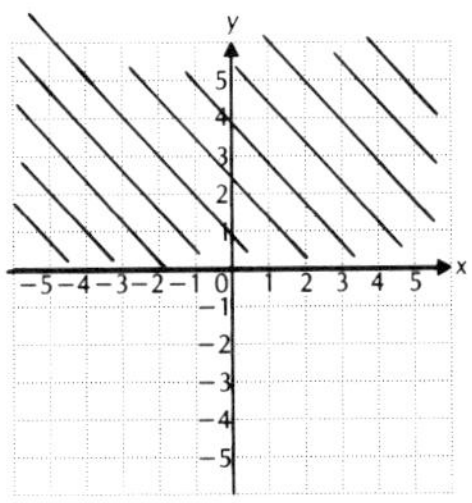

Now graph the intersection of the graphs.

$x + 2y \leqslant 12$

$2x + y \leqslant 12$

$x \geqslant 0$

$y \geqslant 0$

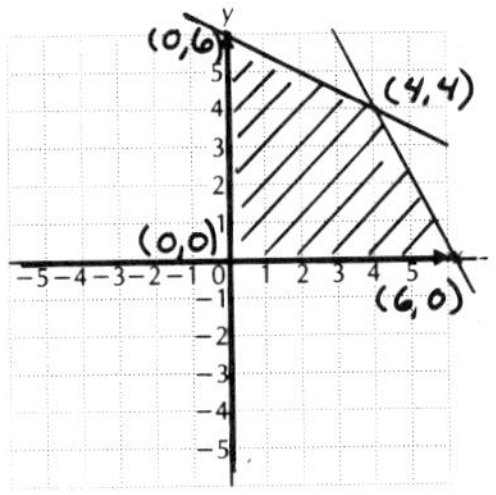

We find vertex (0, 0) by solving the system

$x = 0$

$y = 0.$

We find vertex (0, 6) by solving the system

$x + 2y = 12$

$x = 0.$

We find vertex (4, 4) by solving the system

$x + 2y = 12$

$2x + y = 12.$

We find vertex (6, 0) by solving the system

$2x + y = 12$

$y = 0.$

26. $8x + 5y \leqslant 40$

$x + 2y \leqslant 8$

$x \geqslant 0$

$y \geqslant 0$

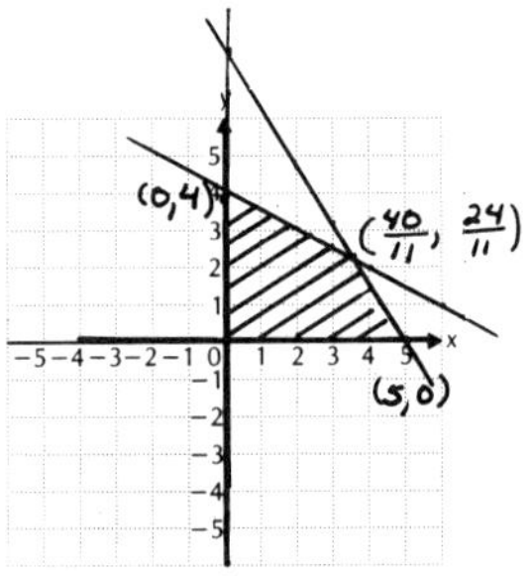

27\. Graph this system: $3x + 4y \geqslant 12$

$5x + 6y \leqslant 30$

$1 \leqslant x \leqslant 3$

Graph the line $3x + 4y = 12$. The y-intercept is (0, 3); the x-intercept is (4, 0). Plot these points and draw the line solid since the inequality is $\geqslant$. Since the origin is not a solution of $3x + 4y \geqslant 12$, shade the half-plane not containing (0, 0).

$3x + 4y \geqslant 12$

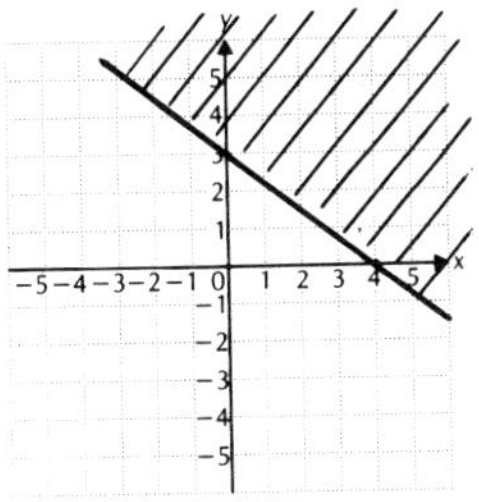

Graph the line $5x + 6y = 30$. The y-intercept is (0, 5); the x-intercept is (6, 0). Plot these points and draw the line solid since the inequality is $\leqslant$. Since the origin is a solution of $5x + 6y \leqslant 30$, shade the half-plane containing (0, 0).

$5x + 6y \leqslant 30$

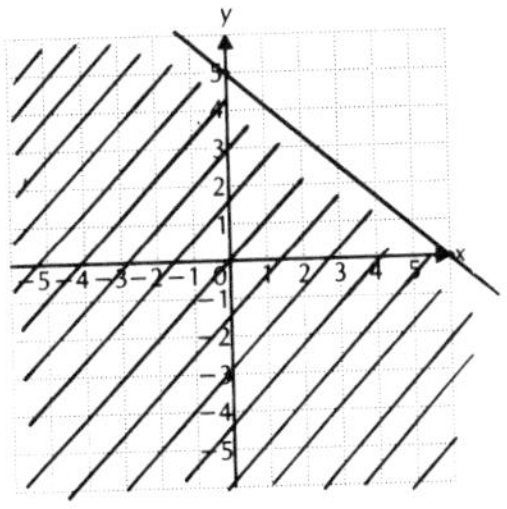

Graph the inequality $1 \leqslant x \leqslant 3$. Graph $1 \leqslant x$ and $x \leqslant 3$ separately and then graph the intersection.

$1 \leqslant x$

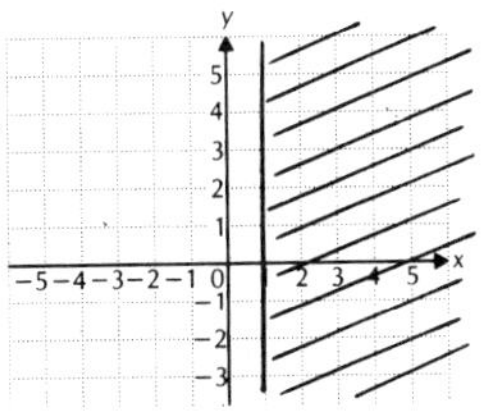

$x \leqslant 3$

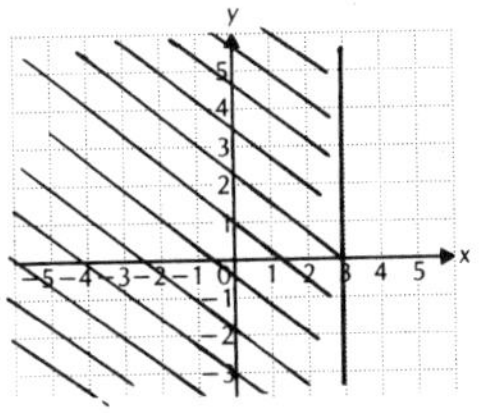

$1 \leqslant x \leqslant 3$

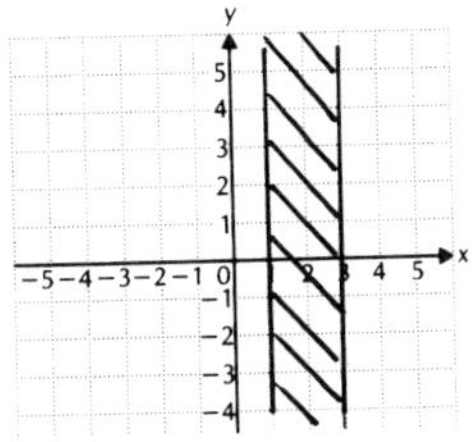

27\. (continued)

Now graph the intersection of the graphs.

$3x + 4y \geqslant 12$

$5x + 6y \leqslant 30$

$1 \leqslant x \leqslant 3$

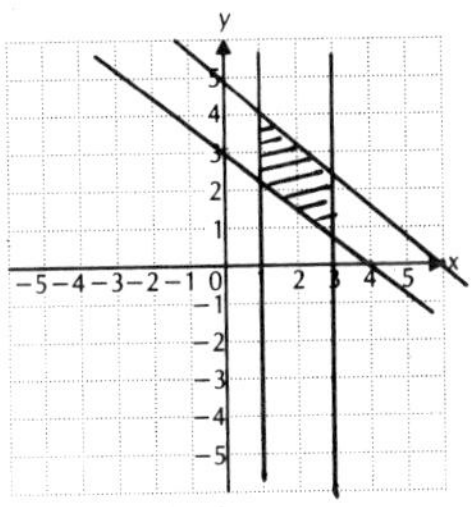

We find the vertex $\left(1, \frac{9}{4}\right)$ by solving the system

$3x + 4y = 12$

$x = 1.$

We find the vertex $\left(1, \frac{25}{6}\right)$ by solving the system

$5x + 6y = 30$

$x = 1.$

We find the vertex $\left(3, \frac{5}{2}\right)$ by solving the system

$5x + 6y = 30$

$x = 3.$

We find the vertex $\left(3, \frac{3}{4}\right)$ by solving the system

$3x + 4y = 12$

$x = 3.$

28\. $y - 2x \geqslant 3$

$y - 2x \leqslant 5$

$6 \leqslant y \leqslant 8$

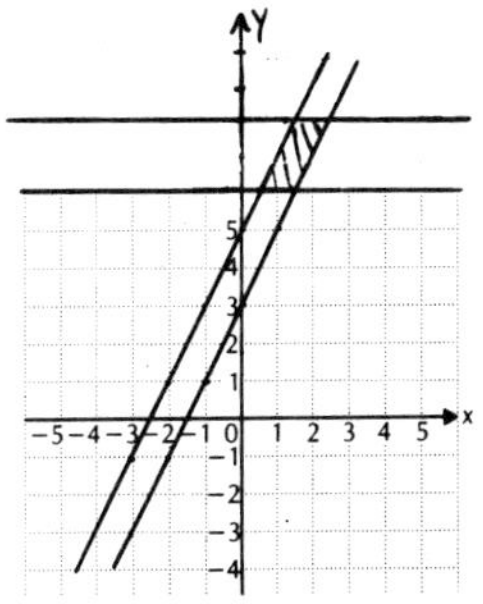

The vertices are:

$\left(\frac{3}{2}, 6\right)$

$\left(\frac{1}{2}, 6\right)$

$\left(\frac{5}{2}, 8\right)$

$\left(\frac{3}{2}, 8\right)$

29\. Graph this system: $y \geqslant x^2 - 2$

$y \leqslant 2 - x^2$

Graph the equation $y = x^2 - 2$. A few solutions are (0, -2), (1, -1), (-1, -1), (2, 2), and (-2, 2). Plot these points and draw the graph solid since the inequality is $\geqslant$. Since for any point above the graph y is greater than $x^2 - 2$, we shade above $y = x^2 - 2$.

$y \geqslant x^2 - 2$

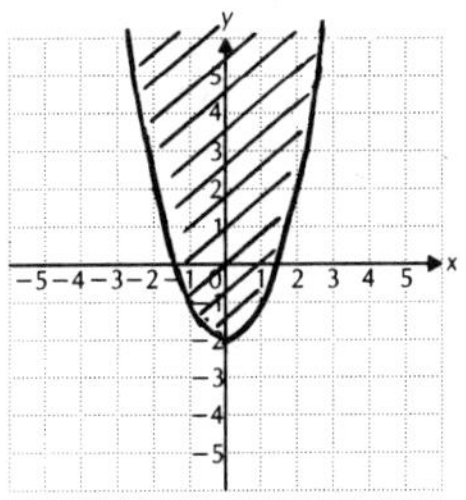

29. (continued)

Graph the equation $y = 2 - x^2$. A few solutions are (0, 2), (1, 1), (-1, 1), (2, -2), and (-2, -2). Plot these points and draw the graph solid since the inequality is $\leqslant$. Since for any point below the graph y is less than $2 - x^2$, we shade below $y = 2 - x^2$.

$y \leqslant 2 - x^2$

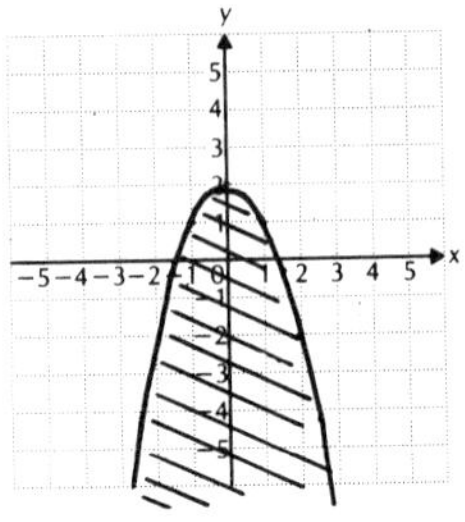

Now graph the intersection of the graphs.

$y \geqslant x^2 - 2$
$y \leqslant 2 - x^2$

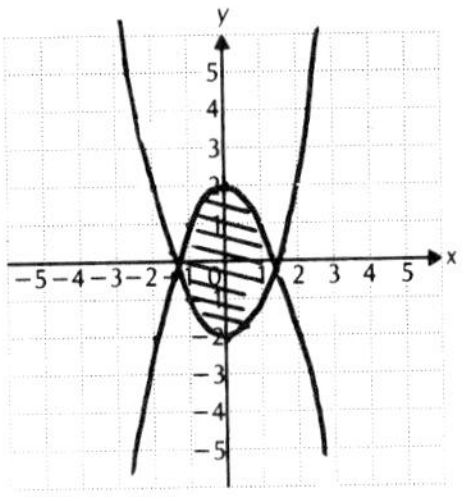

30. $x \geqslant 2y^2 - 1$
$x < y^2$

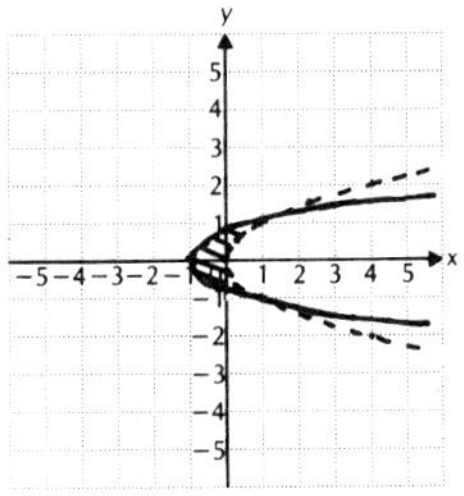

31. Graph this system: $y < x + 1$
$y \geqslant x^2$

Graph the line $y = x + 1$. The y-intercept is (0, 1); the x-intercept is (-1, 0). Plot these points and draw the line dashed since the inequality is <. Since for any point below the line y is less than x + 1, we shade the half-plane below the line.

$y < x + 1$

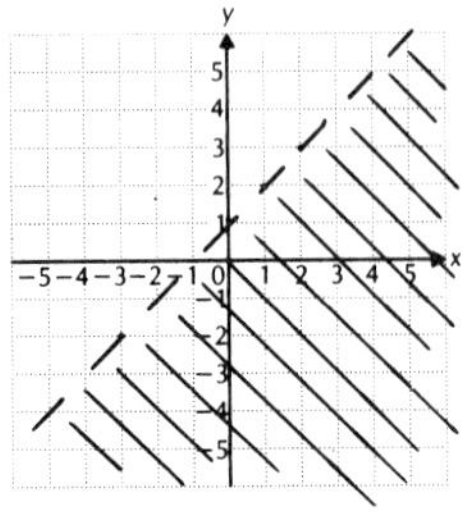

31. (continued)

Graph the equation $y = x^2$. A few solutions are (0, 0), (1, 1), (-1, 1), (2, 4), and (-2, 4). Plot these points and draw the graph solid since the inequality is $\geqslant$. Since for any point above the graph y is greater than x^2, we shade above $y = x^2$.

$y \geqslant x^2$

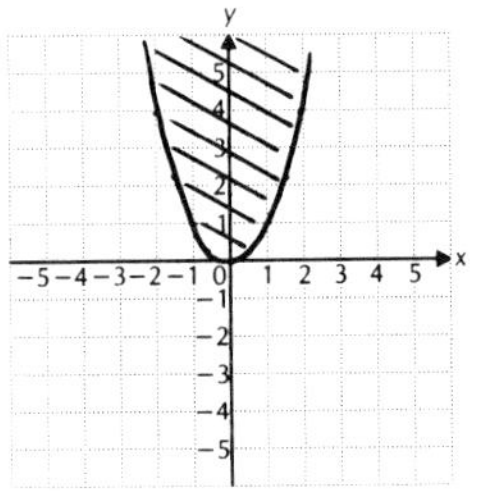

Now graph the intersection of the graphs.

$y < x + 1$
$y \geqslant x^2$

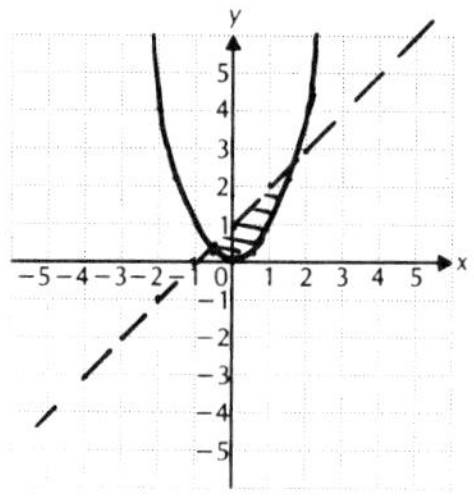

32. $y \leqslant -x^2 + 5$
$y > \frac{1}{2}x^2 - 1$

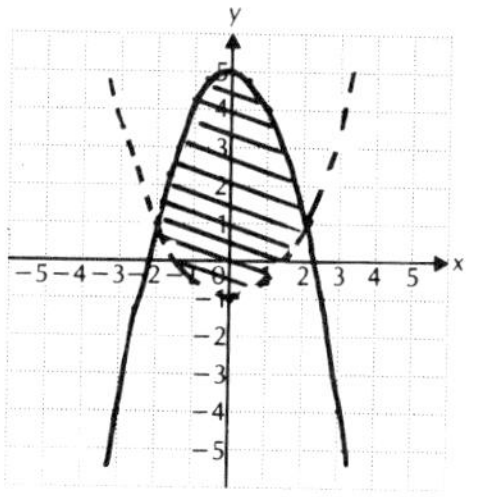

Exercise Set 10.11

1. The graph of the domain is as follows:

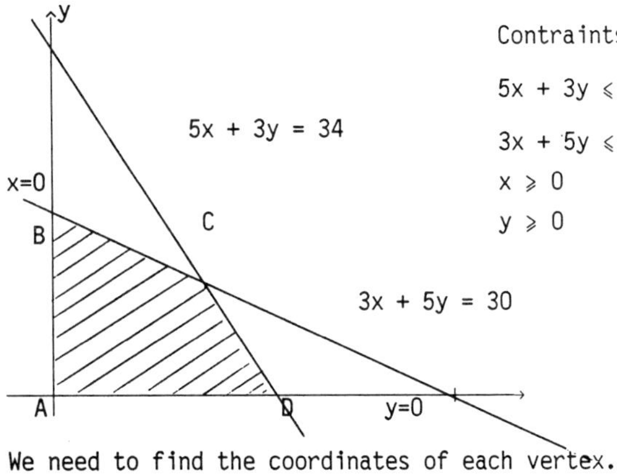

Contraints:

$5x + 3y \leqslant 34$
$3x + 5y \leqslant 30$
$x \geqslant 0$
$y \geqslant 0$

We need to find the coordinates of each vertex.

Vertex A: (0, 0)

Vertex B:

We solve the system $3x + 5y = 30$ and $x = 0$. The coordinates of point B are (0, 6).

1. (continued)

Vertex C:

We solve the system $5x + 3y = 34$ and $3x + 5y = 30$. The coordinates of point C are (5, 3).

Vertex D:

We solve the system $5x + 3y = 34$ and $y = 0$.

The coordinates of point D are $\left(\frac{34}{5}, 0\right)$.

We compute the value of F for each vertex.

Vertex	$F(x, y) = 4x + 28y$
A(0, 0)	$4 \cdot 0 + 28 \cdot 0 = 0$
B(0, 6)	$4 \cdot 0 + 28 \cdot 6 = 168$
C(5, 3)	$4 \cdot 5 + 28 \cdot 3 = 104$
$D\left(\frac{34}{5}, 0\right)$	$4 \cdot \frac{34}{5} + 28 \cdot 0 = 27\frac{1}{5}$

The maximum value of F is 168 when $x = 0$ and $y = 6$.

The minimum value of F is 0 when $x = 0$ and $y = 0$.

2. Maximum value of G is 92.8, when $x = 0$ and $y = 5.8$.

Minimum value of G is 0, when $x = 0$ and $y = 0$.

3. The graph of the domain is as follows:

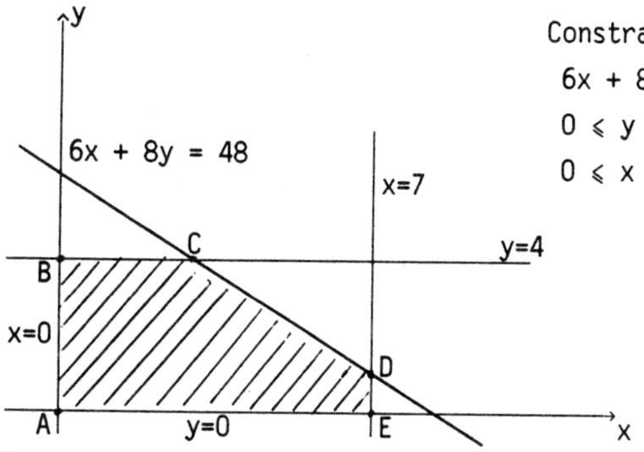

Constraints:

$6x + 8y \leqslant 48$

$0 \leqslant y \leqslant 4$

$0 \leqslant x \leqslant 7$

We need to find the coordinate of each vertex.

Vertex A: (0, 0)

Vertex B:

We solve the system $x = 0$ and $y = 4$. The coordinates of point B are (0, 4).

Vertex C:

We solve the system $6x + 8y = 48$ and $y = 4$.

The coordinates of point C are $\left(\frac{8}{3}, 4\right)$.

Vertex D:

We solve the system $6x + 8y = 48$ and $x = 7$.

The coordinates of point D are $\left(7, \frac{3}{4}\right)$.

Vertex E:

We solve the system $x = 7$ and $y = 0$. The coordinates of point E are (7, 0).

3. (continued)

We compute the value of P for each vertex.

Vertex	$P(x, y) = 16x - 2y + 40$
A(0, 0)	$16 \cdot 0 - 2 \cdot 0 + 40 = 40$
B(0, 4)	$16 \cdot 0 - 2 \cdot 4 + 40 = 32$
$C\left(\frac{8}{3}, 4\right)$	$16 \cdot \frac{8}{3} - 2 \cdot 4 + 40 = 74\frac{2}{3}$
$D\left(7, \frac{3}{4}\right)$	$16 \cdot 7 - 2 \cdot \frac{3}{4} + 40 = 150\frac{1}{2}$
E(7, 0)	$16 \cdot 7 - 2 \cdot 0 + 40 = 152$

The maximum value of P is 152 when $x = 7$ and $y = 0$.

The minimum value of P is 32 when $x = 0$ and $y = 4$.

4. Maximum value of Q is 124 when $x = 3$ and $y = 0$. Minimum value of Q is 40 when $x = 0$ and $y = 4$.

5. We organize the information in a table.

Type	Number of points for each	Number answered	Total points for type
A	4	$5 \leqslant x \leqslant 10$	$4x$
B	7	$3 \leqslant y \leqslant 10$	$7y$
Total		$x + y \leqslant 18$	$4x + 7y$
		No more than 18 can be answered.	This is the total score on the test.

We have used x to represent the number of items of Type A and y to represent the number of items of Type B. Suppose the total score on the test is T. We have T as a function of two variables x and y: $T(x, y) = 4x + 7y$.

Let us consider the domain of the function. We know these things about x and y.

Constraints: $5 \leqslant x \leqslant 10$

$3 \leqslant y \leqslant 10$

$x + y \leqslant 18$

The graph of the domain is as follows:

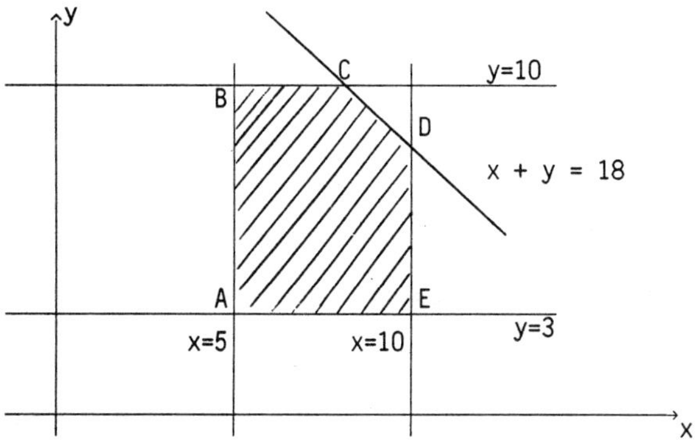

We now find the coordinates of each vertex.

5. (continued)

Vertex A:

We solve the system x = 5 and y = 3.
The coordinates of point A are (5, 3).

Vertex B:

We solve the system x = 5 and y = 10.
The coordinates of point B are (5, 10).

Vertex C:

We solve the system x + y = 18 and y = 10.
The coordinates of point C are (8, 10).

Vertex D:

We solve the system x + y = 18 and x = 10.
The coordinates of point D are (10, 8).

Vertex E:

We solve the system x = 10 and y = 3.
The coordinates of point E are (10, 3).

We compute the value of T for each vertex.

Vertex	T(x, y) = 4x + 7y
A(5, 3)	$4 \cdot 5 + 7 \cdot 3 = 41$
B(5, 10)	$4 \cdot 5 + 7 \cdot 10 = 90$
C(8, 10)	$4 \cdot 8 + 7 \cdot 10 = 102$
D(10, 8)	$4 \cdot 10 + 7 \cdot 8 = 96$
E(10, 3)	$4 \cdot 10 + 7 \cdot 3 = 61$

The maximum test score is 102 when 8 questions of Type A and 10 questions of Type B are answered.

6. Maximum score of 425 when 5 questions of Type A and 15 of Type B are answered.

7. We organize the information in a table.

Bank	Amount invested	Interest rate	Interest earned
X	$2000 \leqslant x \leqslant 14{,}000$	6%	6%x
Y	$0 \leqslant y \leqslant 15{,}000$	$6\frac{1}{2}$%	$6\frac{1}{2}$%y
Total	$x + y \leqslant 22{,}000$		6%x + $6\frac{1}{2}$%y
	No more than \$22,000 will be invested.		This is the total interest earned.

We have used x to represent the amount invested in Bank X and y to represent the amount invested in Bank Y. Suppose the total interest earned is I. We have I as a function of two variables: I = 0.06x + 0.065y.

Let us consider the domain of the function. We know these things about x and y.

Constraints: $2000 \leqslant x \leqslant 14{,}000$

$0 \leqslant y \leqslant 15{,}000$

$x + y \leqslant 22{,}000$

7. (continued)

The graph of the domain is as follows:

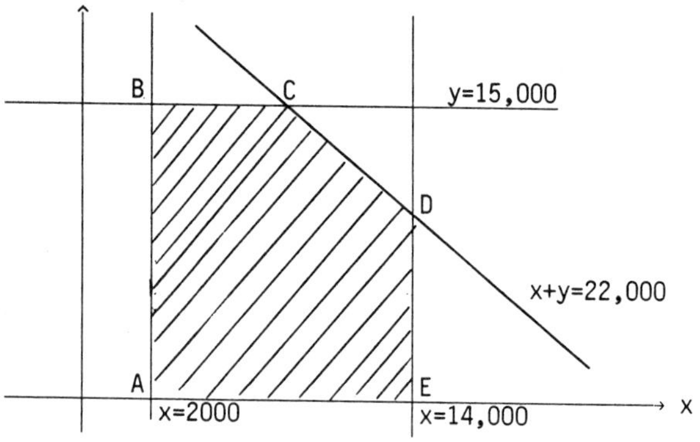

We now find the coordinates of each vertex.

Vertex A:

We solve the system x = 2000 and y = 0.
The coordinates of point A are (2000, 0).

Vertex B:

We solve the system x = 2000 and y = 15,000.
The coordinates of point B are (2000, 15,000).

Vertex C:

We solve the system x + y = 22,000 and y = 15,000. The coordinates of point C are (7000, 15,000).

Vertex D:

We solve the system x + y = 22,000 and x = 14,000. The coordinates of point D are (14,000, 8000).

Vertex E:

We solve the system x = 14,000 and y = 0.
The coordinates of point E are (14,000, 0).

We compute the value of I for each vertex.

Vertex	I(x, y) = 0.06x + 0.065y
A(2000, 0)	\$ 120
B(2000, 15,000)	1095
C(7000, 15,000)	1395
D(14,000, 8000)	1360
E(14,000, 0)	840

The maximum interest income is \$1395 when \$7000 is invested in Bank X and \$15,000 is invested in Bank Y.

8. Maximum income of \$3110 when \$22,000 is invested in corporate bonds and \$18,000 is invested in municipal bonds.

9. We organize the information in a table.

Suit	Number made	Cutting time per suit	Sewing time per suit
Knit	x	2	4
Worsted	y	4	2

Suit	Total cutting time	Total sewing time	Profit per suit	Total profit
Knit	2x	4x	34	34x
Worsted	4y	2y	31	31y
Total	$2x+4y \leqslant 20$	$4x+2y \leqslant 16$		34x + 31y (Total profit from knit and worsted)

We have used x to represent the number of knit suits made per day and y to represent the number of worsted suits made per day. We let P represent the total profit. We have P as a function of two variables: $P = 34x + 31y$.

Let us consider the domain of the function. We know these things about x and y.

Constraints: $2x + 4y \leqslant 20$ (or $x + 2y \leqslant 10$)

$4x + 2y \leqslant 16$ (or $2x + y \leqslant 8$)

$x \geqslant 0$, $y \geqslant 0$ — The number of suits made cannot be negative.

The graph of the domain is as follows:

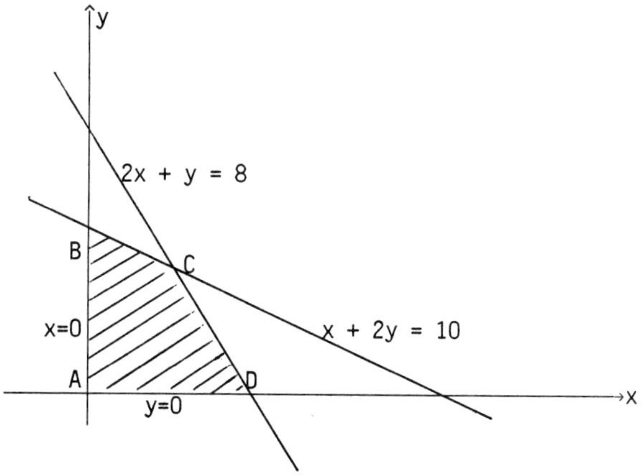

We now find the coordinates of each vertex.

Vertex A: (0, 0)

Vertex B:

We solve the system $x = 0$ and $x + 2y = 10$. The coordinates of point B are (0, 5).

Vertex C:

We solve the system $x + 2y = 10$ and $2x + y = 8$. The coordinates of point C are (2, 4).

Vertex D:

We solve the system $2x + y = 8$ and $y = 0$. The coordinates of point D are (4, 0).

9. (continued)

We compute the value of P for each vertex.

Vertex	$P(x, y) = 34x + 31y$
A(0, 0)	$34\cdot 0 + 31\cdot 0 = 0$
B(0, 5)	$34\cdot 0 + 31\cdot 5 = 155$
C(2, 4)	$34\cdot 2 + 31\cdot 4 = 192$
D(4, 0)	$34\cdot 4 + 31\cdot 0 = 136$

The maximum profit per day is \$192 when 2 knit suits and 4 worsted suits are made.

10. Maximum profit is \$2520 when 125 batches of Smello and 187.5 batches of Roppo are made.

11. We organize the information in a table.

Plane	Number of Planes	Passengers		
		First	Tourist	Economy
P-1	x	40x	40x	120x
P-2	y	80y	30y	40y

Let x represent the number of P-1 planes and y represent the number of P-2 planes. Then $40x + 80y$ represents the total number of first-class passengers, $40x + 30y$ represents the total number of tourist passengers, and $120x + 40y$ represents the total number of economy-class passengers.

The total cost <u>per mile</u> for the x P-1 planes is 12,000x. The total cost <u>per mile</u> for the y P-2 planes is 10,000y. The cost function we wish to minimize is

$$C(x, y) = 12{,}000x + 10{,}000y$$

We know the following constraints about x and y.

$40x + 80y \geqslant 2000$ or $x + 2y \geqslant 50$

$40x + 30y \geqslant 1500$ or $4x + 3y \geqslant 150$

$120x + 40y \geqslant 2400$ or $3x + y \geqslant 60$

$x \geqslant 0$ or $x \geqslant 0$

$y \geqslant 0$ or $y \geqslant 0$

This system of inequalities describes the domain of C.

The graph of the domain is as follows:

11. (continued)

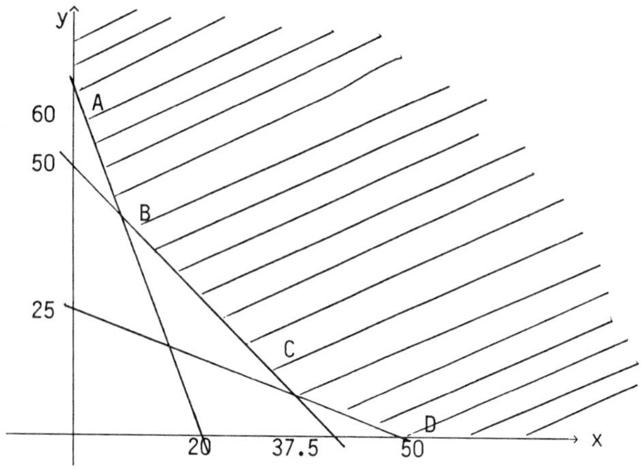

We next find the coordinates of each vertex.

Vertex A:

We solve the system x = 0 and 3x + y = 60. The coordinates of point A are (0, 60).

Vertex B:

We solve the system 4x + 3y = 150 and 3x + y = 60. The coordinates of point B are (6, 42).

Vertex C:

We solve the system x + 2y = 40 and 4x + 3y = 150. The coordinates of point C are (30, 10).

Vertex D:

We solve the system y = 0 and x + 2y = 50. The coordinates of point D are (50, 0).

We compute the cost for each ordered pair.

Vertex	C(x, y) = 12,000x + 10,000y
A(0, 60)	$600,000
B(6, 42)	492,000
C(30, 10)	460,000
D(50, 0)	600,000

The minimum cost per mile occurs when 30 P-1 planes and 10 P-2 planes are used.

12. Minimum: $483,333

$\frac{40}{3}$ P2 airplanes, $\frac{70}{3}$ P3 airplanes

Exercise Set 11.1

1. $$x^2 - y^2 = 0$$
$$(x + y)(x - y) = 0$$
$$x + y = 0 \quad \text{or} \quad x - y = 0$$
$$y = -x \quad \text{or} \quad y = x$$

The graph consists of two intersecting lines.

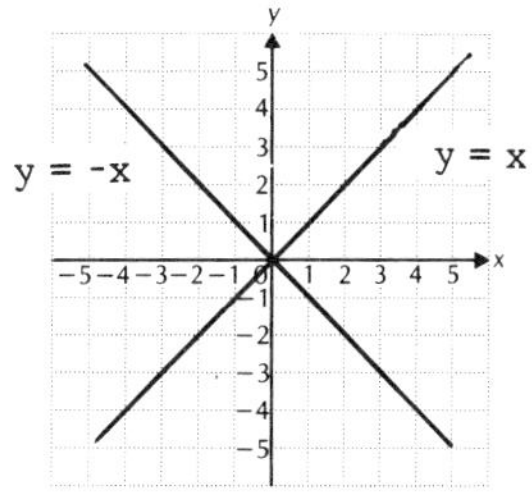

2.

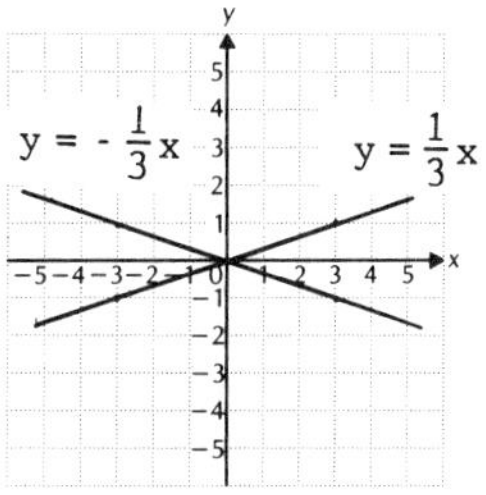

3. $$3x^2 + xy - 2y^2 = 0$$
$$(3x - 2y)(x + y) = 0$$
$$3x - 2y = 0 \quad \text{or} \quad x + y = 0$$
$$y = \frac{3}{2}x \quad \text{or} \quad y = -x$$

The graph consists of two intersecting lines.

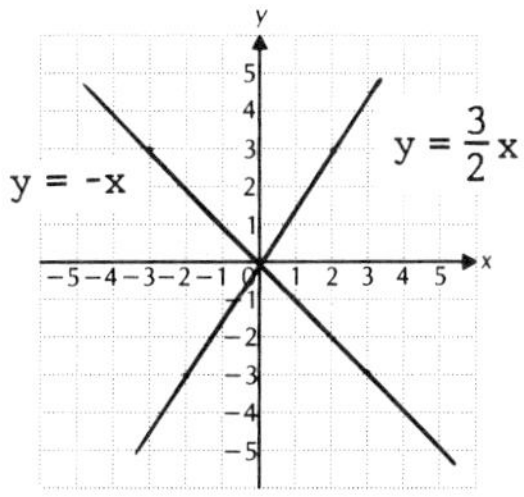

4.

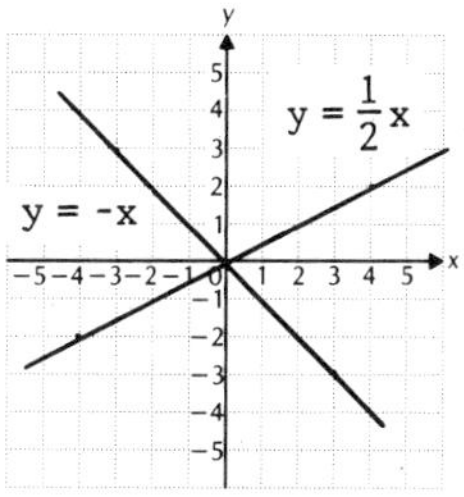

5. $2x^2 + y^2 = 0$

The expression $2x^2 + y^2$ is not factorable in the real-number system. The only real-number solution of the equation is (0, 0).

6. $5x^2 + y^2 = -3$

Since squares of numbers are never negative, the left side of the equation can never be negative. The equation has no real-number solution, hence has no graph.

7. $(x - h)^2 + (y - k)^2 = r^2$ (Standard form)

$(x - 0)^2 + (y - 0)^2 = 7^2$ (Substituting 0 for h, 0 for k, and 7 for r)

$x^2 + y^2 = 49$

8. $(x + 2)^2 + (y - 7)^2 = 5$

9. $$(x + 1)^2 + (y + 3)^2 = 4$$
$[x - (-1)]^2 + [y - (-3)]^2 = 2^2$ (Standard form)

Center: (-1, -3), Radius: 2

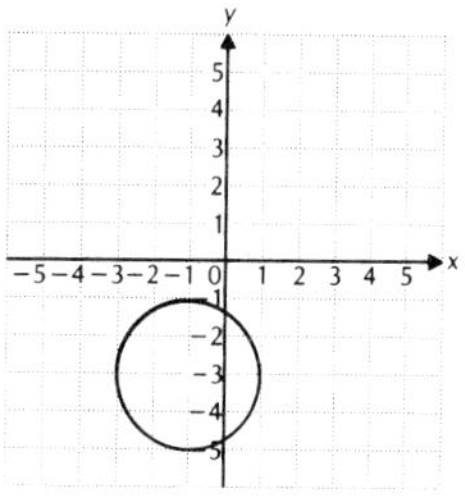

10.

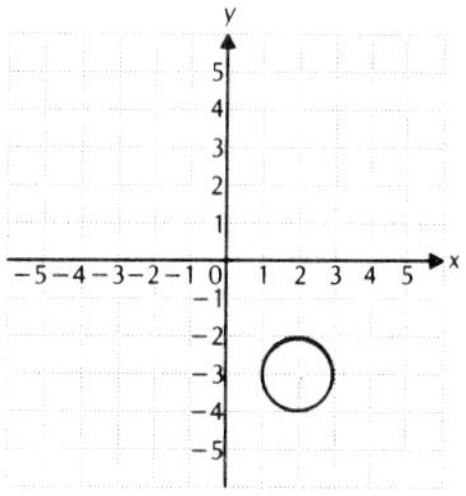

Center: (2, -3)
Radius: 1

11. $$(x - 8)^2 + (y + 3)^2 = 40$$
$$(x - 8)^2 + [y - (-3)]^2 = (\sqrt{40})^2 = (2\sqrt{10})^2$$

Center: (8, -3), Radius: $2\sqrt{10}$

12. Center: (-5, 1), Radius: $5\sqrt{3}$

13. $$x^2 + y^2 + 8x - 6y - 15 = 0$$
$$(x^2 + 8x) + (y^2 - 6y) = 15$$
$$(x^2 + 8x + 16) + (y^2 - 6y + 9) = 15 + 16 + 9$$
$$(x + 4)^2 + (y - 3)^2 = 40$$
$$[x - (-4)]^2 + (y - 3)^2 = (2\sqrt{10})^2$$

Center: (-4, 3), Radius: $2\sqrt{10}$

14. Center: $\left(-\frac{25}{2}, -5\right)$, Radius: $\frac{\sqrt{677}}{2}$

15.
$$x^2 + y^2 - 4x = 0$$
$$(x^2 - 4x + 4) + y^2 = 4$$
$$(x - 2)^2 + (y - 0)^2 = 2^2$$

Center: (2, 0), Radius: 2

16. Center: (0, -5), Radius: 10

17.
$$x^2 + y^2 + 8.246x - 6.348y - 74.35 = 0$$
$$(x^2 + 8.246x) + (y^2 - 6.348y) = 74.35$$
$$(x^2+8.246x+16.999)+(y^2-6.348y+10.074) = 74.35 + 16.999 + 10.074$$
$$(x + 4.123)^2 + (y - 3.174)^2 = 101.423$$
$$[x - (-4.123)]^2 + (y - 3.174)^2 = (10.071)^2$$

Center: (-4.123, 3.174), Radius: 10.071

18. Center: (-12.537, -5.002), Radius: 13.044

19.
$$9x^2 + 9y^2 = 1$$
$$x^2 + y^2 = \frac{1}{9}$$
$$(x - 0)^2 + (y - 0)^2 = \left(\frac{1}{3}\right)^2$$

Center: (0, 0), Radius: $\frac{1}{3}$

20. Center: (0, 0), Radius: $\frac{1}{4}$

21. Since the center is (0, 0), we have
$(x - 0)^2 + (y - 0)^2 = r^2$ or $x^2 + y^2 = r^2$

The circle passes through (-3, 4). We find r by substituting -3 for x and 4 for y.

$$(-3)^2 + 4^2 = r^2$$
$$9 + 16 = r^2$$
$$25 = r^2$$
$$5 = r$$

Then $x^2 + y^2 = 25$ is an equation of the circle.

22. $(x - 3)^2 + (y + 2)^2 = 64$

23. If the circle with center (2, 4) is tangent to the x-axis, the radius is 4. We substitute 2 for h, 4 for k, and 4 for r.

$$(x - h)^2 + (y - k)^2 = r^2$$
$$(x - 2)^2 + (y - 4)^2 = 4^2 = 16$$

24. $(x + 3)^2 + (y + 2)^2 = 9$

25. Label the drawing with additional information and lettering.

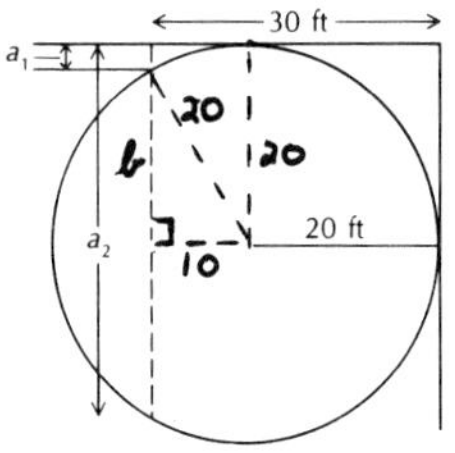

Find b using the Pythagorean Theorem.

$$b^2 + 10^2 = 20^2$$
$$b^2 + 100 = 400$$
$$b^2 = 300$$
$$b = 10\sqrt{3}$$
$$b \approx 17.32$$

Find a_1:

$a_1 = 20 - b \approx 20 - 17.32 \approx 2.68$ ft

Find a_2:

$a_2 = 2b + a_1 \approx 2(17.32) + 2.68 \approx 37.32$ ft

26. $(x - 1)^2 + (y - 2)^2 = 41$

27. a)

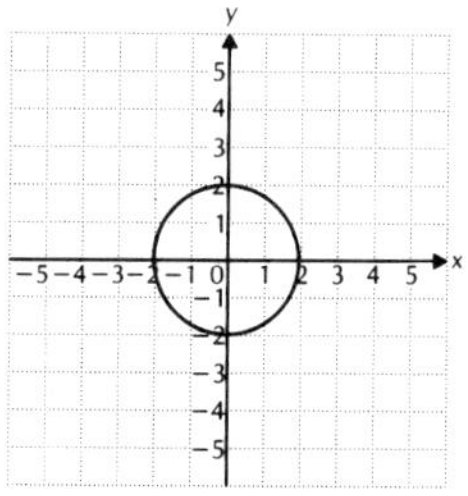

The relation is not a function. It does not pass the vertical line test.

b)
$$x^2 + y^2 = 4$$
$$y^2 = 4 - x^2$$
$$y = \pm\sqrt{4 - x^2}$$

c)

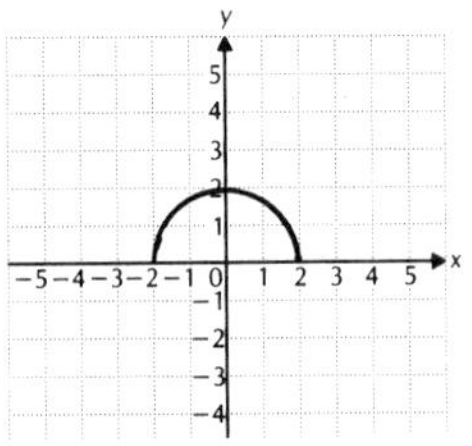

This relation is a function. It does pass the vertical line test.

Domain: $\{x \mid -2 \leqslant x \leqslant 2\}$

Range: $\{y \mid 0 \leqslant y \leqslant 2\}$

d)

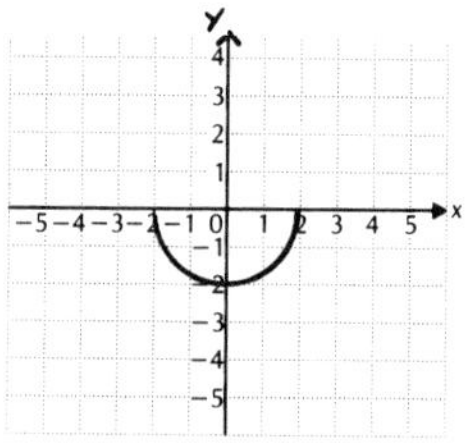

This relation is a function. It does pass the vertical line test.

Domain: $\{x \mid -2 \leqslant x \leqslant 2\}$

Range: $\{y \mid -2 \leqslant y \leqslant 0\}$

28. $(x - h)^2 - (y - k)[-(y - k)] = r^2$, so

$$(x - h)^2 + (y - k)^2 = r^2$$

29.

$x^2 + y^2 = 1$		
$0^2 + (-1)^2$	1	(Substituting)
1		

The point (0, -1) lies on the unit circle.

30. Yes

31.

$x^2 + y^2 = 1$		
$(\sqrt{2} + \sqrt{3})^2 + 0^2$	1	(Substituting)
$2 + 2\sqrt{6} + 3 + 0$		
$5 + 2\sqrt{6}$		

The point $(\sqrt{2} + \sqrt{3}, 0)$ does not lie on the unit circle.

32. No

33.

$x^2 + y^2 = 1$		
$\cos^2\theta + \sin^2\theta$	1	(Substituting)
1		

The point $(\cos\theta, \sin\theta)$ lies on the unit circle.

34. Yes

35.

$x^2 + y^2 = 1$	
$\tan^2\theta + \sec^2\theta$	1
	$\sec^2\theta - \tan^2\theta$

The point $(\tan\theta, \sec\theta)$ does not lie on the unit circle.

36. No

Exercise Set 11.2

1. $\frac{x^2}{4} + \frac{y^2}{1} = 1$

$\frac{x^2}{2^2} + \frac{y^2}{1^2} = 1$

The center of the ellipse is (0, 0); a = 2 and b = 1.

Two of the vertices are (-2, 0) and (2, 0). These are also the x-intercepts. The other two vertices are (0, -1) and (0, 1). These are also the y-intercepts.

Since a > b, we find c using $c^2 = a^2 - b^2$.

$c^2 = a^2 - b^2 = 4 - 1 = 3$

Thus $c = \sqrt{3}$.

1. (continued)

The foci are on the x-axis. They are $(-\sqrt{3}, 0)$ and $(\sqrt{3}, 0)$.

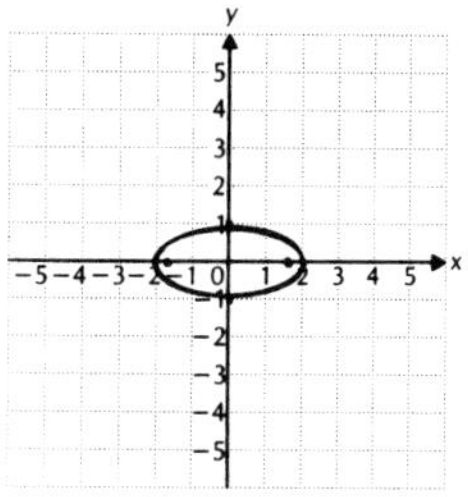

2. V: (-1, 0), (1, 0), (0, -2), (0, 2)

F: $(0, -\sqrt{3})$, $(0, \sqrt{3})$

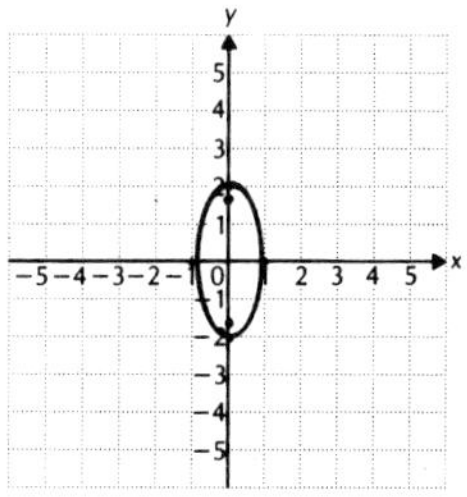

3. $16x^2 + 9y^2 = 144$

$\frac{x^2}{9} + \frac{y^2}{16} = 1$ $\left[\text{Multiplying by } \frac{1}{144}\right]$

$\frac{x^2}{3^2} + \frac{y^2}{4^2} = 1$

The center of the ellipse is (0, 0); a = 3 and b = 4.

Two of the vertices are (-3, 0) and (3, 0). These are also the x-intercepts. The other two vertices are (0, -4) and (0, 4). These are also the y-intercepts.

Since b > a, we find c using $c^2 = b^2 - a^2$.

$c^2 = b^2 - a^2 = 16 - 9 = 7$

Thus $c = \sqrt{7}$.

The foci are on the y-axis. They are $(0, -\sqrt{7})$ and $(0, \sqrt{7})$.

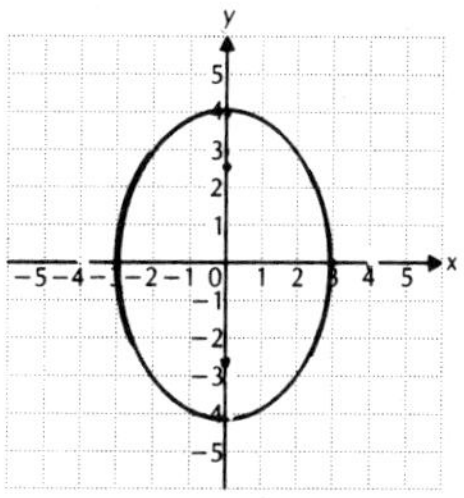

4. V: (-4, 0), (4, 0), (0, -3), (0, 3)

F: $(-\sqrt{7}, 0)$, $(\sqrt{7}, 0)$

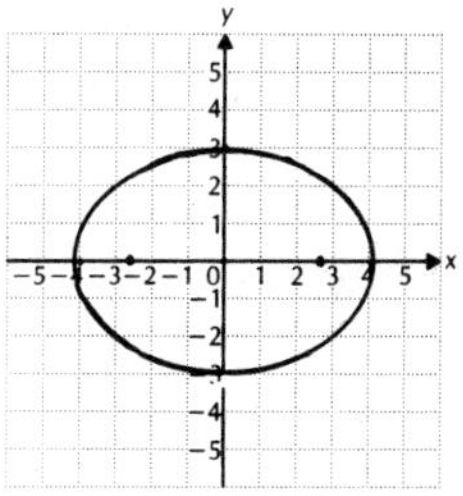

5. $2x^2 + 3y^2 = 6$

$$\frac{x^2}{3} + \frac{y^2}{2} = 1 \qquad \left[\text{Multiplying by } \frac{1}{6}\right]$$

$$\frac{x^2}{(\sqrt{3})^2} + \frac{y^2}{(\sqrt{2})^2} = 1$$

The center of the ellipse is (0, 0); $a = \sqrt{3}$ and $b = \sqrt{2}$.

Two of the vertices are $(-\sqrt{3}, 0)$ and $(\sqrt{3}, 0)$. These are also the x-intercepts. The other two vertices are $(0, -\sqrt{2})$ and $(0, \sqrt{2})$. These are also the y-intercepts.

Since a > b, we find c using $c^2 = a^2 - b^2$.

$c^2 = a^2 - b^2 = 3 - 2 = 1$

Thus c = 1.

The foci are on the x-axis. They are (-1, 0) and (1, 0).

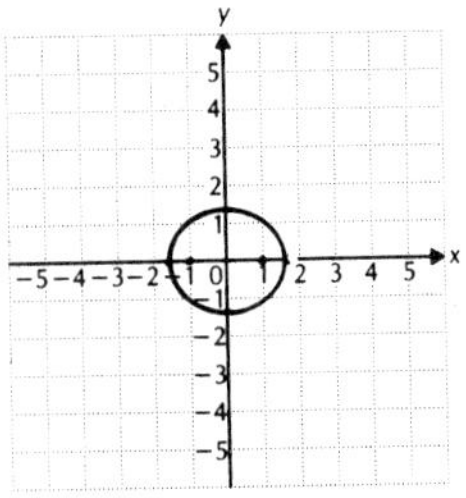

6. V: $(-\sqrt{7}, 0)$, $(\sqrt{7}, 0)$

$(0, -\sqrt{5})$, $(0, \sqrt{5})$

F: $(-\sqrt{2}, 0)$, $(\sqrt{2}, 0)$

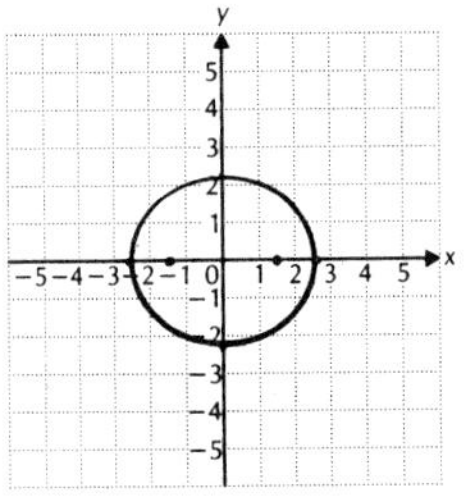

7. $4x^2 + 9y^2 = 1$

$$\frac{x^2}{\frac{1}{4}} + \frac{y^2}{\frac{1}{9}} = 1$$

$$\frac{x^2}{\left(\frac{1}{2}\right)^2} + \frac{y^2}{\left(\frac{1}{3}\right)^2} = 1$$

The center of the ellipse is (0, 0); $a = \frac{1}{2}$ and $b = \frac{1}{3}$.

Two of the vertices are $\left(-\frac{1}{2}, 0\right)$ and $\left(\frac{1}{2}, 0\right)$. These are also the x-intercepts. The other two vertices are $\left(0, -\frac{1}{3}\right)$ and $\left(0, \frac{1}{3}\right)$. These are also the y-intercepts.

Since a > b, we find c using $c^2 = a^2 - b^2$.

$c^2 = a^2 - b^2 = \frac{1}{4} - \frac{1}{9} = \frac{5}{36}$

Thus $c = \frac{\sqrt{5}}{6}$.

7. (continued)

The foci are on the x-axis. They are $\left(-\frac{\sqrt{5}}{6}, 0\right)$ and $\left(\frac{\sqrt{5}}{6}, 0\right)$.

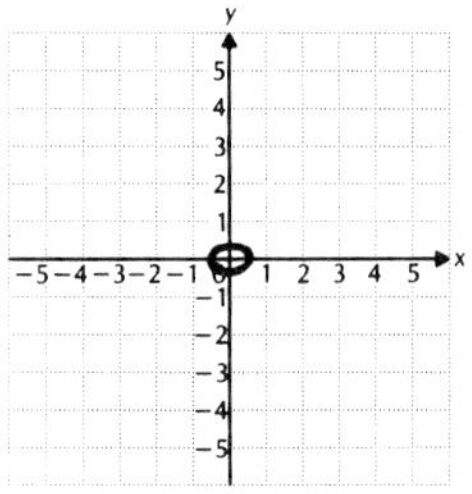

8. V: $\left(-\frac{1}{5}, 0\right)$, $\left(\frac{1}{5}, 0\right)$,

$\left(0, -\frac{1}{4}\right)$, $\left(0, \frac{1}{4}\right)$

F: $\left(0, -\frac{3}{20}\right)$, $\left(0, \frac{3}{20}\right)$

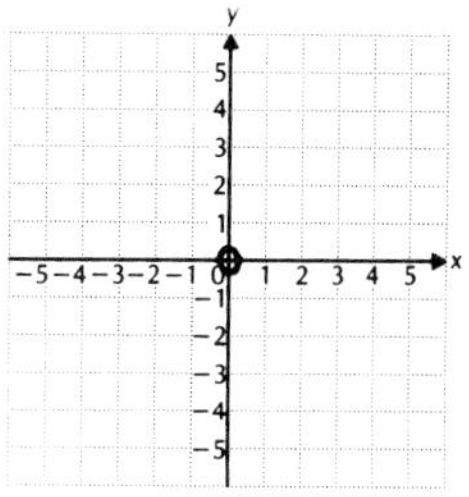

9. $$\frac{(x-1)^2}{4} + \frac{(y-2)^2}{1} = 1$$

$$\frac{(x-1)^2}{2^2} + \frac{(y-2)^2}{1^2} = 1$$

The center is (1, 2); a = 2 and b = 1. Since a > b, we find c using $c^2 = a^2 - b^2$.

$c^2 = a^2 - b^2 = 4 - 1 = 3$

Thus $c = \sqrt{3}$.

Consider the ellipse $\frac{x^2}{2^2} + \frac{y^2}{1^2} = 1$.

The center is (0, 0); a = 2, b = 1, and $c = \sqrt{3}$.

The vertices are (-2, 0), (2, 0), (0, -1), and (0, 1).

The foci are $(-\sqrt{3}, 0)$ and $(\sqrt{3}, 0)$.

The vertices and foci of the translated ellipse are found by translation in the same way in which the center has been translated.

The vertices are

(-2 + 1, 0 + 2), (2 + 1, 0 + 2), (0 + 1, -1 + 2), and (0 + 1, 1 + 2).

or

(-1, 2), (3, 2), (1, 1), and (1, 3).

The foci are

$(-\sqrt{3} + 1, 0 + 2)$ and $(\sqrt{3} + 1, 0 + 2)$

or

$(1 - \sqrt{3}, 2)$ and $(1 + \sqrt{3}, 2)$.

9. (continued)

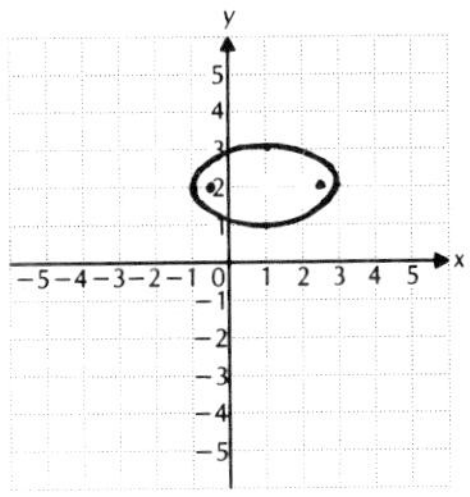

10. C: (1, 2)

V: (0, 2), (2, 2),
(1, 0), (1, 4)

F: $(1, 2 - \sqrt{3})$,
$(1, 2 + \sqrt{3})$

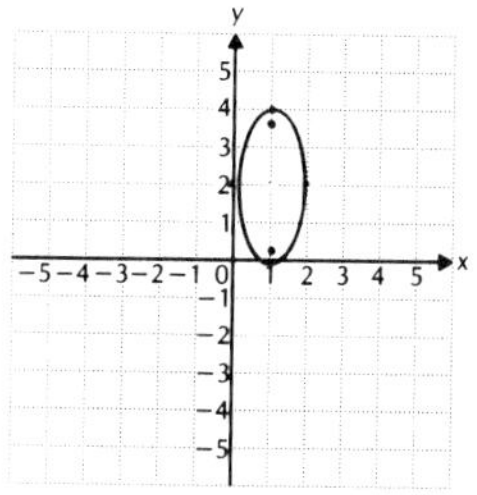

11. $$\frac{(x + 3)^2}{25} + \frac{(y - 2)^2}{16} = 1$$

$$\frac{[x - (-3)]^2}{5^2} + \frac{(y - 2)^2}{4^2} = 1$$

The center is (-3, 2); a = 5 and b = 4. Since a > b, we find c using $c^2 = a^2 - b^2$.

$c^2 = a^2 - b^2 = 25 - 16 = 9$

Thus c = 3.

Consider the ellipse $\frac{x^2}{5^2} + \frac{y^2}{4^2} = 1$.

The center is (0, 0); a = 5, b = 4, and c = 3.

The vertices are (-5, 0), (5, 0), (0, -4) and (0, 4).

The foci are (-3, 0) and (3, 0).

The vertices and foci of the translated ellipse are found by translation in the same way in which the center has been translated.

The vertices are

(-5 - 3, 0 + 2), (5 - 3, 0 + 2), (0 - 3, -4 + 2), and (0 - 3, 4 + 2)

or

(-8, 2), (2, 2), (-3, -2), and (-3, 6).

The foci are

(-3 - 3, 0 + 2) and (3 - 3, 0 + 2)

or

(-6, 2) and (0, 2).

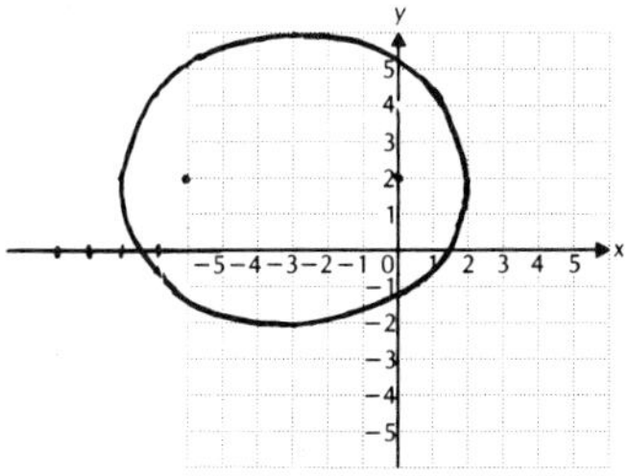

12. C: (2, -3)

V: (-3, -3), (7, -3),
(2, -7), (2, 1)

F: (-1, -3), (5, -3)

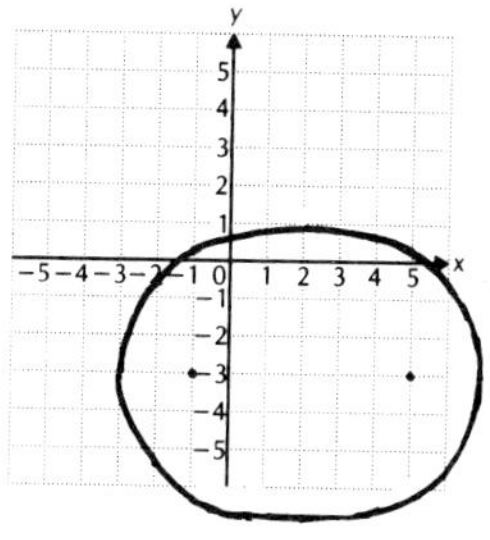

13. $3(x + 2)^2 + 4(y - 1)^2 = 192$

$$\frac{(x + 2)^2}{64} + \frac{(y - 1)^2}{48} = 1 \qquad \left[\text{Multiplying by } \frac{1}{192}\right]$$

$$\frac{(x + 2)^2}{8^2} + \frac{(y - 1)^2}{(4\sqrt{3})^2} = 1$$

The center is (-2, 1); a = 8 and b = $4\sqrt{3}$. Since a > b, we find c using $c^2 = a^2 - b^2$.

$c^2 = a^2 - b^2 = 64 - 48 = 16$

Thus c = 4.

Consider the ellipse $\frac{x^2}{8^2} + \frac{y^2}{(4\sqrt{3})^2} = 1$

The center is (0, 0); a = 8, b = $4\sqrt{3}$, and c = 4.

The vertices are (-8, 0), (8, 0), $(0, -4\sqrt{3})$ and $(0, 4\sqrt{3})$.

The foci are (-4, 0) and (4, 0).

The vertices and foci of the translated ellipse are found by translation in the same way in which the center has been translated.

The vertices are

$(-8 - 2, 0 + 1)$, $(8 - 2, 0 + 1)$, $(0 - 2, -4\sqrt{3} + 1)$ and $(0 - 2, 4\sqrt{3} + 1)$

or

(-10, 1), (6, 1), $(-2, 1 - 4\sqrt{3})$, and $(-2, 1 + 4\sqrt{3})$.

The foci are

(-4 - 2, 0 + 1) and (4 - 2, 0 + 1)

or

(-6, 1) and (2, 1).

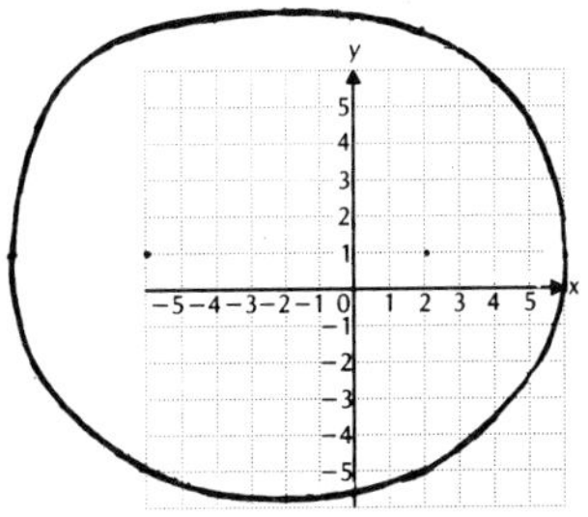

14. C: (5, 5)

V: $(5 - 4\sqrt{3}, 5)$, $(5 + 4\sqrt{3}, 5)$,
(5, -3), (5, 13)

F: (5, 1), (5, 9)

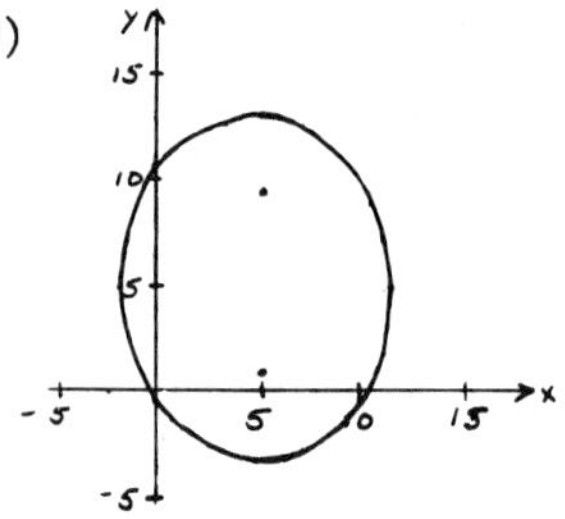

15.

$$4x^2 + 9y^2 - 16x + 18y - 11 = 0$$
$$(4x^2 - 16x) + (9y^2 + 18y) = 11$$
$$4(x^2 - 4x) + 9(y^2 + 2y) = 11$$
$$4(x^2 - 4x + 4) + 9(y^2 + 2y + 1) = 11 + 16 + 9$$
$$4(x - 2)^2 + 9(y + 1)^2 = 36$$
$$\frac{(x - 2)^2}{9} + \frac{(y + 1)^2}{4} = 1$$
$$\frac{(x - 2)^2}{3^2} + \frac{[y - (-1)]^2}{2^2} = 1$$

The center is (2, -1); $a = 3$ and $b = 2$. Since $a > b$, we find c using $c^2 = a^2 - b^2$.

$c^2 = a^2 - b^2 = 9 - 4 = 5$

Thus $c = \sqrt{5}$.

Consider the ellipse $\frac{x^2}{3^2} + \frac{y^2}{2^2} = 1$.

The center is (0, 0); $a = 3$, $b = 2$, and $c = \sqrt{5}$.

The vertices are (-3, 0), (3, 0), (0, -2), and (0, 2).

The foci are $(-\sqrt{5}, 0)$ and $(\sqrt{5}, 0)$.

The vertices and foci of the translated ellipse are found by translation in the same way in which the center has been translated.

The vertices are

(-3 + 2, 0 - 1), (3 + 2, 0 - 1), (0 + 2, -2 - 1), and (0 + 2, 2 - 1)

or

(-1, -1), (5, -1), (2, -3), and (2, 1).

The foci are

$(-\sqrt{5} + 2, 0 - 1)$ and $(\sqrt{5} + 2, 0 - 1)$

or

$(2 - \sqrt{5}, -1)$ and $(2 + \sqrt{5}, -1)$.

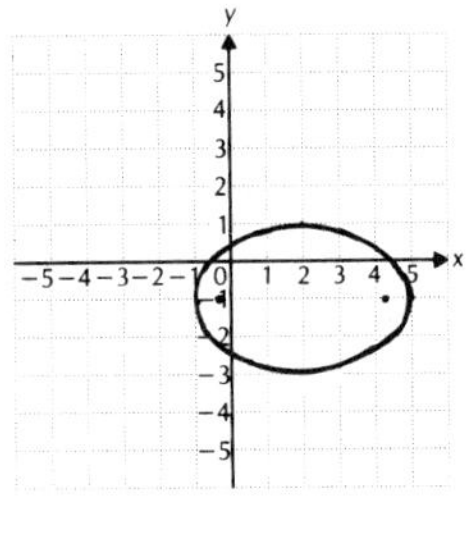

16. C: (5, -2)

V: (3, -2), (7, -2),
$(5, -2 - \sqrt{2})$, $(5, -2 + \sqrt{2})$

F: $(5 - \sqrt{2}, -2)$, $(5 + \sqrt{2}, -2)$

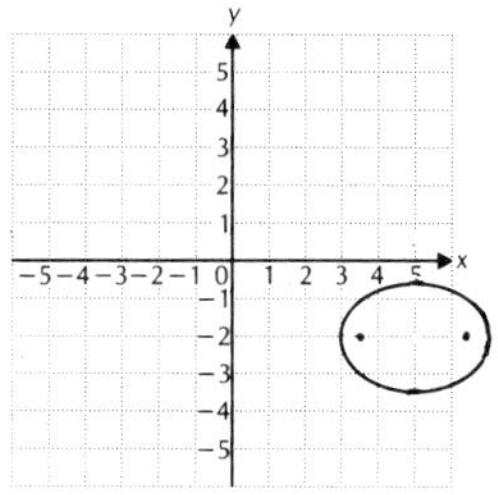

17.

$$4x^2 + y^2 - 8x - 2y + 1 = 0$$
$$(4x^2 - 8x) + (y^2 - 2y) = -1$$
$$4(x^2 - 2x) + (y^2 - 2y) = -1$$
$$4(x^2 - 2x + 1) + (y^2 - 2y + 1) = -1 + 4 + 1$$
$$4(x - 1)^2 + (y - 1)^2 = 4$$
$$\frac{(x - 1)^2}{1^2} + \frac{(y - 1)^2}{2^2} = 1$$

The center is (1, 1); $a = 1$ and $b = 2$. Since $b > a$, we find c using $c^2 = b^2 - a^2$.

$c^2 = b^2 - a^2 = 4 - 1 = 3$

Thus $c = \sqrt{3}$.

Consider the ellipse $\frac{x^2}{1^2} + \frac{y^2}{2^2} = 1$.

The center is (0, 0); $a = 1$, $b = 2$, and $c = \sqrt{3}$.

The vertices are (-1, 0), (1, 0), (0, -2), and (0, 2).

The foci are $(0, -\sqrt{3})$ and $(0, \sqrt{3})$.

The vertices and foci of the translated ellipse are found by translation in the same way in which the center has been translated.

The vertices are

(-1 + 1, 0 + 1), (1 + 1, 0 + 1), (0 + 1, -2 + 1), and (0 + 1, 2 + 1)

or

(0, 1), (2, 1), (1, -1), and (1, 3).

The foci are

$(0 + 1, -\sqrt{3} + 1)$ and $(0 + 1, \sqrt{3} + 1)$

or

$(1, 1 - \sqrt{3})$ and $(1, 1 + \sqrt{3})$.

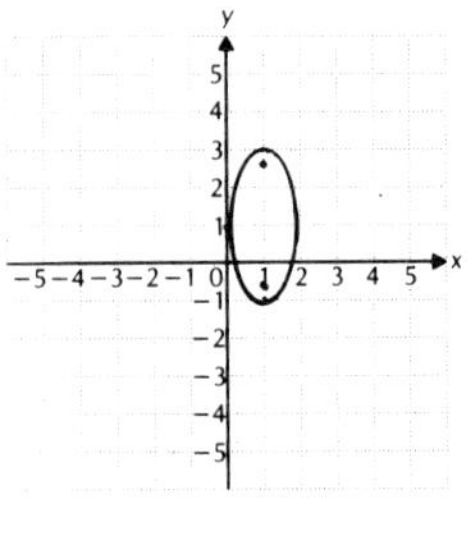

18. C: (-3, 1)

V: (-5, 1), (-1, 1)
(-3, -2), (-3, 4)

F: $(-3, 1 - \sqrt{5})$,
$(-3, 1 + \sqrt{5})$

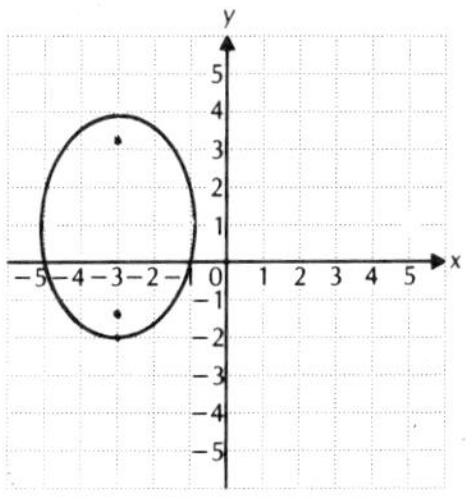

19. $4x^2 + 9y^2 - 16.025x + 18.0927y - 11.346 = 0$

$$(4x^2 - 16.025x) + (9y^2 + 18.0927y) = 11.346$$

$$4(x^2 - 4.00625x) + 9(y^2 + 2.0103y) = 11.346$$

$$4(x^2 - 4.00625x + 4.0125098) + 9(y^2 + 2.0103y + 1.0103265) = 11.346 + 16.0500392 + 9.0929385$$

$$4(x - 2.003125)^2 + 9(y + 1.00515)^2 = 36.4889777$$

$$\frac{(x - 2.003125)^2}{9.122244425} + \frac{(y + 1.00515)^2}{4.054330856} = 1$$

Consider the ellipse $\frac{x^2}{(3.020305)^2} + \frac{y^2}{(2.013537)^2} = 1$.

The center is (0, 0). The vertices are (-3.020305, 0), (3.020305, 0), (0, -2.013537), and (0, 2.013537).

The center of the translated ellipse is (2.003125, -1.00515).

The vertices of the translated ellipse are:
(-3.020305 + 2.003125, 0 - 1.00515),
(3.020305 + 2.003125, 0 - 1.00515),
(0 + 2.003125, -2.013537 - 1.00515),
(0 + 2.003125, 2.013537 - 1.00515)
or
(-1.01718, -1.00515), (5.02343, -1.00515),
(2.003125, -3.018687), (2.003125, 1.008387).

20. C: (-3.0035, 1.002)

V: (-3.0035, -1.97008), (-3.0035, 3.97408)
(-1.02211, 1.002), (-4.98489, 1.002)

21. Graph the vertices and sketch the axes of the ellipse.

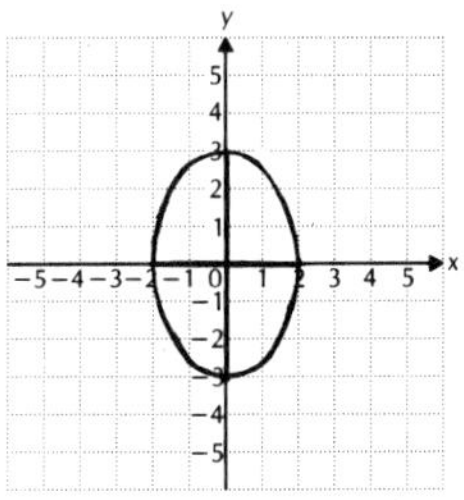

21. (continued)

The intersection of the axes, which is (0, 0), is the center of the ellipse. Note that a = 2 and b = 3. Now find an equation of the ellipse (in standard form).

$$\frac{(x - h)^2}{a^2} + \frac{(y - k)^2}{b^2} = 1$$

$$\frac{(x - 0)^2}{2^2} + \frac{(y - 0)^2}{3^2} = 1 \quad \text{(Substituting)}$$

$$\frac{x^2}{4} + \frac{y^2}{9} = 1$$

22. $x^2 + \frac{y^2}{16} = 1$

23. Graph the vertices and sketch the axes of the ellipse.

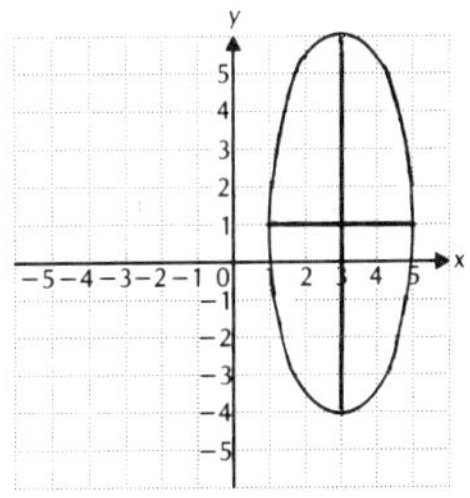

The intersection of the axes, which is (3, 1), is the center of the ellipse. Note that a = 2 and b = 5. Now find an equation of the ellipse (in standard form).

$$\frac{(x - h)^2}{a^2} + \frac{(y - k)^2}{b^2} = 1$$

$$\frac{(x - 3)^2}{2^2} + \frac{(y - 1)^2}{5^2} = 1 \quad \text{(Substituting)}$$

$$\frac{(x - 3)^2}{4} + \frac{(y - 1)^2}{25} = 1$$

24. $\frac{(x + 1)^2}{4} + \frac{(y - 2)^2}{9} = 1$

25. Graph the center, (-2, 3), and sketch the axes of the ellipse. The major axis of length 4 is parallel to the y-axis. The minor axis of length 1 is parallel to the x-axis.

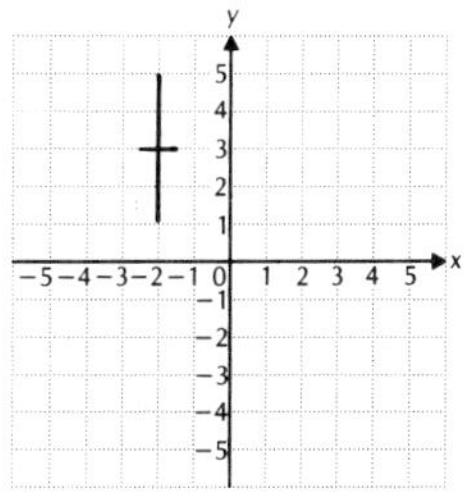

Note that $a = \frac{1}{2}$ and b = 2. Find an equation (in standard form) of the ellipse.

25. (continued)

$$\frac{(x - h)^2}{a^2} + \frac{(y - k)^2}{b^2} = 1$$

$$\frac{[x - (-2)]^2}{\left(\frac{1}{2}\right)^2} + \frac{(y - 3)^2}{2^2} = 1 \qquad \text{(Substituting)}$$

$$\frac{(x + 2)^2}{\frac{1}{4}} + \frac{(y - 3)^2}{4} = 1$$

26. $\frac{x^2}{9} + \frac{5y^2}{484} = 1$

27. a)

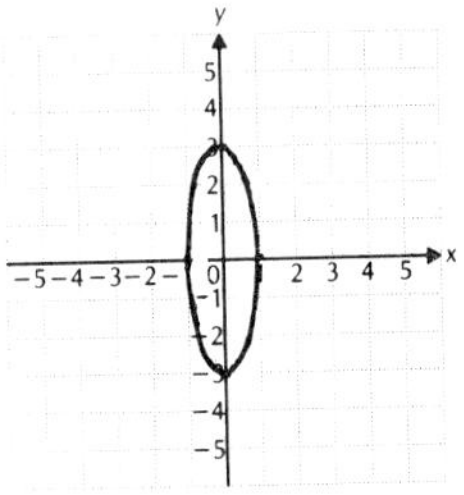

The relation is not a function. It does not pass the vertical line test.

b) $9x^2 + y^2 = 9$

$$y^2 = 9 - 9x^2$$

$$y = \pm\sqrt{9 - 9x^2}$$

$$y = \pm 3\sqrt{1 - x^2}$$

c)

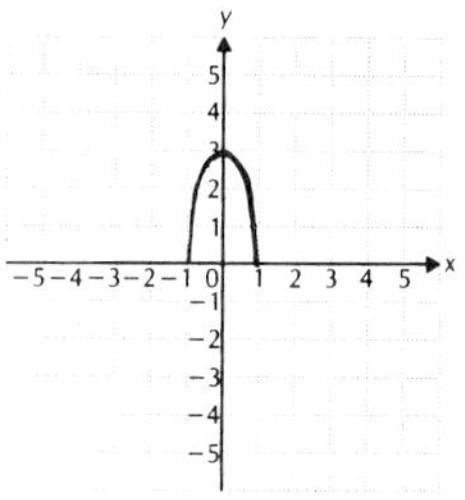

The relation is a function. It does pass the vertical line test.

Domain: $\{x \mid -1 \leqslant x \leqslant 1\}$

Range: $\{y \mid 0 \leqslant y \leqslant 3\}$

d)

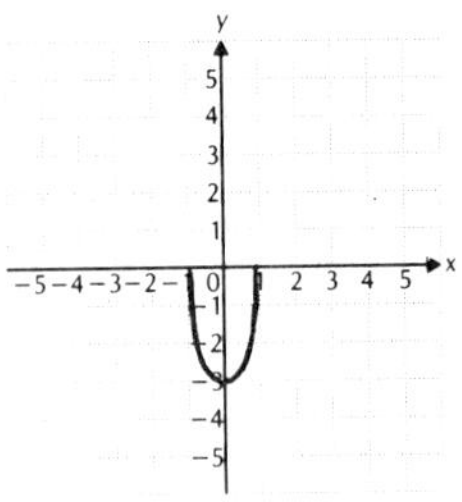

The relation is a function. It does pass the vertical line test.

Domain: $\{x \mid -1 \leqslant x \leqslant 1\}$

Range: $\{y \mid -3 \leqslant y \leqslant 0\}$

28. Circle with center (0, 0) and radius = a.

29.

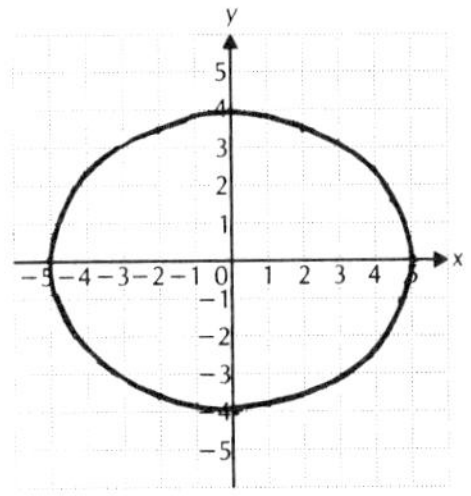

$$\frac{x^2}{25} + \frac{y^2}{36} = 1$$

30. 2×10^6 mi

31. $x = a \cos t$, so $\frac{x^2}{a^2} = \cos^2 t$

$y = b \sin t$, so $\frac{y^2}{b^2} = \sin^2 t$

Adding gives us

$$\frac{x^2}{a^2} + \frac{y^2}{b^2} = \cos^2 t + \sin^2 t$$

$$\frac{x^2}{a^2} + \frac{y^2}{b^2} = 1$$

The above equation is an ellipse centered at the origin.

Exercise Set 11.3

1. $$\frac{x^2}{9} - \frac{y^2}{1} = 1$$

$$\frac{(x - 0)^2}{3^2} - \frac{(y - 0)^2}{1^2} = 1$$

The center is (0, 0); a = 3 and b = 1.

The vertices are (-3, 0) and (3, 0).

Since $c^2 = a^2 + b^2$, $c = \sqrt{a^2 + b^2} = \sqrt{9 + 1}$
$= \sqrt{10}$.

The foci are $(-\sqrt{10}, 0)$ and $(\sqrt{10}, 0)$.

The asymptotes are $y = -\frac{1}{3}x$ and $y = \frac{1}{3}x$.

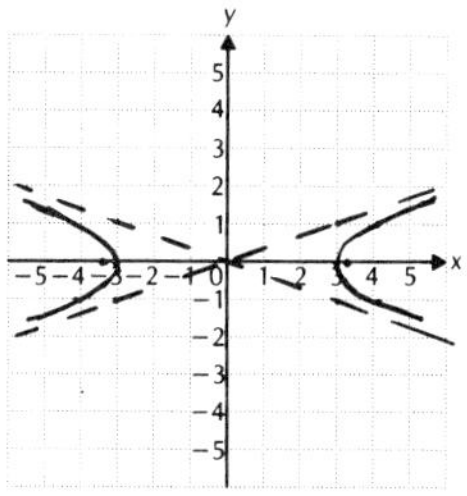

2. C: (0, 0)

V: (-1, 0), (1, 0)

F: $(-\sqrt{10}, 0)$, $(\sqrt{10}, 0)$

A: $y = 3x$, $y = -3x$

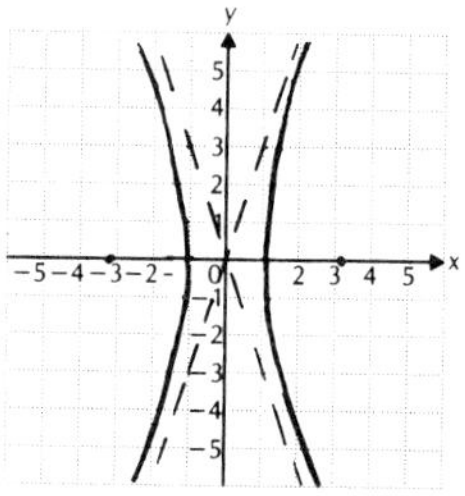

3. $$\frac{(x - 2)^2}{9} - \frac{(y + 5)^2}{1} = 1$$

$$\frac{(x - 2)^2}{3^2} - \frac{[y - (-5)]^2}{1^2} = 1$$

The center is (2, -5); a = 3 and b = 1.

The transverse axis is parallel to the x-axis.

Since $c^2 = a^2 + b^2$, $c = \sqrt{a^2 + b^2} = \sqrt{9 + 1}$
$= \sqrt{10}$.

3. (continued)

Consider the hyperbola $\frac{x^2}{3^2} - \frac{y^2}{1^2} = 1$.

The center is (0, 0); a = 3, b = 1, and c = $\sqrt{10}$.

The vertices are (-3, 0) and (3, 0).

The foci are $(-\sqrt{10}, 0)$ and $(\sqrt{10}, 0)$.

The asymptotes are $y = -\frac{1}{3}x$ and $y = \frac{1}{3}x$.

The vertices, foci, and asymptotes of the translated hyperbola are found by translation in the same way in which the center has been translated.

The vertices are (-3 + 2, 0 - 5) and (3 + 2, 0 - 5), or (-1, -5) and (5, -5).

The foci are $(-\sqrt{10} + 2, 0 - 5)$ and $(\sqrt{10} + 2, 0 - 5)$, or $(2 - \sqrt{10}, -5)$ and $(2 + \sqrt{10}, -5)$.

The asymptotes are

$$y - (-5) = -\frac{1}{3}(x - 2) \text{ and } y - (-5) = \frac{1}{3}(x - 2)$$

$$\text{or } y = -\frac{1}{3}x - \frac{13}{3} \text{ and } y = \frac{1}{3}x - \frac{17}{3}.$$

Graph the hyperpola.

First draw the rectangle which has (-3 + 2, 0 - 5), (3 + 2, 0 - 5), (0 + 2, -1 - 5), and (0 + 2, 1 - 5) or (-1, -5), (5, -5), (2, -6), and (2, -4) as midpoints of its four sides. Then draw the asymptotes and finally the branches of the hyperbola outward from the vertices toward the asymptotes.

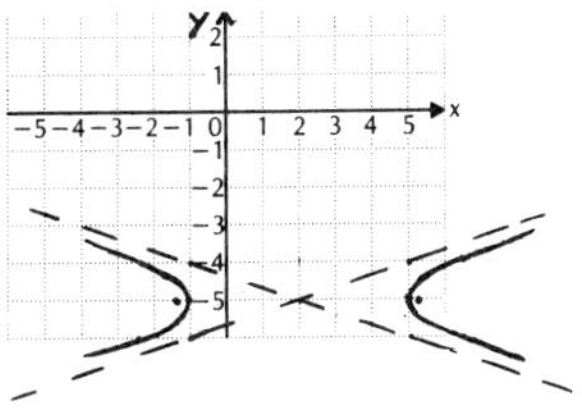

4. C: (2, -5)

V: (1, -5), (3, -5)

F: $(2 - \sqrt{10}, -5)$, $(2 + \sqrt{10}, -5)$

A: y = -3x + 1, y = 3x - 11

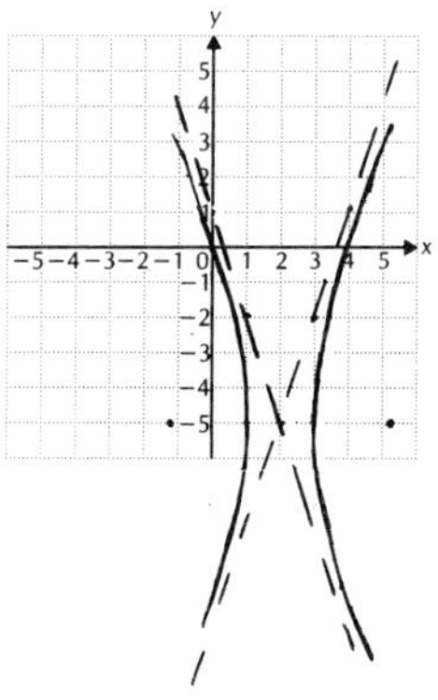

5. $$\frac{(y + 3)^2}{4} - \frac{(x + 1)^2}{16} = 1$$

$$\frac{[y - (-3)]^2}{2^2} - \frac{[x - (-1)]^2}{4^2} = 1$$

The center is (-1, -3); a = 4 and b = 2.

The transverse axis is parallel to the y-axis.

Since $c^2 = a^2 + b^2$, $c = \sqrt{a^2 + b^2} = \sqrt{16 + 4} = 2\sqrt{5}$.

Consider the hyperbola $\frac{y^2}{2^2} - \frac{x^2}{4^2} = 1$.

The center is (0, 0); a = 4, b = 2, and c = $2\sqrt{5}$.

The vertices are (0, -2) and (0, 2).

The foci are $(0, -2\sqrt{5})$ and $(0, 2\sqrt{5})$.

The asymptotes are $y = -\frac{1}{2}x$ and $y = \frac{1}{2}x$.

The vertices, foci, and asymptotes of the translated hyperbola are found by translation in the same way in which the center has been translated.

The vertices are (0 - 1, -2 - 3) and (0 - 1, 2 - 3) or (-1, -5) and (-1, -1).

The foci are $(0 - 1, -2\sqrt{5} - 3)$ and $(0 - 1, 2\sqrt{5} - 3)$ or $(-1, -3 - 2\sqrt{5})$ and $(-1, -3 + 2\sqrt{5})$.

The asymptotes are

$$y - (-3) = -\frac{1}{2}[x - (-1)] \text{ and}$$

$$y - (-3) = \frac{1}{2}[x - (-1)] \text{ or } y = -\frac{1}{2}x - \frac{7}{2}$$

$$\text{and } y = \frac{1}{2}x - \frac{5}{2}.$$

Graph the hyperbola.

First draw the rectangle which has (0 - 1, -2 - 3), (0 - 1, 2 - 3), (4 - 1, 0 - 3), and (-4 - 1, 0 - 3) or (-1, -5), (-1, -1), (3, -3), and (-5, -3) as midpoints of its four sides. Then draw the asymptotes and finally the branches of the hyperbola outward from the vertices toward the asymptotes.

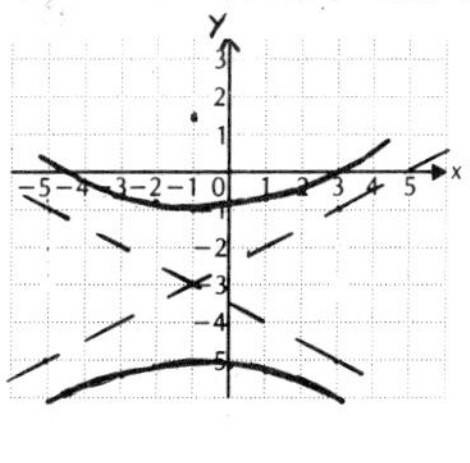

6. C: (-1, -3)

V: (-1, -8), (-1, 2)

F: $(-1, -3 - \sqrt{41})$, $(-1, -3 + \sqrt{41})$

A: $y = -\frac{5}{4}x - \frac{17}{4}$, $y = \frac{5}{4}x - \frac{7}{4}$

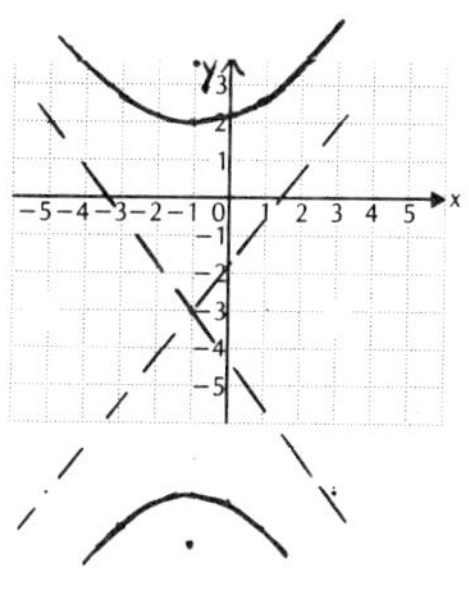

7.
$$x^2 - 4y^2 = 4$$
$$\frac{x^2}{4} - \frac{y^2}{1} = 1$$
$$\frac{(x - 0)^2}{2^2} - \frac{(y - 0)^2}{1^2} = 1$$

The center is (0, 0); a = 2 and b = 1.
The vertices are (-2, 0) and (2, 0).

Since $c^2 = a^2 + b^2$, $c = \sqrt{a^2 + b^2} = \sqrt{4 + 1} = \sqrt{5}$.

The foci are $(-\sqrt{5}, 0)$ and $(\sqrt{5}, 0)$.

The asymptotes are $y = -\frac{1}{2}x$ and $y = \frac{1}{2}x$.

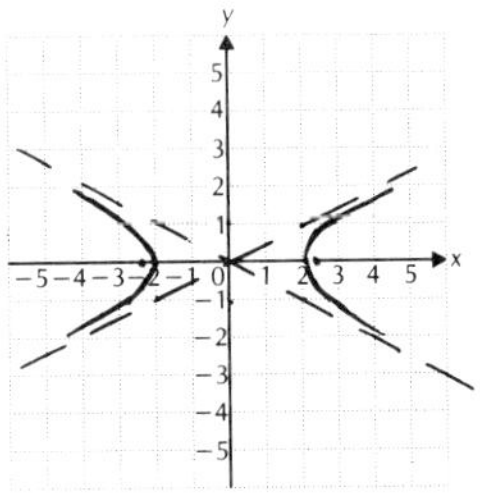

8. C: (0, 0)

V: (-1, 0), (1, 0)

F: $(-\sqrt{5}, 0)$, $(\sqrt{5}, 0)$

A: y = -2x, y = 2x

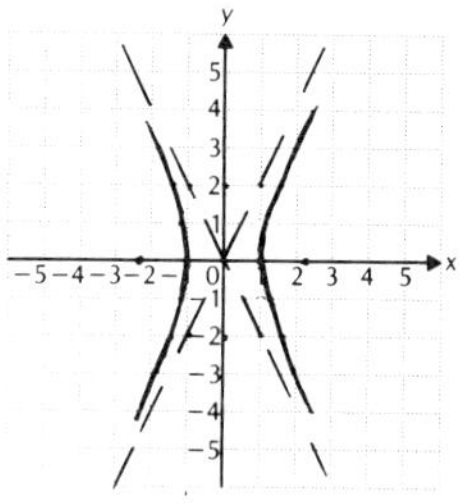

9.
$$4y^2 - x^2 = 4$$
$$y^2 - \frac{x^2}{4} = 1$$
$$\frac{(y - 0)^2}{1^2} - \frac{(x - 0)^2}{2^2} = 1$$

The center is (0, 0); a = 2 and b = 1.
The vertices are (0, -1) and (0, 1).

Since $c^2 = a^2 + b^2$, $c = \sqrt{a^2 + b^2} = \sqrt{4 + 1} = \sqrt{5}$.

The foci are $(0, -\sqrt{5})$ and $(0, \sqrt{5})$.

The asymptotes are $y = -\frac{1}{2}x$ and $y = \frac{1}{2}x$.

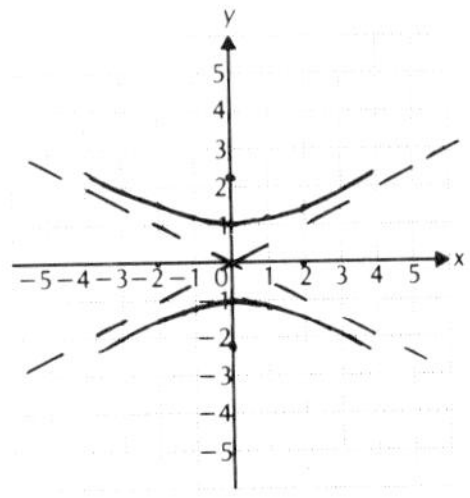

10. C: (0, 0)

V: (0, -2), (0, 2)

F: $(0, -\sqrt{5})$, $(0, \sqrt{5})$

A: y = -2x, y = 2x

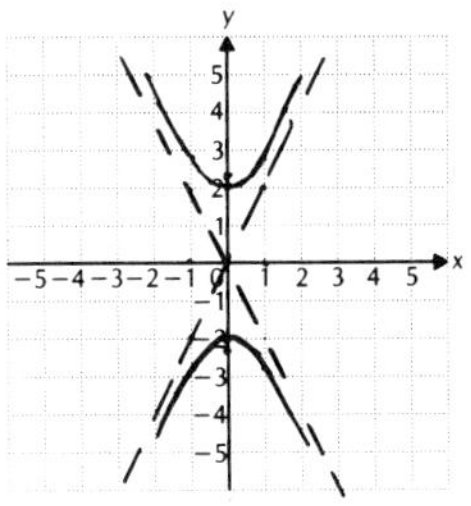

11.
$$x^2 - y^2 = 2$$
$$\frac{x^2}{2} - \frac{y^2}{2} = 1$$
$$\frac{(x - 0)^2}{(\sqrt{2})^2} - \frac{(y - 0)^2}{(\sqrt{2})^2} = 1$$

The center is (0, 0); $a = \sqrt{2}$ and $b = \sqrt{2}$.

The vertices are $(-\sqrt{2}, 0)$ and $(\sqrt{2}, 0)$.

Since $c^2 = a^2 + b^2$, $c = \sqrt{a^2 + b^2} = \sqrt{2 + 2}$
$= \sqrt{4} = 2$.

The foci are (-2, 0) and (2, 0).

The asymptotes are y = -x and y = x.

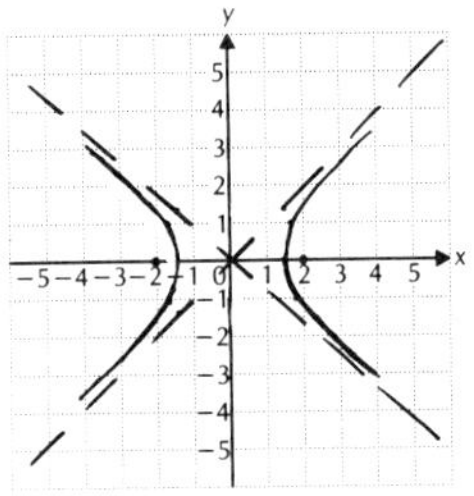

12. C: (0, 0)

V: $(-\sqrt{3}, 0)$, $(\sqrt{3}, 0)$

F: $(-\sqrt{6}, 0)$, $(\sqrt{6}, 0)$

A: y = -x, y = x

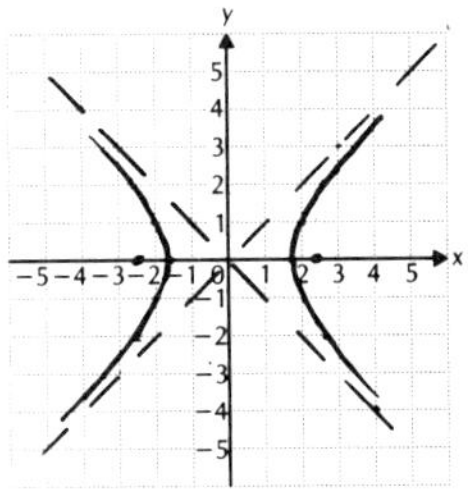

13.
$$x^2 - y^2 = \frac{1}{4}$$
$$\frac{x^2}{\frac{1}{4}} - \frac{y^2}{\frac{1}{4}} = 1$$
$$\frac{(x - 0)^2}{\left(\frac{1}{2}\right)^2} - \frac{(y - 0)^2}{\left(\frac{1}{2}\right)^2} = 1$$

The center is (0, 0); $a = \frac{1}{2}$ and $b = \frac{1}{2}$.

The vertices are $\left(-\frac{1}{2}, 0\right)$ and $\left(\frac{1}{2}, 0\right)$.

Since $c^2 = a^2 + b^2$, $c = \sqrt{a^2 + b^2} = \sqrt{\frac{1}{4} + \frac{1}{4}}$
$= \sqrt{\frac{1}{2}} = \frac{\sqrt{2}}{2}$.

13. (continued)

The foci are $\left(-\frac{\sqrt{2}}{2}, 0\right)$ and $\left(\frac{\sqrt{2}}{2}, 0\right)$.

The asymptotes are $y = -x$ and $y = x$.

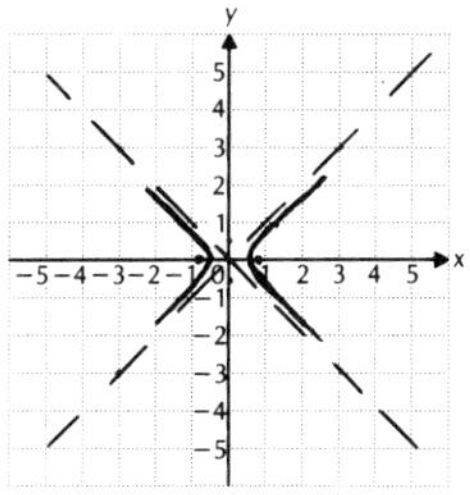

14. C: $(0, 0)$

V: $\left(-\frac{1}{3}, 0\right)$, $\left(\frac{1}{3}, 0\right)$

F: $\left(-\frac{\sqrt{2}}{3}, 0\right)$, $\left(\frac{\sqrt{2}}{3}, 0\right)$

A: $y = -x$, $y = x$

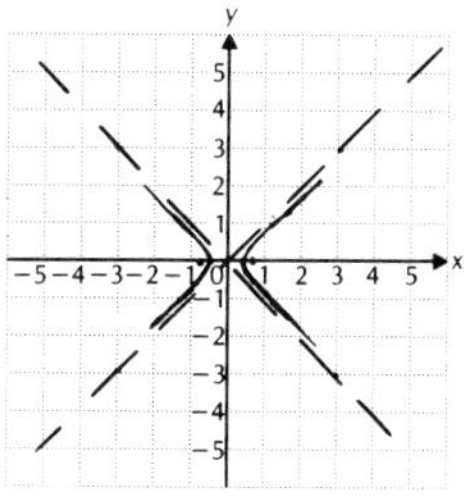

15.

$$x^2 - y^2 - 2x - 4y - 4 = 0$$
$$(x^2 - 2x) - (y^2 + 4y) = 4$$
$$(x^2 - 2x + 1) - (y^2 + 4y + 4) = 4 + 1 - 4$$
$$(x - 1)^2 - (y + 2)^2 = 1$$
$$\frac{(x - 1)^2}{1^2} - \frac{[y - (-2)]^2}{1^2} = 1$$

The center of the hyperbola is $(1, -2)$; $a = 1$ and $b = 1$.

The transverse axis is parallel to the x-axis.

Since $c^2 = a^2 + b^2$, $c = \sqrt{a^2 + b^2} = \sqrt{1 + 1} = \sqrt{2}$.

Consider the hyperbola $\frac{x^2}{1^2} - \frac{y^2}{1^2} = 1$

The center is $(0, 0)$; $a = 1$, $b = 1$, and $c = \sqrt{2}$.
The vertices are $(-1, 0)$ and $(1, 0)$.
The foci are $(-\sqrt{2}, 0)$ and $(\sqrt{2}, 0)$.
The asymptotes are $y = -x$ and $y = x$.

The vertices, foci, and asymptotes of the translated hyperbola are found in the same way in which the center has been translated.

The vertices are

$(-1 + 1, 0 - 2)$ and $(1 + 1, 0 - 2)$

or

$(0, -2)$ and $(2, -2)$.

The foci are

$(-\sqrt{2} + 1, 0 - 2)$ and $(\sqrt{2} + 1, 0 - 2)$

or

$(1 - \sqrt{2}, -2)$ and $(1 + \sqrt{2}, -2)$.

15. (continued)

The asymptotes are

$y - (-2) = -(x - 1)$ and $y - (-2) = x - 1$

or

$y = -x - 1$ and $y = x - 3$.

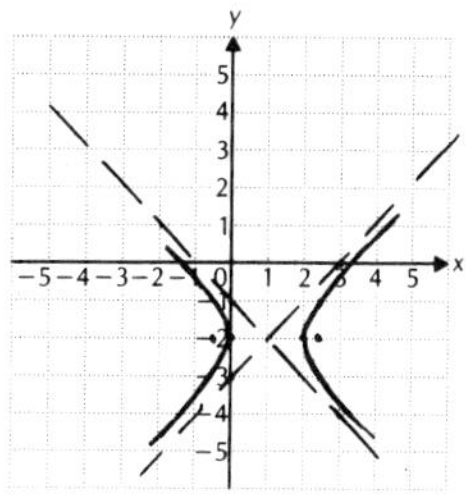

16. C: $(-1, -2)$

V: $(-2, -2)$, $(0, -2)$

F: $(-1 - \sqrt{5}, -2)$,
$(-1 + \sqrt{5}, -2)$

A: $y = -2x - 4$, $y = 2x$

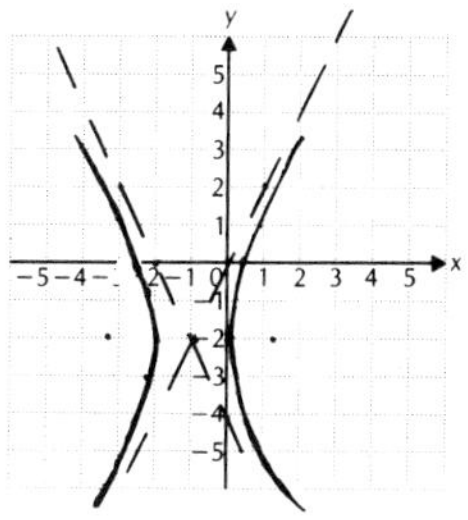

17.

$$36x^2 - y^2 - 24x + 6y - 41 = 0$$
$$(36x^2 - 24x) - (y^2 - 6y) = 41$$
$$36\left(x^2 - \frac{2}{3}x\right) - (y^2 - 6y) = 41$$
$$36\left(x^2 - \frac{2}{3}x + \frac{1}{9}\right) - (y^2 - 6y + 9) = 41 + 4 - 9$$
$$36\left(x - \frac{1}{3}\right)^2 - (y - 3)^2 = 36$$
$$\frac{\left(x - \frac{1}{3}\right)^2}{1} - \frac{(y - 3)^2}{36} = 1$$
$$\frac{\left(x - \frac{1}{3}\right)^2}{1^2} - \frac{(y - 3)^2}{6^2} = 1$$

The center of the hyperbola is $\left(\frac{1}{3}, 3\right)$; $a = 1$ and $b = 6$.

The transverse axis is parallel to the x-axis.

Since $c^2 = a^2 + b^2$, $c = \sqrt{a^2 + b^2} = \sqrt{1 + 36} = \sqrt{37}$.

Consider the hyperbola $\frac{x^2}{1^2} - \frac{y^2}{6^2} = 1$.

The center is $(0, 0)$; $a = 1$, $b = 6$, and $c = \sqrt{37}$.
The vertices are $(-1, 0)$ and $(1, 0)$.
The foci are $(-\sqrt{37}, 0)$ and $(\sqrt{37}, 0)$.
The asymptotes are $y = -6x$ and $y = 6x$.

The vertices, foci, and asymptotes of the translated hyperbola are found in the same way in which the center has been translated.

17. (continued)

The vertices are

$\left(-1 + \frac{1}{3}, 0 + 3\right)$ and $\left(1 + \frac{1}{3}, 0 + 3\right)$

or

$\left(-\frac{2}{3}, 3\right)$ and $\left(\frac{4}{3}, 3\right)$.

The foci are

$\left(-\sqrt{37} + \frac{1}{3}, 0 + 3\right)$ and $\left(\sqrt{37} + \frac{1}{3}, 0 + 3\right)$

or

$\left(\frac{1}{3} - \sqrt{37}, 3\right)$ and $\left(\frac{1}{3} + \sqrt{37}, 3\right)$.

The asymptotes are

$y - 3 = -6\left(x - \frac{1}{3}\right)$ and $y - 3 = 6\left(x - \frac{1}{3}\right)$

or

$y = -6x + 5$ and $y = 6x + 1$.

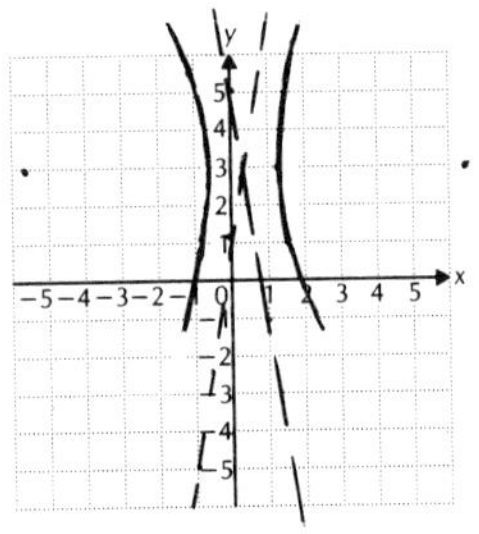

18. C: $(-3, 1)$

V: $\left(-3 + \frac{4\sqrt{2}}{3}, 1\right)$

$\left(-3 - \frac{4\sqrt{2}}{3}, 1\right)$

F: $\left(-3 + \frac{2\sqrt{26}}{3}, 1\right)$,

$\left(-3 - \frac{2\sqrt{26}}{3}, 1\right)$

A: $y = -\frac{3}{2}x - \frac{7}{2}$,

$y = \frac{3}{2}x + \frac{11}{2}$

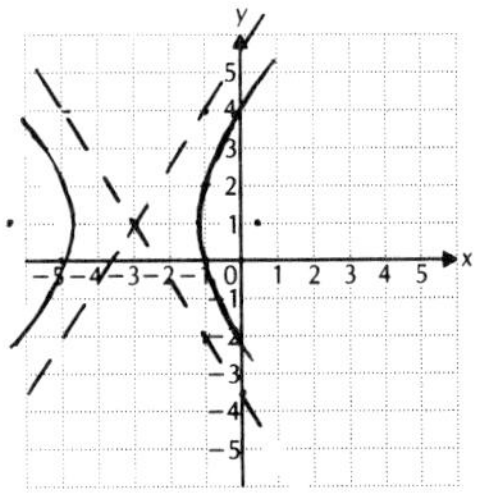

19. $xy = 1$

Since $1 > 0$, the branches of the hyperbola lie in the first and third quadrants. The coordinate axes are its asymptotes.

19. (continued)

x	y
$\frac{1}{4}$	4
$\frac{1}{2}$	2
1	1
$\frac{3}{2}$	$\frac{2}{3}$
3	$\frac{1}{3}$
4	$\frac{1}{4}$

x	y
$-\frac{1}{4}$	-4
$-\frac{1}{2}$	-2
-1	-1
$-\frac{3}{2}$	$-\frac{2}{3}$
-3	$-\frac{1}{3}$
-4	$-\frac{1}{4}$

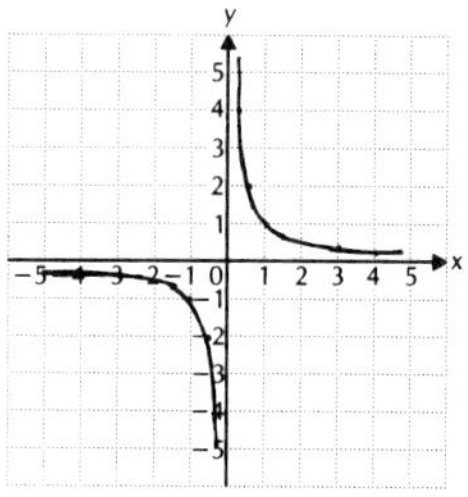

20.

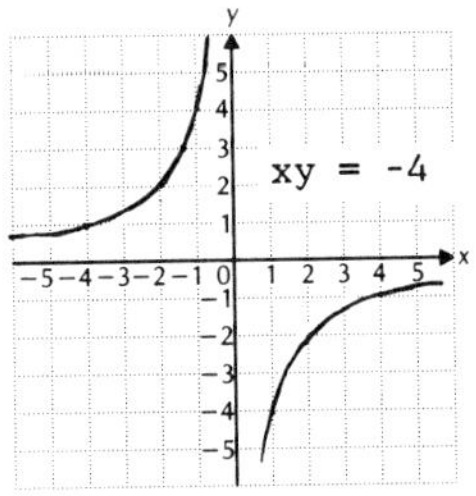

21. $xy = -8$

Since $-8 < 0$, the branches of the hyperbola lie in the second and fourth quadrants. The coordinates axes are its asymptotes.

x	y
1	-8
2	-4
4	-2
8	-1

x	y
-1	8
-2	4
-4	2
-8	1

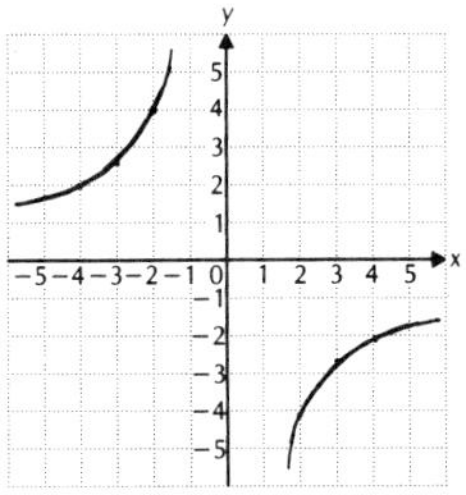

22.

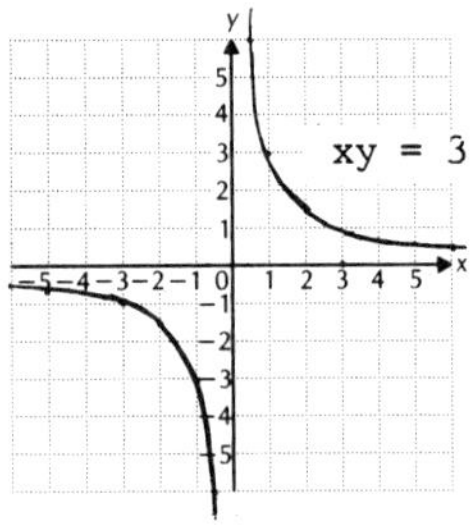

23.

$$x^2 - y^2 - 2.046x - 4.088y - 4.228 = 0$$

$$(x^2 - 2.046x) - (y^2 + 4.088y) = 4.228$$

$$(x^2 - 2.046x + 1.046529) - (y^2 + 4.088y + 4.177936) = 4.228 + 1.046529 - 4.177936$$

$$(x - 1.023)^2 - (y + 2.044)^2 = 1.096593$$

$$\frac{(x - 1.023)^2}{(1.04718)^2} - \frac{[y - (-2.044)]^2}{(1.04718)^2} = 1$$

$$(\sqrt{1.096593} \approx 1.04718)$$

23. (continued)

The center is (1.023, -2.044); a = b = 1.04718.

Consider the hyperbola $\frac{x^2}{(1.04718)^2} - \frac{y^2}{(1.04718)^2} = 1$.

The center is (0, 0); a = b = 1.04718.

The vertices are (-1.04718, 0) and (1.04718, 0).

The asymptotes are y = -x and y = x.

The vertices and asymptotes of the translated hyperbola are found by translation in the same way in which the center has been translated.

The vertices are

(-1.04718 + 1.023, 0 - 2.044) and
(1.04718 + 1.023, 0 - 2.044)

or

(-0.02418, -2.044) and (2.07018, -2.044).

The asymptotes are

y - (-2.044) = -(x - 1.023) and
y - (-2.044) = x - 1.023

or

y = -x - 1.021 and y = x - 3.067.

24. $x^2 - \frac{y^2}{3} = 1$

25. If the asymptotes are $y = \pm \frac{3}{2}x$, then a = 2 and b = 3.

The center of the hyperbola is (0, 0), the intersection of the asymptotes.

The vertices are (-2, 0) and (2, 0); thus the transverse axis is on the x-axis.

The standard form of the equation of the hyperbola must be in the form $\frac{x^2}{a^2} - \frac{y^2}{b^2} = 1$.

The equation is $\frac{x^2}{4} - \frac{y^2}{9} = 1$.

26. a) No

b) $y = \pm \frac{1}{2}\sqrt{x^2 - 4}$

c) Yes, Domain: $\{x \mid x \leqslant -2 \text{ or } x \geqslant 2\}$
Range: $\{y \mid y \geqslant 0\}$

d) Yes, Domain: $\{x \mid x \leqslant -2 \text{ or } x \geqslant 2\}$
Range: $\{y \mid y \leqslant 0\}$

27. See answer section in text.

Exercise Set 11.4

1. $x^2 = 8y$

$x^2 = 4 \cdot 2 \cdot y$ (Writing $x^2 = 4py$)

Vertex: (0, 0)

Focus: (0, 2) [(0, p)]

Directrix: y = -2 (y = -p)

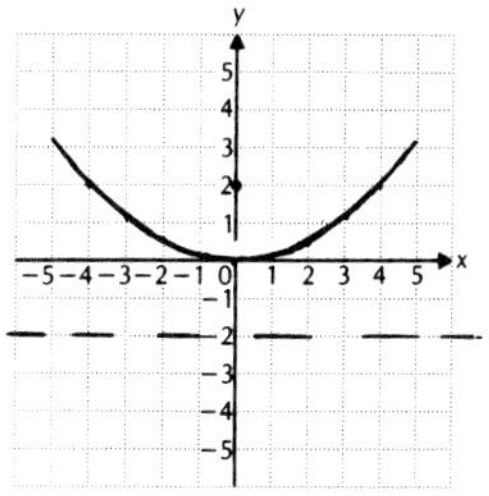

2. V: (0, 0)

F: (0, 4)

Directrix: y = -4

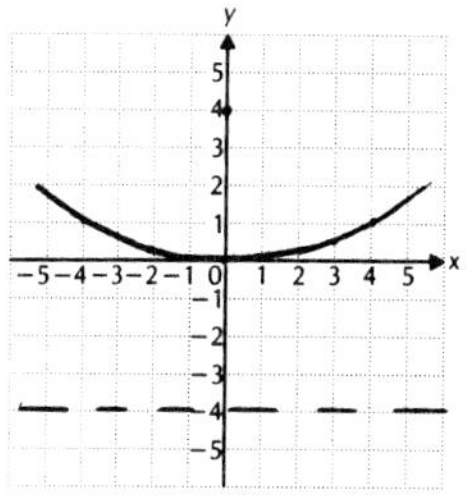

3. $y^2 = -6x$

$y^2 = 4\left(-\frac{3}{2}\right)x$ (Writing $y^2 = 4px$)

Vertex: (0, 0)

Focus: $\left(-\frac{3}{2}, 0\right)$ [(p, 0)]

Directrix: $x = -\left(-\frac{3}{2}\right) = \frac{3}{2}$ (x = -p)

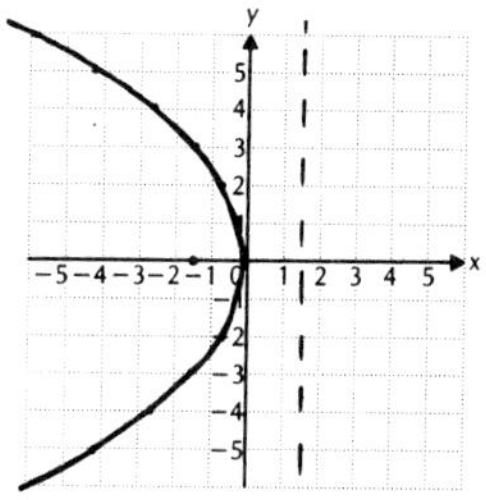

4. V: (0, 0)

F: $\left(-\frac{1}{2}, 0\right)$

Directrix: $x = \frac{1}{2}$

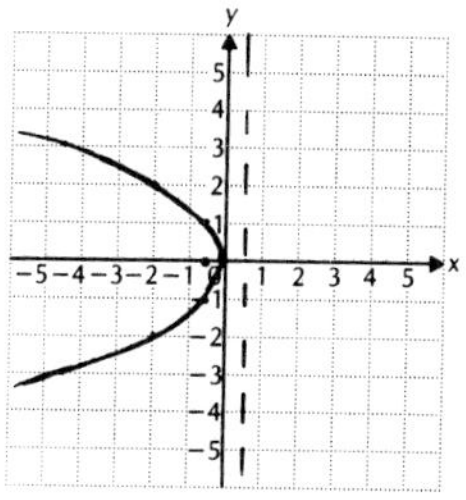

5. $x^2 - 4y = 0$

$x^2 = 4y$

$x^2 = 4 \cdot 1 \cdot y$ (Writing $x^2 = 4py$)

Vertex: (0, 0)

Focus: (0, 1) [(0, p)]

Directrix: y = -1 (y = -p)

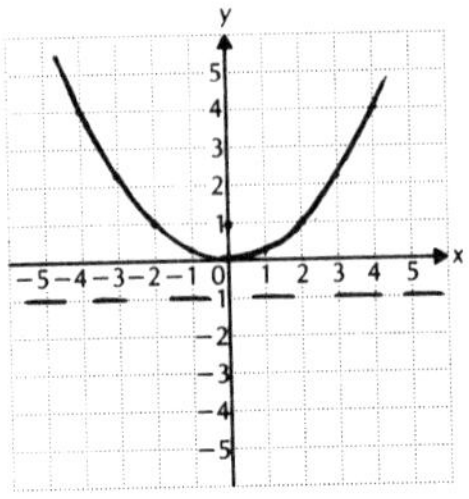

6. V: (0, 0)

F: (-1, 0)

Directrix: x = 1

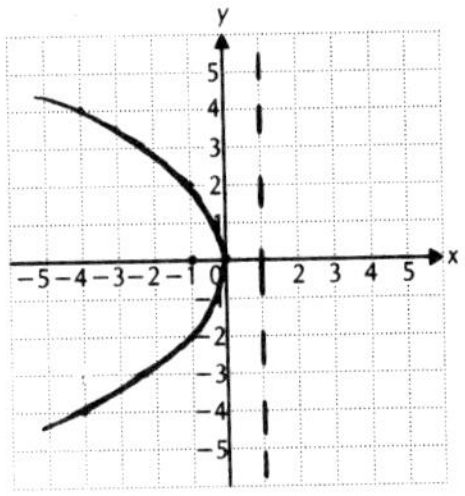

7. $y = 2x^2$

$x^2 = \frac{1}{2} y$

$x^2 = 4 \cdot \frac{1}{8} \cdot y$ (Writing $x^2 = 4py$)

Vertex: (0, 0)

Focus: $\left(0, \frac{1}{8}\right)$ [(0, p)]

Directrix: $y = -\frac{1}{8}$ (y = -p)

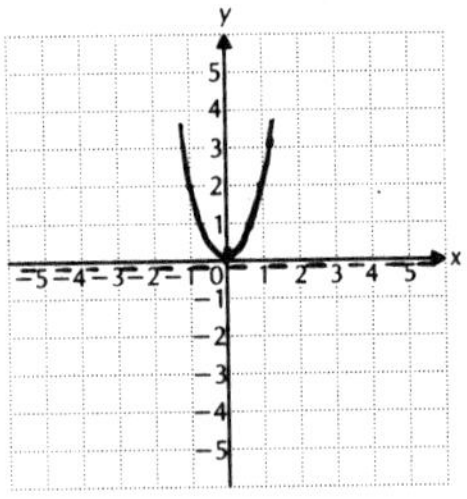

8. V: (0, 0)

F: $\left(0, \frac{1}{2}\right)$

Directrix: $y = -\frac{1}{2}$

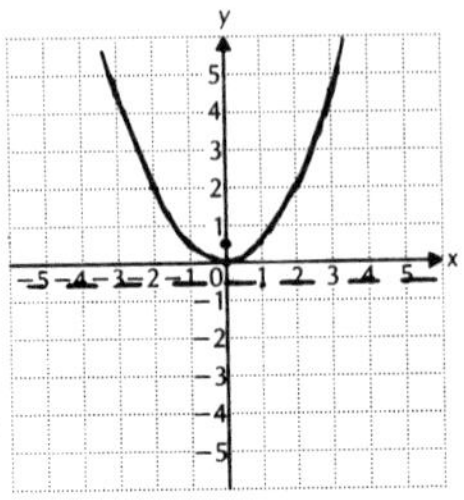

9. It helps to sketch a graph of the focus and the directrix.

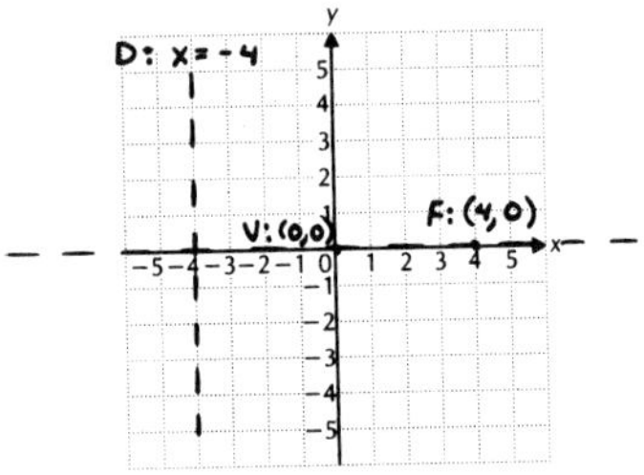

The focus, (4, 0), is on the axis of symmetry, the x-axis. The vertex is (0, 0), and p is 4 - 0, or 4. The equation is of the type

$y^2 = 4px$

$y^2 = 4 \cdot 4 \cdot x$ (Substituting 4 for p)

$y^2 = 16x$

10. $x^2 = y$

11. It is helpful to sketch a graph of the focus and the directrix.

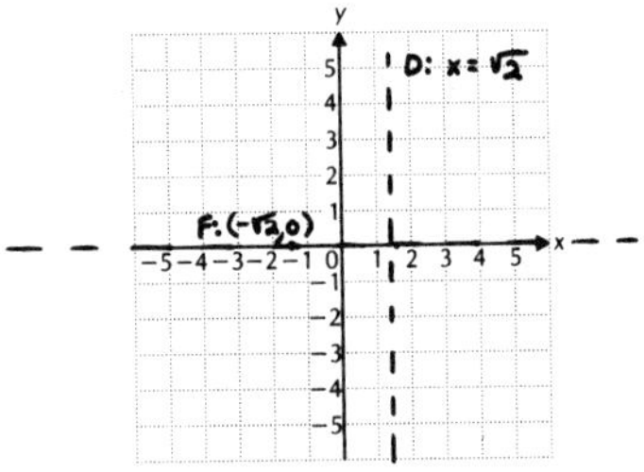

The focus, $(-\sqrt{2}, 0)$, is on the axis of symmetry, the x-axis. The vertex is (0, 0), and $p = -\sqrt{2} - 0$, or $-\sqrt{2}$. The equation is of the type

$y^2 = 4px$

$y^2 = 4(-\sqrt{2})x$ (Substituting $-\sqrt{2}$ for p)

$y^2 = -4\sqrt{2}\,x$

12. $x^2 = -4\pi y$

13. It is helpful to sketch a graph of the focus and the directrix.

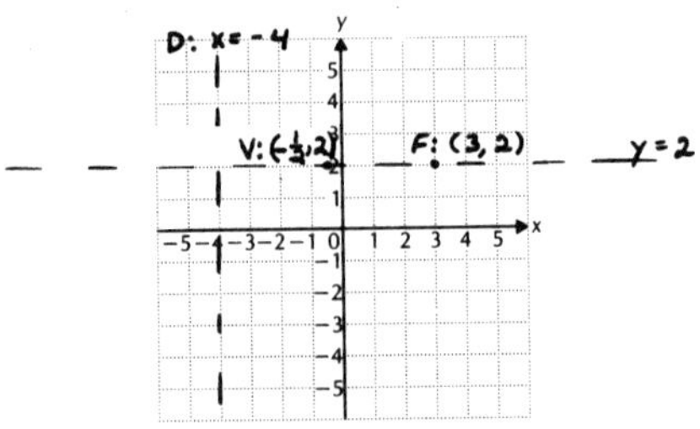

13. (continued)

The focus, (3, 2), is on the axis of symmetry, y = 2, which is parallel to the x-axis. The vertex is $\left(-\frac{1}{2}, 2\right)$, and $p = 3 - \left(-\frac{1}{2}\right)$, or $\frac{7}{2}$. The equation is of the type

$(y - k)^2 = 4p(x - h)$

$(y - 2)^2 = 4 \cdot \frac{7}{2}\left[x - \left(-\frac{1}{2}\right)\right]$ (Substituting)

$(y - 2)^2 = 14\left(x + \frac{1}{2}\right)$

14. $(x + 2)^2 = 12y$

15. $(x + 2)^2 = -6(y - 1)$

$[x - (-2)]^2 = 4\left(-\frac{3}{2}\right)(y - 1)$ $[(x - h)^2 = 4p(y-k)]$

Vertex: (-2, 1) [(h, k)]

Focus: $\left(-2, 1 + \left(-\frac{3}{2}\right)\right)$, or $\left(-2, -\frac{1}{2}\right)$

[(h, k + p)]

Directrix: $y = 1 - \left(-\frac{3}{2}\right) = \frac{5}{2}$ (y = k - p)

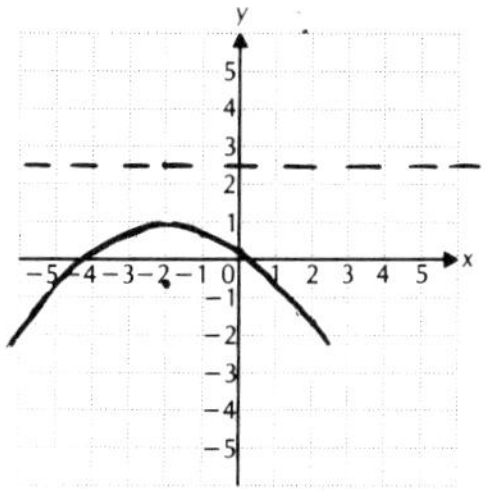

16. V: (-2, 3)

F: (-7, 3)

Directrix: x = 3

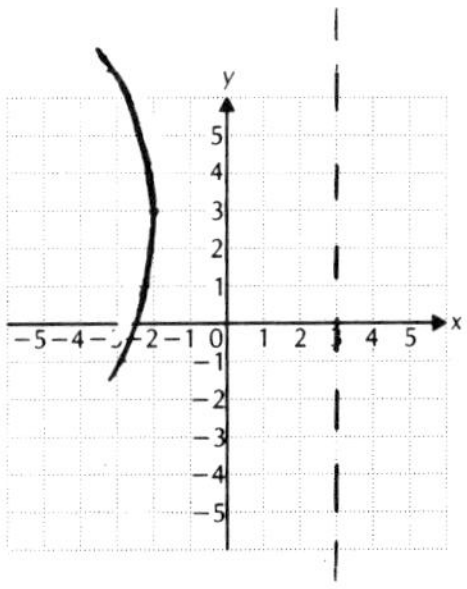

17. $x^2 + 2x + 2y + 7 = 0$

$x^2 + 2x = -2y - 7$

$(x^2 + 2x + 1) = -2y - 7 + 1 = -2y - 6$

$(x + 1)^2 = -2(y + 3)$

$[x - (-1)]^2 = 4\left(-\frac{1}{2}\right)[y - (-3)]$

$[(x - h)^2 = 4p(y - k)]$

Vertex: (-1, -3) [(h, k)]

Focus: $\left(-1, -3 + \left(-\frac{1}{2}\right)\right)$, or $\left(-1, -\frac{7}{2}\right)$ [(h, k+p)]

Directrix: $y = -3 - \left(-\frac{1}{2}\right) = -\frac{5}{2}$ (y = k - p)

17. (continued)

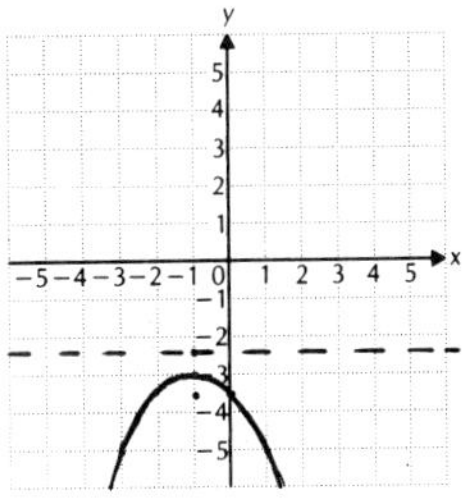

18. V: (7, -3)

F: $\left(\frac{29}{4}, -3\right)$

Directrix: $x = \frac{27}{4}$

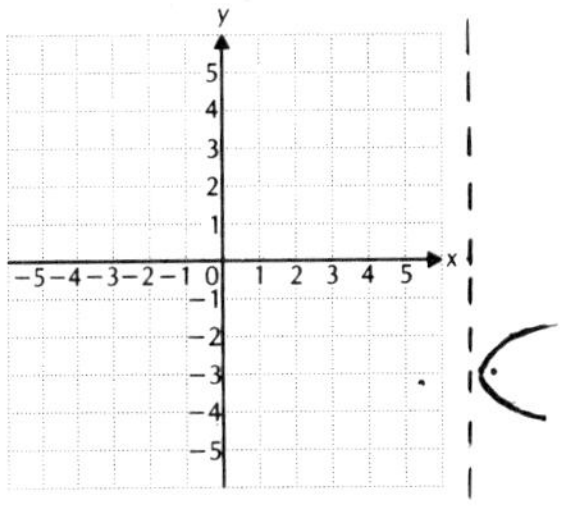

19. $x^2 - y - 2 = 0$

$x^2 = y + 2$

$(x - 0)^2 = 4 \cdot \frac{1}{4} \cdot [y - (-2)]$

$[(x - h)^2 = 4p(y - k)]$

Vertex: (0, -2) [(h, k)]

Focus: $\left(0, -2 + \frac{1}{4}\right)$, or $\left(0, -\frac{7}{4}\right)$ [(h, k + p)]

Directrix: $y = -2 - \frac{1}{4} = -\frac{9}{4}$ (y = k - p)

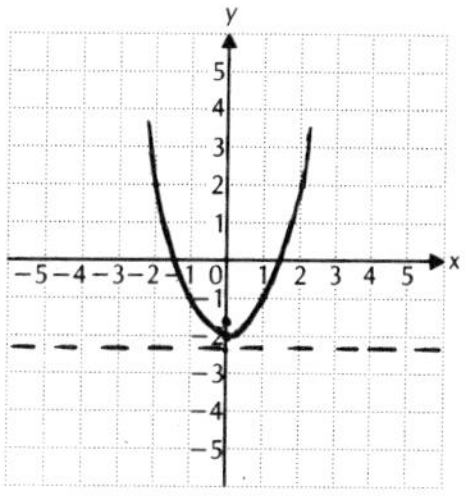

20. V: (2, -2)

F: $\left(2, -\frac{3}{2}\right)$

Directrix: $y = -\frac{5}{2}$

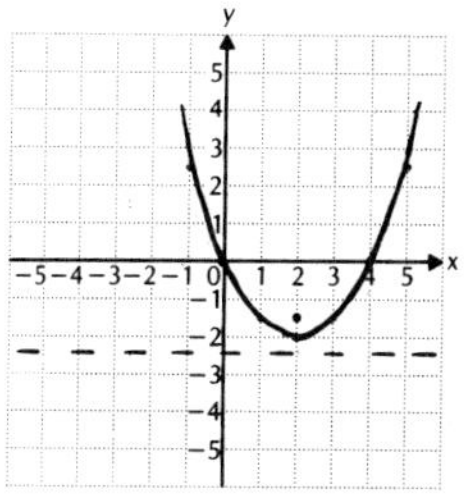

21. $y = x^2 + 4x + 3$

$y - 3 = x^2 + 4x$

$y - 3 + 4 = x^2 + 4x + 4$

$y + 1 = (x + 2)^2$

$4 \cdot \frac{1}{4} \cdot [y - (-1)] = [x - (-2)]^2$

$[(x - h)^2 = 4p(y - k)]$

Vertex: (-2, -1) [(h, k)]

21. (continued)

Focus: $\left(-2, -1 + \frac{1}{4}\right)$, or $\left(-2, -\frac{3}{4}\right)$ $[(h, k + p)]$

Directrix: $y = -1 - \frac{1}{4} = -\frac{5}{4}$ $(y = k - p)$

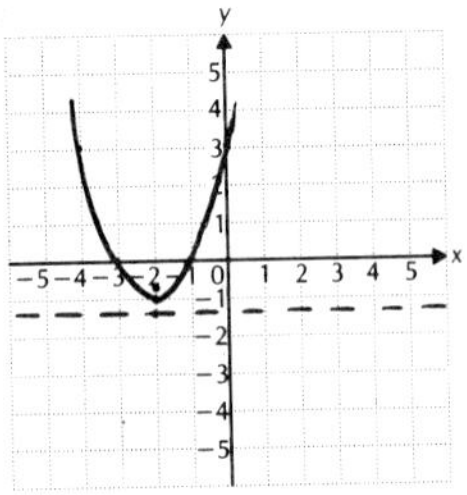

22. V: $(-3, 1)$

F: $\left(-3, \frac{5}{4}\right)$

Directrix: $y = \frac{3}{4}$

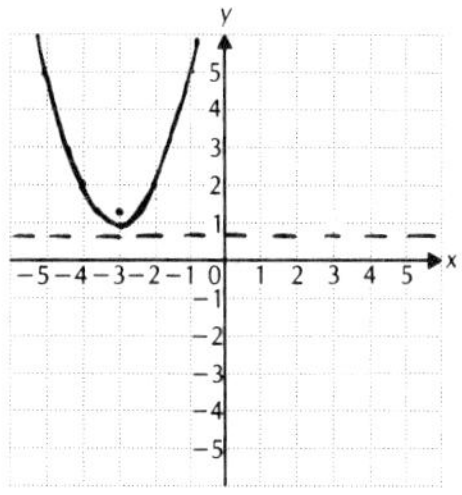

23. $4y^2 - 4y - 4x + 24 = 0$

$y^2 - y - x + 6 = 0$ $\left(\text{Multiplying by } \frac{1}{4}\right)$

$y^2 - y = x - 6$

$y^2 - y + \frac{1}{4} = x - 6 + \frac{1}{4}$

$\left(y - \frac{1}{2}\right)^2 = x - \frac{23}{4}$

$\left(y - \frac{1}{2}\right)^2 = 4 \cdot \frac{1}{4}\left(x - \frac{23}{4}\right)$

$[(y - k)^2 = 4p(x - h)]$

V: $\left(\frac{23}{4}, \frac{1}{2}\right)$ $[(h, k)]$

F: $\left(\frac{23}{4} + \frac{1}{4}, \frac{1}{2}\right)$, or $\left(6, \frac{1}{2}\right)$ $[(h + p, k)]$

D: $x = \frac{23}{4} - \frac{1}{4} = \frac{22}{4}$, or $\frac{11}{2}$ $(x = h - p)$

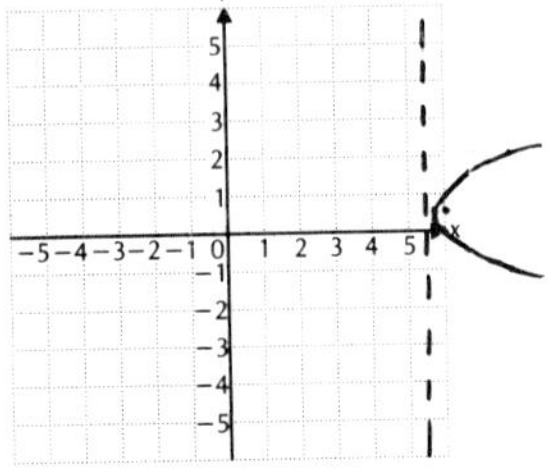

24. V: $\left(-\frac{17}{4}, -\frac{1}{2}\right)$

F: $\left(-4, -\frac{1}{2}\right)$

D: $x = -\frac{9}{2}$

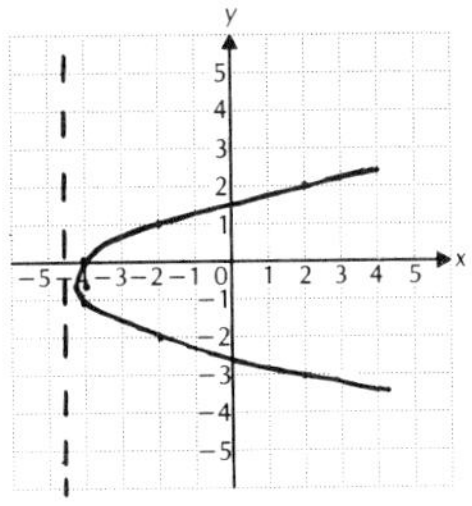

25. $x^2 = 8056.25y$

$x^2 = 4(2014.0625)y$ $(x^2 = 4py)$

V: $(0, 0)$

F: $(0, 2014.0625)$ $[(0, p)]$

D: $y = -2014.0625$ $(y = -p)$

26. V: $(0, 0)$

F: $(-1911.47, 0)$

D: $x = 1911.47$

27.

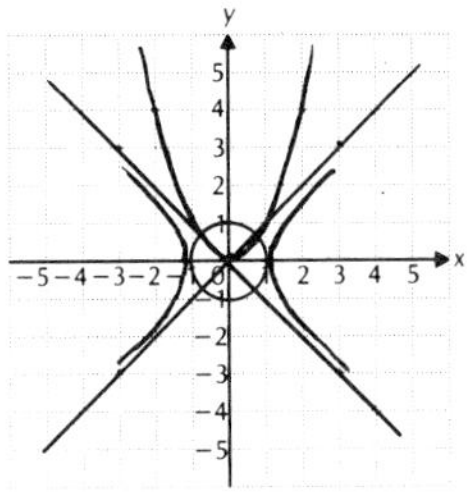

The graph of $x^2 - y^2 = 0$ is the union of the lines $x = y$ and $x = -y$. The others are respectively, a hyperbola, a circle, and a parabola.

28.

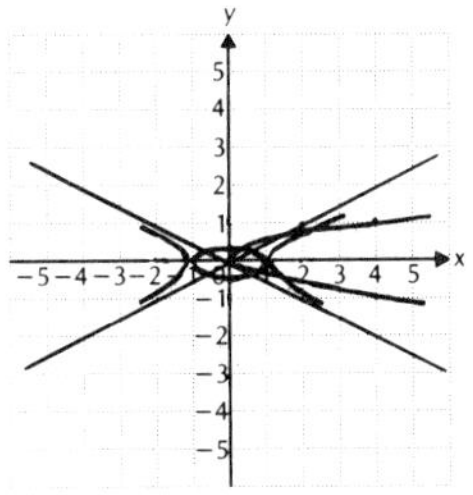

The graph of $x^2 - 4y^2 = 0$ is the union of the lines $x = 2y$ and $x = -2y$. The others are respectively, a hyperbola, an ellipse, and a parabola.

29. If the line of symmetry is parallel to the y-axis and the vertex (h, k) is $(-1, 2)$, then the equation is of the type

$(x - h)^2 = 4p(y - k)$

Solve for p substituting $(-1, 2)$ for (h, k) and $(-3, 1)$ for (x, y).

$[-3 - (-1)]^2 = 4p(1 - 2)$

$4 = -4p$

$-1 = p$

The equation of the parabola is

$[x - (-1)]^2 = 4(-1)(y - 2)$

or

$(x + 1)^2 = -4(y - 2)$

30. a) No; b) No, unless $p = 0$

31. $\begin{vmatrix} y - k & x - h \\ 4p & y - k \end{vmatrix} = 0$

$(y - k)^2 - (x - h)(4p) = 0$

$(y - k)^2 = 4p(x - h)$

32. 10, 11.6, 16.4, 24.4, 35.6 ft

Exercise Set 11.5

1. $x^2 + y^2 = 25$ (1)

$y - x = 1$ (2)

First solve equation (2) for y.

$y - x = 1$

$y = x + 1$

Then substitute x + 1 for y in equation (1) and solve for x.

$x^2 + y^2 = 25$

$x^2 + (x + 1)^2 = 25$

$x^2 + x^2 + 2x + 1 = 25$

$2x^2 + 2x - 24 = 0$

$x^2 + x - 12 = 0$

$(x + 4)(x - 3) = 0$

$x + 4 = 0$ or $x - 3 = 0$

$x = -4$ or $x = 3$

Now substitute these numbers into the linear equation and solve for y.

If $x = -4$, then $y = -4 + 1 = -3$.

If $x = 3$, then $y = 3 + 1 = 4$.

The pairs (-4, -3) and (3, 4) check, hence are solutions.

2. (-8, -6), (6, 8)

3. $y^2 - x^2 = 9$ (1)

$2x - 3 = y$ (2)

Substitute 2x - 3 for y in equation (1) and solve for x.

$y^2 - x^2 = 9$

$(2x - 3)^2 - x^2 = 9$

$4x^2 - 12x + 9 - x^2 = 9$

$3x^2 - 12x = 0$

$x^2 - 4x = 0$

$x(x - 4) = 0$

$x = 0$ or $x - 4 = 0$

$x = 0$ or $x = 4$

Now substitute these numbers into the linear equation and solve for y.

3. (continued)

If $x = 0$, $y = 2\cdot 0 - 3 = -3$.

If $x = 4$, $y = 2\cdot 4 - 3 = 5$.

The pairs (0, -3) and (4, 5) check, hence are solutions.

4. (1, -7), (-7, 1)

5. $4x^2 + 9y^2 = 36$ (1)

$3y + 2x = 6$ (2)

First solve equation (2) for y.

$3y + 2x = 6$

$3y = 6 - 2x$

$y = \frac{6 - 2x}{3}$

Then substitute $\frac{6 - 2x}{3}$ for y in equation (1) and solve for x.

$4x^2 + 9y^2 = 36$

$4x^2 + 9\left(\frac{6 - 2x}{3}\right)^2 = 36$

$4x^2 + 9\left(\frac{36 - 24x + 4x^2}{9}\right) = 36$

$4x^2 + 36 - 24x + 4x^2 = 36$

$8x^2 - 24x = 0$

$x^2 - 3x = 0$

$x(x - 3) = 0$

$x = 0$ or $x - 3 = 0$

$x = 0$ or $x = 3$

Now substitute these numbers into the linear equation and solve for y.

If $x = 0$, $y = \frac{6 - 2\cdot 0}{3} = 2$.

If $x = 3$, $y = \frac{6 - 2\cdot 3}{3} = 0$.

The pairs (0, 2) and (3, 0) check, hence are solutions.

6. (2, 0), (0, 3)

7. $y^2 = x + 3$ (1)

$2y = x + 4$ (2)

First solve equation (2) for x.

$2y = x + 4$

$2y - 4 = x$

Then substitute 2y - 4 for x in equation (1) and solve for y.

$y^2 = x + 3$

$y^2 = (2y - 4) + 3$

$y^2 = 2y - 1$

$y^2 - 2y + 1 = 0$

$(y - 1)(y - 1) = 0$

7. (continued)

$y - 1 = 0$ or $y - 1 = 0$

$y = 1$ or $y = 1$

Now substitute 1 for y into the linear equation and solve for x.

If $y = 1$, $x = 2\cdot 1 - 4 = -2$.

The pair $(-2, 1)$ checks, hence is the solution.

8. $(1, 1)$, $(2, 4)$

9. $x^2 + 4y^2 = 25$ (1)

$x + 2y = 7$ (2)

First solve equation (2) for x.

$x + 2y = 7$

$x = 7 - 2y$

Next substitute $7 - 2y$ for x in equation (1) and solve for y.

$$x^2 + 4y^2 = 25$$
$$(7 - 2y)^2 + 4y^2 = 25$$
$$49 - 28y + 4y^2 + 4y^2 = 25$$
$$8y^2 - 28y + 24 = 0$$
$$2y^2 - 7y + 6 = 0$$
$$(2y - 3)(y - 2) = 0$$

$2y - 3 = 0$ or $y - 2 = 0$

$2y = 3$ or $y = 2$

$y = \frac{3}{2}$ or $y = 2$

Now substitute these numbers into the linear equation and solve for x.

If $y = \frac{3}{2}$, $x = 7 - 2 \cdot \frac{3}{2} = 4$.

If $y = 2$, $x = 7 - 2\cdot 2 = 3$.

The pairs $\left(4, \frac{3}{2}\right)$ and $(3, 2)$ check, hence are solutions.

10. $(3, 5)$, $\left(-\frac{5}{3}, -\frac{13}{3}\right)$

11. $x^2 - xy + 3y^2 = 27$ (1)

$x - y = 2$ (2)

First solve equation (2) for x.

$x - y = 2$

$x = y + 2$

Next substitute $y + 2$ for x in equation (1) and solve for x.

$$x^2 - xy + 3y^2 = 27$$
$$(y + 2)^2 - (y + 2)y + 3y^2 = 27$$
$$y^2 + 4y + 4 - y^2 - 2y + 3y^2 = 27$$
$$3y^2 + 2y - 23 = 0$$

11. (continued)

Using the quadratic formula, we find that

$y = \frac{-1 - \sqrt{70}}{3}$ or $y = \frac{-1 + \sqrt{70}}{3}$.

Now substitute these numbers into the linear equation and solve for x.

If $y = \frac{-1 - \sqrt{70}}{3}$, then $x = \frac{-1 - \sqrt{70}}{3} + 2$, or $\frac{5 - \sqrt{70}}{3}$.

If $y = \frac{-1 + \sqrt{70}}{3}$, then $x = \frac{-1 + \sqrt{70}}{3} + 2$, or $\frac{5 + \sqrt{70}}{3}$.

The pairs $\left(\frac{5 - \sqrt{70}}{3}, \frac{-1 - \sqrt{70}}{3}\right)$ and $\left(\frac{5 + \sqrt{70}}{3}, \frac{-1 + \sqrt{70}}{3}\right)$ check and are solutions.

12. $\left(\frac{11}{4}, -\frac{9}{8}\right)$, $(1, -2)$

13. $3x + y = 7$ (1)

$4x^2 + 5y = 56$ (2)

First solve equation (1) for y.

$3x + y = 7$

$y = 7 - 3x$

Next substitute $7 - 3x$ for y in equation (2) for y and solve for x.

$$4x^2 + 5y = 56$$
$$4x^2 + 5(7 - 3x) = 56$$
$$4x^2 + 35 - 15x = 56$$
$$4x^2 - 15x - 21 = 0$$

Using the quadratic formula, we find that

$x = \frac{15 - \sqrt{561}}{8}$ or $x = \frac{15 + \sqrt{561}}{8}$.

Now substitute these numbers into the linear equation and solve for y.

If $x = \frac{15 - \sqrt{561}}{8}$, $y = 7 - 3\left(\frac{15 - \sqrt{561}}{8}\right)$, or $\frac{11 + 3\sqrt{561}}{8}$.

If $x = \frac{15 + \sqrt{561}}{8}$, $y = 7 - 3\left(\frac{15 + \sqrt{561}}{8}\right)$, or $\frac{11 - 3\sqrt{561}}{8}$.

The pairs $\left(\frac{15 - \sqrt{561}}{8}, \frac{11 + 3\sqrt{561}}{8}\right)$ and $\left(\frac{15 + \sqrt{561}}{8}, \frac{11 - 3\sqrt{561}}{8}\right)$ check and are the solutions.

14. $\left[-3, \frac{5}{2}\right]$, (3, 1)

15. Let x and y represent the numbers.

$x + y = 14$ (1)

$x^2 + y^2 = 106$ (2)

First solve equation (1) for y.

$x + y = 14$

$y = 14 - x$

Then substitute 14 - x for y in equation (2) and solve for x.

$$x^2 + y^2 = 106$$
$$x^2 + (14 - x)^2 = 106$$
$$x^2 + 196 - 28x + x^2 = 106$$
$$2x^2 - 28x + 90 = 0$$
$$x^2 - 14x + 45 = 0$$
$$(x - 9)(x - 5) = 0$$

$x - 9 = 0$ or $x - 5 = 0$

$x = 9$ or $x = 5$

Next substitute these numbers into the linear equation and solve for y.

If $x = 9$, $y = 14 - 9$, or 5.

If $x = 5$, $y = 14 - 5$, or 9.

We now check. The sum of 5 and 9 is 14. The sum of their squares, 25 + 81, is 106. The numbers are 5 and 9.

16. 7, 8

17.

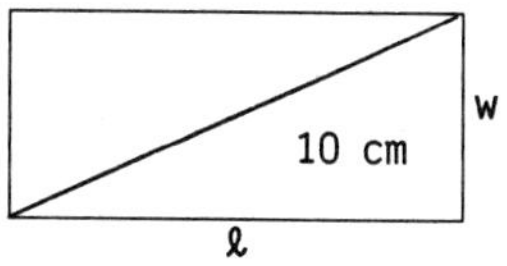

$2\ell + 2w = 28$ (Perimeter)

$\ell^2 + w^2 = 100$ (Using the Pythagorean Theorem)

Solve the linear equation for ℓ:

$2\ell + 2w = 28$

$\ell + w = 14$

$\ell = 14 - w$

Substitute 14 - w for ℓ into $\ell^2 + w^2 = 100$ and solve for w.

$$\ell^2 + w^2 = 100$$
$$(14 - w)^2 + w^2 = 100$$
$$196 - 28w + w^2 + w^2 = 100$$
$$2w^2 - 28w + 96 = 0$$
$$w^2 - 14w + 48 = 0$$
$$(w - 6)(w - 8) = 0$$

$w - 6 = 0$ or $w - 8 = 0$

$w = 6$ or $w = 8$

17. (continued)

Now substitute these numbers into the linear equation and solve for ℓ.

If $w = 6$, then $\ell = 14 - 6$, or 8.

If $w = 8$, then $\ell = 14 - 8$, or 6.

The perimeter is $2\cdot 8 + 2\cdot 6$, or 28. The diagonal is $\sqrt{8^2 + 6^2}$, or 10. The numbers check, so the answer is $\ell = 8$ cm and $w = 6$ cm.

18. 1 m, 2 m

19.

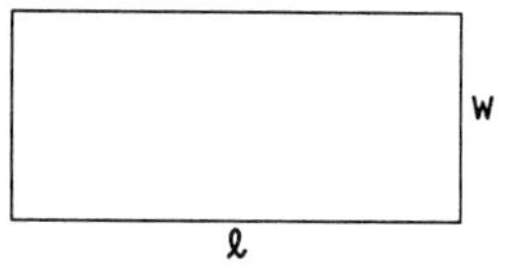

$2\ell + 2w = 18$, or $\ell + w = 9$ (Perimeter)

$\ell w = 20$ (Area)

Solve the linear equation for ℓ.

$\ell + w = 9$

$\ell = 9 - w$

Substitute 9 - w for ℓ into $\ell w = 20$ and solve for w.

$$\ell w = 20$$
$$(9 - w)w = 20$$
$$9w - w^2 = 20$$
$$0 = w^2 - 9w + 20$$
$$0 = (w - 5)(w - 4)$$

$w - 5 = 0$ or $w - 4 = 0$

$w = 5$ or $w = 4$

Now substitute these numbers into the linear eqution and solve for ℓ.

If $w = 5$, $\ell = 9 - 5$, or 4.

If $w = 4$, $\ell = 9 - 4$, or 5.

The perimeter is $2\cdot 5 + 2\cdot 4$, or 18. The area is $5\cdot 4$ or 20. The numbers check, so the dimensions are 5 in. by 4 in.

20. 1 yd, 2 yd

21. $x^2 + y^2 = 19{,}380{,}510.36$ (1)

$27{,}942.25x - 6.125y = 0$ (2)

First solve equation (2) for x.

$$27{,}942.25x - 6.125y = 0$$
$$27{,}942.25x = 6.125y$$
$$x = 0.000219y$$

Then substitute 0.000219y for x in equation (1) and solve for y.

21. (continued)

$$(0.000219y)^2 + y^2 = 19{,}380{,}510.36$$
$$0.000000047y^2 + y^2 = 19{,}380{,}510.36$$
$$1.000000047y^2 = 19{,}380{,}510.36$$
$$y^2 = 19{,}380{,}509.45$$
$$y = \pm\ 4402.33$$

Now substitute these values for y into the linear equation and solve for x.

If $y = -4402.33$, $x = 0.000219(-4402.33)$
$= -0.96411$.

If $y = 4402.33$, $x = 0.000219(4402.33) = 0.96411$.

The pairs (-0.96411, -4402.33) and (0.96411, 4402.33) are the solutions.

22. (45, 785), (785, 45)

23. Solve the system: $2L + 2W = P$
$LW = A$

See text, p. A-54.

24. There is no number x such that

$$\frac{x^2}{a^2} - \frac{\left[\frac{b}{a}x\right]^2}{b^2} = 1,$$ because the left side simplifies to $\frac{x^2}{a^2} - \frac{x^2}{a^2}$ which is 0.

25. $(x - h)^2 + (y - k)^2 = r^2$

If (2, 4) is a point on the circle, then
$(2 - h)^2 + (4 - k)^2 = r^2$.

If (3, 3) is a point on the circle, then
$(3 - h)^2 + (3 - k)^2 = r^2$.

Thus

$$(2 - h)^2 + (4 - k)^2 = (3 - h)^2 + (3 - k)^2$$
$$4 - 4h + h^2 + 16 - 8k + k^2 =$$
$$9 - 6h + h^2 + 9 - 6k + k^2$$
$$-4h - 8k + 20 = -6h - 6k + 18$$
$$2h - 2k = -2$$
$$h - k = -1$$

If the center (h, k) is on the line $3x - y = 3$, then $3h - k = 3$.

Solving the system

$h - k = -1$

$3h - k = 3$

we find that $(h, k) = (2, 3)$.

Find r, substituting (2, 3) for (h, k) and (2, 4) for (x, y). We could also use (3, 3) for (x, y).

25. (continued)

$$(x - h)^2 + (y - k)^2 = r^2$$
$$(2 - 2)^2 + (4 - 3)^2 = r^2$$
$$0 + 1 = r^2$$
$$1 = r^2$$
$$1 = r$$

The equation of the circle whose center is (2, 3) and whose radius is 1 is $(x - 2)^2 + (y - 3)^2 = 1$.

26. $(x + 1)^2 + (y + 3)^2 = 100$

Exercise Set 11.6

1. $x^2 + y^2 = 25$ (1)
$y^2 = x + 5$ (2)

Substitute $x + 5$ for y^2 in equation (1) and solve for x.

$$x^2 + y^2 = 25$$
$$x^2 + (x + 5) = 25$$
$$x^2 + x - 20 = 0$$
$$(x + 5)(x - 4) = 0$$
$$x + 5 = 0 \quad \text{or} \quad x - 4 = 0$$
$$x = -5 \quad \text{or} \quad x = 4$$

Next substitute these numbers for x in equation (2) and solve for y.

If $x = -5$, $y^2 = -5 + 5 = 0$. Thus $y = 0$.

If $x = 4$, $y^2 = 4 + 5 = 9$. Thus $y = \pm\ 3$.

The pairs (-5, 0), (4, 3), and (4, -3) check and are the solutions.

2. (0, 0), (1, 1)

3. $x^2 + y^2 = 9$ (1)
$x^2 - y^2 = 9$ (2)
$2x^2 = 18$ Adding
$x^2 = 9$
$x = \pm\ 3$

Substitute these values for x in either equation (1) or (2) and solve for y. Here we use equation(1).

$$x^2 + y^2 = 9$$
$$(\pm\ 3)^2 + y^2 = 9$$
$$9 + y^2 = 9$$
$$y^2 = 0$$
$$y = 0$$

The pairs (-3, 0) and (3, 0) check and are the solutions.

4. (0, 2), (0, -2)

5. $x^2 + y^2 = 25$ (1)

$xy = 12$ (2)

First solve equation (2) for y.

$xy = 12$

$y = \frac{12}{x}$

Then substitute $\frac{12}{x}$ for y in equation (1) and solve for x.

$$x^2 + y^2 = 25$$
$$x^2 + \left(\frac{12}{x}\right)^2 = 25$$
$$x^2 + \frac{144}{x^2} = 25$$
$$x^4 + 144 = 25x^2$$
$$x^4 - 25x^2 + 144 = 0$$
$$(x^2 - 16)(x^2 - 9) = 0$$

$x^2 - 16 = 0$ or $x^2 - 9 = 0$

$x^2 = 16$ or $x^2 = 9$

$x = \pm 4$ or $x = \pm 3$

Substitute these numbers into equation (2) and solve for y.

If $x = -4$, then $y = \frac{12}{-4} = -3$.

If $x = 4$, then $y = \frac{12}{4} = 3$.

If $x = -3$, then $y = \frac{12}{-3} = -4$.

If $x = 3$, then $y = \frac{12}{3} = 4$.

The pairs (-4, -3), (4, 3), (-3, -4), and (3, 4) check, hence are solutions.

6. (-5, 3), (-5, -3), (4, 0)

7. $x^2 + y^2 = 4$ (1)

$16x^2 + 9y^2 = 144$ (2)

Equation (1) is a circle with center (0, 0) and radius 2. Equation (2) is an ellipse with center (0, 0) and vertices (-3, 0), (3, 0), (0, -4), and (0, 4). The graphs have no points of intersection. Also solving algebraically we find the system has no real solutions.

8. (0, 5), (0, -5)

9. $x^2 + y^2 = 16$ (1)

$y^2 - 2x^2 = 10$ (2)

Rewrite equation (2).

$x^2 + y^2 = 16$ (1)

$-2x^2 + y^2 = 10$ (2)

9. (continued)

Using the addition method, multiply equation (2) by -1 and add the result to equation (1).

$x^2 + y^2 = 16$ (1)

$2x^2 - y^2 = -10$ (2)

$$3x^2 = 6$$
$$x^2 = 2$$
$$x = \pm\sqrt{2}$$

Thus if $x = \sqrt{2}$, $x^2 = 2$, and if $x = -\sqrt{2}$, $x^2 = 2$, so substituting $\sqrt{2}$ or $-\sqrt{2}$ for x in equation (2) we have

$$y^2 - 2x^2 = 10$$
$$y^2 - 2(\pm\sqrt{2})^2 = 10$$
$$y^2 - 4 = 10$$
$$y^2 = 14$$
$$y = \pm\sqrt{14}$$

The pairs $(\sqrt{2}, \sqrt{14})$, $(\sqrt{2}, -\sqrt{14})$, $(-\sqrt{2}, \sqrt{14})$, and $(-\sqrt{2}, -\sqrt{14})$ check and are the solutions.

10. $(3, \sqrt{5})$, $(-3, \sqrt{5})$, $(3, -\sqrt{5})$, $(-3, -\sqrt{5})$

11. $x^2 + y^2 = 5$ (1)

$xy = 2$ (2)

Solve equation (2) for y.

$xy = 2$

$y = \frac{2}{x}$

Substitute $\frac{2}{x}$ for y in equation (1) and solve for x.

$$x^2 + y^2 = 5$$
$$x^2 + \left(\frac{2}{x}\right)^2 = 5$$
$$x^2 + \frac{4}{x^2} = 5$$
$$x^4 + 4 = 5x^2$$
$$x^4 - 5x^2 + 4 = 0$$
$$(x^2 - 4)(x^2 - 1) = 0$$

$x^2 - 4 = 0$ or $x^2 - 1 = 0$

$x^2 = 4$ or $x^2 = 1$

$x = \pm 2$ or $x = \pm 1$

Substitute these values for x in $y = \frac{2}{x}$ and solve for y.

If $x = -2$, $y = \frac{2}{-2} = -1$.

If $x = 2$, $y = \frac{2}{2} = 1$.

If $x = -1$, $y = \frac{2}{-1} = -2$.

If $x = 1$, $y = \frac{2}{1} = 2$.

11. (continued)

The pairs (-2, -1), (2, 1), (-1, -2), and (1, 2) check and are the solutions.

12. (4, 2), (-4, -2), (2, 4), (-2, -4)

13. $x^2 + y^2 = 13$ (1)

$xy = 6$ (2)

Solve equation (2) for y.

$xy = 6$

$y = \frac{6}{x}$

Substitute $\frac{6}{x}$ for y in equation (1) and solve for x.

$$x^2 + y^2 = 13$$
$$x^2 + \left(\frac{6}{x}\right)^2 = 13$$
$$x^2 + \frac{36}{x^2} = 13$$
$$x^4 + 36 = 13x^2$$
$$x^4 - 13x^2 + 36 = 0$$
$$(x^2 - 9)(x^2 - 4) = 0$$

$x^2 - 9 = 0$ or $x^2 - 4 = 0$

$x^2 = 9$ or $x^2 = 4$

$x = \pm 3$ or $x = \pm 2$

Substitute these values for x in $y = \frac{6}{x}$ and solve for y.

If $x = -3$, $y = \frac{6}{-3} = -2$.

If $x = 3$, $y = \frac{6}{3} = 2$.

If $x = -2$, $y = \frac{6}{-2} = -3$.

If $x = 2$, $y = \frac{6}{2} = 3$.

The pairs (-3, -2), (3, 2), (-2, -3), and (2, 3) check and are the solutions.

14. (2, 2), (-2, -2), (4, 1), (-4, -1)

15. $x^2 + y^2 + 6y + 5 = 0$ (1)

$x^2 + y^2 - 2x - 8 = 0$ (2)

Using the addition method, multiply equation (2) by -1 and add the result to equation (1).

$x^2 + y^2 + 6y + 5 = 0$ (1)

$-x^2 - y^2 + 2x + 8 = 0$ (2)

$2x + 6y + 13 = 0$ (3)

Solve equation (3) for x.

$$2x + 6y + 13 = 0$$
$$2x = -6y - 13$$
$$x = \frac{-6y - 13}{2}$$

15. (continued)

Substitute $\frac{-6y - 13}{2}$ for x in equation (1) and solve for y.

$$x^2 + y^2 + 6y + 5 = 0$$
$$\left(\frac{-6y - 13}{2}\right)^2 + y^2 + 6y + 5 = 0$$
$$\frac{36y^2 + 156y + 169}{4} + y^2 + 6y + 5 = 0$$
$$36y^2 + 156y + 169 + 4y^2 + 24y + 20 = 0$$
$$40y^2 + 180y + 189 = 0$$

Using the quadratic formula, we find that $y = \frac{-45 \pm 3\sqrt{15}}{20}$. Substitute $\frac{-45 \pm 3\sqrt{15}}{20}$ for y in $x = \frac{-6y - 13}{2}$ and solve for x.

If $y = \frac{-45 + 3\sqrt{15}}{20}$, then $x = \frac{-6\left(\frac{-45 + 3\sqrt{15}}{20}\right) - 13}{2} = \frac{5 - 9\sqrt{15}}{20}$.

If $y = \frac{-45 - 3\sqrt{15}}{20}$, then $x = \frac{-6\left(\frac{-45 - 3\sqrt{15}}{20}\right) - 13}{2} = \frac{5 + 9\sqrt{15}}{20}$.

The pairs $\left(\frac{5 + 9\sqrt{15}}{20}, \frac{-45 - 3\sqrt{15}}{20}\right)$ and $\left(\frac{5 - 9\sqrt{15}}{20}, \frac{-45 + 3\sqrt{15}}{20}\right)$ check and are the solutions.

16. (2, 1), (-2, -1)

17. $18.465x^2 + 788.723y^2 = 6408$ (1)

$106.535x^2 - 788.723y^2 = 2692$ (2)

$$125x^2 = 9100$$
$$x^2 = 72.8$$
$$x = \pm 8.53$$

Substitute these values for x in either equation (1) or (2) and solve for y. Here we use equation (1).

$$18.465x^2 + 788.723y^2 = 6408$$
$$18.465(\pm 8.53)^2 + 788.723y^2 = 6408$$
$$1344.252 + 788.723y^2 = 6408$$
$$788.723y^2 = 5063.748$$
$$y^2 = 6.42019$$
$$y = \pm 2.53$$

The pairs (-8.53, -2.53), (-8.53, 2.53), (8.53, -2.53) and (8.53, 2.53) check and are the solutions.

18. (400, 1.43), (400, -1.43), (-400, 1.43), (-400, -1.43)

19. Let x and y represent the numbers.

$xy = 156 \quad (1)$

$x^2 + y^2 = 313 \quad (2)$

Solve equation (1) for y.

$xy = 156$

$y = \frac{156}{x}$

Substitute $\frac{156}{x}$ for y in equation (2) and solve for x.

$$x^2 + y^2 = 313$$

$$x^2 + \left(\frac{156}{x}\right)^2 = 313$$

$$x^2 + \frac{24,336}{x^2} = 313$$

$$x^4 + 24,336 = 313x^2$$

$$x^4 - 313x^2 + 24,336 = 0$$

$$(x^2 - 144)(x^2 - 169) = 0$$

$x^2 - 144 = 0$ or $x^2 - 169 = 0$

$x^2 = 144$ or $x^2 = 169$

$x = \pm 12$ or $x = \pm 13$

Substitute these numbers in $y = \frac{156}{x}$ and solve for y.

If $x = 12$, then $y = \frac{156}{12} = 13$.

If $x = -12$, then $y = \frac{156}{-12} = -13$.

If $x = 13$, then $y = \frac{156}{13} = 12$.

If $x = -13$, then $y = \frac{156}{-13} = -12$.

The pairs (12, 13), (-12, -13), (13, 12), and (-13, -12) check. The numbers are 12, 13 and -12, -13.

20. 6, 10 and -6, -10

21. $\ell w = \sqrt{3}$ (Area)

$\ell^2 + w^2 = 2^2$ (Using the Pythagorean Theorem)

Solve $\ell w = \sqrt{3}$ for w.

$\ell w = \sqrt{3}$

$w = \frac{\sqrt{3}}{\ell}$

Substitute $\frac{\sqrt{3}}{\ell}$ for w in $\ell^2 + w^2 = 4$ and solve for ℓ.

$$\ell^2 + w^2 = 4$$

$$\ell^2 + \left(\frac{\sqrt{3}}{\ell}\right)^2 = 4$$

$$\ell^2 + \frac{3}{\ell^2} = 4$$

$$\ell^4 + 3 = 4\ell^2$$

$$\ell^4 - 4\ell^2 + 3 = 0$$

$$(\ell^2 - 3)(\ell^2 - 1) = 0$$

21. (continued)

$\ell^2 - 3 = 0$ or $\ell^2 - 1 = 0$

$\ell^2 = 3$ or $\ell^2 = 1$

$\ell = \pm\sqrt{3}$ or $\ell = \pm 1$

Since the length of the rectangle is positive, we only consider $\sqrt{3}$ and 1. Substitute these values in $\ell w = 2$ and solve for w.

$\ell w = \sqrt{3}$ | $\ell w = \sqrt{3}$

$\sqrt{3} \cdot w = \sqrt{3}$ | $1 \cdot w = \sqrt{3}$

$w = 1$ | $w = \sqrt{3}$

If the dimensions are $\sqrt{3}$ and 1, the area is $\sqrt{3} \cdot 1$, or $\sqrt{3}$, and the diagonal is $\sqrt{(\sqrt{3})^2 + 1^2}$, or 2. The values check. The dimensions are $\sqrt{3}$ m by 1 m.

22. 1 m, $\sqrt{2}$ m

23. Let x represent the length of a side of one peanut bed and y represent the length of a side of the other peanut bed. We translate to a system of equations.

$x^2 + y^2 = 832 \quad (1)$

$x^2 - y^2 = 320 \quad (2)$

$2x^2 = 1152 \quad$ (Adding)

$x^2 = 576$

$x = \pm 24$

Since the length of a side of the bed must be positive, we only consider 24. Substitute 24 for x in equation (1) and solve for y.

$$x^2 + y^2 = 832$$

$$(24)^2 + y^2 = 832$$

$$576 + y^2 = 832$$

$$y^2 = 256$$

$$y = \pm 16$$

Again we only consider the positive length. The areas are 24^2, or 576, and 16^2, or 256. The sum of the areas is 576 + 256, or 832. The difference of the areas is 576 - 256, or 320. The values check. The lengths of the beds are 24 ft and 16 ft.

24. \$125, 6%

25. $(x - h)^2 + (y - k)^2 = r^2$ (Standard form)

Substitute (4, 6), (-6, 2), and (1, -3) for (x, y).

$$(4 - h)^2 + (6 - k)^2 = r^2 \quad (1)$$
$$(-6 - h)^2 + (2 - k)^2 = r^2 \quad (2)$$
$$(1 - h)^2 + (-3 - k)^2 = r^2 \quad (3)$$

Thus

$(4 - h)^2 + (6 - k)^2 = (-6 - h)^2 + (2 - k)^2$

or $5h + 2k = 3$

$(4 - h)^2 + (6 - k)^2 = (1 - h)^2 + (-3 - k)^2$

or $h + 3k = 7$

We solve the system $\begin{array}{l} 5h + 2k = 3 \\ h + 3k = 7 \end{array}$ for (h, k).

Solving we get $h = -\frac{5}{13}$ and $k = \frac{32}{13}$. Substituting these values in equation (1), (2), or (3), we find that $r^2 = \frac{5365}{169}$.

The equation of the circle is

$\left(x + \frac{5}{13}\right)^2 + \left(y - \frac{32}{13}\right)^2 = \frac{5365}{169}$.

26. $(x - 10)^2 + (y + 3)^2 = 100$

Exercise Set 12.1

1. $a_n = 3n + 1$

$a_1 = 3\cdot 1 + 1 = 4$ $\quad a_4 = 3\cdot 4 + 1 = 13$

$a_2 = 3\cdot 2 + 1 = 7$ $\quad a_{10} = 3\cdot 10 + 1 = 31$

$a_3 = 3\cdot 3 + 1 = 10$ $\quad a_{15} = 3\cdot 15 + 1 = 46$

2. 0, 0, 0, 6; 504; 2184

3. $a_n = \frac{1}{n}$

$a_1 = \frac{1}{1} = 1$ $\quad a_4 = \frac{1}{4}$

$a_2 = \frac{1}{2}$ $\quad a_{10} = \frac{1}{10}$

$a_3 = \frac{1}{3}$ $\quad a_{15} = \frac{1}{15}$

4. $-1, \frac{1}{2}, -\frac{1}{3}, \frac{1}{4};$ $\frac{1}{10};$ $-\frac{1}{15}$

5. $a_n = \frac{n}{n+1}$

$a_1 = \frac{1}{1+1} = \frac{1}{2}$ $\quad a_4 = \frac{4}{4+1} = \frac{4}{5}$

$a_2 = \frac{2}{2+1} = \frac{2}{3}$ $\quad a_{10} = \frac{10}{10+1} = \frac{10}{11}$

$a_3 = \frac{3}{3+1} = \frac{3}{4}$ $\quad a_{15} = \frac{15}{15+1} = \frac{15}{16}$

6. $1, -\frac{1}{2}, \frac{1}{4}, -\frac{1}{8};$ $-\frac{1}{512};$ $\frac{1}{16{,}384}$

7. - 14. Answers may vary.

7. 1, 3, 5, 7, 9, ...

$a_1 = 1$

$a_2 = 2\cdot 2 - 1 = 3$

$a_3 = 2\cdot 3 - 1 = 5$

$a_4 = 2\cdot 4 - 1 = 7$

$a_5 = 2\cdot 5 - 1 = 9$

$a_n = 2n - 1$ (General term)

8. 3^n

9. $\frac{2}{3}, \frac{3}{4}, \frac{4}{5}, \frac{5}{6}, \frac{6}{7}, \ldots$

$a_1 = \frac{1+1}{1+2} = \frac{2}{3}$

$a_2 = \frac{2+1}{2+2} = \frac{3}{4}$

$a_3 = \frac{3+1}{3+2} = \frac{4}{5}$

$a_4 = \frac{4+1}{4+2} = \frac{5}{6}$

$a_5 = \frac{5+1}{5+2} = \frac{6}{7}$

9. (continued)

$a_n = \frac{n+1}{n+2}$ (General term)

10. $\sqrt{2n}$

11. $\sqrt{3}, 3, 3\sqrt{3}, 9, 9\sqrt{3}, \ldots$

$a_1 = \sqrt{3}$

$a_2 = (\sqrt{3})^2 = \sqrt{3}\,\sqrt{3} = 3$

$a_3 = (\sqrt{3})^3 = \sqrt{3}\,\sqrt{3}\,\sqrt{3} = 3\sqrt{3}$

$a_4 = (\sqrt{3})^4 = \sqrt{3}\,\sqrt{3}\,\sqrt{3}\,\sqrt{3} = 9$

$a_5 = (\sqrt{3})^5 = \sqrt{3}\,\sqrt{3}\,\sqrt{3}\,\sqrt{3}\,\sqrt{3} = 9\sqrt{3}$

$a_n = (\sqrt{3})^n$ (General term)

12. $n(n + 1)$

13. log 1, log 10, log 100, log 1000, ...

$a_1 = \log 10^{1-1} = \log 10^0 = \log 1$, or 0

$a_2 = \log 10^{2-1} = \log 10^1 = \log 10$, or 1

$a_3 = \log 10^{3-1} = \log 10^2 = \log 100$, or 2

$a_4 = \log 10^{4-1} = \log 10^3 = \log 1000$, or 3

$a_n = \log 10^{n-1}$, or $n - 1$ (General term)

14. $2n$

15. $a_1 = 4$ $\quad a_{k+1} = 1 + \frac{1}{a_k}$

$a_2 = 1 + \frac{1}{4} = \frac{5}{4}$

$a_3 = 1 + \frac{1}{\frac{5}{4}} = 1 + \frac{4}{5} = \frac{9}{5}$

$a_4 = 1 + \frac{1}{\frac{9}{5}} = 1 + \frac{5}{9} = \frac{14}{9}$

16. $16, 4, 2, \sqrt{2}$

17. $a_1 = 64$ $\quad a_{k+1} = \sqrt{a_k}$

$a_2 = \sqrt{64} = 8$

$a_3 = \sqrt{8} = 2\sqrt{2}$

$a_4 = \sqrt{2\sqrt{2}}$

18. e^Q, Q, ln Q, ln ln Q

19. 1, 2, 3, 4, 5, 6, 7, ...

$S_7 = 1 + 2 + 3 + 4 + 5 + 6 + 7 = 28$

20. -8

21. 2, 4, 6, 8, ...
$S_5 = 2 + 4 + 6 + 8 + 10 = 30$

22. 1.4636

23. $\sum_{k=1}^{5} \frac{1}{2k} = \frac{1}{2\cdot 1} + \frac{1}{2\cdot 2} + \frac{1}{2\cdot 3} + \frac{1}{2\cdot 4} + \frac{1}{2\cdot 5}$
$= \frac{1}{2} + \frac{1}{4} + \frac{1}{6} + \frac{1}{8} + \frac{1}{10}$

24. $\frac{1}{3} + \frac{1}{5} + \frac{1}{7} + \frac{1}{9} + \frac{1}{11} + \frac{1}{13}$

25. $\sum_{k=0}^{5} 2^k = 2^0 + 2^1 + 2^2 + 2^3 + 2^4 + 2^5$
$= 1 + 2 + 4 + 8 + 16 + 32$, or 63

26. $\sqrt{7} + 3 + \sqrt{11} + \sqrt{13}$

27. $\sum_{k=7}^{10} \log k = \log 7 + \log 8 + \log 9 + \log 10$

28. $\pi + 2\pi + 3\pi + 4\pi$

29. $\sum_{k=1}^{4} \frac{1}{k^2} = \frac{1}{1^2} + \frac{1}{2^2} + \frac{1}{3^2} + \frac{1}{4^2}$
$= 1 + \frac{1}{4} + \frac{1}{9} + \frac{1}{16}$
$= \frac{144}{144} + \frac{36}{144} + \frac{16}{144} + \frac{9}{144}$
$= \frac{205}{144}$

30. $\sum_{k=1}^{4} \frac{k}{2^k}$, $\frac{13}{8}$

31. $\sum_{k=1}^{6} (-1)^k 2^k = -2 + 4 - 8 + 16 - 32 + 64 = 42$

32. $\sum_{k=1}^{5} \frac{(-1)^{k+1}\cdot k}{2^k}$, $\frac{9}{32}$

33. $a_n = \frac{1}{2n} \log 1000^n$
$a_1 = \frac{1}{2\cdot 1} \log 1000^1 = \frac{1}{2} \cdot 3 = \frac{3}{2}$
$a_2 = \frac{1}{2\cdot 2} \log 1000^2 = \frac{1}{4} \cdot 6 = \frac{3}{2}$
$a_3 = \frac{1}{2\cdot 3} \log 1000^3 = \frac{1}{6} \cdot 9 = \frac{3}{2}$
$a_4 = \frac{1}{2\cdot 4} \log 1000^4 = \frac{1}{8} \cdot 12 = \frac{3}{2}$
$a_5 = \frac{1}{2\cdot 5} \log 1000^5 = \frac{1}{10} \cdot 15 = \frac{3}{2}$

34. $i, -1, -i, 1, i$

35. $a_n = \sin \frac{n\pi}{2}$
$a_1 = \sin \frac{1\cdot\pi}{2} = \sin \frac{\pi}{2} = 1$
$a_2 = \sin \frac{2\cdot\pi}{2} = \sin \pi = 0$
$a_3 = \sin \frac{3\cdot\pi}{2} = -1$
$a_4 = \sin \frac{4\cdot\pi}{2} = \sin 2\pi = 0$
$a_5 = \sin \frac{5\cdot\pi}{2} = 1$

36. 0, 0.693, 1.792, 3.178, 4.787

37. $a_n = \mathrm{Cos}^{-1} (-1)^n$
$a_1 = \mathrm{Cos}^{-1} (-1)^1 = \mathrm{Cos}^{-1} (-1) = \pi$
$a_2 = \mathrm{Cos}^{-1} (-1)^2 = \mathrm{Cos}^{-1} 1 = 0$
$a_3 = \mathrm{Cos}^{-1} (-1)^3 = \mathrm{Cos}^{-1} (-1) = \pi$
$a_4 = \mathrm{Cos}^{-1} (-1)^4 = \mathrm{Cos}^{-1} 1 = 0$
$a_5 = \mathrm{Cos}^{-1} (-1)^5 = \mathrm{Cos}^{-1} (-1) = \pi$

38. $|\cos x|, |\cos x|, |\cos x|, |\cos x|, |\cos x|$

39. $a_n = \left(1 + \frac{1}{n}\right)^n$
$a_1 = \left(1 + \frac{1}{1}\right)^1 = 2$
$a_2 = \left(1 + \frac{1}{2}\right)^2 = (1.5)^2 = 2.25$
$a_3 = \left(1 + \frac{1}{3}\right)^3 = (1.333333)^3 = 2.370370$
$a_4 = \left(1 + \frac{1}{4}\right)^4 = (1.25)^4 = 2.441406$
$a_5 = \left(1 + \frac{1}{5}\right)^5 = (1.2)^5 = 2.488320$
$a_6 = \left(1 + \frac{1}{6}\right)^6 = (1.166666)^6 = 2.521626$
$\sum_{n=1}^{6} \left(1 + \frac{1}{n}\right)^n = 2 + 2.25 + 2.370370 + 2.441406 + 2.488320 + 2.521626$
$= 14.071722$

40. 0.414214, 0.317837, 0.267949, 0.236068, 0.213422, 0.196262; 1.645752

41. $a_1 = 2$, $a_{k+1} = \sqrt{1 + \sqrt{a_k}}$
$a_2 = \sqrt{1 + \sqrt{2}} = 1.553774$
$a_3 = \sqrt{1 + \sqrt{1.553774}} = 1.498834$
$a_4 = \sqrt{1 + \sqrt{1.498834}} = 1.491398$
$a_5 = \sqrt{1 + \sqrt{1.491398}} = 1.490378$

41. (continued)

$a_6 = \sqrt{1 + \sqrt{1.490378}} = 1.490238$

The sum of the first six terms is 9.524622.

42. 2, 1.5, 1.416667, 1.414216, 1.414214, 1.414214; 9.159311

Exercise Set 12.2

1. 2, 6, 10, ...

Note that $a_1 = 2$, $d = 4$, and $n = 12$.

$a_n = a_1 + (n - 1)d$ (Theorem 1)

$a_{12} = 2 + (12 - 1)4 = 2 + 44 = 46$

2. $a_{11} = 0.57$

3. Substitute 106 for a_n, 2 for a_1, and 4 for d. Then solve for n.

$a_n = a_1 + (n - 1)d$ (Theorem 1)

$106 = 2 + (n - 1)4$

$106 = 2 + 4n - 4$

$108 = 4n$

$27 = n$

Thus 106 is the 27th term.

4. 33rd

5. Substitute 5 for a_1, 6 for d, and 17 for n. Then solve for a_{17}.

$a_n = a_1 + (n - 1)d$ (Theorem 1)

$a_{17} = 5 + (17 - 1)6 = 5 + 96 = 101$

6. $a_{20} = -43$

7. Substitute 33 for a_n, 8 for n, and 4 for d. Then solve for a_1.

$a_n = a_1 + (n - 1)d$ (Theorem 1)

$33 = a_1 + (8 - 1)4$

$33 = a_1 + 28$

$5 = a_1$

8. $d = \frac{9}{5}$

9. Substitute 5 for a_1, -3 for d, and -76 for a_n. Then solve for n.

$a_n = a_1 + (n - 1)d$ (Theorem 1)

$-76 = 5 + (n - 1)(-3)$

$-81 = -3n + 3$

$-84 = -3n$

$28 = n$

10. n = 39

11. We know that $a_{17} = -40$ and $a_{28} = -73$. Then we would add d 11 times to get from -40 to -73. That is,

$11d = -73 - (-40)$

$11d = -33$

$d = -3$

Since $a_{17} = -40$, we subtract -3 sixteen times to get to a_1. Thus

$a_1 = -40 - 16(-3) = -40 + 48 = 8.$

The first five terms of the sequence are 8, 5, 2, -1, and -4.

12. 624 ft

13. 5, 8, 11, 14, ...

We note that $a_1 = 5$, $d = 3$, and $n = 20$.

$S_n = \frac{n}{2}[2a_1 + (n - 1)d]$ (Theorem 3)

$S_{20} = \frac{20}{2}[2\cdot5 + (20 - 1)3] = 10[10 + 57] = 670$

14. $S_1 = 375$

15. The series is $1 + 3 + 5 + \ldots + 99$, so $a_1 = 1$, $a_n = 99$, and $n = 50$.

$S_n = \frac{n}{2}(a_1 + a_n)$ (Theorem 2)

$S_{50} = \frac{50}{2}(1 + 99) = 25\cdot100 = 2500$

16. 735

17. $S_n = \frac{n}{2}[2a_1 + (n - 1)d]$ (Theorem 3)

$S_{20} = \frac{20}{2}[2\cdot2 + (20 - 1)5]$ (Substituting)

$= 10[4 + 95]$

$= 990$

18. 34,036

19. $\sum_{k=1}^{16} (7k - 76)$

$a_1 = 7\cdot1 - 76 = 7 - 76 = -69$

$a_{16} = 7\cdot16 - 76 = 112 - 76 = 36$

$n = 16$

$S_n = \frac{n}{2}(a_1 + a_n)$ (Theorem 2)

$S_{16} = \frac{16}{2}(-69 + 36)$ (Substituting)

$= 8(-33)$

$= -264$

20. 432

21. The series is 30 + 29 + 28 + ... + 1.

$a_1 = 30$

$a_{30} = 1$

$n = 30$

$S_n = \frac{n}{2}(a_1 + a_n)$ (Theorem 2)

$S_{30} = \frac{30}{2}(30 + 1) = 15\cdot31 = 465$

There are 465 poles in the pile.

22. $49.60

23. 4, m_1, m_2, m_3, m_4, 13

We look for m_1, m_2, m_3, and m_4 such that 4, m_1, m_2, m_3, m_4, 13 is an arithmetic sequence. In this case $a_1 = 4$, $n = 6$, and $a_6 = 13$. We use Theorem 1.

$a_n = a_1 + (n - 1)d$

$13 = 4 + (6 - 1)d$

$9 = 5d$

$1\frac{4}{5} = d$

Thus,

$m_1 = a_1 + d = 4 + 1\frac{4}{5} = 5\frac{4}{5}$

$m_2 = m_1 + d = 5\frac{4}{5} + 1\frac{4}{5} = 6\frac{8}{5} = 7\frac{3}{5}$

$m_3 = m_2 + d = 7\frac{3}{5} + 1\frac{4}{5} = 8\frac{7}{5} = 9\frac{2}{5}$

$m_4 = m_3 + d = 9\frac{2}{5} + 1\frac{4}{5} = 10\frac{6}{5} = 11\frac{1}{5}$

24. -1, 1, 3

25. 1 + 3 + 5 + 7 + ... + n

$S_n = \frac{n}{2}[2a_1 + (n - 1)d]$ (Theorem 3)

25. (continued)

Note that $a_1 = 1$ and $d = 2$.

$S_n = \frac{n}{2}[2\cdot1 + (n - 1)2] = \frac{n}{2}(2 + 2n - 2)$

$= \frac{n}{2}\cdot 2n$

$= n^2$

26. 3, 5, 7

27. $P(x) = x^4 + 4x^3 - 84x^2 - 176x + 640$ has at most 4 zeros because P(x) is of degree 4. By the rational root theorem, the possible zeros are

±1, ±2, ±4, ±5, ±8, ±10, ±16, ±20, ±32, ±40, ±64, ±80, ±128, ±160, ±320, ±640

We find two of the zeros using synthetic division.

2	1	4	-84	-176	640
		2	12	-144	-640
	1	6	-72	-320	0

-4	1	6	-72	-320
		-4	-8	320
	1	2	-80	0

Also by synthetic division we determine that ±1 and -2 are not zeros. Therefore we determine that d = 6 in the arithmetic sequence.

Possible arithmetic sequences:

a) -4 2 8 14

b) -10 -4 2 8

c) -16 -10 -4 2

The solution cannot be a) because 14 is not a possible zero. Checking -16 by syntnetic division we find that -16 is not a zero. Thus b) is the only arithmetic sequence which contains all four zeros. The zeros are -10, -4, 2, and 8.

28. 16 means, $d = \frac{49}{17}$

29. The sides are a, a + d, and a + 2d. By the Pythagorean property, we get $a^2 + (a + d)^2 = (a + 2d)^2$. Solving for d, we get $d = \frac{a}{3}$. Thus the sides are a, $a + \frac{a}{3}$, and $a + \frac{2a}{3}$, or a, $\frac{4a}{3}$, and $\frac{5a}{3}$ which are in the ratio 3:4:5.

30. $V_n = 5200 - 512.5n$

31. $V_n = 5200 - 512.5n$

$V_0 = 5200 - 512.5(0) = \$5200$
$V_1 = 5200 - 512.5(1) = \$4687.50$
$V_2 = 5200 - 512.5(2) = \$4175$
$V_3 = 5200 - 512.5(3) = \$3662.50$
$V_4 = 5200 - 512.5(4) = \$3150$
$V_5 = 5200 - 512.5(5) = \$2637.50$
$V_6 = 5200 - 512.5(6) = \$2125$
$V_7 = 5200 - 512.5(7) = \$1612.50$
$V_8 = 5200 - 512.5(8) = \$1100$

32. \$8760, \$7961.77, \$7163.54, \$6365.31, \$5567.08, \$4768.85, \$3970.62, \$3172.39, \$2374.16, \$1575.93

33. $S_n = \frac{n}{2}(a_1 + a_n)$

$S_{10} = \frac{10}{2}(8760 + 1575.93)$

$= 5(10{,}335.93)$

$= \$51{,}679.65$

Exercise Set 12.3

1. $\frac{8}{243}, \frac{4}{81}, \frac{2}{27}, \ldots$

Note that $a_1 = \frac{8}{243}$, $r = \frac{\frac{4}{81}}{\frac{8}{243}} = \frac{4}{81} \cdot \frac{243}{8} = \frac{3}{2}$, and $n = 10$.

$a_n = a_1 r^{n-1}$ (Theorem 4)

$a_{10} = \frac{8}{243}\left(\frac{3}{2}\right)^{10-1} = \frac{2^3}{3^5} \cdot \frac{3^9}{2^9} = \frac{3^4}{2^6} = \frac{81}{64}$

2. $a_5 = 1250$

3. 5, 10, 20, ...

Note that $a_1 = 5$, $r = \frac{10}{5} = 2$, and $n = 8$. We use the formula in Theorem 5.

$S_8 = \frac{a_1 - a_1 r^8}{1 - r} = \frac{5 - 5\cdot 2^8}{1 - 2}$

$= \frac{5 - 5\cdot 256}{-1} = \frac{5 - 1280}{-1}$

$= \frac{-1275}{-1} = 1275$

4. $S_6 = \frac{21}{2}$

5. $\frac{1}{18}, -\frac{1}{6}, \frac{1}{2}, \ldots$

Note that $a_1 = \frac{1}{18}$, $r = \frac{-\frac{1}{6}}{\frac{1}{18}} = -\frac{1}{6} \cdot \frac{18}{1} = -3$, and $n = 7$. We use the formula in Theorem 5.

5. (continued)

$S_7 = \frac{a_1 - a_1 r^7}{1 - r} = \frac{\frac{1}{18} - \frac{1}{18}(-3)^7}{1 - (-3)}$

$= \frac{\frac{1}{18} + \frac{2187}{18}}{4} = \frac{2188}{18} \cdot \frac{1}{4} = \frac{547}{18}$

6. $-\frac{171}{32}$

7. $\sum_{k=1}^{6} \left(\frac{1}{2}\right)^{k-1}$ (Geometric series)

$a_1 = \left(\frac{1}{2}\right)^{1-1} = \left(\frac{1}{2}\right)^0 = 1$

$a_2 = \left(\frac{1}{2}\right)^{2-1} = \left(\frac{1}{2}\right)^1 = \frac{1}{2}$

$r = \frac{\frac{1}{2}}{1} = \frac{1}{2}$

$S_n = \frac{a_1 - a_1 r^n}{1 - r}$ (Theorem 5)

$S_6 = \frac{1 - 1\left(\frac{1}{2}\right)^6}{1 - \frac{1}{2}} = \frac{1 - \frac{1}{64}}{\frac{1}{2}} = \frac{\frac{63}{64}}{\frac{1}{2}} = \frac{63}{32}$

8. 2046

9. \$800, \$800(1.08), \$800(1.08)^2, \$800(1.08)^3, ...

The amount owed at the end of the second year is the third term of this geometric sequence. The ratio is 1.08.

$a_n = a_1 r^{n-1}$ (Theorem 4)

$a_3 = 800(1.08)^{3-1} = 800(1.08)^2 = \$933.12$

10. \$1360.49

11. The first rebound is 4 ft, the second is 1 ft, the third is $\frac{1}{4}$ ft, and so on. Find the sixth term in the geometric sequence 4, 1, $\frac{1}{4}$, ...

$a_n = a_1 r^{n-1}$ (Theorem 4)

$a_6 = 4\left(\frac{1}{4}\right)^{6-1} = 4\left(\frac{1}{4}\right)^5 = \frac{4^1}{4^5} = \frac{1}{4^4} = \frac{1}{256}$

Thus the ping-pong ball rebounds $\frac{1}{256}$ ft on the sixth rebound. This problem could also be worked by finding the seventh term in the geometric sequence 16, 4, 1, $\frac{1}{4}$, ...

12. 161,051

13. \$1000, \$1000(1.08), \$1000(1.08)², ...

a_1 = \$1000, $r = 1.08$, $n = 5$

$$S_n = \frac{a_1 - a_1 r^n}{1 - r} \quad \text{(Theorem 5)}$$

$$S_5 = \frac{1000 - 1000(1.08)^5}{1 - 1.08}$$

$$= \frac{1000 - 1469.32808}{-0.08} = \frac{-469.32808}{-0.08}$$

= \$5866.60

14. \$1664.54

15. $\frac{a_n}{a_{n+1}} = r$, so $\frac{a_n^2}{a_{n+1}^2} = \frac{a_n}{a_{n+1}} \cdot \frac{a_n}{a_{n+1}} = r^2$.

Hence $a_1^2, a_2^2, \ldots$ is geometric with ratio r^2.

16. $\frac{a_{n+1}}{a_n} = r$, so $\frac{a_{n+1}^{-3}}{a_n^{-3}} = r^{-3}$

Thus, $a_1^{-3}, a_2^{-3}, a_3^{-3}, \ldots$ is geometric with ratio r^{-3}.

17. $$\frac{a_n}{a_{n+1}} = r$$

$$\ln \frac{a_n}{a_{n+1}} = \ln r$$

$$\ln a_n - \ln a_{n+1} = \ln r$$

Thus, $\ln a_1, \ln a_2, \ln a_3, \ldots$ is an arithmetic sequence.

18. $a_{n+1} - a_n = d$ ($a_1, a_2, a_3, \ldots$ is arithmetic)

$\frac{5^{a_{n+1}}}{5^{a_n}} = 5^{a_{n+1} - a_n} = 5^d$, a constant

Hence $5^{a_1}, 5^{a_2}, 5^{a_3}, \ldots$ is a geometric sequence.

19. p, G, q form a geometric sequence.

Hence $G = pr$ and $q = Gr = pr \cdot r = pr^2$.

$\sqrt{pq} = \sqrt{p \cdot pr^2} = \sqrt{p^2 r^2} = pr = G$

Thus, $G = \sqrt{pq}$.

20. a) 6, b) $2\sqrt{3}$, c) $\sqrt{\frac{1}{6}}$, d) $\sqrt{3}$

21. To find the thickness when doubled 20 times, we find the twenty-first term in the geometric sequence.

0.01, 0.02, 0.04, 0.08, ...

where $a_1 = 0.01$ and $r = 2$.

$a_n = a_1 r^{n-1}$ (Theorem 4)

$a_{21} = 0.01(2^{20})$

$= 0.01(1{,}048{,}576)$

$= 10{,}485.76$

The resulting thickness is about 10,486 inches.

22. $3100 \frac{45}{128}$ ft

Exercise Set 12.4

1. 5, 10, 20, 40, ...

$r = \frac{10}{5} = 2$

The sum does not exist because $r = 2$, so $|r| > 1$.

2. Yes

3. $6, 2, \frac{2}{3}, \frac{2}{9}, \ldots$

$r = \frac{2}{6} = \frac{1}{3}$

The sum does exist because $r = \frac{1}{3}$, so $|r| < 1$.

4. No

5. 1, 0.1, 0.01, 0.001, 0.0001, ...

$r = \frac{0.1}{1} = 0.1$

The sum does exist because $r = 0.1$, so $|r| < 1$.

6. Yes

7. $1, -\frac{1}{5}, \frac{1}{25}, -\frac{1}{125}, \ldots$

$$r = \frac{-\frac{1}{5}}{1} = -\frac{1}{5}$$

The sum does exist because $r = -\frac{1}{5}$, so $|r| < 1$.

8. No

9. $4 + 2 + 1 + \ldots$ (Geometric series)

$a_1 = 4$ and $r = \frac{2}{4} = \frac{1}{2}$

$S_\infty = \frac{a_1}{1 - r}$ (Theorem 7)

$S_\infty = \frac{4}{1 - \frac{1}{2}} = \frac{4}{\frac{1}{2}} = 8$

10. $\frac{49}{4}$

11. $25 + 20 + 16 + \ldots$ (Geometric series)

$a_1 = 25$ and $r = \frac{20}{25} = \frac{4}{5}$

$S_\infty = \frac{a_1}{1 - r}$ (Theorem 7)

$S_\infty = \frac{25}{1 - \frac{4}{5}} = \frac{25}{\frac{1}{5}} = 125$

12. 48

13. $\sum_{k=1}^{\infty} \frac{1}{2^{k-1}}$ (Sum of infinite geometric sequence)

$a_1 = \frac{1}{2^{1-1}} = \frac{1}{1} = 1$

$a_2 = \frac{1}{2^{2-1}} = \frac{1}{2^1} = \frac{1}{2}$

$r = \frac{\frac{1}{2}}{1} = \frac{1}{2}$

$S_\infty = \frac{a_1}{1 - r}$ (Theorem 7)

$S_\infty = \frac{1}{1 - \frac{1}{2}} = \frac{1}{\frac{1}{2}} = 2$

14. $\frac{16}{3}$

15. $\sum_{k=1}^{\infty} 16(0.1)^{k-1}$ (Sum of infinite geometric sequence)

$a_1 = 16(0.1)^{1-1} = 16 \cdot 1 = 16$

$a_2 = 16(0.1)^{2-1} = 16(0.1) = 1.6$

$r = \frac{1.6}{16} = 0.1$

$S_\infty = \frac{a_1}{1 - r}$ (Theorem 7)

$S_\infty = \frac{16}{1 - 0.1} = \frac{16}{0.9} = \frac{160}{9}$, or $17\frac{7}{9}$

16. 10

17. $\$1000(1.08)^{-1} + \$1000(1.08)^{-2} + \$1000(1.08)^{-3} + \ldots$ (Geometric series)

$a_1 = 1000(1.08)^{-1}$ and $r = (1.08)^{-1}$

$S_\infty = \frac{a_1}{1 - r}$ (Theorem 7)

$S_\infty = \frac{1000(1.08)^{-1}}{1 - (1.08)^{-1}} = \frac{\frac{1000}{1.08}}{1 - \frac{1}{1.08}}$

$= \frac{\frac{1000}{1.08}}{\frac{0.08}{1.08}} = \frac{1000}{1.08} \cdot \frac{1.08}{0.08} = \frac{1000}{0.08} = \$12{,}500$

18. \$4166.67

19. $0.777\overline{7} = 0.7 + 0.07 + 0.007 + 0.0007 + \ldots$

Note that $a_1 = 0.7$ and $r = 0.1$

$S_\infty = \frac{a_1}{1 - r}$ (Theorem 7)

$S_\infty = \frac{0.7}{1 - 0.1} = \frac{0.7}{0.9} = \frac{7}{9}$

20. $\frac{8}{15}$

21. $0.2121\overline{21} = 0.21 + 0.0021 + 0.000021 + \ldots$

Note that $a_1 = 0.21$ and $r = 0.01$.

$S_\infty = \frac{a_1}{1 - r}$ (Theorem 7)

$S_\infty = \frac{0.21}{1 - 0.01} = \frac{0.21}{0.99} = \frac{21}{99} = \frac{7}{33}$

22. $\frac{29}{45}$

23. $5.1515\overline{15} = 5 + 0.1515\overline{15}$

First we consider the repeating part.

$0.1515\overline{15} = 0.15 + 0.0015 + 0.000015 + \ldots$

$a_1 = 0.15$ and $r = 0.01$

$S_\infty = \frac{a_1}{1 - r} = \frac{0.15}{1 - 0.01} = \frac{0.15}{0.99} = \frac{15}{99} = \frac{5}{33}$

Thus

$5.1515\overline{15} = 5 + \frac{5}{33} = 5\frac{5}{33}$, or $\frac{170}{33}$

24. $\frac{4121}{9990}$

25. The total effect on the economy is
$\$8{,}000{,}000{,}000 + \$8{,}000{,}000{,}000(0.85) + \$8{,}000{,}000{,}000(0.85)^2 + \$8{,}000{,}000{,}000(0.85)^3 + \ldots$
which is a geometric series with
$a_1 = 8{,}000{,}000{,}000$ and $r = 0.85$.

25. (continued)

$S_\infty = \frac{a_1}{1 - r}$ (Theorem 7)

$S_\infty = \frac{8,000,000,000}{1 - 0.85} = \frac{8,000,000,000}{0.15}$

$= \$53,333,333,333.33$

26. $\$9.4(10^{11})$

27. The total number of people who will buy the product can be expressed as a geometric series.

$5,000,000(0.4) + 5,000,000(0.4)^2 +$
$5,000,000(0.4)^3 + 5,000,000(0.4)^4 + \ldots$

Note that $a_1 = 5,000,000(0.4)$ and $r = 0.4$.

$S_\infty = \frac{a_1}{1 - r}$ (Theorem 7)

$S_\infty = \frac{5,000,000(0.4)}{1 - 0.4} = \frac{2,000,000}{0.6} \approx 3,333,333$

3,333,333 represents what % of the population?

$3,333,333 = x \cdot 5,000,000$

$\frac{3,333,333}{5,000,000} = x$

$0.666\overline{6} = x$

$66\frac{2}{3}\% = x$

28. 24 ft

Exercise Set 12.5

1. $n^2 < n^3$

$1^2 < 1^3,\ 2^2 < 2^3,\ 3^2 < 3^3,\ 4^2 < 4^3,\ 5^2 < 5^3$

2. $1^2 - 1 + 41$ is prime, $2^2 - 2 + 41$ is prime, etc.

3. A polygon of n sides has $\frac{n(n - 3)}{2}$ diagonals.

A polygon of 3 sides has $\frac{3(3 - 3)}{2}$ diagonals.

A polygon of 4 sides has $\frac{4(4 - 3)}{2}$ diagonals., etc.

4. The sum of the angles of a polygon of 3 sides is $(3 - 2)\cdot 180°$., etc.

5. See answer section in text.

6. S_n: $4 + 8 + 12 + \ldots + 4n = 2n(n + 1)$

S_1: $4 = 2\cdot 1\cdot(1 + 1)$

S_k: $4 + 8 + 12 + \ldots + 4k = 2k(k + 1)$

S_{k+1}: $4 + 8 + 12 + \ldots + 4k + 4(k + 1) = 2(k + 1)(k + 2)$

6. (continued)

1) Basis step: S_1 is true by substitution.

2) Induction step: Assume S_k. Deduce S_{k+1}. Starting with the left side of S_{k+1}, we have

$\underbrace{4 + 8 + 12 + \ldots + 4k} + 4(k + 1)$

$= 2k(k + 1) + 4(k + 1)$ (By S_k)

$= (k + 1)(2k + 4)$

$= 2(k + 1)(k + 2)$

7. See answer section in text.

8. S_n: $3 + 6 + 9 + \ldots + 3n = \frac{3n(n + 1)}{2}$

S_1: $3 = \frac{3(1 + 1)}{2}$

S_k: $3 + 6 + 9 + \ldots + 3k = \frac{3k(k + 1)}{2}$

S_{k+1}: $3 + 6 + 9 + \ldots + 3k + 3(k + 1) = \frac{3(k + 1)(k + 2)}{2}$

1) Basis step: S_1 is true by substitution.

2) Induction step: Assume S_k. Deduce S_{k+1}. Starting with the left side of S_{k+1}, we have

$\underbrace{3 + 6 + 9 + \ldots + 3k} + 3(k + 1)$

$= \frac{3k(k + 1)}{2} + 3(k + 1)$ (By S_k)

$= \frac{3k(k + 1) + 6(k + 1)}{2}$

$= \frac{(k + 1)(3k + 6)}{2}$

$= \frac{3(k + 1)(k + 2)}{2}$

9. See answer section in text.

2) Induction step: Assume S_k. Deduce S_{k+1}.

$\frac{1}{1\cdot 2} + \frac{1}{2\cdot 3} + \ldots + \frac{1}{k(k + 1)} = \frac{k}{k + 1}$ (S_k)

$\frac{1}{1\cdot 2} + \frac{1}{2\cdot 3} + \ldots + \frac{1}{k(k+1)} + \frac{1}{(k+1)(k+2)} = \frac{k}{k+1} + \frac{1}{(k+1)(k+2)}$

[Adding $\frac{1}{(k+1)(k+2)}$ on both sides]

$= \frac{k(k + 2) + 1}{(k + 1)(k + 2)}$

$= \frac{k^2 + 2k + 1}{(k + 1)(k + 2)}$

$= \frac{(k + 1)(k + 1)}{(k + 1)(k + 2)}$

$= \frac{k + 1}{k + 2}$

10. S_n: $2 + 4 + 8 + \ldots + 2^n = 2(2^n - 1)$

S_1: $2 = 2(2 - 1)$

S_k: $2 + 4 + 8 + \ldots + 2^k = 2(2^k - 1)$

S_{k+1}: $2 + 4 + 8 + \ldots + 2^k + 2^{k+1} = 2(2^{k+1} - 1)$

1) Basis step: S_1 is true by substitution.

2) Induction step: Assume S_k. Deduce S_{k+1}.
Starting with the left side of S_{k+1}, we have

$$\underbrace{2 + 4 + 8 + \ldots + 2^k}_{} + 2^{k+1}$$
$$= 2(2^k - 1) + 2^{k+1} \quad \text{(By } S_k\text{)}$$
$$= 2^{k+1} - 2 + 2^{k+1}$$
$$= 2 \cdot 2^{k+1} - 2$$
$$= 2(2^{k+1} - 1)$$

11. S_n: $n < n + 1$

S_1: $1 < 1 + 1$

S_k: $k < k + 1$

S_{k+1}: $k + 1 < k + 2$

1) Basis step: S_1 is true by substitution.

2) Induction step: Assume S_k. Deduce S_{k+1}.

$k < k + 1$ (S_k)

$k + 1 < k + 1 + 1$ (Adding 1)

$k + 1 < k + 2$

12. S_n: $2 \leqslant 2^n$

S_1: $2 \leqslant 2^1$

S_k: $2 \leqslant 2^k$

S_{k+1}: $2 \leqslant 2^{k+1}$

1) Basis step: S_1 is true by substitution.

2) Induction step: Assume S_k. Deduce S_{k+1}.

$2 \leqslant 2^k$ (S_k)

$2 \cdot 2 \leqslant 2^k \cdot 2$ (Multiplying by 2)

$2 < 2 \cdot 2 \leqslant 2^{k+1}$ $(2 < 2 \cdot 2)$

$2 \leqslant 2^{k+1}$

13. S_n: $3^n < 3^{n+1}$

S_1: $3^1 < 3^{1+1}$

S_k: $3^k < 3^{k+1}$

S_{k+1}: $3^{k+1} < 3^{k+2}$

1) Basis step: S_1 is true by substitution.

2) Induction step: Assume S_k. Deduce S_{k+1}.

$3^k < 3^{k+1}$ (S_k)

$3^k \cdot 3 < 3^{k+1} \cdot 3$ (Multiplying by 3)

$3^{k+1} < 3^{k+1+1}$

$3^{k+1} < 3^{k+2}$

14. S_n: $2n \leqslant 2^n$

S_1: $2 \cdot 1 \leqslant 2^1$

S_k: $2k \leqslant 2^k$

S_{k+1}: $2(k + 1) \leqslant 2^{k+1}$

1) Basis step: S_1 is true by substitution.

2) Induction step: Assume S_k. Deduce S_{k+1}.

$2k \leqslant 2^k$ (S_k)

$2 \cdot 2k \leqslant 2 \cdot 2^k$ (Multiplying by 2)

$4k \leqslant 2^{k+1}$

Since $1 \leqslant k$, $k + 1 \leqslant k + k$, or $k + 1 \leqslant 2k$.
Then $2(k + 1) \leqslant 4k$.

Thus $2(k + 1) \leqslant 4k \leqslant 2^{k+1}$, so $2(k + 1) \leqslant 2^{k+1}$.

15. S_n: $1^3 + 2^3 + 3^3 + \ldots + n^3 = \dfrac{n^2(n + 1)^2}{4}$

S_1: $1^3 = \dfrac{1^2(1 + 1)^2}{4}$

S_k: $1^3 + 2^3 + 3^3 + \ldots + k^3 = \dfrac{k^2(k + 1)^2}{4}$

S_{k+1}: $1^3 + 2^3 + 3^3 + \ldots + k^3 + (k + 1)^3 = \dfrac{(k + 1)^2(k + 2)^2}{4}$

1) Basis step: S_1 is true by substitution.

2) Induction step: Assume S_k. Deduce S_{k+1}.
Add $(k + 1)^3$ on both sides of S_k.
Then simplify the right side.

$$1^3 + 2^2 + 3^3 + \ldots + k^3 + (k + 1)^3$$
$$= \frac{k^2(k + 1)^2}{4} + (k + 1)^3$$
$$= \frac{k^2(k + 1)^2 + 4(k + 1)^3}{4}$$
$$= \frac{(k + 1)^2[k^2 + 4(k + 1)]}{4}$$
$$= \frac{(k + 1)^2(k^2 + 4k + 4)}{4}$$
$$= \frac{(k + 1)^2(k + 2)^2}{4}$$

16. S_n: $\frac{1}{1\cdot2\cdot3} + \frac{1}{2\cdot3\cdot4} + \frac{1}{3\cdot4\cdot5} + \dots + \frac{1}{n(n+1)(n+2)} = \frac{n(n+3)}{4(n+1)(n+2)}$

S_1: $\frac{1}{1\cdot2\cdot3} = \frac{1(1+3)}{4\cdot2\cdot3}$

S_k: $\frac{1}{1\cdot2\cdot3} + \frac{1}{2\cdot3\cdot4} + \dots + \frac{1}{k(k+1)(k+2)} = \frac{k(k+3)}{4(k+1)(k+2)}$

S_{k+1}: $\frac{1}{1\cdot2\cdot3} + \frac{1}{2\cdot3\cdot4} + \dots + \frac{1}{k(k+1)(k+2)} + \frac{1}{(k+1)(k+2)(k+3)}$

$= \frac{(k+1)(k+1+3)}{4(k+1+1)(k+1+2)} = \frac{(k+1)(k+4)}{4(k+2)(k+3)}$

1) Basis step: S_1 is true by substitution.

2) Induction step: Assume S_k. Deduce S_{k+1}.
Add $\frac{1}{(k+1)(k+2)(k+3)}$ on both sides of S_k and simplify the right side.

Only the right side is shown here.

$$\frac{k(k+3)}{4(k+1)(k+2)} + \frac{1}{(k+1)(k+2)(k+3)}$$
$$= \frac{k(k+3)(k+3)+4}{4(k+1)(k+2)(k+3)}$$
$$= \frac{k^3+6k^2+9k+4}{4(k+1)(k+2)(k+3)}$$
$$= \frac{(k+1)^2(k+4)}{4(k+1)(k+2)(k+3)}$$
$$= \frac{(k+1)(k+4)}{4(k+2)(k+3)}$$

17. See answer section in text.

18. S_n: $a_1 + (a_1 + d) + \dots + [a_1 + (n-1)d] = \frac{n}{2}[2a_1 + (n-1)d]$

S_1: $a_1 = \frac{1}{2}[2a_1 + (1-1)d]$

S_k: $a_1 + (a_1 + d) + \dots + [a_1 + (k-1)d] = \frac{k}{2}[2a_1 + (k-1)d]$

S_{k+1}: $a_1 + (a_1 + d) + \dots + [a_1 + (k-1)d] + [a_1 + kd] = \frac{k+1}{2}[2a_1 + kd]$

1) Basis step: S_1 is true by substitution.

2) Induction step: Assume S_k. Deduce S_{k+1}.
Starting with the left side of S_{k+1}, we have

$$\underbrace{a_1 + (a_1 + d) + \dots + [a_1+(k-1)d]} + [a_1 + kd]$$
$$= \frac{k}{2}[2a_1 + (k-1)d] + [a_1 + kd] \quad \text{(By } S_k\text{)}$$

18. (continued)

$$= \frac{k[2a_1 + (k-1)d]}{2} + \frac{2[a_1 + kd]}{2}$$
$$= \frac{2ka_1 + k(k-1)d + 2a_1 + 2kd}{2}$$
$$= \frac{2a_1(k+1) + k(k-1)d + 2kd}{2}$$
$$= \frac{2a_1(k+1) + (k-1+2)kd}{2}$$
$$= \frac{2a_1(k+1) + (k+1)kd}{2}$$
$$= \frac{k+1}{2}[2a_1 + kd]$$

19. See answer section in text.

20. S_n: $\cos n\pi = (-1)^n$

S_1: $\cos \pi = (-1)^1$

S_k: $\cos k\pi = (-1)^k$

S_{k+1}: $\cos[(k+1)\pi] = (-1)^{k+1}$

1) Basis step: S_1 is true by substitution.

2) Induction step: Assume S_k. Deduce S_{k+1}.
Starting with the left side of S_{k+1}, we have

$$\cos[(k+1)\pi] = \cos(k\pi + \pi)$$
$$= \cos k\pi \cos \pi - \sin k\pi \sin \pi$$
$$= [\cos k\pi](-1) - [\sin k\pi]\cdot 0$$
$$= (-1)\cos k\pi$$
$$= (-1)(-1)^k \quad \text{(By } S_k\text{)}$$
$$= (-1)^{k+1}$$

21. See answer section in text.

22. S_n: $|\sin(nx)| \leqslant n|\sin x|$

S_1: $|\sin x| \leqslant |\sin x|$

S_k: $|\sin(kx)| \leqslant k|\sin x|$

S_{k+1}: $|\sin(k+1)x| \leqslant (k+1)|\sin x|$

1) Basis step: S_1 is true by substitution.

2) Induction step: Assume S_k. Deduce S_{k+1}.
Starting with the left side of S_{k+1}, we have

$$|\sin(k+1)x|$$
$$= |\sin(kx + x)|$$
$$= |\sin kx \cos x + \cos kx \sin x|$$
$$\leqslant |\sin kx \cos x| + |\cos kx \sin x|$$
$$= |\sin kx||\cos x| + |\cos kx||\sin x|$$
$$\leqslant k|\sin x||\cos x| + |\cos kx||\sin x| \quad \text{(By } S_k\text{)}$$
$$= |\sin x|(k|\cos x| + |\cos kx|)$$
$$\leqslant |\sin x|(k+1),$$

since $|\cos x| \leqslant 1$ and $|\cos kx| \leqslant 1$

23. See answer section in text.

24. S_2: $\log_a (b_1b_2) = \log_a b_1 + \log_a b_2$

S_k: $\log_a (b_1b_2...b_k) = \log_a b_1 + \log_a b_2 + ... + \log_a b_k$

$\log_a (b_1b_2...b_{k+1})$ (Left side of S_{k+1})

$= \log_a (b_1b_2...b_k) + \log_a b_{k+1}$ (By S_2)

$= \log_a b_1 + \log_a b_2 + ... + \log_a b_k + \log_a b_{k+1}$

25. See answer section in text.

26. $S_1 = \overline{z^1} = \overline{z}^1$

$S_k = \overline{z^k} = \overline{z}^k$

$\overline{z^k} \cdot \overline{z} = \overline{z}^k \cdot \overline{z}$ (Multiplying both sides of S_k by $\overline{z}$)

$\overline{z^k \cdot z} = \overline{z}^{k+1}$

$\overline{z^{k+1}} = \overline{z}^{k+1}$

27. See answer section in text.

28. S_2: $\overline{z_1z_2} = \overline{z_1} \cdot \overline{z_2}$

S_k: $\overline{z_1z_2...z_k} = \overline{z}_1\overline{z}_2...\overline{z}_k$

Starting with the left side of S_{k+1}, we have

$\overline{z_1z_2...z_kz_{k+1}} = \overline{z_1z_2...z_k} \cdot \overline{z_{k+1}}$ (By S_2)

$= \overline{z}_1\overline{z}_2...\overline{z}_k \cdot \overline{z}_{k+1}$ (By S_k)

29. See answer section in text.

30. S_1: 3 is a factor of $1^3 + 2\cdot1$

S_k: 3 is a factor of $k^3 + 2k$, i.e. $k^3 + 2k = 3\cdot m$

S_{k+1}: 3 is a factor of $(k + 1)^3 + 2(k + 1)$

Consider:

$(k + 1)^3 + 2(k + 1)$

$= k^3 + 3k^2 + 5k + 3$

$= (k^3 + 2k) + 3k^2 + 3k + 3$

$= 3m + 3(k^2 + k + 1)$ (A multiple of 3)

31. See answer section in text.

Exercise Set 12.6

1. ${}_4P_3 = 4\cdot3\cdot2 = 24$ (Theorem 9, Formula 1)

2. 2520

3. ${}_{10}P_7 = 10\cdot9\cdot8\cdot7\cdot6\cdot5\cdot4 = 604{,}800$ (Theorem 9, Formula 1)

4. 720

5. Without repetition: ${}_5P_5 = 5\cdot4\cdot3\cdot2\cdot1 = 120$ (Theorem 8)

With repetition: $5^5 = 3125$ (Theorem 12)

6. 24; 256

7. Line: ${}_5P_5 = 5\cdot4\cdot3\cdot2\cdot1 = 120$ (Theorem 8)

Circle: $(5 - 1)! = 4! = 4\cdot3\cdot2\cdot1 = 24$ (Theorem 11)

8. 5040; 720

9. DIGIT

There are 2 I's, 1 D, 1 G, and 1 T for a total of 5.

$P = \frac{5!}{2!\cdot1!\cdot1!\cdot1!}$ (Theorem 10)

$= \frac{5!}{2!} = \frac{5\cdot4\cdot3\cdot2!}{2!} = 5\cdot4\cdot3 = 60$

10. 360

11. There are only 9 choices for the first digit since 0 is excluded. There are also 9 choices for the second digit since 0 can be included and the first digit cannot be repeated. Because no digit is used more than once there are only 8 choices for the third digit, 7 for the fourth, 6 for the fifth, 5 for the sixth and 4 for the seventh. By the Fundamental Counting Principle the total number of permutations is

$9\cdot9\cdot8\cdot7\cdot6\cdot5\cdot4$, or 544,320.

Thus 544,320 7-digit phone numbers can be formed.

12. 2880

13. $a^2b^3c^4 = a\cdot a\cdot b\cdot b\cdot b\cdot c\cdot c\cdot c\cdot c$

There are 2 a's, 3 b's, and 4 c's, for a total of 9.

Thus $P = \frac{9!}{2!\cdot3!\cdot4!}$ (Theorem 10)

$= \frac{9\cdot8\cdot7\cdot6\cdot5\cdot4!}{2\cdot1\cdot3\cdot2\cdot1\cdot4!} = \frac{9\cdot8\cdot7\cdot6\cdot5}{2\cdot3\cdot2} = 1260$

14. 60

15. The number of distinct circular arrangements of 13 objects (King Arthur plus 12 knights) is $(13 - 1)!$ (Theorem 11).

$(13 - 1)! = 12! = 12\cdot11\cdot10\cdot9\cdot8\cdot7\cdot6\cdot5\cdot4\cdot3\cdot2\cdot1$

$= 479{,}001{,}600$

16. 6

17. a) ${}_5P_5 = 5!$ (Theorem 8)

$= 5\cdot4\cdot3\cdot2\cdot1 = 120$

b) There are 5 choices for the first coin and 2 possibilities (head or tail) for each choice. This results in a total of 10 choices for the first selection.

There are 4 choices (no coin can be used more than once) for the second coin and 2 possibilities (head or tail) for each choice. This results in a total of 8 choices for the second selection.

There are 3 choices for the third coin and 2 possibilities (head or tail) for each choice. This results in a total of 6 choices for the third selection.

Likewise there are 4 choices for the fourth selection and 2 choices for the fifth selection.

Using the Fundamental Counting Principle we know there are

$10\cdot8\cdot6\cdot4\cdot2$, or 3840

ways the coins can be lined up.

18. a) 24, b) 384

19. ${}_{52}P_4 = 52\cdot51\cdot50\cdot49$ (Theorem 9, Formula 1)

$= 6,497,400$

20. 254,251,200

21. $P = \dfrac{24!}{3!\cdot5!\cdot9!\cdot4!\cdot3!}$ (Theorem 10)

$= \dfrac{24\cdot23\cdot22\cdot21\cdot20\cdot19\cdot18\cdot17\cdot16\cdot15\cdot14\cdot13\cdot12\cdot11\cdot10\cdot9!}{3\cdot2\cdot1\cdot5\cdot4\cdot3\cdot2\cdot1\cdot4\cdot3\cdot2\cdot1\cdot3\cdot2\cdot1\cdot9!}$

$= 23\cdot11\cdot7\cdot19\cdot17\cdot8\cdot15\cdot14\cdot13\cdot12\cdot11\cdot10$ (Simplifying)

$= 16,491,024,950,400$

22. 20,951,330,400

23. MATH:

${}_4P_4 = 4\cdot3\cdot2\cdot1 = 24$ (Theorem 8)

BUSINESS: 1 B, 1 U, 3 S's, 1 I, 1 N, 1 E, a total of 8.

$P = \dfrac{8!}{1!\cdot1!\cdot3!\cdot1!\cdot1!\cdot1!}$ (Theorem 10)

$= \dfrac{8!}{3!} = \dfrac{8\cdot7\cdot6\cdot5\cdot4\cdot3!}{3!} = 8\cdot7\cdot6\cdot5\cdot4 = 6720$

23. (continued)

PHILOSOPHICAL: 2 P's, 2 H's, 2 I's, 2 L's, 2 O's, 1 S, 1 C, and 1 A, a total of 13

$P = \dfrac{13!}{2!\cdot2!\cdot2!\cdot2!\cdot2!\cdot1!\cdot1!\cdot1!}$ (Theorem 10)

$= \dfrac{13\cdot12\cdot11\cdot10\cdot9\cdot8\cdot7\cdot6\cdot5\cdot4\cdot3\cdot2\cdot1}{2\cdot1\cdot2\cdot1\cdot2\cdot1\cdot2\cdot1\cdot2\cdot1\cdot1\cdot1\cdot1}$

$= 13\cdot12\cdot11\cdot5\cdot9\cdot4\cdot7\cdot3\cdot5\cdot2\cdot3$ (Simplifying)

$= 194,594,400$

24. 720, 2520, 4,989,600

25. There are 80 choices for the number of the county, 26 choices for the letter of the alphabet, and 9999 choices for the number that follows the letter. By the Fundamental Counting Principle we know there are $80\cdot26\cdot9999$, or 20,797,920 possible license plates.

26. a) 120, b) 625, c) 24, d) 6

27. ${}_nP_5 = 7\cdot{}_nP_4$

$\dfrac{n!}{(n-5)!} = 7\cdot\dfrac{n!}{(n-4)!}$ (Theorem 9)

$\dfrac{n!}{7(n-5)!} = \dfrac{n!}{(n-4)!}$ $\left[\text{Multiplying by } \dfrac{1}{7}\right]$

$7(n-5)! = (n-4)!$ (The denominators must be the same.)

$7(n-5)! = (n-4)(n-5)!$

$7 = n - 4$ $\left[\text{Multiplying by } \dfrac{1}{(n-5)!}\right]$

$11 = n$

28. 8

29. ${}_nP_5 = 9\cdot{}_{n-1}P_4$

$\dfrac{n!}{(n-5)!} = 9\cdot\dfrac{(n-1)!}{(n-5)!}$ (Theorem 9)

$n! = 9(n-1)!$ [Multiplying by $(n-5)!$]

$\dfrac{n!}{(n-1)!} = 9$ $\left[\text{Multiplying by } \dfrac{1}{(n-1)!}\right]$

$\dfrac{n(n-1)!}{(n-1)!} = 9$

$n = 9$

30. 11

Exercise Set 12.7

1. ${}_9C_5 = \dbinom{9}{5} = \dfrac{9!}{5!(9-5)!}$ (Theorem 13)

$= \dfrac{9!}{5!\cdot4!} = \dfrac{9\cdot8\cdot7\cdot6\cdot5!}{5!\cdot4!}$

$= \dfrac{9\cdot8\cdot7\cdot6}{4\cdot3\cdot2\cdot1} = 126$

2. 91

3. $\binom{50}{2} = \frac{50!}{2!(50-2)!}$ (Theorem 13)

$= \frac{50!}{2!\cdot 48!} = \frac{50\cdot 49\cdot 48!}{2!\cdot 48!}$

$= \frac{50\cdot 49}{2\cdot 1} = 1225$

4. 9880

5. ${}_nC_3 = \binom{n}{3} = \frac{n!}{3!(n-3)!}$ (Theorem 13)

$= \frac{n(n-1)(n-2)(n-3)!}{3!(n-3)!}$

$= \frac{n(n-1)(n-2)}{6}$

6. $\frac{n(n-1)}{2}$

7. ${}_{23}C_4 = \binom{23}{4} = \frac{23!}{4!(23-4)!}$ (Theorem 13)

$= \frac{23!}{4!\cdot 19!} = \frac{23\cdot 22\cdot 21\cdot 20\cdot 19!}{4!\cdot 19!}$

$= \frac{23\cdot 22\cdot 21\cdot 20}{4\cdot 3\cdot 2\cdot 1} = 8855$

8. 36; 72

9. ${}_{10}C_6 = \binom{10}{6} = \frac{10!}{6(10-6)!}$ (Theorem 13)

$= \frac{10!}{6!\cdot 4!} = \frac{10\cdot 9\cdot 8\cdot 7\cdot 6!}{6!\cdot 4!}$

$= \frac{10\cdot 9\cdot 8\cdot 7}{4\cdot 3\cdot 2\cdot 1} = 210$

10. 330

11. Since two points determine a line and no three of these 8 points are collinear, we need to find out the number of combinations of 8 points taken 2 at a time, ${}_8C_2$.

${}_8C_2 = \binom{8}{2} = \frac{8!}{2!(8-2)!}$ (Theorem 13)

$= \frac{8\cdot 7\cdot 6!}{2\cdot 1\cdot 6!}$

$= \frac{8\cdot 7}{2} = 28$

Thus 28 lines are determined.

Since three noncollinear points determine a triangle and no four of these 8 points are coplanar, we need to find out the number of combinations of 8 points taken 3 at a time, ${}_8C_3$.

${}_8C_3 = \binom{8}{3} = \frac{8!}{3!(8-3)!}$ (Theorem 13)

$= \frac{8\cdot 7\cdot 6\cdot 5!}{3\cdot 2\cdot 1\cdot 5!}$

$= \frac{8\cdot 7\cdot 6}{3\cdot 2} = 56$

Thus 56 triangles are determined.

12. 21; 35

13. ${}_{10}C_7 \cdot {}_5C_3 = \binom{10}{7}\cdot\binom{5}{3}$ (Using the Fundamental Counting Principle)

$= \frac{10!}{7!(10-7)!} \cdot \frac{5!}{3!(5-3)!}$ (Theorem 13)

$= \frac{10\cdot 9\cdot 8\cdot 7!}{7!\cdot 3!} \cdot \frac{5\cdot 4\cdot 3!}{3!\cdot 2!}$

$= \frac{10\cdot 9\cdot 8}{3\cdot 2\cdot 1} \cdot \frac{5\cdot 4}{2\cdot 1} = 120\cdot 10 = 1200$

14. 112

15. ${}_{58}C_6 \cdot {}_{42}C_4$ (Using the Fundamental Counting Principle)

$= \binom{58}{6}\cdot\binom{42}{4}$

$= \frac{58!}{6!(58-6)!} \cdot \frac{42!}{4!(42-4)!}$ (Theorem 13)

$= \frac{58\cdot 57\cdot 56\cdot 55\cdot 54\cdot 53\cdot 52!}{6!\cdot 52!} \cdot \frac{42\cdot 41\cdot 40\cdot 39\cdot 38!}{4!\cdot 38!}$

$= \frac{58\cdot 57\cdot 56\cdot 55\cdot 54\cdot 53}{6\cdot 5\cdot 4\cdot 3\cdot 2\cdot 1} \cdot \frac{42\cdot 41\cdot 40\cdot 39}{4\cdot 3\cdot 2\cdot 1}$

$= (29\cdot 19\cdot 14\cdot 11\cdot 9\cdot 53)\cdot(21\cdot 41\cdot 10\cdot 13)$

$= (40{,}475{,}358)(111{,}930)$

$= 4{,}530{,}406{,}820{,}940$

16. $\binom{63}{8}\binom{37}{12}$

17. In a 52-card deck there are 4 aces and 48 cards that are not aces.

${}_4C_3 \cdot {}_{48}C_2 = \binom{4}{3}\binom{48}{2}$ (Using the Fundamental Counting Principle)

$= \frac{4!}{3!(4-3)!} \cdot \frac{48!}{2!(48-2)!}$ (Theorem 13)

$= \frac{4!}{3!\cdot 1!} \cdot \frac{48!}{2!\cdot 46!}$

$= \frac{4\cdot 3!}{3!\cdot 1!} \cdot \frac{48\cdot 47\cdot 46!}{2!\cdot 46!}$

$= \frac{4}{1} \cdot \frac{48\cdot 47}{2}$

$= 4512$

18. 103,776

19. The total number of subsets of a set with 7 members is 2^7, or 128 (Theorem 14).

20. 2^6, or 64

21. The total number of subsets of a set with 26 members is 2^{26}, or 67,108,864.

22. 2^{24}, or 16,777,216

23. $_{52}C_5 = \binom{52}{5} = \frac{52!}{5!(52-5)!}$ (Theorem 13)

$= \frac{52!}{5!\cdot 47!}$

$= \frac{52\cdot 51\cdot 50\cdot 49\cdot 48\cdot 47!}{5!\cdot 47!}$

$= \frac{52\cdot 51\cdot 50\cdot 49\cdot 48}{5\cdot 4\cdot 3\cdot 2\cdot 1}$

$= 13\cdot 17\cdot 10\cdot 49\cdot 24$

$= 2{,}598{,}960$

24. 635,013,559,600

25. $_8C_3 = \binom{8}{3} = \frac{8!}{3!(8-3)!} = \frac{8!}{3!\cdot 5!}$

$= \frac{8\cdot 7\cdot 6\cdot 5!}{3\cdot 2\cdot 1\cdot 5!}$

$= 8\cdot 7 = 56$

26. $\binom{n}{4} = \frac{n(n-1)(n-2)(n-3)}{24}$

27. For each parallelogram 2 lines from each group of parallel lines are needed.

Two lines can be selected from the group of 5 parallel lines in $_5C_2$ ways and the other two lines can be selected from the group of 8 parallel lines in $_8C_2$ ways.

Using the Fundamental counting Principle, it follows that the number of possible parallelograms is

$_5C_2 \cdot {_8C_2} = \binom{5}{2}\binom{8}{2}$ (Using the Fundamental Counting Principle)

$= \frac{5!}{2!(5-2)!}\cdot\frac{8!}{2!(8-2)!}$ (Theorem 13)

$= \frac{5\cdot 4\cdot 3!}{2\cdot 1\cdot 3!}\cdot\frac{8\cdot 7\cdot 6!}{2\cdot 1\cdot 6!}$

$= \frac{5\cdot 4}{2}\cdot\frac{8\cdot 7}{2}$

$= 10\cdot 28 = 280$

Thus 280 parallelograms can be formed.

28. $\binom{n}{r} = \frac{n!}{r!(n-r)!} = \frac{n!}{(n-r)!r!}$

$= \frac{n!}{(n-r)![n-(n-r)]!}$

$= \binom{n}{n-r}$

29. $\binom{n+1}{3} = 2\cdot\binom{n}{2}$

$\frac{(n+1)!}{3!(n+1-3)!} = 2\cdot\frac{n!}{2!(n-2)!}$ (Theorem 13)

$\frac{(n+1)(n)(n-1)(n-2)!}{3!(n-2)!} = \frac{n(n-1)(n-2)!}{(n-2)!}$

$\frac{(n+1)(n)(n-1)}{6} = n(n-1)$

$(n+1)(n)(n-1) = 6n(n-1)$

$n+1 = 6$

$n = 5$

30. 4

31. $\binom{n+2}{4} = 6\cdot\binom{n}{2}$

$\frac{(n+2)!}{4!(n+2-4)!} = 6\cdot\frac{n!}{2!(n-2)!}$ (Theorem 13)

$\frac{(n+2)!}{4!(n-2)!} = 6\cdot\frac{n!}{2!(n-2)!}$

$\frac{(n+2)!}{4!} = 6\cdot\frac{n!}{2!}$

$4!\cdot\frac{(n+2)!}{4!} = 4!\cdot 6\cdot\frac{n!}{2!}$

$(n+2)! = 72\cdot n!$

$(n+2)(n+1)n! = 72\cdot n!$

$(n+2)(n+1) = 72$

$n^2 + 3n + 2 = 72$

$n^2 + 3n - 70 = 0$

$(n+10)(n-7) = 0$

$n + 10 = 0$ or $n - 7 = 0$

$n = -10$ or $n = 7$

The only solution is 7 since we cannot have a set of -10 objects.

32. 6

Exercise Set 12.8

1. 100 surveyed
   57 wore either glasses or contacts
   43 wore neither glasses nor contacts

   Using Principle P,

   P(wearing either glasses or contacts) $= \frac{57}{100}$, or 0.57

   P(wearing neither glasses nor contacts) $= \frac{43}{100}$, or 0.43

2. 0.18, 0.24, 0.23, 0.23, 0.12; Opinions might vary, but it would seem that people tend not to pick the first or last numbers.

3. 1044 Total letters in the three paragraphs
78 A's occurred
140 E's occurred
60 I's occurred
74 O's occurred
31 U's occurred

Using Principle P,

P(the occurrence of letter A) = $\frac{78}{1044} \approx 0.075$

P(the occurrence of letter E) = $\frac{140}{1044} \approx 0.134$

P(the occurrence of letter I) = $\frac{60}{1044} \approx 0.057$

P(the occurrence of letter O) = $\frac{74}{1044} \approx 0.071$

P(the occurrence of letter U) = $\frac{31}{1044} \approx 0.030$

4. 0.367

5. 1044 Total letters
383 Total vowels
661 Total consonants

Using Principle P,

P(the occurrence of a consonant) = $\frac{661}{1044} \approx 0.633$

6. Z, 0.999

7. Let D represent the number of deer in the preserve. If there are D deer in the preserve and 318 of them are tagged, then the probability that a deer is tagged is $\frac{318}{D}$. Later, 168 deer are caught of which 56 were tagged. The ratio of deer tagged to deer caught is $\frac{56}{168}$. We assume the two ratios are the same. We solve the proportion for D.

$$\frac{318}{D} = \frac{56}{168}$$

$$168 \cdot 318 = D \cdot 56$$

$$53{,}424 = 56D$$

$$954 = D$$

Thus we estimate that there are 954 deer in the preserve.

8. 287

9. There are 52 equally likely outcomes.

10. $\frac{1}{13}$

11. Since there are 52 equally likely outcomes and there are 13 ways to obtain a heart, by Principle P we have

P(drawing a heart) = $\frac{13}{52} = \frac{1}{4}$.

12. $\frac{1}{4}$

13. Since there are 52 equally likely outcomes and there are 4 ways to obtain a 4, by Principle P we have

P(drawing a 4) = $\frac{4}{52} = \frac{1}{13}$.

14. $\frac{1}{2}$

15. Since there are 52 equally likely outcomes and there are 26 ways to obtain a black card, by Principle P we have

P(drawing a black card) = $\frac{26}{52} = \frac{1}{2}$.

16. $\frac{2}{13}$

17. Since there are 52 equally likely outcomes and there are 8 ways to obtain a 9 or a king (four 9's and four kings), we have, by Principle P,

P(drawing a 9 or a king) = $\frac{8}{52} = \frac{2}{13}$.

18. $\frac{2}{7}$

19. Since there are 14 equally likely ways of selecting a marble from a bag containing 4 red marbles and 10 green marbles, we have, by Principle P,

P(selecting a green marble) = $\frac{10}{14} = \frac{5}{7}$.

20. 0

21. There are 14 equally likely ways of selecting any marble from a bag containing 4 red marbles and 10 green marbles. Since the bag does not contain any white marbles, there are 0 ways of selecting a white marble. By Principle P, we have

P(selecting a white marble) = $\frac{0}{14} = 0$.

22. $\frac{11}{4165}$

23. The number of ways of drawing 4 cards from a deck of 52 cards is ${}_{52}C_4$. Now 13 of the 52 cards are hearts, so the number of ways of drawing 4 hearts is ${}_{13}C_4$. Thus,

$$\text{P(getting 4 hearts)} = \frac{{}_{13}C_4}{{}_{52}C_4} = \frac{\frac{13!}{4!\cdot 9!}}{\frac{52!}{4!\cdot 48!}}$$

$$= \frac{13\cdot 12\cdot 11\cdot 10\cdot 9!}{4!\cdot 9!} \cdot \frac{4!\cdot 48!}{52\cdot 51\cdot 50\cdot 49\cdot 48!}$$

$$= \frac{13\cdot 12\cdot 11\cdot 10}{52\cdot 51\cdot 50\cdot 49} = \frac{11}{4165}$$

24. $\frac{60}{143}$

25. The number of ways of selecting 4 people from a group of 15 is ${}_{15}C_4$. Two men can be selected in ${}_8C_2$ ways, and 2 women can be selected in ${}_7C_2$ ways. By the Fundamental Counting Principle, the number of ways of selecting 2 men and 2 women is ${}_8C_2 \cdot {}_7C_2$. Thus,

$$P(\text{2 men and 2 women are chosen}) = \frac{{}_8C_2 \cdot {}_7C_2}{{}_{15}C_4}$$

$$= \frac{\frac{8!}{2!6!} \cdot \frac{7!}{2!5!}}{\frac{15!}{4!11!}} = \frac{\frac{8\cdot7\cdot6!}{2\cdot1\cdot6!} \cdot \frac{7\cdot6\cdot5!}{2\cdot1\cdot5!}}{\frac{15\cdot14\cdot13\cdot12\cdot11!}{4\cdot3\cdot2\cdot1\cdot11!}}$$

$$= \frac{28\cdot21}{15\cdot7\cdot13} = \frac{28}{65}$$

26. $\frac{5}{36}$

27. On each die there are 6 possible outcomes. The outcomes are paired so there are 6·6, or 36 possible ways in which the two can fall. The pairs that total 3 are (1, 2) and (2, 1). Thus there are 2 possible ways of getting a total of 3, so the probability is $\frac{2}{36}$, or $\frac{1}{18}$.

28. $\frac{1}{36}$

29. On each die there are 6 possible outcomes. The outcomes are paired so there are 6·6, or 36 possible ways in which the two can fall. There is only 1 way of getting a total of 12, the pair (6, 6), so the probability is $\frac{1}{36}$.

30. $\frac{245}{1938}$

31. The number n of ways of getting 6 coins from a bag containing 20 coins is ${}_{20}C_6$. Three nickels can be selected in ${}_6C_3$ ways since the bag contains 6 nickels. Two dimes can be selected in ${}_{10}C_2$ ways since the bag contains 10 dimes. One quarter can be selected in ${}_4C_1$ ways since the bag contains 4 quarters. By the Fundamental Counting Principle the number of ways of selecting 3 nickels, 2 dimes, and 1 quarter is ${}_6C_3 \cdot {}_{10}C_2 \cdot {}_4C_1$. Thus,

P(getting 3 nickels, 2 dimes, and 1 quarter)

$$= \frac{{}_6C_3 \cdot {}_{10}C_2 \cdot {}_4C_1}{{}_{20}C_6} = \frac{\frac{6!}{3!3!} \cdot \frac{10!}{2!8!} \cdot \frac{4!}{1!3!}}{\frac{20!}{6\cdot14!}}$$

$$= \frac{\frac{6\cdot5\cdot4}{3\cdot2\cdot1} \cdot \frac{10\cdot9}{2\cdot1} \cdot \frac{4}{1}}{\frac{20\cdot19\cdot18\cdot17\cdot16\cdot15}{6\cdot5\cdot4\cdot3\cdot2\cdot1}} = \frac{20\cdot45\cdot4}{19\cdot17\cdot8\cdot15} = \frac{30}{323}$$

32. $\frac{9}{19}$

33. The roulette wheel contains 38 equally likely slots. Eighteen of the 38 slots are colored red. Thus, by Principle P,

$$P(\text{the ball falls in a red slot}) = \frac{18}{38} = \frac{9}{19}$$

34. $\frac{18}{19}$

35. The roulette wheel contains 38 equally likely slots. Only 1 slot is numbered 00. Then, by Principle P,

$$P(\text{the ball falls in the 00 slot}) = \frac{1}{38}.$$

36. $\frac{1}{38}$

37. The roulette wheel contains 38 equally likely slots, 2 of which are numbered 00 and 0. Thus, using Principle P,

$$P\begin{bmatrix}\text{the ball falling in either} \\ \text{the 00 or the 0 slot}\end{bmatrix} = \frac{2}{38} = \frac{1}{19}.$$

38. $\frac{9}{19}$

39. $${}_{52}C_5 = \frac{52!}{5!47!} = \frac{52\cdot51\cdot50\cdot49\cdot48\cdot47!}{5\cdot4\cdot3\cdot2\cdot1\cdot47!}$$
$$= 26\cdot17\cdot10\cdot49\cdot12$$
$$= 2{,}598{,}960$$

40. a) 4, b) $\frac{4}{2{,}598{,}960} \approx 1.54 \times 10^{-6}$

41. Consider a suit

A K Q J 10 9 8 7 6 5 4 3 2

A straight flush can be any of the following combinations in the same suit.

K Q J 10 9
Q J 10 9 8
J 10 9 8 7
10 9 8 7 6
9 8 7 6 5
8 7 6 5 4
7 6 5 4 3
6 5 4 3 2
5 4 3 2 A

Remember a straight flush does not include A K Q J 10 which is a royal flush.

a) Since there are 9 straight flushes per suit, there are 36 straight flushes in all 4 suits.

b) Since 2,598,960, or ${}_{52}C_5$, poker hands can be dealt from a standard 52-card deck and 36 of those hands are straight flushes, the probability of getting a straight flush is $\frac{36}{2{,}598{,}960}$, or 0.0000139.

42. a) $13 \cdot 48 = 624$, b) $\frac{624}{2,598,960} \approx 2.4 \times 10^{-4}$